Charged Particle Beams

Charged Particle Beams

STANLEY HUMPHRIES, Jr.

*Department of Electrical
and Computer Engineering
University of New Mexico
Albuquerque*

A WILEY-INTERSCIENCE PUBLICATION

John Wiley & Sons, Inc.

NEW YORK / CHICHESTER / BRISBANE / TORONTO / SINGAPORE

03704555

PHYSICS

Library of Congress Cataloging in Publication Data:

Humphries, Stanley.
    Charged particle beams / Stanley Humphries, Jr.
       p.   cm.
    "A Wiley-Interscience publication."
    Includes bibliographical references.
    ISBN 0-471-60014-8
    1. Particle beams. I. Title.
  QC786.H86 1990   89-29218
  539.7'3—dc20   CIP

Printed in the United States of America

10 9 8 7 6 5 4 3 2 1

To Sandy, Colin, and Courtney
for the joyful years of the past and the ones to come

Preface

Charged Particle Beams is the product of a two-term course sequence that I teach on accelerator technology and beam physics at the University of New Mexico and at Los Alamos National Laboratory. The material for the two terms divides neatly into the dynamics of single charged particles and the description of large groups of particles—the collective behaviour of beams. My first book, *Principles of Charged Particle Acceleration*, covered single particle topics such as linear transfer matrices and the operation of accelerators. The new book is an introduction to charged-particle-beam physics.

In writing *Charged Particle Beams*, my goal was to create a unified description that would be valuable to a broad audience—accelerator designers, accelerator users, industrial engineers, and physics researchers. I organized the material to provide beginning students with the background to understand advanced literature and to use accelerators effectively. This book can serve as an independent reference. Combining *Charged Particle Beams* with *Principles of Charged Particle Acceleration* gives a programmed introduction to the complete field of particle acceleration.

I began my research on particle beams about ten years ago with a background in plasma physics. This change in direction involved a painful process of searching for material, learning from experts, and seeking past insights. Although I found excellent advanced references on specialized areas, no single work covered the topics necessary to understand high-power accelerators and high-brightness beams. The difficulties I faced encouraged me to write *Charged Particle Beams*. The book succinctly describes the basic ideas behind modern beam applications such as stochastic cooling, high-brightness injectors, and the free-electron laser.

Another motivation for the book was to share insights I gained working on applications of collective physics. I believe that students face a particularly difficult problem studying collective particle behavior. When teaching other areas of physics and engineering, we can usually provide students with a definite sequence of steps to solve problems. In contrast, collective physics seldom allows exact solutions or cut-and-dried methods. The complexity of problems involving a multiplicity of interacting particles makes collective physics a science of approximation. The correct answer to a problem depends more on the insights and simplifications involved in setting up equations than on the elegance of the mathematical methods applied to their solution. The ability to discard extraneous material while preserving the essential physics usually comes only after years of experience. I organized *Charged Particle Beams* to accelerate students on this path. The book clearly lists the approximations used for theoretical models and explains their motivation. The book concentrates on the process of creating physical models; the resulting equations are solved the quickest way possible, usually with numerical methods. Because of the availability of personal computers, I believe that an emphasis on numerical methods represents a modern way of doing physics. With available mathematical software, students and experimentalists can quickly solve theoretical problems that previously required pages of painstaking derivations.

No work that attempts to cover an area as diversified as beam physics can be complete. The ideal book is never finished—many potentially valuable references languish indefinitely as lecture notes waiting for the "final" additions. I will apologize in advance for some practical limitations of *Charged Particle Beams*. The book is primarily a text for students—the purpose is to explain how things work. It is not a history of beam science or a detailed review of current literature. As a result, the work contains few reference citations. Given the time and space, I would have enjoyed crediting the contributions of researchers who built the field of beam physics. As a penance for the omission, I have scrupulously avoided citing my own works. I have also included a list of valuable sources for advanced reading at the end of the book.

The most hazardous task in writing a survey book is choosing topics. My criterion was that the subjects should illustrate useful applications of beam physics. Inevitably, they lean toward my own interests and those of laboratories where I have worked. I realize that many readers will feel that I have omitted critical subjects and I propose the following resolution. If readers have important topics relevant to their work that should be included and if (1) they are associated with an institution proximate to hiking trails with a minimum of 500 m elevation change and (2) this institution serves bakery-quality croissants and bran muffins at meetings, then I suggest that they invite me for intensive discussions. If this procedure is followed faithfully, I can guarantee that the subject will appear in future editions.

I was fortunate to have abundant help creating this book, both direct and indirect. Richard Cooper of Los Alamos National Laboratory made outstanding contributions. He applied his legendary proofreading ability to the entire manuscript. Along with innumerable mechanical corrections, his suggestions on technical points and emphasis were invaluable.

The creation of this book was supported, in part, by a sabbatical leave from the Department of Electrical and Computer Engineering at the University of New Mexico. David Woodall, former Chairman of the Department of Chemical and Nuclear Engineering at the University of New Mexico, suggested the idea of the accelerator course sequence. I am grateful for his support during the development of the courses. My decision to write books based on the courses was inspired by R. Bruce Miller's pioneering text on high-power beams and by John Lawson's classic reference on advanced beam physics.

Several people contributed expert advice on specific sections of the book. Commentators included Charles Roberson of the Office of Naval Research, Kevin O'Brien of Sandia National Laboratories, John Creedon of Physics International Company, Brendan Godfrey of the Air Force Weapons Laboratory, Edward Lee of Lawrence Berkeley Laboratory, William Herrmannsfeldt of the Stanford Linear Accelerator Center, and Carl Ekdahl of Los Alamos National Laboratory. I would also like to thank A. V. Tollestrup of Fermi National Accelerator Laboratory for permission to paraphrase his article (coauthored by G. Dugan) on "Elementary Stochastic Cooling."

I want to express appreciation to the students in my beam physics course at the University of New Mexico and at the Los Alamos Graduate Center. They struggled with ambiguous problem sets, handwritten course notes, and less-than-perfect presentations. Through their contributions, I clarified and expanded the material through the years. Los Alamos National Laboratory supported the courses since their inception. I want to thank Robert Jameson and Alan Wadlinger of the Accelerator Technology Division for their encouragement. The efforts of the Instructional Television Center of UNM made it feasible to present classes at Los Alamos. I have also taught the material in short course format. I am grateful to Thomas Roberts and Stanley Pruett for organizing a course at the Strategic Defense Command.

There is a new tendency among textbook authors to thank their software. Although I feel that I showed adequate appreciation when I paid for it, I must make one exception. I used SCIMATH for almost all the numerical calculations in the book. This program, created by Robert Kares of Berkeley Research Associates and Spindrift Software, is an eminently useful and versatile tool for scientists that deserves much wider recognition.

My professional associations in accelerator science contributed significantly to the material in this book. I have worked closely with the Heavy Ion Fusion Accelerator Research Group at Lawrence Berkeley Laboratory for several years. I want to thank Henry Rutkowski, Thomas Fessenden, Denis Keefe, and Edward Lee for their suggestions on the book and for providing the opportunity to work in the innovative field of accelerator inertial fusion.

The long-term support of Charles Roberson of the Office of Naval Research has been critical for advanced accelerator research at the University of New Mexico. I would like to thank my colleague, Edl Schamiloglu, for supervising the

ONR Project on High Current Betatrons at UNM during my sabbatical leave. The University has also received generous research support from Groups CLS-7 and P-14 of the Los Alamos National Laboratory. I am grateful to Roger Bangerter and the late Kenneth Riepe who initiated the UNM program on vacuum arc plasma sources. I would also like to thank Carl Ekdahl—much of the material in this book evolved from our spirited discussions (arguments to the casual observer) on high-current beam physics.

During the composition of this book, I had the opportunity to participate in diverse research programs on high-power accelerators. I would like to thank Ralph Genuario and George Fraser of Physics International Company, Sidney Putnam of Pulse Sciences Incorporated, Robert Meger of the Naval Research Laboratory, Martin Nahemow of the Westinghouse Research and Development Center, Richard Adler of Northstar Research Corporation, Daniel Sloan of CH2M-Hill, Kenneth Moses of Jaycor, and R. Bruce Miller of Titan Technologies.

I would like to acknowledge two meetings that I attended during the creation of the book. The first is the NATO Workshop on High Brightness Beams in Pitlochry, Scotland. I express my appreciation to Anthony Hyder for organizing this workshop, generally acknowledged as one of mankind's supreme conference achievements. I have enjoyed participating in the American Physical Society Accelerator Summer Schools organized by Melvin Month. These schools attract accelerator researchers from the full spectrum of beam applications and provide a unique service to the community. I appreciate the opportunity to teach at the Fermilab School and look forward to presenting a course at Harvard. Perhaps this acknowledgment will impel to an invitation to Geneva.

It is the prerogative of every textbook author to have a long-suffering family anticipating his return. My family refused to honor this obligation, making it difficult for me to think up a final paragraph for the preface. My wife, Sandra, my son, Colin, and my daughter, Courtney, were so busy with their own activities that they willingly excused me for two years. Now that I am finished, I look forward to expanded blocks of time on their calendars.

STANLEY HUMPHRIES, JR.

Albuquerque, New Mexico
January 1990

Contents

Charged
Particle Beams

1

Introduction

1.1. CHARGED PARTICLE BEAMS

A *charged particle beam* is a group of particles that have about the same kinetic energy and move in about the same direction. Usually, the kinetic energies are much higher than the thermal energies of particles at ordinary temperatures. The high kinetic energy and good directionality of charged particles in beams make them useful for applications. Although we often associate accelerators with the large machines of high-energy physics, charged particle beams have continually expanding applications in many branches of research and technology. Recent active areas include flat-screen cathode-ray tubes, synchrotron light sources, beam lithography for microcircuits, thin-film technology, production of short-lived medical isotopes, radiation processing of food, and free-electron lasers.

The importance of accelerators for applications in research and industry sometimes overshadows beam physics as an intellectual discipline in its own right. The theory of charged particle beams is much more than a tool to design machines—it is one of the richest and most active areas of classical physics. In our study of charged particle beams, we shall gain a comprehensive understanding of applied electromagnetism and collective physics.

Despite the practical importance and underlying unity of beam physics, the field has not yet achieved a strong identity like plasma physics. Although there are many specialized review papers and texts, few general works cover the full range of beam processes. There are several reasons for fragmentation in the field. Accelerator scientists are largely goal-oriented, concentrating on the theory and technology to solve the problem at hand. Each large accelerator has its own mission and its own group of scientists. Because of the broad range of required

1

beam parameters, different accelerators use a diversity of technologies that often have little in common. Although there are large differences in technology, we shall see that a few basic principles underlie the design of all accelerators and beam transport devices. As the problems of accelerators become more challenging and beam applications become more sophisticated, it is increasingly important for accelerator scientists to share their insights and expertise. In recent years, there have been several efforts to emphasize the unity of the field and to promote communication among researchers. In the United States, examples include the Particle Accelerator Conference with its steadily increasing attendance from all areas of accelerator research, the U.S. Particle Accelerator School and its educational publications, and the recently formed American Physical Society Division of Accelerator Physics.

This book was written to guide students entering accelerator science and to provide researchers with a comprehensive reference. It contains a unified treatment of beam physics at an introductory level. This book and a previous one, *Principles of Charged Particle Acceleration*, provide a bridge to carry students to advanced work in specialized fields of accelerator science and beam theory. *Principles of Charged Particle Acceleration* reviews the fundamentals of single-particle dynamics. That book describes how accelerators work, from small low-current devices to the largest machines of high-energy and high-power research. The present book concentrates on problems of beam physics, the acceleration and control of large numbers of charged particles. The range of topics is extensive, with reference material for designers and users of all types of accelerators.

In this section, we begin by reviewing properties of charged particles and particle beams. Section 1.2 discusses some of the problems of collective physics and outlines the organization of the book. Section 1.3 summarizes theoretical results from *Principles of Charged Particle Accelerators* that will be useful for many of the derivations. The goal of beam theory is to describe how the multitude of particles in a beam interact with one another. For this purpose, we need not consider the internal structure of charged particles. Usually, it is sufficient to represent a particle as a point entity with two properties: charge, q, and rest mass, m_0. We assume that the particle characteristics are constant during acceleration and transport. In this book, we will not examine the effects of finite particle dimensions and quantum properties such as spin. Except for specialized applications, these properties have little effect on the formation and acceleration of beams.

Much of the material in this book applies to any charged particle, from the elementary particles of high-energy physics research to hyper-velocity charged clusters. Familiar applications usually involve one of two types of particles: electrons or ions. The electron is an elementary particle with the following characteristics:

$$q_e = -e = -1.602 \times 10^{-19} \, C$$
$$m_e = 9.109 \times 10^{-31} \, kg. \qquad (1.1)$$

We shall apply MKS units exclusively throughout the book with the exception of the electron volt, a useful unit for the energy of individual particles.

Ions are composite particles. An ion is an atom missing one or more electrons or, in the case of negative ions, with an extra electron attached. The following quantities characterize an ion:

A: the atomic mass number, equal to the total number of protons and neutrons in the nucleus.

Z: the atomic number, equal to the number of electrons in the neutral atom.

Z^*: the charge state of the ion, equal to the number of electrons removed from or added to the atom.

The proton is the simplest ion—it is a hydrogen atom with its single electron removed. The proton is an elementary particle with charge and mass:

$$q_p = +e = +1.602 \times 10^{-19}\,C,$$
$$m_p = 1.673 \times 10^{-27}\,kg. \tag{1.2}$$

We denote the charge and mass of other ions as

$$q_i = Z^* q_p = Z^*(1.602 \times 10^{-19})\,C, \tag{1.3}$$

$$m_i \cong A m_p = A(1.673 \times 10^{-27})\,kg. \tag{1.4}$$

The *rest energy* of a particle equals the rest mass multiplied by the square of the speed of light. The rest energy of an electron is

$$m_e c^2 = 8.19 \times 10^{-14}\,J.$$

If the kinetic energy of a particle approaches or exceeds its rest energy, we must use relativistic equations of motion. The MKS energy unit of joules is not convenient for individual charged particles. The standard energy unit in beam physics is the *electron volt* (eV). One electron volt equals the change in kinetic energy of an electron or proton that crosses a potential difference of 1 V, or

$$1\,eV = 1.6 \times 10^{-19}\,J. \tag{1.5}$$

The electron rest energy in electron volts is

$$m_e c^2 = 5.11 \times 10^5\,eV = 0.511\,MeV. \tag{1.6}$$

When an electron accelerates through a potential difference of 5.11×10^5 V, the kinetic energy equals the rest energy. Electrons are relativistic when they have kinetic energy above about 100 keV (10^5 eV). The Newtonian equations of motion are approximately correct for electron beams with kinetic energy below this level. The proton rest mass exceeds the electron mass by a factor of 1843—the proton rest energy is correspondingly higher:

$$m_pc^2 = 938\,\text{MeV}. \tag{1.7}$$

Because of the high rest energy, we can use Newtonian dynamics to predict the motion of ions in many applications.

   Although single charged particles may be useful for some physics experiments, we need large numbers of energetic particles for most applications. A flux of particles is a *beam* when the following two conditions hold:

1. The particles travel in almost the same direction.
2. The particles have a small spread in kinetic energy.

A beam is an ordered flow of charged particles. A disordered set of particles, such as a thermal plasma, is not a beam. Figure 1.1 illustrates the difference between a beam and a plasma. The relationship between a charged particle beam and a plasma is analogous to the relationship between a laser and a light bulb. The photons from a laser are directed and monochromatic. The degree of order in a flow of particles is called *coherence*. A high level of coherence is essential for most applications. For example, the minimum spot size of a scanning electron microscopic depends on the parallelism of the electrons in the beam.

   Several quantities are useful to characterize charged particle beams, including (1) type of particle, (2) average kinetic energy, (3) current, (4) power, (5) pulse length, (6) transverse dimension, (7) parallelism, and (8) energy spread. The parameters of charged particle beams for applications extends over a remarkable, range. Table 1.1 gives estimates of high and low values for beam properties. No other field of engineering or applied physics extends over such a broad parameter space.

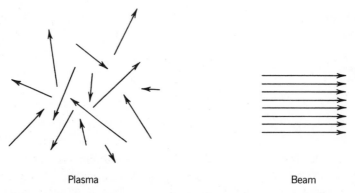

Plasma   Beam

Figure 1.1   Particle orbits in a plasma compared with a beam. The arrows represent velocity vectors.

TABLE 1.1   Charged Particle Beam Parameters for Applications

Property	Lower Limit	Upper Limit	Range
Mass	m_e $(9.1 \times 10^{-31}\,\mathrm{kg})$	$238 m_p$ $(4.0 \times 10^{-25}\,\mathrm{kg})$	10^6
Charge	e $(1.6 \times 10^{-19}\,\mathrm{C})$	$\sim 100 e$ $(1.6 \times 10^{-19}\,\mathrm{C})$	10^2
Kinetic energy	$\leqslant 1\,\mathrm{eV}$	$10^{12}\,\mathrm{eV}$	10^{12}
Current	$10^{-9}\,\mathrm{A}$	$10^6\,\mathrm{A}$	10^{15}
Power	$< 1\,\mathrm{W}$	$10^{12}\,\mathrm{W}$	$> 10^{12}$
Pulse length	$< 10^{-10}\,\mathrm{s}$	Continuous	$> 10^{10}$
Dimension	$10^{-6}\,\mathrm{m}$	$> 1\,\mathrm{m}$	$> 10^6$
Angular divergence	$10^{-6}\,\mathrm{rad}$	1 rad	10^6

In conventional accelerators, particle mass spans the range from electrons to the heavy ions used for nuclear physics and accelerator inertial fusion. The mass of a uranium ion is 238 times that of the proton, or 440,000 times the electron mass. The charge state of particles in most accelerators is $q = \pm e$. Heavy-ion accelerators are an exception. In these machines, highly stripped ions ($Z^* > 50$) result when a medium-energy beam passes through a thin foil. The multiply-charged ions then accelerate to high kinetic energy in a linear accelerator.

The kinetic energy of beams for applications spans about 12 orders of magnitude. At the low end, we shall encounter energies less than 1 eV when we study electron emission from a thermionic cathode. The current achievements of high-energy-physics accelerators define the high end of the energy spectrum, about 1 TeV ($10^{12}\,\mathrm{eV}$).

The beam current in present devices spans an even broader range, $\sim 10^{15}$. Ion and electron microprobes have a current of about $1\,\mathrm{nA} = 10^{-9}\,\mathrm{A}$. Despite the low flux of such beams, we must apply collective beam theory to predict the minimum spot size. At the other extreme, pulsed ion or electron diodes generate beams with current exceeding $1\,\mathrm{MA} = 10^6\,\mathrm{A}$.

To characterize beam power, we must distinguish between average power and peak power. Many accelerators have a pulsed duty cycle. The highest peak power, over $10^{12}\,\mathrm{W}$ for $\sim 50\,\mathrm{ns}$, occurs in experiments on inertial fusion. At the low end, commercial devices such as CRT tubes operate continuously at power levels below 1 W. Continuous machines define the upper limit on beam pulse length. At the lower limit, resonant accelerators generate trains of very short pulses. Pulse durations may be less than $100\,\mathrm{ps} = 10^{-10}\,\mathrm{s}$.

The maximum transverse dimension of charged particle beams is immense if we include astrophysical jets as examples. In conventional applications, industrial sheet beam irradiators create the largest beams, about 2 m in length. Scanning electron microscopes generate small beam spots less than 1 μm in diameter. The parallelism of orbits in beams also has a wide range. Accelerators under development for defense applications have a requirement on angular divergence of about 1 μrad. At the other extreme, intense pinched electron beams may have a divergence angle approaching 1 rad, with a spread in longitudinal kinetic energy comparable to the directed energy.

1.2. METHODS AND ORGANIZATION

The central issue in beam physics is the solution of collective problems involving large numbers of particles. The orbits of the particles depend on electric and magnetic fields. The fields, in turn, result partly from contributions of the beam particles. Therefore, the field values depend on the positions and velocities of all particles. An exact prediction of beam behavior demands the simultaneous calculation of every particle orbit. The challenge is formidable—a low-current beam may contain more than 10^{10} particles. Clearly, exact solutions are impossible, even with the most powerful computers.

Collective physics is a science of approximation. Predictions involve insight and experience—to solve problems, we must eliminate unnecessary material but preserve the essential processes. Beam physics can be difficult for beginning students because there are no cut-and-dried methods. Each calculation demands a careful analysis and a reduction with simplifying assumptions. One goal of this book is to give students insights to resolve collective problems. The material of the book was organized with this goal in mind:

1. The order of topics is from the simplest to the most complex. Ideally, the reader should follow the text from the beginning to the end. The early chapters give background material necessary to understand advanced subjects like beam instabilities.

2. In collective problems, the initial analysis and reduction is as important as the correct mathematical solution of the equations. The best mathematical methods are useless if the statement of the problem is not physically correct. Therefore, we shall concentrate on setting up problems, carefully listing all limiting conditions. After defining the governing equations, we shall apply straightforward mathematical methods to find a solution.

3. A frustrating problem in many advanced works on beam physics is that the derivations often have missing steps. These leaps may be obvious to the author but are obscure to nonexperts. To avoid this difficulty, we shall follow all stages of derivations at the expense of some repetition.

This book is an introductory text. It does not address sophisticated methods of mathematical analysis, the history of beam physics, or the vast range of advanced literature. The references in Appendix 1 are a good starting point for further reading on advanced topics.

The material in Chapters 1–6 is the foundation for later chapters. Chapter 2 is a capsule summary of collective physics with an emphasis on charged-particle-beam theory. Collective physics organizes information about the motion of large numbers of particles. Rather than calculate the orbits of individual particles, we try to identify general trends in behavior. The best way to organize information about particles is to plot orbit vectors in phase space. The theorem of conserved particle density in phase space leads to the fundamental equation of collective physics, the Boltzmann equation. From this relationship, we derive moment equations that describe the conservation of particles, momentum, and energy in large groups.

The introductory accelerator theory of *Principles of Charged Particle Acceleration* concentrates on the orbits of single particles or on laminar beams where all orbits are similar. In Chapter 3, we remove this limitation and study beams where the particles have random spreads in direction and energy. Real beams always have such a diversity of orbits—to design accelerators, we must understand the limitations set by beam imperfections. Chapter 3 defines emittance, a quantity that characterizes the parallelism of beams. The principle of emittance conservation has extensive applications to accelerators and beam optics systems.

Chapter 4 discusses consequences of beam emittance in low-current beams with small space-charge forces. The first three sections define the transport parameters of a beam and review transport theory. This theory is useful for the design of beam-transport systems. Section 4.4 reviews imperfections in charged particle lenses and how they contribute to the growth of beam emittance. The final two sections discuss the importance of beam emittance in storage rings and beam colliders. We shall study methods that circumvent the principle of phase-volume conservation to produce beams with low spreads in direction and energy.

Chapter 5 discusses equilibrium effects of beam-generated electric and magnetic fields. The chapter introduces the idea of self-consistent calculations. In this chapter we follow the motion of beam particles in fields that depend on the instantaneous position of all other particles. The Child derivation is the prototype calculation of a self-consistent beam equilibrium. It leads to the Child limit, a constraint on the current density from a beam extractor. Chapter 6 uses the expressions for the fields generated by equilibrium beams to calculate one-dimensional current flow in several practical cases. The chapter also introduces the KV distribution, a starting point for self-consistent models of two-dimensional equilibria.

Chapters 7 and 8 introduce methods to create beams, while Chapters 9–12 discuss beam transport and acceleration. Chapter 7 deals with electron and ion guns at low to medium current. Sections 7.1–7.3 review design techniques for guns. Section 7.4 discusses electron sources, while Sections 7.5 and 7.6 review ion sources and ion extraction from plasmas. The final two sections describe methods to generate large-area, high-current ion beams.

Chapter 8 is devoted to high-power pulsed electron and ion diodes. These devices use pulsed power technology to generate beams with very high current. Section 8.1 discusses the motion of electrons in crossed electric and magnetic fields—the resulting equations are also useful for conventional devices like the magnetron. The next two sections review the generation of pulsed electron beams. Sections 8.4 and 8.5 discuss two important processes for diode technology, magnetic insulation and plasma erosion. The final four sections cover methods to create pulsed ion beams with current density far beyond the Child limit.

In Chapter 9, we begin the study of beam transport. This chapter discusses the effect of space charge and emittance on beams in conventional accelerators. The beams in these devices are paraxial—particle orbits make small angles with respect to the axis. Sections 9.1–9.3 derive envelope equations for beams in several focusing systems. These equations, based on transverse force balance, are important first-order design tools for beam optics systems. Section 9.4 applies the equations to define the maximum beam current in accelerators. Section 9.5 describes multiple-beam transport, a method to circumvent current limitations. The final section reviews limitations on beam power set by axial space-charge forces.

Chapters 10–12 describe methods to control high-power electron and ion beams. Chapter 10 concentrates on high-current electron beams in high-vacuum regions—the material is useful for applications such as microwave tubes. Solenoid lenses are effective for containment of low-energy electron beams—the first three sections describe electron motion and linear beam propagation in axial magnetic fields. Section 10.5 describes methods to focus relativistic beams with thin foils or meshes. As background, Section 10.4 summarizes the scattering and energy loss of electrons passing through matter. Section 10.6 derives the charge and current distributions induced by beams in surrounding metal structures— the relationships are important for subsequent calculations of beam stability. Sections 10.7–10.9 treat the steering and focusing of high-current electron beams in curved transport systems.

In Chapter 11, we turn our attention to high-current ion beams. Moderate energy ions have low velocity—for the same current and energy, an ion beam has higher space charge than an electron beam. Therefore, it is difficult to transport high-flux ion beams through vacuum. High-current ion beams must be neutralized—the addition of low-energy electrons reduces the beam-generated electric fields. Sections 11.1–11.3 describe methods to add electrons to ion beams. Section 11.4 reviews focal limits on neutralized ion beams, while Section 11.5 describes methods to control and to accelerate high-flux ion beams.

Chapter 12 discusses the propagation of electron beams through plasmas. We shall review the properties of plasmas that affect their response to all types of pulsed beams. Sections 12.1 and 12.2 introduce two basic plasma quantities, the Debye length and the plasma frequency. Section 12.3 applies the theory of plasma oscillations to describe the transverse motion of an electron beam in an ion column. Sections 12.4–12.6 concentrate on plasma responses to pulsed electron beams. Sections 12.7–12.9 review the properties of beam equilibria in plasmas and processes that limit beam current and propagation length.

Chapters 13 and 14 cover instabilities, spontaneous departures from equilibrium driven by the free energy of beams. The theoretical description of instabilities is a challenge—we must handle time-dependent effects of beam-generated electric and magnetic fields with self-consistent methods. The field of beam instabilities is broad. It would take an entire chapter just to list the diverse processes covered in the literature. Instead, we shall concentrate on a few important examples. The detailed discussions illustrate methods and insights that have application to the full range of instability calculations.

Most beam instabilities involve the transfer of axial kinetic energy to undesired random motions. We use the term *transverse instability* when the energy contributes to transverse particle motion. These disturbances can lead to increased emittance or a sweeping motion of the beam. Sections 13.2–13.4 review background material. Section 13.2 classifies transverse beam oscillations in focusing systems. Section 13.3 summarizes the effects of wall resistance and a spread of particle momentum on coherent oscillations. Section 13.4 reviews the theory of transverse resonant modes in accelerator cavities. These modes can effectively couple the beam kinetic energy to transverse oscillations. The other sections in the chapter describe collective instabilities for a broad range of accelerators.

Chapter 14 discusses longitudinal instabilities. Here, the kinetic energy of a monoenergetic beam couples to an axial velocity spread. The resulting momentum dispersion can interfere with beam containment and focusing. Section 14.2 derives reference expressions for axial electric fields in perturbed beams. The other sections contain descriptions of specific instabilities.

In Chapter 15, we shall study an application of charged particle beams, the generation of electromagnetic radiation. The description of radiation sources intimately involves beam theory. Besides reviewing some practical microwave devices, the chapter introduces material of general interest. Section 15.2 covers the use of resonant cavities as impedance transformers. Cavities convert beam energy at high voltage and low current to microwave energy at low voltage and high current. Section 15.3 describes axial beam bunching, a process critical to the operation of the klystron and RF accelerators. The other sections cover several ways to convert the energy of a beam to electromagnetic radiation: the inverse diode, the klystron, the traveling-wave tube, and the magnetron. To conclude the book, Sections 15.7 and 15.8 introduce the theory of the free-electron laser.

1.3. SINGLE-PARTICLE DYNAMICS

In the following chapters, we shall study the behavior of ordered groups of charged particles. Although much of the material is introduced as needed, the reader must have a good preliminary knowledge of single-particle dynamics. The companion book, *Principles of Charged Particle Acceleration*, provides the necessary prerequisites. In this section, we shall summarize important background equations for later reference. The symbol [CPA] appears throughout this book to reference sections of *Principles of Charged Particle Acceleration* with relevant supplementary material.

A. Particle Dynamics

We shall construct theories of collective beam behavior by summing over the orbits of many individual charged particles. We use equations of motion to predict single-particle orbits. Four quantities specify the status of a charged particle: the rest mass, m_0; the charge, q; the vector position, $\mathbf{x}$; and the vector velocity, $\mathbf{v}$. In most of the derivations in following chapters, q and m_0 are constant while $\mathbf{x}$ and $\mathbf{v}$ change. The velocity causes a change in position:

$$dx/dt = \mathbf{v}. \tag{1.8}$$

Although the rest mass of a particle is constant, the special theory of relativity states that the inertia of a particle observed in a frame of reference depends on the magnitude of its speed in that frame. In Cartesian coordinates, the particle speed is

$$v = (v_x^2 + v_y^2 + v_z^2)^{1/2}. \tag{1.9}$$

Relativistic dynamics uses the special function of v:

$$\gamma = \frac{1}{[1 - (v/c)^2]^{1/2}}, \tag{1.10}$$

where c is the speed of light:

$$c = 2.998 \times 10^8 \text{ m/s}. \tag{1.11}$$

The quantity γ is always greater than unity because the observed speed of a particle can never equal or exceed c. Another useful parameter is the ratio of the particle speed to c:

$$\beta = v/c. \tag{1.12}$$

Substituting Eq. (1.12) into Eq. (1.10) gives:

$$\gamma = (1 - \beta^2)^{-1/2}, \tag{1.13}$$

$$\beta = (1 - 1/\gamma^2)^{1/2}. \tag{1.14}$$

The inertia of a particle is proportional to γ. The apparent mass is:

$$m = \gamma m_0. \tag{1.15}$$

The particle momentum, a vector quantity, equals

$$\mathbf{p} = \gamma m_0 \mathbf{v}. \tag{1.16}$$

In response to a force $\mathbf{F}$, the momentum changes as

$$dp/dt = d(\gamma m_0 \mathbf{v})/dt = \mathbf{F}. \tag{1.17}$$

If the force is a known function of $\mathbf{x}$ and $\mathbf{v}$, we can use numerical methods to calculate the orbit of a particle by a simultaneous solution of Eqs. (1.8) and (1.17). In relativistic dynamics, the total energy of a particle is $mc^2 = \gamma m_0 c^2$. The kinetic energy equals the total energy minus the rest energy:

$$T = (\gamma - 1)m_0 c^2. \tag{1.18}$$

Newtonian dynamics describes the motion of low-energy particles when

$$T \ll m_0 c^2. \tag{1.19}$$

In the nonrelativistic limit, $\gamma \cong 1$. Here, the equations of particle dynamics are

$$d\mathbf{x}/dt = \mathbf{v}, \tag{1.20}$$

$$m \cong m_0, \tag{1.21}$$

$$\mathbf{p} \cong m_0 \mathbf{v}, \tag{1.22}$$

$$d\mathbf{p}/dt \cong m_0 \, d\mathbf{v}/dt = \mathbf{F}, \tag{1.23}$$

$$T \cong m_0 v^2/2. \tag{1.24}$$

Equations (1.20)–(1.24) have a simpler form than the relativistic equivalents—it is usually easier to find analytic solutions to nonrelativistic problems.

Sometimes, we can use modified Newtonian equations to describe the transverse motion of relativistic particles in beams. In most beams, the transverse velocities of particles are much smaller than the axial velocities. We shall consistently use the coordinate z as the average direction of beam motion; therefore, $v_x, v_y \ll v_z$. For transverse particle motion with no acceleration in z, the total particle energy is almost constant. If we take γ as a constant, Eqs. (1.8) and (1.17) become

$$dx/dt = v_x, \tag{1.25}$$

$$dp_x/dt = \gamma m_0 \, dv_x/dt = F_x. \tag{1.26}$$

Similar equations hold for the y direction. Equations (1.25) and (1.26) have the form of Newtonian equations of motion with a modified mass γm_0.

B. Electromagnetic Forces

The motion of charged particles in accelerators depends almost entirely on electromagnetic forces. The Lorentz force expression for a particle with charge q

and velocity **v** is

$$\mathbf{F} = q(\mathbf{E} + \mathbf{v} \times \mathbf{B}).\tag{1.27}$$

The values of the electric and magnetic fields in Eq. (1.27) are evaluated at the instantaneous position of the particle.

We calculate electric and magnetic fields from the Maxwell equations using known distributions of charge and current densities. The charge density ρ is a scalar quantity with units of coulombs per cubic meter. The current density **j** is a vector quantity with units of amperes per square meter. Several sources can contribute to the net charge density, including charges deposited on electrodes by external power supplies ($\rho_{applied}$), displaced charges in dielectric materials ($\rho_{dielectric}$), and the charge of beam particles moving freely through vacuum (ρ_{space}). The total current density may have contributions from applied currents in magnets ($\mathbf{j}_{applied}$), from atomic currents in ferromagnetic materials ($\mathbf{j}_{ferromagnetic}$), and from beam particles ($\mathbf{j}_{space}$). The Maxwell equations are

$$\nabla \cdot \mathbf{E} = (\textstyle\sum \rho)/\varepsilon_0 \tag{1.28}$$
$$= (\rho_{applied} + \rho_{dielectric} + \rho_{space})/\varepsilon_0,$$

$$\nabla \times \mathbf{E} = -\partial \mathbf{B}/\partial t, \tag{1.29}$$

$$\nabla \cdot \mathbf{B} = 0, \tag{1.30}$$

$$\nabla \times \mathbf{B} = \mu_0(\textstyle\sum \mathbf{j}) + (1/c^2)\,\partial \mathbf{E}/\partial t \tag{1.31}$$
$$= \mu_0(\mathbf{j}_{applied} + \mathbf{j}_{ferromagnetic} + \mathbf{j}_{space}) + (1/c^2)\,\partial \mathbf{E}/\partial t.$$

The summation symbols denote the sum of all contributions to the charge and current densities. The constants in Eqs. (1.28) and (1.31) are

$$\varepsilon_0 = 8.85 \times 10^{-12}, \tag{1.32}$$

$$\mu_0 = 1.26 \times 10^{-6}, \tag{1.33}$$

and

$$c = 1/(\mu_0 \varepsilon_0)^{1/2}. \tag{1.34}$$

Often, the portion of the electric field created by charges in dielectric materials is *linearly* proportional to the total electric field in the material. In this case Eq. (1.28) becomes,

$$\nabla \cdot \mathbf{E} = \rho_{free}/\varepsilon = (\rho_{applied} + \rho_{space})/\varepsilon, \tag{1.35}$$

where ε is a constant that depends on the material, $\varepsilon \geqslant \varepsilon_0$. The symbol ρ_{free}

represents all contributions to the charge density except the charges bound in the dielectric. We can make similar definitions for the effect of atomic currents in ferromagnetic materials. In the special case that field components arising from dielectric and ferromagnetic materials are linearly proportional to the total fields in the materials, we can write the Maxwell equations in an alternative form:

$$\nabla \cdot \mathbf{E} = \rho_{\text{free}}/\varepsilon, \tag{1.36}$$

$$\nabla \times \mathbf{E} = -\partial \mathbf{B}/\partial t, \tag{1.37}$$

$$\nabla \cdot \mathbf{B} = 0, \tag{1.38}$$

$$\nabla \times \mathbf{B} = \mu \mathbf{j}_{\text{free}} + \varepsilon\mu\, \partial \mathbf{E}/\partial t. \tag{1.39}$$

The constant μ depends on the magnetic properties of the material; here, $\mathbf{j}_{\text{free}}$ is the total current density from all sources except the ferromagnetic material. We can characterize dielectrics in terms of the relative dielectric constant:

$$\varepsilon_r = \varepsilon/\varepsilon_0, \tag{1.40}$$

and magnetic materials in terms of the relative magnetic permeability:

$$\mu_r = \mu/\mu_0. \tag{1.41}$$

C. Coordinate Transformations

In derivations of the following chapters, it is useful to change between different frames of reference. We often define two special frames. The *stationary frame* is the rest frame of the physical devices that accelerate and confine a beam. In this frame, the beam moves at average velocity $v_z = \beta_z c$. The *beam rest frame* moves at velocity $\beta_z c$ relative to the stationary frame. In this frame, particles are at rest if the beam has no axial velocity spread.

The Lorentz transformations relate position and velocity between two frames of reference in relative motion. Suppose that we determine the position and velocity of a particle in a frame that we consider stationary—the measured quantities are (x, y, z, v_x, v_y, v_z). Consider the viewpoint of an observer who moves relative to our frame at speed βc in the $+z$ direction. The observer measures the position and velocity of the particle as x', y', z', v'_x, v'_y, v'_z). The Lorentz transformations give relationships between the quantities measured in the two frames of reference:

$$x' = x, \tag{1.42}$$

$$y' = y, \tag{1.43}$$

$$z' = \gamma(z - \beta ct), \tag{1.44}$$

$$t' = \gamma(t - \beta cz/c^2), \tag{1.45}$$

$$v'_x = \gamma v_{x'} \tag{1.46}$$

$$v'_y = \gamma v_{y'} \tag{1.47}$$

$$v'_z = (v_z - \beta c)/(1 - v_z\beta/c), \tag{1.48}$$

where

$$\gamma = (1 - \beta^2)^{-1/2}.$$

Equations (1.42)–(1.48) hold if we define the origin of time so that $z = z'$ at $t = t' = 0$.

Some derivations in later chapters are simplified by transforming electric and magnetic fields between frames in relative motion. Suppose we measure the quantities $(E_x, E_y, E_z, B_x, B_y, B_z)$ in a frame that we consider stationary. In a frame in relative motion, the measured field components are $(E'_x, E'_y, E'_z, B'_x, B'_y, B'_z)$. If the frame moves at relative velocity $\mathbf{v} = v_z\mathbf{z} = \beta c\mathbf{z}$, the electric and magnetic fields are related by

$$E'_x = \gamma(E_x - \beta cB_y), \tag{1.49}$$

$$E'_y = \gamma(E_y + \beta cB_x), \tag{1.50}$$

$$E'_z = E_z, \tag{1.51}$$

$$B'_x = \gamma(B_x + \beta E_y/c^2), \tag{1.52}$$

$$B'_y = \gamma(B_y - \beta E_x/c^2), \tag{1.53}$$

$$B'_z = B_z. \tag{1.54}$$

D. Transfer Matrices

Most beam-transport devices, such as charged particle lenses and bending magnets, apply transverse forces that are linearly proportional to the distance of a particle from a preferred axis. Suppose a device produces a linear transverse force in the x direction over an axial length. We want to compare the particle orbit at the exit of the device to the orbit at the entrance. To specify the orbit of the particle in the x direction, we must give its position x and velocity v_x. The convention in charged particle optics is to represent particle orbits in terms of their angle relative to the main axis, rather than the transverse velocity. In the limit that $v_x \ll v_z$, the angle is

$$x' = dx/dz \cong v_x/v_z. \tag{1.55}$$

We can symbolize the entrance orbit as a vector, $[x_0, x_0']$. The exit vector is $[x_1, x_1']$. If the x-directed forces in the device are linear, then we can express the exit vector as a linear combination of the entrance vector components:

$$x_1 = a_{11}x_0 + a_{12}x_0', \tag{1.56}$$

$$x_1' = a_{21}x_0 + a_{22}x_0.$$

In matrix notation, the relationship is

$$\begin{bmatrix} x_1 \\ x_1' \end{bmatrix} = \begin{bmatrix} a_{11} & a_{12} \\ a_{21} & a_{22} \end{bmatrix} \begin{bmatrix} x_0 \\ x_0' \end{bmatrix}. \tag{1.57}$$

The quantities a_{mn} depend on the distribution of forces. The focusing effect of any one-dimensional linear device is specified by the four numbers a_{mn}. The matrix of Eq. 1.57 is the *transfer matrix* of the device. Without acceleration, the determinant of the transfer matrix equals unity:

$$\det \mathbf{A} = a_{11}a_{22} - a_{12}a_{21} = 1. \tag{1.58}$$

If a particle travels through linear device A and then through device B, the final orbit vector is

$$\begin{bmatrix} x_2 \\ x_2' \end{bmatrix} = \begin{bmatrix} b_{11} & b_{12} \\ b_{21} & b_{22} \end{bmatrix} \begin{bmatrix} a_{11} & a_{12} \\ a_{21} & a_{22} \end{bmatrix} \begin{bmatrix} x_0 \\ x_0' \end{bmatrix}, \tag{1.59}$$

or

$$x_2 = \mathbf{B}(\mathbf{A}\,x_0) = (\mathbf{B}\,\mathbf{A})x_0 = \mathbf{C}x_0. \tag{1.60}$$

The quantity $\mathbf{C}$ is the matrix product of $\mathbf{B}$ times $\mathbf{A}$. The orbit vector transformation from any combination of one-dimensional focusing elements is a single transfer matrix, the product of the individual matrices of all the elements.

The particle orbit vector for a two-dimensional focusing system is $\mathbf{x} = [x, x', y, y']$. A 4×4 matrix represents the effect of a general linear focusing element or system. In many practical devices, such as quadrupole lens arrays, the forces in the x and y directions are independent. Then, we can calculate motion in x and y separately using individual 2×2 matrices.

E. Periodic Focusing Systems

Most high-energy accelerator systems use quadrupole lens arrays for focusing. The transverse forces in a quadrupole array must vary periodically along the axis for net focusing in both the x and y directions. A general periodic focusing system consists of a repeated set of focusing elements and drift spaces. The smallest periodic subset is called a *focusing cell*. For example, the cell of a quadrupole lens array may consist of a focusing lens, a defocusing lens, and intervening drift spaces. If transverse forces are linear, each element of the cell has a transfer matrix. We can find the transfer matrix **M** for the complete cell by multiplication of the matrices of individual elements.

If x_0 is an orbit vector at the entrance to a focusing cell, the orbit at the entrance to the next cell is $x_1 = Mx_0$. At the end of n cells, the orbit vector is $x_n = M^n x_0$—the notation means the matrix multiplication of **M** by itself n times. Through analysis of the matrix power operation, we can show that if $x_0' = 0$ the orbit displacements at the cell entrances follow the equation

$$x_n = x_0 \cos(n\mu_0). \tag{1.61}$$

The quantity μ_0 is the *vacuum phase advance* per cell. It depends on the variation of forces in the focusing cell devices through the equation

$$\mu_0 = \cos^{-1}(\mathrm{Tr}\,M/2). \tag{1.62}$$

An orbital instability occurs in the periodic system when

$$|\det M/2| > 1. \tag{1.63}$$

With a continuous linear focusing force, all particles oscillate harmonically in the transverse direction at ω_b, the *betatron frequency*. The particles trace out harmonic curves as they move axially. The axial wavelength of the orbit traces is the *betatron wavelength*

$$\lambda_b = \beta c(2\pi/\omega_b). \tag{1.64}$$

In a periodic system with axially varying forces, particles do not follow perfect sinusoidal orbits. Nonetheless, Eq. (1.61) shows that the locus of particle displacements *at the cell boundaries* is a harmonic curve. If we disregard small-scale motions within each cell, we can define an effective betatron wavelength in periodic systems. If the cell has length L, then the actual orbit intersects the curve:

$$x(z) = x_0 \cos(\mu_0 z/L). \tag{1.65}$$

The effective betatron wavelength in a periodic focusing system is related to the phase advance per cell by

$$\lambda_b = L(2\pi/\mu_0). \tag{1.66}$$

F. Phase Dynamics

Despite differences in geometry, all radio-frequency accelerators use a traveling electromagnetic wave to accelerate charged particles. For ion acceleration, the axial component of the electric field on the axis has the form

$$E_z(z, t) = - E_0(z) \sin [k(z)z - \omega t].$$ (1.67)

Figure 1.2 shows the axial variation of the electric field of Eq. (1.67) at time $t = 0$. The frequency ω is constant through the length of the accelerator, while the wave number $k(z)$ may vary. In the following discussion, we assume that the magnitude of the electric field E_0 is constant in z. The wave can accelerate particles to high energy only if they stay within the region of accelerating electric field. In other words, the particles must remian at about the same phase of the accelerating wave. This means that the wave phase velocity must increase to match the velocity of the accelerating particles.

Figure 1.2 defines the phase of a particle with respect to a traveling wave. A particle with zero phase, $\phi = 0$, sees no axial electric field. The wave accelerates particles with phase in the range $0 < \phi < \pi$ and decelerates particles in the phase range $0 > \phi > - \pi$. We can define conditions where a particle stays at a constant phase. A particle with this property is a *synchronous particle*—its phase is the *synchronous phase*, ϕ_s. Figure 1.2 shows that the synchronous particle experiences a constant axial electric field, $E_{zs} = E_0 \sin \phi_s$. Limiting attention to nonrelativistic ions, the synchronous particle velocity changes as

$$dv_s(z)/dt = (qE_0 \sin \phi_s)/m_0.$$ (1.68)

The accelerating structure must vary along its length so that the wavenumber is

$$k(z) = \omega/v_s(z).$$ (1.69)

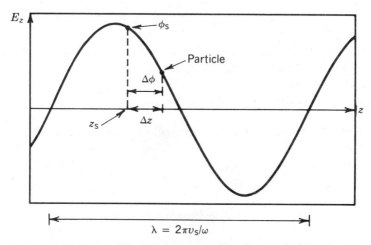

Figure 1.2 Axial variation of the longitudinal electric field of a traveling wave at a given time. The figure shows the definition of particle phase.

An accelerator has a synchronous particle if Eqs. (1.68) and (1.69) hold. If a synchronous particle exists, we can show that under some conditions nonsynchronous particles have stable oscillations about the synchronous particle position z_s. This is an essential requirement for beam acceleration because we can never create a singular distribution of perfectly synchronized particles. Let z and v be the axial position and velocity of a nonsynchronous particle. We define the small quantities

$$\Delta z = z - z_s, \tag{1.70}$$

$$\Delta v = v - v_s,$$

and

$$\Delta \phi = \phi - \phi_s.$$

Inspection of Fig. 1.2 shows that

$$\Delta \phi / 2\pi = -\Delta z / \lambda = -k\Delta z / 2\pi = -\Delta z / (2\pi v_s / \omega). \tag{1.71}$$

The quantity λ is the local wavelength of the traveling wave. The instantaneous acceleration of the nonsynchronous particle is

$$d^2 z / dt^2 = dv/dt = (qE_0 \sin \phi)/m_0. \tag{1.72}$$

We can combine Eqs. (1.70)–(1.72) into the general phase equations for nonrelativistic particles:

$$\phi = \phi_s - \omega \Delta z / v_s, \tag{1.73}$$

$$d^2 \Delta z / dt^2 = (qE_0 \sin \phi)/m_0 - d^2 z_s / dt^2$$
$$= (qE_0/m_0)(\sin \phi - \sin \phi_s). \tag{1.74}$$

Numerical solutions of Eqs. (1.73) and (1.74) describe the axial motion of particles when E_0 and v_s vary with z. If E_0 and v_s are almost constant during an axial oscillation of a nonsynchronous particle, then we can combine Eqs. (1.73) and (1.74) to give the familiar nonlinear differential equation:

$$d^2 \phi / dt^2 \cong -(\omega q E_0 / m_0 v_s)(\sin \phi - \sin \phi_s). \tag{1.75}$$

For small oscillations about the phase of the synchronous particle, $\Delta \phi \ll \phi_s$, Eq. (1.75) reduces to

$$d^2 (\Delta \phi) / dt^2 \cong -(\omega q E_0 / m_0 v_s)(\cos \phi_s) \Delta \phi. \tag{1.76}$$

The axial oscillations of nonsynchronous particles are stable if $\cos \phi_s > 0$. The conditions for synchronized particles acceleration are $\cos \phi_s > 0$ and $\sin \phi_s > 0$, or

$$0 < \phi < \pi/2 \quad [\text{acceleration}], \tag{1.77}$$

A traveling wave can also decelerate particles—this is the basis for many microwave devices and the free-electron laser. The conditions for synchronized deceleration are $\cos \phi_s > 0$ and $\sin \phi_s < 0$, or

$$0 > \phi > -\pi/2 \quad [\text{deceleration}]. \tag{1.78}$$

2

Phase-Space Description of Charged Particle Beams

This chapter introduces theoretical tools for application in the rest of this book. We shall review methods to predict the average behavior of large numbers of particles. The emphasis is on beams, where the particles have high kinetic energy, have good directionality, and may be relativistic.

Section 2.1 discusses the representation of particle orbits in phase space. For nonrelativistic particles, phase space is a six-dimensional space with axes in space (x, y, z) and velocity (v_x, v_y, v_z). At a given time, a point in phase space represents the complete parameters of a particle orbit. As time evolves, the orbit point of a particle traces out a trajectory $[x(t), y(t), z(t), v_x(t), v_y(t), v_z(t)]$. We adopt the phase space viewpoint because plots of multiple-particle trajectories are more orderly than the familiar orbit plots in conventional space, $[x(t), y(t), z(t)]$.

Section 2.2 introduces the distribution function, a method to organize information on large numbers of particle orbit points in phase space. The distribution function is a record of particle orbit coordinates—it changes with time as particles move. This section defines both the discrete distribution function and the continuous distribution function. The discrete function is the foundation for computer simulations, while the continuous function is the basis for analytic theories of collective behavior.

Section 2.3 gives a general discussion of self-consistent-orbit calculations with beam-generated forces. The section concentrates on numerical methods as an illustration of a strategy to advance orbits and fields simultaneously. We divide time into small steps. At each step, we advance the orbits and calculate the resulting fields. The process converges toward the actual solution if the forces acting on particles are almost constant over the time step. The section reviews a useful numerical technique for orbit calculations, the leapfrog method.

Section 2.4 applies the leapfrog method to prove the theorem of phase-volume conservation. The theorem states that the volume occupied by a collection of orbit points in phase space is constant in time when there are no collisions. Here, the term "collision" means a short-range force that varies over a length comparable to the spacing between particles. An equivalent statement of the theorem is that the distribution function is constant if we follow the orbit of a particle. Phase-volume conservation is the fundamental principle of beam physics. From it we derive the basic equations and the principle of emittance conservation (Chapters 3 and 4).

Sections 2.5 and 2.6 illustrate the use of the distribution function. Section 2.5 discusses macroscopic quantities. These are the measurable properties of a collection of particles, such as density and average velocity. We shall see how to calculate these quantities as velocity-space averages over the distribution function. Section 2.6 reviews the properties of a specific function, the Maxwell distribution. The distribution describes a collection of particles in thermal equilibrium—we shall encounter it several times in following chapters. For reference, the section gives several macroscopic averages of the Maxwell distribution.

Section 2.7 derives the collisionless Boltzmann equation. The equation describes the evolution of particle orbit points in phase space in the fluid approximation. The equation follows from the demonstration in Section 2.4 that the phase fluid is incompressible. The Vlasov equation is a special case of the collisionless Boltzmann equation where only electric and magnetic forces act on particles. The Vlasov equation is the fundamental relationship of beam physics.

For self-consistent calculations, we must find the electric and magnetic fields associated with a distribution of particles. Section 2.8 shows how to calculate the space-charge density and current density of a beam from the distribution function. We shall also review the basis of the phase-fluid model and the validity limits on the Vlasov equation. The models assume that field variations are smooth—variations of the field are small over a length comparable to the distance between particles.

Section 2.9 reviews computer simulations of beams. The material illustrates the physical meanings of distribution functions and the Vlasov equation. Simulations are concrete and easily visualized examples of collective-particle motion. They help us to see the distribution function as a practical method of particle bookkeeping rather than an abstract theoretical concept. We shall review the particle-in-cell method, the basis for simulations of high-current beams.

We derive the moment equations in Section 2.10 by taking velocity averages over the collisionless Boltzmann equation. The process generates simple and useful equations by removing detailed information about the velocity distribution. We shall use the moment equations extensively, particularly in treating macroscopic beam instabilities where the orbits of all particles move in a similar way. Section 2.11 discusses the consequences of velocity dispersion in a beam distribution. We can represent the effect of a velocity spread as an average pressure force. Because we often think of pressure in terms of collisional particles such as gases, we must carefully review the nature of the pressure force in a collisionless distribution.

To conclude the chapter, Section 2.12 describes the properties of relativistic particle distributions, a subject critical to charged particle beams. We must define phase space in terms of position and momentum rather than velocity, $[x, y, z, p_x, p_y, p_z]$. If we adopt this convention, we can preserve the principle of phase-volume conservation and derive a modified Vlasov equation for relativistic particles.

2.1. PARTICLE TRAJECTORIES IN PHASE SPACE

The motion of a single charged particle is usually portrayed in three-dimensional geometric space, or *configuration space*. The particle follows a trajectory described by the vector $(x(t), y(t), z(t))$. The position vector is calculated from the velocity $(v_x(t), v_y(t), v_z(t))$—the velocity reflects the action of applied forces. The configuration-space representation is not an effective way to portray the behavior of large numbers of particles. For example, Fig. 2.1 shows trajectories of particles in a beam confined by a nonlinear transverse force with a distribution of amplitudes and phases. In configuration space, the motion appears disordered. From an inspection of the confused picture, it is difficult to envision general methods to predict the behavior of the beam as it moves downstream. The only option appears to be direct solution of a large number of individual orbits.

In order to develop theoretical tools for the description of large numbers of particles, we need a more ordered picture of particle dynamics. We can improve the representation by representing particle trajectories in a six-dimensional space with axes in both space and velocity (x, y, z, v_x, v_y, v_z). This mathematical space is called *phase space*. At a particular time, each particle in a beam is represented as a single point in phase space. Although it is impossible to display all six phase-space dimensions simultaneously, we can gain insight into phase-space dynamics by viewing the projection of particle motion in a single direction (x, v_x). Then, the evolution of particle orbits can be displayed on a two-dimensional plot.

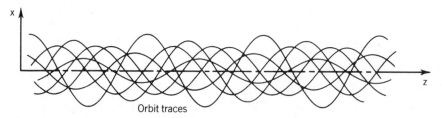

Figure 2.1   The orbits of particles in a beam portrayed in configuration space. The particles move in a nonlinear transverse force with a mixture of amplitudes, phases, and oscillation frequencies.

Throughout this book, we will use two-dimensional plots to gain physical intuition before addressing general results for six-dimensional phase space.

When plotted in phase space, the trajectories of large numbers of particles have a high degree of order if forces vary smoothly in space. Without collisions, two particles that start out with similar phase-space coordinates will always be neighbors. This follows from the fact that the particles have about the same initial position and velocity and are influenced by similar forces. Another consequence is that the trajectories of particles acted on by smooth forces never cross in phase space. As two trajectories approach, the forces acting on both particles becomes almost identical. The implication is that the trajectories of particles in phase space are *laminar*. The trajectories follow nonintersecting streamlines, as shown in Fig. 2.2*a*. Particle orbits plotted in configuration space clearly do not have this property (Fig. 2.1). Laminar phase flow is the foundation for theories of collective behavior. Application of the laminar flow condition leads to equations for large numbers of particles similar to those that describe fluids (Section 2.7). Because of the adherence to fluid equations, the phase-space coordinates for a collection of particles is often called a *phase fluid*.

The criterion for laminar flow is that the forces acting on particles are smooth. A force is *smooth* when the scale length for spatial force variations is long compared with the distance between particles. Similarly, if the force depends on velocity, the velocity scale for variations of a smooth force is larger than the velocity difference of adjacent particles in phase space. The applied forces in accelerators are usually smooth. For example, transverse focusing forces vary over scale lengths comparable to a beam radius, ~ 0.01 m. In contrast, the spacing between particles in a 1-A, 100-keV electron beam of radius 0.01 m is only about 10^{-3} cm.

Short-range forces between particles are generally called collisions. When particles collide with a background or with one another there is a discrete change

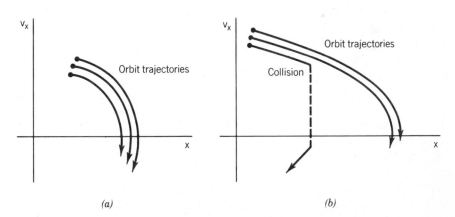

(a)   (b)

Figure 2.2 Representation of particle motions in phase space. (*a*) Laminar phase-space trajectories of particle-orbit vectors with no collisions. (*b*) Effect of a collision on the phase-space position of a particle-orbit vector.

in phase-space coordinates. Two particles that are phase-space neighbors may be widely separated in velocity space after one particle suffers a collision. Figure 2.2*b* shows phase-space trajectories in the presence of collisions. The trajectories are no longer laminar. In most charged-particle-beam accelerators and transport systems, the effect of collisions is small. Changes in phase trajectories from collisions usually take place slowly compared with the motion of particles under the action of long-range forces.

To gain an intuition for phase-space trajectories, we shall review some examples. To begin, consider the coordinate trajectories of particles accelerated by a constant axial electric field E_z. We shall neglect transverse motions of particles and concentrate on the $z - v_z$ phase-space plane. The axial position and velocity of nonrelativistic particles are given by the parametric equations

$$z(t) = z_0 + v_{z0}t + (qE_z/m_0)t^2/2, \qquad (2.1)$$

$$v_z(t) = v_{z0} + (qE_z/m_0)t. \qquad (2.2)$$

Equation (2.1) and (2.2) represent a parabola in phase space. Figure 2.3*a* illustrates some orbits of protons in an electric field of 10^5 V/m. Note that trajectories are laminar with no crossings. Adjacent particles are always localized in phase space even though they undergo substantial acceleration.

As a second example, consider particles acted on by a spatially varying transverse focusing force of the form

$$F_x = -ax. \qquad (2.3)$$

The linear force variation of Eq. (2.3) is important in charged-particle-beam applications. Almost all accelerator focusing devices exert approximately linear transverse forces. The orbits of nonrelativistic particles are described by

$$x(t) = x_0 \cos(\omega t + \phi), \qquad (2.4)$$

$$v_x(t) = -x_0 \omega \sin(\omega t + \phi), \qquad (2.5)$$

where

$$\omega = (a/m_0)^{1/2}. \qquad (2.6)$$

The phase-space trajectories of particle orbits are ellipses (Fig. 2.3*b*) The phase-space rotation frequency ω is the same for all particles. Individual trajectories may vary in amplitude x_0 or phase ϕ. A collection of particles with different oscillation amplitudes follows a nested set of elliptical trajectories. A set of particles with the same amplitude but different phase follow each other along the same ellipse. The trajectories neither overtake nor cross one another.

Analysis of relativistic particles is performed in a phase space with axes of spatial coordinates and momentum (x, y, z, p_x, p_y, p_z), rather than velocity. Velocity is not a useful quantity to characterize relativistic orbits, since all

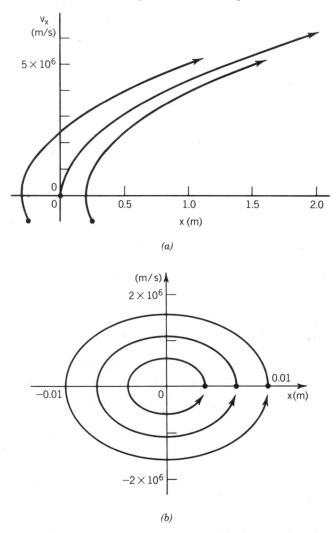

Figure 2.3 Representation of particle motions in phase space. (*a*) Longitudinal motion ($z-v_z$) of protons accelerated by an axial electric field, $E_z = 10^5$ V/m. (*b*) Transverse motion ($x-v_x$) of particles in a linear focusing force: 300-keV protons with a betatron wavelength of 0.3 m.

particles travel at approximately the speed of light when their kinetic energy is comparable to or greater than their rest energy. Representation of relativistic particle distributions in $(\mathbf{x}, \mathbf{p})$ phase space leads to fluidlike equations for collective behavior (Section 2.12). To illustrate the phase dynamics of high-energy particles, we shall consider the relativistic harmonic oscillator. Here, the change of kinetic energy of a particle moving in the linear force of Eq. (2.3) is comparable to or greater than the rest energy—we must account for a change in the relativistic mass factor γ. We should note that this derivation is seldom applicable to the transverse oscillation of charged particles in beams. In a typical beam, transverse particle motion accounts for only a small fraction of the particle's kinetic energy and the quantity γ is almost constant.

The relativistic equations of motion are

$$dx/dt = p_x/\gamma m_0, \tag{2.7}$$

$$dp_x/dt = -ax. \tag{2.8}$$

Momentum is related to γ by

$$(\gamma m_0 c^2)^2 = (cp_x)^2 + (m_0 c^2)^2. \tag{2.9}$$

Equation (2.7) can be written as

$$\frac{dx}{dt} = \frac{p_x}{m_0[1 + p_x^2/m_0^2 c^2]^{1/2}}. \tag{2.10}$$

We shall rewrite Eqs. (2.8) and (2.10) in dimensionless form to illustrate scaling parameters for the solution. Assume a maximum amplitude for particle oscillations x_0 and define the dimensionless variable $X = x/x_0$. For relativistic particles, a good choice for the scaling velocity is c, the speed of light. The time scale of interest is roughly x_0/c—we define a dimensionless time variable $\tau = t/(x_0/c)$. Finally, inspection of Eq. (2.10) shows that a good choice for the dimensionless momentum is $P_x = p_x/m_0 c$. The reduced equations of motion are

$$dX/d\tau = P_x/[1 + P_x^2]^{1/2}, \tag{2.11}$$

$$dP_x/d\tau = -\alpha X, \tag{2.12}$$

where

$$\alpha = ax_0^2/m_0 c^2. \tag{2.13}$$

The single governing parameter α is equal to twice the particle potential energy at x_0 divided by the rest energy. The quantity α indicates the regime of particle dynamics—motion is relativistic when $\alpha > 1$.

Figures 2.4a and 2.4b show numerical solutions for configuration space orbits. The quantity $X(\tau)$ is plotted for $\alpha = 0.5$ and 5.0. As expected, motion is nearly sinusoidal when α is small. At high values of α, the orbit approaches a sawtooth function. This behavior results from the fact that the particle travels close to the velocity of light most of the time. The velocity is small compared with c only at the turning points. In contrast to nonrelativistic particles, the transverse oscillation frequency of relativistic particles in a linear force is amplitude dependent. This property can reduce the growth of resonant instabilities in high-current relativistic beams. Figure 2.4c shows the nonelliptical phase-space trajectory of the relativistic particle with $\alpha = 5$.

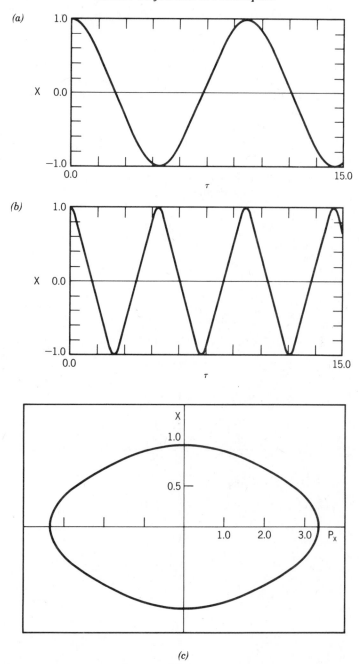

Figure 2.4 Oscillation of a relativistic particle in a linear force. (*a*) Variation of normalized transverse position with normalized time, $X(\tau)$, for $\alpha = ax_0^2/m_0c^2 = 0.5$. (*b*) Plot of $X(\tau)$ for $\alpha = 5.0$. (*c*) Phase space plot of a particle-orbit-vector trajectory. Normalized position versus normalized momentum, $X - P_x$, for $\alpha = 5.0$.

2.2. DISTRIBUTION FUNCTIONS

Collective physics predicts the behavior of systems with multiple interacting components. The main task is keeping track of large numbers of items. Usually, it is impossible to monitor individual components—the amount of information is overwhelming. Two procedures are commonly used to reduce large data sets to a comprehensible level:

1. Identification of trends in behavior or other relationships among components.
2. Development of methods of record-keeping so that information on group behavior can be extracted efficiently.

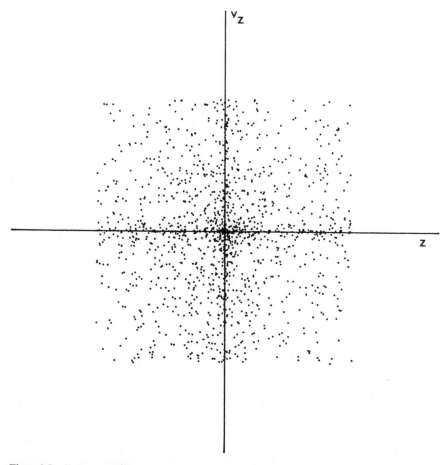

Figure 2.5 Computer-generated one-dimensional distribution—2000 particle-orbit vectors. Random distribution in position and velocity with the weighting function $f(x, v_x) \sim \cos(\pi x/2x_{max})\cos(\pi v_x/2v_{xmax})$.

In the first procedure, components are assigned to groups with similar properties. This makes it easier to predict average properties by studying group behavior. We have already made a start in this direction by adopting the phase-space description of particle dynamics. Particles in a particular phase-space region all behave about the same way.

Regarding the second technique, the function of record-keeping in charged particle physics is served by the *distribution function*. The distribution function is a database of information on particles from which predictions of average behavior can be extracted. It is a time-resolved record of how particles occupy phase space. The distribution function is more than a simple list of properties— the function itself obeys a fluid equation and is constrained by conservation principles. As a result, we can often study the evolution of the distribution function to predict the behavior of a beam rather than follow the orbits of individual particles.

To visualize distribution functions, we will concentrate on particle motion in a single direction represented by a two-dimensional phase space. We begin by studying nonrelativistic particles in z-v_z space. Section 2.12 treats relativistic particle beams. At time t, four quantities represent all information about a particle: q, m_0, $z(t)$, and $v_z(t)$. Assume that all particles in a collection have the same charge q and rest mass m_0. In nonrelativistic beams, these quantities are constant in time. Therefore, all information about a collection of particles at time t is contained in a plot of the coordinate points $[z, v_z]$. Figure 2.5 represents such a plot. The collection of coordinate points in the figure is called the *particle distribution* at time t.

Another way to represent information about the particle distribution is to divide the phase-space region of interest into boxes with dimensions Δz and Δv_z (Fig. 2.6). The process is called defining a *mesh* in phase space. Given a mesh, we can define the *discrete distribution function F*. The function is simply the number of particle coordinates in each box. The discrete distribution appears in computer simulations of beam dynamics. In a computer memory, the discrete function has the form of a table of integer numbers. For a two-dimensional distribution, the discrete function is an $M \times N$ matrix, where MN is the total number of phase-space boxes defined. The full six-dimensional phase-space representation of a beam is stored in a computer in a six-dimensional matrix.

For one-dimensional motion, the discrete distribution function is denoted

$$F(z, v_z, t) = F(m\Delta z, n\Delta v_z),$$

where z and v_z are the coordinates at the center of a box. Note that F is a function of time, since particles may move in and out of a box if they travel along the z axis or undergo acceleration. The discrete distribution function for the outlined box in Fig. 2.6 is $F = 5$.

To understand the utility of the discrete distribution function, we must ask what information F contains about the beam. If Δz and Δv_z are made small, the boxes may contain the coordinates of a single particle or none at all. In the limit

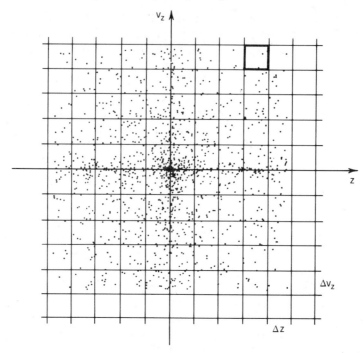

Figure 2.6 Phase space mesh to define the discrete distribution function, F. The function has the value $F = 5$ in the outlined area element.

that Δz and Δv_z approach zero, the function F gives detailed information on the velocity and position of every particle. In principle, a knowledge of F at time t is sufficient to predict the beam behavior exactly at all following times. In the opposite limit, where Δz and Δv_z are large, each box contains many particle coordinates. On such a coarse mesh, the function F tells how many particles have position within a range $\pm \Delta z/2$ about z and velocity $\pm \Delta v_z/2$ about v_z. Although the discrete distribution function on a coarse mesh may contain useful information about average beam properties, it cannot predict each particle orbit in detail.

A large amount of information is required to define F on a fine mesh. The amount of information may be too extensive to obtain or to analyze. Furthermore, most applications do not call for such detailed knowledge. For example, we saw in Section 2.1 that particles with phase-space coordinates that are close follow almost the same trajectories when applied forces are smooth. For this condition, a mesh division with dimension Δz comparable to the distance for force variations may be sufficient.

The boxes on a coarse mesh contain many particles. When forces vary smoothly, the average behavior of particles in the box can be determined by calculating the orbit of a single test particle. This simplification substantially reduces the number of calculations to predict the evolution of the distribution. The test-particle approximation is the basis of computer simulations. The criterion for an accurate analysis is that Δz and Δv_z are small enough to resolve major physical processes. For example, if a beam is subject to a filamentation instability, where the fastest growth occurs at a wavelength of 0.01 m, then the dimension of the box along z should probably be smaller than 1 mm.

When particle motions along the three coordinate axes are not independent, the definition of the discrete distribution function must be extended. For example, when a solenoid lens is combined with a quadrupole transport system, motions in the x and y directions are coupled. The discrete distribution function, denoted $F(x, v_x, y, v_y)$, is referenced to a mesh with box dimensions $(\Delta x, \Delta v_x, \Delta y, \Delta v_y)$. The function F gives the number of particles within a range $(\pm \Delta x/2, \pm \Delta v_x/2, \pm \Delta y/2, \pm \Delta v_y/2)$ of the point (x, v_x, y, v_y) in the four-dimensional phase space.

Although discrete functions are ideal for computer calculations, analytic treatments of collective beam physics are usually carried out in terms of fluid models. The discrete nature of particles is ignored and the distribution in phase space is approximated as a continuum. The fluid description is valid when the distribution has a high density in phase space. The term *high density* implies that there are many particle orbit points in a phase-space box. Accordingly, we can define a quantity that describes the local density of particle coordinates in phase space, independent of the dimensions of the mesh:

$$f(z, v_z) = \lim_{\Delta z, \Delta v_z \to 0} [F(z, v_z)/\Delta z \Delta v_z]. \qquad (2.14)$$

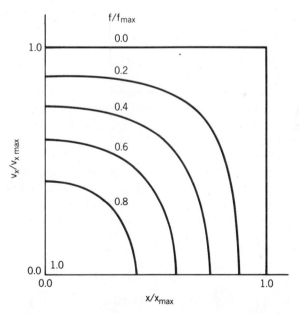

Figure 2.7 Phase space plot of contour lines of the normalized continuous distribution function, $f(x, v_x)/f(0, 0) = \cos(\pi x/2x_{max}) \cos(\pi v_x/2v_{x\,max})$. The function $f(x, v_x)$ is the continuous approximation to the distribution of Fig. 2.5. (One quadrant included.)

The function f is called the *continuous distribution function*, or simply the *distribution function.*

The continuous distribution function has properties that are analogous to the space-charge density ρ in electrostatics. When there are many charged particles in a volume of interest, ρ can be treated as a continuous spatial function, leading to the differential form of the Poisson equation. Similarly, the function f varies smoothly in space and time in the limit of high phase-space density. We will see in Section 2.7 that a fluid equation describes the time evolution of f. This means that an extensive body of knowledge on fluid dynamics can be applied to the analysis of beam behavior. For one-dimensional motion, the continuous distribution function is often represented by a contour plot. Figure 2.7 shows a contour plot of the continuous limit of the discrete function of Fig. 2.6. When there is a mixture of particle species (such as a charge-neutralized beam), each species is represented by a separate distribution function.

2.3. NUMERICAL CALCULATION OF PARTICLE ORBITS WITH BEAM-GENERATED FORCES

As mentioned in Chapter 1, the major challenge in collective physics is the solution of the self-consistent problem. We would like to predict the evolution of a charged particle distribution in which the forces generated by the particles themselves play a role in the dynamics. We shall have an opportunity in following chapters to study several analytic approaches to the problem. These methods are often indirect—they may involve geometric simplifications, recognition of trends in particle orbits, or application of symmetry principles. In this section, we will defer analytic methods and concentrate, instead, on direct numerical solutions. This approach is the most straightforward way to reduce the self-consistent problem to a manageable level. Also, we shall gain an insight into the basis of finite difference solutions of charged particle orbits. These concepts will be applied in Section 2.4 to derive the principle of phase-volume conservation. The section concludes with an illustration of a specific finite difference method, the leapfrog algorithm. This algorithm finds extensive application in computer simulations of beams and plasmas.

Consider the motion of nonrelativistic particles in a beam. When forces between the particles are small compared with externally applied forces, the solution is conceptually simple. Here, we can solve the equations of motion for each particle independently to find the position and velocity. The nonrelativistic equations for the nth particle are

$$d\mathbf{x}_n/dt = \mathbf{v}_n, \tag{2.15}$$

$$d\mathbf{v}_n/dt = \mathbf{F}_e[\mathbf{x}_n(t), \mathbf{v}_n(t), t]. \tag{2.16}$$

The quantity $\mathbf{F}_e$ in Eq. (2.16) is an externally applied force that may depend on time and the current position and velocity of the specific particle. The calculation of Eqs. (2.15) and (2.16) is performed over an interval for all particles to give an accurate prediction of the final distribution.

Calculating the evolution of a beam distribution is more difficult when the particles interact. If the electric or magnetic fields generated by the beam are strong, then the force on a specific particle depends on the location of all other particles in the beam. As a result, it is impossible to determine a test particle orbit directly. The orbit depends on the time-dependent positions and velocities of the other particles in the beam. In turn, these orbits depend on the motion of the test particle. The problem is circular—we cannot find the exact orbits of a collection of particles without a foreknowledge of the solution.

One approach to self-consistent solutions is to use a stepwise approximation—we advance all orbits simultaneously over a small time increment Δt. The resulting orbit predictions converge to the actual solutions if the forces generated by particles satisfy two conditions: (1) they vary over long distance scales compared with the spacing between particles and (2) they change slowly compared with Δt. The conditions imply that the spatial variation of the forces does not change greatly between t and $t + \Delta t$.

A stepwise method to advance a particle distribution with self-consistent forces consists of the following operations:

1. The positions and velocities of all particles $(\mathbf{x}_i, \mathbf{v}_i)$ are specified at an initial time $t = 0$.

2. The velocity of each particle is advanced through the first time step according to

$$\mathbf{v}_n(\Delta t) \cong \mathbf{v}_n(0) + \mathbf{F}[\mathbf{x}_n(0), \mathbf{x}_i(0), 0]\Delta t. \tag{2.17}$$

The notation in Eq. (2.17) suggests that the total force $\mathbf{F}$ depends on the location of all other beam particles at time $t = 0$ and the position of the test particle.

3. The values of $\mathbf{x}_n$ are estimated at $t = \Delta t$ through the equation

$$\mathbf{x}_n(\Delta t) \cong \mathbf{x}_n(0) + \mathbf{v}_n(0)\Delta t. \tag{2.18}$$

4. Since the forces are almost constant over Δt, the positions predicted by Eq. (2.18) will be close to the actual positions of interacting particles. Given the new positions, the electric and magnetic fields of the beam can be estimated at time Δt. This leads to an approximation for the total force at the position of each particle,

$$\mathbf{F}(\mathbf{x}_n, \mathbf{x}_i, \Delta t).$$

5. Steps 2 and 3 are repeated to predict the approximate particle positions and velocities at $2\Delta t$, $3\Delta t$, and so forth The process continues until the particle distribution advances to the final time.

As in any numerical calculation, the procedure involves compromises. A short time step improves the accuracy but increases the number of operations. On the other hand, a large time step leads to inaccuracies that may mask important physical processes. The goal is to extract the maximum amount of information through the minimum number of calculations. The following strategies reduce beam calculations to a level that a computer can accommodate:

1. The physical processes of the problem are analyzed to find Δt small enough to resolve critical beam behavior, but large enough to avoid the generation of needless information.

2. Further analysis is performed to find the minimum number of particle orbits to represent changes in the beam distribution. Representation of average beam behavior by a reduced number of test particles is the basis of computer simulations (Section 2.9).

3. Numerical methods are applied that yield the highest accuracy for a given Δt.

Concerning the third item, we shall study a particular procedure in the remainder of this section. The *leapfrog method* gives accurate predictions of particle orbits with few operations—it is well suited to simulations that involve large numbers of particles. To simplify the discussion, assume that particles move only in the x direction and that the force is a function of position and time, not velocity. The second condition applies, for example, to electric forces in the static limit. The equation of motion for the nth particle in a distribution is

$$\frac{d^2 x_n}{dt^2} = \frac{F_x(x_n, x_i, t)}{m_0}. \tag{2.19}$$

The total force at time t depends on the position of the particle $x_n(t)$ and the positions of all other particles $x_i(t)$. Equation (2.19) separates into two first-order differential equations:

$$\frac{dv_n}{dt} = \frac{F(x_n, x_i, t)}{m_0}, \tag{2.20}$$

$$\frac{dx_n}{dt} = v_n. \tag{2.21}$$

The differential equations are symmetric—the change in position depends on velocity, while the change in velocity depends on position.

Before addressing the simultaneous solution of Eqs. (2.20) and (2.21), it is useful to review the solution of a first-order differential equation of the type

$$\frac{dy}{dt} = F(t). \tag{2.22}$$

The exact change in y over the interval Δt can be written as

$$y(t + \Delta t) - y(t) = \int_{t}^{t+\Delta t} F(t')dt'. \tag{2.23}$$

For a numerical solution, we must find an approximation for the integral on the right-hand side of Eq. (2.23). One choice is

$$\int_{t}^{t+\Delta t} F(t')dt' \cong F(t)\Delta t. \tag{2.24}$$

Equation (2.24) is the basis of the Eulerian difference method [CPA, 115]. The drawback of this approach is that it introduces errors proportional to the first power of the interval Δt. Figure 2.8a illustrates the origin of the error. The rectangle $F(t)\Delta t$ is a poor approximation to the area under the curve.

An improved estimate results if the value of the function is taken at the intermediate time, $t + \Delta t/2$. Here, we have

$$\int_{t}^{t+\Delta t} F(t')dt' \cong F(t + \Delta t/2)\Delta t. \tag{2.25}$$

Inspection of Fig. 2.8b shows that Eq. (2.25) is a better approximation than Eq. (2.24) because first-order errors cancel—the solution is accurate to order Δt^2. The general class of difference methods that use Eq. (2.25) are called *time-centered methods* because the value of F is taken at the midpoint of the interval. The algorithm to advance y is

$$y(t + \Delta t) \cong y(t) + F(t + \Delta t/2)\Delta t. \tag{2.26}$$

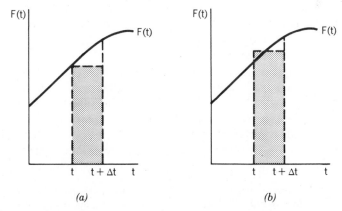

(a)   *(b)*

Figure 2.8 Approximations to the integral $\int F(t)\,dt$ for the numerical solution of the first-order differential equation. (a) Eulerian method. (b) Time-centered method.

Time-centering is simple when F is a known function of time. In the more general case, where F depends on y and t, we must use approximations to estimate $F(y(t + \Delta t/2), t + \Delta t/2)$. One such method, the two-step integration procedure, is described in [CPA, Section 7.6]. The leapfrog method is an alternative that involves fewer calculations—it is well suited to electrostatic particle simulations. At the start of such a simulation, we usually have the set of particle positions and velocities at $t = 0$, $x_i(0)$ and $v_i(0)$. Suppose, instead, that we know the positions at $t = 0$ and the velocities at the displaced time $\Delta t/2$. In other words, the initial quantities are the set $x_i(0)$ and $v_i(\Delta t/2)$.

The first step of the leapfrog method is to advance the positions according to

$$x_n(\Delta t) = x_n(0) + v_n(\Delta t/2)\Delta t. \tag{2.27}$$

Note that Eq. (2.27) has a time-centered form. Once we know the particle positions at $t = \Delta t$, we can find the electric field at time Δt. The field is used to advance the velocities according to

$$v_n(3\Delta t/2) = v_n(\Delta t/2) + F[x_i(\Delta t)]\Delta t/m_0. \tag{2.28}$$

Equation (2.28) is also time-centered. The process extends over additional time intervals, with the velocity and position advancing ahead of each other. Equations (2.27) and (2.28) define the leapfrog method. Despite its simplicity, it achieves second-order accuracy in Δt. The main problem is finding accurate values of $v_i(\Delta t/2)$ to start the calculation if $x_i(0)$ and $v_i(0)$ are known. Often, more complex difference methods are applied over the first half time step.

2.4. CONSERVATION OF PHASE-SPACE VOLUME

Conservation of the phase-space volume occupied by a particle distribution is a fundamental theorem of collective physics.[1] The conservation principle leads to the Boltzmann equation (Section 2.7) and to the fluid moment equations (Section 2.10). Furthermore, the theorem is the basis for the principle of emittance conservation, discussed in Chapter 3.

Figure 2.9 illustrates the physical meaning of phase-volume conservation. Again, we use a two-dimensional phase space for an illustration—the general theorem applies to hypervolumes in a six-dimensional phase space. At time t_1, a set of adjacent coordinate points in phase space represents a collection of particles with similar orbit properties. A boundary, B_1, circumscribes the coordinate points. In the two-dimensional representation of Fig. 2.9, the

---

[1] The principle of the incompressibility of a phase fluid is oten called Liouville's theorem. This is not strictly correct—Liouville's theorem refers to the conservation of the number of possible macrostates of a system of N particles plotted in a $6N$-dimensional phase space. For a complete explanation, see H. Goldstein, *Classical Mechanics*, Addison-Wesley, New York, 1950, p. 266.

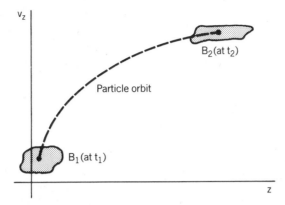

Figure 2.9 Illustration of phase-volume conservation—the boundary around the included group of particle-orbit vectors changes shape with time but has the same area. The dashed line shows the orbit-vector trajectory of a particle at the center of the group.

boundary defines a phase-space area—in the full six-dimensional phase space, the boundary surrounds a hypervolume. At time t_2, the particle positions and velocities have changed. The dashed line in Fig. 2.9 indicates the phase-space motion of a particle orbit vector near the center of the set as time evolves. We know from the properties of phase-space trajectories that the trajectories at t_2 will still be near each other if there are no collisions. The set can be circumscribed by another boundary, B_2. The theorem of phase-volume conservation states that the area (or hypervolume) enclosed by B_2 is equal to the original area inside B_1.

It is important to define carefully the validity conditions for conservation of phase volume. Many important procedures in beam physics, such as beam cooling, depend on violation of the theorem. The following conditions must be satisfied:

1. The particle distribution is dense—the number of particles in a phase-volume element of interest is large. In the high-density limit, the phase-fluid concept is valid, and the continuous distribution function f provides an accurate description.

2. Forces on particles vary smoothly in space and time—there are no collisions. The implication is that proximate particles in phase space always remain close.

3. Although forces may vary with time and position, frictional forces that depend on particle velocity are excluded.

In the following discussion, we will limit attention to nonrelativistic particle motion. This limit applies to low-energy particles or the transverse motion of relativistic particles in the paraxial limit. Section 2.12 extends the analysis to relativistic particles.

If the phase volume occupied by a group of particle orbit vectors remains constant, then the phase-space density in the vicinity of the particles is a conserved quantity. Adopting the phase-fluid viewpoint, conservation of phase volume implies that the fluid is incompressible. By definition, the continuous distribution function equals the density of orbit vectors. The condition of phase-space incompressibility is equivalent to

$$f[x(t), v_x(t), y(t), v_y(t), z(t), v_z(t), t] = \text{constant.} \qquad (2.29)$$

The function is evaluated at the phase-space location $x(t), v_x(t), y(t), v_y(t)$, $z(t), v_z(t)$—the coordinates correspond to the center of the volume element under consideration. Equation 2.29 shows that f remains constant as we follow a particle trajectory—it does not imply that f is constant at all locations in phase space.

We shall prove the theorem of phase-volume conservation for the special case of a two-dimensional distribution with particle orbit coordinates x and v_x. The extension to six dimensions is straightforward. To begin, consider a collection of orbit points contained within a rectangular boundary at $t = 0$ (Fig. 2.10a). The area contains many orbit points, so its boundary can be identified unambiguously as the distribution evolves. If we can prove that the area circumscribed by the boundary remains constant, we can conclude that the area of any bounded region remains constant, since any shape can be approximated by a number of small rectangles.

We can find changes in the shape of the phase-shape region by analyzing the motion of individual particles in the region. Section 2.3 discussed finite difference methods to calculate particle orbits by advancing in small time steps Δt. We will concentrate on the leapfrog method that sequentially advances position and velocity. In this context, the spatial coordinates of a particle orbit are defined at times t, $t + \Delta t$, $t + 2\Delta t$, ..., while the velocity coordinates are defined at $t + \Delta t/2$, $t + 3\Delta t/2$, ...

The first step is to find the change in the spatial positions of particles at time $t = t + \Delta t$. Equation (2.27) implies that

$$x_i(t + \Delta t) = x_i(t) + v_{xi}(t + \Delta t/2)\Delta t, \qquad (2.30)$$

where i is the index number of particles within the group. The half-step velocities can be expressed in terms of deviations from the average velocity of particles in the region:

$$v_{xi}(t + \Delta t/2) = \langle v_x(t + \Delta t/2)\rangle + \Delta v_{xi}(t + \Delta t/2). \qquad (2.31)$$

Equation (2.30) becomes

$$x_i(\Delta t) = x_i(0) + \langle v_x(\Delta t/2)\rangle \Delta t + \Delta v_{xi}(\Delta t/2)\Delta t. \qquad (2.32)$$

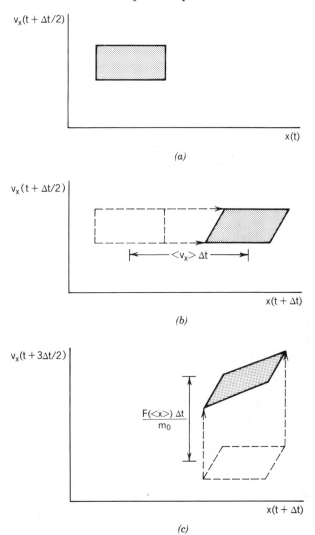

Figure 2.10 Illustration of phase-volume conservation by the leapfrog method to advance particle positions and velocities. (*a*) Boundary around a group of neighboring particles in phase space, with positions defined at time t and velocities defined at time $t + \Delta t/2$. (*b*) Advancing the particle positions to time $t + \Delta t$ using the velocities at $t + \Delta t/2$. (*c*) Advancing the particle velocities to time $t + 3\Delta t/2$ using forces calculated from the positions at time $t + \Delta t$.

The second term on the right-hand side of Eq. (2.32) gives an equal displacement for all particles in the region. Although the rectangle moves, there is no change in its shape or area. Figure 2.10*b* illustrates the effect of the second term. Particles with $v_{xi} = \langle v_x \rangle$ suffer no displacement relative to the center of the rectangle. Particles with $\Delta v_{xi} > 0$ move a relative distance proportional to Δv_{xi} in the $+x$

direction. Similarly, particles with negative velocity compared to the average are retarded. The net effect of the third term in Eq. (2.32) is to distort the shape of the phase-space region from a rectangle to a trapezoid. Note that the phase-space area occupied by the particles does not change—the rectangle and trapezoid have the same base and height.

The second step in the difference method is to advance the velocities of the particles from time $(t + \Delta t/2)$ to $(t + 3\Delta t/2)$. The forces are calculated from the values of the particle positions at time $(t + \Delta t)$. Following Eq. (2.28),

$$v_i(t + 3\Delta t/2) = v_i(t + \Delta t/2) + [F(x_i(t + \Delta t), t + \Delta t]\Delta t/m_0. \qquad (2.33)$$

The force F is a function of the particle position and time. It also depends implicitly on the positions of all other particles in the beam and charges on external electrodes. We assume that F varies smoothly with the position of the particle x_i. As a result, variation of the force over the small phase-space region of Fig. 2.10a at time $t + \Delta t$ can be approximated by a Taylor series expansion. The expansion is performed about the average position of particles in the region, $\langle x(t + \Delta t) \rangle$:

$$F[x_i(t + \Delta t), t + \Delta t] \cong F[\langle x(t + \Delta t) \rangle, t + \Delta t] + \frac{\partial F}{\partial x}\Delta x_i(t + \Delta t), \qquad (2.34)$$

where

$$x_i(t + \Delta t) = \langle x(t + \Delta t) \rangle + \Delta x_i(t + \Delta t).$$

Substituting Eq. (2.34) in Eq. (2.33), we find that the equation to advance the velocity has a form similar to Eq. (2.32):

$$v_i(t + 3\Delta t/2) = v_i(t + \Delta t/2) + \{F[\langle x(t + \Delta t) \rangle, t + \Delta t]/m_0\}\Delta t$$

$$+ \left(\frac{\partial F(t + \Delta t)}{\partial x}\right)\Delta x_i(t + \Delta t)\Delta t. \qquad (2.35)$$

When Eq. (2.35) is applied to the particles in the region, the second term on the right-hand side contributes a uniform displacement of orbit coordinates along the velocity axis. The third term distorts the boundary of the region from a trapezoid to diamond (Fig. 2.10c). Again, it is easy to verify that this transform-ation preserves the area of the figure. In following time steps, the diamond may change to another diamond, a trapezoid, or back to a rectangle. In any case, the area of the figures is the same.

The preceding derivation, although simple, has broad application. A phase-space region of any shape can be divided into a set of rectangular segments. As time advances, the shapes of individual segments change, but their net area remains constant. Adjacent segments always share the same boundary particles. In the smooth force limit, the segments cannot overlap or exchange particles. Figure 2.11 illustrates these facts. In the figure, an applied force of the form $F_x =$

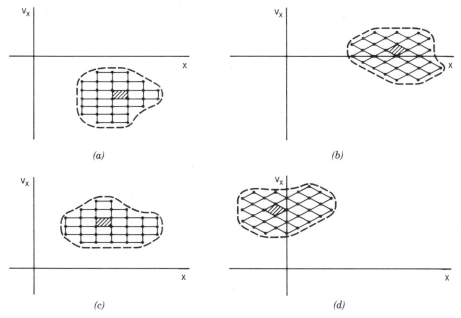

Figure 2.11 Conservation of phase space area for a one-dimensional particle distribution in a linear focusing force. The bounded area is divided into rectangles. The four figures show the calculated evolution of the rectangles with time as the particles advance through three-eighths of a betatron oscillation. The shaded region marks a particular group of particles.

$- kx$ acts on the particles. The particles initially occupy an irregular phase-space area. Plots of the evolution of the rectangular boundaries at three later times are shown. Note that the boundaries between segments are always straight lines. The shapes of segments change, but the areas are constant. As a result, the area of the complete distribution is constant.

Conservation of phase-space area (or volume) occupied by a distribution results from the fact that individual particle orbits advance through linear transformations. Motion of a group of particles during a time step is equivalent to the transformation of the orbit coordinates in phase space. As an illustration, take a particle orbit represented by the vector

$$\mathbf{X}_i(t) = [\Delta x_i(t), \Delta v_i(t + \Delta t/2)]. \tag{2.36}$$

We will neglect changes of the position or velocity centroids of the group—we know that these changes do not modify the phase area. The first orbit vector transformation of the leapfrog method can be represented by the matrix equation

$$\mathbf{X}'_i(t) = \mathbf{A} \cdot \mathbf{X}_i(t), \tag{2.37}$$

where

$$\mathbf{X}'_i(t) = [\Delta x_i(t + \Delta t), \Delta v_i(t + \Delta t/2)]$$

and

$$A = \begin{bmatrix} 1 & \Delta t \\ 0 & 1 \end{bmatrix}.$$

The second step is:

$$X_i(t + \Delta t) = B \cdot X_i'(t), \qquad (2.38)$$

where

$$B = \begin{bmatrix} 1 & 0 \\ (\partial F/\partial x)\Delta t & 1 \end{bmatrix}.$$

The net transformation of position and velocity over a time step is

$$X_i(t + \Delta t) = C \cdot X_i(t), \qquad (2.39)$$

where

$$C = \begin{bmatrix} 1 & \Delta t \\ (\partial F/\partial x)\Delta t & 1 + (\partial F/\partial x)\Delta t^2 \end{bmatrix}.$$

Equation 2.39 is a linear transformation of particle orbit parameters. Note that the determinant of the transformation matrix C is unity. Under this condition, the theory of linear algebra states that two-dimensional transformations preserve elements of area. In other words, if two particles are initially separated in position by Δx and in velocity by Δv_x and if individual positions and velocities are modified by a linear transformation, then

$$(\Delta x \Delta v_x)_{\text{final}} = (\Delta x \Delta v_x)_{\text{initial}}. \qquad (2.40)$$

Extension of the derivations to six-dimensional phase-space orbit vectors leads to the conclusion that the hypervolume occupied by a collection of particles is conserved over each time step.

The condition of conservation of phase volume is often written in the following notation:

$$\frac{D f(x, v_x, y, v_y, z, v_z, t)}{Dt} = 0. \qquad (2.41)$$

The symbol D/Dt denotes the *convective derivative*. In fluid dynamics, convection denotes moving along with fluid. For example, if we follow the orbit represented by the dashed line in Fig. 2.9, then the phase-space density f remains constant near the particle.

To conclude the section, it is important to note circumstances when the derivation leading to Eq. (2.29) is invalid. The effect of collisions is obvious;

particles can move instantaneously from one phase-space segment to another. Consequently, the phase-space density inside a particular segment can change in time. Collisions sometimes result from strong interactions between two particles in the distribution. In addition, collisions can occur between particles in the distribution and an external distribution of particles. For example, a beam passing through neutral gas can exchange momentum with the background.

In contrast to single-particle collisions, friction is a smooth force.[2] It causes gradual changes in particle orbits. Usually, a frictional force can be represented as a function of the particle velocity v:

$$\mathbf{F}_f = -k\mathbf{v}^n \tag{2.42}$$

where n is a number. Friction may result from the interaction of a beam with an external resistive structure or from the cumulative effect of many weak collisions. Friction leads to a net loss of energy from a beam. Generally, the beam particles transfer energy to an external medium in the form of heat. The following examples illustrate some instances where friction is important to describe the evolution of a charged particle beam:

1. Ions passing through a diffuse neutral gas background interact primarily with the electrons of the atoms. Displacement of the light electrons absorbs energy from the massive ions. Interaction with large numbers of electrons results in a cumulative decrease of longitudinal ion kinetic energy.

2. When a pulsed beam propagates through a conducting pipe, it induces return currents in the wall (Section 10.6). If the wall is resistive, the return current deposits thermal energy in the wall that is ultimately supplied by deceleration of the beam.

3. Particles may transfer energy to an external structure by radiation. A familiar example is synchrotron radiation of electrons in synchrotrons and storage rings.

To illustrate how friction modifies the phase-volume conservation theorem, assume that a spatially uniform, velocity-dependent force acts on a beam. We will concentrate on the axial motion of particles contained in the phase-space area illustrated in Fig. 2.12. The force can be represented as

$$F(v_{zi}) = F(\langle v_z \rangle) + [\partial F(\langle v_z \rangle)/\partial v_z]\Delta v_{zi}. \tag{2.43}$$

The quantity $F(v_{zi})$ is the force acting on a single particle and $F(\langle v_z \rangle)$ is the average force acting on all particles in the region. Inserting Eq. (2.42) gives

$$F(v_{zi}) = -k\langle v_z \rangle^n[1 + n\Delta v_{zi}/\langle v_z \rangle]. \tag{2.44}$$

[2]This is strictly true only if we ignore shot noise.

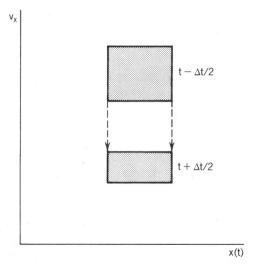

Figure 2.12   Effect of a frictional force on a bounded phase space region of orbit vectors. As the particles decelerate, the area of the region decreases.

Consider advancing the velocity with the leapfrog method. If the $\langle v_z \rangle$ is initially positive, the constant term in the brackets of Eq. (2.44) leads to a net slowing of particles in the segment with no change in the segment area. On the other hand, the term proportional to Δv_{zi} modifies the segment area. If the exponent n in Eq. (2.42) is greater than zero, the retarding frictional force on the top of the segment exceeds the force on the bottom. Consequently, the top shifts down more than the bottom—the area of the rectangle shrinks. The implication is that frictional forces can compress a phase fluid, raising the phase-space density. This fact is the physical basis of beam cooling discussed in Section 4.6.

An example will clarify the relationship between strong collisions resulting in velocity diffusion and weak collisions that approximate a frictional force. Both effects violate conservation of volume. A steady-state proton beam with kinetic energy in the 200-MeV range decelerates in a solid density metal target (Fig. 2.13). Transverse and longitudinal motions are independent, since there are no long-range electric or magnetic fields—we shall discuss only motion along the z axis. The beam initially has infinite axial extent and a small spread in axial velocity (A). As shown in Section 10.4, the energy loss rate from collisions with atomic electrons is small at high energy. Therefore, interactions of the beam with the target are initially dominated by strong nuclear collisions. These collisions cause *energy straggling*, a spread in the longitudinal velocity of the protons. The phase-space area occupied by the beam particles moves to lower velocity because of the cumulative effect of inelastic collisions and expands in area because of the random nature of the nuclear collisions (B). At lower energy, the rate of energy transfer to atomic electrons becomes the dominant energy loss mechanism. All

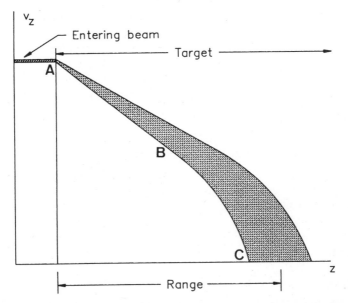

Figure 2.13   Axial phase space distribution $(z-v_z)$ of an energetic, steady-state ion beam slowing down in a solid target. (A) The entering beam has a narrow velocity spread. (B) Inelastic nuclear collisions with target nuclei reduce the average beam velocity and broaden the velocity spread. (C) Small-angle collisions with atomic electrons bring the ions to rest at the end of their range with zero velocity spread.

ions slow down almost uniformly through the summation of many weak collisions. At low energy, the force is frictional and the phase area occupied by protons decreases (C). Ultimately, the area shrinks to zero as the protons come to rest in the target.

Finally, we should note that the velocity-dependent forces associated with magnetic fields do not cause a change in the phase-space volume occupied by a particle distribution. In contrast to frictional forces, the magnetic force is perpendicular to the particle velocity. The magnetic force extracts no energy from the particle. We can prove that the phase area of a distribution is conserved in the presence of magnetic forces by invoking a result from electromagnetic field theory. It is always possible to find a frame of reference in which a given magnetic field vanishes.[3] In this frame, only electric fields are present. The electric force does not depend on velocity; therefore, the phase volume is conserved under the action of the force. In a nonrelativistic transformation between frames, the volume of differential phase-space elements in preserved—the volume conservation principle must hold in both frames.

[3] See, for instance, J. D. Jackson, *Classical Electrodynamics, 2nd Edition* Wiley, New York, 1975, Section 11.10.

2.5. DENSITY AND AVERAGE VELOCITY

When we measure the properties of a particle distribution, we often use instruments that sample spatial variations of the beam but cannot resolve details of the velocity distribution. For example, simple probes to measure particle flux at a point usually give no information about the velocities of individual particles that constitute the flow. Velocity averages over a distribution are called *macroscopic quantities* or *moments of the distribution function.* Two such quantities that are useful for charged particle beams are the particle density and average velocity. Other quantities, such as the velocity spread about a mean, are sometimes useful. The source terms for the Maxwell equations, charge density and current density, are also macroscopic quantities (Section 2.8).

We calculate macroscopic quantities by taking velocity-weighted sums over a discrete distribution function (or integrals over a continuous distribution function). The resulting quantities are functions only of the spatial coordinates. There are three motivations to define velocity-averaged quantities:

1. We must have expressions for the charge and current densities to develop self-consistent theories of beam evolution in response to electromagnetic forces.

2. Velocity-averaged quantities can be compared with measurements to check the validity of a theory.

3. Fluid equations directly describe the evolution of macroscopic quantities (Section 2.10). Although the equations provide an incomplete description of a beam, they are often easy to solve.

The density of a distribution of particles equals the number of particles per volume at a location. We will take some time to discuss density expressions in both discrete and continuous forms. To begin, consider the calculation for a discrete distribution function. We shall refer to the one-dimensional distribution illustrated in Fig. 2.14. We divide the region occupied by particles into elements with area $\Delta x \Delta v_x$. The discrete density function $N(x)$ is the number of particles between $x - \Delta x/2$ and $x + \Delta x/2$ divided by Δx. It is easy to find $N(x)$ using a computer that stores the positions and velocities of the particles. The computer loops through the particle array, counting the number of particles within Δx. The process ignores information about the particle velocity. Figure 2.14 indicates that the computer performs a sum of particles over all phase area elements at position x. In terms of the function $F(x, v_x, y, v_y, z, v_z)$, the symbolic representation of the process in three dimensions is

$$N(x, y, z) = \sum_{\text{all } v_x} \sum_{\text{all } v_y} \sum_{\text{all } v_z} F(x, v_x, y, v_y, z, v_z)/\Delta x \, \Delta y \, \Delta z. \qquad (2.45)$$

Section 2.2 showed that a continuous distribution function $f(x, v_x, y, v_y, z, v_z)$ is useful when a beam contains many particles. The function f equals the density of particles in phase space. Similarly, we can define a configuration-space density

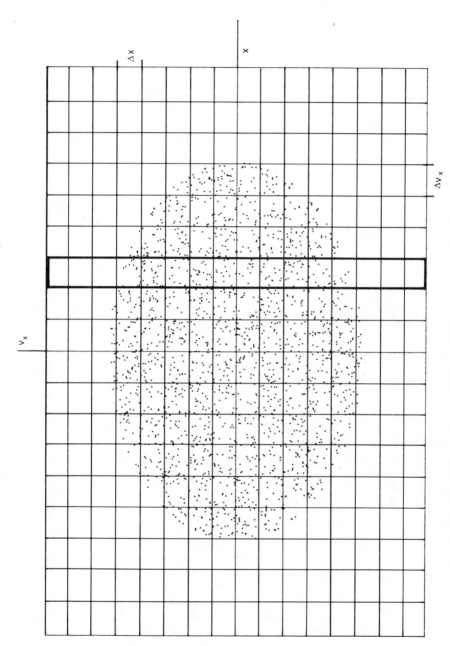

Figure 2.14 Phase-space orbit vectors of a one-dimensional distribution containing 2000 particles with a uniform random distribution over an elliptical region. Phase space is divided into elements to compute the density, $n(\mathbf{x})$. The outlined area indicates a sum over velocities.

$n(x, y, z)$ as

$$n(x, y, z) \cong N(x, y, z)/\Delta x\, \Delta y\, \Delta z, \qquad (2.46)$$

in the limit that $\Delta x, \Delta y, \Delta z \to 0$. The units of $n(x, y, z)$ are particles per cubic meter. We can write an analog for Eq. (2.45) in continuous form. Again, consider a one-dimensional distribution. The relation between the discrete and continuous distribution function is

$$F(x, v_x) \cong \int_{v_x - \Delta v_x/2}^{v_x + \Delta v_x/2} dv_x \int_{x - \Delta x/2}^{x + \Delta x/2} dx\, f(x, v_x). \qquad (2.47)$$

If we extend the integration limits on the right-hand side of Eq. (2.47) over all velocities, then the left-hand side equals $N(x)$. Substituting in Eq. (2.46), the continuous density function is

$$n(x) = \frac{1}{\Delta x} \int_{-\infty}^{+\infty} dv_x \int_{x - \Delta x/2}^{x + \Delta x/2} dx\, f(x, v_x) \cong \int_{-\infty}^{+\infty} dv_x f(x, v_x). \qquad (2.48)$$

The three-dimensional generalization of Eq. (2.48) is

$$n(x, y, z) \cong \int_{-\infty}^{+\infty} dv_x \int_{-\infty}^{+\infty} dv_y \int_{-\infty}^{+\infty} dv_z\, f(x, v_x, y, v_y, z, v_z). \qquad (2.49)$$

The average velocity of a one-dimensional discrete distribution is calculated by taking a weighted sum over the distribution. Again, we divide phase space into elements with dimensions Δx and Δv_x. The average velocity at position x is the sum over all elements at x of v_x times the probability that a particle is in the element. The probability that a particle is in the element at (x, v_x) is

$$P(x, v_x) = F(x, v_x)/N(x). \qquad (2.50)$$

The resulting expression for the average velocity is

$$\langle v_x(x) \rangle = [1/N(x)] \sum_{\text{all } v_x} v_x F(x, v_x). \qquad (2.51)$$

The extension to a one-dimensional continuous distribution function is straightforward. The fraction of particles at x with velocity in the range $v_x - \Delta v_x/2$ to $v_x + \Delta v_x/2$ is

$$\text{fraction} \cong f(x, v_x)\Delta v_x \bigg/ \int_{-\infty}^{+\infty} f(x, v_x)\, dv_x. \qquad (2.52)$$

The average velocity is defined as

$$\langle v_x(x) \rangle = \int_{-\infty}^{+\infty} dv_x\, v_x\, f(x, v_x) \bigg/ \int_{-\infty}^{+\infty} dv_x\, f(x, v_x). \tag{2.53}$$

Note that the denominator in Eq. (2.53) is equal to the density, $n(x)$. The three-dimensional extension of Eq. (2.53) is

$$\langle v_x(x, y, z) \rangle = \int_{-\infty}^{+\infty} dv_x \int_{-\infty}^{+\infty} dv_y \int_{-\infty}^{+\infty} dv_z\, v_x\, f(x, v_x, y, v_y, z, v_z)/n(x, y, z). \tag{2.54}$$

The following section on the Maxwell distribution illustrates a specific application of Eqs. (2.49) and (2.54).

2.6. MAXWELL DISTRIBUTION

The Maxwell velocity distribution appears in all branches of collective physics. Advanced kinetic theory shows that groups of energetic particles that interact through collisions ultimately approach a Maxwell distribution. Particles in an isotropic Maxwell distribution are in *thermal equilibrium*. They have a spread in kinetic energy. Charged particle beams are nonisotropic and usually almost monoenergetic. In a sense, a goal of beam technology is to create non-Maxwellian distributions and to preserve them over time scales set by the application. Nevertheless, there are many applications of the Maxwell distribution in beam physics. An example is the distribution of electrons from a thermionic cathode. In another application, we will study the thermalization of low-energy electrons injected to neutralize high-current ion beams in Chapter 11.

The distribution of nonrelativistic particles in thermal equilibrium can be expressed as the product of the spatial density function times a function of velocity:

$$f(x, v_x, y, v_y, z, v_z) = n(x, y, z)\, g(v_x, v_y, v_z). \tag{2.55}$$

If particle motions are closely coupled in all three directions, the velocity distribution is isotropic. Here, the velocity function is given by

$$g(v_x, v_y, v_z) = A \exp[-m(v_x^2 + v_y^2 + v_z^2)/2kT]. \tag{2.56}$$

The argument of the exponent is proportional to particle kinetic energy divided by kT. The quantity kT therefore characterizes the kinetic energy spread of the distribution. The quantity T is the temperature of the distribution and k is Boltzmann's constant,

$$k = 1.38 \times 10^{-23}\ \text{J/K}.$$

If a collection of particles is in thermal equilibrium, kT is constant over all regions of space. Sometimes, the velocity distribution at a point may approximate the form of Eq. (2.56), but kT may vary in space. Then, the distribution is called a *local Maxwellian distribution*. In chemistry and fluid mechanics, the temperature T is specified in degrees Kelvin. In beam physics, the temperature in degrees Kelvin is not a convenient unit; usually, we express kT as a single quantity in units of joules or electron volts.

We can find the normalization constant A in Eq. (2.56) by setting

$$\int_{-\infty}^{+\infty} dv_x \int_{-\infty}^{+\infty} dv_y \int_{-\infty}^{+\infty} dv_z \, g(v_x, v_y, v_z) = 1. \qquad (2.57)$$

Using the relation

$$\int_{-\infty}^{+\infty} dv_x \exp\left(\frac{-mv_x^2}{2kT}\right) = \left(\frac{2\pi kT}{m}\right)^{1/2}, \qquad (2.58)$$

the Maxwell velocity function is

$$g(v_x, v_y, v_z) = \left[\frac{m}{2\pi kT}\right]^{3/2} \exp\left[-\frac{m(v_x^2 + v_y^2 + v_z^2)}{2kT}\right]. \qquad (2.59)$$

The quantity $g(v_x)dv_x$, the probability that the particle has velocity in the interval dv_x, is plotted in Fig. 2.15.

We can apply Eq. (2.59) to find velocity averages that will be useful in following sections. As discussed in Section 2.5, the average speed projected in the x

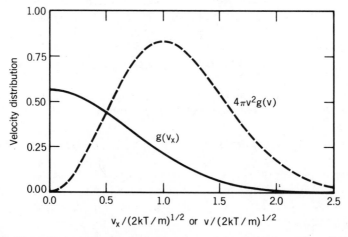

Figure 2.15 Normalized velocity distribution functions, $g(v_x)$ and $g(v)$, for a Maxwell distribution. The functions plotted have an integral of unity.

direction is

$$\langle |v_x| \rangle = \int_{-\infty}^{+\infty} dv_y \int_{-\infty}^{+\infty} dv_z \int_{-\infty}^{+\infty} dv_x \, v_x \, g(v_x, v_y, v_z). \tag{2.60}$$

Substituting Eq. (2.59) for g, we find that

$$\langle |v_x| \rangle = (2kT/\pi m)^{1/2}. \tag{2.61}$$

Similarly, we can find the average particle kinetic energy projected along an axis by taking the distribution integral weighted by $mv_x^2/2$. The result is

$$\langle mv_x^2/2 \rangle = kT/2. \tag{2.62}$$

The average particle speed (independent of direction) is a useful quantity for isotropic distributions. The speed is

$$v = (v_x^2 + v_y^2 + v_z^2)^{1/2}. \tag{2.63}$$

Inspection of Eq. (2.59) shows that $g(v) \sim \exp(-mv^2/2kT)$. In taking averages over the distribution of particle speed, we must remember that the fraction of particles with speed v in the range $v - dv/2$ to $v + dv/2$ is $4\pi v^2 g(v)\,dv$.

The quantity $4\pi v^2 dv$ is the volume of a spherical shell of thickness dv in velocity space. The normalized distribution function in terms of speed is

$$g(v) = [m/2\pi kT]^{3/2} \exp(-mv^2/2kT). \tag{2.64}$$

Figure 2.15 shows a plot of $4\pi v^2 g(v)$, the probability that a particle has speed in the interval dv. Equation (2.64) gives the average speed as

$$\langle v \rangle = (8kT/\pi m)^{1/2}, \tag{2.65}$$

and the average total kinetic energy as

$$\langle mv^2/2 \rangle = 3kT/2. \tag{2.66}$$

The Maxwell distribution is often applied to systems where the velocity distribution is nonisotropic. Consider, for example, a nonrelativistic electron beam extracted from a thermionic cathode. At the cathode, the electron distribution is Maxwellian in the transverse direction with a temperature equal to that of the cathode T_c. If the electrons are accelerated and transported in an ideal accelerator, they all gain an equal increment of axial velocity v_0. If we observe the beam in a frame moving at v_0, we find that the distribution is still isotropic and Maxwellian with temperature T_c. On the other hand, suppose that the beam is separated and compressed into axial bunches for injection into an RF accelerator

(Section 15.3). As we shall see in Section 3.8, compression of the beam into small bunches causes an increase in the spread of axial kinetic energy. If the forces are solely in the axial direction, the kinetic energy spread in the x and y directions remains equal to $kT_c/2$. Under these conditions, it is useful to approximate the beam distribution as Maxwellian with different temperatures in the axial and transverse directions:

$$g(v_x, v_y, v_z) \sim \exp(-mv_x^2/2kT_x)\exp(-mv_y^2/2kT_y)\exp(-mv_z^2/2kT_z). \quad (2.67)$$

For the thermionic cathode example, $kT_x = kT_y = kT_c$ and $kT_z > kT_c$.

A common assumption used in beam theory is that the particles have a Maxwell distribution when observed in the beam rest frame. The transformed distribution observed in the stationary frame of the accelerator is called a *displaced Maxwellian distribution*. The distribution has a kinetic energy spread superimposed on an ordered axial velocity. As an example, consider a nonrelativistic ion beam extracted from a plasma source with ion temperature T_i (Section 7.6). The beam is axially bunched passing through a radio-frequency quadrupole accelerator. The beam emerges from the accelerator with kinetic energy E_0. We can represent the exit beam distribution in the stationary frame as

$$g(v_x, v_y, v_z) \sim \exp\left[\frac{-m_i(v_x^2 + v_y^2)}{2kT_i}\right]\exp\left[\frac{-m_i(v_z - v_0)^2}{2kT_i'}\right], \quad (2.68)$$

where $v_0 = (2E_0/m_i)^{1/2}$. Because of the axial bunching, T_i' is larger than T_i.

2.7. COLLISIONLESS BOLTZMANN EQUATION

Section 2.4 showed that the convective derivative of the continuous distribution function equals zero:

$$Df/Dt = 0. \quad (2.69)$$

Equation (2.69) means that the density of orbit vectors of a collisionless group of particles is constant in a frame of reference moving with the vectors through phase space. If we view the collection of vector points as a fluid, Eq. (2.69) states that the fluid is locally incompressible. In fluid dynamics, relationships like Eq. (2.69) are called *Lagrangian equations*—they are defined in a frame of reference that moves with the fluid.

Often, the Lagrangian viewpoint is not the most convenient form. Usually, we would like to describe variations of fluid quantities in a fixed frame through which the fluid moves. For example, in beam physics we reference equations to the frame of the stationary accelerator. Furthermore, there are circumstances where a Lagrangian frame is undefined. For example, theories of neutralized beams (Chapter 11) require a description of intermixed fluids moving at different velocities. *Eulerian equations* specify the evolution of fluid quantities in a stationary frame. In this section, we shall convert Eq. (2.69) to an Eulerian form. The derivation leads to one of the most important equations of collective physics, the collisionless Boltzmann equation. A form of the Boltzmann equation that includes only electromagnetic forces, the *Vlasov equation*, is often encountered in charged particle beam physics.

Figure 2.9 shows a region of phase space occupied by a one-dimensional beam distribution. The function $f(x, v_x, t)$ represents the density of orbit vector points at a particular time. Suppose we measure f near an orbit vector point $[x_0(t_0), v_{x0}(t_0)]$ at time t_0. Equation (2.69) states that the measurements always give the same value if we follow the orbit vector as it moves with time, or

$$f[x_0(t), v_{x0}(t), t] = f[x_0(t_0), v_{x0}(t_0), t_0].$$

Here, we seek an alternate equation that prescribes how f varies in time at a constant position in phase space, (x, v_x). The advantage is that the resulting equation describes quantities that we can measure. Figure 2.16 shows how a detector could determine $f(x, v_x)$ at a stationary point in phase space. A sheet beam, uniform in y and z, moves past the detector. Transverse velocity components v_x are small compared with the axial velocity v_0. Any motion of the detector in the x direction must be slow compared with the velocities of beam particles—the detector effectively occupies a stationary point on the position axis. The aperture admits particles only in a small range of position Δx near x. The detector can be inclined at an angle, v_x/v_0, so that only particles within a

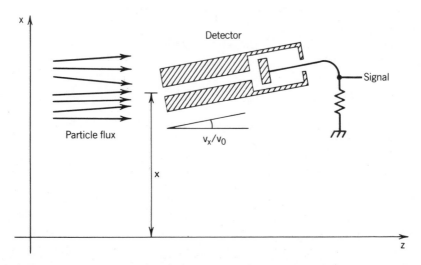

Figure 2.16 Schematic drawing of a detector to measure the transverse distribution function of a paraxial beam. The detector moves in postion and rotates in angle to admit particles with different values of x and v_x.

range Δv_x near v_x can enter. The output signal of the detector is proportional to the entering flux, which is in turn proportional to $f(x, v_x, t)$. We can calculate the absolute value of f by analyzing the properties of the aperture. For a steady-state beam, the detector can be moved and tilted to map the entire distribution function.

In a distribution that changes with time, orbit vector points flow past a fixed location. If f is constant over all x and v_x, then the phase-space density observed at the location is constant in time. Here, an equal number of points replace orbit vectors that leave the region near the specified location. Temporal variations of f (observed at a location) occur if there are nonzero gradients of the distribution function in x or v_x. For example, consider a point (x_0, v_{v0}). If v_{x0} is positive, particles with $x < x_0$ replace particles that start near x_0 at t_0. Suppose that f has a negative spatial gradient at x_0 and t_0. Then, the phase-space density observed at the point increases with time,

$$f[x_0(t), v_{x0}(t), t] > f[x_0(t_0), v_{x0}(t_0), t_0].$$

We shall develop a quantitative expression for the time variation of f with the help of Fig. 2.17. The solid line shows $f(x, v_x, t)$ at time t in a region near the point (x, v_x). The function has a negative gradient. At a later time, $t + \Delta t$, all particles shift to the right a distance $\Delta x = v_x \Delta t$. A second line represents the modified distribution. Although f retains its original value at the new position of the particles, the value of f at the stationary point (x, v_x) drops by an amount

$$\Delta f_x(x, v_x) = -\Delta x \left(\frac{\partial f}{\partial x}\bigg|_{x,v_x} \right) = -v_x \Delta t \left(\frac{\partial f}{\partial x}\bigg|_{x,v_x} \right). \tag{2.70}$$

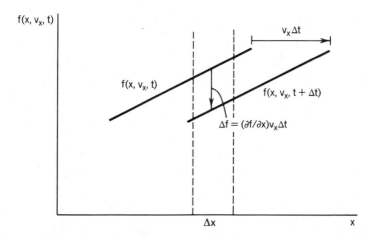

Figure 2.17   Distribution function with a spatial gradient plotted at two times. The positive spatial gradient leads to a decrease of f at a point with time.

The partial derivative notation refers to a variation of f along a path in the x direction.

Similarly, we can write the change of f at the point (x, v_x) that results from a gradient of f in velocity. Suppose that near (x, v_x) the phase-space density is higher for higher v_x. In the presence of a decelerating force, the high-velocity particles shift to a velocity near v_x. As a result, $f(x, v_x, t)$ rises with time. The change in velocity produced by an acceleration a_x acting over an interval Δt is $v_x = a_x \Delta t$. The change in f resulting from a gradient of f in velocity and a force can be written:

$$\Delta f_{v_x}(x, v_x) = - \Delta v_x \left(\frac{\partial f}{\partial v_x} \bigg|_{x,v_x} \right) = - a_x \Delta t \left(\frac{\partial f}{\partial v_x} \bigg|_{x,v_x} \right). \tag{2.71}$$

The symbol $\partial f / \partial t$ denotes the change in f with time at a constant location in phase space. The combined effects of position and velocity gradients give the following equation:

$$\frac{\partial f}{\partial t} = - v_x \frac{\partial f}{\partial x} - a_x \frac{\partial f}{\partial v_x}. \tag{2.72}$$

Equation (2.72) implies that the total change in the continuous distribution function from all causes—the passage of time, motion of the phase fluid, and fluid acceleration—equals zero. The full form of Eq. (2.72) for motion in three dimensions is

$$\frac{D f(x, v_x, y, v_y, z, v_z)}{Dt} = \frac{\partial f}{\partial t} + v_x \frac{\partial f}{\partial x} + v_y \frac{\partial f}{\partial y} + v_z \frac{\partial f}{\partial z}$$

$$+ a_x \frac{\partial f}{\partial v_x} + a_y \frac{\partial f}{\partial v_y} + a_z \frac{\partial f}{\partial v_z} = 0. \tag{2.73}$$

Equation (2.73) is the *collisionless Boltzmann equation*. The equation is equivalent to the principle of the conservation of phase volume. Therefore, Eq. (2.73) has the same validity limitations. Forces on the particles must vary smoothly in space and time. Collisions and friction are excluded. Equation (2.73) is often written in vector notation as

$$\frac{\partial f}{\partial t} + \mathbf{v} \cdot \frac{\partial f}{\partial \mathbf{x}} + \mathbf{a} \cdot \frac{\partial f}{\partial \mathbf{v}} = 0. \tag{2.74}$$

The definitions of the vector derivative and dot product notations are evident from a comparison of Eq. (2.74) with Eq. (2.73).

Equation (2.74) takes a special form when f corresponds to a group of nonrelativistic charged particles accelerated by electric and magnetic fields. The acceleration is given by the Lorentz expression:

$$\mathbf{a} = q(\mathbf{E} + \mathbf{v} \times \mathbf{B})/m_0. \tag{2.75}$$

If the particles have charge q and rest mass m_0, then Eq. (2.74) becomes

$$\frac{\partial f}{\partial t} + \mathbf{v} \cdot \frac{\partial f}{\partial \mathbf{x}} + \frac{q}{m_0}(\mathbf{E} + \mathbf{v} \times \mathbf{B}) \cdot \frac{\partial f}{\partial \mathbf{v}} = 0. \tag{2.76}$$

Equation (2.76) is the *Vlasov equation*, a fundamental relationship in plasma physics. The Vlasov equation is often applied to the stability analysis of charged particle beams.

If collisions occur, the phase-space density of orbit vectors may change, even in a Lagrangian reference frame. Collisions cause a random walk of particles through velocity space. Usually, collisions result in a decrease of phase-space density with time as a distribution diffuses to fill a large phase volume. For particle densities typical of most charged-particle-beam applications, long-range forces dominate the motions of particles, while collisions perturb the distribution. The effect of weak collisions is often added symbolically to Eq. (2.74) as a term that contributes a small variation to f at a point:

$$\frac{\partial f}{\partial t} + \mathbf{v} \cdot \frac{\partial f}{\partial \mathbf{x}} + \mathbf{a} \cdot \frac{\partial f}{\partial \mathbf{v}} = \frac{\partial f_c}{\partial t}. \tag{2.77}$$

Analyses of discrete-particle interactions and particle migration in velocity space lead to special forms of Eq. (2.77) such as the Fokker–Planck equation.

2.8. CHARGE AND CURRENT DENSITIES

Solution of the Vlasov equation for a charged particle beam depends on the net electric and magnetic fields in the accelerator or transport region. At low current, the field contributions of beam particles are small—electric and magnetic fields arise mainly from external charges and currents. Furthermore, presence of the beam has little effect on the distribution of external charges and currents. The implication is that we can treat the electric and magnetic field as given functions that do not depend on the beam motion.

When we proceed beyond the single-particle limit to collective effects, we include beam contributions to the electromagnetic fields. Calculations of the fields and particle orbits must be self-consistent, as discussed in Section 2.3. In this section, expressions that relate the electric and magnetic fields to the distribution function are summarized. Integrals over the distribution function give the source terms of the Maxwell equations. For reference, the Maxwell equations are

$$\nabla \cdot \mathbf{E} = \rho/\varepsilon_0 = (\rho_b + \sum \rho_e)/\varepsilon_0, \tag{2.78}$$

$$\nabla \times \mathbf{E} = -\partial \mathbf{B}/\partial t, \tag{2.79}$$

$$\nabla \cdot \mathbf{B} = 0, \tag{2.80}$$

$$\nabla \times \mathbf{B} = \mu_0 \mathbf{j}_b + \mu_0 \sum \mathbf{j}_e + (1/c^2)\, \partial \mathbf{E}/\partial t. \tag{2.81}$$

Equation 2.78 states that electric fields arise from a generating function, the charge density $\rho(x, y, z, t)$. The charge density is a scalar function with units of coulombs per cubic meter. The second form on the right-hand side of Eq. (2.78) emphasizes that different types of charge may contribute to the electric field. The charge density includes contributions for the beam ρ_b and a summation over all other types of external changes on electrodes and in dielectrics. The beam charge is called *space charge*, since it exists in free space. Equation (2.79) is Faraday's law—electric fields also result from changing magnetic flux. Equation (2.81) implies that spatially distributed currents generate magnetic fields. The current density $\mathbf{j}(x, y, z, t)$ has units amperes per square meter and it is a vector quantity. Magnetic fields may result from beam current in space, $\mathbf{j}_b$, or external currents. Equation (2.81) also shows that magnetic fields can arise from time-varying electric fields through the displacement current

$$\mathbf{j}_d = \varepsilon_0 \partial \mathbf{E}/\partial t.$$

In many charged-particle-beam applications, the electric field is static. Even if this is not true, static field equations sometimes provide a good approximation. The *quasistatic* limit is valid if the fields vary slowly compared with the time scale for light to cross a characteristic dimension of the beam or accelerator. In this limit, the effects of displacement current and electric fields generated by induction are small. In the static limit, the equations for electric fields are decoupled from the magnetic fields:

$$\nabla \cdot \mathbf{E} = \rho/\varepsilon_0, \tag{2.82}$$

$$\nabla \times \mathbf{E} = 0. \tag{2.83}$$

We can define the electrostatic potential $\phi(x, y, z)$ when Eqs. (2.82) and (2.83) hold. The potential is a scalar, related to the electric field by

$$\mathbf{E} = -\nabla \phi. \tag{2.84}$$

It is usually easier to find a solution for ϕ rather than to solve for $\mathbf{E}$ directly, because it is easier to identify scalar boundary conditions. Once ϕ is known, Eq. (2.84) determines all the electric field components. The combination of Eqs. (2.82) and (2.84) yield the Poisson equation

$$\nabla^2 \phi(x, y, z) = -\rho(x, y, z)/\varepsilon_0. \tag{2.85}$$

Equation (2.85) appears extensively in derivations in this book.

Although a continuous space-charge density ρ does not occur in reality, it is often a good approximation when a beam contains many particles. We can understand limits on the validity of the continuous approximation by reference to Fig. 2.18. Figure 2.18a shows a collection of stationary, randomly distributed charged particles. Figure 2.18b shows the exact axial electric field along the dashed line of Fig. 2.18a. Coulomb's law determines the field—the electric field at the position of particle i is

$$\mathbf{E}_i = \sum_{j \neq i} \frac{q_j \mathbf{r}_{ji}}{\varepsilon_0 |\mathbf{r}_j - \mathbf{r}_i|^2}, \qquad (2.86)$$

where $\mathbf{r}_{ji}$ is a unit vector pointing from particle j to particle i, while $\mathbf{r}_j$ and $\mathbf{r}_i$ are the position vectors of the two particles. The sum is taken over all other particles.

Note in Fig. 2.18 that field fluctuations occur over a scale length comparable with the spacing between particles. We neglect these small-scale variations when we solve the Maxwell equations with a continuous charge distribution. The dashed line in Fig. 2.18b represents the continuous solution. The fluctuations of electric field may change the velocity of a particle while leaving a phase-space neighbor unaffected. Therefore, the fluctuations have much the same effect as collisions—close encounters between charged particles are usually called *Coulomb collisions*. The electric field fluctuations are small when a beam contains many particles. The smooth force approximation is well justified for almost all useful charged particle beams. Usually, the general motion of beams is well

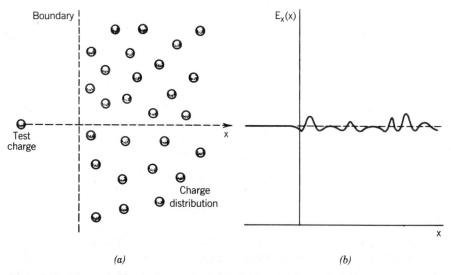

(a)   *(b)*

Figure 2.18 Macroscopic and microscopic electric field variations (a) The orbit of a test particle (dashed line) through randomly distributed charged particles. (b) Variation of the axial electric field along the test particle orbit. Solid line: Exact field. Dashed line: Macroscopic field, discrete charges replaced by a uniform density.

described by the Maxwell equations with Coulomb collisions added as a perturbation.

For a single type of charged particle in a region of interest, the charge density equals the charge of each particle, q, multiplied by number of particles per cubic meter, $n(x, y, z)$:

$$\rho(x, y, z) = qn(x, y, z). \tag{2.87}$$

Inserting Eq. (2.49), we find that the charge density can be calculated directly from the continuous distribution function as

$$\rho(x, y, z, t) \cong q \int_{-\infty}^{+\infty} dv_x \int_{-\infty}^{+\infty} dv_y \int_{-\infty}^{+\infty} dv_z \, f(x, v_x, y, v_y, z, v_z, t). \tag{2.88}$$

Equation (2.88) suggests an approach to the solution of the Vlasov equation with self-consistent electric fields. We apply the Vlasov equation to advance the distribution function through a time step, giving the charge and current densities of the beam at time t. We then combine the beam contributions to the field with external charge and current components to find the total electric field at time t. The information is then inserted into the Vlasov equation to advance the distribution function another time step. Repetition of the process leads to a prediction of the evolution of the beam distribution.

When more than one type of charged particle is present, the charge density is a sum of contributions from individual components:

$$\rho = \rho_1 + \rho_2 + \rho_3 + \cdots. \tag{2.89}$$

The most common application of Eq. (2.89) in beam physics is to neutralized beams, which consist of a mixture of ions and electrons. If the ions have charge state Z_i, Eq. (2.89) takes the form

$$\rho = -en_e + Z_i en_i. \tag{2.90}$$

In well-neutralized beams, the charge densities are almost balanced, $\rho_e \cong -\rho_i$. Here, the net charge is small and electric fields are almost completely canceled. The procedure to find the self-consistent evolution of multiple charge species is similar to the approach for a single species. Each type of particle is represented by separate distribution functions and Vlasov equations. Changes in the individual distributions couple through the net electric and magnetic fields.

The current density $\mathbf{j}$ is the charge crossing an area of $1\,\text{m}^2$ per second. The current density vector points in the direction of the average flow and the unit area is normal to the flow. For a single species, the current density equals the charge density multiplied by the average velocity,

$$\mathbf{j}(x, y, z) = qn(x, y, z)\langle v(x, y, z)\rangle. \tag{2.91}$$

We can modify Eq. (2.54) to give an expression for **j** in terms of the continuous distribution function:

$$\mathbf{j}(x, y, z, t) \cong q \int_{-\infty}^{+\infty} dv_x \int_{-\infty}^{+\infty} dv_y \int_{-\infty}^{+\infty} dv_z \, \mathbf{v} \, f(x, v_x, y, v_y, z, v_z, t). \qquad (2.92)$$

Equation (2.92) can be combined with the Maxwell and Vlasov equations to incorporate magnetic fields in the self-consistent solutions. When multiple species are present, the current density is the vector sum of components. For example, in a system with electrons and ions,

$$\mathbf{j} = \mathbf{j}_e + \mathbf{j}_i = -en_e \langle \mathbf{v}_e \rangle + Z_i n_i \langle \mathbf{v}_i \rangle. \qquad (2.93)$$

2.9. COMPUTER SIMULATIONS

In the prediction of the collective behavior of beams, numerical simulations are an alternative to analytic solutions of the Vlasov or fluid equations. The use of simulations in beam physics has expanded rapidly with advances in computer capabilities. There are two motivations to discuss simulations in this section. First, they are an important design tool for modern accelerator technology. Second, understanding computer simulations serves an educational purpose. The methods used in simulations provide a good example of the application of phase-space dynamics. Calculations of density, directed velocity, and other functions in simulations are literal and easily understood applications of Eqs. (2.88) and (2.92).

To view computer simulations in perspective, we must examine the differences between an ideal simulation and the real calculations carried out on existing computers. The ideal simulation replicates the collective interactions of a beam by calculating accurately the orbits of every particle in the beam. To carry out a solution of the equations of motion, the ideal simulation evaluates electromagnetic fields at the position of each particle. The fields include contributions from other beam particles. The ideal simulation describes a beam exactly without the simplifications required for tractable analytic theories. In principle, we need not worry about interpreting results and deriving validity limits. It is tempting to believe that a technique exists that can eliminate the work and worry of collective physics. This desire is reflected in a prevalent attitude that computer simulations are the final arbiters of theoretical debates—computer simulations are sometimes called *numerical experiments*.

In reality, computer simulations can never precisely model beams of practical interest. Real simulations require approximations, exclusion of noncritical processes, and other acts of judgment. For example, we must apply initial conditions and boundary conditions to finite intervals and spatial regions— these conditions often profoundly affect the results. The generation of good simulations demands as much on preliminary analysis of the input and interpretation of the results as an analytic theory. We can easily show why simulations can never replicate beam experiments. Consider, for example, a description of the electrons in a 1 cm length of a 10-A, 10-keV electron beam. The segment contains about 10^{10} particles. The orbit vector of each particle, $(\mathbf{x}, \mathbf{v})$, requires six floating point numbers, or about 30 bytes of random access memory. The total memory required for 10^{10} particles is a large number—300,000 megabytes.

Storage is not the only problem—the ideal simulation would demand a huge amount of computer time. Suppose we advance 10^{10} particles. The direct application of Eq. (2.86) involves 10^{10} sequences of floating point operations for each particle, or 10^{20} sequences for each time step. A computer that operates at 100 MFlops (10^8 floating point operations per second) would take more than 10^{12} sec, or 32,000 years to advance one step. These numbers make it clear that even the most powerful computer cannot create a literal representation of a beam. For practical calculations, we must simplify the process in two ways:

1. Reduce the number of particle vectors in the computer memory.

2. Reduce the number of operations to calculate electromagnetic fields at each time step.

We can accomplish the first goal by applying the insights of Sections 2.2 and 2.4. We shall represent the motions of many particles by a single test particle, or *computational particle*. We know that forces that vary smoothly in $\mathbf{x}$ and $\mathbf{v}$ preserve the continuity of particle distributions. Suppose we divide phase space into small regions (Fig. 2.19). Although the shapes of the regions change with

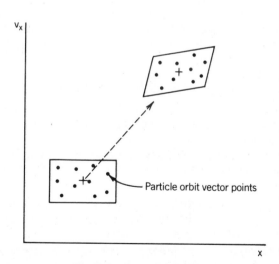

Figure 2.19 The physical basis of computer simulations of collective particle motions. The behavior of a group of particles adjacent in phase space is represented by the trajectory of a test-particle-orbit vector.

time, they enclose the same particles. We need not calculate the orbits of all particles inside a region, since they all move in about the same way. For each region, it is sufficient to determine the orbit of one particle at the centroid. A typical computer simulation follows the orbits of 10^3–10^5 computational particles—each computational particle represents a region that contains 10^3–10^8 real particles. At each time step, the computational particles are advanced by single-particle equations of motion. The fields at the new time are calculated with the assumption that all particles in a region move with the computational particle. Another viewpoint is that the computational particle advances as a single particle, but is assigned multiple charges when the electromagnetic fields are calculated.

Regarding the field calculation, we could estimate the electric field by applying Eq. (2.86) to the computational particles. The process still consumes a considerable amount of the time—the number of operations scales as the square of the number of computational particles. Furthermore, the resulting field function includes nonphysical fluctuations that result from the discrete nature of the computational particles. We must remove these field variations to avoid spurious collisional effects. The *particle-in-cell* (PIC) method greatly reduces the number of calculations and also brings about field smoothing.

The PIC method for electric and magnetic field calculations consists of the following procedure. Configuration space is divided into small regions called *cells*. Discrete charge-density and current-density arrays are defined with values in each cell. The charge density in a cell at time t equals the product of the number of computational particles in the cell times the charge per computational particle divided by the cell volume. The current density is the average vector velocity of computational particles multiplied by their charge divided by the cell area. Sometimes, computational particles are given a mathematical width to generate density functions with smoother variations. Depending on its position, a fraction of a particle may be assigned to a cell. This procedure is called the *cloud-in-cell* method.

The space-charge and current-density functions are combined with a finite difference solution of the Maxwell equations to generate electric and magnetic field values. The PIC procedure involves one pass through the array of computational particles to assign them to cells, followed by a field computation. The number of mathematical operations is linearly proportional to the number of particles and the number of cells. In a simulation involving thousands of particles, the PIC process takes much less time than a direct evaluation of the forces between particles.

We shall illustrate the PIC method by computing a quasistatic electric field from a particle distribution using the Poisson equation. The cubic cells have width Δ and volume Δ^3 (Fig. 2.20). Points at the cell centers define a three-dimensional Cartesian mesh. The index numbers i, j, and k specify a mesh point with location $x = i\Delta$, $y = j\Delta$, and $z = k\Delta$. We define a discrete function $\Phi(i,j,k)$. The value of the function is equal to that of the electrostatic potential ϕ at a mesh

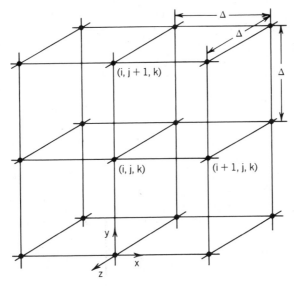

Figure 2.20 Spatial mesh for the finite difference solution of the Poisson equation. Cartesian mesh with uniform spacing in x, y, and z.

point:

$$\Phi(i,j,k) = \phi(i\Delta, j\Delta, k\Delta). \tag{2.94}$$

We can fit the values of $\Phi(i,j,k)$ with a smooth curve to define $\phi(x, y, z)$. Gradients of ϕ yield the components of electric field.

The finite-difference form of the Poisson equation is [CPA, p. 67]

$$\Phi(i,j,k) \cong [\Phi(i+1,j,k) + \Phi(i-1,j,k) + \Phi(i,j+1,k) + \Phi(i,j-1,k)$$
$$+ \Phi(i,j,k+1) + \Phi(i,j,k-1)]/6 - \rho(i,j,k)\Delta^3/\varepsilon_0. \tag{2.95}$$

The final term is the space-charge density at the point $(i\Delta, j\Delta, k\Delta)$. The quantity $\rho(i,j,k)\Delta^3$ equals the number of computational particles within a cell of volume Δ^3 centered at mesh point (i,j,k), multiplied by the charge per computational particle. When $\rho\Delta^3$ is known, Eq. (2.95) can be solved by a variety of methods including successive overrelaxation [CPA, p. 55].

An electrostatic computer simulation using the PIC method for field calculations consists of the following operations at each time step:

1. At time t, the electric field at the location of each computational particle is interpolated from the function $\Phi(i,j,k)$. The field is used to advance the vectors **x** and **v** for the particles to $t + \Delta t$ using an accurate difference scheme (Section 2.3).

2. The function $\rho(i,j,k)\Delta^3$ is evaluated at the mesh points by assigning the charge of the computational particles according to their position.

3. The Poisson equation is solved to find electric fields at $t + \Delta t$, and the process is repeated.

The procedure continues until the beam advances to the desired final state. We shall have an opportunity to study one-dimensional computer simulations in Section 12.4 and 13.1.

The charge assignment process is the heart of a PIC simulation. Table 2.1 contains a listing of a subroutine to calculate the discrete density and average velocity function (or equivalently, the charge and current densities) of a set of computational particles for a one-dimensional simulation. Two arrays, x[j] and v[j], define the distribution function of computational particles. The routine incorporates the following assumptions:

1. There are NPart computational particles with positions x_i in a region of width L between $x = 0$ and $x = L$.

2. The discrete density function n_i is defined at (NMesh + 1) points uniformly

TABLE 2.1   Density and Average Velocity of a Distribution[a]

```
{Set the density and velocity arrays equal to zero}
  for i:= 0 to Nmesh do
    begin
      Dens[i]:= 0.0;
      Vav[i]:= 0.0;
    end;

{Define factors to eliminate redundant floating point operations}
  NormFact: = Dens0*L;
  HalfDelta:= Delta/2.0;

{Sweep through all computational particles and assign them to mesh points, and find the
sum of particle velocities in a cell}
  for j:= 1 to NPart do
    begin
      index:= int((x[j] + HalfDelta)/Delta);
      Dens[index]:= Dens[index] + 1.0;
      Vav[index]:= Vav[index] + v[j];
    end;

{Normalize the results}
  for i:= 0 to NMesh do
    begin
      Vav[i]:= Vav[i]/Dens[i];
      Dens[i]:= Dens[i]*NormFact;
    end;
```

[a]Written in PASCAL, simple point assignment.

spaced a distance Delta = L/NMesh apart. The location of mesh point i is $x_i = i*$Delta.

3. If particles are uniformly spaced, the beam has line density Dens0 (particles/m). Each computational particle represents Dens0*L/NPart real particles.

The routine determines if a particle is within a distance $\pm \Delta/2$ of a mesh point and assigns the density carried by the computational particle to the corresponding cell. The approach to find the average velocity array, Vav[i], is similar.

2.10. MOMENT EQUATIONS

In beam theory, the investigation of velocity space instabilities and collective phenomena like Landau damping (Section 14.4) requires a complete solution of the Vlasov equation. Although the equation has a simple form, its solution is usually difficult. Sometimes, we can describe motion of a beam without a detailed description of its velocity distribution. It is often sufficient to approximate a beam velocity distribution with a simple form such as a delta function (cold beam) or a Maxwell distribution.

In this section, we shall derive a set of reduced equations from the Vlasov equation by taking weighted averages over velocity space. We shall limit attention to low-energy distributions described by the nonrelativistic Boltzmann equation of Section 2.7—the more complex equations that result from relativistic distributions are of limited use. In classical mechanics, weighted averages over a mass distribution are called *moments*; hence, the equations we shall derive are called *moment equations*. The moment equations are useful even though they reveal no information about the evolution of the velocity distribution. They involve measurable quantities, such as particle density, average velocity, and temperature—their application is helpful to develop physical insight into collective phenomena.

A velocity moment taken over a particle distribution function has the following general form:

$$\mathbf{M}_n(x, y, z, t) = \int dv_x \int dv_y \int dv_z \mathbf{v}^n f(x, v_x, y, v_y, z, v_z, t). \tag{2.96}$$

We have already encountered a version of Eq. (2.96) in Section 2.8. The moment corresponding to $n = 0$ is a scalar quantity, the particle density $n(x, y, z)$. The moment for $n = 1$ is the average velocity of the particles at a location multiplied by the density. We shall derive the moment equations by applying the operation of Eq. (2.96) to all terms in the collisionless Boltzmann equation [Eq. (2.73)] for different values of n. The resulting equations apply to charged particle beams when the force in Eq. (2.73) is the Lorentz force. To simplify notation, the calculations are performed for a one-dimensional distribution $f(x, v_x)$.

The $n = 0$ moment over the Boltzmann equation has the form

$$\int_{-\infty}^{+\infty} dv_x \frac{\partial f}{\partial t} + \int_{-\infty}^{+\infty} dv_x v_x \frac{\partial f}{\partial x} + \int_{-\infty}^{+\infty} dv_x a_x \frac{\partial f}{\partial v_x} = 0. \tag{2.97}$$

We shall analyze the terms of Eq. (2.97) separately. In the first term, we can move the partial derivative operation outside the integral. The result is

$$\int_{-\infty}^{+\infty} dv_x \frac{\partial f}{\partial t} = \frac{\partial}{\partial t} \int_{-\infty}^{+\infty} dv \, f(x, v_x) = \frac{\partial n(x)}{\partial t}. \tag{2.98}$$

The final form on the right-hand side of Eq. (2.98) proceeds from Eq. (2.48).

We also move the spatial partial derivative operation outside the integral in the second term of Eq. (2.97). Multiplying and dividing the term by $\int dv_x f$ gives

$$\int_{-\infty}^{+\infty} dv_x v_x \frac{\partial f}{\partial x} = \frac{\partial}{\partial x} \left[\left(\int dv_x f \right) \left(\frac{\int dv_x v_x f}{\int dv_x f} \right) \right]. \tag{2.99}$$

The integrals in Eq. (2.99) extend over all velocity space. The term in parentheses is the average velocity in the x direction [Eq. (2.53)]. The multiplying factor equals the particle density. The second term thus assumes the form

$$\frac{\partial}{\partial x} [n(x) \langle v_x(x) \rangle]. \tag{2.100}$$

To simplify the third term, we note that Eq. (2.73) is invalid if the force depends on v_x. Therefore, the acceleration in Eq. (2.97) varies only with x. Moving the acceleration outside the velocity integral leads to a modified form for the third term:

$$\int_{-\infty}^{+\infty} dv_x a_x \frac{\partial f}{\partial v_x} = a_x \int_{-\infty}^{+\infty} dv_x \frac{\partial f}{\partial v_x} = a_x f \Big|_{-\infty}^{+\infty}. \tag{2.101}$$

If a beam has finite kinetic energy density, the distribution function must drop to zero as $v_x \to \pm \infty$. Therefore, the third term equals zero.

In summary, the $n = 0$ moment equation is

$$\frac{\partial n}{\partial t} + \frac{\partial}{\partial x} (n \langle v_x \rangle) = 0. \tag{2.102}$$

Equation (2.102) describes the changes in particle density with time at a particular location. We can understand the physical meaning of the equation by expanding

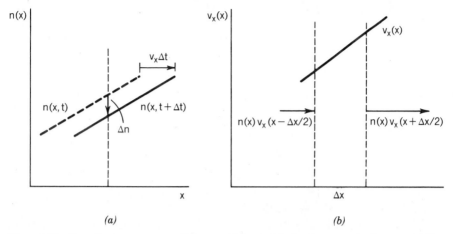

Figure 2.21 Contributions to temporal changes of the particle density at a point in space. (a) The density changes when particles have an average directed velocity and a spatial density gradient. (b) The density changes when there is a gradient of the average particle velocity. Here, density decreases at a point because more particles leave through the right-hand boundary than enter through the left-hand boundary.

the x derivative:

$$\frac{\partial n(x)}{\partial t} = - n(x) \frac{\partial \langle v_x(x) \rangle}{\partial x} - \langle v_x(x) \rangle \frac{\partial n(x)}{\partial x}. \tag{2.103}$$

Changes in $n(x)$ at a point occur when there are spatial gradients of the density or of the average particle velocity. Figure 2.21a illustrates the effect of a density gradient. The spatial distribution of particles is shown at times t and $t + \Delta t$ for a positive value of $\langle v_x \rangle$. The density changes by an amount

$$\Delta n = - \left(\frac{\partial n}{\partial x} \right) \langle v_x \rangle \Delta t. \tag{2.104}$$

Figure 2.21b shows the effect of a velocity gradient on the number of particles contained in a differential element Δx centered at x. The average velocity at each side of the element is

$$\langle v_x(x \pm \Delta x/2) \rangle \cong \langle v_x(x) \rangle \pm \frac{\partial \langle v_x(x) \rangle}{\partial x} \frac{\Delta x}{2}. \tag{2.105}$$

The change in the number of particles in the element over time Δt equals the difference in particle flux crossing the boundaries multiplied by Δt. The particle flux equals the product of the density and average velocity. The rate that particles enter the left-hand boundary is

$$\Delta n(\text{left}) \cong n(x)\langle v_x(x)\rangle - n(x)\frac{\partial\langle v_x(x)\rangle}{\partial x}\frac{\Delta x}{2}. \tag{2.106}$$

The change in the number of particles in the element resulting from the velocity gradient is $(\partial n/\partial t)\Delta x = \Delta n(\text{left}) - \Delta n(\text{right})$, or

$$\Delta n = -n\left(\frac{\partial\langle v_x\rangle}{\partial x}\right)\Delta t. \tag{2.107}$$

Adding the effects of a density gradient [Eq. (2.104)] and a gradient of average velocity [Eq. (2.107)] leads to Eq. (2.102).

The three-dimensional extension of Eq. (2.102) is written in vector notation as:

$$\frac{\partial n}{\partial t} = -\nabla\cdot(n\langle \mathbf{v}\rangle). \tag{2.108}$$

The quantity $\langle \mathbf{v}\rangle$ is the average velocity vector. We recognize the right-hand side of Eq. (2.108) as the divergence of the particle flux [CPA, pg. 46]. The equation states that the time rate of change in the number of particles in a volume equals the particle flux into or out of the region. Equation (2.108) is called the *continuity equation*—it expresses conservation of the total number of particles in a system. If particles are created or absorbed, we must add source or sink terms to the equation.

We shall derive another equation from the collisionless Boltzmann equation by taking moments with $n = 1$ [Eq. (2.96)]. The one-dimensional form is

$$\int_{-\infty}^{+\infty} dv_x v_x \frac{\partial f}{\partial t} + \int_{-\infty}^{+\infty} dv_x v_x^2 \frac{\partial f}{\partial x} + \int_{-\infty}^{+\infty} dv_x a_x v_x \frac{\partial f}{\partial v_x} = 0. \tag{2.109}$$

Again, we shall deal with each term in sequence. From the preceding discussion, we recognize that the first term equals

$$\frac{\partial(n\langle v_x\rangle)}{\partial t}. \tag{2.110}$$

Rearrangement of the second term gives

$$\frac{\partial}{\partial x}(n\langle v_x\rangle^2). \tag{2.111}$$

Charged particle beams usually have a highly directed velocity with a small spread about the mean. It is convenient to write the velocity of particles in terms of a deviation from the average:

$$v_x = \langle v_x \rangle + \delta v_x.$$

By the definition of the average velocity, the average of deviations about $\langle v_x \rangle$ must be zero, $\langle \delta v_x \rangle = 0$. Therefore, the quantity $\langle v_x^2 \rangle$ in Eq. (2.109) equals

$$\langle v_x^2 \rangle = \langle v_x \rangle^2 + \langle \delta v_x^2 \rangle. \qquad (2.112)$$

We can again extract the acceleration from the velocity integral in the third term. We can modify the expression by noting that

$$v_x \frac{\partial f}{\partial v_x} = \frac{\partial}{\partial v_x}(f v_x) - f. \qquad (2.113)$$

The third term of Eq. (2.109) becomes

$$a_x \left[\int_{-\infty}^{+\infty} dv_x \frac{\partial}{\partial v_x}(f v_x) - \int_{-\infty}^{+\infty} dv_x f \right]. \qquad (2.114)$$

The integrand of the first integral in brackets is an exact differential. Again, if the average velocity of the distribution is finite, $(f v_x)$ must approach zero as $v_x \rightarrow \pm \infty$. Only the second term in Eq. (2.114) is nonzero; the integral $\int dv_x f$ equals the particle density.

If we combine Eqs. (2.110), (2.112), and (2.114) and multiply all terms by the particle rest mass, we find the following form for the $n = 1$ moment equation:

$$\frac{\partial}{\partial t}(m_0 n \langle v_x \rangle) = -\frac{\partial}{\partial x}\underbrace{[m_0 n \langle v_x \rangle^2]}_{\{1\}} - \frac{\partial}{\partial x}\underbrace{[m_0 n \langle \delta v_x^2 \rangle]}_{\{2\}} + \underbrace{n F_x}_{\{3\}}. \qquad (2.115)$$

The quantity F_x is the force on individual particles, $F_x = m_0 a_x$. Equation (2.115) expresses conservation of momentum at spatial positions within the particle distribution.

The left-hand side of Eq. (2.115) is the time rate of change of momentum per unit volume, $m_0 n \langle v_x \rangle$, at a location. The terms on the right-hand side contribute to the momentum change. We should note that the average momentum at a point can change by two processes:

1. An applied force may accelerate particles in the volume element.

2. Particles may leave the element and be replaced by new particles with a different average momentum.

The first process is familiar from single-particle dynamics. The second process is unique to collective physics—it can occur only if a system contains many particles. In the Eulerian viewpoint, it is unimportant which specific particles are at a location. Instead, we want to find the average properties of whatever particles occupy the point at a particular time. Although this contention may seem straightforward, we shall find in later chapters that it sometimes leads to results that challenge our intuition.

Term 3 on right-hand side of Eq. (2.115) represents the acceleration of particles in a volume element. A force F_x causes a time rate of change momentum per volume equal to $n(x)F_x(x)$. Terms 1 and 2 in Eq. (2.115) represent changes of momentum per volume resulting from particle migration. Term 1 describes momentum convection by a directed velocity. We can see this by noting that the flux of momentum across a surface normal to the x direction equals the momentum per particle multiplied by the number of particles that cross the surface per unit area and time. The mathematical expression of this statement is

$$\text{Momentum flux} = [m_0 \langle v_x \rangle][n \langle v_x \rangle]. \qquad (2.116)$$

Equation (2.116), combined with the arguments explaining Eq. (2.102), shows that the change in momentum resulting from convection in a differential length element is proportional to the gradient of $m_0 n \langle v_x \rangle^2$. Term 2 in Eq. (2.115) represents the migration of momentum associated with spatial variations of random velocity components. This term is often represented by ficticious forces, *pressure* and *viscosity*. We shall discuss the effects of velocity spread in Section 2.11.

We usually write Eq. (2.115) in a form that gives the time rate of change of the average velocity at a point:

$$\frac{\partial \langle v_x \rangle}{\partial t} = - \langle v_x \rangle \frac{\partial \langle v_x \rangle}{\partial x} - \frac{1}{n} \frac{\partial (n \langle \delta v_x^2 \rangle)}{\partial x} + \frac{F_x}{m_0}. \qquad (2.117)$$

The derivation of Eq. (2.117) involves the expansion of Eq. (2.115) and substitution from the continuity equation [Eq. 2.102)].

The condition $\langle \delta v_x^2 \rangle = 0$ defines a *cold beam*. The three-dimensional extension of Eq. (2.117) for a cold beam with electromagnetic forces is

$$\frac{\partial \langle \mathbf{v} \rangle}{\partial t} = -(\langle \mathbf{v} \rangle \cdot \nabla) \langle \mathbf{v} \rangle + \frac{q(\mathbf{E} + \langle \mathbf{v} \rangle \times \mathbf{B})}{m_0}. \qquad (2.118)$$

The vector notation of the convective term has the following form in Cartesian coordinates:

$$(\langle \mathbf{v} \rangle \cdot \nabla \langle \mathbf{v} \rangle = \mathbf{x} \left[\langle v_x \rangle \frac{\partial \langle v_x \rangle}{\partial x} + \langle v_y \rangle \frac{\partial \langle v_x \rangle}{\partial y} + \langle v_z \rangle \frac{\partial \langle v_x \rangle}{\partial z} \right]$$

$$+ \mathbf{y} \left[\langle v_x \rangle \frac{\partial \langle v_y \rangle}{\partial x} + \langle v_y \rangle \frac{\partial \langle v_y \rangle}{\partial y} + \langle v_z \rangle \frac{\partial \rangle v_y \rangle}{\partial z} \right] \qquad (2.119)$$

$$+ \mathbf{z} \left[\langle v_x \rangle \frac{\partial \langle v_z \rangle}{\partial x} + \langle v_y \rangle \frac{\partial \langle v_z \rangle}{\partial y} + \langle v_z \rangle \frac{\partial \langle v_z \rangle}{\partial z} \right].$$

The terms in Eq. (2.119) represent convection of momentum in three possible directions.

It is possible to generate higher-order moment equations from the collisionless Boltzmann equation by choosing $n > 1$. The equation for $n = 2$ describes heat transfer in a collisionless particle distribution. There is, in fact, an infinite set of moment equations. Higher-order equations give more detailed information about the velocity-space evolution of the distribution. Ultimately, a complete set of moment equations yields the same information about the distribution as a direct solution of Eq. (2.73). High-order moments of the Vlasov equation are sometimes applied to describe transport phenomena in plasmas. For charged-particle-beam theory, it is seldom necessary to extend the set beyond the momentum equation.

2.11. PRESSURE FORCE IN COLLISIONLESS DISTRIBUTIONS

In this section, we shall discuss how velocity spreads affect momentum balance among collisionless particles. Velocity dispersion leads to a force density that acts on volume elements of distributions. For a one-dimensional beam with uniform properties in the y and z directions, Eq. (2.115) shows that the force per volume resulting from random velocity components about a mean is

$$-\frac{\partial}{\partial x}[m_0 n \langle \delta v_x^2 \rangle]. \tag{2.120}$$

In Eq. (2.120), m_0 is the particle mass, n is the density, and δv_x is difference of the velocity of a particle from the mean, $\delta v_x = v_x - \langle v_x \rangle$. Equation (2.120) implies that momentum may be transferred to or from a volume element when there is a gradient in either the particle density or the mean squared velocity spread, $\langle \delta v^2 \rangle$.

The volumetric force of Eq. (2.120) is often called the *pressure force*. This term sometimes confuses people who are unfamiliar with collective physics. Most of us have a strong intuitive view of pressure from gas dynamics. The force arises when gas atoms with random velocity components collide with one another—we feel that the gas atoms push against each another. If this is true, how can we define pressure for collisionless distributions?

The key to understanding pressure is to remember that a volumetric force is synonymous with a time rate of change of momentum per volume. Section 2.10 pointed out that momentum density can change as a result of both applied forces and particle migration. An observer measuring macroscopic properties of a beam cannot distinguish between the two processes. To the observer, momentum change through particle migration appears as a force acting at a location. In the macroscopic view of the observer, the pressure term of Eq. (2.120) acts as a force per volume. In the microscopic view, the collisionless particles do not interchange momentum. There is no actual transfer of momentum between particles. Instead, gradients in $n \langle \delta v_x^2 \rangle$ effect changes in the momentum density at a location by particle transport.

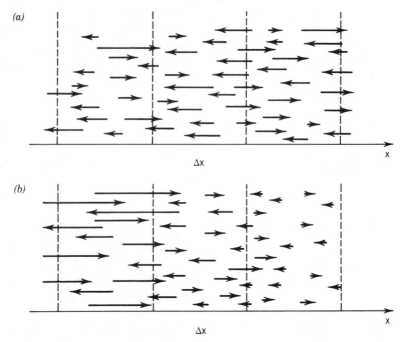

Figure 2.22 Changes in the average particle momentum at a point in space resulting from a velocity spread. Velocity vectors in figure represent a beam with uniform density and zero average velocity in the x direction. (*a*) With a spatially uniform velocity spread, $\langle v_x^2 \rangle$, the average velocity in the center cell remains equal to zero. (*b*) With a negative gradient of velocity spread, more particles enter through the left-hand boundary than through the right-hand boundary. The average momentum in the $+x$ direction increases with time.

Figure 2.22 illustrates the meaning of Eq. (2.120) for a one-dimensional distribution. Figure 2.22*a* shows two boundaries defining differential length elements—the quantities n and $\langle \delta v^2 \rangle$ are equal on both sides of each boundary. Assume that the particles in the elements have no average drift velocity, $\langle v_x \rangle = 0$. The velocity dispersion causes momentum flux across the boundaries. From the discussion of Section 2.10, the flux across a boundary equals the average of individual particle momenta weighted by their velocity:

$$\text{Boundary momentum flux} = \pm m_0 n \langle \delta v_x^2 \rangle. \tag{2.121}$$

Because n and $\langle \delta v_x^2 \rangle$ are uniform, the magnitudes of flux in the positive and negative directions are equal. Therefore, there is no change in momentum density in the volume element between the boundaries. In the macroscopic view, there is no apparent volumetric force.

   Figure 2.22*b* shows the magnitude of flux across the boundaries when there is a negative gradient in $m_0 n \langle \delta v_x^2 \rangle$. The fraction of particles with momentum in the positive x direction increases in the central element—the average momentum in the element assumes a positive, nonzero value. In the macroscopic view, an apparent volumetric force in the $+x$ direction acts on the element. We can apply the technique described in Section 2.10 to estimate the momentum fluxes on the boundary. A Taylor's expansion of $m_0 n \langle \delta v_x^2 \rangle$ about the center of the length element leads to Eq. (2.120).

We shall identify the quantity $m_0 n \langle \delta v_x^2 \rangle$ as the particle *pressure* in x direction:

$$p_x = m_0 n \langle \delta v_x^2 \rangle. \tag{2.122}$$

In MKS units, pressure has dimensions newtons per square meter. The expression of Eq. (2.120) can be recast as:

$$- \partial p_x / \partial x. \tag{2.123}$$

The pressure notation can be extended to distributions with three-dimensional variations. If particle motions are decoupled in x, y, and z, then we can define individual terms for the pressure:

$$p_x = m_0 n \langle \delta v_x^2 \rangle, \qquad p_y = m_0 n \langle \delta v_y^2 \rangle, \qquad p_z = m_0 n \langle \delta v_z^2 \rangle. \tag{2.124}$$

Incorporating the definitions of Eqs. (2.123) and (2.124), the momentum equation for nonrelativistic charged particles [Eq. 2.117)] becomes

$$\frac{\partial \langle \mathbf{v} \rangle}{\partial t} = -(\langle \mathbf{v} \cdot \nabla \rangle \langle \mathbf{v} \rangle - \frac{\nabla \cdot \mathbf{p}}{n m_0} + \frac{q(\mathbf{E} + \langle \mathbf{v} \rangle \times \mathbf{B})}{m_0}. \tag{2.125}$$

The pressure vector for decoupled motion is

$$\mathbf{p} = p_x \mathbf{x} + p_y \mathbf{y} + p_z \mathbf{z}. \tag{2.126}$$

In Cartesian coordinates, the expansion of the divergence operator is

$$\nabla \cdot \mathbf{p} = \partial p_x / \partial x + \partial p_y / \partial y + \partial p_z / \partial z. \tag{2.127}$$

When particle motions in the x, y, and z directions are not decoupled, momentum flow becomes more complex. The momentum equation involves terms like $m_0 n \langle (v_x v_y) \rangle$ —pressure must be defined as a tensor. The cross terms represent collisionless viscosity. Although viscosity is a useful concept in gas dynamics and plasma physics, it is seldom required for charged-particle-beam theory.

Another quantity, the total pressure, is useful if the distribution of particle velocity is isotropic (uniform in the three dimensions). The total pressure is defined by

$$\begin{aligned} p &= m_0 n (\langle \delta v_x^2 \rangle + \langle \delta v_y^2 \rangle + \langle \delta v_z^2 \rangle)/3 \\ &= (p_x + p_y + p_z)/3 = p_x. \end{aligned} \tag{2.128}$$

A global temperature T can be defined for an isotropic Maxwellian velocity distribution. Both temperature and pressure depend on the velocity spread. Comparison of Eqs. (2.128) and (2.66) shows that

$$p = nkT. \qquad (2.129)$$

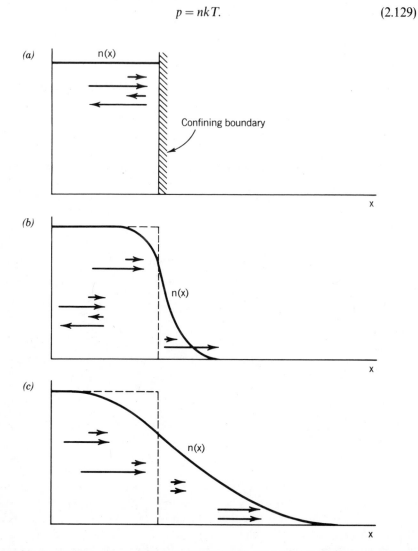

Figure 2.23 Pressure force in a collisionless particle distribution—free expansion at a boundary. Figure plots density as a function of position with arrows to represent velocity vectors. (*a*) Initial state with particles confined at the boundary. Uniform density for $x < 0$, average momentum equals zero at all positions. (*b*) Shortly after removal of boundary. The density decreases in region $x < 0$ as particles move into free space with no reflection. Also, there is an average momentum in the $+x$ direction. (*c*) Later, particles drift into the region $x > 0$. In the macroscopic viewpoint, apparent acceleration by a pressure force causes the increase in average momentum. Particles separate according to the magnitude of their velocity, resulting in cooling of the distribution.

Two examples serve to clarify the relationships between the macroscopic expressions for the pressure force and the microscopic view of single-particle motions. First, consider the free expansion of a one-dimensional particle distribution. In gas dynamics, a free expansion occurs when a high-pressure gas is released by a punctured diaphragm. In the macroscopic viewpoint, the pressure gradient drives the gas outward. As it expands, the gas cools. Collisionless particle distributions have similar behavior. Figure 2.23 shows the free expansion of collisionless particles. A localized force initially confines the particles at a boundary (Fig. 2.23a)—the force is strong enough to reverse the velocity of all particles at the boundary. In the macroscopic view, the confining force balances the pressure force.

The confining force disappears instantaneously at time $t = 0$. The collisionless particles drift into the region $x > 0$ according to their initial velocities. Fast particles move farther than slow particles. Plots of the density and average velocity as a function of x are shown for a short delay in Fig. 2.23b and for a long delay in Fig. 2.23c. Because of particle migration, length elements near the boundary have a positively directed average momentum that grows in time. In the macroscopic view, the length elements appear to accelerate in response to a force pointing in the $+x$ direction. At later times, drift motion separates the fast and slow particles, reducing the velocity spread in downstream length elements. This process lowers the temperature, or velocity spread, in a length element. As particles spread out in space, the apparent acceleration of length elements decreases. In the microscopic view, this effect results from the reduction of gradients of n and $\langle \delta v^2 \rangle$.

As a second example, we shall derive the variation of density of a group of charged particles in equilibrium subject to a one-dimensional electric force. The positively charged particles have an isotropic Maxwell velocity distribution. We shall apply the results to thermionic emission in Section 6.2 and to plasma sheaths in Section 12.1. Assume that the particle temperature T is independent of position. The electric field is zero in the region $x \leqslant 0$, and has variation $-E_x(x)$ in the region $x > 0$. The electric force confines particles near a boundary at $x = 0$. Since there are no applied forces in the region $x \leqslant 0$, the particle density has the uniform value n_0.

Since the particles are in equilibrium, the average velocity is zero everywhere, $\langle v_x \rangle = 0$. Equation (2.115) reduces to

$$-(kT)[dn(x)/dx] + eE_x(x)n(x) = 0. \tag{2.130}$$

Equation (2.130) has the solution,

$$n(x) = n_0 \exp\left[\int^x dx' eE_x(x')/kT\right]. \tag{2.131}$$

If electrostatic potential at the boundary equals zero, $\phi(0) = 0$, then Eq. (2.131) assumes the familiar form,

$$n(x) = n_0 \exp[-e\phi(x)/kT]. \tag{2.132}$$

The same equation results from a microscopic analysis. The velocity distribution at $x = 0$ is

$$n(x, v_x) = n_0[m_0/2\pi kT]^{1/2} \exp[-m_0 v_x^2/2kT]. \tag{2.133}$$

The density is lower at a location $x > 0$ because only a fraction of the particles have a kinetic energy high enough to overcome the electrostatic potential energy, $m_0 v_x^2/2 \geqslant e\phi$. The fraction of particles that can reach position x with electrostatic potential $\phi(x)$ is

$$n(x)/n_0 = \int_{(2e\phi/m_0)^{1/2}}^{\infty} dv_x \exp(-m_0 v_x^2/2kT) \bigg/ \int_0^{\infty} dv_x \exp(-m_0 v_x^2/2kT)$$

$$= \exp[-e\phi(x)/kT]. \tag{2.134}$$

Equation (2.134) is identical to Eq. (2.133). Note also that the relative velocity distribution in Eq. (2.134) is independent of x—this property of the Maxwell distribution justifies the assumption of constant temperature.

In beam theory, particle motions in the transverse directions (x, y) are usually separable from those in the axial direction (z). When a charged particle beam has a velocity spread in the transverse direction, focusing forces must be applied to confine the beam to a limited cross-sectional area. Although we sometimes use the terms "beam pressure" or "beam temperature," the common name for velocity dispersion in beam theory is *emittance*. Chapters 3 and 4 give detailed discussions of beam emittance, including the apparent expansion force associated with transverse velocity spread (Section 3.5).

2.12. RELATIVISTIC PARTICLE DISTRIBUTIONS

The derivations of phase-volume conservation in Section 2.4 and the collisionless Boltzmann equation Section 2.7 were limited to nonrelativistic particles. This approximation is adequate for plasma physics and sometimes applies to high-energy charged particle beams. For example, transverse particle motion in a paraxial beam is well described by Newtonian dynamics with an adjusted particle mass, $m = \gamma m_0$. Nonetheless, there are many applications for relativistically correct equations. One example is the axial motion of high-energy electrons in the RF fields of a resonant accelerator [CPA, Chapter 13]. When observed in the accelerator frame, all electrons travel close to the speed of light, independent of their kinetic energy. Longitudinal forces cause changes of momentum rather than velocity.

We must revise our definition of phase space for relativistic distributions—it is impractical to organize particle orbit parameters in terms of their velocity. For example, a one-dimensional axial distribution of relativistic particles would appear simply as a narrow line near $v_z = c$. A 10-MeV electron beam with a ± 250-keV spread in kinetic energy has a relative velocity spread of only $\pm 0.005\%$. The electrons do have a significant spread in axial momentum because of variations in the relativistic mass. This suggests organizing relativistic particle distributions in a phase space of position **x** and momentum **p** rather than velocity.

It is easy to recast the collisionless Boltzmann equation in terms of momentum. Again, we shall use a one-dimensional distribution as an example. Consider axial particle motion in terms of the variables z and p_z. The relativistic equations of motion are

$$dp_z/dt = F_z \qquad (2.135)$$

and

$$dz/dt = p_z/\gamma m_0, \qquad (2.136)$$

where

$$\gamma = [1 + (p_z/m_0 c)^2]^{1/2}. \qquad (2.137)$$

There are two differences from the nonrelativistic equations of motion. First, the force equation [Eq. (2.135)] describes the time rate of change of momentum rather than velocity. Second, the position equation [Eq. (2.136)] is more complex.

Using the reasoning of Section 2.4, we can show that the (z, p_z) phase-space area occupied by a group of collisionless particles remains constant in time. Figure 2.24a shows an area element in (z, p_z) phase space. To begin, imagine how

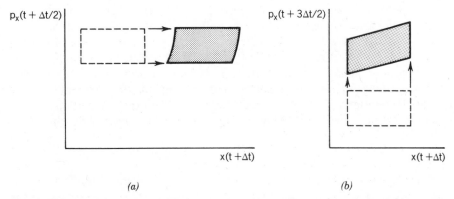

$p_x(t + \Delta t/2)$

$x(t + \Delta t)$

(a)

$p_x(t + 3\Delta t/2)$

$x(t + \Delta t)$

(b)

Figure 2.24 Phase-volume conservation for a relativistic one-dimensional distribution (z, p_z) illustrated by the leapfrog method. (a) Change of a bounded phase space region over a time step. Momenta at $t + \Delta t/2$ advance positions to time $t + \Delta t$. (b) Forces calculated from positions at time $t + \Delta t$ advance momenta to time $t + 3\Delta t/2$.

the element changes in an interval Δt when there are no forces. Axial velocity causes a shift of particle positions along the z axis. In contrast to a nonrelativistic beam, the position shift is not linearly proportional to the difference in momentum from the mean value. An element that is initially rectangular does not change to a trapezoid. Nonetheless, the width of each slice of the element does not change in Δt. Figure 2.24a shows that the area of the element is constant although the shape changes. Modification of the element over Δt by forces follows the same reasoning used in Section 2.4. We can expand the forces in a Taylor's series about the mean position. When plotted in (z, p_z) space, a rectangular phase area subjected to a force over a time Δt is distorted to a trapezoid with equal area (Fig. 2.24b).

Given that a phase volume in (z, p_z) space occupied by a group of collisionless relativistic particles is constant in time, we can derive a modified collisionless Boltzmann equation following the development of Section 2.7. We define a continuous relativistic distribution function, $f(z, p_z)$. The quantity $f(z, p_z)\Delta z \Delta p_z$ represents the number of particles in a beam located in region of phase space of area $\Delta z \Delta p_z$ near z, p_z. The mathematical expression of conservation of phase volume is

$$\frac{Df(z, p_z)}{Dt} = \frac{\partial f}{\partial t} + \left(\frac{p_z}{\gamma m_0}\right)\frac{\partial f}{\partial z} + F_z(z)\frac{\partial f}{dp_z}. \tag{2.138}$$

The three-dimensional form of the Vlasov equation becomes

$$\frac{\partial f}{\partial t} + \left(\frac{\mathbf{p}}{\gamma m_0}\right)\cdot\frac{\partial f}{\partial \mathbf{x}} + q(\mathbf{E} + \mathbf{v} \times \mathbf{B})\cdot\frac{\partial f}{\partial \mathbf{p}} = 0. \tag{2.139}$$

It is disappointing to find that the relativistic collective equations are less symmetric and are more difficult to solve than the corresponding Newtonian equations. Generally, we hope that physical formulations become simpler as we approach closer to the truth. The factor of γ in the denominator of the second term of the Vlasov equation upsets the symmetry. The derivation of moment equations from the Vlasov equation (Section 2.10) becomes difficult in the general relativistic limit. As a result, there is little motivation to describe relativistic beams in terms of averaged quantities and moment equations except in the limit that $\gamma \cong$ constant. Theoretical treatments of relativistic distributions usually assume a small spread in kinetic energy and proceed directly from the Vlasov equation.

3

Introduction to Beam Emittance

Emittance is a measure of the parallelism of a beam—it allows us to compare the quality of beams for applications. This chapter describes techniques to find emittance and shows how to use the quantity to predict the evolution of beams. The emittance is a conserved quantity in an ideal focusing system. We can gauge imperfections in transport systems by measuring emittance growth.

Beams with random components of transverse velocity have a spread in angle relative to the axis of propagation. Section 3.1 shows how the angular dispersion limits the ability to focus beams. Section 3.2 defines emittance for beams where particle motions are independent in the x and y directions. We can find the quantity by plotting orbit vector points in a modified phase space called trace space with axes of position and transverse angle, $[x, x']$. Emittance is proportional to the area filled by the points. Section 3.3 illustrates the physical meaning of emittance by discussing methods to measure the quantity. Section 3.4 extends the emittance definition to beams where transverse motions are not separable. The section introduces some new quantities, including the *brightness*, a function of the emittance. Brightness quantifies the maximum focused power flux of a beam.

Section 3.5 incorporates the emittance into a simple expression for the effective transverse force resulting from the pressure, or velocity dispersion, of a beam. Section 3.6 applies the expression to a practical problem, the expansion or compression of a beam with nonzero emittance in a drift region. We can use the results to find the maximum propagation length or the minimum focal spot size for a beam. Section 3.7 shows how imperfect focusing systems can cause emittance growth. We shall investigate the effects of lenses where the magnitude of the focusing force does not increase linearly from the axis. Section 3.8 applies the principle of emittance conservation to derive some useful equations that relate the velocity dispersion of a beam to a given change in its dimensions.

3.1. LAMINAR AND NONLAMINAR BEAMS

As an introduction to emittance, we shall review the subject of order and disorder in beams. In particular, we shall find how disorder limits the transport properties of beams. Beams with good parallelism are easier to transport than beams with large random transverse velocity components. Ordered beams can focus to a small spot size. We shall first discuss the properties of ideal ordered beams and then review the limitations set by disorder.

The ideal charged particle beam has laminar particle orbits. Orbits in a laminar beam flow in layers (or *laminae*) that never intersect. A laminar beam satisfies two conditions:

1. All particles at a position have identical transverse velocities. If this is not true, the orbits of two particles that start at the same position could separate and later cross each other.

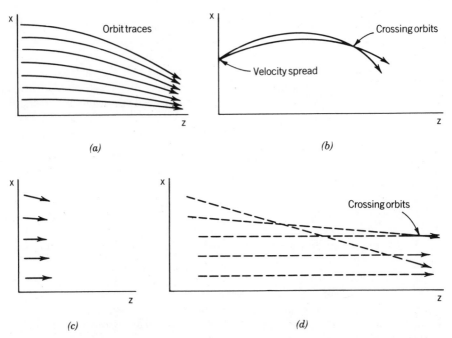

Figure 3.1 Motion of beam particles viewed in configuration space. (*a*) Particle orbits in a laminar beam. (*b*) Crossing orbits, where two particles at the same point have different transverse velocities. (*c*) Nonlaminar distribution, where peripheral particles have excess inward velocity. (*d*) Downstream projection of particle orbits for the nonlaminar distribution.

2. The magnitude of the transverse particle velocity is linearly proportional to the displacement from the axis of beam symmetry.

Some examples will illustrate the implications of the conditions. Figure 3.1 shows particle orbits of a beam viewed in configuration space. Orbits in Fig. 3.1a are laminar. The beam of Fig. 3.1b is nonlaminar because two particles at the same position have different transverse velocities—condition 1 is not satisfied. As the beam propagates through a linear focusing system, the orbits separate and later cross one another. Figures 3.1c and 3.1d illustrate a circumstance where condition 2 is violated. At an initial position (Fig. 3.1c), particles near the axis have small transverse velocity, while displaced particles have a large inward-

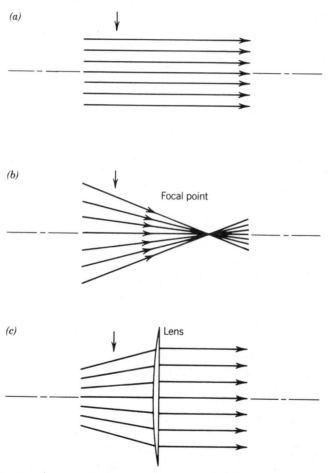

Figure 3.2 Configuration-space view of particle orbits in a laminar beam. (*a*) Ideal parallel beam. (*b*) Converging laminar beam, where orbits pass through a common focus. (*c*) Diverging laminar beam converted to a parallel beam by a linear lens.

directed velocity. The transverse velocity is not linearly proportional to displacement. Figure 3.1d shows the state of orbits at a downstream location. The peripheral particles have moved toward the axis, crossing the inner particle orbits. Particles at a location may have multiple values of transverse velocity.

Figure 3.2 illustrates some examples of orbits in laminar beams. The beam of Fig. 3.2a is parallel—all particles have zero transverse velocity. There are no orbits that cross in such a beam. The parallel beam propagates an infinite distance with no change in its width. As a second example, Fig. 3.2b shows a converging laminar beam. Since the transverse velocities are proportional to displacement, particle orbits define similar triangles, which converge to a point. After passing through the singularity at the focal point, the particles follow uniformly diverging orbits. Figure 3.2c shows a diverging beam focused by the forces of an electric field lens. If the lens forces are linearly proportional to displacement from the symmetry axis, the lens maintains the laminar flow of the

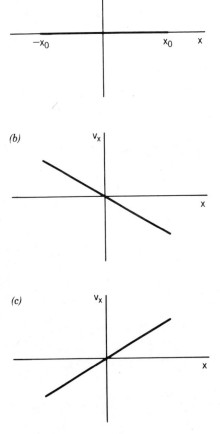

Figure 3.3 Laminar-beam orbit-vector distributions viewed in transverse phase space. (*a*) Distribution for a parallel beam of width $\pm x_0$. (*b*) Distribution for a converging beam. (*c*) Distribution for a diverging beam.

beam. We can always convert a converging or diverging beam to a parallel beam by using a lens of the proper focal length.

Figure 3.3 depicts the three particle distributions of Fig. 3.2 as orbit vector plots in phase space. Although the three configuration-space plots differ, the phase-space representations are similar. The phase-space vector plot of a laminar beam is always a straight line of zero thickness. The condition that the particle distribution has zero thickness proceeds from condition 1; the line straightness is a consequence of condition 2. The distribution of a laminar beam propagating through a transport system with ideal linear focusing elements is a straight line with variable length.

An ideal lens can focus a laminar beam to a point of zero dimension. Figure 3.4a illustrates focusing by a lens of focal length f in configuration space—the parallel incident beam has a halfwidth x_0. After deflection, the particles converge to a point a distance f from the lens. Figure 3.4b is a phase-space view of the same process. The incident beam distribution is a straight line of length $2x_0$ aligned along the x axis. The lens displaces the distribution in the v_x direction while preserving the projected length along the x axis. The velocity displacement has a maximum value of $(x_0/f)v_z$ at the beam edge. During subsequent transit through the drift region of length f, the orbit vectors converge toward $x = 0$. The orientation of the distribution changes until it aligns with the v_x axis at the focal point. At this point, the distribution has dimension equal to zero in x and a halfwidth along v_x of $\pm (x_0/f)v_z$.

Particles in a nonlaminar beam have a random distribution of transverse velocities at a location. Particles at the position x have different values of v_x and a spread in directions. Because of the disorder of a nonlaminar beam, it is impossible to aim all particles from a location in the beam toward a common point. Lenses can influence only the average motion of particles. Focal spot

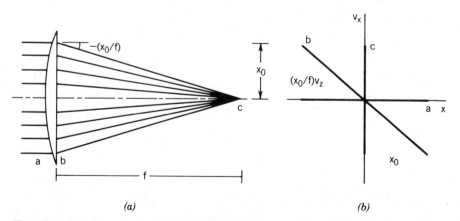

(a) *(b)*

Figure 3.4 Focusing a laminar beam. (*a*) Configuration-space view of particle-orbit traces. (*b*) Snapshots of orbit-vector distributions in phase space at the positions (*a*), (*b*), and (*c*) marked in part *a*.

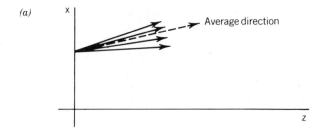

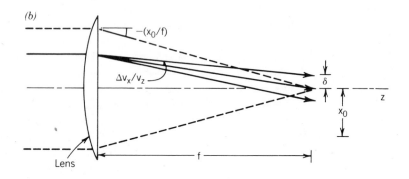

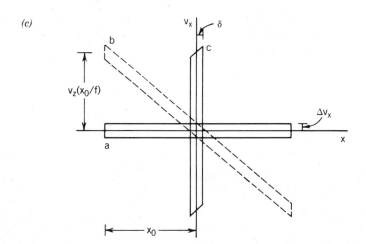

Figure 3.5 Properties of nonlaminar beams. (*a*) Configuration-space view of particle orbits—particles at the same point move in different directions. (*b*) Configuration-space view of the transverse focusing of a nonlaminar beam. Incident beam has an equal spread in angle at all positions. (*c*) Snapshots of orbit-vector distributions in phase space at positions (*a*), (*b*), and (*c*) marked in part *b*.

limitations are a major concern for a wide variety of applications, from electron microscopy to ion-beam inertial fusion.

Figure 3.5a shows orbits at a location of a nonlaminar beam and the definition of the average direction. Figure 3.5b is a configuration-space view of the focusing process for nonlaminar beams. The beam portrayed has a uniform spread of transverse velocity, $\pm \Delta v_x$, at all positions in the cross section. The beam entering the lens is almost parallel—the average transverse velocity at all positions equals zero. Individual particles have inclination angles in the range $\pm \Delta v_x/v_z$. Passing through the lens, particles acquire a directed convergence angle that sums with the angular spread. The best focus occurs at approximately a distance f from lens. Figure 3.5b shows that the orbit vectors at the focus spread over a nonzero spatial width. The halfwidth is

$$\delta \cong \Delta v_x f/v_z. \tag{3.1}$$

Equation (3.1) reveals that the width of the focal spot is proportional to the transverse velocity spread of the incident beam.

Figure 3.5c illustrates the same process viewed in phase space. The incident beam distribution is no longer a line; it occupies a rectangular region in phase space with dimensions $2x_0$ and $2\Delta v_x$. The lens shifts the rectangle along the velocity axis—the maximum displacement is

$$\Delta v_0 = v_z(x_0/f). \tag{3.2}$$

In the drift region, the rectangle rotates to an upright orientation. The beam has a finite width at the focal point. In the limit of small velocity spread, $\Delta v_x \ll \Delta v_0$, conservation of phase-space area implies that the width of the focal spot is given by Eq. (3.1). The phase-space area occupied by the orbits of the incident parallel beam is

$$A_p = 4x_0\Delta v_x. \tag{3.3}$$

At the focal point, the height of the beam along the velocity axis is almost equal to Δv_0 [(Eq. (3.2)]. For a halfwidth δ, the phase area at the focal point is

$$A_p = 4\delta v_z(x_0/f). \tag{3.4}$$

Setting the expressions of Eqs. (3.3) and (3.4) equal leads to Eq. (3.1).

We can apply similar consideration to particle motion in the axial direction. In particular, we can define the longitudinal equivalent of laminar flow—all particles at an axial location are monoenergetic. Just as a spread in transverse angles limits the focal spot of a beam, a spread in axial velocity limits the ability to compress the beam to short pulses. Axial focusing of beams is called *bunching*. The process is used, for example, to match beams into RF accelerators. As with laminar beams, monoenergetic beams can converge to pulses of arbitrarily small duration.

To illustrate bunching, consider a pulsed monoenergetic beam with axial length z_0. Initially, all particles have the same axial velocity v_0. The pulse length of the beam (the time to pass a given point) is $\Delta t_p = z_0/v_0$. The particles pass through a buncher, an element that imparts a velocity change linearly proportional to the distance from the beam midpoint. Particles at the midpoint have no change in velocity, particles at the beam head decelerate by an amount $-\Delta v_0$, while particles at the tail accelerate by $+\Delta v_0$. Peak bunching occurs when the tail particles overtake particles at the midpoint. The propagation time from the buncher to the point of peak convergence is

$$\tau = z_0/(2\Delta v_0). \tag{3.5}$$

The corresponding drift distance is

$$L = z_0 v_0/(2\Delta v_0). \tag{3.6}$$

We can apply an approach similar to that used for transverse focusing to find the minimum bunched pulse length for a beam with a longitudinal velocity spread. We assume that the velocity spread Δv_z is small compared with the velocity shift introduced by the buncher, Δv_0. At the point of peak compression, some of the particles at the tail of the beam travel a distance $(v_0 + \Delta v_0 + \Delta v_z)\tau$ during transit from the buncher to the focal point, while others travel only a distance $(v_0 + \Delta v_0 - \Delta v_z)\tau$. The quantity τ is given in Eq. (3.5). Similarly, particles at the beam centroid can shift a maximum distance $\pm \Delta v_z \tau$ during transit. As a result of the velocity spread, the beam has a nonzero length at the bunching point given by

$$\Delta v_z \tau = z_0(\Delta v_z/\Delta v_0). \tag{3.7}$$

The corresponding pulse length of the bunched beam is

$$\Delta t_b = (z_0/v_0)/(\Delta v_0/\Delta v_z). \tag{3.8}$$

The compression ratio, defined as the ratio of the initial to the reduced pulse length, is

$$\Delta t_p/\Delta t_b = \Delta v_0/\Delta v_z. \tag{3.9}$$

A high compression ratio is possible under two circumstances: either the incident beam has small longitudinal velocity spread or the buncher introduces a large velocity shift. We have seen that strong transverse focusing has analogous requirements: either the beam has a small spread in angle or the lens has a low f/number.

3.2. EMITTANCE

Section 2.4 proved that applied forces and beam-generated forces acting over large length scales compared with the interparticle spacing preserve the phase-space volume of a distribution. Nonetheless, noncollisional processes in accelerators can warp the shape of the distribution, enlarging the effective phase volume. Although the net phase volume occupied by a beam may be constant, nonlinear field components and fringing fields of focusing lenses and steering magnets can stretch and distort the distribution. To designate the quality of a beam or an application, we must adopt a figure of merit based on the *effective* volume occupied by the distribution. This quantity is the *emittance*.

To understand the motivation behind the definition of emittance, we must review properties of particle distributions acted on by smooth forces. Section 3.1 showed that the orbit vector points of a one-dimensional sheet beam fill an area in x-v_x space. The smaller the phase area occupied by the beam, the better the quality of the beam. Here, the term quality implies focusability or parallelism. The minimum phase-space volume of a distribution is determined by the characteristics of the beam injector. Processes that increases the phase volume are undesirable.

In principle, a modified beam distribution can be restored to its original state by reversing the orbits of all the individual particles. In practice, we cannot hope to control the orbits of individual particles in a beam that may contain more than 10^{10} particles. From the macroscopic viewpoint, some modifications of distributions are irreversible—it is impossible to sort out and compensate changes by applying broad-scale macroscopic forces. In this context, modification of a

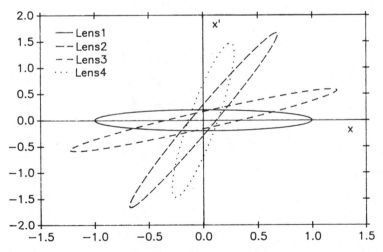

Figure 3.6 Trace-space view of focusing of a nonlaminar beam by an array of thin linear lenses. Four lenses with $\mu_0 = 80°$. Note that the process is reversible—the area enclosed by the distribution and shape of the boundary are conserved.

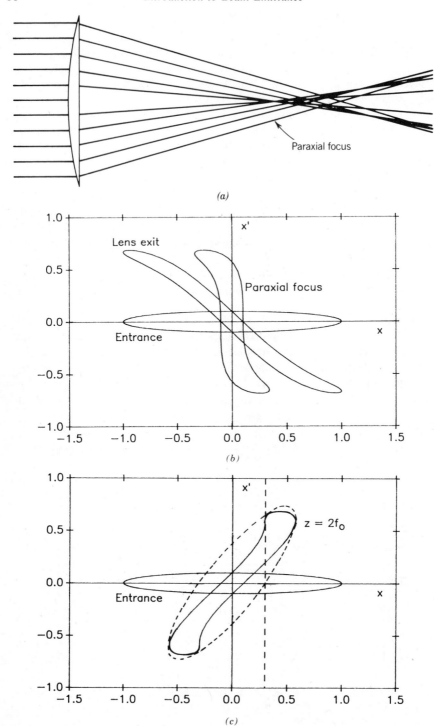

(a)

(b)

(c)

distribution is reversible if the process preserves not only the volume but also the continuity of the distribution. As an example of a reversible process, when a sheet beam that occupies a rectangle in x-v_x phase space propagates through a series of linear lenses and drift spaces, the rectangle changes to a series of parallelograms with equal area. Here, we can always find a combination of linear optical elements that will restore the distribution to its original shape. The linear focusing system acts reversibly. To illustrate this fact, Fig. 3.6 shows calculated shapes of a distribution in a uniform thin lens array. Note that the lens focal length and drift distance give a phase advance of $80°$ — the distribution returns to its initial shape after nine lenses.

Irreversible processes change distributions in such a way that they cannot be restored by macroscopic forces. Figure 3.7 illustrates such a process. A sheet beam passes through a nonlinear lens. The transverse force of the lens is *soft*; the lens force is weak at large displacement compared with a linear lens. As a result, the lens underfocuses particles on the periphery. Figure 3.7a shows particle orbits in configuration space, while Fig. 3.7b illustrates the beam distribution at two points: the lens exit and the point of maximum convergence. Downstream from the focal point (Fig. 3.7c), the distribution folds back on itself. There are positions in the beam, shown by dashed lines, where the particles have two different values of transverse velocity. Here, we can see that it is impossible to restore the distribution to a straight line with any optical component that generates forces that are functions only of position. Such an element could not separate the two overlapping velocity components.

Note that the phase-space area filled by the beam in Fig. 3.7c is unchanged. Nonetheless, the distorted distribution surrounds regions of unoccupied phase space—the *effective* area of the distribution is larger. If we sought to focus all particles in the distribution of Fig. 3.7c to a spot with an ideal linear lens, the relevant phase-space area is that inside a boundary surrounding all particles. Figure 3.7c shows such a boundary as a dotted line.

Emittance is an empirical quantity that characterizes the effective phase volume (or area) of a beam distribution, including the effects of irreversible processes. As opposed to the actual phase volume, emittance is an inexact figure of merit. As a result, its value involves approximations and conventions that may vary with the application, sometimes leading to confusion. We express emittance in terms of position and transverse angle rather than transverse velocity since the inclination of particle orbits can usually be measured directly. We denote the angle a particle makes with the beam axis of symmetry as $x' = dx/dz$ and $y' = dy/dz$. For paraxial beams, the relationship between the inclination angles and transverse velocities is

$$x' \cong v_x/v_z, \qquad y' = v_y/v_z. \tag{3.10}$$

Figure 3.7 Beam focusing by a nonlinear lens with soft force. (*a*) Configuration-space view of orbits with a parallel incident beam. (*b*) Trace-space orbit vector distributions at the lens entrance, lens exit, and paraxial focal point. (*c*) Trace-space orbit vector distribution beyond the focal point.

The coordinates (x, x', y, y') are usually treated as functions of z rather than time t. They describe the trace of a particle orbit along the axial direction, $[x(z), y(z)]$, rather than the time-dependent position $[x(t), y(t)]$. Hence, the space defined by the coordinates is called *trace space*. We can represent trace-space distributions for both relativistic and nonrelativistic beams. The conversions of trace coordinates to phase-space coordinates for relativistic beams, (x, p_x), is

$$p_x = x' p_z. \tag{3.11}$$

We can plot the distributions of paraxial beams in trace space rather than phase space. Equations (3.10) and (3.11) show that the plots contain the same information for known values of v_z or p_z. Emittance is the effective volume (or area) occupied by a distribution in trace space. To specify the volume, we must designate the distribution boundary. To begin, we shall limit our attention to a simple distribution with an unambiguous boundary. Assume that the orbit vectors of a one-dimensional beam uniformly fill a well-defined region with a sharp boundary (Fig. 3.8a). We circumscribe the distribution with an ellipse, shown as a dashed line in Fig. 3.8a. The curve surrounds the minimum area that contains all the orbit vector points. We define emittance as the area of the ellipse divided by π:

$$\varepsilon_x = \iint dx\, dx'/\pi. \tag{3.12}$$

The subscript x indicates that orbit parameters are measured in the x direction. The following section generalizes emittance for two-dimensional beams.

Sometimes, an ellipse circumscribed around a distribution is upright. In other words, the major and minor axes are aligned with the x-x' coordinate axes (Fig. 3.8b). Section 4.1 shows that an upright distribution ellipse corresponds to a

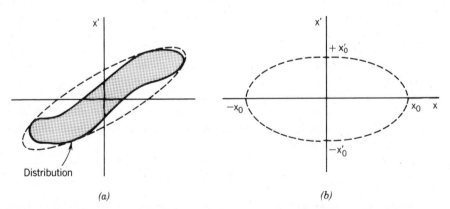

Figure 3.8 Definition of emittance. (a) Uniform orbit-vector distribution inside a boundary, surrounded by a minimum-area ellipse. (b) Upright trace-space ellipse—the enclosed emittance equals $x_0 x_0'$ π-m-rad.

beam with an envelope parallel to the axis of propagation. If x_0 and x_0' are the ellipse dimensions, then Eq. (3.12) reduces to

$$\varepsilon_x = x_0 x_0'. \tag{3.13}$$

In some works, the emittance is taken as the area of the ellipse without the divisor of π. When quoting an emittance value, the recommended procedure to show which emittance convention has been used is to include a symbol in the units. The modern standard is to include the factor of $1/\pi$. To designate this convention, we append the symbol π to the units. For example, if the distribution of a parallel beam fills an ellipse with $x_0 = 1$ cm and $x_0' = 30$ mrad, then the emittance in MKS units should be given as

$$\varepsilon_x = 3 \times 10^{-4} \pi\text{-m-rad.}$$

Note that the symbol π is a flag attached to the dimensions to signify the standard emittance convention. The specification does not mean that the dimensions have been multiplied by the number π. The dimensional units are chosen for convenience; units of π-cm-mrad are often encountered. In the older convention, where emittance equals the ellipse area, the beam emittance for the given parameters is quoted as

$$\varepsilon_x = 9.4245 \times 10^{-4} \text{m-rad.}$$

The π symbol does not appear in the unit declaration.[1]

In a transport system with no acceleration, the emittance is proportional to the effective phase volume of the beam. Focusing and steering elements preserve the continuity of the distribution if the forces are linear. If this is true, the emittance ellipse has a constant area throughout the length of the system. Emitance has an important physical interpretation—it is a conserved quantity when a beam is subject to reversible processes. In contrast, irreversible processes distort and convolute the boundaries of beam distributions, even if they preserve a constant phase-space area. When the distribution is distorted, the ellipse must enclose a larger area containing empty regions of phase space. Irreversible processes lead to *emittance growth* of a distribution. Usually, emittance growth is undesirable since it degrades the parallelism of a beam.

The final emittance of a beam represents the sum of the intrinsic emittance from the source and emittance growth during acceleration. Figure 3.9 illustrates sources of intrinsic emittance of a low-current-density electron beam emerging from an injector. The device (Fig. 3.9a) generates a parallel sheet beam of kinetic energy eV_0. The thermionic electron source operates at temperature T_c. If the combined applied and beam-generated electric fields are purely linear, then the average orbit inclination angle equals zero at all positions, $\langle x'(x) \rangle = 0$. Nonethe-

---

[1] If readers feel that this convention is backward, they are not alone.

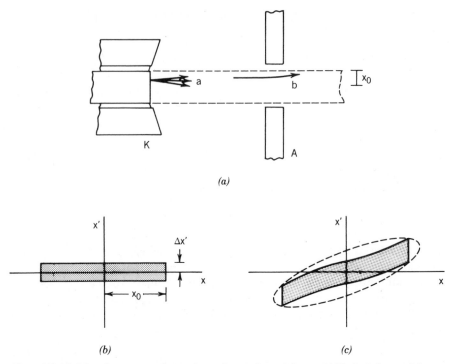

(a)

(b)   (c)

Figure 3.9 Origins of nonzero emittance in an electron beam injector. (a) Typical electron injector with a cathode (K) (consisting of an electron source and a focusing electrode) and an anode (A). Lines at (a) represent divergence of thermal electrons from the source. Line at (b) shows nonlinear deflection by fringing electric fields. (b) Trace-space distribution of orbit vectors, electrons emerging from the source. (c) Trace-space distribution of orbit vectors for electrons emerging from the anode aperture, including thermal divergence plus effects of distortion and defocusing at the anode.

less, the beam has nonzero emittance because of the thermal motions of electrons leaving the source. If the source electrons have a Maxwell distribution (Section 2.6), the average thermal velocity in the x direction is $\langle v_x \rangle = (2kT/\pi m)^{1/2}$. The resulting trace-space distribution (Fig. 3.9b) is rectangular with an angular spread of $\langle v_x \rangle/v_z$, or

$$\Delta x' \sim [kT/\pi(mv_z^2/2)]. \tag{3.14}$$

The transverse electric force in injectors always has a nonlinear component. Usually, the force is stronger near electrodes—particles near the beam envelope are overfocused. The resulting distorted trace-space distribution is shown in Fig. 3.9c. The emittance ellipse shown includes all particle orbits at the injector exit. The total area reflects contributions from both the thermal velocity spread and distortions from optical errors.

Emittance is defined in terms of an elliptical boundary because this curve plays an important role in linear focusing systems. For example, we saw in Section 2.1 that particles subject to a continuous linear focusing force follow elliptical trajectories in both phase space and trace space. Section 4.1 shows that an elliptical distribution that enters a system with discrete linear optical elements maintains an elliptical shape. Ellipses have special significance for periodic linear focusing systems. In such a system, there is a special elliptical trace-space boundary called the *acceptance* (Section 4.3). An incident beam distribution that fits within the acceptance exhibits minimal envelope oscillations and propagates through the focusing system without striking physical boundaries. In this context, the elliptical acceptance boundary sets an upper limit on the emittance of the incident beam.

3.3. MEASUREMENT OF EMITTANCE

In Section 3.2, we defined emittance for an idealized distribution with a sharp boundary in trace space. In this section, we shall discuss how to calculate emittance for the type of distributions encountered in experiments. We apply the *root-mean-square emittance*, or *RMS emittance*, when distributions have diffuse boundaries. We shall first summarize methods for measuring the distribution of particles in position and inclination angle. We shall concentrate on techniques for low-energy beams. Such measurements are made near the entrance of an accelerator to gauge the operation of particle sources and extractors. The input-beam emittance sets a limit on the quality of the ouput beam from an accelerator. Emittance diagnostics for low-energy beams are simple devices. Because of the low-power density, it is usually possible to place the analyzer in the path of the beam. Devices that stop beams are called *destructive* or *interceptive diagnostics*— they are usually inserted and removed from the beamline by vacuum translators during accelerator operation.

To infer emittance, we must measure particle displacement in both position and angle. Measurements of spatial variations of charged particles beams are straightforward. Moving collectors are used for continuous beams. The spatial profile of pulsed beams is measured with arrays of collectors or imaging detectors, such as phosphor sheets, fast scintillators, and channel electron multiplier arrays. Information on the distribution in angle must be obtained indirectly. There are no simple detectors that measure the angle of a particle orbit directly. Instead, we must convert an angular distribution into a related spatial distribution. The conversion occurs when particles are allowed to drift freely after passing through a slit or aperture.

Figure 3.10*a* illustrates the free-space propagation of particles emerging from a narrow slit. The particles have an angular spread $\Delta\theta$ normal to the slit. As they drift they move transversely according to their angle. If D is the distance between the aperture and the detector, a particle with incident angle x' moves a transverse distance

$$\Delta x = x'D. \qquad (3.15)$$

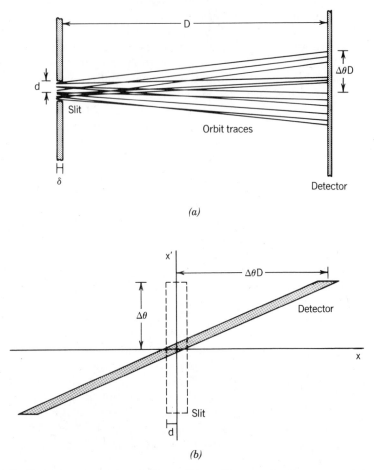

Figure 3.10 Free-space propagation of a beam with a uniform angular divergence emerging from a slit. (*a*) Configuration-space view of orbits. (*b*) Trace-space view of orbit-vector distributions at the slit and at the detector.

Figure 3.10*b* shows a trace-space representation of the collimation and drift. The spatial profile of particles at the detector gives a direct measurement of the incident angular distribution if the following conditions are met.

1. The width of the particle profile at the detector must be large compared with the dimension of the aperture, or

$$\Delta\theta \, D \gg d. \tag{3.16}$$

If Eq. (3.16) is true, inspection of Fig. 3.10*b* shows that the beam distribution stretches to a thin line at the detector—the spatial distribution is almost independent of the aperture geometry.

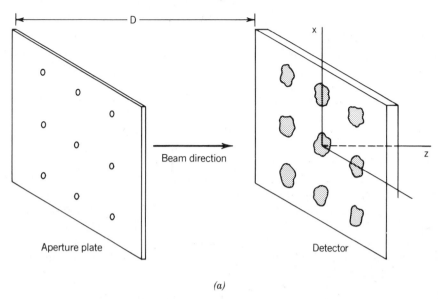

(a)

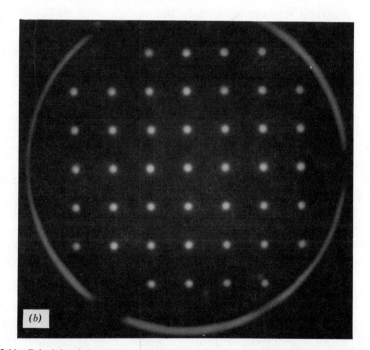

Figure 3.11 Principle of the pepperpot emittance diagnostic. (*a*) Schematic view of a pepperpot, showing aperture and images of projected beam particles. (*b*) Pepperpot data, time-integrated photograph of 250-keV electrons from a needle cathode striking a phosphor screen. (courtesy K. Zieher, University of New Mexico.)

2. Particles must follow ballistic trajectories to the detector—space-charge forces must be small. We review criteria for this condition in Section 5.4.

3. The aperture should admit all particles over its width, independent of their angle. Unfortunately, aperture plates must have a nonzero thickness and vignetting can occur—the plate intercepts a portion of the particles with large angle. If the plate has thickness δ and the slit has width d, then vignetting is small if

$$\delta \, \Delta\theta \ll d. \tag{3.17}$$

The *pepperpot*, illustrated in Fig. 3.11a, is a common aperture plate geometry used with an imaging detector. The front plate has a two-dimensional array of apertures. The apertures are far enough apart so that beam projections from adjacent holes do not interfere. Figure 3.11b shows photographic data from a pepperpot measurement of a 300-keV, 250-A pulsed electron beam. Cerenkov light emission from a glass plate generated the beam image. Analysis of the light distribution in the photograph gives a variety of information, illustrated schematically in Fig. 3.12:

1. An integral of the intensity of a beam spot gives the relative current density of the beam (Fig. 3.12a). Values from the different apertures can be combined to construct a current density profile of the beam in the aperture plane.

2. Measurements of the displacement of the spot centroid from the position of the corresponding aperture gives information on beam aiming in the aperture plane (Fig. 3.12b). We can find whether the beam is converging or diverging.

3. A scan averaged over the y direction gives the relative distribution in x' at the position (x, y) (Fig. 3.12c). Among other information, this process yields a value of the average divergence angle, $\Delta\theta_x(x_0)$. A similar analysis gives $\Delta\theta_y(y_0)$.

4. We can estimate complete four-dimensional trace-space distributions by combining scans of images from all apertures in the pepperpot. Figure 3.12d shows the portion of an aperture projection subtended by a detector with angular resolution $\Delta x'$ and $\Delta y'$. The dimensions of the aperture define the spatial resolution, Δx and Δy. If the aperture is at position (x_0, y_0) and the detector position corresponds to angles (x'_0, y'_0), then the detector signal is proportional to the trace-space density multiplied by the entrance phase volume of the detector. If we denote the continuous distribution function in trace space as $g(x, x', y, y')$, then the signal is proportional to

$$S(x, x', y, y') \sim g(x_0, x'_0, y_0, y'_0) \Delta x \, \Delta x' \, \Delta y \, \Delta y'. \tag{3.18}$$

Although we carry out the analysis at discrete locations, we can infer the complete continuous distribution function by interpolation.

Often, it is unnecessary to perform a complete analysis to find $g(x, x', y, y')$. For example, if particle motions are separable in x and y, the relative distribution of x'

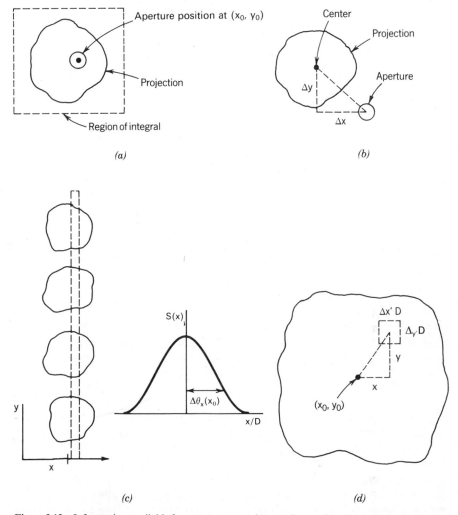

Figure 3.12 Information available from pepperpot emittance diagnostic—beam projections in the detector plane. (*a*) Spatial integral of spot signal intensities (region of solid line) gives the spatial profile of beam current. (*b*) Displacement of the spot center shows aiming errors and the convergence or divergence of the beam. (*c*) A measurement of the relative intensity along ζ_x (integrated over y') gives the angular divergence at x, $\Delta\theta_x(x)$, and the emittance, ε_x. (*d*) A full analysis of the spot intensity along ζ_x and ζ_y leads to the angular distribution function, $g(x, x', y, y')$ at (x, y).

should be the same at all values of y. Here, we can generate an emittance plot from scans in x' at one location in y.

Figure 3.13 illustrates the wire scan, an alternative method to measure trace-space distributions when x and y motions are decoupled. A detector with an analyzing slit moves over the cross section of a beam to sample different values of x. A wire beam collector moves within the detector to sample different values of

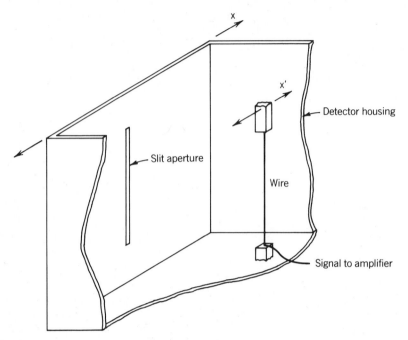

Figure 3.13 Scanned wire emittance diagnostic. The entire detector moves sideways to sample different values of x while the wire translates within the assembly to give the distribution in x' at x.

x'. Averaging along y and y' occurs automatically because of the extended geometry of the slit and wire. Wire scans are feasible only for steady-state or continuously pulsed beams. Figure 3.14 shows data from such a measurement for a diverging beam.

The detectors that we have discussed generate extensive information on diffuse distributions—usually, we represent this information through contour plots. Often, we want a single number that characterizes the quality of a beam for quick comparisons. The RMS emittance, calculated from the full distribution measurement, is such a quantity. For the *upright* distribution of a parallel beam, the following equation defines the RMS emittance:

$$\varepsilon_x = 4\left[\frac{\int dx \int dx' (x - \langle x \rangle)^2 g(x, x')}{\int dx \int dx' g(x, x')}\right]^{1/2}$$
$$\times \left[\frac{\int dx \int dx' (x' - \langle x' \rangle)^2 g(x, x')}{\int dx \int dx' g(x, x')}\right]^{1/2}$$
$$= 4\Delta x_{\text{rms}} \, \Delta x'_{\text{rms}}. \tag{3.19}$$

Again, $g(x, x')$ is the continuous trace-space distribution function. The quantity

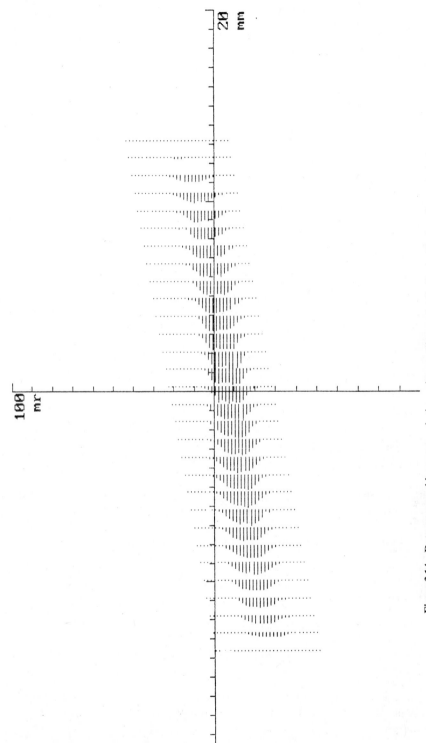

Figure 3.14 Data generated by a scanned wire emittance diagnostic. Ordinate: Motion along x', abscissa: motion along x. Signal strength of current collected on wire displayed as horizontal length of lines. Skewness and distortion of distribution reflect the optics of the simple planar injector. (Courtesy H. Rutkowski, Lawrence Berkeley Laboratory.)

TABLE 3.1

```
function RMSEMIT (var PartDist : Distribution;
                  var Npart : PartIndex): real;
        {Returns RMS emittance in pi-m-rad.
        Upright or skewed distributions.
        NPart is the number of particles
        PartDist is a particle distribution consisting
          of an array of particle records.
        Each particle record consists of two real
          quantities:
                x (position, m)
                v (angle, radians)}
var
  xav, vav, skewav : real;
  j : PartIndex;
  xsqav, vsqav : real;
begin
        {−Find xav and vav---}
  xav:= 0.0;
  vav:= 0.0;
  for j:= 1 to Npart do
    with PartDist[j] do
      begin
        xav:= xav + x;
        vav:= vav + v
      end;
  xav:= xav/Npart;
  vav:= vav/Npart;
        {−Find skewav---}
  skewav:= 0.0;
  for j:= 1 to Npart do
    with PartDist[j] do
        skewav:= skewav + (x-xav)*(v-vav);
  skewav:= skewav/Npart;
        {−Compute average of squares---}
  xsqav:= 0.0;
  vsqav:= 0.0;
  for j:= 1 to Npart do
    with PartDist[j] do
      begin
        xsqav:= xsqav + (x-xav)*(x-xav);
        vsqav:= vsqav + (v-vav)*(v-vav);
      end;
  xsqav:= xsqav/Npart;
  vsqav:= vsqav/Npart;
        {−Compute emittance---}
  RMSEmit:= 4.0*sqrt (xsqav*vsqav − skewav*skewav)
end; {RMSEMIT}
```

Δx_{rms} is the RMS beam width and $\Delta x'_{\text{rms}}$ is the RMS divergence angle:

$$\Delta x'_{\text{rms}} = \left[\frac{\int dx \int dx' (x' - \langle x' \rangle)^2 g(x, x')}{\int dx \int dx' g(x, x')} \right]^{1/2} \qquad (3.20)$$
$$= \langle (x' - \langle x' \rangle)^2 \rangle.^{1/2}$$

We introduce the factor of 4 so that Eq. (3.19) gives the correct emittance when applied to an ideal distribution, a uniformly filled ellipse. We can prove the result by evaluating the integrals in Eq. (3.19) for a uniform value of $g(x, x')$ between the limits

$$-[1 - (x/x_0)^2] \leqslant x' \leqslant [1 - (x/x_0)^2],$$
$$-x_0 \leqslant x \leqslant x_0.$$

In experimental measurements, we cannot expect that the location of diagnostics will correspond to a beam waist, a position where the beam envelope is parallel to the axis. Generally, the trace-space distribution is skewed as in Fig. 3.14. Then, we must use the following corrected version of Eq. (3.19)[2]:

$$\varepsilon_x = 4[\langle (x - \langle x \rangle)^2 \rangle \langle (x' - \langle x' \rangle)^2 \rangle - \langle (x - \langle x \rangle)(x' - \langle x' \rangle) \rangle^2]^{1/2}. \qquad (3.21)$$

Table 3.1 illustrates the application of Eq. (3.21). The table lists a computer subroutine to evaluate the RMS emittance for a beam distribution generated by a one-dimensional computer simulation. The distribution is stored as an array of particles; each particle is a record consisting of two real values, the position x and angle x'.

3.4. COUPLED BEAM DISTRIBUTIONS, LONGITUDINAL EMITTANCE, NORMALIZED EMITTANCE, AND BRIGHTNESS

In this section, we shall discuss several topics related to emittance, including normalized emittance and brightness. To begin, we shall extend definitions of emittance to include cases where particle motions in different directions are coupled. In previous discussions of emittance, we limited attention to one-dimensional beams or we assumed that motions in the x and y directions were independent. Transverse particle motions in many practical focusing systems are not separable—therefore, we must extend the emittance definition to include four-dimensional trace-space volumes. Furthermore, when axial motion couples to tranverse motion, we must deal with trace-space volumes in a six-dimensional space.

[2]See, for instance, C. Lejevne and J. Aubert, "Emittance and Brightness: Definitions and Measurements," in *Applied Charged Particle Optics*, A. Septier, Ed., Academic Press, New York, 1980, p. 159.

A.   Coupled Transverse Beam Distributions

The focusing system of a high-energy-particle accelerator consists mainly of quadrupole lenses and dipole bending magnets. In these optical elements, particle motions in the x and y directions are independent. Motion is not separable in a variety of other focusing devices. Some common examples are solenoidal magnetic lenses, liquid metal lenses, or cylindrical electrostatic lenses in acceleration columns [CPA, Chapter 6].

Consider a paraxial beam in a focusing system where x and y motions couple but transverse motion is independent of axial motion. Emittances in the x and y directions are no longer separately conserved quantities. Instead, the total four-dimensional trace-space volume in (x, x', y, y') is constant in the absence of acceleration. The four-dimensional extension of emittance is called *hyperemittance*. The definition is similar to those for the two-dimensional quantity. For example, suppose a parallel beam has a uniform distribution in four-dimensional trace space with a sharp boundary. If the distribution fits into the four-dimensional ellipsoid

$$\left(\frac{x}{x_0}\right)^2 + \left(\frac{x'}{x_0'}\right)^2 + \left(\frac{y}{y_0}\right)^2 + \left(\frac{y'}{y_0'}\right)^2 = 1, \tag{3.22}$$

then the hyperemittance (ε_4) is

$$\varepsilon_4 = V_4/\pi^2 = x_0 x_0' y_0 y_0' \quad \pi^2\text{-m}^2\text{rad}^2. \tag{3.23}$$

The symbol V_4 in Eq. (3.23) represents the four-dimensional volume occupied by the distribution. If the distribution ellipsoid is not upright, we must calculate the orientation and volume of the ellipsoid by more advanced methods. We can easily extend the equations for RMS emittance (Section 3.3) to coupled distributions. Finally, if particle motions in the x and y directions are independent, we can identify separate emittances, ε_x and ε_y. The hyperemittance is the product of the two-dimensional emittances:

$$\varepsilon_4 = \varepsilon_x \varepsilon_y. \tag{3.24}$$

The definition of emittance presents some problems when we deal with cylindrical beams. We transport and focus such beams with lenses that exert only radial forces. We would like to define a quantity ε_r that characterizes the velocity spread in the r direction and the radial pressure force of the beam. Unfortunately, the simple geometric interpretation of emittance in terms of a trace-space plot (Section 3.2) does not hold for motion in a curved coordinate system—the size of a differential element in configuration space varies with position by a factor proportional to $2\pi r dr$. Also, motion in r and θ is not strictly separable.

Nonetheless, we shall adopt an approximate quantity ε_r for use in the paraxial ray equation. We define the quantity so that it gives the correct value of radial pressure force (Section 3.5). We set $\pi \varepsilon_r$ equal to the area in r-r' space circumscribed by the orbit of a particle that reaches the envelope with $v_r = 0$. This quantity has the dimensions of emittance and represents the effect of a spread in v_r. For diffuse beams, we take the envelope radius as an RMS radius. To avoid confusion, we must remember that ε_r is an approximate quantity that does not follow directly from the theoretical framework used to define $\varepsilon_x, \varepsilon_y$, and ε_4.

B.   Longitudinal Emittance

Early researchers in particle-beam physics created the quantity emittance to describe the transverse motion of low-energy-electron beams in cathode ray tubes and electron microscopes. These continuous beams are generated by electrostatic accelerators—they are uniform in z with a single value of v_z. Therefore, there are no axial forces that arise from velocity spreads. In contrast, longitudinal motions play an important role in modern high-energy accelerators such as RF linacs. Conservation of the effective longitudinal phase-space volume, (z, p_z), is an important principle with many applications. In extending emittance to the axial direction, we recognize that angles and orbit traces are undefined. Therefore, we must plot orbit vectors to represent distributions directly in z-p_z space. In RF accelerators, we measure and plot longitudinal distributions relative to a point of constant phase of the accelerating wave. Here, common distribution coordinates are ϕ, the phase position of the particle relative to the wave, and ΔT, the difference in kinetic energy from the average value.

Figure 3.15 illustrates a method to construct a complete axial distribution plot for a pulse of H^- ions emerging from the buckets of an RF linac [CPC, Chap. 13]. A short laser pulse irradiates an axial slice of a beam bunch at a phase ϕ. The tuned laser radiation detaches the loosely bound electron of the negative ion, creating an energetic neutral hydrogen atom, H°. A magnet deflects the main beam, while the atoms continue forward. The time variation of atomic flux is measured at the end of a drift length. The integral of the signal gives the relative beam density at ϕ. The time variation of the signal gives the velocity dispersion and hence the relative distribution of beam kinetic energy error, ΔT. The diagnostic measures the full axial distribution by varying the laser delay with respect to the bunch.

Sometimes, longitudinal and transverse particle motions couple. For example, the forces in most focusing elements are energy dependent—transverse deflections depend on the axial velocity. This effect, *chromatic aberration*, leads to a increase in a focal spot size when beams have a substantial energy spread. Other examples of transverse–longitudinal coupling include instabilities that transfer longitudinal energy to transverse motion (Chapter 13), orbital resonances in circular machines, and intense electron beams where the transverse kinetic energy is comparable to the longitudinal energy. When there is strong coupling, we cannot apply the conservation of transverse emittances or even hyperemittance. The only conserved quantity is the total six-dimensional phase volume. Here, we

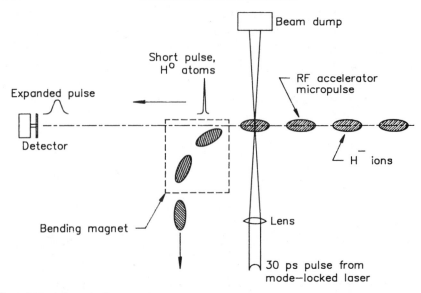

Figure 3.15 Schematic diagram of an apparatus to measure the longitudinal distribution of H^- ions from an RF accelerator. A synchronized laser pulse dissociates the negative ions to H^0 at a position in the micropulse. The shape of the pulse of atoms after a downstream drift gives the axial velocity distribution about the mean at a point in the beam. The distribution is measured over many shots by changing the timing of the laser pulses relative to the ion micropulses. (Adapted from W. B. Cottingame, J. H. Cortez, W. W. Higgins, O. R. Sander, and D. P. Sandoval, "Longitudinal Emittance Measurements on the ATS," in *Proc. 1986 Linear Accelerator Conf., Standford Linear Accelerator Center.*)

must quote the emittance in terms of a boundary around the full distribution in (x, p_x, y, p_y, z, p_z) space.

C. Normalized Emittance

Acceleration reduces emittance. The transverse momentum of particles may remain constant while the axial momentum increases, leading to a reduction in x'. We shall find it useful to designate an alternative quantity that remains constant during acceleration, the *normalized emittance*. With the effects of acceleration removed, changes in the normalized emittance signal degradation of beam quality resulting from nonlinear forces or beam perturbations. Also, normalized emittance is useful to show improvements in beam quality from beam-cooling methods (Section 4.6).

Although the *trace-space volume* of a beam decreases during acceleration, we know that the *phase-space volume* stays constant in a linear focusing system. The transverse momenta are related to the inclination angles by

$$p_x = x'(\beta\gamma)(m_0 c). \tag{3.25}$$

We calculated normalized emittance from the boundary of the distribution of orbit vector points in a modified trace space, $[x, \beta\gamma x']$. The correction factor $(\beta\gamma)$ ensures that the normalized emittance is invariant when the beam accelerates. The normalized emittance of a relativistic paraxial beam is

$$\varepsilon_{nx} = (\beta\gamma)\varepsilon_x = (\text{Area in } x\text{-}p_x \text{ space})/\pi m_0 c \quad \pi\text{-m-rad.} \tag{3.26}$$

Note the use of the symbol π is the dimensions to signify that the quantity includes the factor of $1/\pi$. For nonrelativistic beams, Eq. (3.26) has the form

$$\varepsilon_{nx} = \beta\varepsilon_x = v_z\varepsilon_x/c = (\text{Area in } x\text{-}v_x \text{ space})/\pi c \quad \pi\text{-m-rad.} \tag{3.27}$$

D. Brightness

The quantity *brightness* was adopted from conventional optics where it characterizes the quality of light sources. In charged-particle-beam applications, beam brightness is the current density per unit solid angle in the axial direction. Bright beams have high current density and good parallelism.

Figure 3.16 illustrates the meaning of brightness for a cylindrical charged particle beam. The beam has current I, average radius Δr, and average divergence

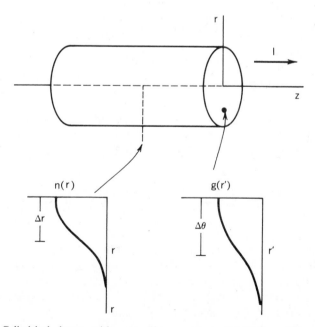

Figure 3.16 Cylindrical beam with nonuniform current distribution. Figure shows the root-mean-squared radius (Δr) and divergence angle ($\Delta\theta$).

angle $\Delta\theta$. In the limit $\Delta\theta \ll 1$, the brightness is

$$B \cong \frac{I}{(\pi\Delta r^2)(\pi\Delta\theta^2)} \quad \text{A/m}^2\text{-rad}^2. \tag{3.28}$$

We can rewrite Eq. (3.28) in terms of the average current density j_b:

$$B \cong j_b/\pi\Delta\theta^2. \tag{3.29}$$

When beams have Cartesian symmetry in the transverse direction, we can write an expression for brightness in terms of the emittances. As an example, suppose a beam has an ideal distribution. Orbit vectors uniformly fill an ellipse with axes (x_0, x_0') and (y_0, y_0'). The associated brightness is

$$B \cong \frac{I}{(\pi x_0 x_0')(\pi y_0 y_0')} = \frac{I}{\pi^2 \varepsilon_x \varepsilon_y} \quad \frac{\text{A}}{\text{m}^2\text{-rad}^2}. \tag{3.30}$$

The following approximation holds for isotropic beams with average emittance ε:

$$B \cong I/\pi^2\varepsilon^2. \tag{3.31}$$

Note that if ε is constant, the beam brightness is also a conserved quantity.

We can appreciate the significance of brightness by reviewing the example of nonlaminar beam focusing from Section 3.1. A parallel cylindrical beam with angular divergence $\Delta\theta$ carries current I. The beam passes through a lens of focal length f. The minimum size of the focal spot is about $\delta \cong \Delta\theta f$—the average current density at the focus is

$$j_{max} = I/\pi\delta^2 = I/\pi(\Delta\theta f)^2. \tag{3.32}$$

Taking R as the useful lens radius and beam radius at the lens, we can write Eq. (3.32) as

$$j_{max} \cong (\pi/4)[I/(\pi R\Delta\theta)^2(f/2R)^2] \sim B/(f/\text{number})^2. \tag{3.33}$$

The f/number depends on the focusing system—the highest power density results are achieved with the lowest f/number optics. For a given optical system, the power density at the focus is proportional to the beam brightness.

Like the emittance, brightness is not constant during beam acceleration. We can define a conserved quantity, the *normalized brightness*, in analogy with the normalized emittance. The relativistic expression is

$$B_n = I/\pi^2\varepsilon_n^2 = B/(\beta\gamma)^2. \tag{3.34}$$

3.5. EMITTANCE FORCE

We derived equations for the effective force resulting from particle velocity dispersions in Section 2.11. In this section, we shall express the transverse pressure force of a paraxial beam in terms of its emittance. We shall also review the principles of beam envelope equations [CPA, Chapter 7]—Chapter 9 develops this topic in detail. In this section, we shall limit attention to modifications of the envelope equation to include the effects of emittance.

An ideal laminar beam with parallel orbits propagates indefinitely with no change in radius. In contrast, a beam with nonzero emittance expands—some of the particles are aimed outward. To maintain a constant radius for a beam with emittance, focusing forces must be applied to reverse the outwardly directed particles. In a sense, we can view nonzero emittance in terms of an outward force that balances the focusing force to maintain a constant radius beam. In this section, we shall calculate the effective emittance force by seeking the focusing force that guarantees radial force balance.

To simplify the discussion, we adopt the following assumptions:

1. The cylindrical beam has azimuthal symmetry.

2. All particle orbits are contained within a maximum radius R.

3. The beam is paraxial. This condition implies that the axial distance for a substantial change in the envelope radius is much larger than R. This condition allows us to ignore axial forces and to describe the beam in the infinite length limit.

We recognize that only the radial motions of particles are important, since there are no forces in the θ direction. An azimuthal velocity spread makes only a small modification to the centrifugal force.

Suppose a linear, axicentered force contains the beam—the force varies in z over scale lengths long compared with the envelope radius R. We write the linear focusing force as

$$F_r(r) = -F_0(r/R_0). \qquad (3.35)$$

If no other forces act on the beam, the orbit vector points of individual particles follow ellipses in trace space as the particles perform radial oscillations. The oscillation frequency for all particles is

$$\omega_r = \sqrt{F_0/\gamma m_0 R_0}, \qquad (3.36)$$

when $v_r \ll v_z$. In Eq. (3.36), m_0 is the particle rest mass and $(\gamma - 1)m_0c^2$ is the kinetic energy. A plot of the boundary beam orbit is shown in Fig. 3.17. The family of trace-space orbits is a nested set of ellipses that fill the area inside the boundary orbit. If the beam is in radial equilibrium, particle orbit vector points are distributed uniformly in phase along the ellipses.

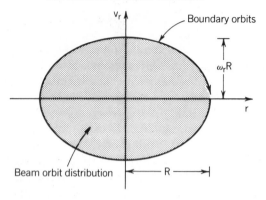

Figure 3.17 Phase space orbit-vector trajectories for particle motion in a linear, radial force.

By the definition of Section 3.4, the radial emittance of the beam equals the production of the maximum displacement and angle of the boundary orbit:

$$\varepsilon_r = R v_{ro}/v_z = \omega_r R^2/v_z. \tag{3.37}$$

Solving Eq. (3.37) for the oscillation frequency gives

$$\omega_r = \varepsilon_r v_z/R^2. \tag{3.38}$$

We can equate the expressions of Eqs. (3.36) and (3.38) to find the focusing force needed to balance emittance on the beam envelope:

$$F_r(R) = - F_0(R/R_0) = - \varepsilon_r^2(\gamma m_0 v_z^2)/R^3. \tag{3.39}$$

The expression on the right-hand side of Eq. (3.39) is the effective emittance force.

In the special case we have considered with linear focusing forces and no beam-generated forces, Eq. (3.39) guarantees radial force balance at all positions in the beam cross section. Generally, this is not true. In practice, we cannot measure detailed characteristics of beam motion over the whole cross section. Without a complete knowledge of transverse motion, a common approach is to seek conditions for force balance only on the beam boundary to generate a rough global prediction of beam behavior. Envelope equations summarize this information.

To construct a paraxial envelope equation, suppose that changes in the envelope radius and focusing force occur over long time scales compared with the betatron frequency ω_r In the quasistatic limit, the following approximate equation describes changes in the envelope radius of a beam with nonzero emittance subject to a linear focusing force:

$$\frac{d^2(\gamma m_0 R)}{dt^2} = - F_0\left(\frac{R}{R_0}\right) + \frac{\varepsilon_r^2(m_0 v_z^2)}{R^3}. \tag{3.40}$$

It is usually more informative to calculate the *envelope* trace of a paraxial beam, $R(z)$, rather than temporal variations of the radius, $R(t)$. Accordingly, we convert Eq. (3.40) from a force equation to a trace equation by exchanging axial derivatives for all time derivatives. We use the chain rule for derivatives:

$$\frac{d}{dt} \to v_z\left(\frac{d}{dz}\right). \tag{3.41}$$

Equation (3.41) is valid for paraxial beams where γ and v_z are uniform over the beam cross section at an axial location. For a constant velocity beam, the resulting equation is

$$R'' = -\frac{F_0(R/R_0)}{[\gamma m_0(\beta\gamma)^2]} + \frac{\varepsilon_r^2}{R^3} \tag{3.42}$$

When the beam distribution does not have a sharp boundary, the emittance in Eq. (3.42) can be taken as an RMS value and R can be associated with the RMS beam radius. Note that additional terms must appear in Eq. (3.42) when the beam accelerates—the procedure to convert a force equation to a trace equation in the presence of acceleration is presented in [CPA, Chap. 7]. Section 3.6 discusses the application of Eq. (3.42) to beam expansion in a drift region.

3.6. NONLAMINAR BEAMS IN DRIFT REGIONS

We can apply Eq. (3.42) to study propagation of a beam with nonzero emittance in a field-free drift region. We retain only the emittance term on the right-hand side. The equation

$$R'' = \varepsilon^2/R^3, \tag{3.43}$$

governs the envelope radius. Figure 3.18a shows the geometry of the calculation. Define the axial coordinate so that the beam envelope is parallel to the axis at $z = 0$, or $R'(0) = 0$. A position where a beam has a minimum radius and zero envelope angle ($R' = 0$) is called a *beam waist*—let the radius at the waist be R_0. Since the outward emittance force is the only radial force, the envelope radius increases as the beam moves downstream.

To solve Eq. (3.43), we multiply both sides by $2R'$. The left-hand side, $2R'R''$, is a perfect differential of the quantity $(R')^2$. We can integrate both sides of Eq. (3.43) from the origin to a position z. Applying the boundary conditions at $z = 0$, we find that

$$(R')^2 = \varepsilon^2(1/R_0^2 - 1/R^2). \tag{3.44}$$

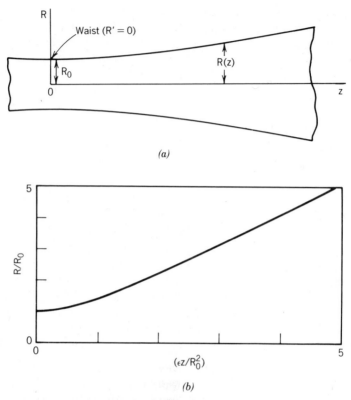

Figure 3.18 Free-space expansion of a cylindrical beam with nonzero emittance. (*a*) Geometry for calculation. (*b*) Calculated axial variation of the normalized envelope radius.

Manipulation and integration of Eq. (3.44) gives an expression for $R(z)$:

$$R(z)^2 = R_0^2 + \varepsilon^2 z^2 / R_0^2. \tag{3.45}$$

The solution is plotted in Fig. 3.18*b*. At $z = 0$, the envelope has zero slope. At long distances from the origin ($z \gg R_0/\varepsilon$), $R(z)$ approaches a straight line with slope

$$R' \cong \varepsilon / R_0. \tag{3.46}$$

According to Eq. (3.13), the quantity ε/R_0 equals the maximum orbit angle of particles in an upright elliptical distribution at $z = 0$. Note that the beam is parallel to the axis at this point. After the particles drift downstream, the envelope expansion angle approaches the angle of particles with the maximum inclination.

We can modify Eq. (3.45) to describe converging or diverging beams that have a waist point at any position z_0:

$$R(z)^2 = R_0^2 + \varepsilon^2 (z - z_0)^2 / R_0^2. \tag{3.47}$$

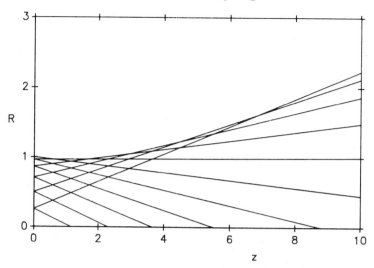

Figure 3.19 Axial expansion of beam from a waist point. Configuration-space plot of selected orbits on the boundary of a beam with an elliptical distribution ($R_0 = 1$, $R'_0 = 0.2$).

Imagine a beam traveling in the $+z$ direction that converges to a waist at $z = 0$ and then expands. In the microscopic view, particles follow ballistic orbits through the convergence point without interaction. In the macroscopic view, we say a repulsive emittance force reflects the beam when it converges. The force is proportional to $1/R^3$. Figure 3.19 clarifies the relationship between the microscopic and macroscopic descriptions. Orbits on the periphery on an elliptical distribution are plotted in configuration space for a focused beam. Inspection of Fig. 3.19 shows how the summation of a large number of straight line orbits leads to the envelope curve of Eq. (3.45).

Figure 3.20 illustrates the trace-space behavior of a freely expanding beam. We take an ideal elliptical distribution of orbit vector points. At the waist point (curve A), the axes of the ellipse are parallel to the trace coordinates. The inclination angle of the beam envelope (R') equals the angle of individual particles on the envelope. Inspection of curve A shows that particles at $r = R$ have zero inclination angle. As the beam moves downstream, free particle motion results in the deformation of the ellipse shown as B. The envelope radius, indicated by a dashed line, increases. For $z > 0$, particles with positive inclination angle move to the boundary; therefore, the envelope angle appears to increase. At long distance from the waist (curve C), the particles on the periphery are those with almost the maximum inclination angle. Therefore, the envelope angle approaches the value of Eq. (3.46).

We can apply Eq. (3.47) to a practical problem, the field-free propagation of a beam through a tube. We want to find the maximum length pipe that the beam can traverse without losses. Figure 3.21 shows the geometry—the tube has length L and radius R_i. A beam with emittance ε fills the tube entrance—the

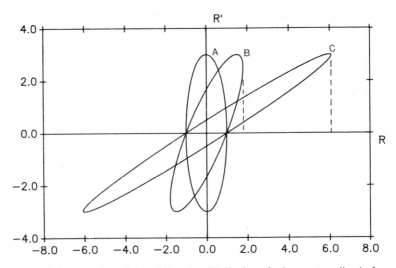

Figure 3.20   Trace-space view of the orbit-vector distribution of a beam expanding in free space $(R_0 = 1, R'_0 = 3)$. (*A*) Beam waist. (*B*) Downstream from waist—dashed line shows envelope radius. (*C*) Farther downstream.

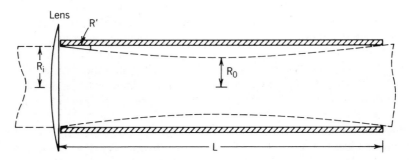

Figure 3.21   Geometry—propagation of a cylindrical beam with nonzero emittance through a tube. The beam enters through a thin lens.

envelope radius is R_i at $z = 0$. From the symmetry of Eq. (3.47), we expect that the best solution has a beam that converges at the pipe entrance ($z = 0$), reaches a waist at the midpoint ($z = L/2$), and expands to radius R_i at the exit ($z = L$).

We assume that a focusing lens at the entrance allows us to adjust the input convergence angle. Different angles give different values of the waist radius R_0. We seek an maximum tube length by expressing L as a function of R_0 and then setting dL/dR_0 equal to zero. Substituting $(z - z_0) = L/2$ in Eq. (3.47), we find

$$L^2 = (4/\varepsilon^2)(R_0^2 R_i^2 - R_0^4). \tag{3.48}$$

Taking the derivative, the following value of R_0 gives the maximum value of L:

$$R_0 = R_i/\sqrt{2}. \tag{3.49}$$

The maximum allowed tube length is

$$L = \sqrt{2}(R_i^2/\varepsilon). \tag{3.50}$$

The envelope angle at the pipe entrance for the optimum solution is

$$R' = -\varepsilon \left[\frac{1}{R_0^2} - \frac{1}{R_i^2} \right]^{1/2} = -\frac{\varepsilon}{R_i}. \tag{3.51}$$

As an application example, suppose we want to inject a low-energy neutral particle beam through the shield of a fusion reactor. The atomic beam fills a pipe of radius $R_i = 0.02$ m. The beam orbit vector distribution fits in an ellipse with a maximum angular deflection of $\Delta R_i' = 10^{-2}$ rad (0.6°). The corresponding emittance is $\varepsilon = R_i R_i' = 2 \times 10^{-4}$ π-m-rad. At the center of the pipe, the waist envelope radius is $R_0 = 0.0141$ m. Equation (3.50) implies that the length of the pipe is 2.0 m. Inserting values in Eq. (3.51) gives the injection envelope angle as $R_i' = -10^{-2}$ rad. The focal length of lens to match a parallel beam into the tube should be $f = R_i/R_i' = 2.0$ m.

3.7. NONLAMINAR BEAMS IN LINEAR FOCUSING SYSTEMS

This section examines the propagation of nonlaminar beams through the multielement optics systems used in accelerators. Accelerator transport systems combine electric or magnetic field lenses, bending elements, and drift spaces to steer beams and to confine them about an axis. All transport systems accept particles only within a limited range of displacement and inclination angle from the main axis. We must make certain that all particles in the beam can travel through the system without striking a boundary. We display the allowed particle orbits through a trace-space boundary called the *acceptance*. This section introduces the topics of acceptance and beam matching. Section 4.1 presents general mathematical methods to transform trace-space distributions through linear focusing systems, while Section 4.4 describes the effects of nonlinear optics.

We shall limit the discussion to beam propagation at constant energy through combinations of drift spaces and linear lenses. Again, we use one-dimensional distributions in (x, x') for illustration. Furthermore, we limit attention to elliptical distributions—we shall find that the acceptance boundaries of linear focusing systems are often elliptical. To begin, we shall review the relationship between the shape and orientation of the distribution ellipse and the properties of the beam. Figure 3.22*a* shows a beam distribution that is narrow along x and wide along x'.

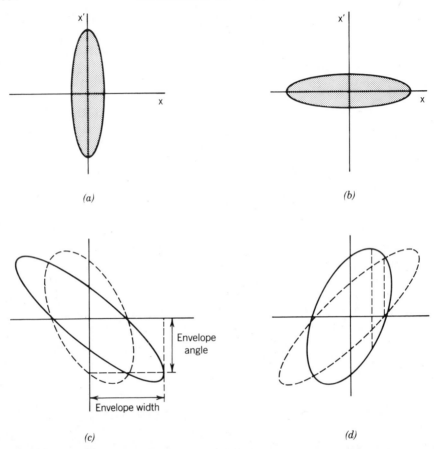

Figure 3.22 Examples of orbit-vector distributions in trace-space. (*a*) Hot distribution—beam at a focal point (waist). (*b*) Cold distribution of an approximately parallel beam with the same emittance as part *a*. (*c*) Skewed elliptical distribution representing a converging beam. Dashed ellipse shows distribution at a latter time. (*d*) Skewed elliptical distribution representing a diverging beam. Dashed lines emphasize that the angular divergence at a particular relative position in the beam decreases as the beam expands.

The beam is at a focal point. Such a distribution is sometimes called a *hot distribution*—there is a large spread in angle and transverse velocity. In contrast, Fig. 3.22*b* represents a *cold distribution*—the angular width is small. The distributions of Figs. 3.22*a* and 3.22*b* have the same emittance—linear optical elements can convert one distribution into the other. Depending on the application, focusing systems can produce beams with small spatial dimension or small angular spread.

The upright ellipses of Figs. 3.22*a* and 3.22*b* represent beam waists. The average angle at all positions of the beam is zero—the beam neither converges nor diverges and the beam envelope is parallel to the *z* axis. Generally, distributions are inclined, as in Figs. 3.22c and 3.22d. Particles in the ellipse of Fig. 3.22c have a negative average angle for $x > 0$ and a positive angle for $x < 0$. Therefore, the beam converges toward the axis—the envelope angle is negative. The dashed line shows the same distribution after drifting a distance downstream—the spatial dimension is smaller while the angular spread at each position inside the beam is larger. The beam of Fig. 3.22d diverges. Again, the dashed line shows how a drift transforms the distribution. The spatial width increases while the angular spread *at each position* shrinks. We say that the beam *cools* as it expands. The inclination or skewness of the distribution ellipse determines the envelope convergence or divergence angle.

Transfer matrix algebra is applied to calculate the effects of linear optical elements on the orbit parameters of individual particles, $[x_n(z), x'_n(z)]$. Section 4.1 reviews the topic while [CPA, Chapter 8] gives a detailed discussion of matrix methods. As an example, if a particle enters a drift space of length d with orbit vector $[x_{0n}, x'_{0n}]$, then the exit vector is

$$\begin{bmatrix} x_n \\ x'_n \end{bmatrix} = \mathbf{M}_d \begin{bmatrix} x_{0n} \\ x'_{0n} \end{bmatrix}. \tag{3.52}$$

The transfer matrix for the drift space is

$$\mathbf{M}_d = \begin{bmatrix} 1 & d \\ 0 & 1 \end{bmatrix}. \tag{3.53}$$

The transfer matrix of a thin lens of focal length f is

$$\mathbf{M}_f = \begin{bmatrix} 1 & 0 \\ -1/f & 1 \end{bmatrix}. \tag{3.54}$$

A simple method to find the effect of a lens or drift space on a distribution ellipse is to apply Eq. (3.52) to a large number of orbit vectors on the boundary. Figure 3.23a depicts the calculation for a free drift length. In Section 4.1, we shall prove that the output distribution is another ellipse with the same area. As expected, the diverging beam distribution cools. Figure 3.23b demonstrates the effect of a thin focusing lens that changes the angle of particles but not their position. The net effect is to modify the orientation of the distribution ellipse. The lens of Fig. 3.23b transforms a diverging beam into a converging beam.

The *beam telescope* of Fig. 3.24 illustrates the effect of combined optical elements on a nonlaminar beam. The function of the device is to focus a beam from an accelerator to a small spot a long distance from the lens. We saw in Section 3.1 that the angular divergence of the beam at the lens limits the minimum size of the focal spot. If the output beam from the accelerator were focused directly by a small-diameter lens, the focal spot would be large. To reduce the spot size, the telescope first expands the beam and then focuses the cooled

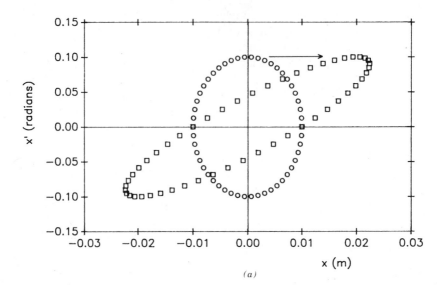

(a)

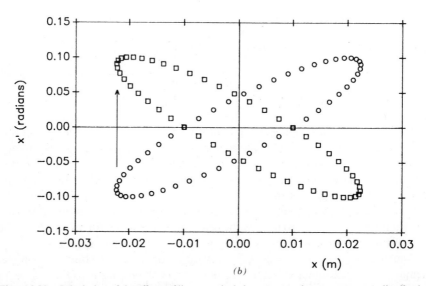

(b)

Figure 3.23   Calculation of the effects of linear optical elements on a beam trace-space distribution by the projection of the position and angle of several particles on the beam boundary. (a) Effect of a drift length. Circles represent boundary-particle-orbit vectors at a beam waist. Squares show orbit vectors a distance $L = 0.2$ m downstream. (b) Effect of a lens with $f = 0.2$ m. Circles represent boundary orbit vectors of a diverging beam at the lens entrance, squares show orbit vectors at the lens exit.

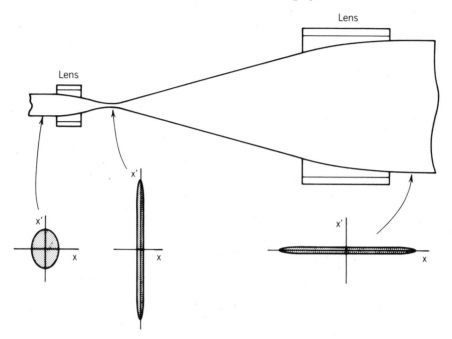

Figure 3.24   Configuration-space and trace-space views of beam motion through a one-dimensional beam telescope.

beam with a large-diameter lens. Figure 3.24 contains trace-space distributions at various points in the telescope. The beam is parallel at the entrance. The first lens expands the beam diameter by focusing particles through a point. The final lens directs particles in the average direction of the focal point. The particles converge at a distant point with a focal spot size limited by the angular divergence of the distribution at the final lens.

We shall introduce the idea of beam matching with the example of propagation in a continuous linear focusing force. Assume the force is uniform in z and linear in x:

$$F_x = - Ax. \tag{3.55}$$

In the paraxial limit, individual particles follow orbits described by

$$x_n(z) = x_{0n} \sin (k_x z + \phi_n), \tag{3.56}$$

$$x'_n(z) = x_{0n} k_x \cos (k_x z + \phi_n), \tag{3.57}$$

where

$$k_x = (A/\gamma m_0)^{1/2}/\beta c.$$

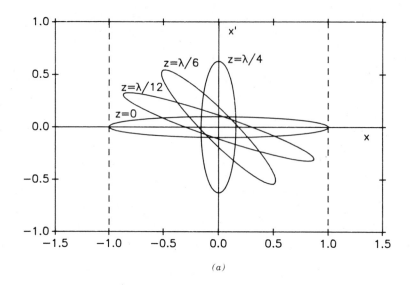

(a)

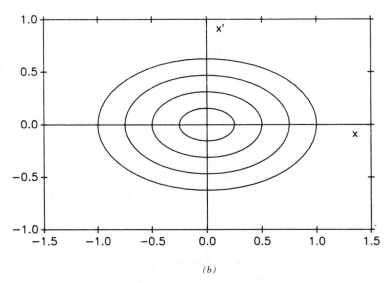

(b)

Figure 3.25 Evolution of orbit-vector distributions in trace-space for a beam moving in a continuous, linear focusing force. (*a*) Beam mismatched at the entrance ($z = 0$). Initially, the envelope width decreases and the angular divergence increases. Envelope oscillations occur at the betatron wavelength, λ. (*b*) Contour lines for a distribution matched to the focusing force.

Equations (3.56) and (3.57) are the parametric equations of ellipses in trace space with dimensions x_{0n} and $x_{0n}k_x$ along the axes. Although the amplitude and phase of individual particle orbits vary, all particle orbits rotate in trace space at the same frequency.

Suppose we pick an initial distribution with a boundary of arbitrary shape, such as the ellipse of Fig. 3.25a. We can use Eqs. (3.56) and (3.57) to predict the boundary of the ellipse at any axial position z. Figure 3.25a shows results of the calculation—the ellipse rotates, completing a circuit after one betatron wavelength, $\lambda = 2\pi/k_x$. The envelope width of the beam, marked by a dashed line in Fig. 3.25a, corresponds to the maximum spatial extension. The envelope width oscillates as the beam propagates. Here, we say that the distribution is *mismatched* to the focusing force.

A matched beam exhibits minimal envelope oscillations. The advantage of matching a beam with a given emittance to a focusing system is that the beam has the smallest possible spatial width as it propagates. This characteristic is important in large accelerators—lenses with small bore diameters cost less. We can easily find a matched distribution when focusing is performed by a continuous force [Eq. (3.55)]. Take x_0 as the maximum allowed width of the beam. A particle with oscillation amplitude x_0 *defines a trace-space ellipse:*

$$\left[\frac{x}{x_0}\right]^2 + \left[\frac{x'}{x_0 k_x}\right]^2 = 1. \tag{3.58}$$

We choose a distribution boundary where particle orbit vectors are uniformly distributed along the ellipse of Eq. (3.58)—Fig. 3.25b shows the matched distribution boundary. We specify, further, that the distribution enclosed by the boundary has uniform density contours along the curves of Eqs. (3.56) and (3.57). The distribution density does not change as the beam propagates, even though individual particles rotate in trace space. The departure of particles at any point $[x, x']$ in the distribution is balanced by the arrival of an equal number of particles. The distribution is *matched* to the focusing system—the envelope width is constant.

The matched beam defined by Eq. (3.58) is in equilibrium. Even though individual particles move, macroscopic properties of the beam do not change with time as the beam propagates. Note that the ellipses of Eqs. (3.56) and (3.57) are curves of constant transverse total energy. The total energy W_x is the sum of transverse kinetic energy plus potential energy:

$$W_x = (\gamma m_0 \beta^2 c^2/2)(x'^2 + k_x^2 x^2). \tag{3.59}$$

The condition for a stationary distribution is that the continuous distribution function $f(x, x')$ is constant along the ellipses. This is equivalent to saying f must be a function of the constant of motion for a beam in equilibrium.

For four-dimensional distributions, $f(x, x', y, y')$, there are two constants of the motion. If motions in the x and y directions are uncoupled, then the constants

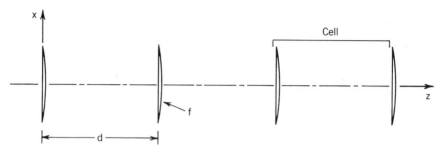

Figure 3.26 An array of thin, linear lenses. By definition, a focusing cell extends from the entrance of a lens to the entrance of the following lens.

are the total energy in each direction, W_x and W_y. Any function $f\ (W_x, W_y)$ represents a potential beam equilibrium that automatically satisfies the Vlasov equation when $\partial f/\partial t = 0$. An example is a uniformly filled matched ellipse:

$$f(W_x, W_y) = N g_1(W_x) g_2(W_y), \tag{3.60}$$

where N is the total number of particles and

$$
\begin{aligned}
g_1(W_x) &= 1/(\pi k_x x_0^2) &\quad (W_x \leqslant k_x x_0^2/2) \\
&= 0 &\quad (W_x > k_x x_0^2/2).
\end{aligned}
\tag{3.61}
$$

Another example is the truncated Maxwell distribution,

$$
\begin{aligned}
f(W_x, W_y) &= A\exp(-B_x W_x^2)\exp(-B_y W_y^2) &\quad (W_x \leqslant k_x x_0^2/2 \text{ and } W_y \leqslant k_y y_0^2/2) \\
&= 0 &\quad (W_x > k_x x_0^2/2 \text{ or } W_y > k_y y_0^2/2).
\end{aligned}
\tag{3.62}
$$

Next, we shall extend the concept of matched beams from a continuous focusing force to a periodic focusing system. As an example, we shall use a one-dimensional beam in the thin lens array of Fig. 3.26. The *cell* is the fundamental unit in a periodic focusing system. In the thin lens array, a cell consists of one lens with focal length f and one drift space of length d. Figure 3.27 shows a mismatched beam in the thin lens array. The envelope width varies from cell to cell. A distribution is matched to a periodic system if it has identical properties at the boundaries of all focusing cells. Although the envelope of the beam varies passing through a cell, the width has the same value at cell boundaries. A matched beam exhibits minimal envelope oscillations.

Section 4.3 describes analytic methods to find matched distributions. In this section, we shall concentrate on a simple mathematical method to find the matched distribution of any periodic system. The method consists of the following steps:

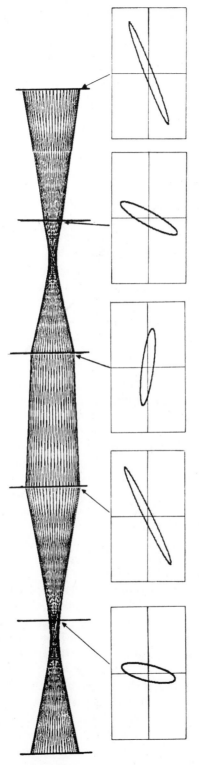

Figure 3.27 Propagation of a mismatched beam through an array of thin lenses. Configuration-space view of 250 particle orbits on the distribution boundary and trace-space views at lens entrances. $\mu_0 = 60°$.

1. Choose a focusing cell boundary—the boundary can be any location that divides the focusing system into the smallest possible repeating sections. As an example, we take the boundary of the lens array at the entrance to the lens.

2. Find the total ray transfer matrix for a focusing cell. The transformation matrix for any combination of linear elements is a single matrix resulting from multiplication of the matrices of individual components. The transfer matrix for the focusing cell consisting of a thin lens following by a drift distance is

$$\mathbf{M}_{fd} = \begin{bmatrix} 1 - d/f & d \\ -1/f & 1 \end{bmatrix}. \tag{3.63}$$

3. Take a test orbit that enters the focusing system at the chosen cell boundary, such as $(x, x')_0 = (x_0, 0)$. Advance the orbit from cell to cell using the transfer matrix. The result is a set of orbit vectors $(x, x')_i$.

4. Plot the locus or orbit vectors on a trace-space diagram. The orbit vectors trace an ellipse as long as the phase advance per focusing cell is not a rational number.

5. The ellipse created is a boundary for a matched distribution. We can adjust the size of the curve to achieve the desired envelope width at the cell boundary.

The method relies on the fact that the test orbit intersects the cell boundary with all possible phase values for orbits on the distribution periphery as long as the vacuum phased advance μ_0 has an irrational value. Figure 3.28 shows a numerical calculation of a matched distribution in the thin lens array and the corresponding configuration-space orbits. Note that the matched ellipse at the entrance to the lens is skewed. Inspection of the configuration-space orbits shows that the matched beam must be diverging when it enters the first lens.

Acceptance refers to the region in trace space accessible for particle transport. We can define acceptance for optical systems with multiple elements or for single devices. For example, Fig. 3.29a shows the acceptance of a slit of width 2δ. The aperture transmits particle orbits with any inclination angle as long as their position lies between $\pm \delta$. Figure 3.29b shows the acceptance of a continuous focusing system [defined by Eqs. (3.56) and (3.57)] with boundaries at $\pm x_w$. Entering particles with orbit vectors in the shaded ellipse have oscillation amplitudes smaller than x_w. Similarly, Fig. 3.29c shows the acceptance diagram for a thin lens array with wall boundaries $\pm x_w$ in the planes of the lenses.

The thick aperture of Fig. 3.29d is an interesting example of an acceptance calculation. The aperture has width $2x_w$ and length L. Orbit vectors at the entrance must have positions in the range $-x_w < x < x_w$. In contrast to the thin aperture, the thick aperture constrains the allowed angles at the entrance. Inspection of Fig. 3.29d shows that entering particles with $x = 0$ pass through the aperture if their angle is in the range $-x_w/L < x' < x_w/L$. Particles that enter at

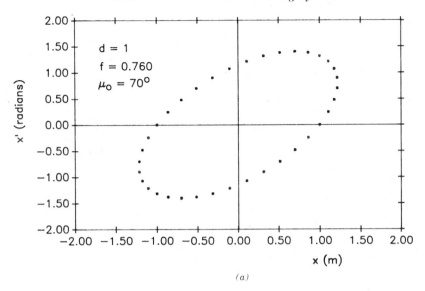

$$d = 1$$
$$f = 0.760$$
$$\mu_0 = 70°$$

(a)

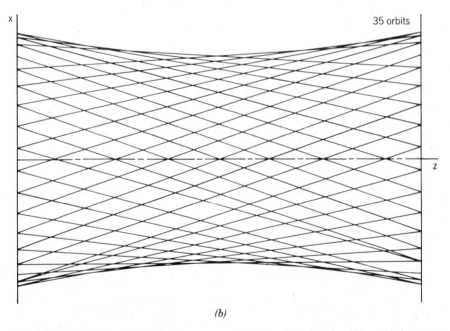

35 orbits

(b)

Figure 3.28 Derivation of a matched trace-space distribution boundary for a linear thin lens array ($f = 0.76$ m, $d = 1$ m, $\mu_0 = 70°$). The cell boundary is at the lens entrance. (a) Trace-space distribution—the points show orbit vectors of a test particle on the distribution boundary at the entrance to 35 lenses. (b) Superimposed configuration-space trajectories of the test particle show the matched beam envelope.

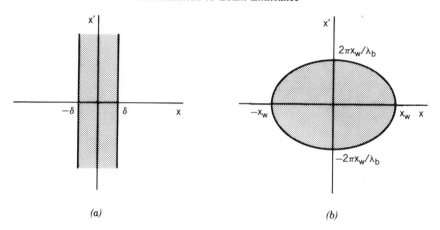

(a)   *(b)*

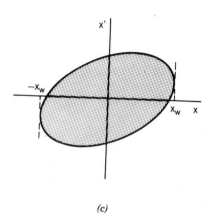

(c)

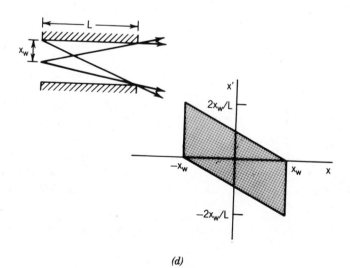

(d)

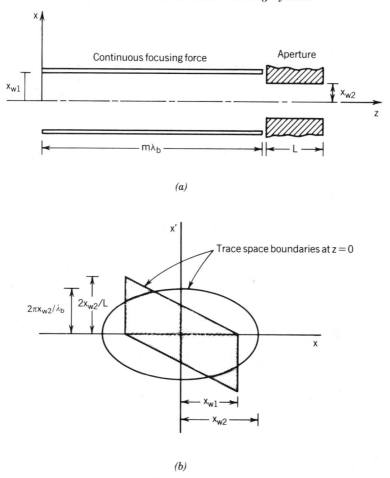

(a)

(b)

Figure 3.30 Combined acceptance of a continuous focusing force followed by a thick aperture. (a) Geometry. (b) Acceptance diagram. The shapes represent the acceptances of individual elements projected to $z = 0$. The shaded area is the combined acceptance.

$+ x_w$ can have no positive angle—the allowed negative angle is $- 2x_w/L$. The acceptance region is the trapezoid of Fig. 3.29d.

We can apply the following procedure to find the acceptance of complex optical systems with many elements. We project the acceptance regions for all individual elements back to the entrance of the system. To make the projection, we first find the total transfer matrix of all elements between the entrance and the element in question. Next, we transform several vectors on the

Figure 3.29 Examples of acceptance diagrams for one-dimensional beams. (a) Thin aperture. (b) Continuous focusing force. (c) Thin lens array—lens entrance. (d) Thick aperture. Inset shows geometry.

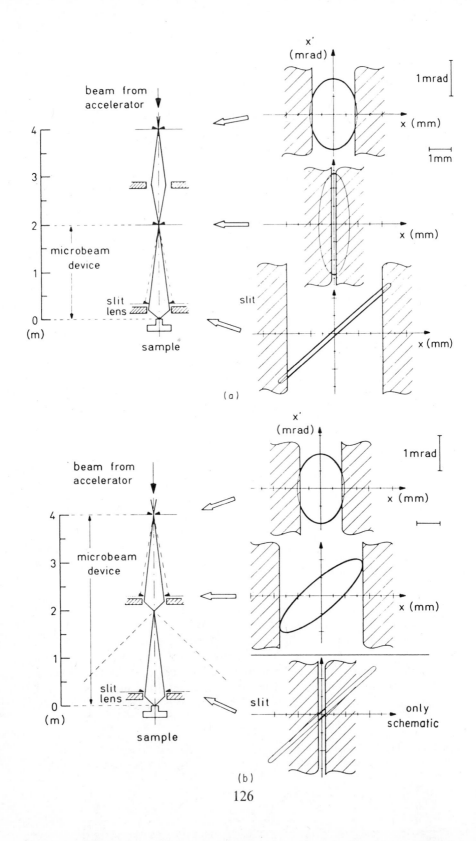

(a)

(b)

126

element acceptance boundary back to the entrance to define a modified boundary. Finally, we combine the projected acceptances to create a global entrance acceptance diagram. The allowed input beam emittance corresponds to the maximum area ellipse that can fit inside the combined boundaries. Figure 3.30 illustrates the combined acceptance for a region of continuous focusing force followed by a thick aperture of length L. The beam performs an integral number of betatron oscillations in the continuous force region.

The ion microprobe is a good example of transport of a nonlaminar beam through linear optical elements. The goal is to focus an ion beam to a very small spot size (~ 1–$5\,\mu$m) on a surface. Analysis of emissions stimulated by the beam gives information on the local surface composition. The measurement usually requires an accumulated charge of 10–100 nC. To probe a single point in 100 sec, the beam current must be in the range 0.1–1 nA. High-energy probe ion beams ($\geqslant 1$ MeV) are usually generated by Van de Graaff accelerators. The emittance of beams from these machines is usually in the range 10^{-6}–10^{-5} m-rad.

We have seen that the best strategy to achieve a small focal spot is to expand the beam to a large diameter at the final focusing lens. The maximum lens diameter is limited by spherical aberration. This effect restricts the beam envelope convergence angle to less than about 0.05 rad. The emittance of a 5-μm beam at the focal spot must be in the range

$$\varepsilon \leqslant (5 \times 10^{-6}\,\text{m})(5 \times 10^{-2}\,\text{rad}) = 2.5 \times 10^{-7}\,\text{m-rad}. \tag{3.64}$$

The value of Eq. (3.64) is much lower than the source emittance. The only way to achieve the low emittance value is to collimate the beam and accept only a fraction of the incident particles. As a result, the accelerator must generate a continuous current of 10 nA or more.

Figure 3.31 shows a scale drawing of two approaches for the optics of an ion microprobe. Either particles with high angular divergence or particles with high displacement are discarded, leaving behind a low emittance beam. In Fig. 3.31a, the beam from the accelerator is projected on a narrow collimator that accepts only the fraction of the initial distribution with small displacement. The beam is subsequently cooled by expansion and then focused by a lens with the shortest focal length allowed by spherical aberration. The geometry of Fig. 3.31b illustrates another method to select particles. The beam from the accelerator is strongly focused and expanded. An aperture in the plane of the final focusing lens selects particles from the cooled distribution with the convergence angle set by lens aberrations.

Figure 3.31 Acceptance diagrams for emittance selection in an ion microprobe. System geometry and trace-space plots at selected locations. (a) Single-stage focusing. Emittance selection at a narrow aperture at the entrance, then focusing of the narrow image to the target. (b) Double-stage focusing with beam cooling by expansion. Transverse dimensions not to scale. (From "High-energy ion microprobes," in *Applied Charged Particle Optics*, A. Septier, Ed. Used by permission, Academic Press.)

3.8. COMPRESSION AND EXPANSION OF
NONLAMINAR BEAMS

We can control the dimensions of a charged particle beam by changing the magnitude of focusing forces. This process is often applied to compress or to bunch beams for transfer between different types of accelerators. The velocity spread of nonlaminar beams limits the dimension change for a given change in focusing force. In this section, we shall apply the principle of conservation of emittance to beam compression.

In the derivations, we shall find it useful to use the equivalent *beam temperature*, a quantity proportional to the mean squared velocity dispersion (Section 2.11). The relationship between volume and temperature in a medium, be it a gas, a plasma, or a charged particle beam, is called an *equation of state*. We can apply emittance conservation to find equations of state for beams that are thermally isolated. In other words, there are no mechanisms, such as collisions, where the beam exchanges energy with external systems. Sometimes, we can find a simple equation of state that circumvents the need for higher-order moment equations (Section 2.10).

To begin, we shall study compression of a one-dimensional beam normal to the direction of propagation. The term compression denotes reduction of the spatial dimension of the beam by increasing the focusing force. We should note that the results are directly applicable to beam expansions where the force decreases. To simplify the model, we adopt the following assumptions:

1. The focusing force is almost continuous along the z direction.

2. Particles sense a slowly varying focusing force compared with their betatron frequency.

3. The beam is almost in transverse equilibrium—the magnitude of the emittance force is approximately equal to the focusing force.

Consider beam particles with uniform axial velocity v_z in a linear focusing force. If $v_x \ll v_z$, we can use an equation of motion with a nonrelativistic form. The following equation describes the transverse motion of particles:

$$d^2x/dt^2 = -\omega(t)^2 x. \tag{3.65}$$

The quantity $\omega(t)^2$ is proportional to the transverse force. The condition that the forces change slowly over a betatron period is

$$(d\omega/dt)/\omega^2 \ll 1. \tag{3.66}$$

In the limit of Eq. (3.66), we can show by direct substitution [CPA, p. 336] that the solution of Eq. (3.65) is

$$x(t) \cong A(t) \sin [\Phi(t)]. \tag{3.67}$$

where

$$\Phi = \Phi_0 + \int \omega(t)dt \qquad (3.68)$$

and

$$\omega A^2 = \text{constant}. \qquad (3.69)$$

Equation (3.67) has the following implications. Particle oscillations at time t are almost harmonic with frequency equal to $\omega(t)$. Also, the particle trace-space trajectory is close to an ellipse with dimensions

$$x_0(t) \cong A(t), \qquad (3.70)$$

$$x_0'(t) = v_{x0}(t)/v_z \cong \omega(t)A(t)/v_z. \qquad (3.71)$$

Taking the product of Eqs. (3.70) and (3.71), we find that the area of the trace-space ellipse is constant, although the relative shape may change.

Suppose that the orbit of Eqs. (3.70) and (3.71) defines the boundary of a matched beam distribution. Initially, particles have trace-space trajectories that follow nested ellipses inside the boundary orbit. Since all orbits have a solution of the form of Eq. (3.67), the trajectories remain within the boundary orbit as the focusing force changes. Therefore, the boundary trajectory encloses a distribution ellipse with constant area:

$$x_0(t)x_0'(t) = \text{constant}. \qquad (3.72)$$

Equations (3.72) is a statement that emittance is conserved when linear focusing forces vary slowly.

Extending the development of Section 2.11, we define the transverse beam temperature as an average over the beam velocity distribution:

$$kT_x = \gamma m_0 \langle v_x^2 \rangle/2. \qquad (3.73)$$

The root-mean-squared velocity must be proportional to the velocity width of the distribution, or

$$[v_{x0}(t)]^2 \sim \langle v_x(t)^2 \rangle \sim T_x(t). \qquad (3.74)$$

We can write Eq. (3.72) in the alternative form

$$v_{x0}(t) \sim 1/x_0(t). \qquad (3.75)$$

Combining Eqs. (3.74) and (3.75), the transverse temperature and beam width are related by the equation

$$T_x(t)/T_0 = [x_0(0)/x_0(t)]^2. \qquad (3.76)$$

In an ideal one-dimensional compression, particle orbits change only in the x-direction—the quantities $\langle v_y^2 \rangle$ and $\langle v_z^2 \rangle$ remain constant. Since the beam dimensions in the y and z directions do not change during the compression, the beam volume V is proportional to $x_0(t)$. Rewriting Eq. (3.76) in terms of the volume gives the equation of state for a one-dimensional compression:

$$T_x(t)/T_x(0) = [V(0)/V(t)]^2. \tag{3.77}$$

We can extend the reasoning that led to Eq. (3.77) to a two-dimensional compression. Assume that beam focusing is performed by forces that vary slowly in the axial direction and are separable in the x and y directions. Furthermore, forces in the two directions have equal magnitude and the beam distribution is initially isotropic, $\langle v_x(0)^2 \rangle = \langle v_y(0)^2 \rangle$. For a slow compression, conservation of emittance leads to the relationship,

$$\langle v_x^2(t) \rangle \sim 1/x_0(t)^2, \qquad \langle v_y^2(t) \rangle \sim 1/y_0(t)^2. \tag{3.78}$$

The assumption of symmetry implies that

$$x_0(t) = y_0(t) \tag{3.79}$$

and

$$\langle v_x(t)^2 \rangle = \langle v_y(t)^2 \rangle. \tag{3.80}$$

Applying Eqs. (3.79) and (3.80), we can write the transverse beam temperature as

$$T_\perp(t) \sim [\langle v_x(t)^2 \rangle + \langle v_y(t)^2 \rangle] \sim 1/x_0(t)y_0(t). \tag{3.81}$$

The following equation of state holds for a symmetric two-dimensional compression:

$$T_\perp(t)/T_\perp(0) = [V(0)/V(t)]. \tag{3.82}$$

Equation (3.82) holds when particles gain energy in the x and y directions but not in the axial direction. The equation is also valid if x and y motions are coupled, as in a rising solenoidal magnetic field.

We can apply the same arguments to the symmetric three-dimensional compression of an isotropic beam. The net beam temperature has proportionality

$$T(t) \sim [\langle v_x(t)^2 \rangle + \langle v_y(t)^2 \rangle + \langle v_z(t)^2 \rangle]. \tag{3.83}$$

The principle of emittance conservation has the form

$$[\langle v_x(t)^2 \rangle \langle v_y(t)^2 \rangle \langle v_z(t)^2 \rangle]^{1/2} \sim 1/x_0(t)y_0(t)z_0(t). \tag{3.84}$$

Equations (3.83) and (3.84) imply the equation of state

$$T(t)/T(0) = [V(0)/V(t)]^{2/3} \sim 1/x_0(t). \tag{3.85}$$

Comparison of Eqs. (3.77), (3.82), and (3.85) leads to a general form for the equation of state of isotropic beams:

$$T(t)/T(0) = [V(0)/V(t)]^{2/\gamma}. \tag{3.86}$$

The quantity γ is the number of degrees of freedom, equal to 1, 2, or 3. Equation (3.86) is a familiar relationship in gas dynamics—it applies to collision-dominated particle distributions in thermal isolation. We can see that it also describes collisionless particle distributions.

We seldom encounter ideal spherical compressions in charged-particle-beam applications. Beams are focused and bunched, but usually the dimension changes are widely different in the transverse and axial directions. When axial particle motion is decoupled, we can define separate transverse and longitudinal temperatures and invoke conservation of emittance to derive two separate equations of state in terms of the width and length of the beam. Sometimes, we can derive a hybrid equation of state when axial motion couples to transverse motion. For example, suppose that a beam enters a region where the transverse force increases along the axis. Although the forces are predominantly normal to the axis, there is a small component of force in the z direction. As particle orbits compress, some of the extra kinetic energy transfers to random axial motion about a mean velocity. If the coupling is strong, we expect that the distribution viewed in the beam rest frame approaches an isotropic state.

To illustrate how coupling modifies the equation of state, suppose that forces compress a beam in the two transverse directions. Although the axial length of the beam is constant, there are mechanisms that couple energy to an axial velocity spread. For simplicity, we take the velocity distribution as isotropic in three dimensions. In each direction, conservation of emittance implies that

$$\begin{aligned} \langle v_x(t)^2 \rangle &\sim 1/x_0(t)^2, \\ \langle v_y(t)^2 \rangle &\sim 1/y_0(t)^2, \\ \langle v_z(t)^2 \rangle &\sim 1/z_0(t)^2. \end{aligned} \tag{3.87}$$

The product of Eqs. (3.87) is

$$[\langle v_x(t)^2 \rangle \langle v_y(t)^2 \rangle \langle v_z(t)^2 \rangle]^{1/2} \sim 1/x_0(t)y_0(t)z_0(t). \tag{3.88}$$

With the definition of beam temperature from Eq. (3.83), Eq. (3.88) leads to the volume dependence as Eq. (3.85):

$$T(t)/T(0) = [V(0)/V(t)]^{2/3}. \tag{3.89}$$

On the other hand, the scaling of temperature with the transverse dimension of the beam is different from Eq. (3.85) because of the modified dependence of volume on dimension:

$$V(t) \sim 1/x_0(t)^2. \tag{3.90}$$

Equation (3.90) implies that

$$T(t)/T(0) = [x_0(0)/x_0(t)]^{4/3}. \tag{3.91}$$

We will use Eq. 3.91 in Section 11.4 to study details of neutralized beam focusing.

4

Beam Emittance—Advanced Topics

In this chapter, we continue our study of emittance by discussing several critical areas for beam applications. The first three sections outline the linear transformation theory of beam distributions. Here, we extend the method of transfer matrices to deal with particle distributions. Transfer matrix theory simplifies the description of single-particle motion in linear optical systems. A 2×2 (or 4×4) matrix summarizes the effect of any linear focusing element. Similarly, transport theory leads to matrices that describe the action of an optical element on an entire beam distribution.

Section 4.1 reviews that special nature of beam distributions with elliptical trace-space boundaries. We shall find that a linear optical element transforms an elliptical distribution to another ellipse. The ellipse area (or distribution emittance) is constant, but the shape may change. The section introduces the transport parameters α, β, and γ. These quantities specify the geometry of an elliptical distribution. We derive equations that give the change in the transport parameters of a distribution passing through an optical element in terms of the element's transfer matrix. Section 4.2 presents an alternative derivation of the transport parameters from the properties of particle orbits in a circular accelerator with a periodic focusing system. Section 4.3 illustrates beam matching, an application of transport theory. Often, we need specific distributions for injection into accelerators and beam transport devices. Transport theory leads to systematic methods to choose matching lenses that convert a given beam to a desired distribution.

Section 4.4 discusses periodic focusing systems with nonlinear forces. This topic is important for intense beam transport because beam-generated forces are inevitably nonlinear. Although the emittance of mismatched beams grows in nonlinear periodic focusing systems, we shall find that the emittance of matched beams is constant. The section presents methods to calculate matched distributions and to compress beams in focusing channels with arbitrary variations of transverse force.

Section 4.5 reviews the importance of low emittance for beams in storage rings for high-energy-physics research, particularly colliding-beam experiments. The quality of a beam in a storage ring is expressed in terms of luminosity. The rate of reactions for high-energy-physics experiments is proportional to luminosity—a high value of luminosity demands a stored beam with low emittance. Section 4.6 reviews methods for beam cooling, the reduction of beam emittance. The techniques, which depend on the long-term storage of beams, have application mainly to circular accelerators. Beam cooling circumvents the principle of phase-volume conservation. Understanding cooling methods will give us insight into the limitations of fluid models for beams.

4.1. LINEAR TRANSFORMATIONS OF ELLIPTICAL DISTRIBUTIONS

Transfer matrix theory [CPA, Chap. 8] succinctly describes the motion of *single* particles through complex linear optical systems. When we treat nonlaminar beams, we must deal with large numbers of particles with different orbit characteristics. In this section, we shall extend the method of transfer matrices to a form that advances an entire elliptical distribution rather than a single-particle orbit. If we know the properties of the input distribution ellipse and the ray transfer matrix for an extended system of linear elements, we can predict the output distribution directly. This approach is much more efficient than the calculation of multiple orbits in Section 3.7.

The theory of linear transformation of distribution ellipses is often called *distribution transport theory* since it forms the basis for the well-known computer code TRANSPORT—the quantities that characterize ellipses are called the *transport parameters*. Distribution transport theory has many applications in accelerator design:

1. Estimation of macroscopic beam properties, such as the envelope width and convergence angle, within complex optical systems.

2. Calculation of matched beams for periodic focusing systems.

3. Prediction of the distribution at the entrance to an accelerator or transport system necessary to obtain an exit distribution required by an application.

4. Calculation of optical elements for a matched beam transfer between periodic focusing systems with different focal properties.

Following a brief review of transfer matrices for single-particle orbits, this section introduces distribution transport theory for collections of discrete optical elements. Section 4.2 addresses periodic focusing systems and beam matching.

A.   Linear Transformations of Single-Particle Orbits

Transfer matrix theory describes particle motion relative to a known main equilibrium orbit. The first-order theory employs two assumptions:

1. Particle motions are paraxial. The inclination angles are small and all particles have approximately the same axial velocity v_z at an axial location.

2. Transverse focusing forces vary linearly with displacement from the main axis and are independent of the transverse velocity.

Particle orbits are characterized by displacement from the main axis and inclination angle. In Cartesian coordinates, a particle orbit at some axial position is specified by a set of four quantities that we write as a vector:

$$\mathbf{x} = [x, x', y, y'].\tag{4.1}$$

The quantities x and y are the displacement from the main axis, while x' and y' are the inclination angles.

Charged-particle-beam focusing elements generate localized regions of transverse forces. Examples include bending magnets, quadrupole lenses, and drift lengths. Some focusing elements, such as acceleration gaps, also have axial forces. If the transverse forces are linear, we display the effect of an optical element on an orbit vector by

$$\mathbf{x}_1 = \mathbf{M}_1 \mathbf{x}_0.\tag{4.2}$$

The quantity $\mathbf{M}_1$ is a 4×4 matrix called a *transfer matrix*. If there are no accelerating forces in the element, the transfer matrix has the property:

$$\det \mathbf{M} = 1,\tag{4.3}$$

where det designates the determinant operation.

Equation 4.2 follows the standard rules for the multiplication of a 4×4 matrix times a four-component vector. We denote the individual components of $\mathbf{M}$ as m_{ij}—i is the row index and j is the column index. Equation (4.1) is a linear transformation if the components of the matrix are constants with no dependence on x, x', y, or y'. Here, the terms of the output orbit vector are linear combinations of the terms of the input vector, such as

$$x_1 = m_{11}x_0 + m_{12}x'_0 + m_{13}y_0 + m_{14}y'_0.\tag{4.4}$$

When a particle travels through two sequential optical elements, the net change of the orbit vector is

$$\mathbf{x}_2 = \mathbf{M}_2 \mathbf{x}_1 = \mathbf{M}_2(\mathbf{M}_1 \mathbf{x}_0).\tag{4.5}$$

The quantity $\mathbf{x}_0$ is the input vector, $\mathbf{M}_1$ and $\mathbf{M}_2$ are the transfer matrices of the elements, and $\mathbf{x}_1$ and $\mathbf{x}_2$ are the vectors at the exits of the two elements. We can write the result as

$$\mathbf{x}_2 = (\mathbf{M}_2 \cdot \mathbf{M}_1)\mathbf{x}_0. \tag{4.6}$$

Evaluation of the quantity in parentheses follows the standard rules of matrix multiplication. If there is a third optical element. the cumulative effect of the system is given by

$$\mathbf{x}_3 = (\mathbf{M}_3(\mathbf{M}_2 \cdot \mathbf{M}_1))\mathbf{x}_0. \tag{4.7}$$

Note that the matrix multiplication operation is not commutative; multiplications must be carried out in the order that the particle traverses the elements. The input vector is multiplied by the matrix for the first element, then the second element, and so forth.

The forces in a charged particle transport system are often separable in the x and y directions. Here, the transverse force along x does not depend on y or y'. Applications where forces are separable include storage rings with dipole bending magnets and quadrupole focusing magnets, electrostatic quadrupole arrays for low-energy ion beam transport, and radio-frequency quadrupole ion accelerators. The transport matrix reduces to the simple form

$$\mathbf{M} = \begin{bmatrix} m_{11} & m_{12} & 0 & 0 \\ m_{21} & m_{22} & 0 & 0 \\ 0 & 0 & m_{33} & m_{34} \\ 0 & 0 & m_{43} & m_{44} \end{bmatrix}. \tag{4.8}$$

Orbit vectors in the x and y directions are advanced independently by 2×2 matrices, a subset of the matrix of Eq. (4.8):

$$\mathbf{M} = \begin{bmatrix} m_{11} & m_{12} \\ m_{21} & m_{22} \end{bmatrix}. \tag{4.9}$$

We will concentrate on 2×2 matrices for the remainder of this section; the extension to 4×4 matrices is straightforward. For reference, the determinant of a 2×2 matrix is

$$\det \mathbf{M} = m_{11}m_{22} - m_{21}m_{12}. \tag{4.10}$$

We shall illustrate transfer matrices with two simple examples—the thin lens and the field-free drift space. We can often approximate more complex elements, such as a quadrupole lens, with combinations of these two elements. The matrix for a thin lens with focal length f is

$$\mathbf{M}(\text{lens}) = \begin{bmatrix} 1 & 0 \\ -1/f & 1 \end{bmatrix}. \tag{4.11}$$

Equation 4.11 shows that the lens changes the angle of a particle orbit by a factor proportional to the displacement but it does not affect the displacement. The matrix of a field-free drift region of length D is

$$\mathbf{M}(\text{drift}) = \begin{bmatrix} 1 & D \\ 0 & 1 \end{bmatrix}. \tag{4.12}$$

The matrix of Eq. (4.12) modifies the displacement of an orbit vector by a factor proportional to the inclination angle but leaves the angle unchanged.

Transfer matrix theory is well suited to periodic focusing systems. A periodic system consists of a series of identical elements or collections of elements called cells. The most common periodic system in accelerator science is the quadrupole lens array [CPA, Section 8.7]. Quadrupole lenses must be arranged in focusing–defocusing combinations to ensure containment in both the x and y directions. We denote an array with a focusing lens in the x direction followed by a defocusing lens as FD. Because a quadrupole lens that focuses in the x direction defocuses in the y direction, the designation for the y direction is DF. In circular accelerators, quadrupole arrays usually consist of a focusing lens, a drift space to accommodate a bending magnet, a defocusing lens, and another drift length. This geometry is called a $FODO$ system, where the O denotes "open." The cell of the FODO system consists of two lenses and two drift lengths.

We can represent the effect of each optical element in a periodic focusing cell by a transfer matrix. Matrix multiplication of the elements of the cell gives a matrix $\mathbf{M}$ that represents the total effect of the cell. When forces are separable, two 2×2 matrices, $\mathbf{M}_x$ and $\mathbf{M}_y$, advance orbits through a cell independently in the x and y directions. Passage through n identical focusing cells transforms an incident orbit vector according to

$$x_n = (\mathbf{M}_x^n)x_0, \tag{4.13}$$

$$y_n = (\mathbf{M}_y^n)y_0. \tag{4.14}$$

The symbol $\mathbf{M}_x^n$ in Eq. (4.13) represents matrix multiplication of $\mathbf{M}_x$ by itself n times.

B. Elliptical Distributions

We have seen in Chapter 3 that beam distributions enclosed by elliptical boundaries play an important role in accelerators with linear focusing systems.

Trace-space ellipses have geometric properties that lead to a compact-beam transport theory. A linear transformation always transforms an elliptical distribution into another ellipse. Furthermore, with no acceleration, a linear transformation does not change the area of a distribution ellipse—the beam emittance is constant. If the optical element has transfer matrix $\mathbf{M}$, emittance conservation holds if

$$\det \mathbf{M} = 1. \tag{4.15}$$

Section 3.7 showed that distribution ellipses at different positions in a linear system may be either skewed or upright. To begin, we shall write an equation for an upright ellipse. Suppose the elliptical distribution has dimensions X_0 and X_0' at a beam waist. The phase-space equation for the boundary is

$$(x/X_0)^2 + (x'/X_0')^2 = 1. \tag{4.16}$$

Multiplying both sides of Eq. (4.16) by $X_0 X_0'$ gives a standard form for elliptical boundaries that emphasizes conservation of emittance ε:

$$(X_0'/X_0)x^2 + (X_0/X_0')x'^2 = X_0 X_0' = \varepsilon. \tag{4.17}$$

The beam represented by Eq. (4.17) has zero envelope angle.

The envelope angle is nonzero for converging or diverging beams. Here, the distribution ellipse is tipped. We can find the general form for an inclined ellipse by applying a coordinate rotation to an upright ellipse. The following equation represents the upright ellipse in a coordinate system (X, Y):

$$aX^2 + bY^2 = 1. \tag{4.18}$$

We express the curve in terms of coordinates X' and Y' rotated by an angle $-\theta$. The coordinates are related by

$$X' = X \cos \theta + Y \sin \theta, \tag{4.19}$$

$$Y' = -X \sin \theta + Y \cos \theta. \tag{4.20}$$

Substituting Eqs. (4.19) and (4.20) into Eq. (4.18) gives

$$(a \cos^2 \theta + b \sin^2 \theta)X'^2 + (2[a - b] \cos \theta \sin \theta)X'Y'$$
$$+ (a \sin^2 \theta + b \cos^2 \theta)Y'^2 = 1. \tag{4.21}$$

In the transformed coordinate system, the ellipse has the mathematical form

$$AX'^2 + 2BX'Y' + CY'^2 = 1. \tag{4.22}$$

where A, B, and C are constants. Later, we will use the fact that the equation for a tipped ellipse, Eq. (4.22), always contains a cross product term proportional to XY.

C. Linear Transformations of Ellipses

We shall now derive the general form for a skewed distribution ellipse by applying a linear transformation to an upright ellipse. This is equivalent to finding the effect of a linear optical system on the input elliptical distribution, since any combination of optical elements can be represented by a single transfer matrix. We can generate any distribution ellipse by applying a linear transformation to an upright ellipse with the same area. We take the ellipse of Eq. (4.17) with dimensions X_0 and X_0'. Individual orbit vectors on the distribution boundary transform according to

$$x_1 = m_{11}x_0 + m_{12}x_0' \tag{4.23}$$

and

$$x_1' = m_{21}x_0 + m_{22}x_0', \tag{4.24}$$

where x_0 is the input vector to the system and x_1 is the output vecvor. The inverse forms of Eqs. (4.23) and (4.24) are

$$x_0 = m_{22}x_1 - m_{12}x_1' \tag{4.25}$$

and

$$x_0' = -m_{21}x_1 + m_{11}x_1'. \tag{4.26}$$

We let the input vectors lie on the upright ellipse of Eq. (4.17). To find the equation of the elliptical boundary at the output of the optical system, we substitute Eqs. (4.25) and (4.26) into Eq. (4.17). Performing the operation and collecting terms, the trace-space boundary of the output beam is

$$\gamma x_1^2 + 2\alpha x_1 x_1' + \beta x_1'^2 = \varepsilon. \tag{4.27}$$

We recognize that Eq. (4.27) has the form of a tipped ellipse. The values of the coefficients in Eq. (4.27) are given by the parameters of the initial upright ellipse and the properties of the optical system:

$$\gamma = (X_0'/X_0)m_{22}^2 + (X_0/X_0')m_{21}^2, \tag{4.28}$$

$$\alpha = -(X_0'/X_0)m_{12}m_{22} - (X_0/X_0')m_{11}m_{21}, \tag{4.29}$$

and

$$\beta = (X_0'/X_0)m_{12}^2 + (X_0/X_0')m_{11}^2. \tag{4.30}$$

The quantities α, β and γ are called the *transport parameters* (or sometimes Twiss parameters). Combined with the emittance ε they specify the distribution at the output of the optical system.

D. Significance of Transport Parameters

We shall adopt Eq. (4.27) as the general mathematical form for a distribution ellipse:

$$\gamma x^2 + 2\alpha x x' + \beta x'^2 = \varepsilon. \tag{4.31}$$

The parameters α, β, and γ take on different values as the beam moves through an optical system. The values give the size and orientation of the distribution ellipse. To find the relationship between the transport parameters and the beam properties, we assume that the beam has constant kinetic energy and emittance. To begin, we recognize that the distribution is upright when α equals zero, since there is no cross-product term. The condition $\alpha = 0$ implies that the beam is at a waist point. Comparison of Eqs. (4.17) and (4.31) (without the xx' term) shows that the spatial and angular halfwidths of the beam at a waist point are

$$X_0 = (\varepsilon/\gamma)^{1/2} \tag{4.32}$$

and

$$X'_0 = (\varepsilon/\beta)^{1/2}. \tag{4.33}$$

Taking the product of Eqs. (4.32) and (4.3) and remembering that $\varepsilon = X_0 X'_0$, we find that

$$\gamma\beta = 1. \tag{4.34}$$

Applying the above condition, Eqs. (4.32) and (4.33) can be rewritten

$$X_0 = (\beta\varepsilon)^{1/2} \tag{4.35}$$

and

$$X'_0 = (\gamma\varepsilon)^{1/2}. \tag{4.36}$$

For an upright ellipse, the quantity β gives the spatial width of the beam envelope and γ determines the angular width.

    Figure 4.1 shows the association between α, β, and γ of a skewed ellipse and the physical properties of the beam. We shall first cite the relationships, and then prove them mathematically. The halfwidth of the beam distribution is equal to the point of maximum extension of the distribution boundary along the spatial dimension. The expression is identical to that for an upright ellipse:

$$x_{\max} = \sqrt{\beta\varepsilon}. \tag{4.37}$$

Similarly, the maximum extent in the angular direction is

$$x'_{\max} = \sqrt{\gamma\varepsilon}. \tag{4.38}$$

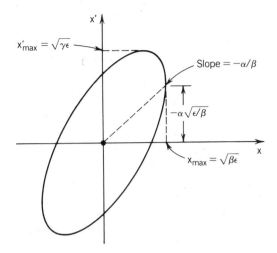

Figure 4.1 Trace-space distribution with a skewed elliptical boundary. Relationship of transport parameters (α, β, γ) to the ellipse geometry.

The quantity α determines the envelope angle of the beam, $d(x_{max})/dz$. The envelope angle equals the orbit inclination of peripheral particles at x_{max}. Figure 4.1 indicates that

$$d(x_{max}/dz) = -\alpha\sqrt{\varepsilon/\beta}. \tag{4.39}$$

We shall prove Eqs. (4.37) and (4.39) and leave the proof of Eq. (4.38) to the reader. To expedite the derivations, we shall first find a relationship between the transport parameters equivalent to emittance conservation. Consider the expression $\gamma\beta - \alpha^2$. By direct substitution from Eqs. (4.28)–(4.30), we can show that the expression reduces to

$$\gamma\beta - \alpha^2 = (m_{11}m_{22} - m_{12}m_{21}) \tag{4.40}$$

for any choice of the transfer matrix. Therefore, Eq. (4.40) holds for all distribution ellipses with the same value of ε. The quantity on the right-hand side of Eq. (4.40) is the determinant of the transfer matrix; it equals unity for a transformation that conserves emittance. The result is

$$\gamma\beta - \alpha^2 = 1. \tag{4.41}$$

In beam physics, the quantity $\gamma\beta - \alpha^2$ is called the Courant–Snyder invariant. Applying the quadratic formula, we can solve Eq. (4.31) for x:

$$x = -(\alpha x'/\gamma) \pm (\alpha^2 x'^2 - \gamma\beta x'^2 - \gamma\varepsilon)^{1/2}/\gamma. \tag{4.42}$$

Applying the Courant–Snyder condition, (Eq. (4.42) reduces to

$$x = -(\alpha x'/\gamma) \pm (\varepsilon \gamma - x'^2)^{1/2}/\gamma. \tag{4.43}$$

Inspection of Fig. 4.1 shows that the periphery of the beam x_{max} is defined by the condition $dx/dx' = 0$. Setting the derivative of Eq. (4.43) equal to zero and applying the Courant–Snyder condition implies that the orbit angle at the periphery is

$$x'(\text{at } x_{max}) = (\varepsilon/\beta)^{1/2}\alpha, \tag{4.44}$$

verifying Eq. (4.39). Substituting Eq. (4.44) into Eq. (4.43) gives Eq. (4.37).

E. Transformations of Beam Distributions

In a linear optical system, the transfer matrix $\mathbf{M}$ advances the orbit properties of a single particle through optical elements. We would like to find a similar operation that advances the properties of an entire distribution through the optical system. We take the distribution as elliptical with an area and shape given by values of ε, α, β, and γ at the entrance and exit of the system. We can write the transformation operation symbolically as

$$(\alpha_0, \beta_0, \gamma_0) = \mathbf{A(M)} \cdot (\alpha_1, \beta_1, \gamma_1). \tag{4.45}$$

The quantity $\mathbf{A}$ is an operator that depends on the transport system matrix $\mathbf{M}$. The boundary of the initial distribution is given by the equation

$$\gamma_0 x_0^2 + 2\alpha_0 x_0 x_0' + \beta_0 x_0'^2 = \varepsilon. \tag{4.46}$$

For a known matrix $\mathbf{M}$, we can transform the coordinates of points on the boundary by Eqs. (4.25) and (4.26). Substituting into Eq. (4.46) and collecting terms yields the following equation for the distribution ellipse at the exit of the optical system:

$$
\begin{aligned}
&x_1^2[\gamma_0 m_{22}^2 - 2\alpha_0 m_{21} m_{22} + \beta_0 m_{21}^2] \\
&+ 2x_1 x_1'[-\gamma_0 m_{12} m_{22} + \alpha_0 (m_{11} m_{22} + m_{12} m_{21}) - \beta_0 m_{11} m_{21}] \\
&+ x_1'^2[\gamma_0 m_{12}^2 - 2\alpha_0 m_{11} m_{12} + \beta_0 m_{11}^2] = \varepsilon.
\end{aligned}
\tag{4.47}
$$

We can identify the transport parameters of the exit ellipse as

$$\gamma_1 = \gamma_0 m_{22}^2 - 2\alpha_0 m_{21} m_{22} + \beta_0 m_{21}^2, \tag{4.48}$$

$$\alpha_1 = -\gamma_0 m_{12} m_{22} + \alpha_0 (1 - 2m_{12} m_{21}) - \beta_0 m_{11} m_{21},$$

$$\beta_1 = \gamma_0 m_{12}^2 - 2\alpha_0 m_{11} m_{12} + \beta_0 m_{11}^2.$$

To reduce the expression for α_1, we invoked the condition $\det \mathbf{M} = 1$. We can represent the transformation operation in the matrix form:

$$
\begin{bmatrix} \gamma_1 \\ \alpha_1 \\ \beta_1 \end{bmatrix} = \begin{bmatrix} m_{22}^2 & -2m_{21}m_{22} & m_{21}^2 \\ -m_{12}m_{22} & 1+2m_{12}m_{21} & -m_{11}m_{21} \\ m_{12}^2 & -2m_{11}m_{12} & m_{11}^2 \end{bmatrix} \begin{bmatrix} \gamma_0 \\ \alpha_0 \\ \beta_0 \end{bmatrix}. \tag{4.49}
$$

F.   Focusing a Beam with Nonzero Emittance

In a multielement transport system, Eq. (4.49) can lead to complex expressions. We shall illustrate implications of Eq. (4.49) with a simple example—focusing a nonlaminar beam. Suppose a parallel beam enters a thin lens followed by a drift space. We want to find the distance to the best focus and the size of the smallest focal spot. Previously, we found an approximate solution (Section 3.1) valid in the limit of small emittance. With distribution transport theory, we can find an exact solution for a beam with high emittance.

Multiplying the matrices of Eqs. (4.11) and (4.12) gives a total ray transfer matrix for a thin lens of focal length f followed by a drift length d:

$$
\mathbf{M} = \begin{bmatrix} 1 - d/f & d \\ -1/f & 1 \end{bmatrix}. \tag{4.50}
$$

Substituting the components of Eq. (4.50) in Eq. (4.49), we find:

$$
\begin{bmatrix} \gamma_1 \\ \alpha_1 \\ \beta_1 \end{bmatrix} = \begin{bmatrix} 1 & 2/f & 1/f^2 \\ -d & 1-2d/f & (1-d/f)/f \\ d^2 & -2d(1-d/f) & (1-d/f)^2 \end{bmatrix} \begin{bmatrix} \gamma_0 \\ \alpha_0 \\ \beta_0 \end{bmatrix}. \tag{4.51}
$$

The condition that the input beam is parallel is equivalent to $\alpha_0 = 0$. The ellipse of the incident beam is upright with dimensions x_0 and x_0'—the emittance is $\varepsilon = x_0 x_0'$. The exit transport parameters are

$$
\gamma_1 = \gamma_0 + \beta_0/f^2, \tag{4.52}
$$

$$
\alpha_1 = -\gamma_0 d + (1 - d/f)\beta_0/f, \tag{4.53}
$$

$$
\beta_1 = \gamma_0 d^2 + \beta_0(1 - d/f)^2. \tag{4.54}
$$

We shall review some of the implications Eqs. (4.52)–(4.54). First, we verify that the results agree with our previous treatment of focusing for small emittance. In this limit, the incident emittance is small enough or the lens is strong enough so that beam halfwidth at the focus is much less than x_0. An equivalent condition is that the angular width of the input distribution is small compared with the envelope deflection angle caused by the lens:

$$x_0' \ll x_0/f. \tag{4.55}$$

Furthermore, we expect that the focus occurs at a distance from the lens close to the focal length, $d \cong f$. With these conditions, Eq. (4.54) reduces to $\beta_1 \cong \gamma_0 f^2$. The focal spot size is

$$x_1 = (\varepsilon \beta_1)^{1/2} = (\gamma_0 x_0 x_0' f)^{1/2} = x_0' f. \tag{4.56}$$

The predicted focal spot size agrees with Eq. (3.1).

When the limit of Eq. (4.55) does not hold, the distance of best focus does not equal the lens focal length. We find the point where the envelope width is a minimum by setting the derivative of β_1 with respect to d equal to zero. Performing the operation on Eq. (4.54), we find that the best focus occurs at a distance from the lens

$$d = \frac{\beta_0 f}{\gamma_0 + \beta_0/f^2}. \tag{4.57}$$

In the limit of small emittance, we can show that $\gamma_0 \ll 1/f$ and $\beta_0 \gg f$. Inserting these conditions, Eq. (4.57) reduces to

$$d \cong \frac{f}{1 + (x_0' f/x_0)^2}. \tag{4.58}$$

Equation (4.58) implies that the focal point of a nonlaminar beam is closer to the lens than the focus of a zero emittance beam. Figure 3.19 illustrates the basis of this result.

The envelope angle of the beam must be zero at the best focal point. We can confirm the existence of a beam waist at the focal point specified by Eq. (4.57) by substituting the value of d in Eq. (4.53). We find that α_1 is identically equal to zero. To find the beam width at the focal point, we must find the transport parameter β_1. Inserting Eq. (4.57) into Eq. (4.54) gives

$$\beta_1 = \frac{\beta_0}{1 + (\beta_0/f)^2}. \tag{4.59}$$

After some algebraic manipulation, we find the exact expression for the beam halfwidth at the focal spot:

$$x_1 = (\beta_1 \varepsilon)^{1/2} = \frac{x_0' f}{[1 + (x_0' f/x_0)^2]^{1/2}}. \tag{4.60}$$

4.2. TRANSPORT PARAMETERS FROM PARTICLE ORBIT THEORY

In this section, we shall study an alternative derivation of the transport parameters. Section 4.1 used transfer matrix theory to introduce α, β, and γ. In this view, optical elements make discrete changes in orbit vectors—distributions are defined at the boundaries between elements. Here, we shall derive the transport parameters by studying the properties of particle orbits focused by continuous forces that vary periodically in the axial direction. This approach is often applied to circular accelerators, such as synchrotrons, where a beam circulates many times through the identical focusing cells. The derivation introduces some ideas that are often used in circular accelerator theory such as the machine ellipse and the β function.

To develop the theory, we make three assumptions:

1. Transverse forces vary linearly with displacement from the main axis.

2. Focusing forces vary smoothly along the direction of beam propagation.

3. Acceleration takes place slowly compared with the particle revolution time in the accelerator—we use the condition of constant kinetic energy.

Since most circular accelerators use quadrupole focusing, we shall limit the treatment to motion in a single dimension, (x, x').

Consider a particle focused by a transverse force $F_x(z)$. The force function repeats periodically over a cell distance L. The following form for the force displays the linear variation:

$$F_x(x) = - \gamma m_0 \beta^2 c^2 k(z) x. \qquad (4.61)$$

The periodic function ($k(z)$ has the property

$$k(z) = k(z + L) = k(z + 2L) = \cdots. \qquad (4.62)$$

We can represent the transverse component of paraxial orbits by an equation with nonrelativistic form:

$$x'' = d^2 x / dz^2 = - k(z) x. \qquad (4.63)$$

If k is constant, Eq. (4.63) describes a simple harmonic oscillation. For variable $k(z)$, we shall assume that the solution for $x(z)$ has the form

$$x(z) = \sqrt{\varepsilon} \sqrt{\beta(z)} \cos [\Phi_x(z) + \phi]. \qquad (4.64)$$

The quantity ($\varepsilon^{1/2}$) is an arbitrary amplitude factor. Later, we shall associate ε

with the beam emittance enclosed if the orbit represents a particle on the boundary ellipse of the beam distribution. The functions $\beta(z)$ and $\Phi_x(z)$ are periodic over length L. The quantity ϕ is an arbitrary phase factor.

We can justify Eq. (4.64) and find a relationship between $\beta(z)$, $k(z)$, and $\Phi_x(z)$ by substituting the form into Eq. (4.63). Evaluating derivatives and eliminating the constant factor $\varepsilon^{1/2}$, Eq. (4.63) becomes

$$-\sin\left[\Phi_x + \phi\right]\left[\frac{\beta'\Phi_x'}{\beta^{1/2}} + \beta^{1/2}\Phi_x''\right]$$

$$+\cos\left[\Phi_x + \phi\right]\left[\frac{\beta''}{2\beta^{1/2}} - \frac{\beta'^2}{4\beta^{3/2}} - \beta^{1/2}(\Phi_x')^2 + k\beta^{1/2}\right] = 0. \qquad (4.65)$$

Equation (4.65) holds at all positions in the system and for all values of ϕ; therefore, the sine and cosine terms are independently equal to zero. The sine terms imply the following equation:

$$\Phi_x''/\Phi_x' = -\beta'/\beta. \qquad (4.66)$$

Equation (4.66) has the solution $\Phi_x'\beta = A$, where A is a constant. We will make the specific choice $A = 1$. With this choice, we shall find that $\beta(z)$ has a physical meaning similar to that of the transport parameter β derived in Section 4.1. For this choice, the following relationship holds:

$$\Phi_x(z) = \int_0^z \frac{dz'}{\beta(z')}. \qquad (4.67)$$

Collecting the cosine terms, we find a nonlinear differential equation that determines the β function in terms of applied force:

$$\beta'' - \frac{(\beta')^2}{2\beta} - \frac{2}{\beta} + 2k(z)\beta = 0. \qquad (4.68)$$

We can find an expression for the orbit angle by taking the axial derivative of Eq. (4.64) and applying the relationship $\Phi_x' = 1/\beta$:

$$x'(z) = \left[\frac{\varepsilon}{\beta}\right]^{1/2}\left[\frac{\beta'\cos(\Phi_x + \phi)}{2} - \sin(\Phi_x + \phi)\right]. \qquad (4.69)$$

We can simplify the expression further by defining a quantity α in terms of the axial derivative of the $\beta(z)$ function:

$$\beta'(z) = -2\alpha(z). \qquad (4.70)$$

Substituting Eq. (4.70), Eq. (4.69) becomes

$$x'(z) = -\left[\frac{\varepsilon}{\beta}\right]^{1/2}\left[\alpha(z)\cos(\Phi_x + \phi) + \sin(\Phi_x + \phi)\right]. \qquad (4.71)$$

We can write the orbit angle in an alternative form by using the trigonometric relationship

$$A\cos\psi + B\sin\psi = (A^2 + B^2)^{1/2}\cos[\psi - \tan^{-1}(A/B)].$$

Equation (4.71) becomes

$$x'(z) = \sqrt{\varepsilon\gamma(z)}\cos[\chi(z) + \phi], \qquad (4.72)$$

where

$$\chi(z) = \Phi_x(z) - \tan^{-1}[1/\alpha(z)].$$

The quantity $\gamma(z)$ is defined by the relationship

$$\gamma(z) = [1 + \alpha(z)^2]/\beta(z). \qquad (4.73)$$

We assume that the focusing forces and the functions $\alpha(z)$, $\beta(z)$, and $\gamma(z)$ vary periodically over length L. To understand the physical implications of the functions, imagine a plot of the locus of orbit vectors in a plane for a particle confined in a circular accelerator. At a specific location z_0, a particle defines a new orbit with each revolution. The locus of orbit vectors at z_0 generates a curve in (x, x'). The curve is given by Eqs. (4.69) and (4.72) with the parameters $\alpha(z_0)$, $\beta(z_0)$, and $\gamma(z_0)$. The two equations are parametric equations in the variable ϕ. We can derive an explicit expression for the curve by eliminating ϕ and combining Eqs. (4.69) and (4.72). The algebraic manipulations are complex[1]; we shall simply quote the final resuts:

$$\gamma(z_0)x^2 + 2\alpha(z_0)xx' + \beta(z_0)x'^2 = \varepsilon. \qquad (4.74)$$

Equation (4.74) is a trace-space ellipse of area $\pi\varepsilon$—note that the form is identical to Eq. (4.31). The curve of Eq. (4.74) could represent the boundary of a beam distribution for a group of particles uniformly spread over all values of ϕ. Here, the quantities $\beta(z_0)$, $\gamma(z_0)$, and $\alpha(z_0)$ have the same relation to the beam width, angular width, and envelope angle as the transport parameters of Section 4.1. The defining equation for γ[Eq. (4.73)] states the Courant–Snyder

---

[1] See, for example, K. Brown and R. Servranckx, "Optics Modules for Circular Accelerator Design" in *Charged Particle Optics*, S. O. Schriber and L. S. Taylor, Eds., North-Holland, Amsterdam, 1987, p. 480.

condition. Furthermore, the beam envelope angle is given by

$$dx_{max}/dz = d(\sqrt{\varepsilon\beta})/dz = \beta'\sqrt{\varepsilon/\beta}/2 = -\alpha\sqrt{\varepsilon/\beta}. \tag{4.75}$$

Equation (4.75) agrees with Eq. (4.39).

A distribution bounded by the curve of Eq. (4.74) is matched to the focusing system of the accelerator. If particles are uniformly spread over all values of phase, the distribution is identical each time the particles revolve around the machine. The matched shape is called the *machine ellipse*. The shape and

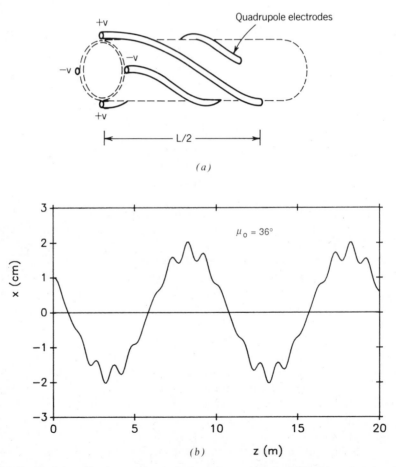

Figure 4.2 Particle orbits in a twisted electrostatic quadrupole field with a focusing force $F_x(z) \sim \sin(2\pi z/L)$. (a) Schematic view of the electrode geometry. (b) Numerical calculation of a particle orbit for $\mu_0 = 36°$. (c) Matched distribution ellipse at the system entrance ($z = 0$)—circles show vectors of a particle orbit with $x = 0.01$ and $x' = 0$ at $z = 0, L, 2L, 3L, \ldots$. (d) Matched distribution at $z = L/4$.

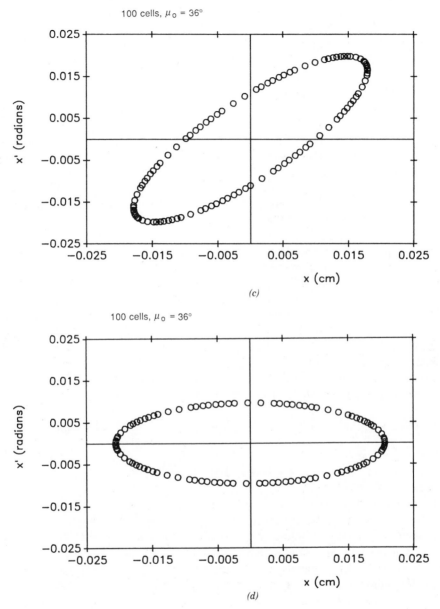

Figure 4.2 *(Continued)*

orientation of the machine ellipse varies, moving from position to position around the machine. We can find the machine ellipse at any point in a circular accelerator by creating numeric solutions of Eq. (4.68) for known spatial variations of force over a large number of particle revolutions. For example, suppose that we construct the machine ellipses at the injection point of a storage ring. The distribution is matched to the ring and has envelope oscillations of minimum value if it fits into a machine ellipse.

Figure 4.2 illustrates the matched ellipse of a periodic focusing system that repeats indefinitely. The transverse force varies continuously according to

$$F_x(z) \sim \sin(2\pi z/L). \tag{4.76}$$

A focusing cell has length L. The force of Eq. (4.76) can be produced with the helical quadrupole electrodes of Fig. 4.2α. Figure 4.2b illustrates a numerical solution for a particle orbit with a phase advance per cell of 36°. Figures 4.2c and 4.2d show the matched ellipses at $z = 0$ and $z = L/4$. The curves were generated by following the trajectory of a single orbit through many focusing cells. Characteristics of the matched beam are evident in Figs. 4.2c and 4.2d. At the point $z = 0$ the force is changing from defocusing to focusing—it has zero amplitude. At this point the envelope of the matched beam diverges. The beam reaches a maximum diameter waist at the midpoint of the focusing region.

The function $\beta(z)$ gives the envelope width of the beam at all points of the focusing system through Eq. (4.37). Knowledge of this function is important to design the lens arrays, or *focussing lattice*, of a circular accelerator. For example, in many machines the envelope width must meet criteria at different points in the ring. Consider a collider, a synchrotron with counter rotating beams. The envelope must be small at the beam intersection point to achieve an acceptable interaction rate (Section 4.5). Sections of the accelerator with small envelope are called *low-β regions*. In another application, we can use Eq. (4.67) to find the total phase advance for particle orbits circulating around a circular accelerator in terms of the β function. To avoid orbital resonances, the integral of Eq. (4.67) must be a noninteger number for all operating regimes of the machine.

4.3.   BEAM MATCHING

A beam matched to a periodic focusing system has envelope oscillations of minimum amplitude. Furthermore, we shall find in Section 4.4 that the emittance growth caused by lens nonlinearities is smallest for a matched beam. We outlined a numerical method to find matched-beam distributions in Section 3.7. In this section, we shall study analytic methods that use distribution transport theory. These methods lead to algebraic solutions to the important problem of beam matching between periodic focusing systems with different properties. An example is the transfer of a beam in a high-intensity ion accelerator from a radio-frequency quadrupole to a drift-tube linac. We can find the matched distribution in either focusing system—the challenge is to find optical elements that convert the output distribution from one system into a matched input distribution for the next system. Transport theory is an organized approach to the matching problem.

We can represent the transformation between two focusing systems symbolically as

$$(\alpha_0, \beta_0, \gamma_0) \longrightarrow (\alpha_1, \beta_1, \gamma_1). \tag{4.77}$$

The transport parameters $(\alpha_0, \beta_0, \gamma_0)$ represent the output distribution from the upstream system, while $(\alpha_1, \beta_1, \gamma_1)$ gives the desired matched distribution in the downstream system. We want to find a combination of linear focusing devices that accomplishes the transformation. The combined optical elements constitute the maching system. We can represent the effect of the matching system on particle orbits with the transfer matrix **M**.

To begin, we shall derive an alternative expression for **M** in terms of the input and output transport parameters, $(\alpha_0, \beta_0, \gamma_0)$ and $(\alpha_1, \beta_1, \gamma_1)$. The matching system has linear transverse forces that vary over its length. Following the development of Section 4.2, the displacement and angle of a particle orbit are

$$x(z) = \sqrt{\varepsilon \beta(z)} \cos \left[\Phi_x(z) + \phi \right] \tag{4.78}$$

and

$$x'(z) = -\sqrt{\varepsilon/\beta(z)} \{ \alpha(z) \cos \left[\Phi_x(z) + \phi \right] + \sin \left[\Phi_x(z) + \phi \right] \}. \tag{4.79}$$

The function $\Phi_x(z)$ is the phase of the orbit as the particle passes through the system. We assign the phase a value of zero at the system entrance:

$$\Phi_{x0} = 0. \tag{4.80}$$

Similarly, we define the phase at the exit of the matching system as

$$\Phi_{x1} = \mu. \tag{4.81}$$

We say that μ is the total phase advance of the orbit in the matching section.
Expanding the trigonometric functions in Eqs. (4.78) and (4.79), we find that

$$x = \sqrt{\varepsilon \beta} (\cos \Phi_x \cos \phi - \sin \Phi_x \sin \phi) \tag{4.82}$$

and

$$x' = -\sqrt{\varepsilon/\beta} (\alpha \cos \Phi_x \cos \phi - \alpha \sin \Phi_x \sin \phi + \sin \Phi_x \cos \phi + \cos \Phi_x \sin \phi). \tag{4.83}$$

Using Eq. (4.80), the orbit vectors at the entrance are

$$x_0 = \sqrt{\varepsilon \beta_0} \cos \phi \tag{4.84}$$

and

$$x_0' = -\sqrt{\varepsilon/\beta_0} (\alpha_0 \cos \phi + \sin \phi). \tag{4.85}$$

Substituting Eq. (4.81) into Eqs. (4.82) and (4.83), the exit orbit vector is given by

$$x_1 = \sqrt{\varepsilon\beta_1}\,(\cos\mu\cos\phi - \sin\mu\sin\phi), \tag{4.86}$$

$$\begin{aligned}x_1' = -\sqrt{\varepsilon/\beta_1}\,(\alpha_1\cos\mu\cos\phi - \alpha_1\sin\mu\sin\phi \\ + \sin\mu\cos\phi + \cos\mu\sin\phi).\end{aligned} \tag{4.87}$$

We can eliminate the quantities $\cos\phi$ and $\sin\phi$ in Eqs. (4.84) through (4.87) to find expressions for the output orbit vector in terms of the input vector and the transport parameters at the entrance and exit of the matching section:

$$x_1 = x_0\sqrt{\beta_1/\beta_0}\,(\cos\mu + \alpha_0\sin\mu) + (x_0'\sqrt{\beta_0\beta_1})\sin\mu, \tag{4.88}$$

$$\begin{aligned}x_1' = (x_0/\sqrt{\beta_0\beta_1})[(1 + \alpha_0\alpha_1)\sin\mu + (\alpha_1 - \alpha_0)\cos\mu] \\ + (x_0'\sqrt{\beta_0/\beta_1})(\cos\mu - \alpha_1\sin\mu).\end{aligned} \tag{4.89}$$

By inspecting Eqs. (4.88) and (4.89), we can find the terms of the transfer matrix of the matching section $\mathbf{M}$:

$$m_{11} = (\beta_1/\beta_0)^{1/2}(\cos\mu + \alpha_0\sin\mu), \tag{4.90}$$

$$m_{12} = (\beta_0\beta_1)^{1/2}\sin\mu, \tag{4.91}$$

$$m_{21} = -[(1 - \alpha_0\alpha_1)\sin\mu + (\alpha_1 - \alpha_0)\cos\mu]/(\beta_0\beta_1)^{1/2}, \tag{4.92}$$

$$m_{22} = (\beta_0\beta_1)^{1/2}(\cos\mu - \alpha_1\sin\mu). \tag{4.93}$$

Equations (4.90) through (4.93) are the starting point for beam-matching calculations. The choice of μ, the known values $(\alpha_0, \beta_0, \gamma_0)$, and the desired values $(\alpha_1, \beta_1, \gamma_1)$ uniquely determine the components of the transfer matrix. The problem resolves into a search for a combination of optical elements to give the known elements of $\mathbf{M}$. The procedure is complex if simultaneous matching in the x and y directions is required.

In this section, we shall limit the discussion to a simple but useful application, how to find the matched distributions of a periodic focusing system. In terms of the above derivation, we identify the matching optics as one cell of the periodic system. The cell connects identical upstream and downstream cells. The condition for a matched beam means that the input and output distributions are identical, or $\beta_1 = \beta_0 = \beta$, $\gamma_1 = \gamma_0 = \gamma$, $\alpha_1 = \alpha_1 = \alpha$. Using Eqs. (4.90) through (4.93), the transfer matrix for a cell has the form

$$\mathbf{M} = \begin{bmatrix} \cos\mu + \alpha\sin\mu & \beta\sin\mu \\ -\gamma\sin\mu & \cos\mu - \alpha\sin\mu \end{bmatrix}. \tag{4.94}$$

As a specific example of matching, we shall study the FODO geometry, consisting of alternating focusing and defocusing quadrupole lenses separated by a drift distance. We must address motion in both the x and y directions.

Figure 4.3a shows the discrete quadrupole lenses that comprise the focusing system. Each lens is rotated 90° relative to the previous one. Figure 4.3b illustrates an alternative optical system that is often used to model the FODO array—thin lenses of focal length $\pm f$ separated by a length D. The quantity D represents the total distance between quadrupole lens centers, including the lens and drift lengths. The choice of the cell boundary depends on where we want to know the matched distribution. To begin, suppose that the focusing cell starts and ends at the center of a quadrupole lens and that this lens focuses in the x direction. Figure 4.3c represents a cell. Particles first travel through half a lens, focusing in the x direction but defocusing in y. We represent the effect of the half lens by a thin lens of focal length $2f$—the associated transfer matrices are

$$\mathbf{M}_x(\text{half lens}) = \begin{bmatrix} 1 & 0 \\ -1/2f & 1 \end{bmatrix} \tag{4.95}$$

and

$$\mathbf{M}_y(\text{half lens}) = \begin{bmatrix} 1 & 0 \\ +1/2f & 1 \end{bmatrix}. \tag{4.96}$$

The next element is a drift path of length D—it has the same effect in both the x and y directions:

$$\mathbf{M}_x(\text{drift}) = \mathbf{M}_y(\text{drift}) = \begin{bmatrix} 1 & D \\ 0 & 1 \end{bmatrix}. \tag{4.97}$$

The next element is a full length lens, defocusing in x and focusing in y:

$$\mathbf{M}_x(\text{lens}) = \begin{bmatrix} 1 & 0 \\ +1/f & 1 \end{bmatrix} \tag{4.98}$$

and

$$\mathbf{M}_y(\text{lens}) = \begin{bmatrix} 1 & 0 \\ -1/f & 1 \end{bmatrix}. \tag{4.99}$$

Another drift length and a half lens complete the focusing cell.

We find the transfer matrix for a cell by multiplication of the matrices of the individual components. The results are

$$\mathbf{M}_x = \begin{bmatrix} (1 - D^2/2f^2) & 2D(1 + D/2f) \\ -(D/2f^2)(1 - D/2f) & (1 - D^2/2f^2) \end{bmatrix} \tag{4.100}$$

and

$$\mathbf{M}_y = \begin{bmatrix} (1 - D^2/2f^2) & 2D(1 - D/2f) \\ -(D/2f^2)(1 + D/2f) & (1 - D^2/2f^2) \end{bmatrix}. \tag{4.101}$$

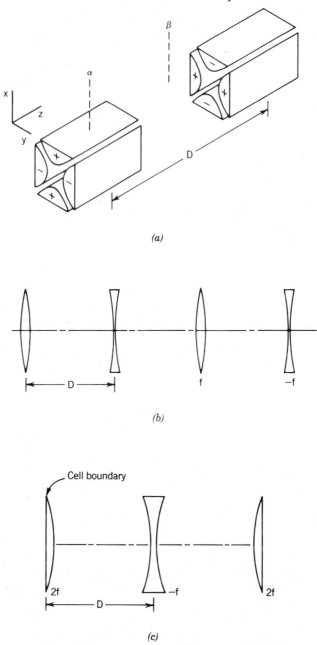

(a)

(b)

(c)

Figure 4.3 Ion beam focusing by a quadrupole lens array. (*a*) Electrostatic quadrupoles—electrode geometries for a *FODO* system. (*b*) Approximate model for focusing in a uniform *FODO* system in either *x* or *y*. Array of thin lenses with alternating focal length, $\pm f$. (*c*) Definition of a focusing cell—the cell boundary is at the center of a lens (dashed line α in part *a* relative to the *x* direction). (*d*) Definition of a cell with the boundary at the midpoint between two lenses (dashed line β in part *a*).

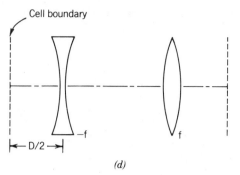

(d)

Figure 4.3 *(Continued)*

Comparison of Eqs. (4.100) and (4.101) with Eq. (4.94) gives the parameters of the matched ellipse at the cell boundary and the phase advance per cell:

$$\cos \mu = (1 - D^2/2f^2), \tag{4.102}$$

$$\alpha_x = \alpha_y = 0, \tag{4.103}$$

$$\beta_x = 2D(1 + D/2f)/\sin \mu, \tag{4.104}$$

$$\beta_y = 2D(1 - D/2f)/\sin \mu, \tag{4.105}$$

$$\gamma_x = 1/\beta_x, \tag{4.106}$$

and

$$\gamma_y = 1/\beta_y. \tag{4.107}$$

We can rewrite the expressions for the β parameters entirely in terms of μ by recognizing that

$$\sin (\mu/2) = [(1 - \cos \mu)/2]^{1/2} = D/2f. \tag{4.108}$$

The final expressions are

$$\beta_x = (2D)[1 + \sin (\mu/2)]/\sin \mu, \tag{4.109}$$

$$\beta_y = (2D)[1 - \sin (\mu/2)]/\sin \mu. \tag{4.110}$$

The results imply that the beam envelope should be parallel to the axis at the center of the quadrupole lens since $\alpha_x = \alpha_y = 0$. At this point, the ratio of the β parameters in the x and y directions is

$$\frac{\beta_x}{\beta_y} = \frac{1 + \sin (\mu/2)}{\sin \mu}. \tag{4.111}$$

If the injected beam has equal emittance in both directions, $\varepsilon_x = \varepsilon_y = \varepsilon$, then the ratio of the envelope widths of the beam at the center of the lens is

$$\frac{x_0}{y_0} = \left[\frac{1 + \sin(\mu/2)}{\sin \mu}\right]^{1/2}. \tag{4.112}$$

As in all FD systems, the envelope width is larger in the focusing direction than in the defocusing direction. As an example, for a phase advance of $\mu = 60°$ the ratio of the envelope widths is $x_0/y_0 = 1.316$.

Figure 4.4 shows plots of the beam envelope in the x and y directions to illustrate the physical meaning of Eqs. (4.102) through (4.111). The width in the y direction is maximum in the second lens. As an exercise, the reader should derive the transport parameters of a matched beam distribution in the same FODO channel, but with the focusing cell boundary at the center of the first drift space. Figure 4.3d illustrates the cell. The results are

$$\beta_x = \beta_y \tag{4.113}$$

and

$$\alpha_x = -\alpha_y. \tag{4.114}$$

Assuming emittances in the x and y directions are equal, the envelope widths are equal and the envelope angles have equal magnitude but opposite sign. The beam converges in x and diverges in y.

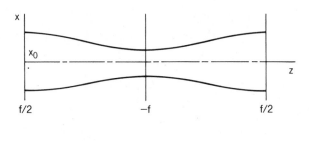

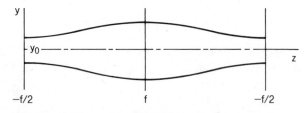

Figure 4.4   Envelope projections in the x and y directions for a matched beam in a *FODO* system. Cell boundary at the center of a lens (focusing in x and defocusing in y).

4.4. NONLINEAR FOCUSING SYSTEMS

The force in a nonlinear lens is not linearly proportional to distance from a symmetry axis. Usually, accelerator designers take care to minimize nonlinearities in charged particle focusing systems. There is an important reason for this precaution in strong focusing circular accelerators. In these devices, it is essential to avoid orbit resonance instabilities that occur when the machine circumference equals an integral number of betatron wavelengths. Elimination of resonances is difficult in a nonlinear optics system since the betatron wavelength depends on the amplitude of transverse motion.

On the other hand, nonlinear lenses have potential advantages for beam transport in linear accelerators. In these devices, a spread in betatron wavelength can reduce many of the instabilities described in Chapters 13 and 14. Furthermore, we shall see (Section 10.9) that some inherently nonlinear focusing arrays, such as magnetic cusps, have stability advantages for high-current beam transport. Nonetheless, nonlinear lenses are seldom used in linacs. This reflects, in part, a general belief that nonlinear focusing forces inevitably lead to beam emittance growth. In this section, we shall review the properties of nonlinear focusing systems and discuss their advantages and disadvantages. We shall see, for example, that emittance growth occurs only when a beam is mismatched to a nonlinear focusing channel. We can always find matched distributions that propagate with no change in emittance even for focusing systems with highly nonlinear lenses.

A. Nonlinear Lens

To begin, we shall review the properties of individual nonlinear lenses. Figure 4.5 shows that a nonlinear lens cannot focus a beam to point, even if the beam is laminar. The force of the lens in Fig. 4.5 exceeds the linear value at large

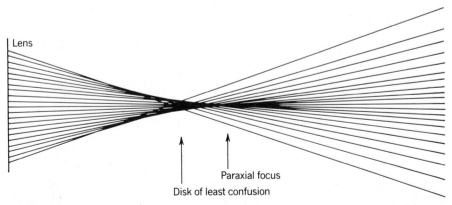

Figure 4.5 Focusing by a nonlinear lens with spherical aberration. Configuration-space orbits for an incident laminar beam.

displacement, resulting in overfocusing of peripheral particles. There is no downstream location where the particle orbits meet. A related problem occurs in imaging applications—a nonlinear lens cannot relay a point-focused beam to another point. This property is called *spherical aberration*, where aberration implies a deviation from ideal behavior. The term spherical comes from light optics—the problem arises if the surfaces of lenses are spherical sections rather than paraboloids.

We can represent the transverse force of a lens with cylindrical symmetry in the form

$$F(r) = ar + br^3. \qquad (4.115)$$

There is no r^2 term because of the symmetry—terms with order higher than r^3 are generally unimportant. The cubic term is responsible for spherical aberration. An alternative form of Eq. (4.115) emphasizes the cubic term as a perturbation to a linear lens:

$$F(r) = Ar[1 + \eta r^2]. \qquad (4.116)$$

The quantity η is an error parameter—a lens is almost linear if

$$\eta \ll 1/r_0^2. \qquad (4.117)$$

The quantity r_0 in Eq. (4.117) is the maximum radius of the beam or the radius of

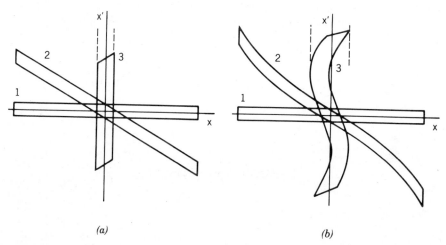

(a)   *(b)*

Figure 4.6 Trace-space views—focusing a beam with a rectangular distribution (uniform angular dispersion at all transverse positions). (1) Lens entrance, (2) lens exit, (3) position of best focus. *(a)* Ideal linear lens—dashed lines show beam width a focal point. *(b)* Nonlinear lens with spherical aberration—dashed lines shows beam width at point of maximum convergence.

the lens aperture. Equation (4.116) shows that focusing errors depend strongly on radius. Spherical aberration limits the useful bore of charged particle lenses.

Figure 4.6 shows trace-space plots of a parallel beam focused by linear and nonlinear lenses. The linear lens tips the beam distribution so that it converges to an emittance-limited focal spot. In contrast, the nonlinear lens distorts the shape of the distribution. There is no downstream location where the entire distribution aligns with the r' axis. The focal point for paraxial particles occurs at a different axial location than the focus of envelope particles. The beam reaches its minimum width at an intermediate location—the profile at this location is called the disk of least confusion.

We represent the departure of a lens from ideal focusing properties by the *spherical aberration coefficient* C_s. To illustrate the definition of C_s, consider a lens with a paraxial focal length of f and a maximum unsable radius of r_0. The quantity α is the approximate envelope convergence angle,

$$\alpha \cong -r_0/f. \tag{4.118}$$

Because of the error in the lens deflection angle, particles on the beam envelope pass the paraxial focal point at a distance $\sim \alpha \eta r_0^2$ from the axis. Therefore, the radius of the disk of least confusion δ is close to

$$\delta \leqslant \eta r_0^3 = (\eta/f^3)\alpha^3. \tag{4.119}$$

Equation (4.119) shows that the spot size is proportional to the cube of the convergence angle for a given focal length lens. Accordingly, we define the spherical aberration coefficient in terms of the disk of least confusion and the convergence angle:

$$\delta = C_s \alpha^3/4. \tag{4.120}$$

In practical lenses the cubic error of the focusing force is small—following the example of light optics, we might hope to correct the errors by designing multielement lenses with positive and negative values of η. Unfortunately, this approach is impossible for cylindrical electrostatic or magnetic lenses. All such lenses, whether electrostatic or magnetic, have positive values of η. The transverse force invariably increases more rapidly near the electrodes or coils that generate the fields. For common lenses, the only options to reduce spherical aberration are to design for highly linear fields and to limit the useful lens aperture.

B. Emittance Growth in an Array of Nonlinear Lenses

Section 3.7 showed that periodic focusing systems with linear lenses preserve both the area and general shape of beam distributions. In particular, an elliptical distribution remains an ellipse, even if the distribution is mismatched to the focusing system. Nonlinear focusing systems do not have this property. A nonlinear lens array distorts the shape of a mismatched distribution, leading to emittance growth.

To illustrate emittance enhancement in a nonlinear focusing system, consider a beam confined by an axially uniform transverse force with radial variation given by Eq. (4.116). Without the cubic term, all particle orbit vectors follow elliptical trace-space trajectories that circulate at the same rate. In other words, particles have the same betatron wavelength. All parts of a mismatched distribution rotate in trace space at the same rate—the distribution returns to its original form after one betatron oscillation.

In contrast, particle oscillations in a nonlinear force are anharmonic. Although all paraxial particles have about the same value of λ_b, the betatron wavelength of peripheral particles is shorter (for $\eta > 0$) because the average focusing force is higher. In trace space, outer portions of the distribution circulate faster than inner parts—as a result, the distribution wraps around itself. Figure 4.7 shows a trace-space view of a mismatched beam propagating in a continuous nonlinear force. Although the initial beam is almost laminar, longitudinal phase mixing ultimately creates disorder.

A similar process happens in periodic focusing systems. Here, we represent the trace-space distribution at focusing cell boundaries—the change in particle orbit vectors between boundaries is given by μ_0, the phase advance per cell. In a linear system, μ_0 is the same for all particles—a mismatched distribution can eventually return to its original orientation and shape after crossing several cells. With nonlinear lenses, μ_0 is larger for peripheral particles, resulting in distortion of the distribution. Figure 4.8 shows a trace-space view of a mismatched beam

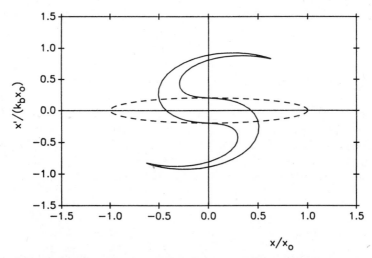

Figure 4.7 Trace-space view of a mismatched beam distribution propagating in a nonlinear transverse force, uniform in the axial direction. The focusing force has the form $F_x(x) = \gamma\beta^2 c^2 m_0 k_b^2 x[1 + 0.2(x/x_0)^2]$, where k_b is the betatron wavenumber for paraxial particles.

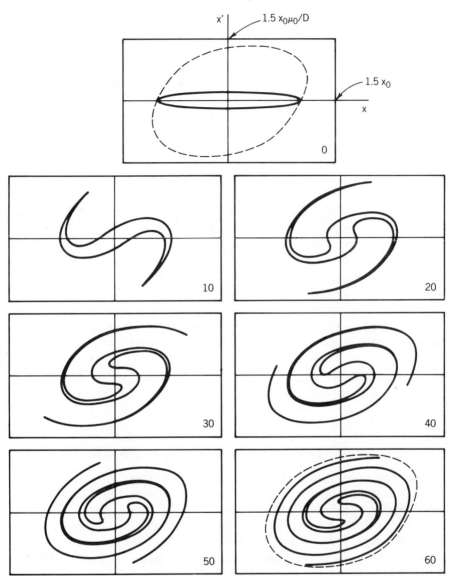

Figure 4.8 Trace-space views of the propagation of a mismatched beam in an array of 60 nonlinear lenses with spherical aberration. Thin lenses with focal length $f(x) = f_0[1 + (x/x_0)^2]$ separated by a distance D. Cell boundary at lens exit. Paraxial phase advance: $\mu_0 = 0.51$ rad. Dashed lines at entrance and exit show trace-space trajectory for particles on the beam periphery.

transported in a array of nonlinear thin lenses. In the example, the focal length of the lenses is

$$f = f_0[1 + \eta(x/x_0)^2], \tag{4.121}$$

where $\eta = 1$. The phase advance per cell for paraxial particles is $\mu_0 = 0.51$. Figure 4.8 shows the state of the distribution propagating through 60 lenses, an axial distance of about $5\lambda_b$.

Figure 4.8 illustrates some characteristics of beam propagation in a nonlinear system:

1. In principle, the distribution distortion illustrated in Fig. 4.8 could be corrected, since the actual area occupied in trace space is constant. In practice, there is no hope of unwinding the final distribution with any combination of conventional lenses. For practical purposes, the process is irreversible.

2. The beam emittance is higher, since the smallest elliptical curve that can surround the final distribution is larger than the boundary of the input distribution.

3. The emittance of the unmatched beam does not grow without bound—the particle orbits are stable. The emittance growth results from phase mixing of transverse oscillations—growth ends when orbits are uniformly distributed in phase.

4. We can find the upper limit on the final emittance by inspection of Fig. 4.8. In the final state, the distribution fills the area circumscribed by the trace-space orbits of particles on the envelope of the mismatched beam.

C. Matched Distributions in Nonlinear Channels

For the beam shown in Fig. 4.8, emittance growth stops when orbit vectors uniformly fill the boundary defined by the peripheral particles. Therefore, we conclude that a distribution that fills the region is matched to the nonlinear channel. If we injected the beam with the matched distribution, the emittance would not increase, although the focusing forces are nonlinear.

From the discussion of Section 3.7, we know that the distribution function of a matched beam has constant value along single-particle orbits in trace space. We can generate contours of a matched distribution simply by plotting the trace-space loci of orbit vectors at the focusing cell boundary. By choosing test particles with different oscillation amplitudes and phases, we can find a family of continuous trace-space curves that correspond to isodensity contours of the distribution function. Figure 4.9 shows the results of such a calculation for a thin lens array with focal length given by Eq. (4.121) and a paraxial phase advance of $\mu_0 = 0.51$. At large displacement, the curves are not elliptical.

All particle orbits in a linear lens array are stable if $\mu_0 < \pi$[CPA, Chap. 8]. In contrast, the orbits of peripheral particles in a nonlinear periodic focusing system

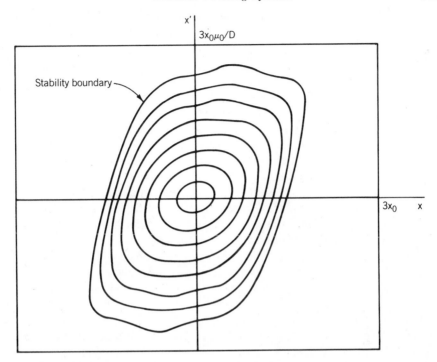

Figure 4.9 Isodensity contours of matched distributions in a lens array with spherical aberration. Thin lenses with focal length $f(x) = f_0[1 + (x/x_0)^2]$ separated by distance D, cell boundary at lens exit. Paraxial phase advance: $\mu_0 = 0.51$ rad.

may be subject to resonant instabilities, even if paraxial particles are stable. This behavior arises because the phase advance per cell is a function of the orbit displacement amplitude. In nonlinear systems, there is a bounded region of stable orbit vectors in trace space. We can find the stability boundary numerically by computing single-particle orbits through several cells and checking for a diverging amplitude. Figure 4.9 shows the stability boundary for the thin lens array. Several distribution function isodensity lines have been included inside the stability boundary. Note the contours near the axis are skewed ellipses, identical to the matched distributions of a linear lens array. The nonlinear forces distort the large amplitude orbits. Particles with orbit vectors outside the boundary are unstable because they have μ_0 less than π.

D. Beam Compression by Nonlinear Forces

We define a compression as any process than reduces the width of a beam while increasing the spread in transverse angle. Although we saw that a single nonlinear lens cannot direct a beam to an emittance-limited focal spot, we shall see that an array of nonlinear optical elements can compress a beam to a minimal spot size.

The validity condition is that the nonlinear elements induce gradual changes in the beam distribution. The term gradual means that the time for a change in the distribution is long compared with the oscillation time of the particles.

To illustrate compression by nonlinear forces, we shall study beam propagation in a converging pipe. The pipe "wall" is a region of localized transverse force that reflects particles that move outward. The transverse force variation is nonlinear—it is zero inside the pipe and rises to a high value at the wall. The idealized pipe is a good approximation for some transport schemes, such as dense plasma channels with embedded magnetic fields to confine intense light ion beams for inertial fusion. A major issue in light ion fusion studies is how much compression can be achieved in a channel.

To simplify the calculation, we shall first address particle orbits in a sheet beam that has mirror reflections at two converging planar walls (Fig. 4.10a). The planes incline at an angle α relative to the symmetry axis. Imagine the orbit of a particle that enters the converging pipe parallel to axis at a distance x_1 from the axis. The variable n represents the number of the orbit reflections from the walls. Figure 4.10 shows the first and second reflections, $n = 1$ and $n = 2$. Note that the transverse velocity of the particle increases with each collision. Since kinetic energy is conserved, longitudinal velocity is converted to transverse velocity. After many collisions, the axial velocity becomes negative—the particle reverses direction. The point of reversal defines a maximum penetration distance into the pipe. To define focal limits, we will calculate the maximum number of reflections N that take place before a particle reaches the turning point and the minimum transverse dimension of the particle orbit x_N at the point of maximum penetration.

Figure 4.10a shows that $y_1 + y_2$ is the axial distance the particle travels between the first and second reflections. Over this length, the halfwidth of the pipe decreases by an amount $(y_1 + y_2)\tan \alpha$. Therefore, the orbit displacements at the first and second reflections are related by

$$x_2 = x_1 - (y_1 + y_2)\tan \alpha. \tag{4.122}$$

The rule of similar triangles constrains the quantities in Eq. (4.122):

$$x_1/y_1 = x_2/y_2. \tag{4.123}$$

We can define additional relationships in terms of θ, the angle between the wall and the particle orbit after the first collision:

$$y_1 = x_1/\tan(\theta + \alpha), \tag{4.124}$$

$$y_2 = x_2/\tan(\theta + \alpha). \tag{4.125}$$

We can combine Eqs. (4.122) through (4.125) and eliminate the quantities y_1 and y_2, arriving at an equation that relates the orbit displacements after the first

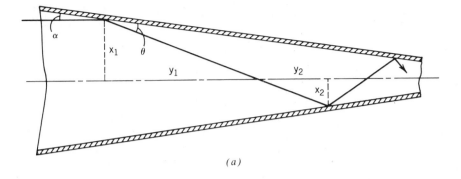

(a)

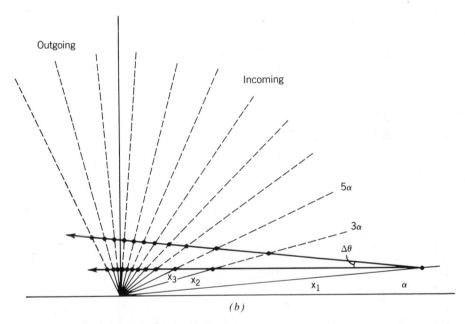

(b)

Figure 4.10 Nonlinear focusing by reflection from a tapered converging pipe. (a) Geometry. (b) Construction to calculate the displacements and angles at particle reflections.

two collisions:

$$\frac{x_2}{x_1} = \frac{\tan(\theta + \alpha) - \tan(\alpha)}{\tan(\theta + \alpha) + \tan(\alpha)} = \frac{\sin(\theta)}{\sin(\theta + 2\alpha)}. \qquad (4.126)$$

For an initially parallel orbit, Fig. 4.10a shows that $\theta = \alpha$ for a specular reflection. Equation (4.126) becomes

$$x_2/x_1 = \sin(\alpha)/\sin(3\alpha). \qquad (4.127)$$

Similarly, after the second reflection that orbit inclination relative to the wall is $\theta + 2\alpha = 3\alpha$. Extending the equations for the first and second reflections, we can show that the displacements on the second and third reflections are related by

$$x_3 = x_2 \sin(3\alpha)/\sin(5\alpha) = x_1 \sin(\alpha)/\sin(5\alpha). \tag{4.128}$$

By induction, the orbit displacement at the nth reflection is

$$x_n = x_1 \sin(\alpha)/\sin[(2n-1)\alpha]. \tag{4.129}$$

Next, suppose a particle enters the pipe with an inclination angle $\Delta\theta$. After a reflection, the angle between the orbit and the wall is

$$\theta = \alpha + \Delta\theta. \tag{4.130}$$

Reviewing the arguments leading to Eq. (4.129), we can show that the displacement at the nth reflection is

$$x_n = x_1 \sin(\alpha + \Delta\theta)/\sin[\Delta\theta + (2n-1)\alpha]. \tag{4.131}$$

The minimum value of the orbit displacement occurs when the denominator equals one, or

$$x_{min} = x_1 \sin(\alpha + \Delta\theta). \tag{4.132}$$

We can find the maximum number of wall reflections from the equation

$$\Delta\theta + [(2N-1)]\alpha = \pi/2, \tag{4.133}$$

or

$$N = [(\pi/2 - \Delta\theta)/\alpha + 1]/2. \tag{4.134}$$

Consider the implications of Eq. (4.132) for an incident beam of particles distributed between $\pm x_1$ with an angular spread $\pm \Delta\theta$. The beam emittance is $\varepsilon \cong x_1 \Delta\theta$. Section 3.1 showed that if we focused such a beam with an ideal linear lens, the minimum spot size is approximately

$$x_{min} \cong x_1 \Delta\theta/(x_1/f). \tag{4.135}$$

The quantity x_1/f is the envelope convergence angle; the highest possible beam convergence angle is about one radian. The minimum spot size using a single ideal single lens is on the order of

$$x_{min} \geq x_1 \Delta\theta. \tag{4.136}$$

A comparison to Eq. (4.132) shows that the same focal spot size could be achieved with a nonlinear converging channel if the beam compression is gradual.

Equation (4.132) implies a relationship to define the term "gradual":

$$\alpha \ll \Delta\theta. \tag{4.137}$$

To produce an emittance-limited focus, the convergence angle must be small compared with the input beam divergence. In the limit of Eq. (4.137), the compression is reversible.

We can easily extend the above derivation to a two-dimensional compression. Imagine a converging pipe with a square or rectangular cross section. Since there are no applied forces inside the pipe, particle motions in the x and y directions are independent. Also, wall reflections do not couple x and y velocities. Therefore, displacements in the x and y directions can be described independently according to Eq. (4.131).

As a final point of interest, we can represent the orbits of particles in a converging pipe by the geometric construction of Fig. 4.10*b*. We draw a set of radial lines in the upper half plane of a polar coordinate system (r, ϕ) at angles α, $3\alpha, 5\alpha, \ldots, (2n-1)\alpha, \ldots$. Again, α is the pipe convergence angle. We next draw an orbit line starting at a point on the $\phi = \alpha$ ray a distance x_1 from the origin. The line inclines at an angle $\Delta\theta$ relative to the axis. We leave it as an exercise to show the following characteristics of the construction:

1. The intersections of the orbit line with the rays represent reflections from the wall. The angle between the line and the intersected ray equals the angle between the orbit and wall before the reflection.

2. The distance from the origin to the intersection of the orbit line with the nth ray equals the beam displacement at the nth collision.

3. The construction gives the history of particle reflections in the pipe. The particle axial velocity is positive for reflections in the range $x > 0$ and negative in the range $x < 0$.

4.5. EMITTANCE IN STORAGE RINGS[2]

Elementary particle physics has been a major motivation for the development of high-energy charged particle accelerators. In the early days of particle research, beams were injected into *fixed targets*, blocks of solid density material at rest in the accelerator frame. Although fixed targets are practical for incident beams of moderate energy ($\leqslant 10$ GeV), the approach is ineffective in the TeV energy range of the largest modern accelerators. The figure of merit in elementary particle research is the net kinetic energy of incident and target particles in the center-of-momentum reference frame[3]—this energy is available to create particles or to

[2] Material adapted from "Elementary Stochastic Cooling," A. V. Tollestrup and G. Dugan, in *Physics of High Energy Particle Accelerators, Conf. Proc. No. 105*, M. Month Ed., (American Physical Society, New York, 1983).

[3] In the center-of-momentum reference frame, the vector sum of the momenta of the target and incident particles equals zero.

liberate the components of composite particles. With a fixed target, most of the incident beam energy transforms to the kinetic energy of reaction products—the energy is wasted. For example, a 1-TeV proton incident on fixed protons in a target provides only 42 GeV of center-of-momentum energy [CPA, p. 541]. Colliding beams provide a solution to this problem. When a 1-TeV proton collides head-on with another 1-TeV proton, the maximum available center-of-momentum energy is 2 TeV.

Generation and storage of colliding beams takes place in large circular accelerators called colliders. Magnetic lenses compress the beams to small diameters at positions in the ring called *interaction regions*. Particle detectors are located near these regions. Present high-energy colliders use beams of protons on protons, protons on antiprotons, electrons on positrons, or protons on electrons. A machine with two proton beams is called a *p–p collider*—it requires two separate rings with opposite polarity magnetic bending fields. Straight sections of the rings intersect at small angles at one or more interaction regions. A p–p̄ collider requires only a single ring. Antiprotons have the same mass as a proton but have a negative charge; hence, p and p̄ beams circulate in opposite directions in the same magnet array.

Although the principle is simple, the practical realization of colliding beams is difficult. The application demands the most complex accelerator systems ever designed. A fixed target has a high density of target particles, giving a good probability for interactions. Therefore, the intensity and emittance of the incident beam are not usually a major concern in fixed target experiments. In contrast, the density of a colliding beam target is infinitesimally small compared with a solid target. To achieve sufficient reactions, the beams must be bright, with the highest possible intensity and lowest emittance. Furthermore, the momentum spread must be small enough to match the longitudinal acceptance of the accelerator.

The emittance requirements for a collider are expressed in terms of the quantity *luminosity*. To understand luminosity, we must first review the concept of reaction *cross sections* and their relationship to the reaction rate. Cross sections describe the probability of reactions between particles. To illustrate the meaning, Fig. 4.11a shows a beam particle aimed toward a target particle. The impact parameter ρ is the projected distance of closest approach. Generally, an interaction occurs only if the incident particle passes close to the target. Depending on the nature of forces between the particles, the reaction probability $P(\rho)$ is a decreasing function of ρ.

The reaction cross section σ is a probability weighted interaction area:

$$\sigma = \int_0^\infty 2\pi\rho \, d\rho \, P(\rho). \tag{4.138}$$

The cross section has the following interpretation. If an incident particle passes through a plane of area $1 \, m^2$ that contains one target particle, then the probability of an interaction is $\sigma(m^2)/1(m^2)$. Colliding charged particles may have a variety of interactions—each reaction has a cross section. The probability

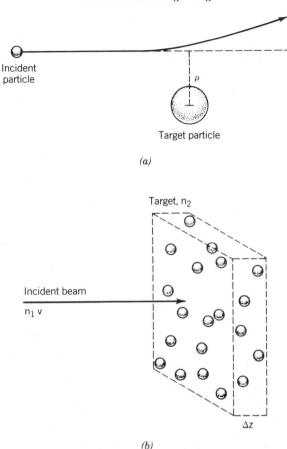

Figure 4.11 Particle collisions. (a) Definition of impact parameter, ρ. (b) Geometric construction to define the collision cross section.

function and the cross section generally depend on the relative kinetic energy of the particles. The cross sections for many atomic processes have values close to the physical area of an atom, $1\,\text{Å}^2$ $(10^{-20}\,\text{m}^2)$. The standard unit for nuclear reactions is the barn, where 1 barn $= 10^{-28}\,\text{m}^2$. Nuclear cross sections are much lower than atomic cross sections because of the small size of the nucleus. Cross sections decrease as the relative velocity of the particles increases—the particles spend less time near each other. A cross-section value typical of high-energy processes is $10^{-37}\,\text{m}^2$. Designers of colliding beam systems face two problems in achieving adequate reaction rates: the beam densities are low and the reaction probabilities are small.

For a known cross section, we can calculate the rate of reactions in the laboratory between colliding beams. The construction of Fig. 4.11b illustrates the method. As an example, suppose that a single proton impinges on a slab of antiprotons with unit area. The slab has thickness Δz and the antiprotons have density n_2. The total convergence velocity in the laboratory frame is $v \sim 2c$. When the proton passes through the slab, the probability of a reaction equals the product of the cross section times the number of antiprotons in the slab:

$$\Delta P = \sigma n_2 \Delta z. \qquad (4.139)$$

The reaction probability for a proton passing through a 1-m length of the target distribution with cross-section area of 1 m^2 is

$$P = \sigma n_2. \qquad (4.140)$$

The total rate of reactions occurring in 1 m^3 of the target is the product of P in Eq. (4.140) times the number of protons incident per second per square meter. The flux of protons incident on the target is $n_1 v$, where n_1 is the beam density. The reaction rate is

$$R(\text{reactions/sec-m}^3) = \sigma n_1 n_2 v. \qquad (4.141)$$

Luminosity L measures the inherent ability of a colliding-beam system to produce reactions. It is proportional to the normalized reaction rate R/σ; therefore, the product of L times σ for a particular reaction gives the total number of events expected. As an example, suppose a collider has counterrotating bunches consisting of N_1 protons and N_2 antiprotons. If each bunch has length Δz and cross-section area A, then the densities are $n_1 = N_1/A\Delta z$ and $n_2 = N_2/A\Delta z$. From Eq. (4.141), the reaction rate during the time the bunches intersect is

$$R = \sigma N_1 N_2 v/(A\Delta z)^2. \qquad (4.142)$$

The total number of reactions during an intersection equals the product of R times the volume $(A\Delta z)$ and the interaction time $(\Delta z/v)$:

$$\text{Reactions/crossing} = \sigma N_1 N_2/A. \qquad (4.143)$$

The *luminosity per crossing* equals the reactions per crossing divided by σ, or

$$L/\text{crossing} = N_1 N_2/A. \qquad (4.144)$$

The *luminosity per second* equals the product of the luminosity per crossing times f, the number of bunch crossings per second:

$$L = N_1 N_2 f/A. \qquad (4.145)$$

We can increase the luminosity by trapping more protons and antiprotons in the ring. Space-charge forces limit the maximum number of particles in a bunch to about $N_1 \sim N_2 \sim 10^{11}$. The size of the collider ring fixes the rotation frequency f. The ring diameter depends on the energy of the particles and the maximum field of the bending magnets. The only free parameter is the transverse area of the beam in the interaction region—the quantity A should be as small as possible. In colliders, focusing magnets compress the beam to a small diameter over the length of the interaction region. Following the discussion of Section 4.2, the beam width depends on the maximum strength of the lenses and the emittance. Low-emittance beams are essential to achieve high luminosity.

To illustrate luminosity requirements, we shall use parameters from the Fermilab TEV I collider. The bunches contain about 10^{11} protons and antiprotons. The rotation frequency in the 6-km ring is $f = 5 \times 10^4 \, \text{s}^{-1}$. The total cross section for all p–p̄ reactions is about 0.1 barn at a center-of-momentum energy of 1 TeV. To avoid overloading detectors, there should be no more than a single event per crossing. Equations (4.143) and (1.145) imply that the maximum luminosity per crossing is approximately $10^{29} \, \text{m}^{-2}$. For the given parameters, A should equal $10^{-7} \, \text{m}^2$. The corresponding luminosity per second of the collider is $L = 5 \times 10^{33} \, \text{m}^{-2} \text{s}^{-1}$. The cross section for a specific reaction, such as the production of W particles, is much smaller than the total cross section. A typical value is $\sigma \sim 1$ nanobarn $= 10^{-37} \, \text{m}^2$. At a luminosity per second of $5 \times 10^{33} \, \text{m}^{-2} \text{s}^{-1}$, it takes 560 h of operation to achieve 1000 reactions. This is a typical time scale for a high-energy physics experimental run. The *integrated luminosity* during the run is $10^{40} \, \text{m}^{-2}$.

Proton beams trapped in circular accelerators have low emittance. It is not difficult to achieve proton bunch areas less than or equal to $10^{-7} \, \text{m}^2$. The main problem in a p–p̄ collider is the generation of low-emittance antiproton beams. Antiprotons result from collisions of high-energy protons with a fixed target. They leave the target with a large spread in angle and kinetic energy. Even though the net efficiency for antiproton production may be as high as a few percent, the resulting beam has a dispersion in angle and energy that is far too large to fit into the acceptance of a circular accelerator. The only solution is to capture only the fraction of antiprotons with low angular divergence within a limited kinetic energy range. To accumulate 10^{11} antiprotons, production and capture extends over long periods of time. Even though the captured beam may fit the acceptance of the accelerator, usually its emittance is too high for the required luminosity. The antiproton beam must be cooled; the next section addresses this topic.

Figure 4.12 shows the efficiency for the generation of forward-directed antiprotons ($x' \leqslant 30$ mrad) by 120-GeV protons. The efficiency is less than 1% and the antiprotons have a broad spread in momentum. The antiproton production target and beam capture system must be designed to maximize p̄ production within the acceptance of the downstream transport system. To begin, we shall discuss methods to reduce the transverse emittance. The angular spread of antiprotons from the target is an inherent property of the generation process—it does not depend on the size of the incident proton beam. With a constant angular divergence, the lowest antiproton emittance results from the smaller

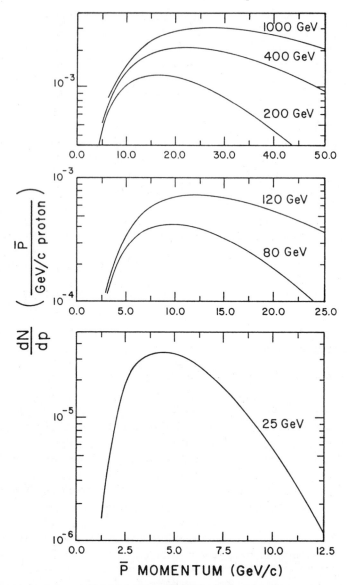

Figure 4.12 Efficiency for the production of antiprotons by the interaction of energetic protons with a tungsten target. Graphs show the probability for the creation of an antiproton within a momentum range with angle (relative to the proton beam axis) less than 30 mrad. Curves show different values of incident proton beam kinetic energy. (Courtesy A. V. Tollestrup, Fermi National Accelerator Laboratory. Used by permission, American Institute of Physics.)

(a)

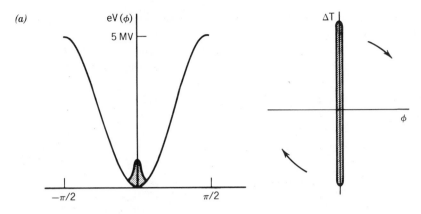

(b)

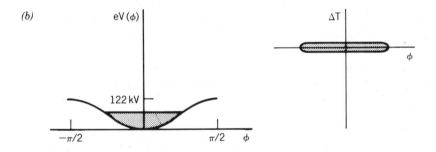

(c)

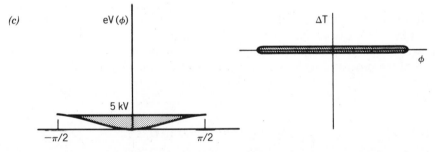

Figure 4.13 Sequence of events in the Debuncher Ring, Fermilab. (Phase diagrams not to scale.) (a) Antiproton bunch enters rings with short pulse length and large energy spread. Particles begin phase oscillation in a strong RF confining force. (b) During one-quarter period of a phase oscillation, the confining RF bucket depth drops from 5 MV to 122 kV to damp the phase oscillation and match the RF bucket to the bunch. (c) The RF bucket depth then drops slowly from 122 to 5 kV—the bunch expands in pulse length while the energy spread decreases.

source diameter. Hence, the proton beam is focused to a small spot size on the target.

Antiprotons diverging from the point source must be formed into a parallel beam for transport and capture in a storage ring. A focusing lens placed one focal length from the target performs the task. To capture a good fraction of the antiprotons, the lens must have a short focal length. The Fermilab lens is a lithium cell 1 cm in radius and 15 cm in length. The liquid metal carries a pulsed axial current of 500 kA—the strong azimuthal magnetic field creates a linear focusing force to capture the particles. The effective focal length is only 0.2 m. The lens and downstream transport system capture antiprotons with an average momentum of 8.9 GeV/c and a momentum spread of ± 2%.

It is necessary to reduce substantially the momentum spread of the antiprotons before they enter the main acceleration ring. The strategy is similar to that applied in the transverse direction—the antiprotons are created in a short pulse, the longitudinal analogy of a small area. The protons that generate the antiprotons are compressed into bunches with a pulse length of only 0.15 ns. The resulting antiproton pulses enter a storage ring called the debuncher. Figure 4.13 illustrates the transformation of the longitudinal beam distribution in the debuncher. At injection (Fig. 4.13a) the pulse has a large energy spread but fills only a small fraction of the ring circumference. The RF accelerating gap of the buncher has a high voltage to define a bucket depth of 5 MV. The beam pulse, injected at a phase of 0°, fills only a small fraction of the bucket length. Since the beam is not in axial equilibrium, it expands axially in the bucket—in phase space, the distribution rotates. If the RF field level were kept constant, the beam would perform phase oscillations. Instead, the phase motion is strongly damped by lowering the RF field. In the Fermilab debuncher, the bucket depth drops from 5 MV to 122 kV in a quarter of a phase oscillation period so that the beam reaches an axial equilibrium. At the end, the beam (Fig. 4.13b) has an extended length and a reduced energy spread. A further reduction is achieved by lowering the bucket depth to 5 kV over a time long compared with a phase oscillation period— Section 3.8 discussed beam cooling by a slow one-dimensional expansion. At the end of the process (Fig. 4.13c), the beam occupies a substantial fraction of the debuncher circumference—the longitudinal energy spread is about ± 0.1%.

4.6. BEAM COOLING

The goal of *cooling* is to reduce the phase-space volume occupied by a beam distribution. Beam-cooling processes can lower velocity spreads in both the transverse and longitudinal directions. Existing methods extend over long periods of time; therefore, applications are limited to high-energy circular accelerators and storage rings. As we mentioned in Section 4.4, a major application of beam cooling is to condition antimatter beams in colliding-beam experiments. In this section, we shall review techniques of cooling and concentrate on a specific example, stochastic cooling.

To cool a beam, we must find ways to circumvent the phase volume-conservation theorem of Section 2.4. We have already discussed one exception that occurs when velocity-dependent forces act on a beam. An example of cooling by frictional force is the reduction of the longitudinal velocity dispersion of an electron beam by synchrotron radiation. Synchrotron radiation results from the transverse acceleration of high-energy electrons confined in a circular accelerator. The electrons emit radiation predominantly in the forward direction, reducing their longitudinal kinetic energy. The radiation power loss per electron depends strongly on kinetic energy—in MKS units, the power radiated by a single electron in a circular orbit is[4]

$$P = 2cE^4 r_e / 3R^2 (m_e c^2)^3 \quad \text{W}. \tag{4.146}$$

In Eq. (4.146), E is the electron kinetic energy, R is the radius of the accelerator, and r_e is the classical electron radius,

$$r_e = e^2 / 4\pi\varepsilon_0 m_e c^2 = 2.818 \times 10^{-15}\,\text{m}. \tag{4.147}$$

Electrons in a storage ring are continuously accelerated to compensate for radiation losses. An electron beam distribution with an initial spread in E rapidly becomes almost monoenergetic. Electrons with energy higher than the mean emit more radiation and lose energy faster, while electrons with low E gain more energy from the acceleration system than they radiate. Synchrotron damping also reduces the transverse emittance of a beam—the acceleration associated with betatron oscillations causes emission of radiation by electrons. The damping time for random electron motions, either transverse or longitudinal, is roughly[5]

$$\tau_d \sim E/P \sim (m_e c^2 / E)^3 (R/r_e)(R/c). \tag{4.148}$$

Equation (4.148) implies that synchrotron cooling of electrons is rapid. For example, the cooling time for a 1-GeV electron beam in a 20-m-diameter ring is only about 20 ms. Synchrotron cooling of ions is insignificant because of the cubic dependence of the cooling time on particle mass.

The reduction in phase volume by friction depends on the conditions that the force varies smoothly in time and space. In reality, friction results from collisional processes. For an accurate description, we must superimpose small scale force variations on the average forces represented by expressions such as Eq. (2.42). Collisions cause diffusion of particle orbit vectors, a process that increases the beam phase volume. Real frictional processes have components that simulta-

---

[4]See, for instance, M. Sands, *The Physics of Electron Storage Rings* (Stanford Linear Accelerator Center, SLAC-121, 1970), p. 98. (Available from National Technical Information Service, Springfield, Virginia 22151.)

[5]See, for instance, F. T. Cole and F. E. Mills, "Increasing the phase-space density of high energy particle beams," *Ann. Rev. Nucl. Sci.* **31**, 295 (1981).

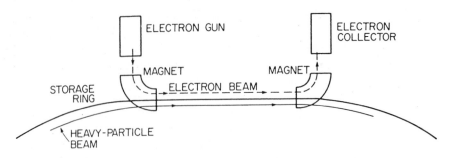

Figure 4.14 Ion beam cooling by energy transfer to a costreaming electron beam. (Courtesy F. Cole, Fermilab. Used by permission, American Institute of Physics.)

neously cool and heat beam distributions. Although friction may initially cool a beam, the discrete nature of the force (*Schottky noise*) limits the process. The discrete nature of synchrotron radiation sets limits on the minimum phase volume achieved by cooling. On the quantum level, synchrotron radiation consists of individual photons.

We must use alternate methods to cool ions in circular accelerators. *Electron beam cooling* is based on the interaction between the high-energy ions and a low-energy electron beam injected into the ring. Remember that the theorem of phase-volume conservation depends on the condition that particles in a distribution are isolated. The theorem need not hold if other particles interact with the beam. Figure 4.14 shows the basis of electron beam cooling of antiprotons in a storage ring. The electron beam is injected into the ring so that it is collinear with the stored antiproton beam. The choice of injector voltage gives an electron beam with precisely the same velocity as the antiprotons—a typical value is 100 kV. The weak magnetic field that bends the low-energy electrons has little effect on the energetic antiprotons. The electron beam has high current (~ 1A) and low beam divergence. After a drift distance, another bending magnet removes the electrons, directing them to an inverse diode (Section 15.1) to recover most of their energy.

In the rest frame of the drifting beams, the transverse and longitudinal energy spreads of the electron beam are much lower than those of the antiproton beam. Hot antiprotons lose thermal energy when they have Coulomb collisions with the cold electrons. The heated electrons carry the excess energy outside the storage ring. Cold electrons replace heated electrons for continuous refrigeration. The energy transfer rates are described by relativistic extensions of the familiar energy transfer formulas of plasma physics.[6]

Experiments on storage rings have demonstrated the principles of electron

[6]See, for instance, G. Schmidt, *Physics of High Temperature Plasmas, 2nd ed.*, Academic Press, New York, 1979, Section 11.3.

beam cooling. The process is most effective for antiprotons with low kinetic energy, partly because the energy transfer rate is higher, and partly because of technological difficulties in generating continuous high-power electron beams. Electron beam cooling of antiprotons is a complex process. Generation of antiprotons must take place at high kinetic energy where the production cross section is large. The particles then must be decelerated to an energy in the range 200–400 MeV for accumulation and cooling. After cooling, the antiprotons are reaccelerated for injection into a collider.

In the remainder of this section, we shall concentrate on stochastic cooling of beams. The word *stochastic* means "involving chance or probability." Stochastic cooling is a statistical process based on the fact that a beam is not a perfect fluid but a collection of discrete particles, albeit a very large number. Information on random variations of the macroscopic properties of the beam associated with discreteness is used to influence specific groups of particles to achieve net cooling.

An underlying assumption of the theorem of conservation of phase volume is that we cannot apply specific forces to individual particles. If we could individually accelerate antiprotons in a collider, we could create any distribution desired. Individual control is impossible because the number of particles in a bunch ranges from 10^8 to 10^{11}. Nonetheless, by recognizing trends in the random distribution of particles, we can sense groups of particles with similar properties. Given sufficient time, applied forces can effect an average emittance reduction.

Figure 4.15a shows a schematic diagram of a stochastic-beam-cooling system for a storage ring. The system lowers the transverse emittance of the beam in the horizontal direction ε_x. A detector, which we shall call the *pickup*, measures the centroid of the beam as it passes. After amplification, the signal from the pickup drives a set of deflection plates, called the *kicker*. The kicker, located across a chord from pickup, applies an accelerating or decelerating electric force in the horizontal direction. The cable and amplifier delays are chosen so that the signal generated in the pickup by a group of particle arrives at the kicker at the same time as the particles.

The pickup cannot resolve the positions of individual particles. Instead, it measures the centroid $\langle x \rangle$ of a group of particles. We denote the minimum number of particles that can be resolved as N_s. If $N_s \rightarrow \infty$, random variations in the centroid approach zero and the pickup generates no signal. For the finite number of particles in a group, we expect that statistical variations cause changes in the centroid. Sometimes there are more particles on the inside than the outside. Then, the centroid shifts inward. Statistical theory predicts that the fractional imbalance of particle between the inside and outside is roughly

$$\Delta N_s/N_s \sim 1/(N_s)^{1/2}. \tag{4.149}$$

Suppose we locate the kicker a distance equal to an odd number of quarter betatron wavelengths from the pickup:

$$D = \lambda_x(2m + 1)/4. \tag{4.150}$$

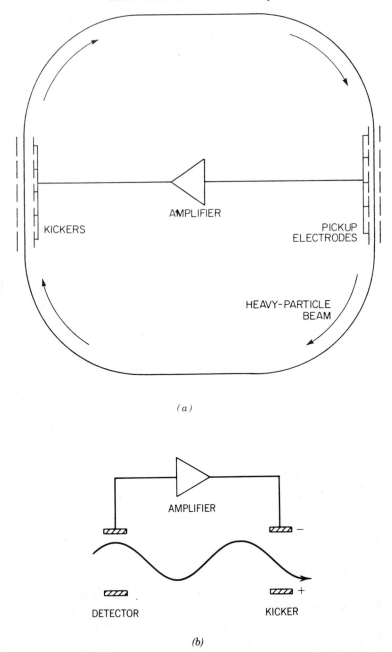

(a)

(b)

Figure 4.15 Stochastic cooling of an ion beam. (*a*) Schematic drawing of apparatus. (*b*) Projected transverse particle orbit between the detector and kicker. Ideal orbit has its maximum displacement at the detector and maximum velocity at the kicker. (Courtesy F. Cole, Fermilab. Used by permission, American Institute of Physics.)

The quantity λ_x is the average betatron wavelength for a group of particles. Figure 4.15b shows a projection of a horizontal particle oscillation for $m = 2$. The condition of Eq. (4.150) means that a positive centroid displacement at the pickup becomes a negative average velocity at the kicker.

For antiprotons, a negative electric field in the kicker decelerates the particles in the transverse direction, thereby reducing the amplitude of the betatron oscillation. Depending on the phase of the particle oscillation, an electric field at the kicker either increases or decreases the betatron oscillation amplitude. If the particle oscillations between the kicker plates are randomly distributed, then the electric field increases the oscillation amplitude for about half the particles and lowers the amplitude of the other half. On the other hand, if the electric field at time t is proportional to the pickup signal at time $[t - \lambda_x(2m + 1)/4\beta c]$, then we can achieve a net reduction in the amplitude of particle betatron oscillations. The pickup signal shows when there will be an excess of particles with inward directed velocity at the kicker.

The damping rate for betatron oscillations, or the cooling rate, depends on the fractional imbalance of particles moving inward and outward at the kicker. According to Eq. (4.149), the procedure works best if the number of particles in the group sensed by the pickup N_s is small. There are two ways we can assure a low value N_s:

1. The total number of particles in the storage ring N is low.

2. The detector–amplifier system has high-frequency response so that it can sense changes in a small fraction of the particles in the ring.

For high beam luminosity, we want to avoid the first option. Ideally, we should design a feedback system with the highest possible frequency response.

If we rotate the pickup and kicker electrodes 90° about the beam axis, the feedback system of Fig. 4.15a will damp vertical betatron oscillations. Stochastic cooling systems can also reduce the longitudinal emittance of a beam. One way to approach longitudinal cooling is to design the optics of the storage ring so that the beam particles spread in the horizontal direction according to their longitudinal energy. High-energy particles move outward compared with low-energy particles. Such sections of the accelerator are called *high dispersion regions*. For this geometry, measurement of the beam centroid in the horizontal direction yields information on the average kinetic energy of a group of particles relative to the mean. The pickup signal drives a kicker with a longitudinal electric field. An outward centroid displacement at the kicker implies that the group of particles has an excess of kinetic energy—the kicker should apply a decelerating force.

In the following discussion, we shall concentrate on stochastic cooling in the transverse direction, or *betatron cooling*. The theory presented gives a good idea of the mechanism of stochastic cooling and motivates some important scaling laws. Nonetheless, for a comprehensive description of topics such as mixing, a more sophisticated approach based on a frequency domain analysis is required.

To begin, imagine how the cooling system of Fig. 4.15a would function if it could sense individual particles. Assume the system responds to a single test particle. The frequency with which the particle passes the detector is $1/T_0$, where T_0 is the time to complete a revolution around the ring. If the average orbit radius of the particle orbit is R, then

$$T_0 = 2\pi R/\beta c. \qquad (4.151)$$

The test particle executes betatron oscillations of wavelength λ. The quantity x_k^0 represents the amplitude of the oscillation. The superscript 0 flags the test particle, while k shows that the particle has completed k revolutions around the ring.

To begin, suppose that the particle has its maximum displacement when it passes the pickup (Fig. 4.15b). Later, we shall add corrections to represent the fact that particles pass the pickup at random phases in their oscillations. If Eq. (4.150) holds, the particle reaches the kicker with a negative velocity:

$$v_k^0 = - x_k^0(\beta c/\lambda). \qquad (4.152)$$

For antiprotons, a negative kicker electric field reduces the transverse velocity. There is a specific value of electric field E_{x0} that will cancel the velocity completely. If the kicker has length D, then E_{x0} is approximately

$$E_{x0} \cong - x_k^0[\gamma m_p(\beta c)^2/qD\lambda]. \qquad (4.153)$$

Equation (4.153) shows that the kicker field is linearly proportional to the particle displacement at the pickup.

We can represent the action of the cooling system on the test particle in one transit by the equation

$$x_{k+1}^0 = x_k^0 - gx_k^0. \qquad (4.154)$$

Equation (4.154) gives the betatron oscillation amplitude of the particle arriving at the detector on turn $k + 1$ as a function of the amplitude on turn k. The quantity g is the system gain, which depends on the signal strength from the pickup, the amplifier gain, and the kicker plate geometry. If a displacement x_k^0 at the kicker results in the electric field of Eq. (4.153), then we take $g = 1$. For this choice, the system cools the orbit of Fig. 4.15b in a single pass.

Stochastic cooling is complicated by two factors that influence actual systems:

1. The detector–amplifier has a finite frequency bandwidth and senses many particles at once.
2. Noise generated in the amplifier distorts the signal applied to kicker.

Regarding the first point, the electronics responds to changes in the beam centroid with a nonzero time resolution that we shall denote T_s. The effective frequency bandwidth of the system W is related to the sampling time by

$$T_s = 1/2W. \tag{4.155}$$

Suppose that the storage ring contains a total of N particles. For effective cooling, we want the detector to sense a small number of particles in the time window T_s. Therefore, it is better if the particles are spread uniformly around the ring rather than bunched. If we assume a uniform beam, the number of particles sampled by detector is

$$N_s = N(T_s/T_0). \tag{4.156}$$

where T_0 is the average particle revolution time [Eq. (4.151)].

Consider a group of N_s particles completing turn k around the ring. The group includes the test particle with betatron oscillation amplitude x_k^0. The feedback signal from the detector is proportional to the average imbalance of particles between the inside and the outside. The average displacement of the beam at the detector is

$$\langle x_k \rangle = \frac{1}{N_s}\left[x_k^0 + \sum_{j=1}^{N_s-1} x_k^j \right]. \tag{4.157}$$

The quantity x_k^j is the amplitude of the betatron oscillation of the jth particle in the group. We must modify Eq. (4.154) to include the fact that the test particle oscillation amplitude passing through the kicker is proportional to the average displacement $\langle x_k \rangle$ rather than to x_k^0. The equation becomes

$$x_{k+1}^0 = x_k^0 - g[\langle x_k \rangle + \xi]. \tag{4.158}$$

We have included the possibility of detector noise in Eq. (4.158) through the quantity ξ. It is a random variable uncorrelated with any beam-dependent quantities; the effect of noise appears as an error in the detector centroid signal that is amplified with the real signal.

Since the betatron oscillation amplitudes of other particles are randomly distributed relative to the test particle, we must construct a statistical theory to describe variations of x^0 averaged over many transits through the system. We shall construct an equation that describes the rate of reduction of the root-mean-squared test particle displacement averaged over many transits. To begin, the change in the square of the test particle displacement in a single transit is

$$\begin{aligned}
\Delta(x_k^0)^2 &= (x_{k+1}^0)^2 - (x_k^0)^2 \\
&= [x_k^0 - g(\langle x_k \rangle + \xi)]^2 - (x_k^0)^2 \\
&= -2g(\langle x_k \rangle + \xi)x_k^0 + g^2(\langle x_k \rangle + \xi)^2. \tag{4.159}
\end{aligned}$$

The second form of Eq. (4.159) comes from Eq. (4.158). Expanding terms and substituting for the beam centroid at the detector from Eq. (4.157) gives the following equation:

$$\Delta(x_k^0)^2 = -2g\{x_k^0(\sum x_k^j)/N_s + (x_k^0)^2/N_s + x_k^0\xi\}$$
$$+ g^2\{[\sum x_k^j + x_k^0]^2/N_s^2 + 2[\sum x_k^j + x_k^0]\xi/N_s + \xi^2\}. \quad (4.160)$$

The sums in Eq. (4.160) are taken over the same range as Eq. (4.157).

Next, we take averages of the terms in Eq. (4.160) over many transits through the feedback system. We apply the condition that the displacement amplitudes of other particles at the detector are uncorrelated with the amplitude of the test particle. In other words, the betatron oscillations become completely mixed each time the group of particles traverses the detector. (We should note that this condition is usually violated in actual systems—we shall discuss the consequences of poor mixing later.) A bar over a quantity denotes an average taken over many transits. The assumption of random oscillations leads to the elimination or modification of many terms on the right-hand side of Eq. (4.160). For example,

$$\overline{\sum x_k^j/N_s} = 0, \quad (4.161)$$

because the displacement of the beam centroid from the main beam axis averages to zero over multiple transits.

We can simplify the notation by defining the root-mean-squared betatron oscillation amplitude of particles in the test group:

$$X^2 = \sum (x_k^j)/N_s. \quad (4.162)$$

We want the test particle orbit to characterize the particles in the group. Let the displacement amplitude of the test particle equal X:

$$X^2 = \overline{(x_k^0)^2}. \quad (4.163)$$

We can write

$$\Delta X^2 = \overline{\Delta(x_k^0)^2}. \quad (4.164)$$

Now, consider the other terms on the right-hand side of Eq. (4.160). We find that

$$\overline{x_k^0 \sum x_k^j} = 0, \quad (4.165)$$

by the condition that particle orbits mix completely between transits. Also, by the assumption that the amplifier noise is uncorrelated with the particle displacement amplitude, we find

$$\overline{x_k^0 \xi} = 0 \qquad (4.166)$$

and

$$\overline{\xi \sum x_k^j} = 0. \qquad (4.167)$$

Finally,

$$\overline{(\sum x_k^j)(\sum x_k^i)} = \overline{\sum (x_k^j)^2} = (N_s - 1)X^2, \qquad (4.168)$$

since the particles are statistically independent and the cross terms cancel.

Collecting all the non zero terms of Eq. (4.160) gives the following relationship for the change in the mean-squared betatron oscillation amplitude of a group of particles in the cooling system:

$$\Delta X^2 = -2gX^2/N_s + g^2[(N_s - 1)X^2/N_s + X^2/N_s + \xi^2]. \qquad (4.169)$$

The quantity ξ^2 is the mean-squared detector noise figure. We must correct Eq. (4.169) to account for the fact that the original test particle passes the pickup and kicker at random phasess in its betatron oscillations. We multiply all terms on the right-hand side by 0.5 because the random mean-squared displacement is half the square of the betatron oscillation amplitude. With the correction, Eq. (4.169) gives the average change of the mean-squared betatron oscillation amplitude in a single transit. We can write an approximate differential equation by remembering that the time per transit is T_0. The rate of change of the mean-square oscillation amplitude is

$$\frac{dX^2}{dt} \cong \frac{\Delta X^2}{T_0} = -\frac{gX^2}{N_s T_0} + \frac{g^2 X^2}{2N_s T_0} + \frac{g^2 \xi^2}{2T_0}. \qquad (4.170)$$

Equation (4.170) is the fundamental relationship of stochastic beam cooling. The first term on the right-hand side represents the coherent response of the system to a horizontal imbalance in the particle distribution at the detector. This is the desired effect that we discussed qualitatively at the beginning of the section. Note that the cooling rate is inversely proportional to N_s and T_0. As expected, cooling is less effective when there are many particles in the sensed bunch. Also, cooling is proportional to the frequency of particle transits through the system $1/T_0$.

The second and third terms on the right-hand side of Eq. (4.170) have opposite signs compared with the first term—they describe effects that heat the distribution. The third term represents electronic noise in the detector circuit—the noise is amplified and applied to the kicker plates. As expected, the noise term does not depend on N_s or X^2, the properties of the particle distribution. The second term in Eq. (4.170) represents statistical granularity, or Schottky noise, in the detector signal that results from the finite number of particles sampled. The granularity is uncorrelated with the detected centroid variations that allow beam cooling.

To understand the implications of Eq. (4.170), we shall first discuss the solution for a perfect amplifier. With no electronic noise ($\xi^2 = 0$), the equation takes the form,

$$dX^2/dt = -(X^2/N_sT_0)g(1 - g/2). \tag{4.171}$$

Equation (4.71) has the solution

$$X^2 = X_0^2\exp(-\lambda t), \tag{4.172}$$

where λ is the cooling rate:

$$\lambda = (1/N_sT_0)g(1 - g/2).$$

The transverse emittance of the beam is proportional to X^2; therefore, ε varies as

$$\varepsilon = \varepsilon_0\exp(-\lambda t). \tag{4.173}$$

Equation (4.172) shows that cooling is most rapid when $g = 1$. Here, the cooling rate is

$$\lambda_{max} = 1/2N_sT_0. \tag{4.174}$$

We can write the peak damping rate in terms of the sampling time T_s and the total number of particles in the ring:

$$\lambda_{max} = (1/2T_0)(T_0/NT_s) = 1/2NT_s = W/N. \tag{4.175}$$

Equation (4.175) illustrates that beam cooling improves with high bandwidth and a low number of contained particles. As an example, take $N = 10^{11}$ and $W = 1\,\text{GHz}$. The e-folding time for emittance reduction is $1/\lambda_{max} = 100\,\text{s}$. In actual systems, cooling of beams takes much longer because of imperfect mixing.

Electronic noise usually has little effect on the initial cooling rate—it limits the final achievable emittance reduction. At late time, we expect that the time derivative on the left-hand side of Eq. (4.170) approaches zero. Setting the terms on the right-hand side equal to zero gives the following value for the minimum mean-squared betatron oscillation amplitude:

$$X_{min}^2 = \left(\frac{g}{2g^2 - g}\right)\frac{\xi^2N}{2WT_0}. \tag{4.176}$$

The term in parentheses equals unity if $g = 1$. Equation (4.176) shows that we can reduce the final beam emittance by lowering the detector noise, increasing the system bandwidth, and reducing the number of particles stored in the ring.

The above conclusions rely on the assumption that the betatron oscillations of particles in a group are completely randomized between transits of the feedback system. Cooling does not take place if particle orbits remain correlated between transits. To show this, imagine a monoenergetic group of particles in a focusing system with perfect linear forces. If the initial distribution of betatron oscillation phase is random, the pickup initially senses an irregular signal. Nonetheless, the displacement fluctuations of the beam are coherent—over long times, the signal repeats periodically. The feedback system senses the initial displacement fluctuations and corrects the particle orbits, slightly reducing the beam emittance. After this there are no more centroid fluctuations to detect and beam cooling ceases. Only the effects of Schottky and detector noise remain—the feedback system subsequently heats the beam.

The success of stochastic cooling depends on the continuous randomization of the particle orbits sampled. In storage rings, randomization results primarily from a spread in the axial momentum of beam particles. Usually, there is no correlation between axial and transverse motions. The mixing process is a continuous redistribution of the sampled betatron oscillations so that particle displacements at the pickup are random. Unfortunately, stochastic cooling makes contradictory demands on the mixing process. Ideally, there should be no mixing when the beam travels from the pickup to the kicker. For effective cooling, a beam displacement at the pickup must appear as an average transverse velocity at the kicker. The requirement is that the betatron wavelengths of the N_s particles in the group must be almost the same to preserve a coherent oscillation. On the other hand, we require that effective mixing occurs when the particles travel from the kicker back to the pickup—the beam distribution at transit $k + 1$ should be unrelated to the distribution at transit k.

In a real system, mixing involves compromises. Information is inevitably lost when particles move from the pickup to the kicker and randomization of the distribution is incomplete after a revolution. A detailed description of stochastic cooling with incomplete mixing requires an advanced frequency domain theory. Instead, we shall concentrate on qualitative arguments to emphasize the main ideas. The pickup system senses a portion of the beam in a storage ring of length $cT_s = c/2W$. Suppose that particles in the ring have a momentum spread Δp. As a result, the localized group of particles spreads axially as it propagates. We define the quantity ΔL as the length of an initially short beam bunch after it makes one revolution around the ring.

We define the *mixing parameter M* in terms of ΔL:

$$M = cT_s/\Delta L. \tag{4.177}$$

The mixing parameter has the following interpretation. Mixing is effective if $M \sim 1$—there is sufficient coherence between the pickup and kicker while the particles are significantly redistributed over a full revolution. In many storage rings, the momentum spread is small, so that $M \gg 1$. Here, there is substantial turn-to-turn coherence and the cooling rate drops.

We can express ΔL in terms of the momentum spread and the properties of the transverse focusing system. The transition gamma factor γ_t is an important characteristic of focusing systems in circular machines. The definition and implications of γ_t are discussed in [CPA, p. 554]. If S is the average path length for particle orbits around a storage ring, then changes in the path length are related to the momentum spread by

$$\Delta S/S = (\Delta p/p)/\gamma_t^2. \tag{4.178}$$

Assume that the antiproton beam in the storage ring is highly relativistic ($\gamma \gg \gamma_t$). Then, all particles travel at about speed of light and variations in transit time around the ring arise primarily from differences in path length. After a transit, a group of particles with momentum spread Δp expands to a length

$$\Delta L = \Delta S = (2\pi R)\Delta p/p\gamma_t^2 = cT_0\Delta p/p\gamma_t^2. \tag{4.179}$$

Substitution of Eq. (4.179) into Eq. (4.177) gives the following expression for the mixing parameter:

$$M = \gamma_t^2(T_s/T_0)/(\Delta p/p) = \gamma_t^2/(2WT_0)(\Delta p/p). \tag{4.180}$$

In a sample of N_s particles, the fraction of the group that remains coherent from turn-to-turn is not cooled. Although coherency reduces cooling, it does not change the effect of Shottky noise. The frequency domain analysis shows that imperfect mixing enhances the beam noise term by a factor of M. The revised cooling equation is

$$\frac{dX^2}{dt} \cong \frac{\Delta X^2}{T_0} = -\frac{2gX^2}{N_sT_0} + \frac{Mg^2X^2}{N_sT_0} + \frac{g^2\xi^2}{T_0}. \tag{4.181}$$

Equation (4.181) implies that the maximum cooling rate occurs when $g = 1/M$. Ignoring amplifier noise, the modified cooling rate is

$$\lambda_{max} = W/MN. \tag{4.182}$$

As an example, consider the parameters of the antiproton debuncher ring at Fermilab. The ring has a radius of $R = 83$ m, corresponding to a antiproton revolution time of 1.7 μs. For typical operation of the ring, $\gamma_t = 13$, $\Delta p/p = 0.002$, and $N = 10^8$. The bandwidth of the cooling system is about $W = 2$ GHz. Insertion of the parameters into Eq. (4.180) shows that mixing is poor, with $M \cong 12$. For the optimum system gain, the cooling rate is $\lambda_{max} = 1.7\,\text{s}^{-1}$. For injection into the next storage ring, the transverse emittance of the injected beam must be reduced to about 3% of its initial value. The process takes about 2 s. During this time, the beam passes through the cooling system about 10^6 times.

5

Introduction to Beam-Generated Forces

In Chapters 3 and 4, we studied methods to describe beams of particles with diverse orbit parameters. In this chapter, we address the second main problem of collective-beam physics, the calculation of orbits when particles exert forces on one another. Collections of charged particles interact in a different way from familiar systems such as molecules in gas. Charge-neutral molecules move freely except for collisions with neighboring particles. Furthermore, a molecule collides with only one other particle at a time. In contrast, charged particles create long-range forces. Large numbers of charged particles interact simultaneously.

Section 5.1 reviews expressions for the electric and magnetic fields generated by beams with simple geometries. Many derivations in this book apply field expressions for cylindrical or sheet beams of infinite length. We shall find that the focusing force from the magnetic field of a relativistic beam almost cancels its repulsive electric force—because of this balance, relativistic beams can propagate at high current.

The main practical implication of beam-generated fields is that they limit transportable beam current. We distinguish two types of current limits—longitudinal and transverse. The longitudinal space-charge limit results from the axial electric fields of a beam. Beam particles decelerate if the energy associated with the space-charge potential in the beam volume is comparable to their kinetic energy. Space-charge fields can completely stop a high-current beam. The defocusing electric forces of a beam define the transverse limit for propagation. The transverse limit occurs when the beam-generated force exceeds the force of the focusing system.

187

Section 5.2 derives the one-dimensional Child law, a longitudinal limit. The relationship gives the maximum current density that can be extracted from a particle source. The Child law derivation also illustrates a method to find the orbits of charged particles with self-consistent space-charge fields in a one-dimensional geometry. Section 5.3 illustrates calculation of a longitudinal limit in a two-dimensional geometry. We derive the maximum current of a cylindrical electron beam in a strong solenoid field.

Section 5.4 gives an example of the effect of transverse beam-generated forces. We study the expansion of high-current beams in drift spaces with no applied force. Like the emittance force (Section 3.6), the space-charge force sets limits on propagation length and minimum focal spot size. Section 5.5 reviews the transverse forces generated by relativistic beams. The properties of relativistic coordinate transformations lead to the processes of Lorentz contraction and time dilation. We shall show that these effects imply the transformation laws of electric and magnetic fields [Eqs. (1.49)–(1.54)] and the balance between the electric and magnetic forces of relativistic beams.

5.1. ELECTRIC AND MAGNETIC FIELDS OF BEAMS

To describe particle motion in high-current beams, we must have expressions for the electric and magnetic fields generated by the particles. In most analytic calculations of collective charged particle dynamics, we use simple geometries that approximate real beams. It is easy to derive field expressions for the sheet and cylindrical beams since the beam density varies in only one dimension. In this section, we shall collect useful formulas for the electric and magnetic fields generated by beams with these geometries.

A. Sheet Beam

Figure 5.1 shows an ideal sheet beam of particles with charge q. The beam extends infinitely in the y and z directions—average particle motion is along the z axis. The particle density $n(x)$ varies only in the x direction—particles are contained between $\pm x_0$. Derivatives in the y and z directions are equal to zero. Equation 1.28 reduces to

$$\frac{\partial E_x}{\partial x} = \frac{qn(x)}{\varepsilon_0}. \tag{5.1}$$

For simplicity, we shall take the particle distribution as symmetric about the axis $x = 0$. Then, the electric field at the axis equals zero:

$$E_x(0) = 0. \tag{5.2}$$

If there is no externally applied field, integration of Eq. (5.1) and application

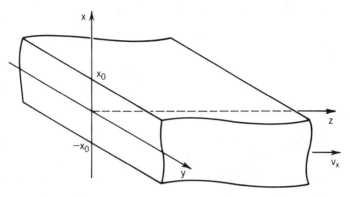

Figure 5.1 Geometry of a sheet beam of infinite width in the y direction propagating in the z direction.

of Eq. (5.2) gives an expression for the electric field in terms of the particle density:

$$E_x(x) = q \int_0^x dx' n(x')/\varepsilon_0. \tag{5.3}$$

Equation (5.3) gives a good approximation for finite length beams if their properties vary over distances long compared with x_0.[1] Here, the gradients of the electric field in the y and z directions are nonzero but much smaller than the gradient in the x direction. Figure 5.2 shows equipotential lines in a cross section of a sheet beam in a rectangular box with finite width in the y direction. The field varies mainly in x except near the edges. The sheet approximation is also useful for annular beams with radii much greater than their thicknesses.

In the one-dimensional approximation, the magnetic field is related to the z-directed current density $j_z(x)$ by

$$\frac{\partial B_y}{\partial x} = \mu_0 j_z(x). \tag{5.4}$$

If the particles in the beam of Fig. 5.1 are monoenergetic and have small transverse velocity components ($v_x, v_y \ll v_z$), then the axial velocity is constant over the beam cross section. The current density is

$$j_z(x) = en(x)v_z. \tag{5.5}$$

---

[1] For relativistic beams, the infinite beam expressions are approximately correct if the beam properties vary over axial distances much greater than x_0 in the beam rest frame. In the stationary frame, the axial length scale is x_0/γ.

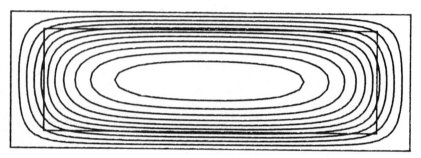

Figure 5.2 Lines of constant beam-generated electrostatic potential for a sheet beam with finite width in the y direction in a rectangular metal chamber. Numerical solution using POISSON code.

Substituting Eq. (5.5) into (5.4) and integrating, we find that

$$B_y(x) = (qv_z\mu_0) \int_0^x dx' n(x').\tag{5.6}$$

Comparison of Eqs. (5.3) and (5.6) gives a relationship between the transverse electric and magnetic fields of paraxial sheet beams:

$$B_y = (\varepsilon_0\mu_0)v_z E_x = (v_z/c^2)E_x.\tag{5.7}$$

Note that Eq. (5.7) holds for any variation of $n(x)$.

We often use the condition that a beam has uniform density to make estimates when we do not have a detailed knowledge of the particle distribution. Suppose that the density has the form

$$n(x) = n_0 \quad (-x_0 \leqslant x \leqslant +x_0),\tag{5.8}$$
$$n(x) = 0 \quad (|x| > x_0).$$

Evaluation of the integrals of Eqs. (5.3) and (5.6) gives

$$E_x = (qn_0/\varepsilon_0)x,\tag{5.9}$$

$$By = (\mu_0 qn_0 v_z)x = (qn_0 v_z/\varepsilon_0 c^2)x.\tag{5.10}$$

Equations (5.9) and (5.10) illustrate an important property of the uniform density beam—the electric and magnetic fields are linear. This property makes it easy to describe the motion of particles through a linear focusing system in the presence of beam-generated fields.

We can also find expressions for the electrostatic potential and magnetic vector potential of a uniform density beam. Integration of the Poisson equation [Eq. (2.85)] gives

$$\phi(x) = \phi(0) - (qn_0/2\varepsilon_0)x^2. \tag{5.11}$$

The quantity $\phi(0)$ is the potential at the center of the beam. The difference in electrostatic potential between the center of the beam and the edge is

$$\Delta\phi_{max} = qn_0 x_0^2/2\varepsilon_0. \tag{5.12}$$

We expect that the condition $q\Delta\phi_{max} \ll m_0 v_z^2/2$ holds for paraxial particle orbits.

The equation that defines the vector potential for a sheet beam is

$$\frac{\partial A_z}{\partial x} = -B_y. \tag{5.13}$$

We calculate A_z by inserting the expression of Eq. (5.10) into Eq. (5.13) and integrating. Adopting the boundary condition $A_z(0) = 0$, the vector potential is

$$A_z(x) = -(\mu_0 qn_0 v_z/2)x^2. \tag{5.14}$$

Often, it is unnecessary or impossible to understand the details of particle motion over the full cross section of a beam. Instead, we use an *envelope equation* that describes the balance of forces only at the periphery, or envelope, of a beam. The beam-generated electric field on the envelope of a sheet beam is given by Eq. (5.9) with $x = x_0$. For a paraxial beam, we can write a useful alternative expression in terms of the net beam current. If v_z is almost constant, the line current per unit length of a sheet beam is

$$J = 2qn_0 v_z x_0 = 2qn_0 \beta c x_0. \tag{5.15}$$

The quantity β is the longitudinal relativistic velocity factor. The envelope electric field is

$$E_{x0} = J/2\varepsilon_0 v_z = J/2\varepsilon_0 \beta c. \tag{5.16}$$

Similarly, the envelope magnetic field is

$$B_{y0} = \mu_0 J/2. \tag{5.17}$$

The beam-generated electric and magnetic forces acting on the envelope of a sheet beam are

$$F_{x0}(\text{electric}) = eJ/2\varepsilon_0\beta c \qquad (5.18)$$

and

$$F_{x0}(\text{magnetic}) = -ev_zB_y = -e\mu_0\beta cJ/2 = -e\beta J/2\varepsilon_0 c. \qquad (5.19)$$

The ratio of magnetic to electric force is

$$F_x(\text{magnetic})/F_x(\text{electric}) = -\beta^2. \qquad (5.20)$$

Equation (5.20) holds at all locations in a paraxial sheet beam, independent of the density distribution. Equation (5.20) has an important implication for the transport of high-current beams. For nonrelativistic particles, such as ions, the beam magnetic force is usually negligible. In contrast, magnetic forces are important in relativistic electron beams. Here, the focusing magnetic force almost balances the electric force, allowing high-current transport.

B. Cylindrical Beams

Figure 5.3 shows an ideal cylindrical beam. The beam has azimuthal symmetry and infinite extent in the direction of propagation. The density drops to zero at radius r_0. Again, the infinite length beam gives good estimates for the fields of beams that vary over distances long compared with r_0/γ. Here, axial derivatives of field quantities are much smaller than derivatives in the radial direction.

Expressions for the fields of cylindrical beams are similar to those for the sheet beam. A uniform cylindrical beam has density n_0 that extends from the axis to a sharp termination at r_0. For a uniform density, the beam-generated fields vary linearly with radius. The radial magnetic force, $-qv_zB_\theta$, has magnitude equal to β^2 times the electrical forces at all radii. We shall show in Section 5.5 that this condition is a consequence of relativistic transformations and holds for all beam geometries.

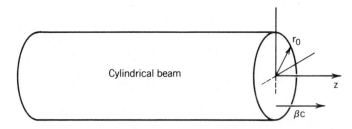

Figure 5.3   Cylindrical beam propagating in the z direction.

C. Summary of Equations

I. Sheet Beam

1. **Varying Density in x Direction, $n(x)$**
Electric field:

$$E_x(x) = q \int_0^x dx' n(x')/\varepsilon_0. \tag{5.3}$$

Magnetic field, paraxial beam:

$$B_y(x) = (qv_z\mu_0) \int_0^x dx' n(x'). \tag{5.6}$$

Relation between electric and magnetic field, paraxial beam:

$$B_y = (\beta/c)E_x \quad (\beta = v_z/c). \tag{5.21}$$

Relation between transverse forces, paraxial beam:

$$F_x(\text{magnetic}) = -\beta^2 F_x(\text{electric}). \tag{5.20}$$

2. **Uniform Density, $n(x) = n_0$ in range $-x_0 \leqslant x \leqslant x_0$**
Electric field:

$$E_x = (qn_0/\varepsilon_0)x. \tag{5.9}$$

Magnetic field, paraxial beam:

$$B_y = (\mu_0 qn_0 v_z)x = (qn_0 v_z/\varepsilon_0 c^2)x. \tag{5.10}$$

Electrostatic potential:

$$\phi(x) = \phi(0) - (qn_0/2\varepsilon_0)x^2. \tag{5.11}$$

Difference in ϕ between axis and x_0:

$$\Delta\phi = qn_0 x_0^2/2\varepsilon_0. \tag{5.12}$$

Vector potential, paraxial beam:

$$A_z(x) = -(\mu_0 qn_0 v_z/2)x^2. \tag{5.14}$$

Introduction to Beam-Generated Forces

3. Envelope Fields and Forces, Paraxial Beam

Total line current density:

$$J = 2qv_z \int_0^{x_0} dx' n(x').$$ (5.15)

Electric field:

$$E_{x0} = J/2\varepsilon_0 v_z = J/2\varepsilon_0 \beta c.$$ (5.16)

Magnetic field:

$$B_{y0} = \mu_0 J/2.$$ (5.17)

Electric force:

$$F_{x0}(\text{electric}) = eJ/2\varepsilon_0 \beta c.$$ (5.18)

Magnetic force:

$$F_{x0}(\text{magnetic}) = -e\mu_0 \beta c J/2 = -e\beta J/2\varepsilon_0 c.$$ (5.19)

II. Cylindrical Beam

1. Varying Density in Radial Direction, $n(r)$

Electric field:

$$E_r(r) = \frac{q}{2\pi\varepsilon_0 r} \int_0^r 2\pi r' dr' n(r').$$ (5.22)

Magnetic field, paraxial beam:

$$B_\theta(r) = \frac{qv_z\mu_0}{2\pi r} \int_0^r 2\pi r' dr' n(r').$$ (5.23)

Relation between electric and magnetic field, paraxial beam:

$$B_\theta(r) = \beta E_r(r)/c.$$ (5.24)

Relation between transverse forces, paraxial beam:

$$F_r(\text{magnetic}) = -\beta^2 F_r(\text{electric}).$$ (5.25)

2. Uniform Density $n(r) = n_0$ in range $0 \leqslant r \leqslant r_0$

Electric field:

$$E_r(r) = (qn_0/2\varepsilon_0)r.$$ (5.26)

Magnetic field, paraxial beam:

$$B_\theta(r) = (qn_0v_z\mu_0/2)r. \tag{5.27}$$

Electrostatic potential:

$$\phi(r) = \phi(0) - (qn_0/4\varepsilon_0)r^2. \tag{5.28}$$

Difference in ϕ between axis and r_0:

$$\Delta\phi = (qn_0/4\varepsilon_0)r_0^2. \tag{5.29}$$

Vector potential, paraxial beam:

$$A_z = -(\mu_0 qn_0 v_z/4)r^2. \tag{5.30}$$

3. Envelope Fields and Forces, Paraxial Beam

Total beam current:

$$I = qv_z \int_0^{r_0} 2\pi r' dr' n(r'). \tag{5.31}$$

Electric field:

$$E_r(r_0) = I/2\pi\varepsilon_0\beta c r_0. \tag{5.32}$$

Magnetic field

$$B_\theta(r_0) = \mu_0 I/2\pi r_0. \tag{5.33}$$

Electric force:

$$F_{r0}(\text{electric}) = eI/2\pi\varepsilon_0\beta c r_0. \tag{5.34}$$

Magnetic force:

$$F_{r0}(\text{magnetic}) = -eI\beta/2\pi\varepsilon_0 c r_0. \tag{5.35}$$

5.2. ONE-DIMENSIONAL CHILD LAW FOR NONRELATIVISTIC PARTICLES

The Child law states the maximum current density that can be carried by charged particle flow across a one-dimensional extraction gap. The limit arises from the longitudinal electric fields of the beam space charge. We shall study the Child law in detail—it is one of the most important results in collective-beam physics. The derivation is significant for two reasons:

1. The Child limit gives the maximum current density from a charged particle extractor. Although the derivation applies to a specialized geometry, the results provide good estimates for a variety of high-power beam devices.

2. The derivation illustrates the calculation of a charged particle equilibrium with self-consistent space-charge fields. In latter chapters, we shall apply similar methods to more complex problems.

The *extraction gap* is the first stage of an accelerator—low-energy charged particles from a source are accelerated to moderate energy ($\sim 10\,\text{keV}$ to $\sim 1\,\text{MeV}$) and formed into a beam. The Child law calculation applies to the one-dimensional gap of Fig. 5.4. A voltage $-V_0$ is applied across a vacuum gap of width d. Charged particles with low kinetic energy enter at the grounded boundary. The particles have rest mass m_0 and carry positive charge $+Ze$—it is easy to modify the treatment for electrons. Particles leave the right-hand boundary with kinetic energy eV_0—we shall assume that the exit electrode is an ideal mesh that defines an equipotential surface while transmitting all particles. The following assumptions simplify the calculation of self-consistent flow:

1. Particle motion is nonrelativistic ($eV_0 \ll m_0 c^2$). We shall discuss relativistic effects in Section 6.5.

2. The source on the left-hand boundary supplies an unlimited flux of particles. Restrictions of flow result entirely from space-charge effects.

3. The transverse dimension of the gap is large compared with d. The only significant components of particle velocity and electric field are in the z direction. We shall develop methods for finite width injectors in Section 7.1.

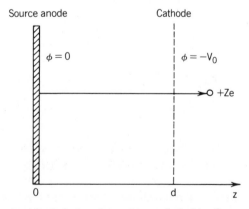

Figure 5.4 Geometry for the calculation of space-charge-limited ion flow across an infinite planar acceleration gap.

4. The transverse magnetic force generated by current across the gap is small compared with the axial electric force. As a result, particles follow straight line trajectories across the gap. This assumption is valid for ion beams, but it is usually violated in high-current relativistic electron beam injectors.

5. Particles flow continuously—the electric fields and space-charge density at all positions in the gap are constant.

The steady-state condition means that the space-charge density $\rho(z)$ is constant in time:

$$\partial\rho/\partial t = 0. \tag{5.36}$$

Combining Eq. (5.36) with the law of conservation of particles [Eq. (2.108)] gives

$$\frac{\partial[Zen(z)v_z(z)]}{\partial z} = 0. \tag{5.37}$$

Equation (5.37) implies that the current density, equal to $Zen(z)v_z(z)$, is the same at all positions in the gap. We will designate the constant value as j_0. The density of particles as a function of position is

$$n(z) = \frac{j_0/Ze}{v_z(z)}. \tag{5.38}$$

If we take the electrostatic potential equal to zero at the particle source, the velocity and potential are related by

$$m_0 v_z^2/2 = - Ze\phi. \tag{5.39}$$

We express the density as a function of ϕ by inserting Eq. (5.39) into Eq. (5.38). We can then substitute the result in the one-dimensional Poisson equation:

$$\frac{d^2\phi}{dz^2} = \frac{-j_0}{\varepsilon_0(-2Ze\phi/m_0)^{1/2}}. \tag{5.40}$$

We can solve Eq. (5.40) with appropriate boundary conditions to find the self-consistent variation of $\phi(z)$. We then substitute into Eq. (5.38) to find the variation of particle density. To review, the steps in the self-consistent equilibrium calculation for this special case are as follows. First, we use conservation of energy and particle flux to express the beam density as a function of the field quantity ϕ. Second, we substitute the expression into the field equation to find ϕ.

We can write Eq. (5.40) more succinctly if we introduce the dimensionless variables $\zeta = z/d$, $\Phi = -\phi/V_0$. The equation becomes

$$d^2\Phi/d^2\zeta = \alpha/\Phi^{1/2}, \tag{5.41}$$

where

$$\alpha = \frac{j_0}{\varepsilon_0(2ZeV_0/m_0)^{1/2}}. \tag{5.42}$$

Two of the boundary conditions are $\Phi(0) = 0$ and $\Phi(1) = 1$. We must have a third boundary condition to define a unique solution. For space-charge-limited flow, we take the condition that the electric field at the source equals zero, or

$$\frac{d\Phi(0)}{d\zeta} = 0. \tag{5.43}$$

We can understand the reason for this condition by inspecting Fig. 5.5. The figure shows a set of possible solutions of Eq. (5.41). If the available current density from the source is low, then a solution like that of curve A results. The potential variation is almost the same as the vacuum solution with no contribution from space charge—the electric field is nearly uniform across the gap. With higher source flux there is more positive space charge in the gap. Curve B shows the potential with space-charge contributions. The average potential in the gap is higher and the electric field at the source is lower. Suppose we continue raising the flux until the electric field at the source approaches zero (curve C). Then, particles with low kinetic energy are just able to leave the source. Higher gap flux is impossible—it would result in a negative electric field at the source that repels entering particles. The flux of particles across the extraction gap saturates at the level where $\Phi'(0) = 0$, independent of further increases in the available flux from

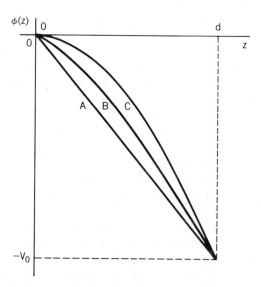

Figure 5.5 Axial variation of electrostatic potential in an infinite planar acceleration gap. (A) Low beam current. (B) Moderate beam current. (C) Space-charged-limited beam current.

the source. At this point, we say that the extraction gap passes from *source-limited flow* to *space-charge-limited flow*. Section 6.2 discusses the physical basis of the transition in detail.

We can solve Eq. (5.41) by multiplying both sides by $2\Phi'$, where $\Phi' = d\Phi/d\zeta$:

$$2\Phi'\Phi'' = 2\alpha\Phi'/\Phi^{1/2}. \tag{5.44}$$

The left-hand side is an exact differential of $(\Phi')^2$. Integrating both sides of Eq. (5.44) from the source to position ζ and applying the boundary conditions gives

$$[\Phi'(\zeta)]^2 = 4\alpha[\Phi(\zeta)]^{1/2}. \tag{5.45}$$

We can rewrite Eq. (5.45) as

$$d\Phi/\Phi^{1/4} = \sqrt{4\alpha}\, d\zeta. \tag{5.46}$$

Both sides of Eq. (5.46) can be integrated—the result is

$$\Phi^{3/4} = 3\sqrt{4\alpha}\,\zeta/4. \tag{5.47}$$

The boundary condition $\Phi(1) = 1$ implies that $\alpha = \frac{4}{9}$. Substituting the definition of α from Eq. (5.42) and solving for the current density gives the well-known Child law for space-charge-limited extraction:

$$j_0 = \frac{4\varepsilon_0}{9}\left[\frac{2Ze}{m_0}\right]^{1/2}\frac{V_0^{3/2}}{d^2}. \tag{5.48}$$

For space-charge-limited flow, the electrostatic potential varies with position as

$$\phi(z) = -V_0(z/d)^{4/3}. \tag{5.49}$$

Equation (5.49) determines curve C of Fig. 5.5.

Equation (5.48) states that for a given gap voltage and geometry, the current density is proportional to the square root of the charge-to-mass ratio of the accelerated particles $(Ze/m_0)^{1/2}$. The allowed current density of electrons is roughly 43 times higher than that of protons. In electron and proton extractors with the same d and $|V_0|$, the densities are the same but the electrons move faster by a factor $(m_p/m_e)^{1/2}$. For reference, the Child limit for nonrelativistic electrons is

$$j_0 = (2.33 \times 10^{-6})V_0^{3/2}/d^2 \quad \text{[electrons]}. \tag{5.50}$$

The quantity V_0 is given in volts. For d expressed in centimeters, the units of Eq. (5.50) are amperes per square centimeters. Similarly, if d is given in meters, j_0

has units of amperes per square meter. The Child limit for ions is

$$j_0 = (5.44 \times 10^{-8})(Z/A)^{1/2} V_0^{3/2}/d^2 \quad \text{[ions]}. \tag{5.51}$$

The quantity Z is the ionization state and A is the atomic number of the ions. Figure 5.6 gives values of $j_0 d^2$ in amperes as a function of V_0 for electrons and some ions. Electron values are plotted only up to 100 kV; relativistic corrections should be added for higher voltages (Section 6.5).

The electric field magnitude limits the maximum current density from an extractor—electrodes break down at fields exceeding ~ 10 MV/m. For space-

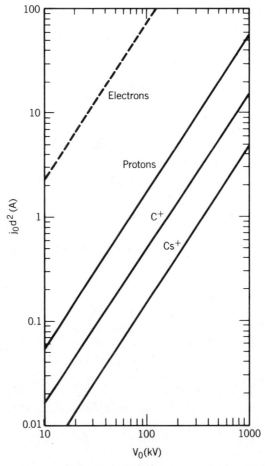

Figure 5.6 Normalized space-charge-limited current densities for several charged particles — j_0/d^2 versus V_0.

charge-limited flow, the electric field varies as

$$E_z = -\frac{d\phi}{dz} = \frac{4V_0 z^{1/3}}{d^{4/3}}.$$ (5.52)

The electric field is highest at the output boundary, $E_z(d) = E_{max} = 4V_0/3d$. We can write the Child law in terms of the maximum electric field:

$$j_0 = \frac{\varepsilon_0}{4}\left[\frac{2Ze}{m_0}\right]^{1/2}\frac{E_{max}^2}{V_0^{1/2}}.$$ (5.53)

Equation (5.53) implies that the maximum current density decreases with increasing applied voltage. For $E_{max} = 10\,MV/m$, the highest current density of a 50-kV proton extractor is roughly $1.37 \times 10^4\,A/m^2$ ($1.37\,A/cm^2$).

Beams extracted from a one-dimensional, space-charged-limited extractor cannot propagate an indefinite distance under vacuum. The space charge of the beams creates electric fields—depending on the geometry of the propagation region, the fields may be strong enough to reverse the direction of the beam. Figure 5.7a shows the space-charge potential of a one-dimensional electron beam propagating between a grounded mesh anode and a grounded collector. In the figure, the distance between the anode and collector is less than or equal to d, the width of the extraction gap. We shall find (Section 6.1) that the maximum space-charge potential between the anode and collector is less than V_0. Although electrons slow down, they still reach the collector. With larger spacing between the anode and collector, the peak potential in the propagation region increases.

We can use a simple construction to find the maximum allowed propagation length. To begin, suppose we locate a plate at potential $-V_0$ a distance d from the anode (Fig. 5.7b). In the region $d \leqslant z \leqslant 2d$, electron orbits are mirror images of those in the extractor. Similarly, the variation of space-charge density and potential are mirror images—all electrons reach the plate. Next, we replace the plate with a transparent mesh and locate a grounded collector at the position $z = 3d$ (Fig. 5.7c). With a perfectly transmitting mesh, the electron flux is the same in all three regions. The electron orbits, density, and electrostatic potential in the region $2d \leqslant z \leqslant 3d$ are identical to those in the extractor. Finally, we recognize that the mesh collects no charge—it can be removed without changing the potential. The solution of Fig. 5.7c is a limiting case. If the spacing between anode and collector exceeds $2d$, the electrostatic potential causes electron reflection (Fig. 5.7d). The reflection plane, where the potential reaches $-V_0$, is called a *virtual cathode*. We conclude that beams generated by a one-dimensional space-charge-limited extractor cannot propagate a distance greater than twice the extractor width.

To complete the discussion of space-charge flow, we shall consider the Child law derivation as an example of a self-consistent equilibrium. Section 2.4 showed that the distribution function for a collisionless beam is constant following a single-particle orbit. If we pick a distribution function that depends on constants

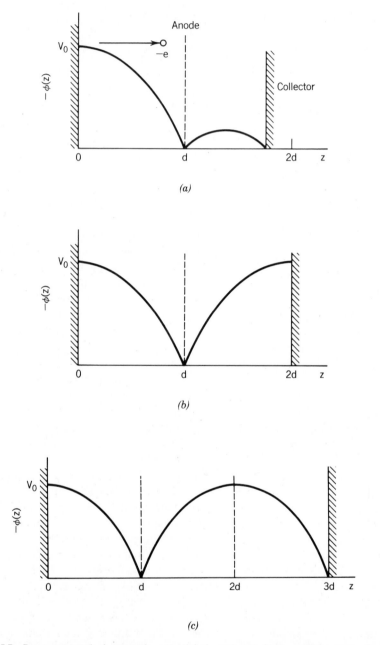

Figure 5.7 Beam-generated electrostatic potential downstream from a planar space-charged-limited acceleration gap with voltage V_0 and spacing d (polarities apply to electrons). (*a*) Grounded collector at a distance smaller than d from the anode mesh. (*b*) Biased collector ($-V_0$) at a distance d from the anode mesh. (*c*) Grounded collector at a distance $2d$ from the anode mesh. (*d*) Formation of a virtual cathode when a grounded collector is moved to a distance greater than $2d$ from the anode mesh.

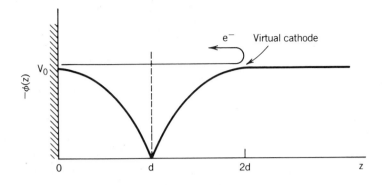

(d)

Figure 5.7 *(Continued)*

of single-particle motion, such as total energy, then that function must be a valid equilibrium solution to the Vlasov equation. For one-dimensional motion, the single constant is the total particle energy. In an extractor, the total energy is the sum of kinetic and potential energy contributions:

$$U = mv_z^2 - Ze\phi(z). \tag{5.54}$$

We have defined the electrostatic potential so that the total energy is zero at the source, $U(0) = 0$. We assume that all particles enter the gap with negligible kinetic energy; therefore, the particle distribution function has the form

$$f(z, v_z) \sim \delta(U) \sim \delta[v_z^2 - 2Ze\phi(z)/m_0], \tag{5.55}$$

where $\delta(U)$ represents the delta function.

For reference, the delta function has the following properties:

$$\int_{-\infty}^{+\infty} dX\, \delta(X) = 1; \tag{5.56}$$

$$\int_{-\infty}^{+\infty} dX\, g(X)\, \delta(X - X_0) = g(X_0); \tag{5.57}$$

$$\int_{-\infty}^{+\infty} dX\, g(X)\frac{d[\delta(X - X_0)]}{dX} = -\frac{dg(X_0)}{dX}. \tag{5.58}$$

We calculate the particle density by taking an integral of the distribution function over velocity space:

$$n(z) = \int_{-\infty}^{+\infty} f(x, v_z)\, dv_z. \tag{5.59}$$

We must be careful to express the integral in the proper form to apply Eq. (5.57). If we set $X = v_z^2$ and $dX = 2v_z dv_z$, then the Eq. (5.59) becomes

$$n(z) \sim \int_{-\infty}^{+\infty} \frac{\delta[X - 2Ze\phi(z)/m_0]\, dX}{X^{1/2}}. \tag{5.60}$$

Comparing Eq. (5.60) with Eq. (5.57) and taking $g(X) = 1/X^{1/2}$, we find that

$$n(x) \sim 1/(2Ze\phi/m_0)^{1/2}. \tag{5.61}$$

Equation (5.61) has the same form as the density function used in Eq. (5.38). The formal method leads to the same expressions for the self-consistent particle equilibrium.

5.3. LONGITUDINAL TRANSPORT LIMITS FOR MAGNETICALLY CONFINED ELECTRON BEAMS

We saw in Section 5.2 that space-charge effects can strongly influence the transport of a one-dimensional high-current beam injected into a vacuum region. The theoretical description of high-current beam propagation can become complex when we address the full three-dimensional problem. Figure 5.8*a* illustrates some of the interactive processes that can take place when a cylindrical beam enters a transport tube through an anode mesh. Some of the particles travel forward while others are reflected at a virtual cathode—all particles are subject to transverse space-charge deflections.

In this section, we limit attention to a simplified transport geometry so that we can develop analytic formulas. We shall study a cylindrical relativistic electron beam in a transport pipe with a strong axial magnetic field. The magnetic field allows electrons to move only in the axial direction—we say that electrons are tied to the field lines. As a result, particle motion is easy to describe. In this section, we calculate limits on propagating beam current—we shall look for conditions where the space-charge potential of the beam is comparable to the electron kinetic energy. We restrict attention to electrons since solenoidal magnetic fields are ineffective for ion beam containment. Section 10.2 discusses conditions for electron confinement in a magnetic field.

Figure 5.8*b* shows the geometry of the calculation. The steady-state beam enters a conducting pipe of infinite length through a foil or mesh. The pipe wall has radius r_w. The mesh and pipe are at ground potential. We assume that the electrons emerge from an electrostatic injector with applied voltage V_0. The relativistic electrons are monoenergetic with $\gamma = 1 + eV_0/m_e c^2$. In the strong magnetic field the electron velocity is directed entirely along the axis, $\beta_z \cong \beta$. We take the current density equal to a uniform value j_0 over the region $0 < r < r_0$. The net beam current is $I_0 = j_0(\pi r_0^2)$.

Upon entering the pipe, the beam passes through a transition region with

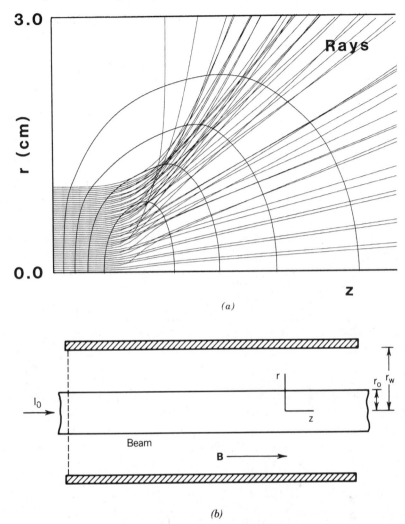

(a)

(b)

Figure 5.8 Injection of a cylindrical beam into a grounded metal pipe. (*a*) Numerical calculation using the EGUN code showing traces of rays and lines of constant electrostatic potential. Incident laminar beam properties: 0.01 m radius, 100 keV energy, 350A current, uniform current density. Equipotential lines at 20, 40, 60, and 80 kV. (*b*) Geometry for the calculation of the limiting current for a beam confined by a strong axial magnetic field.

strong axial components of electric field. We shall use a strategy to avoid detailed calculations of complex particle orbits in this region. We calculate a beam equilibrium in the downstream region far from the entrance mesh ($z \gg r_w$) and match this solution to the input beam distribution by applying conservation principles. In the downstream region, we expect that the cylindrical beam is axially uniform; therefore, axial variations of electric and magnetic field are small.

Introduction to Beam-Generated Forces

It is acceptable to approximate the fields with expressions for an infinite length beam.

We can specify some properties of the downstream beam. For instance, the radius is everywhere equal to r_0 because electrons are tied to magnetic field lines. Also, the axial current of the steady-state beam is constant at all positions. As a result, the current density is uniform throughout the length of the pipe,

$$j_z(r, z) = j_0 \quad (0 \leqslant r \leqslant r_0). \tag{5.62}$$

In the region far from the mesh, the space-charge potential is a function of radius only, $\phi(r)$. By conservation of total energy, the relativistic gamma factor is

$$\gamma(r) = 1 + eV_0/m_ec^2 + e\phi(r)/m_ec^2. \tag{5.63}$$

We can apply Eq. (5.63) to calculate the axial velocity as a function of electrostatic potential ϕ. We then combine the expression with Eq. (5.38) to find an expression for the electron charge density as a function of ϕ, $\rho(\phi)$. Finally, we substitute the charge density in the cylindrical form of the Maxwell equations to determine the self-consistent space-charge potential.

We will add one more simplification to calculate the space-charge potential of the downstream beam—the space charge is approximate uniform:

$$\rho(r) = \rho_0. \tag{5.64}$$

Equation (5.64) is valid if all electrons in the beam have about the same axial velocity. This condition holds for relativistic electrons if

$$-e\phi \ll eV_0. \tag{5.65}$$

For uniform density, the Maxwell equation $\nabla \cdot \mathbf{E} = \rho(r)/\varepsilon_0$ implies that the electric field has the form:

$$
\begin{aligned}
E_r &= (\rho_0/2\varepsilon_0)r & (0 \leqslant r \leqslant r_0), \\
E_r &= (\rho_0/2\varepsilon_0)(r_0^2/r) & (r_0 \leqslant r \leqslant r_w).
\end{aligned}
\tag{5.66}
$$

To find the maximum change in electron kinetic energy, we want to find the electrostatic potential on the axis. The quantity is given by

$$e\phi(0) = \int_0^{r_w} E_r \, dr \tag{5.67}$$

or

$$e\phi(0) = \left[\frac{e\rho_0 \pi r_0^2}{2\pi\varepsilon_0}\right]\left[\int_0^{r_0} \frac{r\,dr}{r_0^2} + \int_{r_0}^{r_w} \frac{dr}{r}\right].$$

Identifying the total beam current as $I_0 = \rho_0 \pi r_0^2 \beta c$, Eq. (5.67) takes the form

$$e\phi(0) = \frac{-eI_0}{4\pi\varepsilon_0 \beta c}\left[1 + 2\ln\left(\frac{r_w}{r_o}\right)\right].\qquad(5.68)$$

In Eq. (5.68), the first term in brackets represents the change in potential across the beam, while the second term gives the potential difference between the edge of the beam and the wall. If we evaluate the constants in the multiplying factor, we find that

$$\phi(0) = \frac{30I_0}{\beta}\left[1 + 2\ln\left(\frac{r_w}{r_o}\right)\right].\qquad(5.69)$$

The potential has units of volts if the beam current is given in amperes. Figure 5.9 shows predictions of Eq. (5.69). A current of 10 kA is typical of high-power linear induction accelerators. Equation (5.69) shows that the corresponding space-charge potential is high, nearly 1 MV. For this reason, induction linac injectors operate at high voltage ($V_0 \geqslant 2\,\mathrm{MV}$).

The assumption of uniform space-charge density in the preceding calculation applies if electron deceleration by space-charge forces is small enough so that $\beta_z \cong 1$. This condition is violated in many intense relativistic beam experiments. We shall next extend our models to account for differences in β_z from the injection value. In a beam with high values of electrostatic potential far from the injection

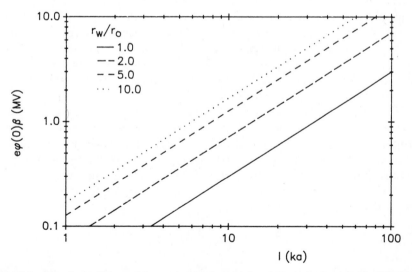

Figure 5.9 Magnitude of the peak beam-generated space-charge electrostatic potential. Cylindrical magnetically confined electron beam of radius r_0 with uniform current density. Normalized potential plotted versus beam current and the normalized wall radius, r_w/r_0.

mesh, the density and average velocity are functions of radial position. Conservation of charge implies that the quantities are related by

$$n_e(r) = j_0/e\beta_z(r). \tag{5.70}$$

Conservation of total energy gives the following relationship between $\phi(r)$ and $\beta_z(r)$:

$$\beta_z = \frac{(\gamma^2 - 1)^{1/2}}{\gamma} = \frac{[2e\phi/m_ec^2 + (e\phi/m_ec^2)^2]^{1/2}}{1 + e\phi/m_ec^2}. \tag{5.71}$$

Substituting from Eq. (5.70), the Poisson equation is

$$\frac{1}{r}\frac{d}{dr}\left[r\frac{d\phi}{dr}\right] = \frac{j_0(1 + e\phi/m_ec^2)}{\varepsilon_0[2e\phi/m_ec^2 + (e\phi/m_ec^2)^2]^{1/2}}. \tag{5.72}$$

Outside the beam, the right-hand side of Eq. (5.72) equals zero.

We can apply the following numerical procedure to find the space-charge-limited current. First, we choose a value of peak electrostatic potential in the range. We then solve Eq. (5.72) numerically. We start at the origin with the conditions $\phi(0) = \phi_0$, $d\phi(0)/dr = 0$ and match the potential and its radial derivative at the boundary, $r = r_o$. We carry out the calculation for different values of j_0 until we find a choice where ϕ approaches 0 at $r = r_w$. The procedure yields the current density and net current for the choice of ϕ_0. We repeat the calculation many times to estimate I_0 as a function of ϕ_0.

To find the space-charge limit, we must find the value of peak potential in the range $0 \leqslant \phi_0 \leqslant V_0$ that gives the highest value of I_0 for a given γ. When $\phi_0 = V_0$, electrons at the center of the beam lose all their kinetic energy. Here, many electrons enter the pipe, but their average axial velocity is low—the net current is small. At the other extreme ($\phi_0 = 0$), few electrons enter, but they move at high velocity. Again, the current is low. We expect that the maximum value of current occurs at an intermediate value of ϕ_0.

We can make a simple estimate of the current limit if the beam radius is small compared with the wall radius. Then, Eq. (5.68) shows that most of the potential drop occurs between the wall and the beam boundary. Electrons in the narrow beam all have about the same kinetic energy—we can take γ and β_z as average values over the beam cross section. If the beam carries a net current I_0, Eq. (5.68) implies that the average space-charge potential in the cylindrical beam is roughly

$$e\Delta\phi \cong \frac{eI_0}{4\pi\varepsilon_0\beta c}\left[1 + 2\ln\left(\frac{r_w}{r_o}\right)\right]. \tag{5.73}$$

Conservation of energy implies that

$$e\Delta\phi \cong m_ec^2(\gamma_0 - \gamma), \tag{5.74}$$

where γ_0 gives the injection energy of the electrons. Combining Eqs. (5.73) and (5.74) and expressing β in terms of γ gives an expression for the current as a function of the average γ factor in the transported beam:

$$\left[\frac{eI_0}{4\pi\varepsilon_0 m_e c^3}\right]\left[1+2\ln\left(\frac{r_w}{r_o}\right)\right]=\frac{(\gamma_0-\gamma)(\gamma-1)^{1/2}}{\gamma}. \tag{5.75}$$

We can find the maximum value of the function on the right-hand side of Eq. (5.75) by setting the derivative with respect to γ equal to zero. The value of γ for the maximum beam current is

$$\gamma=(\gamma_0)^{1/3}. \tag{5.76}$$

Substituting Eq. (5.76) into Eq. (5.75) gives the following formula for the current limit of a narrow beam:

$$I_{\max}=\frac{4\pi\varepsilon_0 m_e c^3/e}{1+2\ln(r_w/r_o)}(\gamma_0^{2/3}-1)^{3/2}. \tag{5.77}$$

The factor $4\pi\varepsilon_0 m_e c^3/e$ on the right-hand side of Eq. (5.77) equals 17.1 kA. As an example, consider the propagation of a 0.01-m-radius beam in a 0.03-m-radius pipe. The injection energy of 1.5 MeV corresponds to $\gamma_0 = 3.94$. The space-charge-limiting current is 9.2 kA. At this value, the space-charge potential of the beam is 1.2 MV—the beam propagates with an average kinetic energy of 0.3 MeV ($\gamma = 1.58$).

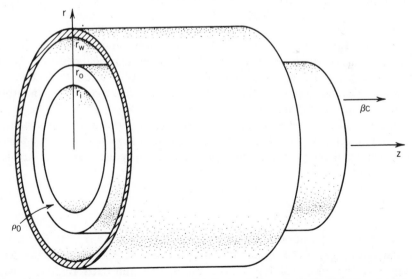

Figure 5.10   Geometry of a uniform-current-density annular beam in a cylindrical pipe of radius r_w.

Longitudinal space-charge effects can be reduced, in principle, by using beams with nonuniform current density. An annular beam can carry more current in equilibrium than a solid beam. The beam illustrated in Fig. 5.10 has uniform charge density ρ_0 distributed between inner radius r_i and outer radius r_o. The electric field inside and outside the beam is given by the following expressions:

$$E_r = \frac{\rho_0}{2\varepsilon_0}\left(r - \frac{r_i^2}{r}\right) \qquad (r_i < r < r_o), \qquad (5.78)$$

$$E_r = \frac{\rho_0}{2\varepsilon_0}\left(\frac{r_o^2 - r_i^2}{r}\right) \qquad (r_o < r < r_w). \qquad (5.79)$$

Following the method we used for the solid beam, we integrate Eqs. (5.78) and (5.79) to find the potential drop between r_i and r_w. The maximum space-charge potential is

$$e\phi(\text{max}) = \frac{\rho_0}{4\pi\varepsilon_0}\left[\pi(r_o^2 - r_i^2) - 2\pi r_i^2 \ln\left(\frac{r_o}{r_i}\right) + 2\pi(r_o^2 - r_i^2)\ln\left(\frac{r_w}{r_o}\right)\right]. \qquad (5.80)$$

Introducing the total current,

$$I_o = \pi(r_o^2 - r_i^2)\rho_0\beta c, \qquad (5.81)$$

Eq. (5.80) becomes

$$e\phi(\text{max}) = \left(\frac{eI_0}{4\pi\varepsilon_0\beta c}\right)\left[1 - \frac{2r_i^2 \ln(r_o/r_i)}{r_o^2 - r_i^2} + 2\ln\left(\frac{r_w}{r_o}\right)\right] \qquad (5.82)$$

Equation (5.82) is identical to Eq. (5.68) for a solid beam except for the second term in brackets on the right-hand side. This term, which depends on the geometry of the annulus, reduces the potential difference across the beam. The potential difference between the beam and wall, represented by the third bracketed term, is unchanged.

The ratio of the space-charge potential drop across an annular beam of relativistic electrons compared to the potential across a uniform-density cylindrical beam carrying the same current is given by the reduction factor:

$$F\left(\frac{r_i}{r_o}\right) = 1 - \frac{2r_i^2 \ln(r_o/r_i)}{r_o^2 - r_i^2}. \qquad (5.83)$$

The function $F(r_i/r_o)$ approaches unity for a solid beam, $r_i/r_o \to 0$. For a thin annular beam ($r_i/r_o \to 1$), the potential drop across the beam approaches zero.

Ignoring the effects of beam instabilities, one approach to high-current ($> 10\,\text{kA}$) electron beam transport is to generate a thin annular beam close to the conducting wall of a transport pipe in a strong magnetic field. The idea is to minimize the potential drops across both the beam and the space between the beam and the wall. For a thin annulus, we can apply arguments similar to those that lead to Eq. (5.77) to derive the following limiting current:

$$I_{max} \cong \frac{4\pi\varepsilon_0 m_e c^3 / e}{2 \ln(r_w / r_o)} (\gamma_0^{2/3} - 1)^{3/2}. \tag{5.84}$$

As an example of Eq. (5.84), suppose we have an annular beam with outer radius $r_o = 0.09$ m in a tube with $r_w = 0.10$ m. At injection into the transport system, the particles have 1.5-MeV kinetic energy. The current limit predicted by Eq. (5.84) is 148 kA compared with the figure of 9.2 kA for the narrow solid beam. Despite the theoretical advantage of annular beams, in practice they are susceptible to instabilities that lead to particle losses. The feasibility of high-current annular beam transport is still an open question.

5.4. SPACE-CHARGE EXPANSION OF A DRIFTING BEAM

In this section, we shift attention from longitudinal field limitations to the effects of transverse electric and magnetic fields on beam transport. Again, we shall introduce the topic with a simple example—in this case, a high-current cylindrical beam that expands under the influence of its own space-charge force in a drift region. We take the beam emittance equal to zero. Section 9.2 describes the complete paraxial ray equation with emittance, space-charge, and external focusing forces included.

We describe a cylindrical, paraxial beam. The paraxial condition means that particle orbits make small angles with the axis—variations in beam properties occur over long distances compared with the beam radius. As a result, we can represent transverse field components using expressions for an infinite length beam. In addition, paraxial orbits have $v_r \ll v_z$. As a result, the beam β and γ are almost constant in r and z if there in no acceleration. Finally, we assume that the zero emittance beam has uniform current density with radius; therefore, the net transverse force is linear. Since orbits are laminar, we need treat only particle motions on the beam envelope. We denote the envelope radius as R.

The beam-generated electric and magnetic forces at the envelope are listed in Eqs. (5.34) and (5.35). The motion of envelope particles in the combined fields follows the equation

$$\frac{d(\gamma m_0 \, dR/dt)}{dt} = \frac{eI_0}{2\pi\varepsilon_0 \beta c \gamma^2} \frac{1}{R}. \tag{5.85}$$

We apply the chain rule of derivatives to convert Eq. (5.85) to a trace equation,

$$\frac{d}{dt} \rightarrow (\beta c) \left(\frac{d}{dz} \right). \tag{5.86}$$

The constant-energy condition means that we can extract factors of β and γ from the derivatives. The result is

$$\frac{d^2R}{dz^2} = \frac{eI_0}{2\pi\varepsilon_0 m_0(\beta\gamma c)^3}\frac{1}{R}.$$

(5.87)

To simplify Eq. (5.87), we define the following dimensionless parameter:

$$K = \frac{eI_0}{2\pi\varepsilon_0 m_0(\beta\gamma c)^3}.$$

(5.88)

$$I_A \equiv \frac{4\pi\varepsilon_0 \, m_e c^3 \beta\gamma}{e} \quad (12.127) \quad \Rightarrow \quad k = \frac{2I_0}{\beta^2\gamma^2 I_A}$$

TABLE 5.1   Values of the Function $F(\chi)$

χ	$F(\chi)$	$\sqrt{\ln(\chi)}$	$F(\chi)/\chi$
1.10000	0.63700	0.30872	0.57910
1.20000	0.90822	0.42699	0.75685
1.30000	1.12090	0.51222	0.86223
1.40000	1.30387	0.58006	0.93133
1.50000	1.46814	0.63676	0.97876
1.60000	1.61933	0.68557	1.01208
1.70000	1.76073	0.72844	1.03572
1.80000	1.89447	0.76667	1.05248
1.90000	2.02200	0.80116	1.06421
2.00000	2.14441	0.83255	1.07220
2.10000	2.26246	0.86136	1.07736
2.20000	2.37678	0.88795	1.08035
2.30000	2.48784	0.91264	1.08167
2.40000	2.59604	0.93566	1.08168
2.50000	2.70169	0.95723	1.08067
2.60000	2.80505	0.97750	1.07887
2.70000	2.90636	0.99662	1.07643
2.80000	3.00579	1.01470	1.07350
2.90000	3.10351	1.03185	1.07018
3.00000	3.19966	1.04815	1.06655
3.10000	3.29436	1.06367	1.06270
3.20000	3.38772	1.07849	1.05866
3.30000	3.47984	1.09267	1.05450
3.40000	3.57079	1.10624	1.05023
3.50000	3.66065	1.11927	1.04590
3.60000	3.74950	1.13178	1.04153
3.70000	3.83738	1.14382	1.03713
3.80000	3.92437	1.15542	1.03273
3.90000	4.01049	1.16661	1.02833
4.00000	4.09582	1.17741	1.02395
4.10000	4.18037	1.18785	1.01960
4.20000	4.26420	1.19795	1.01529
4.30000	4.34734	1.20773	1.01101
4.40000	4.42981	1.21721	1.00678
4.50000	4.51166	1.22641	1.00259
4.60000	4.59290	1.23534	0.99846
4.70000	4.67356	1.24401	0.99438
4.80000	4.75368	1.25244	0.99035
4.90000	4.83326	1.26065	0.98638
5.00000	4.91233	1.26864	0.98247

TABLE 5.1 *(Continued)*

x	$F(x)$	$\sqrt{\ln(x)}$	$F(x)/x$
5.00000	4.91233	1.26864	0.98247
5.50000	5.30070	1.30566	0.96376
6.00000	5.67881	1.33857	0.94647
6.50000	6.04821	1.36814	0.93049
7.00000	6.41008	1.39496	0.91573
7.50000	6.76536	1.41947	0.90205
8.00000	7.11479	1.44203	0.88935
8.50000	7.45901	1.46290	0.87753
9.00000	7.79853	1.48230	0.86650
9.50000	8.13377	1.50043	0.85619
10.00000	8.46512	1.51743	0.84651
10.50000	8.79288	1.53342	0.83742
11.00000	9.11734	1.54851	0.82885
11.50000	9.43874	1.56280	0.82076
12.00000	9.75728	1.57636	0.81311
12.50000	10.07317	1.58925	0.80585
13.00000	10.38656	1.60155	0.79897
13.50000	10.69761	1.61329	0.79242
14.00000	11.00646	1.62452	0.78618
14.50000	11.31322	1.63528	0.78022
15.00000	11.61801	1.64562	0.77453
15.50000	11.92093	1.65555	0.76909
16.00000	12.22207	1.66511	0.76388
16.50000	12.52151	1.67432	0.75888
17.00000	12.81935	1.68322	0.75408
17.50000	13.11564	1.69180	0.74947
18.00000	13.41045	1.70011	0.74503
18.50000	13.70386	1.70815	0.74075
19.00000	13.99590	1.71594	0.73663
19.50000	14.28665	1.72349	0.73265
20.00000	14.57614	1.73082	0.72881
20.50000	14.86442	1.73794	0.72509
21.00000	15.15155	1.74486	0.72150
21.50000	15.43755	1.75159	0.71803
22.00000	15.72247	1.75814	0.71466
22.50000	16.00634	1.76452	0.71139
23.00000	16.28921	1.77073	0.70823
23.50000	16.57109	1.77679	0.70515
24.00000	16.85203	1.78271	0.70217
24.50000	17.13205	1.78848	0.69927
25.00000	17.41117	1.79412	0.69645
25.50000	17.68943	1.79963	0.69370
26.00000	17.96685	1.80502	0.69103
26.50000	18.24345	1.81029	0.68843
27.00000	18.51925	1.81544	0.68590
27.50000	18.79428	1.82049	0.68343
28.00000	19.06856	1.82543	0.68102
28.50000	19.34211	1.83027	0.67867
29.00000	19.61493	1.83502	0.67638
29.50000	19.88707	1.83967	0.67414
30.00000	20.15852	1.84423	0.67195

The dimensionless quantity K is called the *generalized perveance*. The term *perveance* refers to the magnitude of space-charge effects in a beam. Section 7.1 discusses the standard perveance, a dimensional quantity. Equation (5.87) reduces to the form

$$R'' = K/R. \tag{5.89}$$

The prime symbol denotes a derivative with respect to z. The beam-generated forces cause beam expansion—a converging beam reaches a minimum value of envelope radius and then expands. We shall designate the envelope radius at the beam waist as R_m—we choose the axial position of the waist as $z = 0$. The first integral of Eq. (5.87) is

$$R'(z) = (2K)^{1/2}\{\ln[R(z)/R_m]\}^{1/2} = (2K)^{1/2}[\ln(\chi)]^{1/2}. \tag{5.90}$$

The quantity χ is the ratio of the beam envelope radius to the radius at the waist:

$$\chi = R(z)/R_m. \tag{5.91}$$

The variation of envelope radius with distance from the waist is given by

$$z = (R_m/\sqrt{2K})\, F(\chi), \tag{5.92}$$

where

$$F(\chi) = \int_1^\chi \frac{dy}{[\ln(y)]^{1/2}}.$$

Another useful form of the solution is

$$R(z) = (2K)^{1/2}\, z\, [\chi/F(\chi)]. \tag{5.93}$$

Table 5.1 lists the function $F(\chi)$.

We shall work through some examples to understand the significance of the results. To begin, we calculate the expansion of an unneutralized ion beam. Suppose we inject a 200-mA, 300-keV C^+ beam into a vacuum region. The generalized perveance is $K = 2.8 \times 10^{-3}$. The envelope angle is zero at injection—the injection point is a beam waist. The initial radius of the beam is $R_m = 0.01$ m. We want to find the beam radius at a position 0.3 m downstream. Inserting R_m, z, and K in Eq. (5.92), we find $F(\chi) = 2.24$. Table 5.1 shows that this value corresponds to $\chi = 2.1$—the final radius is 0.021 m. At $z = 0.3$ m, Eq. (5.90) predicts that the expansion angle of the envelope is 65 mrad (3.7°).

As a second example, we shall find the maximum distance that a relativistic electron beam can propagate across a vacuum region (Fig. 5.11). Suppose we know the beam energy, current, and entrance radius R_0. We locate a lens at the entrance to the system to adjust the initial convergence angle R_0'. We want to find

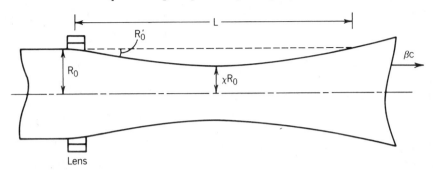

Figure 5.11 Propagation of a high-current relativistic electron beam through a drift region. Incident cylindrical beam has an envelope radius R_0. The converging beam returns to its initial radius after traveling a distance L.

the distance L the beam travels before it expands to its initial radius. Furthermore, we want to find the choice of R'_0 that gives the maximum value of L. From Eq. (5.93), the transit distance is

$$L = (2R_0)[F(\chi)/\chi]/(2K)^{1/2}. \qquad (5.94)$$

Inspection of Table 5.1 shows that the quantity $F(\chi)/\chi$ attains a maximum value of 1.085 at $\chi = 2.35$. The maximum propagation distance is $L_{\max} = 1.53R_0/\sqrt{K}$ if we adjust the entrance envelope angle to $R'_0 = -1.31\sqrt{K}$. As an example, imagine a 100-A, 500-keV electron beam ($\gamma = 1.98$, $\beta = 0.863$) with an initial radius of 0.02 m. The generalized perveance is $K = 2.3 \times 10^{-3}$. The propagation distance is $L_{\max} = 0.63$ m for an injection angle of -63 mrad ($-3.6°$).

As a final example, we calculate the maximum beam current that can propagate to a specified focal spot a given distance from a lens. Here, we know the initial beam radius (R_0), the target radius (R_m), the propagation distance (Z), and the beam energy. We can substitute these quantities into Eq. (5.93) to find the generalized perveance and hence the allowed current of the beam. To illustrate the calculation, we choose parameters relevant to the application of high-power heavy ion beams to inertial fusion. The goal is to transport 10-GeV U^+ beams ($\gamma = 1.048$, $\beta = 0.29$) across a vacuum reactor chamber. The chamber radius may be 10 m, while the radius of the fusion target is only 2.5×10^{-3} m. The total current required for ignition is roughly 50 kA. The current must be divided between a number of beams to stay within the limits set by space-charge effects. If we pick $R_0 = 0.1$ m, the convergence ratio is $\chi = 20$, and the convergence function is $F(\chi) = 14.8$. We find the generalized perveance must be less than 2.7×10^{-5}. Substituting values for the U^+ beam, the current of each beam is less than or equal to 2.8 kA—the total number of beams is about 20.

To conclude, we can use the generalized perveance to compare space-charge forces to the emittance force in a nonlaminar beam. The ratio of forces on the beam envelope is

$$\frac{F(\text{emittance})}{F(\text{self-field})} = \frac{\varepsilon^2}{KR^2}. \tag{5.95}$$

The beam-generated forces dominate beam expansion if

$$K > \varepsilon^2/R^2 \cong \Delta r'^2. \tag{5.96}$$

The quantity $\Delta r'$ is the spread in particle inclination angles. As an example, a 200-mA beam of 300-keV C^+ ions has a generalized perveance of $K = 2.8 \times 10^{-3}$. Beam forces determine the expansion if the angular spread is less than 53 mrad (3.0°).

5.5. TRANSVERSE FORCES IN RELATIVISTIC BEAMS

Section 5.1 showed that there is a close balance between the transverse electric and magnetic forces of a relativistic paraxial beam. The result did not depend on the geometry of the beam or the distribution of charge and current densities. This section will show that the balance between electric and magnetic forces proceeds directly from the properties of relativistic field transformations. To illustrate the application of coordinate transformations in beam theory, we shall study a specific example—propagation of a self-contained ion beam in free space. Under certain conditions, collinear ion and electron flows can achieve radial force equilibrium. As a preface to the example, we shall review basic ideas of beam neutralization. Neutralization is discussed in detail in Chapters 11 and 12.

To begin, we shall list the transformation equations for electromagnetic field quantities. The transformation is between an initial frame of reference and another frame moving at a relative velocity **v**. To simplify notation, we take two coordinate systems with aligned axes. The relative translation velocity is in the z direction:

$$\mathbf{v} = \beta c \mathbf{z}. \tag{5.97}$$

We associate a relativistic gamma factor with the transformation

$$\gamma = 1/(1 - \beta^2)^{1/2}. \tag{5.98}$$

Electric and magnetic fields measured in the two frames of reference have the following relationships[2]:

$$E'_z = E_z, \tag{5.99}$$

---

[2]Adapted from Jackson, *Classical Electrodynamics*, 2nd ed., Wiley, New York, 1975, p. 552.

$$E'_x = \gamma(E_x - \beta c B_y), \tag{5.100}$$

$$E'_y = \gamma(E_y + \beta c B_x), \tag{5.101}$$

$$B'_z = B_z, \tag{5.102}$$

$$B'_x = \gamma(B_x + \beta E_y/c), \tag{5.103}$$

$$B'_y = \gamma(B_y - \beta E_x/c) \tag{5.104}$$

The primed quantities in Eqs. (5.99) through (5.104) are measured in the frame moving at relative velocity $\beta c \mathbf{z}$. In cylindrical coordinates, the transformations are

$$E'_z = E_z, \tag{5.105}$$

$$E'_r = \gamma(E_r - \beta c B_\theta), \tag{5.106}$$

$$E'_\theta = \gamma(E_\theta + \beta c B_r), \tag{5.107}$$

$$B'_z = B_z, \tag{5.108}$$

$$B'_r = \gamma(B_r + \beta E_\theta/c), \tag{5.109}$$

$$B'_\theta = \gamma(B_\theta - \beta E_r/c). \tag{5.110}$$

The example of a long cylindrical beam of paraxial electrons illustrates the implications of Eqs. (5.105) through (5.110). We shall view an axial section of the beam in two specific frames of reference that we shall call the *accelerator frame* and the *beam rest frame*. In the accelerator frame, the electrons move at velocity $\beta c \mathbf{z}$. The rest frame moves at velocity $\beta c \mathbf{z}$—in the rest frame, electrons are stationary. For this choice of reference frames, the transformation γ factor equals the kinetic energy factor of the beam electrons in the accelerator frame.

In the accelerator frame, the beam section has length Δz. The electrons have axial velocity βc and kinetic energy $(\gamma - 1)m_e c^2$. The beam has radius r_0 and carries current I_0. For simplicity, we take a uniform density n_0—it is easy to generalize the results for any radial density variation. The current is related to the density by

$$I_0 = \pi r_0^2 e n_0 \beta c. \tag{5.111}$$

From Section 5.1, we know that the beam has a radial electric field E_r and toroidal magnetic field B_θ. The fields are related by

$$B_\theta = \beta E_r/c. \tag{5.112}$$

The net radial force on the electrons of a relativistic beam is small because magnetic focusing cancels electric repulsion. The net radial force at the envelope is

$$F_r(r_0) = (I_0/2\pi\varepsilon_0\beta c r_0)(1 - \beta^2) = (I_0/2\pi\varepsilon_0\beta c r_0)/\gamma^2. \qquad (5.113)$$

Equation (5.113) shows that magnetic focusing reduces the net radial force by a factor $1/\gamma^2$.

Next, consider the fields of the beam section viewed in the rest frame. The radius of the beam still equals r_0. Because of the Lorentz contraction, the measured axial length of the beam section in the rest frame, $\Delta z'$, is larger than Δz. The section lengths are related by

$$\Delta z' = \gamma\Delta z. \qquad (5.114)$$

As a result, the beam densities in the two frames are

$$n_0' = n_0/\gamma. \qquad (5.115)$$

There is no beam-generated magnetic field in the rest frame since the electrons are stationary:

$$B_\theta' = 0. \qquad (5.116)$$

We can calculate the radial electric field in terms of the density n_0' [Eq. (5.26)]. The electric fields measured in the two frames of reference are related by

$$E_r' = E_r/\gamma. \qquad (5.117)$$

We can show that Eqs. (5.116) and (5.117) are consistent with the general rules for field transformations. Using Eq. (5.112) in Eqs. (5.106) and (5.110) gives

$$E_r' = \gamma(E_r - \beta c B_\theta) = \gamma E_r(1 - \beta^2) = E_r/\gamma. \qquad (5.118)$$

$$B_\theta' = \gamma(B_\theta - \beta E_r/c) \equiv 0. \qquad (5.119)$$

We can derive the conclusion of Eq. (5.113) as a direct implication of the Lorentz transformations [CPA, Section 2.8]. In the rest frame, the only force is electric, $F' = F_e'$. Because of the difference in beam density, the rest frame electric force is a factor $1/\gamma$ smaller than the electric force in the accelerator frame F_e. We can apply time dilation to find the total transverse force F in the accelerator frame in terms of F'. Suppose a force accelerates a beam particle in the transverse direction over a small time increment $\Delta t'$. In the rest frame, the radial particle displacement scales with the interval according to

$$\Delta r' = (F'/m_0)\Delta t'^2/2. \qquad (5.120)$$

Now, imagine the same event observed in the accelerator frame. The equation for radial displacement has a similar form, but the measured interval and particle mass are different.

$$\Delta r = (F/\gamma m_0)\Delta t^2/2. \tag{5.121}$$

The displacements observed in both frames are the same, $\Delta r = \Delta r'$—the intervals are related through the time dilation law, $\Delta t = \gamma \Delta t'$. Therefore, as in Eq. (5.113), the total transverse forces are related by

$$F = F'/\gamma = F_e/\gamma^2. \tag{5.122}$$

Coordinate transformations are useful in discussions of neutralization of high-energy beams. The term *neutralization* refers to mixing positively charged particles with negative particles to reduce or to eliminate beam-generated fields. Beams may be either *space-charge neutralized* or *current neutralized*. In the first process, mixing particles with positive and negative charges lowers the beam electric field. In the second process, the opposing flow of positive and negative charged particles reduces the current in the beam, lowering the magnetic field. Both types of neutralization can take place simultaneously—the division is not definite because the relative importance of electric and magnetic fields depends on the frame of reference.

We can neutralize ion beams with electrons and electron beams with ions. We define the type of beam in terms of the particle that has the highest kinetic energy. First, consider the neutralization of ion beams. If an ion beam has density $n_i(\mathbf{x})$, complete space-charge neutralization takes place if we add low energy electrons with density $n_e(\mathbf{x}) = n_i(\mathbf{x})$. Here, there is no charge density in the beam—hence, there is no beam-generated electric field. If the neutralizing electrons are stationary in the accelerator frame, the beam carries a net current from the motion of the ions—there is a beam-generated magnetic field. On the other hand, if the electrons move at a velocity $v_e = n_i v_i/n_e$, there is no net current. If $n_e = n_i$ and $v_e = v_i$, the ion beam is both space-charge and current neutralized. This often occurs in practice when ions propagate through field-free regions. Because of their small mass, low-energy electrons can follow the ions. If the ions are nonrelativistic, the electron kinetic energy for a matched drift velocity is

$$T_e = m_0 v_e^2/2 = m_0 v_i^2/2 = T_i(m_0/m_i). \tag{5.123}$$

The quantity T_i is the ion kinetic energy. As an example, a 1-MeV C^+ beam can be charge and current neutralized by 45-eV electrons.

We can reduce the electric fields of electron beams with a background of low-energy ions. Again, complete space-charge neutralization results when $n_i(\mathbf{x}) = n_e(\mathbf{x})$. For relativistic electrons, partial space-charge neutralization is often sufficient for beam confinement. Accordingly, we shall define a partial neutralization factor f_e for electron beams:

$$f_e = \langle n_i(\mathbf{x}) \rangle / \langle n_e(\mathbf{x}) \rangle. \tag{5.124}$$

The average of Eq. (5.124) is taken over the beam cross section.

For certain values of f_e, a relativistic electron beam can propagate in a background of stationary ions with constant radius. Such a beam is focused by magnetic forces and defocused by electric forces. We have seen that the magnitude of the transverse magnetic force equals the electric force multiplied by β^2. Adding ions reduces the electric force by an average factor of $1 - f_e$. The approximate condition for radial force balance of a laminar beam is

$$f_e = 1 - \beta^2 = 1/\gamma^2. \tag{5.125}$$

We cannot use ions to neutralize the current of an electron beam. Since the ion density in a practical beam must be comparable to the electron density, current neutralization requires that $v_i \sim v_e$. According to Eq. (5.123), ions that move at the same velocity as electrons have a much higher kinetic energy. Then we must classify the collection of particles as an ion beam rather than an electron beam. Current neutralization may take place when an electron beam travels through a plasma. Here, the background consists of both low-energy electrons and ions. Usually, the plasma electron density is much greater than the beam density. A small reduction in the plasma electron density gives space-charge neutralization. The current is neutralized if the remaining plasma electrons move with a small velocity in the direction opposite to the beam.

We shall conclude by discussing neutralized ion beams that are self-contained. Here, internal fields can provide radial equilibrium for both the ions and electrons, even if they have nonzero emittance. Such a self-contained beam is sometimes called a *plasmoid*. Plasmoids are formations of energetic charged particles that can propagate long distances without expansion. Considerable effort has been expended to generate and understand plasmoids for possible defense applications. We shall use relativistic transformations to investigate the properties of the self-contained beam. We take an intense ion beam with high kinetic energy that travels into free space in the z direction. The beam has current I_i, kinetic energy $(\gamma_i - 1)m_i c^2$, and radius r_0. Without neutralization, space-charge expansion limits the propagation distance (Section 5.4). We can reduce the electric fields by injecting a collinear electron beam with lower kinetic energy. Total cancellation of electric and magnetic fields results if the electron beam has the same density and velocity as the ion beam. The absence of fields is undesirable for beam propagation. Without the beam-generated fields, there is no transverse force to counteract emittance expansion.

Figure 5.12 shows an approach to achieve a self-focused equilibrium for both the ion and electron streams. We inject an electron beam that has a different velocity from the ion beam and a higher density—the beam region has a net negative charge density. The resulting radial electric field can confine ions with

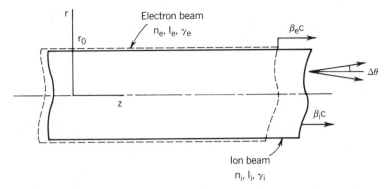

Figure 5.12 Definition of quantities to describe a self-focused ion-electron beam.

nonzero emittance. The beam also carries a net current. With the correct choice of parameters, the resulting magnetic field can balance the effect of the electric field to give electron confinement.

To derive parameters for radial equilibrium, we shall carry out initial calculations in the ion rest frame and then transform the results to the accelerator frame. We denote rest frame quantities with a prime symbol. We know the following characteristics of the ion beam in the accelerator frame: γ_i, I_i, r_0. We also know the angular divergence of the ions $\Delta\theta$. We can write the ion divergence angle in terms of the average transverse velocity in the accelerator frame:

$$\Delta\theta = v_i/\beta_i c, \tag{5.126}$$

where $\beta_i = (1 - 1/\gamma_i^2)^{1/2}$. The accelerator frame density of the ion beam is $n_i = I_i/(e\pi r_0^2\beta_i c)$. We want to find the current I_e and kinetic energy $(\gamma_e - 1)m_0 c^2$ of a collinear electron beam of radius r_0 that provides radial equilibrium for both particle species. For simplicity, we assume that the electron stream is cold—the electrons have no transverse velocity.

The ion rest frame moves at velocity $\beta_i c$ relative to the accelerator frame. The density of stationary ions in the rest frame is

$$n_i' = n_i/\gamma_i. \tag{5.127}$$

Time dilation implies that the ion transverse velocity is

$$v_\perp' = \gamma_i v_\perp. \tag{5.128}$$

We write the rest frame density of electrons in terms of a rest frame neutralization factor:

$$n_e'/n_i' = n_e/n_i = 1/f_e, \tag{5.129}$$

where $f_e < 1$.

We postulate that the electrons move at velocity $\beta'_e c$ in the positive z direction in the rest frame with kinetic energy $(\gamma'_e - 1)m_0 c^2$. From Eq. (5.125), the condition for radial electron equilibrium in the ion rest frame is

$$\gamma'_e = 1/f_e^{1/2}. \tag{5.130}$$

For ion equilibrium, the radial electric force must counteract the effect of the transverse velocities. Chapter 9 covers the balance between emittance and space-charge forces in detail. For this discussion, we estimate an equilibrium criterion by setting the transverse ion kinetic energy equal to the difference in electrostatic potential energy between the axis and the beam envelope. Using Eq. (5.129), we find that

$$\frac{m_i(v'_\perp)^2}{2} \cong \frac{e^2 n'_i r_0^2}{4\varepsilon_0} \frac{1 - f_e}{f_e}. \tag{5.131}$$

The electron velocity in the accelerator frame is given by the relativistic velocity addition law [CPA, Section 2.8],

$$\beta_e = (\beta'_e + \beta_i)/(1 + \beta'_e \beta_i). \tag{5.132}$$

TABLE 5.2   Self-Contained Ion-Electron Beams[a]

f_e	γ_e	I_e(A)	β_e	$\Delta\theta$ (rad)
0.050	15.086	237.048	0.998	1.787E-03
0.100	10.539	118.249	0.995	1.230E-03
0.150	8.497	78.640	0.993	9.761E-04
0.200	7.263	58.827	0.990	8.201E-04
0.250	6.408	46.932	0.988	7.102E-04
0.300	5.766	38.995	0.985	6.264E-04
0.350	5.258	33.319	0.982	5.588E-04
0.400	4.840	29.056	0.978	5.022E-04
0.450	4.487	25.733	0.975	4.533E-04
0.500	4.180	23.067	0.971	4.100E-04
0.550	3.909	20.879	0.967	3.709E-04
0.600	3.666	19.047	0.962	3.348E-04
0.650	3.443	17.487	0.957	3.009E-04
0.700	3.236	16.139	0.951	2.684E-04
0.750	3.040	14.957	0.944	2.367E-04
0.800	2.852	13.905	0.936	2.050E-04
0.850	2.665	12.954	0.927	1.723E-04
0.900	2.473	12.071	0.915	1.367E-04
0.950	2.259	11.212	0.897	9.407E-05

[a] Ion beam parameters: ion species, protons; current, 10 A; kinetic energy, 800 MeV.

The current of the electron stream is

$$I_e = I_i(n_e/n_i)(\beta_e/\beta_i) = I_i(\beta_e/\beta_i)/f_e. \tag{5.133}$$

Table 5.2 lists the implications of Eqs. (5.126) through (5.133) for the transport of a 10-A, 800-MeV proton beam. The beam has radius $r_0 = 0.05$ M. Laboratory frame values of γ_e, β_e, I_e, and $\Delta\theta$ are tabulated as a function of the rest frame neutralization parameter f_e. A value of $f_e \ll 1$ implies that the electron stream density is much higher than the ion beam density. Then, the electron equilibrium requires almost complete cancellation of the electric force by the magnetic force. As a result, the electron beam must have high energy and current. The advantage of this regime is that the ion beam can have a high divergence. This type of equilibrium is not feasible for an electron stream with a significance emittance. At the other extreme ($f_e \lesssim 1$), requirements of the electron stream are less severe, but the allowed ion beam divergence is smaller. As an example, we take $f_e = 0.9$. In the accelerator frame, the electron beam has current 12.1 A and a kinetic energy of only 0.75 MeV. The permissible ion beam divergence is $\Delta\theta = 0.14$ mrad. The electron velocity is 7% higher than the ion velocity.

A fundamental theorem of plasma physics[3] states that a self-contained plasmoid cannot exist. How can we reconcile the above results with this limitation? The resolution is that although the self-contained beam has radial equilibrium, it is not confined in the axial direction. Radial force balance depends on axial slippage between the ion and electron streams. In free space, the velocity difference between ions and electrons in a finite length beam would immediately lead to charge separation and axial electric fields. The fields would decelerate the electron stream, resulting in a loss of radial equilibrium. The self-contained ion beam equilibrium can exist only in a long conducting pipe. The pipe provides a return path for the extra electron current, preventing axial electric fields.

[3]See, for instance, G. Schmidt, *Physics of High Temperature Plasmas, 2nd ed.*, Academic Press, New York, 1979, p. 71.

6

Beam-Generated Forces—Advanced Topics

In this chapter we shall continue the study of beam-generated forces and self-consistent particle flow calculations. For the most part, the chapter concentrates on one-dimensional models. With this limitation, we can find analytic solutions with straightforward mathematics.

Section 6.1 extends the nonrelativistic Child model to describe particles that enter an acceleration gap with an initial velocity. The results are useful for designing multiple-gap ion extractors and high-current ion accelerators. Section 6.2 reviews space-charge-limited flow from a thermionic cathode. We assume that the cathode emits a Maxwell distribution of electrons, some of which cross the extraction gap. Besides its practical value, the model illustrates a self-consistent calculation for particles with a nonsingular distribution. Section 6.3 derives the Child law in spherical rather than planar geometry. We shall use the results to understand the design of high-current electron guns in Chapter 7. Section 6.4 describes the space-charge-limited flow of counterstreaming ions and electrons. Partial space-charge neutralization allows a higher flux of each species—this is an example of current flow enhancement. As a final case of one-dimensional flow, Section 6.5 derives the relativistic Child law. The result is useful for pulsed electron and ion diodes (Chapter 8).

Section 6.6 presents a general calculation of a one-dimensional self-consistent equilibrium for nonrelativistic particles. We shall calculate the transverse density distribution of a infinite-length sheet beam with nonzero emittance in a linear focusing force. Section 6.7 presents methods to extend equilibrium calculations to beams with two- and three-dimensional density variations. For the special case of a uniform-density beam with an ellipsoidal shape, we shall find that electric and magnetic fields vary linearly from the center in all three dimensions. The section introduces the KV distribution function, a starting point for many treatments of beam equilibrium and stability. Velocity integrals over the KV distribution give a beam with elliptical boundaries and uniform density.

6.1. SPACE-CHARGE-LIMITED FLOW WITH AN INITIAL INJECTION ENERGY

We can extend the one-dimensional Child limit to describe acceleration gaps where particles enter with a nonzero energy. The results have application to the gaps of high-current multistage accelerators and the flow of electrons through grid-controlled devices like vacuum triodes. Suppose that we apply a voltage V_0 across a gap of width d—Fig. 6.1 shows polarity conventions for positively charged particles. Particles enter with an initial kinetic energy T_0 and leave the gap with energy $T_0 + qV_0$. The problem is most easily formulated in terms of the *absolute potential*: ϕ—the absolute potential equals zero at the particle source. The kinetic energy of a nonrelativistic particle is related to ϕ by

$$m_0 v^2 / 2 = -q\phi. \tag{6.1}$$

We can combine Eq. (6.1) with the condition of steady-state flux at all positions in the gap to give the Poisson equation in the same form as Eq. (5.40).

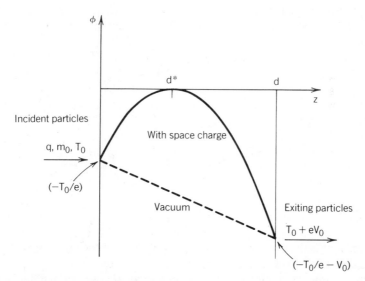

Figure 6.1 Variation of electrostatic potential across a one-dimensional acceleration gap for a space-charge-limited flow of ions with initial kinetic energy T_0. The dashed line shows the potential variation with no ions.

The difference lies in the boundary conditions on ϕ:

$$\phi(z = 0) = \phi_1 = -(T_0/q), \tag{6.2}$$

$$\phi(z = d) = \phi_2 = -(T_0/q + V_0). \tag{6.3}$$

Furthermore, because of their initial energy, particles can cross the gap even if the electric field at the entrance is negative. Figure 6.1 shows that the presence of particles in the gap increases ϕ. The limit on current occurs when the absolute potential reaches zero. We denote the position where ϕ equals zero as d^*. Particles reach this point with infinitesimally small kinetic energy. An increase in the injected flux causes reflection of particles.

We shall follow the method of Section 5.2 to find an expression for the potential as a function of position in the gap for the limiting current. The potential at d^* satisfies the following boundary conditions: $\phi(d^*) = 0$, $d\phi(d^*)/dz = 0$. We can integrate the Poisson equation on each side of d^*. For example, the solution in the region $d^* \leqslant z \leqslant d$ is

$$\tfrac{4}{3}(-\phi)^{3/4} = [4j_0/\varepsilon_0(2q/m_0)^{1/2}]^{1/2}(z - d^*). \tag{6.4}$$

Combination of solutions from both regions leads to the relationship

$$\tfrac{4}{3}[(-\phi_1)^{3/4} + (-\phi_2)^{3/4}] = [4j_0/\varepsilon_0(2q/m_0)^{1/2}]^{1/2}d. \tag{6.5}$$

The quantities ϕ_1 and ϕ_2 are defined in Eqs. (6.2) and (6.3). We can solve Eq. (6.5) for the limiting current density. We express the result in terms of the one-dimensional Child limit [Eq. (5.48)] with a correction factor for the particle injection energy:

$$j_0 = \frac{4\varepsilon_0}{9}\left[\frac{2q}{m_0}\right]^{1/2}\frac{V_0^{3/2}}{d^2}F(\chi). \tag{6.6}$$

The correction factor is

$$F(\chi) = \chi^{3/2}[(1 - 1/\chi)^{3/4} + 1]^2, \tag{6.7}$$

where

$$\chi = -\phi_2/(\phi_1 - \phi_2).$$

The quantity χ is the ratio of the kinetic energy of the particles emerging from the acceleration gap to the change in kinetic energy in the gap.

Figure 6.2 shows a plot of $F(\chi)$. As expected, the solution reduces to the standard Child law when the injection energy approaches zero ($\chi = 1$). The function $F(\chi)$ grows rapidly for increasing χ. Figure 6.2 emphasizes that the

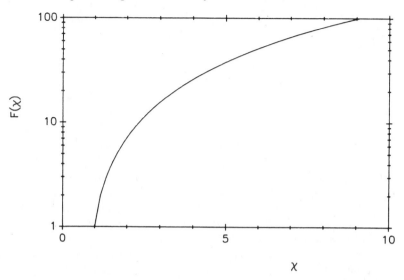

Figure 6.2 Enhancement of space-charge-limited current density, $F(\chi)$, for ions with an initial kinetic energy T_0 crossing an acceleration gap with voltage V_0 as a function of $\chi = -\phi_2/(\phi_1 - \phi_2)$.

longitudinal space-charge limit drops rapidly as particles accelerate. For example, the space-charge limit in a postacceleration gap that doubles the energy of particles ($\chi = 2$) is 7.2 times the Child limit.

6.2. SPACE-CHARGE-LIMITED FLOW FROM A THERMIONIC CATHODE

An inherent assumption in the derivation of space-charge-limited flow of Section 5.2 is that the particle source supplies an unlimited flux of electrons (or ions). This condition is usually true for electron extraction from a dense plasma, but it may not hold for sources like thermionic cathodes. Depending on the surface temperature, a thermionic cathode may generate a current density less than the space-charge-limited value. Then, particle flow is *source limited* rather than *space-charge limited*.

The current density of thermionic cathodes can be highly nonuniform. Small variations in the surface work function result in large variations of current density (Section 7.4). For the generation of high-brightness beams, nonuniform emission is a severe drawback—the asymmetric space-charge forces cause growth of beam emittance in the extractor and postacceleration gaps. To avoid this problem, thermionic cathodes are usually operated in the space-charge-limited mode. When the flux from the source is higher than the Child law value, only a fraction of the flux given by the space-charge limit crosses the extraction gap. As long as the cathode is hot enough so that emission from all points on the surface exceeds the Child limit, the output beam has uniform current density.

The model we developed in Section 6.1 was based on monoenergetic electrons entering the gap. This model cannot explain the selection of particles for space-charge-limited flow—either all the entering electrons cross the gap or all electrons are turned back by a virtual cathode. We can resolve the problem by adding an energy spread to the injected electron distribution. In this section, we shall study space-charge-limited flow in a one-dimensional gap with a thermal spectrum of source electrons. In particular, we shall show how selection of electrons takes place at a virtual cathode to give a Child flux limit, even if the initial energy spread is small compared with the energy gain in the gap.

Before we address the problem of the space-charge flow, we shall review some properties of thermal electrons near a conducting surface. Figure 6.3a shows the geometry of the model. A planar metal cathode at zero potential emits electrons. The electrons have a Maxwell distribution of kinetic energy in the z direction, characterized by the temperature T_e. The space charge of electrons emerging into the half-plane with $z > 0$ creates a negative potential in the region. The associated electric field reflects electrons back to the cathode. The distance an electron travels from the cathode depends on its injection energy.

Following the discussion of Section 2.6, the equilibrium density of electrons in the vacuum region is a function of the electrostatic potential:

$$n(z) = n_0 \exp\left[+ e\phi(z)/kT_e\right]. \tag{6.8}$$

The quantity n_0 is the density adjacent to the cathode. We can substitute Eq. (6.8) into the one-dimensional Poisson equation:

$$\frac{d^2\phi}{dz^2} = \frac{en_0 \exp\left(+ e\phi/kT_e\right)}{\varepsilon_0}. \tag{6.9}$$

If we introduce the following dimensionless variables:

$$\Phi = - e\phi/kT_e, \tag{6.10}$$

$$Z = z/\lambda_d, \tag{6.11}$$

Eq. 6.9 has the simple form

$$\frac{d^2\Phi}{dZ^2} = \exp\left(- \Phi\right). \tag{6.12}$$

The scale length in Eq. (6.11) is

$$\lambda_d = \left[\frac{kT_e\varepsilon_0}{e^2 n_0}\right]^{1/2}. \tag{6.13}$$

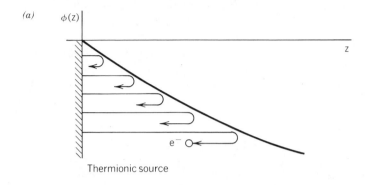

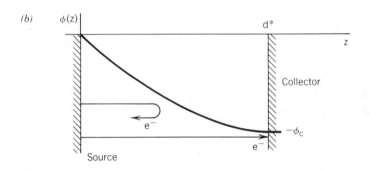

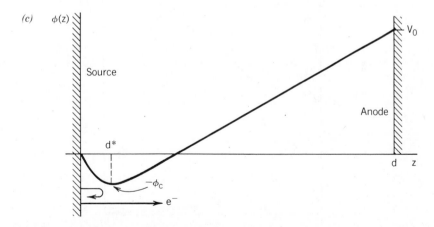

Figure 6.3   Space-charge-limited emission of electrons from a thermionic cathode. (*a*) Variation of electrostatic potential for thermal electrons near a metal surface—the region $z > 0$ is an field-free volume of infinite extent. (*b*) Potential variation with partial collection of thermal electrons by a biased electrode—zero electric field on the electrode surface. (*c*) Potential variation for the extraction of electrons from a thermionic cathode showing the location of the virtual cathode surface.

The quantity λ_d is the *Debye length*, a familiar quantity in plasma physics. It enters all problems that involve interaction of a Maxwell distribution with an electric field. Section 12.1 discusses the significance of λ_d in plasmas

The boundary conditions for the solution of Eq. (6.12) are $\Phi(0) = 0$ and $d\Phi(\infty)/dZ = 0$. The second condition ensures that the electron density approaches zero an infinite distance from the plate. We can integrate Eq. (6.12) after multiplying both sides by $2(d\Phi/dZ)$, and then perform a second integral. The result is

$$Z = \sqrt{2}\,[\exp{(\Phi/2)} - 1]. \qquad (6.14)$$

The normalized electric field at the cathode is $d\Phi(0)/dZ = 2$. Figure 6.4 shows spatial variations of electrostatic potential and density. Note that significant changes in the density take place over the characteristic scale length λ_d. The electrostatic potential decreases monotonically away from the cathode surface; all particles leaving the surface are eventually reflected. Therefore, the particles have a symmetric Maxwell distribution at any position, with half the particles leaving and half returning.

The Debye length for electrons near a high-current thermionic cathode is small. For example, suppose we have a dispenser cathode with a surface temperature of $T_e = 3000°C$ and source-limited current density of $j_s = 10^5\,A/m^2$ $(10\,A/cm^2)$. Electrons leaving the surface have an average velocity (Eq. (2.61)]

$$\langle v_z \rangle = (2kT_e/\pi m_e)^{1/2} = 1.8 \times 10^5\,\text{m/s}. \qquad (6.15)$$

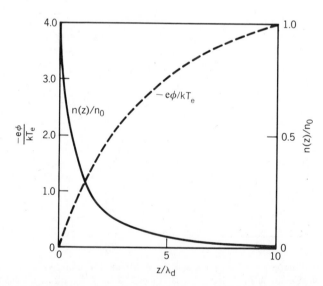

Figure 6.4 Normalized electron density and electrostatic potential near the surface of a thermionic emitter. Electrons have a Maxwell velocity distribution with temperature T_e. The region $z > 0$ is a field-free space of infinite extent. $\lambda_d = [kT_e\varepsilon_0/e^2 n_0]^{1/2}$.

The density at the surface is

$$n_0 = j_s/e\langle v_z \rangle = 3.47 \times 10^{18}\, \text{m}^{-3}. \tag{6.16}$$

Substitution into Eq. (6.13) shows that the electron density extends into the vacuum region a small distance, only $\lambda_d = 2.2\,\mu\text{m}$ ($2.2 \times 10^{-6}\,\text{m}$).

We next add a conducting anode to extract electrons. The presence of the anode makes the theory more complex. Some of the emitted electrons do not return to the cathode—the axial velocity distribution is no longer symmetric. Although the emitted electrons may have a Maxwell distribution, the net distribution is non-Maxwellian and Eq. (6.8) is no longer valid. To understand the self-consistent electron flow, we shall proceed in two steps. First, we place a hypothetical collector plate biased to potential $-\phi_c$ a distance d^* from the cathode (Fig. 6.3b). Emitted electrons either travel to the collector or are reflected back to the cathode. A solution of space-charge flow in the region between the cathode and collector must include two types of electrons—collected and reflected. We shall study the solution in Section 15.1. For now, we shall use two results of the calculation without proof. First, for a given cathode electron temperature and flux, we can find a combination of ϕ_c and d^* such that the electric field at the collector is zero. Second, the full solution shows that $d^* \cong \lambda_d$.

The space-charge flow solution depends on the collected fraction of the electron flux. We denote j_0 as the current density of thermal electrons leaving the cathode and j_c as the collected current density. Collected electrons have kinetic energy in the range:

$$m_e v_z(0)^2 \geqslant e\phi_c. \tag{6.17}$$

From Eq. (2.60), the ratio of current densities is

$$\frac{j_c}{j_0} = \int_{(2e\phi_c/m_e)^{1/2}}^{\infty} dv_z v_z \exp\left(\frac{-mv_z^2}{2kT_e}\right) \Big/ \int_0^{\infty} dv_z v_z \exp\left(\frac{-mv_z^2}{2kT_e}\right),$$

or

$$\frac{j_c}{j_0} = \exp\left(\frac{-e\phi_c}{kT_e}\right). \tag{6.18}$$

Now, suppose we make the collector perfectly transparent and place an anode a distance d from the cathode (Fig. 6.3c). We choose the positive potential of the anode, V_0, so that it attracts all the current that passes through the collector. The anode voltage is just high enough to reduce the electric field at the collector to zero. Since no particles return through the collector, the space-charge solution in the region $0 \leqslant z < d^*$ does not change. We shall concentrate on the region between d^* and d. If $d \gg d^*$, then $V_0 \gg \phi_c$. The thermal energy of electrons emerging from the collector is very small compared with eV_0. Therefore, we have a situation where electrons of approximately zero energy emerge from a surface at zero electric field. The potential difference in the region to the right of the collector must be sufficient to give a space-charge-limited flux:

$$j_c \cong \frac{4\varepsilon_0}{9}\left[\frac{2e}{m_e}\right]^{1/2}\frac{(V_0 + \phi_c)^{3/2}}{(d - d^*)^2}.$$

(6.19)

Finally, we note that we can remove the hypothetical collector without changing the nature of the solution. The collector intercepts no current. Furthermore, since the electric field is zero on both sides, it carries no surface charge. The conclusion is that if $\phi_c \ll V_0$ and $\lambda_d \ll d$, the current density that passes the virtual cathode is approximately equal to the Child limit for a gap with voltage V_0 and spacing d.

6.3. SPACE-CHARGE-LIMITED FLOW IN SPHERICAL GEOMETRY

We can carry out space-charge flow calculations in specialized geometries besides the planar case of Section 5.2. In this section, we shall address charged particle flow between concentric spheres—modifications of the configuration have many applications in beam technology. For example, we use the spherical flow solution in Section 7.2 for the design of high-flux electron guns.

Figure 6.5 illustrates the geometry for ideal spherical flow. The particle source and collector are located on concentric spherical surfaces. The quantity R_s is the source radius, R_c is the collector radius, and r represents a radial position between the two. We can find two types of solutions for inward flow ($R_s > R_c$) and outward

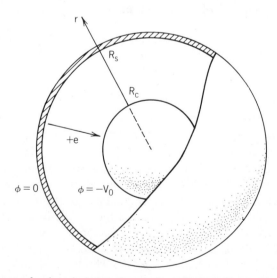

Figure 6.5 Geometry for the calculation of space-charge-limited electron flow between spheres.

flow $(R_s < R_c)$. For convenience we treat a steady-state flow of positive nonrelativistic particles with mass m_0. We take the source potential equal to zero while the collector has bias voltage $-V_0$.

The Poisson equation has the following form for a spherically symmetric potential:

$$\frac{1}{r^2}\frac{d}{dr}\left(r^2\frac{d\phi}{dr}\right) = -\frac{\rho}{\varepsilon_0}. \tag{6.20}$$

The quantity ρ is the space-charge density of ions. Conservation of energy implies that the ion velocity is related to ϕ by

$$v = (-2e\phi/m_e)^{1/2}. \tag{6.21}$$

In equilibrium, the radial current I is constant at all radii. This condition implies that the space-charge density is related to the velocity by

$$\rho = I/4\pi r^2 v. \tag{6.22}$$

Combining Eqs. (6.20), (6.21), and (6.22) gives following self-consistent form for the Poisson equation:

$$\frac{d}{dr}\left(r^2\frac{d\phi}{dr}\right) = -\frac{I}{4\pi\varepsilon_0}\left(\frac{m_0}{-2e\phi}\right)^{1/2}. \tag{6.23}$$

We must solve Eq. (6.23) numerically. As with any numerical calculation, we can generate results of maximum generality if we first rewrite the governing equations in dimensionless form. We define the dimensionless potential and radius as

$$\Phi = -\phi/V_0 \tag{6.24}$$

and

$$R = r/R_s. \tag{6.25}$$

Equation (6.23) takes the form

$$\frac{d}{dR}\left(R^2\frac{d\Phi}{dR}\right) = A\Phi^{-1/2}, \tag{6.26}$$

where

$$A = \left[\frac{I}{4\pi\varepsilon_0 V_0^{3/2}}\left(\frac{m_0}{2e}\right)^{1/2}\right]. \tag{6.27}$$

The boundary conditions for the solution of Eq. (6.26) are $\Phi(1) = 0$, $d\Phi/dR(1) = 0$, and $\Phi(R_c/R_s) = 1$. The parameter A and the aspect ratio of the spherical extractor, R_c/R_s, govern the solution. The current density in a spherical extractor has scaling similar to that of a planar gap. The differences arise from the electric field enhancement of the spherical geometry. Note that Eq. (6.27) implies that the current I does not depend on the absolute value of the source radius R_s. A proportionally smaller spherical extractor has reduced emission surface area but correspondingly higher fields at the source.

Langmuir and Blodgett[1] developed a well-known numerical solution of Eq. (6.26). Following the anticipated scaling, they assumed the following form for the electrostatic potential:

$$\Phi(R)^{3/2} = (9A/4)\alpha(R)^2. \tag{6.28}$$

The function $\alpha(R)$ is the Langmuir function. We can find an equation for α by substituting Eq. (6.28) into Eq. (6.26). Defining the variable $\gamma = \ln(R)$, the equation is

$$3\alpha\frac{d^2\alpha}{d\gamma^2} + \left(\frac{d\alpha}{d\gamma}\right)^2 + 3\alpha\frac{d\alpha}{d\gamma} - 1 = 0. \tag{6.29}$$

The following expression for α, valid near $\gamma = 0$, comes from a series solution of Eq. (6.29):

$$\alpha = \gamma - 0.3\gamma^2 + 0.075\gamma^3 - 0.0143182\gamma^4 + 0.0021609\gamma^5$$
$$- 0.00026791\gamma^6 + \cdots. \tag{6.30}$$

Values for the α function are given in Tables 6.1 and 6.2 for a range of extractor geometries. Table 6.1 applies to inward flow $(R_s > R_c)$, while Table 6.2 holds for outward flow $(R_s < R_c)$.

We can use the α-function tables to calculate the variation of potential between the electrodes of a spherical injector and the total current. We find the total current from Eq. (6.28) by noting that $\Phi(R_c/R_s) = 1$:

$$I = \frac{4\varepsilon_0}{9}\left(\frac{2e}{m_0}\right)^{1/2}\frac{4\pi V_0^{3/2}}{\alpha(R_c/R_s)^2}. \tag{6.31}$$

As an example, suppose there is a converging flow of electrons between spherical electrodes with $R_c/R_s = 0.33$. The applied voltage is 50 kV. Table 6.1 shows that $\alpha(0.33)^2 = 2.512$. The corresponding current from Eq. (6.31) is $I = 130$ A. The current density at the source is

$$j_s = I/4\pi R_s^2. \tag{6.32}$$

If we set the source current equal to 10^4 A/m² (1 A/cm²), then the source radius

[1]I. Langmuir and D. Blodgett, *Phys. Rev.* **24**, 49 (1924).

TABLE 6.1 Langmuir Function versus Normalized
Radius for Converging Beam

$R = r/R_s$	α^2	$R = r/R_s$	α^2
1.0000	0.0000	0.1923	8.636
0.9524	0.0024	0.1852	9.135
0.9091	0.0096	0.1786	10.01
0.8696	0.0213	0.1724	10.73
0.8333	0.0372	0.1667	11.46
0.8000	0.0571	0.1538	13.35
0.7692	0.0809	0.1429	15.35
0.7407	0.1084	0.1333	17.44
0.7143	0.1396	0.1250	19.62
0.6897	0.1740	0.1176	21.89
0.6667	0.2118	0.1111	24.25
0.6250	0.2968	0.1053	26.68
0.5882	0.394	0.1000	29.19
0.5556	0.502	0.0833	39.98
0.5623	0.621	0.0714	51.86
0.5000	0.750	0.0625	64.74
0.4762	0.888	0.0556	78.56
0.4545	1.036	0.0500	93.24
0.4348	1.193	0.0333	178.2
0.4167	1.358	0.0250	279.6
0.4000	1.531	0.0200	395.3
0.3846	1.712	0.0167	523.6
0.3704	1.901	0.0143	663.3
0.3571	2.098	0.0125	813.7
0.3448	2.302	0.0111	974.1
0.3333	2.512	0.0100	1144
0.3125	2.954	0.0083	1509
0.2941	3.421	0.0071	1907
0.2778	3.913	0.0063	2333
0.2632	4.429	0.0056	2790
0.2500	4.968	0.0050	3270
0.2381	5.528	0.0040	4582
0.2273	6.109	0.0033	6031
0.2174	6.712	0.0029	7610
0.2083	7.334	0.0025	9303
0.2000	7.976	0.0020	13015

must be $R_s \geqslant 0.032$ m. Finally, we can use the tables to find α at values of r/R_s between the source and collector. Substituting this information into Eq. (6.28) gives $\Phi(r)$ between the electrodes.

Because Section 7.2 discusses the nature of converging flow, we shall concentrate here on diverging beams. Figure 6.6 shows a plot of the current between spheres and the source current density as a function of R_s/R_c.

TABLE 6.2 Langmuir α-Function versus Normalized Radius for Diverging Beam

$R = r/R_s$	α^2	$R = r/R_s$	α^2
1.0	0.0000	6.5	1.385
1.05	0.0023	7.0	1.453
1.1	0.0086	7.5	1.516
1.15	0.0180	8.0	1.575
1.2	0.0299	8.5	1.630
1.25	0.0437	9.0	1.682
1.3	0.0591	9.5	1.731
1.35	0.0756	10	1.777
1.4	0.0931	12	1.938
1.45	0.1114	14	2.073
1.5	0.1302	16	2.189
1.6	0.1688	18	2.289
1.7	0.208	20	2.378
1.8	0.248	30	2.713
1.9	0.287	40	2.944
2.0	0.326	50	3.120
2.1	0.364	60	3.261
2.2	0.402	70	3.380
2.3	0.438	80	3.482
2.4	0.474	90	3.572
2.5	0.509	100	3.652
2.6	0.543	120	3.788
2.7	0.576	140	3.903
2.8	0.608	160	4.002
2.9	0.639	180	4.089
3.0	0.669	200	4.166
3.2	0.727	250	4.329
3.4	0.783	300	4.462
3.6	0.836	350	4.573
3.8	0.886	400	4.669
4.0	0.934	500	4.829
4.2	0.979	600	4.960
4.4	1.022	800	5.165
4.6	1.063	1000	5.324
4.8	1.103	1500	5.610
5.0	1.141	2000	5.812
5.2	1.178	5000	6.453
5.4	1.213	10000	6.933
5.6	1.247	30000	7.693
5.8	1.280	100000	8.523

a)

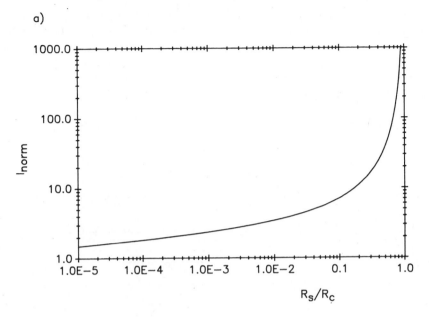

b)

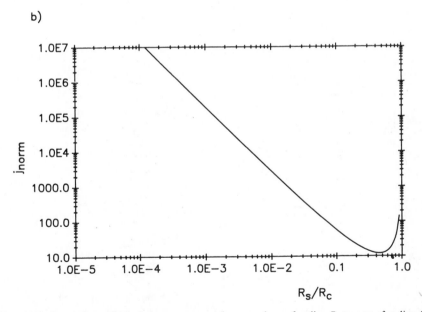

Figure 6.6   Space-charge-limited electron current from a sphere of radius R_s to one of radius R_c $(R_s < R_c)$. (a) $I_{\text{norm}} = 4\pi/\alpha^2$. (b) $j_{\text{norm}} = 1/(\alpha^2(R_s/R_c)^2)$.

The figure emphasizes that the current to a collector of given radius R_c becomes almost independent of the surface area of the source when R_s/R_c approaches zero. Here, the enhanced electric field around the small sphere counteracts the effect of source area.

The weak dependence of the current in a diverging beam on the source geometry gives insight into space-charge flow from an array of sharp points. Figure 6.7a illustrates a cathode for pulsed high-current electron beams. The electrode is covered with an array of fibers, needles, or sharp blades. Upon

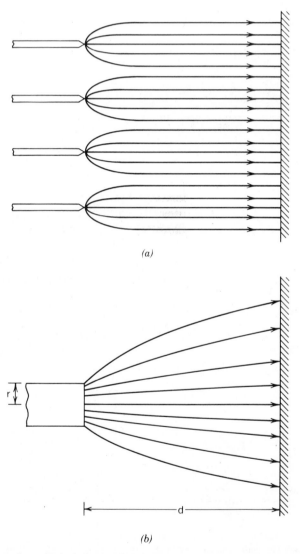

(a)

(b)

Figure 6.7 Space-charge-limited electron flow in complex geometries. (*a*) Electron flow from an array of closely spaced needles. (*b*) Electron flow from a rod cathode to a grid anode.

application of a high voltage, plasma formation occurs at points of electric field enhancement. In contrast to a planar extractor with a uniform emitter, the electrons in Fig. 6.7a emerge from a small fraction of the electrode surface. The high fields around the needles enhance the space-charge-limited electron flow. As electrons move across the gap, individual streams coalesce to form a planar beam. Experiments show that if the gap width is large compared with the distance between emitters, then the net current flow is close to the prediction of Eq. (5.48) for a planar gap with uniform source.

As a second example, imagine electron emission from a narrow rod (Fig. 6.7b)—the rod radius r is small compared with the gap spacing d. Space-charge-limited emission occurs over the surface of the rod. Because of electric field enhancement, emission is particularly strong at the edges. The flow of electrons away from the rod is roughly hemispheric. The total current from such an electron gun is much higher than that predicted from the planar formula [Eq. (5.48)] with area πr^2. In practice, the total current is about

$$I \cong \frac{4\varepsilon_0}{9}\left(\frac{2e}{m_0}\right)^{1/2} V_0^{3/2}\pi[r^2 + (d/2)^2]. \tag{6.33}$$

6.4. BIPOLAR FLOW

The term *bipolar flow* refers to the simultaneous space-charge-limited flow of ions and electrons emitted from opposite sides of an acceleration gap. We shall encounter the bipolar-flow solution when we study intense beam diodes in Chapter 8. In these devices, strong electric fields cause free emission of particles from all exposed electrode surfaces.

To construct a physical model, we shall consider a one-dimensional acceleration gap of width d with applied voltage V_0 (Fig. 6.8a). Particle motion is nonrelativistic, $eV_0 \ll m_e c^2$. We associate the cathode surface with the origin of the z axis and the position of zero electrostatic potential. The cathode and anode can supply an unlimited flux of electrons and ions. The particles leave the cathode and anode with zero velocity. In equilibrium, conservation of charge implies that the electron density equals

$$n_e(\phi) = -(j_e/e)(2e\phi/m_e)^{-1/2}. \tag{6.34}$$

Similarly, the ion density varies as

$$n_i(\phi) = (j_i/e)[2e(V_0 - \phi)/m_e]^{-1/2}. \tag{6.35}$$

In terms of the dimensionless variables $\Phi = \phi/V_0$ and $Z = z/d$, the Poisson equation is

$$\frac{d^2\Phi}{dZ^2} = A\left\{\frac{1}{\Phi^{1/2}} - \frac{1}{(1-\Phi)^{1/2}}\left[\frac{j_i}{j_e}\left(\frac{m_i}{m_e}\right)^{1/2}\right]\right\}, \tag{6.36}$$

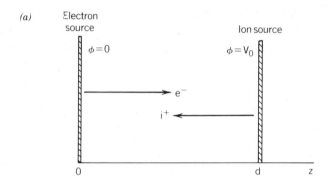

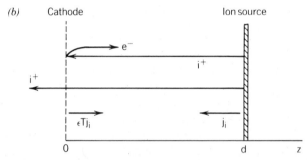

Figure 6.8 Calculation of planar bipolar flow, counterstreaming ions and electrons. (*a*) Space-charge-limited flow of both ions and electrons. (*b*) Space-charge-limited ion flow with secondary emission of electrons at the anode grid.

where

$$A = \frac{j_e d^2}{\varepsilon_0 (2e/m_e)^{1/2} V_0^{3/2}}. \tag{6.37}$$

Again, we multiply both sides of Eq. (6.36) by $2\Phi'$. Integrating the equation from 0 to Z gives the expression

$$\left(\frac{d^2\Phi}{dZ^2}\right)^2 = 4A\left\{\Phi^{1/2} + [(1-\Phi)^{1/2} - 1]\left[\frac{j_i}{j_e}\left(\frac{m_i}{m_e}\right)^{1/2}\right]\right\}. \tag{6.38}$$

The form of Eq. (6.38) guarantees that $d\Phi/dZ = 0$ at $z = 0$. The condition that the axial electric field equals zero at $Z = 1$ implies that

$$\frac{j_1}{j_e} = \sqrt{\frac{m_e}{m_i}}. \tag{6.39}$$

Solution of Eq. (6.36) with the condition of Eq. (6.39) and the boundary conditions $\Phi(0) = 0$, $\Phi(1) = 1$, yields the following expression for space-charge-limited electron current density:

$$j_e = \frac{4\varepsilon_0}{9}\left(\frac{2e}{m_e}\right)^{1/2}\frac{V_0^{3/2}}{d^2}\left[\frac{3}{4}\int_0^1 \frac{d\Phi}{[\Phi^{1/2}+(1-\Phi)^{1/2}-1]^{1/2}}\right]^2. \tag{6.40}$$

Equation (6.40) shows that the electron current density equals the Child law value multiplied by a correction factor, the quantity in brackets. Evaluating the definite integral, we find that

$$j_e(\text{bipolar}) = 1.86\,j_e(\text{Child}). \tag{6.41}$$

Similarly, Eq. (6.39) implies that

$$j_i(\text{bipolar}) = 1.86\,j_i(\text{Child}). \tag{6.42}$$

The results of Eqs. (6.41) and (6.42) have a straightforward physical interpretation. Since the electric field equals zero on both sides of the acceleration gap, Gauss's law implies that the integrals of electron and ion densities are equal. The functional form of the ion density is the mirror image of the electron density. The ion density at the cathode equals the electron density at the anode; therefore, the ratio of current densities is proportional to the exit velocities of the particles. Since both species exit with kinetic energy eV_0, the velocity ratio is inversely proportional to the square root of the mass ratio, leading to Eq. (6.39). The value of 1.86 in Eqs. (6.41) and (6.42) is called an *enhancement factor* for space-charge-limited flow. The increased flux occurs because the electron density partially cancels the ion space charge at the anode. A similar effect occurs for the electron flow.

Other solutions for one-dimensional, self-consistent equilibria of ion and electron flows may be useful in special cases. For example, suppose we have a low-field gap for ion extraction (Fig. 6.8b). The ions, created by an anode source, exit the acceleration gap through a cathode grid of transparency T. Ions in the range 10 keV to 1 MeV have a high probability of ejecting secondary electrons when they strike a metal surface. This process is characterized by the secondary emission coefficient ε, the number of electrons emitted per incident ion. The secondary electrons travel to the anode, attracted by the electric field that accelerates the ions. The ion current density is given by

$$n_i(\phi) = (j_i/e)[2e(V_0 - \phi)/m_e]^{1/2}. \tag{6.43}$$

The electron density equals

$$n_e(\phi) = -(\varepsilon T)(j_i/e)(2e\phi/m_e)^{-1/2}. \tag{6.44}$$

We can solve the Poisson equation with the density expressions of Eqs. (6.43) and (6.44). The electron flow is source limited if $\varepsilon < (m_i/m_e)^{1/2}/T$; therefore, we apply the condition of zero electric field only at the anode. We do not have to specify the cathode electric field to determine a unique solution since the magnitude of the electron flux is no longer a free parameter.

6.5. SPACE-CHARGE-LIMITED FLOW OF RELATIVISTIC ELECTRONS

Single-gap electron extractors have been operated in the megavolt range. These extractors are driven either by electrostatic pulsed power generators or by stacked induction linac cavities. Relativistic effects are important since $eV_0 \geq m_e c^2$. In this section, we shall extend the theory of one-dimensional space-charge-limited electron flow to include relativistic variations of axial velocity.

We must apply the theory with caution. The beam current generated by pulsed power devices is usually in the multikiloampere range—forces from beam-generated magnetic fields may be significant. A strong toroidal magnetic field gives electrons a component of radial velocity that may invalidate the conditions of the one-dimensional model. There are two common situations where the one-dimensional model accurately predicts the electron beam current:

1. We can design high-voltage extractors for moderate current beams ($\leq 1\,\mathrm{kA}$) with applied radial electric fields that counteract focusing by beam-generated magnetic fields. Here, electrons move mainly in the axial direction.

2. We can apply a strong axial magnetic field in the extractor ($B_z \gg B_\theta$). Electrons follow the net magnetic field lines.

Again, we shall consider a one-dimensional gap with applied voltage V_0 and width d. If we set the electrostatic potential ϕ equal to zero at the cathode, conservation of energy gives the following expression for the relativistic gamma factor for electrons in the extractor:

$$\gamma(z) = 1 + e\phi(z)/m_e c^2. \tag{6.45}$$

For electron motion in the z direction, the axial velocity is related to γ by

$$v_z = (c/\gamma)(\gamma^2 - 1)^{1/2}. \tag{6.46}$$

As in Section 5.2, the product of electron density and velocity is constant at all positions. We can express the density in terms of the electron gamma factor:

$$n_e(\gamma) = \frac{\gamma j_e}{ec(\gamma^2 - 1)^{1/2}}. \tag{6.47}$$

The quantity j_e is the electron current density, a constant.
Equation (6.45) implies that

$$\frac{d^2\phi}{dz^2} = \frac{m_ec^2}{e}\frac{d^2\gamma}{dz^2}. \tag{6.48}$$

Substituting Eqs. (6.47) and (6.48) into the one-dimensional Poisson equation, we find an equation for the $\gamma(z)$:

$$\frac{d^2\gamma}{dz^2} = \left[\frac{ej_ed^2}{\varepsilon_0c^3m_e}\right]\frac{\gamma}{(\gamma^2 - 1)^{1/2}}. \tag{6.49}$$

The boundary conditions for Eq. (6.49) corresponding to a space-charge-limited solution are

$$\gamma(0) = 1, \tag{6.50}$$

$$d\gamma(0)/dz = 0, \tag{6.51}$$

$$\gamma(d) = \gamma_0 = 1 + eV_0/m_ec^2. \tag{6.52}$$

Equation (6.51) specifies zero electric field at the cathode. We multiply both sides of Eq. (6.49) by $2(d\gamma/dz)$ and integrate. Following the methods of Sections 6.3 and 6.4, we can express the space-charge-limited current density in terms of a definite integral.

$$j_e = \left[\frac{\varepsilon_0m_ec^3}{ed^2}\right]\frac{G(\gamma_0)^2}{2}. \tag{6.53}$$

The integral is

$$G(\gamma_0) = \int_1^{\gamma_0}\frac{d\xi}{(\xi^2 - 1)^{1/4}}. \tag{6.54}$$

Table 6.3 lists the function G and the quantity j_ed^2 as a function of γ_0 and v_0.
We can check the validity of Eq. (6.53) by comparing its predictions to expressions for current density in limiting cases. We already have results for low-voltage gaps ($eV_0 \ll m_ec^2$) from Section 5.2. We can easily calculate the

TABLE 6.3 Space-Charge Flow in a Relativistic Planar Gap

γ_0	V_0 (MV)	$G(\gamma_0)$	$j_e d^2$ (kA-m^2)
1.978	0.500	1.053	0.754
2.957	1.000	1.709	1.986
3.935	1.500	2.249	3.438
4.914	2.000	2.721	5.031
5.892	2.500	3.146	6.724
6.871	3.000	3.535	8.493
7.849	3.500	3.898	10.323
8.828	4.000	4.238	12.202
9.806	4.500	4.559	14.122
10.785	5.000	4.864	16.078
11.763	5.500	5.156	18.065
12.742	6.000	5.436	20.079
13.720	6.500	5.705	22.117
14.699	7.000	5.965	24.176
15.677	7.500	6.216	26.255
16.656	8.000	6.459	28.351
17.634	8.500	6.696	30.464
18.613	9.000	6.925	32.591
19.591	9.500	7.149	34.732
20.569	10.000	7.368	36.886
21.548	10.500	7.581	39.051
22.526	11.000	7.789	41.227
23.505	11.500	7.993	43.414
24.483	12.000	8.193	45.610
25.462	12.500	8.388	47.815
26.440	13.000	8.580	50.029
27.419	13.500	8.769	52.250
28.397	14.000	8.954	54.480
29.376	14.500	9.136	56.716
30.354	15.000	9.315	58.960
31.333	15.500	9.491	61.210
32.311	16.000	9.664	63.467
33.290	16.500	9.835	65.729
34.268	17.000	10.003	67.997
35.247	17.500	10.169	70.271
36.225	18.000	10.333	72.549
37.204	18.500	10.494	74.833
38.182	19.000	10.653	77.122
39.160	19.500	10.811	79.415
40.139	20.000	10.966	81.713

space-charged-limited flux in the *ultrarelativistic limit*, $eV_0 \gg m_e c^2$. At high gap voltage, injected electrons are quickly accelerated to a velocity close to the speed of light. We can approximate this condition by taking the electron density as a constant, j_e/ec, independent of position. The Poisson equation has the form

$$d^2\phi/dx^2 \cong j_e/\varepsilon_0 c. \tag{6.55}$$

We can solve Eq. (6.55) with the standard boundary conditions to give the result

$$j_e \cong 2\varepsilon_0 c V_0/d^2 \qquad [eV_0 \gg m_e c^2] \tag{6.56}$$

Figure 6.9 plots $j_e d^2$ as a function of V_0 as predicted by Eq. (6.53). The nonrelativistic and the ultrarelativistic predictions are also shown.

In the range of applied voltage $0.5\,\mathrm{MV} < V_0 < 10\,\mathrm{MV}$, the following equation

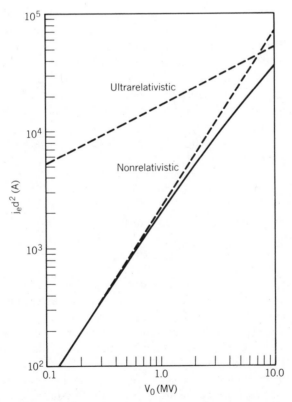

Figure 6.9 Space-charge-limited planar flow of electrons in a relativistic diode—$j_e d^2$ versus V_0. Dashed lines show the predictions of nonrelativistic and ultrarelativistic models.

gives a good approximation for Eq. (6.53):

$$j_e \cong \left[\frac{2e_0 m_e c^3}{ed^2}\right]\left[\left(1 + \frac{eV_0}{m_e c^2}\right)^{1/2} - 0.8471\right]^2. \tag{6.57}$$

As an applications example, suppose we have a 1-MV electron extractor with gap width $d = 0.1$ m. The nonrelativistic Child law predicts a current density of 2.33×10^5 A/m² (23.3 A/cm²). The ultrarelativistic prediction is 5.31×10^5 A/m² (53.1 A/cm²). Inspection of Table 6.3 at $\gamma = 2.96$ shows that $G(2.96) = 1.712$. Substitution in Eq. (6.57) gives a predicted current density of 1.99×10^5 A/m² (19.9 A/cm²), while the approximation of Eq. (6.57) gives 2.07×10^5 A/m² (20.7 A/cm²).

6.6. ONE-DIMENSIONAL SELF-CONSISTENT EQUILIBRIUM

When we derived the self-consistent space-charge flow solutions of Section 5.2, we used a singular distribution—particles were monoenergetic. Although simple distribution functions make calculations easier, they often lead to misleading results when applied in stability analyses. Sometimes, theories based on singular equilibrium distributions predict instabilities that do not occur with more realistic distributions. In this section, we shall solve the general problem of self-consistent transverse beam equilibria in the presence of applied forces, self-forces, and nonzero emittance. The development illustrates how to incorporate more complex distribution functions into an equilibrium theory and how to interpret the results.

We shall adopt some simplifying assumptions to minimize the mathematics of the calculations:

1. We describe a drifting sheet beam with variations only in the x direction. The beam is symmetric about $x = 0$.

2. The transverse focusing force is uniform along the direction of propagation.

3. The beam consists of ions. We shall use nonrelativistic equations of transverse motion and ignore the force of the beam-generated magnetic field.

In the theory of one-dimensional space-charge flow of Section 5.2, only an electric force acted on the particles. To derive a self-consistent solution, we expressed the density in terms of the electrostatic potential and then solved the Poisson equation. In this section, we want to include two forces, the space-charge electric force and a general transverse focusing force. Additional forces may appear in other problems. As the first step, we shall define a generalized potential energy function $U(x)$ that includes contributions from multiple transverse forces. We shall call the function the *confining potential*. Our initial goal is to find an expression for the density of the sheet beam in terms of the confining potential $n(U)$.

We define $U_f(x)$ as the portion of the confining potential associated with a static focusing force $F_f(x)$. The focusing force has odd symmetry about $x = 0$. We exclude even functions since they would lead to net displacements of the beam from the axis of symmetry. The confining potential energy associated with the focusing force is

$$U_f(x) = -\int_0^x F_f(x')dx' + U_f(0). \qquad (6.58)$$

The constant term does not affect the results; therefore, we take $U_f(0) = 0$. The confining potential of a focusing force $U_f(x)$ increases monotonically away from the axis.

We shall also include a confining potential for the defocusing force from the beam-generated electric field $U_s(x)$. The function decreases with displacement. Since the transverse forces add linearly, we can add confining potentials to give a total potential function:

$$U(x) = U_f(x) + U_s(x). \qquad (6.59)$$

In the presence of multiple forces, a particle that moves from position x_1 to x_2 has a change in the kinetic energy

$$\Delta T = U(x_1) - U(x_2). \qquad (6.60)$$

Figure 6.10*a* shows the parabolic confining potential from a linear focusing force. Figure 6.10*b* illustrates the combined confining potentials of a focusing force and a defocusing space-charge force. For the example illustrated, the focusing force exceeds the defocusing force—the total potential function defines a *potential well*. In the potential well, contained particle orbits have stability about the point $x = 0$—a particle with zero kinetic energy at position x oscillates between $\pm x$. When the space-charge force exceeds the focusing force, the total confining potential defines a *potential hill* (Fig. 6.10*c*). Here, particles are repelled from the axis of symmetry. A condition for stable beam transport is that the sum of focusing potentials exceeds the sum of defocusing potentials at all positions in the beam.

Figure 6.10*d* shows the confining potential where the space-charge force equals the focusing force—$U(x)$ is everywhere equal to zero. Since the forces balance at every position, particle orbits in the steady-state beam must be laminar—if this were not true, the beam would expand. The condition for laminar flow is

$$U_s(x) = -U_f(x). \qquad (6.61)$$

For linear focusing forces, Eq. (6.61) can apply only when the space-charge density of the sheet beam is uniform, dropping sharply to zero at the boundary. If

a)

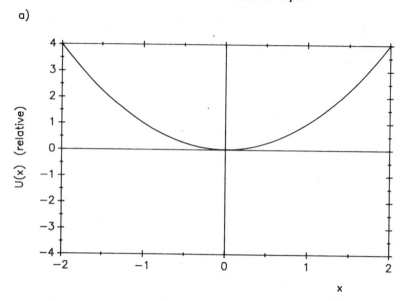

b)

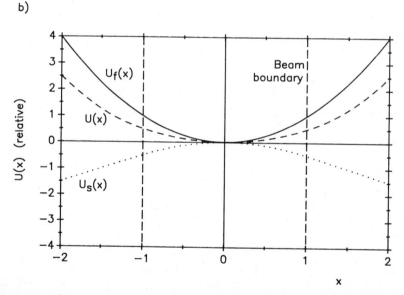

Figure 6.10   Potential energy functions used to calculate focusing and defocusing of particles in a one-dimensional beam. (*a*) Variation of the potential energy function for a focusing force that varies linearly with distance from a symmetry axis. (*b*) Total potential energy function (*U*) with contributions from a linear focusing force (*U*$_f$) and the defocusing space-charge force (*U*$_s$) of a uniform charge-density beam. The dashed lines show the beam boundary. (*c*) Total potential energy function resulting from a linear focusing force and the space charge forces of a high-current, uniform-density beam. (*d*) Total potential energy function resulting from a linear focusing force and the space-charge force of a reduced-current beam.

c)

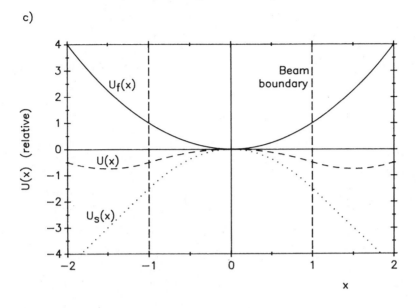

d)

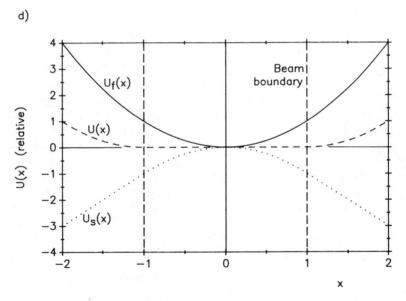

Figure 6.10 (*Continued*)

the beam density is nonuniform, the two forces cannot balance everywhere. Then, an equilibrium exists only if the beam has nonzero emittance.

To understand the nature of nonlaminar beam equilibria, we shall discuss particle orbits in a confining potential well. In equilibrium, the total energy is a constant of particle motion,

$$W_x = m_0 v_x^2/2 + U(x). \tag{6.62}$$

Suppose a sheet beam has envelope dimensions $\pm x_0$. In a nonlaminar beam, particles move back and forth over the cross section. Some of the beam particles reach the envelope—they have a *turning point* equal to x_0. These particles must have zero kinetic energy at x_0, otherwise, the envelope would expand. The total energy of peripheral particles is

$$W_x = U(x_0) = U_0. \tag{6.63}$$

The beam may also contain particles whose orbits do not reach the boundaries. A particle with a turning point $x < x_0$ has total energy less than the maximum value, $W_x < U_0$. For given confining potential U and envelope width x_0, the range of total energy for confined particles is

$$0 \leqslant W_x \leqslant U_0. \tag{6.64}$$

When we know the distribution of particles in W_x and the confining potential, we can find an expression for the particle density as a function of U. We shall illustrate this contention through the example of a monoenergetic beam—all particles have $W_x = U_0$. If we divide the space between $\pm x_0$ into length elements Δx, the particle density in an element is proportional to the number of particles that pass through the element weighted by the relative amount of time they spend there. In a monoenergetic beam, all particles perform identical oscillations between the turning points $\pm x_0$. In equilibrium, the oscillations are uniformly distributed in phase. Furthermore, all particle orbits pass through all volume elements. As a result, the density is inversely proportional to v_x. Equation (6.62) implies that v_x is related to the confining potential by

$$v_x(x) = \sqrt{2[U_0 - U(x)]/m_0}. \tag{6.65}$$

Using Eq. (6.65), we find an expression for the particle density:

$$n(x) = \frac{n_0}{\sqrt{1 - U(x)/U_0}}. \tag{6.66}$$

Figure 6.11 shows the density variation for a parabolic confining potential:

$$U(x) = U_0(x/x_0)^2. \tag{6.67}$$

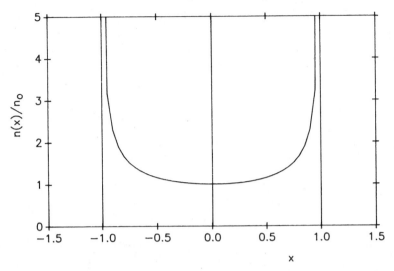

Figure 6.11 Variation of normalized density for a beam with a delta-function distribution in transverse energy in a linear focusing force.

The density has peaks at $\pm x_0$ because particles spend a large fraction of the time there.

We do not expect to observe a discontinuous density function like that of Fig. 6.11 in a real beam. The peaked density function is not consistent with the smooth variation of total confining potential of Eq. (6.67). To generate more realistic density functions, we must use distributions where particles have a spectrum of values of W_x. We shall adopt the methods developed in Section 2.5. For a given distribution function $f(x, v_x)$, the density is

$$n(x) = \int_{-\infty}^{+\infty} dv_x \, f(x, v_x). \tag{6.68}$$

The quantity $f(x, v_x)$ corresponds to an equilibrium if we can express it as a function of W_x, the constant of single-particle motion. For the one-dimensional particle distribution, Eq. (6.68) takes the following form

$$n(U) = \int_{-\sqrt{2(U_0 - U)/m_0}}^{+\sqrt{2(U_0 - U)/m_0}} dv_x f(m_0 v_x^2/2 + U). \tag{6.69}$$

We determine the limits on the velocity integral from Eq. (6.65).

To complete the derivation of a self-consistent equilibrium, we need an equation that relates the confining potential to the particle density. For a sheet beam with space-charge and focusing fields, we can derive such a relationship from the one-dimensional Poisson equation. If U_s is the potential energy associated with electric fields in the beam, then we can write the Poisson equation as

$$\frac{d^2 U_s}{dx^2} = -\frac{e^2 n(U)}{\varepsilon_0}. \tag{6.70}$$

Equation (6.70) leads to the following equation for the total potential:

$$\frac{d^2 U}{dx^2} = \frac{d^2 U_f}{dx^2} - \frac{e^2 n(U)}{\varepsilon_0}. \tag{6.71}$$

The first term on the right-hand side comes from the focusing force—it is a known function of x. If we have an expression for $n(U)$, we can solve Eq. (6.71) with appropriate boundary conditions to find $U(x)$. The confining potential function gives the spatial variations of density, velocity dispersion ($\langle v_x^2 \rangle$), and other quantities.

To illustrate the procedure, we shall follow a specific calculation. We choose a distribution function that allows an analytic solution of Eqs. (6.69) and (6.71). At the same time, the density distribution is physically reasonable—it falls smoothly to zero at the beam boundary. The distribution is uniform over the allowed range of total energy

$$f(x, v_x) = f(W_x) = A, \tag{6.72}$$

where A is a constant. Inserting Eq. (6.72) into Eq. (6.69) gives the following expression for the density:

$$n(U) = 2A \sqrt{2(U_0 - U)/m_0}. \tag{6.73}$$

We can write Eq. (6.73) in terms of the density on axis, $n_0 = n(x = 0)$:

$$n(U) = n_0 \sqrt{1 - U/U_0}. \tag{6.74}$$

We take an applied focusing force that varies linearly with x. The quantity U_{f0} is the magnitude of the confining potential associated with the focusing force at the boundary x_0 or

$$U_{f0} = -\int_0^{x_0} F_f \, dx. \tag{6.75}$$

With the definition of Eq. (6.75), we can express the focusing potential as

$$U_f(x) = U_{f0} (x/x_0)^2. \tag{6.76}$$

Substituting Eqs. (6.76) and (6.74) into Eq. (6.71) gives the following equation:

$$\frac{d^2U}{dx^2} = \frac{2U_{f0}}{x_0^2} - \frac{e^2 n_0}{\varepsilon_0}\left[1 - \frac{U}{U_0}\right]^{1/2}.$$ (6.77)

To simplify Eq. (6.77), we define dimensionless variables:

$$\mathsf{U} = U/U_0,$$ (6.78)

$$\mathsf{X} = x/x_0.$$ (6.79)

In terms of the dimensionless variables, Eq. (6.77) is

$$\frac{d^2\mathsf{U}}{d\mathsf{X}^2} = 2\{F - S\sqrt{1 - \mathsf{U}}\},$$ (6.80)

where

$$F = U_{f0}/U_0,$$ (6.81)

and

$$S = e^2 n_0 x_0^2/2\varepsilon_0 U_0.$$ (6.82)

The two parameters F and S govern the solution of Eq. (6.80). The quantity F is the ratio of the depth of the potential well created by the focusing system to the total confining potential. Inspection of Fig. 6.10c shows that $F \geqslant 1$. The second factor, S, is roughly equal to the ratio of the maximum value of the space-charge potential function to the depth of the total confining potential, $S \sim U_s/U_0$. We recognize that $e^2 n_0 x_0^2/2\varepsilon_0$ is equal to the electrostatic potential energy difference, $e\Delta\phi$, across a beam with uniform density n_0 from 0 to x_0. If the beam density is zero, there is no space charge and the focusing force makes the only contribution to the total confining potential. Then, $F = 1$, and the beam equilibrium represents a balance between emittance and the focusing force. For nonzero S, the space-charge electric fields contribute to the force balance and F must be larger than unity.

We find a relationship between F and S that describes the nature of the equilibrium by solving Eq. (6.80) with the following boundary conditions:

$$\mathsf{U}(\mathsf{O}) = 0,$$ (6.83)

$$\mathsf{U}(1) = 1,$$ (6.84)

$$d\mathsf{U}(0)/d\mathsf{X} = 0.$$ (6.85)

The first condition is a convention—we take $\mathsf{U} = 0$ at the origin. The second condition follows from the definition of the normalized variables [Eq. (6.78)]. The third condition reflects the symmetry of the applied forces about the axis. We can calculate the first integral of Eq. (6.80):

$$\frac{dU}{dX} = \{4FU + \tfrac{8}{3}S[(1 - U)^{3/2} - 1]\}^{1/2}. \tag{6.86}$$

Integration of Eq. (6.86) gives

$$\int_0^1 dY / \{4FY + \tfrac{8}{3}S[(1 - Y)^{3/2} - 1])^{1/2} = 1. \tag{6.87}$$

Figure 6.12a shows S as a function of F determined from Eq. (6.87). When the beam density is small, S approaches zero and F approaches unity. Here, the focusing force is the only contribution to the confining potential, and there is a balance between the focusing force and the emittance force of the beam. When the effect of space charge is high ($S \gg 1$), the curve approaches the condition $S = F$, a laminar flow equilibrium with no emittance. Figure 6.12b plots the focusing, space-charge, and total potentials for an intermediate case with $F = 2.5$ and $S = 1.75$. Emittance and space charge contribute to the force balance. Figure 6.12c shows the beam density for the same parameters.

To complete this section, we shall briefly discuss calculations of self-consistent beam equilibria in two dimensions. The treatment of paraxial beams in axially uniform transverse forces is similar to the one-dimensional theory that we have discussed. One difference is that there are two constants of motion when there are two degrees of freedom. For example, when forces in the x and y directions are independent and decoupled, the total confining potential is the sum of x and y contributions:

$$U(x, y) = U_x(x) + U_y(y). \tag{6.88}$$

The total particle energies in the x and y directions are independent conserved quantities. The two constants of particle motion are

$$W_x = m_0 v_x^2/2 + U_x(x), \qquad W_y = m_0 v_y^2/2 + U_y(y). \tag{6.89}$$

Any function of W_x and W_y is a valid distribution function. For example, the two-dimensional form of a Maxwell distribution is

$$f(x, v_x, y, v_y) = A \exp(-W_x/kT_x)\exp(-W_y/kT_y). \tag{6.90}$$

The equilibrium distribution function depends on the confining potentials and the velocities, $f(W_x, W_y) = f(U_x, v_x, U_y, v_y)$. We can find the density as a function of the confining potentials by taking integrals over velocity:

$$n(U_x, U_y) = \int_{-\infty}^{+\infty} dv_x \int_{-\infty}^{+\infty} dv_y\, f(U_x, v_x, U_y, v_y). \tag{6.91}$$

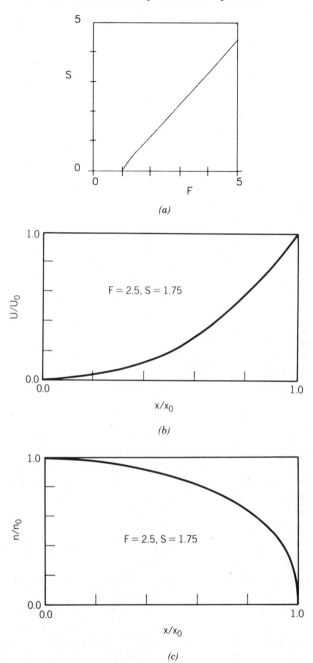

Figure 6.12 Self-consistent equilibrium theory for a sheet beam. (*a*) Space-charge function, $S = e^2 n_0 x_0^2 / 2\varepsilon_0 U_0$, versus the emittance function, $F = U_{f0}/U_0$. (*b*) Variation of the normalized total potential energy function versus normalized position for $F = 2.5$, $S = 1.75$. (*c*) Variation of the normalized density versus normalized position for $F = 2.5$, $S = 1.75$.

When particle motions are coupled in the x and y directions, we must use alternate constants of the motion. An example of coupled motion is the confinement of a cylindrical electron beam in a solenoidal magnetic field. We can find the constants of motion by considering the symmetry of the system. One constant is the total transverse energy. In a uniform cylindrical beam, the transverse energy varies only with radius:

$$W(r) = U_s(r) + \gamma m_0 v_r^2/2 + \gamma m_0 v_\theta^2/2. \tag{6.92}$$

Since there are no forces in the azimuthal direction, the other constant of particle motion is the canonial angular momentum

$$P_\theta = \gamma m_0 r v_\theta - e r A_\theta. \tag{6.93}$$

The quantity A_θ is the vector potential from the applied and beam-generated magnetic fields. Any distribution function of the form $f(W, P_\theta)$ gives a valid equilibrium.

6.7. KV DISTRIBUTION

Many accelerators used FD quadrupole arrays for transport. The derivation of a self-consistent equilibria for beams in such systems is more difficult than the simple one-dimensional model of Section 6.6 for two reasons:

1. Beam properties vary in both the x and y directions. At any axial location, the strength of the focusing system is different in the two directions.

2. The transverse forces of the periodic focusing system vary along the direction of propagation. Also, in accelerators such as synchrotrons and RF linacs, the beam density varies in the axial direction.

General derivations require a solution of the full three-dimensional Poisson equation to find the beam-generated fields. Usually, the fields cannot be represented by closed-form expressions. Iterative numerical processes are necessary to find a self-consistent equilibrium.

To construct analytic models for beam equilibria, we must seek special geometries that lead to simple expressions for beam-generated electric and magnetic fields. In this section, we shall discuss two important components for a description of beams in quadrupole arrays—calculation of beam-generated fields for a three-dimensional beam pulse and determination of beam charge density from a three-dimensional distribution. We shall concentrate on a charge density and distribution function that lead to closed-form field expressions. We shall first study the properties of an ellipsoid with uniform charge density. The configuration has the important properties that the electric fields in the three spatial directions are linear and separable. Since the forces in a quadrupole array are separable in x and y and do not depend on the accelerating forces, we can resolve the self-consistent equilibrium problem for an ellipsoid into three one-dimensional calculations. The critical step to close the self-consistent calculation is to find a distribution function for a nonlaminar beam that gives an uniform ellipsoid of charge when integrated over velocity. The KV distribution, widely used in accelerator theory, fulfills this condition.

Figure 6.13 shows a three-dimensional space-charge ellipsoid. The figure could represent a rest frame view of a microbunch of particles in an RF linear accelerator. We denote the rest frame coordinates as (X, Y, Z). The ellipsoid may have different dimensions in the three directions (X_0, Y_0, Z_0). The space-charge density in the rest frame has a uniform value ρR within a boundary defined by the equation

$$\left[\frac{X}{X_0}\right]^2 + \left[\frac{Y}{Y_0}\right]^2 + \left[\frac{Z}{Z_0}\right]^2 = 1. \tag{6.94}$$

The stationary space-charge creates only electric fields. Following Kapchinskij and Vladimirskij[2], the electric fields inside the ellipsoid are separable and vary linearly with X, Y, and Z. The relative strength of the electric field components depends on form factors M_i that are functions of X_0, Y_0, and Z_0:

$$E_X = (\rho_R/\varepsilon_0)\, M_x(X_0, Y_0, Z_0)X, \tag{6.95}$$

$$E_Y = (\rho_R/\varepsilon_0)\, M_Y(X_0, Y_0, Z_0)Y, \tag{6.96}$$

$$E_Z = (\rho_R/\varepsilon_0)\, M_Z(X_0, Y_0, Z_0)Z. \tag{6.97}$$

The general expression for the form factors is

$$M_i = \frac{X_0 Y_0 Z_0}{2}\int_0^\infty \frac{d\tau}{(A_i^2 + \tau)[(X_0^2 + \tau)(Y_0^2 + \tau)(Z_0^2 + \tau)]^{1/2}} \tag{6.98}$$

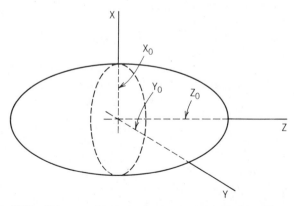

Figure 6.13 Geometry of a space-charge ellipsoid viewed in the beam rest frame.

[2] I. M. Kapchinskij and V. V. Vladimirskij, *Proc. Intl. Conf. High Energy Accelerators*, CERN, Geneva, 1959, p. 274.

where $i = x, y, z$ and $A_i = X_0, Y_0$, or Z_0. The form factors have the property that

$$M_x + M_y + M_z = 1. \tag{6.99}$$

In the special case of a cigar-shaped (prolate) ellipsoid with a circular cross section $(X_0 = Y_0)$, the form factors satisfy the equation

$$M_x = M_y = (1 - M_z)/2. \tag{6.100}$$

In the range $X_0/Z_0 \leqslant 1$, the quantity M_z is

$$M_z = \frac{1 - \Gamma^2}{\Gamma^3} \left[\frac{1}{2} \ln \left(\frac{1 + \Gamma}{1 - \Gamma} \right) - \Gamma \right], \tag{6.101}$$

where

$$\Gamma = (1 - X_0^2/Z_0^2)^{1/2}. \tag{6.102}$$

In the opposite limit of an oblate or saucer-shaped ellipsoid $(X_0/Z_0 > 1)$, the following expressions hold:

$$M_z = \frac{1 + \Gamma^2}{\Gamma^3} [\Gamma - \tan^{-1}(\Gamma)], \tag{6.103}$$

and

$$\Gamma = (X_0^2/Z_0^2 - 1)^{1/2}. \tag{6.104}$$

Usually, we want expressions for the beam-generated forces in the stationary frame. Here, the centroid of the ellipsoid moves axially at velocity βc. In an RF linac, the centroid of the charge bunch corresponds to the position of the synchronous particle z_s. We can use Eqs. (5.105)–(5.110) to transform the rest frame electric fields to electric and magnetic fields in the stationary frame. The net axial beam force in the stationary frame is

$$F_z = \gamma(q\rho_R/\varepsilon_0) M_z(X_0, Y_0, Z_0)(z - z_s), \tag{6.105}$$

where $\gamma = (1 - \beta^2)^{-1/2}$. We must remember to use the *rest frame* dimensions of the ellipsoid to calculate M_z. If the beam has dimensions x_0, y_0, and z_0 in the stationary frame, then we can write the form factor as $M_z(x_0, y_0, \gamma z_0)$. The expression of Eq. (6.105) equals the rest frame electric force multiplied by γ. The factor results from Lorentz contraction of the ellipsoid length; the space-charge density observed in the stationary frame is higher:

$$\rho_0 = \gamma\rho_R. \tag{6.106}$$

We can rewrite Eq. (6.105) in terms of the rest frame charge density:

$$F_z = (q\rho_0/\varepsilon_0) M_z(x_0, y_0, \gamma z_0)(z - z_s). \tag{6.107}$$

We can carry out similar transformations for the transverse field components. We must include the magnetic field that results from motion of the charge in the stationary frame. The Lorentz contraction increases the observed charge and current density, but the balance of electric and magnetic forces reduces the net force. The total transverse forces are

$$F_x = (1/\gamma^2)(q\rho_0/\varepsilon_0) M_x(x_0, y_0, \gamma z_0)x, \tag{6.108}$$

$$F_y = (1/\gamma^2)(q\rho_0/\varepsilon_0) M_y(x_0, y_0, \gamma z_0)y. \tag{6.109}$$

Equations (6.107), (6.108), and (6.109) apply to an ellipsoid in free space. Conducting boundaries close to the beam modify the forces. Section 14.4 discusses field contributions from induced wall charges and currents.

The Kapchinskij–Vladimirskij (KV) distribution is a special equilibrium distribution that gives a uniform particle density within a sharp ellipsoidal boundary. In our discussion of the KV distribution, we shall limit attention to the two-dimensional form in trace space. Transverse motion is decoupled from axial motion. The KV distribution has the mathematical form

$$g_{KV}(x, x', y, y') = A\,\delta\left[\frac{x^2}{x_0^2} + \frac{y^2}{y_0^2} + \frac{x_0^2 x'^2}{\varepsilon_x^2} + \frac{y_0^2 y'^2}{\varepsilon_y^2} - 1\right], \tag{6.110}$$

where δ is the Dirac delta function. The emittances in the x and y directions are

$$\varepsilon_x = x_0 x_0' \tag{6.111}$$

and

$$\varepsilon_y = y_0 y_0'. \tag{6.112}$$

All particles in the distribution of Eq. (6.110) follow trace-space trajectories that lie on the surface of a four-dimensional hyperellipse with axes $x_0, x_0', y_0,$ and y_0'.

We can understand the physical meaning of the KV distribution by first discussing an analogous distribution in one spatial dimension. Consider a sheet beam with particles confined in the x direction by a linear force that is uniform in in the axial direction. Orbits have maximum displacement x_0 and maximum inclination angle x_0'. Particles follow elliptical trace-space trajectories—if they have the same total transverse energy, they all follow the curve

$$\frac{x^2}{x_0^2} + \frac{x'^2}{x_0'^2} - 1 = 0 \tag{6.113}$$

at different phase angles.

A delta function in total transverse energy is a valid equilibrium distribution. The trace-space distribution function for particles that follow the curve of

Eq. (6.113) is

$$g(x, x') = A\delta\left[\frac{x^2}{x_0^2} + \frac{x_0^2 x'^2}{\varepsilon_x^2} - 1\right].$$

(6.114)

The emittance in the x direction is $\varepsilon_x = x_0 x_0'$. We find the particle density by integrating Eq. (6.114) over all angles:

$$n(x) = \int_{-\infty}^{\infty} dx' g(x, x').$$

(6.115)

As before, we must take care to convert Eq. (6.115) to a form where we can apply a standard delta function integral. Making the substitution $\chi = x_0^2 x'^2/\varepsilon_x^2$, the equation becomes

$$n(x) = \frac{A\varepsilon_x}{2x_0} \int_{-\infty}^{\infty} \frac{d\chi}{\sqrt{\chi}} \delta\left[\chi - \left(1 - \frac{x^2}{x_0^2}\right)\right].$$

(6.116)

Applying Eq. (5.57), we find the particle density as

$$n(x) = n_0/[1 - (x/x_0)^2]^{1/2},$$

(6.117)

where $n_0 = A\varepsilon_x/2x_0$. Equation (6.117) is the familiar density expression for a collection of monoenergetic harmonic oscillators. The density is proportional to the relative time spent at a position x as particles follow their trace-space trajectories—its value diverges at the turning points.

The KV distribution is a delta function in net transverse energy in x and y. The nature of the density integral changes when we include motion in both the x and y directions. The density for a two-dimensional distribution is

$$n(x, y) = A \int_{-\infty}^{\infty} dx' \int_{-\infty}^{\infty} dy' g(x, x', y, y').$$

(6.118)

Substituting Eq. (6.110) for $g(x, x', y, y')$, we can follow the method used for the one-dimensional distribution to perform a first integral:

$$n(x, y) = \left(\frac{A\varepsilon_y}{2y_0}\right) \int_{-\infty}^{\infty} dx' \left[1 - \frac{x^2}{x_0^2} - \frac{y^2}{y_0^2} - \frac{x'^2 x_0^2}{\varepsilon_x^2}\right]^{-1/2}.$$

(6.119)

We can also perform a second integral. With the substitutions $\alpha^2 = [1 - (x/x_0)^2 - (y/y_0)^2]$ and $Y = \alpha \sin(x' x_0/\varepsilon_x)$, Eq. (6.119) becomes

$$n(x, y) = \left(\frac{A\varepsilon_y \varepsilon_x}{2y_0 x_0}\right) \int_{-\pi/2}^{\pi/2} dY \frac{\cos Y}{[1 - \sin^2 Y]^{1/2}} = \left(\frac{A\pi\varepsilon_y \varepsilon_x}{2y_0 x_0}\right).$$

(6.120)

We thus find that the two-dimensional KV distribution indeed gives a uniform beam density. The result is different from the one-dimensional calculations because particle orbits corresponding to Eq. (6.110) have a maximum velocity in the y direction when they have a minimum velocity in the x direction. There is no point where particles have zero transverse velocity; therefore, there is no location where the density diverges.

Given the properties of the KV distribution, we can sketch the procedure for a self-consistent equilibrium calculation. Suppose we have linear focusing forces that are axially uniform and separable in the x and y directions. The total force on particles is the sum of applied forces and beam-generated forces. If we have a uniform density beam with an elliptical cross section, the beam forces are also linear. If the net force is linear, we know that particles follow ellipses in trace space and the KV distribution gives a uniform-density elliptical beam. We therefore have a logically consistent structure. To complete the process, we need to choose normalizations to guarantee force balance. Suppose we start with a given beam emittance; in other words, we specify the net transverse energy in the KV distribution. We can find the magnitude of the net forces that gives a particle orbit with dimensions (x_o, x'_0, y_0, y'_0). For a given beam charge density and defocusing beam force, we can then find the applied force that gives the required net transverse force.

We can apply the KV distribution to model space-charge effects in high-energy accelerators with quadrupole focusing. The transverse forces are linear and separable in x and y. The added complication is that the applied forces vary in the axial direction. From the discussion of Sections 4.1 through 4.3, we expect that the orientation of the trace-space distribution and the configuration-space shape of the beam varies along z. In principle, we can extend the analysis of Section 4.2 to find a matched distribution at a cell boundary of a periodic focusing system that includes the beam-generated forces of a KV distribution. Here, the projections of the distribution ellipsoid in the x–x' and y–y' planes may be skewed. Although the KV distribution is helpful in deriving beam equilibria, we must exercise care in using the equilibria as a basis for a stability analysis. The KV distribution is singular and physically unrealistic, leading to predictions for instabilities that often are not observed.

7

Electron and Ion Guns

The first step in the charged particle acceleration process is to extract low-energy particles from a source and to form them into a beam. The particle source and initial acceleration gaps constitute the injector. Although an injector may represent only a small fraction of the cost and size of a high-energy accelerator, it often presents the most difficult physical and technological problems. The particles move slowly in the first acceleration gap, and the space-charge forces are correspondingly strong. The limitations of the injector are often the main constraint on the performance of a large accelerator.

In this chapter, we shall review methods to create beams of electrons and ions. The initial acceleration gaps present a challenging theoretical problem. Both transverse and axial forces are important—we must deal with three-dimensional variations of electric and magnetic fields. Section 7.1 introduces the Pierce design technique for charged particle guns. We shall derive a self-consistent solution for the flow of a finite-width beam. Section 7.2 summarizes analytic techniques to design practical guns for moderate-current beams. Section 7.3 extends the discussion to the highest currents available from conventional guns. Here, we must use numerical techniques to derive the shapes of electrodes. The section reviews the physical principles of ray-tracing computer programs.

The next three sections cover sources of electrons and ions for high-flux beams. Section 7.4 reviews electron sources, including dispenser cathodes and laser-driven photocathodes. Section 7.5 covers the physical basis of ion extraction from free plasmas. We shall see how the properties of particle flow in the plasma and the extraction gap combine to define a shaped plasma emission surface. Successful extraction from a free plasma surface demands a stable, highly uniform

262

plasma source. Section 7.6 gives a brief review of the types of plasma sources that have been used for high-flux ion beams.

Section 7.7 describes electrostatic confinement of a plasma and application of the process to particle extraction from a controlled plasma surface. With this method, the flow of ions or electrons in the extraction gap is independent of the properties of the plasma source. Section 7.8 reviews modifications of ion extractors to achieve large-area, high-current ion beams. The techniques include multiple extraction apertures, accel–decel gaps for plasma neutralization, and multiple-gap extractor designs.

7.1. PIERCE METHOD FOR GUN DESIGN

The analytic derivation of Pierce[1] gives a self-consistent solution for a space-charge-dominated injector. The procedure predicts the shapes of accelerating electrodes to produce a laminar beam with uniform current density. Although the treatment holds only for the special geometry of a sheet beam accelerated through an extraction grid, it gives valuable insights into the design of more complex guns.

We assume that a space-charge-limited injector creates a sheet beam of width $\pm x_0$. The source and extractor electrodes have an electrostatic potential difference of v_0. Particles emerge from the source region with negligible kinetic energy compared with eV_0. We can find an analytic solution with the following limiting assumptions:

1. Particle motion in the extraction gap is nonrelativistic.

2. The force from beam-generated magnetic fields is small.

3. Potentials at the source and extractor electrodes are determined by conducting surfaces—the beam exits the gap through a grid or foil.

We shall use sign conventions for electrons in the following discussion—the extension to ions is straightforward. Conditions 1 and 2 are satisfied if the electron energy is in the range $eV_0 \leqslant 100 \, \text{keV}$. The theory is useful for the electron guns in high-power traveling-wave tubes and klystrons.

Section 5.2 showed how to calculate the space-charge-limited flow of a uniform current density electron beam of infinite cross section. For a finite-width beam, we would like to add boundary effects in a way that preserves uniform current density and laminar flow. Figure 7.1a shows a sheet beam of infinite cross section. If the source is located at $z = 0$ and the extractor electrode at $z = d$, then the potential across the gap varies as

$$\phi(x, y, z)/V_0 = (z/d)^{4/3}, \tag{7.1}$$

[1] J. R. Pierce, *Theory and Design of Electron Beams*, Van Nostrand, Princeton, NJ, 1949.

for space-charge-limited flow. Figure 7.1*b* shows a bounded beam. A source in the lower half plane ($x < 0$) generates electrons.

We seek a geometry where there are no electrons for $x > 0$ (Region 1) but the flow in the lower half plane retains the properties of the infinite sheet beam (Region 2). To satisfy this condition, we invoke the superposition property of electrostatic solutions. Suppose we place a large number of electrodes along the boundary at $x = 0$ (Fig. 7.1*c*) and bias them so that the potential varies as

$$\phi(0, y, z) = V_0(z/d)^{4/3}. \tag{7.2}$$

In other words, the potential on the boundary follows the variation it would have if Region 1 contained an extension of the beam. Therefore, the electric fields in Region 2 are purely axial and have the same variation as those in an infinite-width beam. The biased boundary simulates the effect of the missing portion of the beam in Region 1—particle orbits in Region 2 are laminar and the current density is uniform.

Figure 7.1*d* shows an alternative method to duplicate the infinite beam condition along the boundary. Electrodes in the upper half plane create an electrostatic potential. The potential function in Region 1 must satisfy two conditions at the boundary to match the solution in Region 2:

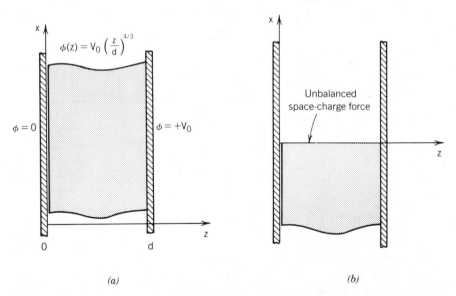

(a)   *(b)*

Figure 7.1 Basis of the Pierce design procedure for a space-charge-dominated electron gun. (*a*) Planar gun with beam of infinite width—the electron source covers the left-hand boundary. (*b*) Electron beam with a sharp boundary—electron source only in region $x < 0$. (*c*) One method to correct electron orbits in a bounded beam by setting electrostatic potential on the boundary equal to $\phi(0, z) = V_0(z/d)^{4/3}$. (*d*) Shaped electrodes to establish correct potential variation along the beam boundary.

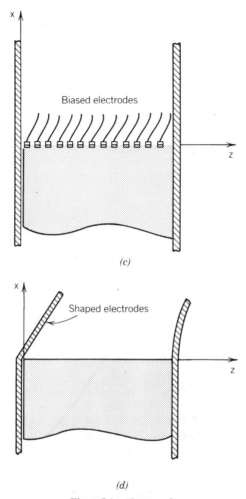

(c)

(d)

Figure 7.1 (*Continued*)

1. The potential must follow Eq. (7.2) at $x = 0$. This condition is equivalent to matching values of the axial electric field at the boundary.

2. The electric field in the x direction must equal zero at the boundary, since there are no transverse electric fields in Region 2, or $\partial\phi/\partial x = 0$.

To find the proper electrode shapes, we must solve the Laplace equation in the upper half plane with specified boundary conditions. For the Cartesian geometry of Fig. 7.1b, there is a quick method to solve such problems using the properties of complex functions. Take the complex variable u as a linear combination of the real coordinate variables,

$$u = z + jx, \qquad (7.3)$$

where $j = \sqrt{-1}$. We can define functions of the complex variable $f(u)$. An *analytic* function of a complex variable varies smoothly and has finite derivatives over the region of interest. Two examples are $f(u) = u^3$ and $f(u) = \exp(-u)$. We can show that any analytic function u satisfies the Laplace equation

$$\frac{\partial^2 f}{\partial z^2} + \frac{\partial^2 f}{\partial x^2} = 0. \tag{7.4}$$

We can verify Eq. (7.4) by using the chain rule of partial derivatives. Rewriting the first derivatives of the Laplacian operator gives

$$\nabla^2 f = \frac{\partial}{\partial z}\left(\frac{\partial f}{\partial u}\frac{\partial u}{\partial z}\right) + \frac{\partial}{\partial x}\left(\frac{\partial f}{\partial u}\frac{\partial u}{\partial x}\right)$$

$$= \frac{\partial}{\partial z}\left(\frac{\partial f}{\partial u}\right) + \frac{\partial}{\partial x}\left(j\frac{\partial f}{\partial u}\right) = 0. \tag{7.5}$$

A second application of the chain rule of derivatives shows that the Laplacian operator applied to $f(u)$ gives a result that is identically zero:

$$\frac{\partial^2 f}{\partial u^2} + (j^2)\frac{\partial^2 f}{\partial u^2} \equiv 0. \tag{7.6}$$

Equation (7.4) implies that the real part of any analytic complex function is a valid form for the electrostatic potential. The function $\phi = \text{Re}(f)$ automatically satisfies the static Maxwell equations with no space charge.

We can generate an infinite set of analytic functions, all giving valid electrostatic solutions. We shall identify a particular function related to the space-charge flow problem:

$$f(u) = V_0\left(\frac{u}{d}\right)^{4/3} = \left(\frac{z + jx}{d}\right)^{4/3}. \tag{7.7}$$

The corresponding electrostatic potential is

$$\phi(x, z) = V_0\,\text{Re}\left(\frac{z + jx}{d}\right)^{4/3}. \tag{7.8}$$

Equation (7.8) satisfies Eq. (7.2) along the boundary at $x = 0$.

To extract the real part of the potential, it is convenient to express the complex function in the polar coordinates:

$$z = \rho\cos\theta, \qquad x = \rho\sin\theta. \tag{7.9}$$

Equation (7.8) can be written in terms of the complex exponential function

$$e^{j\theta} = \cos\theta + j\sin\theta, \tag{7.10}$$

as

$$\frac{\phi}{V_0} = \mathrm{Re}\left(\frac{\rho}{d}e^{j\theta}\right)^{4/3} = \left(\frac{\rho}{d}\right)^{4/3}\mathrm{Re}(e^{4j\theta/3}). \tag{7.11}$$

We use Eq. (7.10) to express the complex expotential in terms of trigonometric functions and then take the real part to derive the result

$$\frac{\phi}{V_0} = \mathrm{Re}\left(\frac{\rho}{d}e^{j\theta}\right)^{4/3} = \left(\frac{\rho}{d}\right)^{4/3}\mathrm{Re}(e^{4j\theta/3}). \tag{7.12}$$

where

$$\rho = (x^2 + z^2)^{1/2}, \qquad \theta = \tan^{-1}(x/z).$$

We can apply the chain rule of derivatives to Eq. (7.12) to verify that $\partial\phi/\partial x = 0$ at $x = 0$.

We can use Eq. (7.12) to find the shape of conducting electrodes in Region 1 that will generate the correct fields. Suppose we take equipotential lines corresponding to two values of potential, ϕ_1 and ϕ_2. If we add conducting plates biased to ϕ_1 and ϕ_2 that follow the lines, the electrostatic solution between the plates is unchanged. We shall choose $\phi_1 = 0$, the source potential, and $\phi_2 = V_0$, the extractor potential. Figure 7.2 shows the shape of the source and extractor electrodes in Region 1 and some intervening equipotential lines. For $x > 0$, the source electrode lies on the curve

$$4\theta/3 = \pi/2. \tag{7.13}$$

Equation (7.13) represents a straight line oriented at 67.5° with respect to the z axis as shown. The line inclines at 22.5° with respect to the electron source. The extractor curve follows the equation

$$(\rho/d)^{4/3}\cos(4\theta/3) = 1. \tag{7.14}$$

Within the beam volume, the source and extractor electrodes follow the straight lines $z = 0$ and $z = d$. We can add another boundary and a set of electrodes at position $x = -2x_0$ without affecting the solution at $x = 0$. The electrodes in the vacuum region are the mirror image of those in Region 1. In a sense, the electrodes outside the beam aim the particles to achieve laminar flow. The most important component is the slanted electrode adjacent to the source. This surface, known as a *focusing* or *Pierce electrode*, bends electric fields to generate focusing forces near the source. The electric force counteracts the defocusing beam-generated forces on the edge of the beam.

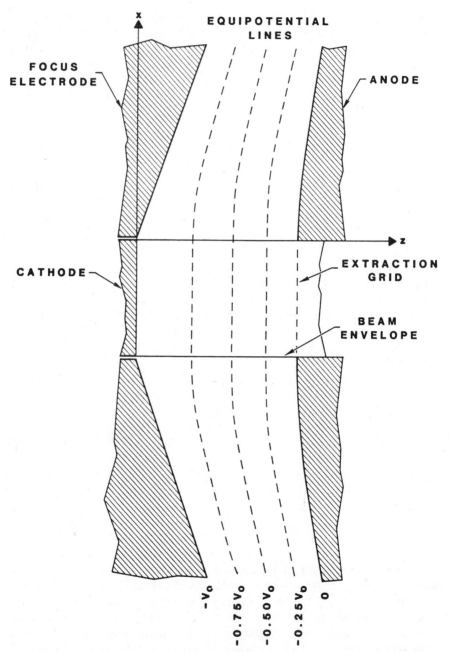

Figure 7.2 Geometry of planar Pierce gun showing calculated equipotential lines within the beam and in vacuum regions.

Many applications require cylindrical electron beams. The design of a cylindrical gun follows the same procedure as a sheet beam gun, although we cannot use the method of complex functions to find an analytic solution. We can apply numerical methods to search for cylindrical electrode shapes that give the variation of potential along a beam boundary at r_0:

$$\phi(r_0, z)/V_0 = (z/d)^{4/3}. \tag{7.15}$$

Figure 7.3 shows a cylindrical gun with focusing electrode and shaped anode. The electrode shapes are close to those for a sheet beam gun when $r_0 \gg d$.

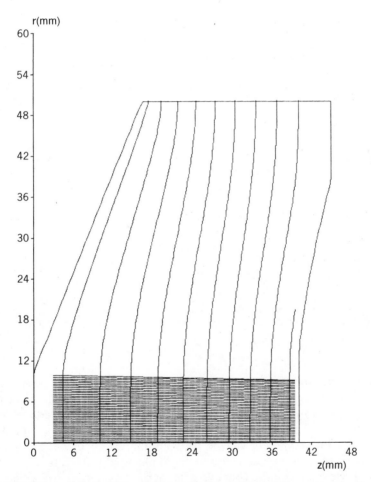

Figure 7.3(a) Numerical calculation of electron flow in a cylindrical Pierce gun using the EGUN code. $V_0 = 50\,\text{kV}$, $I = 4.71$ A. Electrode boundaries, computational rays, and lines of constant electrostatic potential at 2.5, 7.5, 12.5, 17.5, 22.5, 27.5, 32.5, 37.5, 42.5, and 47.5 kV. The magnetic field generated by the beam causes a slight convergence of the electron orbits. Nonrelativistic Child law predicts 5.1 A current.

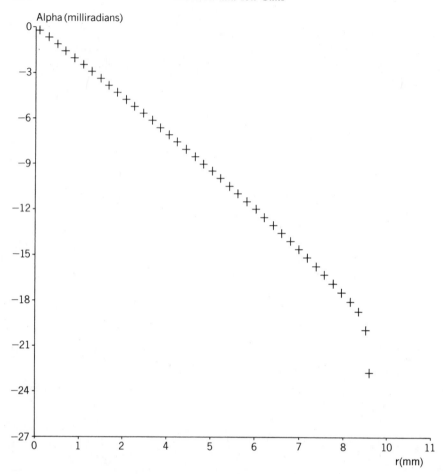

Figure 7.3(b)   Trace-space plot of rays in the anode plane. The output beam is almost laminar with small deviations at large radius resulting from unbalanced electric and magnetic forces.

There is one type of space-charge-limited cylindrical gun that we can design analytically. Figure 7.4 illustrates the *Pierce column*. The design principle is similar to that illustrated in Fig. 7.1c. We locate several ring electrodes along the beam boundary. The rings are biased so that the potential varies roughly as $(z/d)^{4/3}$. The Pierce column can be used for the extraction of space-charge-dominated ion beams at high voltage (≥ 1 MV). Usually, the column of a high-voltage acceleration column contains grading rings. A resistor string biases the rings to impose an even electric field along a stack of insulators. Figure 7.4 shows one method to achieve the proper boundary condition along the beam edge by using reentrant structures to vary the electrode-to-ring spacing. The rings concentrate the axial electric field near the output. The Laplace equation shows

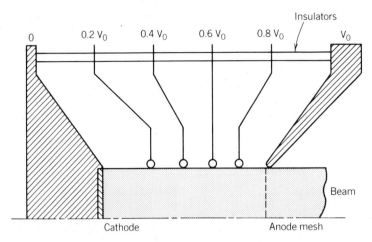

Figure 7.4 Pierce column injector. Electrode spacing establishes the electrostatic potential as $\phi(z) \cong V_0(z/d)^{4/3}$ along the beam envelope.

that there are radial electric fields when there is a gradient of axial electric field. The forces of the radial electric field are just sufficient to balance the space-charge defocusing of the ion beam.

7.2. MEDIUM-PERVEANCE GUNS

Beams exit the planar Pierce gun of Section 7.1 through a conducting mesh or foil anode. This approach is sufficient for low-current or low-duty-cycle guns where the beams have small average power density. Many applications call for high-power-density beams that would quickly melt an anode mesh. These beams must exit the extraction gap through an aperture in the output electrode.

An anode aperture modifies the electric fields in an electron gun. Figure 7.5 shows the distortion of equipotential lines in an acceleration gap near a hole. The radial electric fields defocus exiting electrons. The fields act like an electrostatic lens with negative focal length—the defocusing action is called the *negative lens effect*. Also, the anode aperture reduces the axial electric field at the center of the cathode, leading to depressed beam current density. The change in cathode electric field is small if the diameter of the anode aperture is small compared with the gap width

$$2r_a \ll d. \tag{7.16}$$

The quantity r_a is the radius of the anode aperture. In the limit of Eq. (7.16), the motion of electrons is close to that predicted by the Pierce solution. On

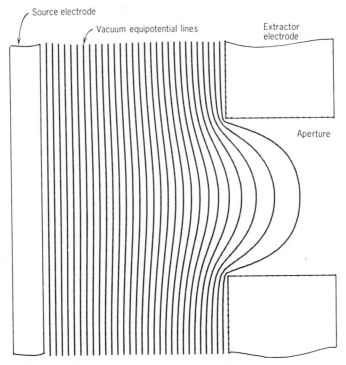

Figure 7.5 Lines of constant electrostatic potential between a planar electrode and an electrode with an extraction aperture. Numerical computation using the POISSON code.

the other hand, if

$$2r_a \geqslant d, \tag{7.17}$$

the field perturbation is strong. Then we must modify the geometry of the gun to achieve an output beam with uniform current density.

We can write the condition of Eq. (7.16) in a form that illustrates the ranges of current and voltage in a cylindrical gun that satisfy the Pierce derivation. Taking the source radius equal to the aperture radius r_a, we apply Eq. (5.48) to estimate the net current from the planar gun:

$$I = \left[\frac{4\varepsilon_0}{9}\right]\left[\frac{2e}{m_e}\right]^{1/2}\left[\frac{\pi r_a^2}{d^2}\right]V_0^{3/2}. \tag{7.18}$$

If we divide both sides of Eq. (7.18) by $V_0^{3/2}$, we find that $(I/V_0^{3/2})$ depends only on the geometry of the extractor and the type of particle. We call this quantity the gun *perveance*,

$$P = I/V_0^{3/2}. \tag{7.19}$$

The unit of perveance is the perv, equal to 1 amp/(volt)$^{3/2}$. If we include the condition of Eq. (7.16), we find that aperture effects are small when the perveance is in the range

$$P \ll \left[\frac{\pi \varepsilon_0}{9} \right] \left[\frac{2e}{m_e} \right]^{1/2} = 0.6 \times 10^{-6}. \tag{7.20}$$

Equation (7.20) implies that the planar Pierce derivation of Section 7.1 applies to apertured cylindrical guns less than 1 μperv. For reference, a 1-μperv, 20-kV electron gun has a current of 2.8 A, while a proton gun has a current of 0.12 A. We should note that limitations on gun perveance are significant only for cylindrical guns with apertures. There is no perveance limit for guns with grids. High perveance also is possible with alternative geometries such as sheet or annular beams.

For cylindrical guns with perveance in the range 1 μperv and above, we must apply a different design procedure. In this section, we shall concentrate on methods for moderate perveance guns, $P \leqslant 1 \mu$perv. These guns usually have the converging geometry of Fig. 7.6. Here, a focusing electrode surrounds a concave cathode. The shaped electrodes produce a converging electron beam that passes through the anode aperture. Compared with a planar gun, the converging gun has several advantages:

1. The aperture diameter can be small because the beam has the minimum radius at the anode.

2. A source with limited current density can generate a high-current beam since the beam width is large at the cathode.

3. The space-charge-limited current for a given aperture area is higher than that of a planar gun since the beam density is smaller near the cathode.

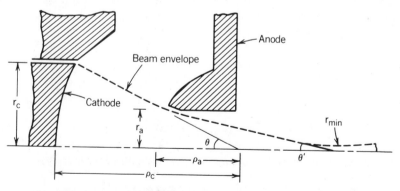

Figure 7.6 Converging gun geometry for a beam with moderate perveance.

4. With converging electrons, it is possible to counter the negative lens effect to generate a parallel output beam.

For small aperture perturbations, we can design a nonrelativistic electron gun by dividing beam motion through the extractor into three roughly independent phases:

1. We treat electron motion from the cathode to the anode using the theory of space-charge-limited flow between sperical electrodes (Section 6.3).
2. We assume that aperture field perturbations are localized near the anode and represent their effect as a thin linear lens with negative focal length.
3. In the propagation region beyond the anode, we treat space-charge flow of the beam using the paraxial theory of Section 5.4.

First, we shall discuss converging electron flow between spherical electrodes—Fig. 7.6 shows the geometry. The cathode and anode are spherical segments referenced to the same center. They have radii of curvature ρ_c and ρ_a. Although the electrodes do not comprise complete spheres, we can apply the results of Section 6.3 by extending the method Section 7.1. We look for a set of electrodes to place outside the beam volume that produce a potential variation along the beam envelope that replicates the spherical flow solution of Section 6.3. Trial-and-error calculations using numerical or analog techniques can be used to find the electrode shapes.

The perveance of a full spherical electron beam is

$$\frac{I_{\text{sphere}}}{V_0^{3/2}} = \frac{4\varepsilon_0}{9} \left[\frac{2e}{m_e} \right]^{1/2} \frac{4\pi}{\alpha^2(\rho_a/\rho_c)}. \tag{7.21}$$

The quantity $\alpha(\rho_a/\rho_c)$ is the Langmuir function for converging flow from Section 6.3. It depends on the radii of curvature of the cathode and anode. The perveance of the electron gun in Fig. 7.6 equals the expression of Eq. (7.21) multiplied by the ratio of the cathode area A_c to the area of a full sphere $4\pi\rho_c^2$. In terms of the coordinate system of Fig. 7.6, a differential surface element of the cathode equals

$$dA_c = 2\pi\rho_c^2 \sin \theta' \, d\theta'. \tag{7.22}$$

Integrating Eq. (7.22) from 0 to θ gives

$$A_c = 4\pi\rho_c^2(1 - \cos \theta)/2 = 4\pi\rho_c^2 \sin^2 (\theta/2). \tag{7.23}$$

If focusing electrodes have the proper shapes, the gun perveance is

$$\frac{I}{V_0^{3/2}} = \frac{4\varepsilon_0}{9} \left[\frac{2e}{m_e} \right]^{1/2} \frac{\sin^2 (\theta/2)}{\alpha^2(\rho_a/\rho_c)}. \tag{7.24}$$

In practical units, Eq. (7.24) becomes

$$P = \frac{29.4 \sin^2(\theta/2)}{\alpha^2(\rho_a/\rho_c)} \quad (\mu\text{perv}). \tag{7.25}$$

After crossing the gap, the beam leaves through the anode aperture. The radial fields near the aperture defocus the electrons. The focal length for the negative lens action of the aperture [CPA, Chap. 6] is roughly

$$f \cong -4V_0/E_a. \tag{7.26}$$

In Eq. (7.26), the quantity V_0 is the gap voltage while E_a is the magnitude of the axial electric near the anode. To estimate the effect, we set E_a equal to the value of electric field without the beam and aperture:

$$E_a \cong V_0(\rho_c/\rho_a)/(\rho_c - \rho_a). \tag{7.27}$$

Inserting Eq. (7.27) into Eq. (7.26) gives the focal length

$$f \cong -4(\rho_a/\rho_c)(\rho_c - \rho_a). \tag{7.28}$$

Passing through the aperture, the beam envelope convergence angle changes from θ to θ', where

$$\theta' = \theta - r_a/f = \theta[1 - \rho_c/4(\rho_c - \rho_a)]. \tag{7.29}$$

Instead of converging toward a point a distance ρ_a from the anode, the beam approaches a point at a distance

$$\rho' = \rho_a[1 - \rho_c/4(\rho_c - \rho_a)]^{-1}. \tag{7.30}$$

In many applications, we want to inject a high-current-density beam of small radius into a magnetic transport channel. For a matched equilibrium, we should locate the magnetic field boundary at a waist of the beam. The beam emerging from the aperture of a converging gun usually has strong space-charge forces and low emittance. We can apply the method of Section 5.4 to find the axial location where the beam reaches a waist. From the gun design, we know the beam current (I), kinetic energy (eV_0), initial radius (r_a), and envelope angle ($-\theta'$). We can modify Eq. (5.90) to describe the minimum beam radius in terms of the envelope angle and beam perveance at the anode:

$$\frac{r_{\min}}{r_a} = \exp\left[\frac{-(3.3 \times 10^{-5})\theta'^2}{(I/V_0)^{3/2}}\right]. \tag{7.31}$$

An example is the best way to illustrate the gun design procedure. Suppose we want to create a 1-A beam of 20-keV electrons. The perveance is only

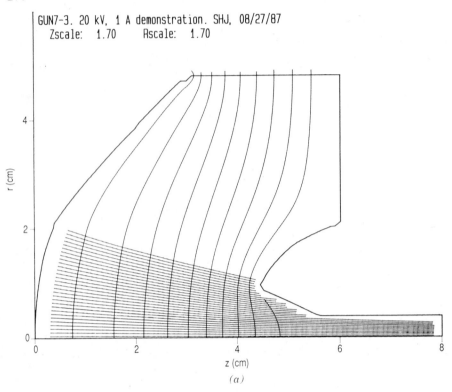

GUN7-3. 20 kV, 1 A demonstration. SHJ, 08/27/87
  Zscale: 1.70   Rscale: 1.70

Figure 7.7 Numerical calculations of converging gun properties using the EGUN code. Figures show electrodes, computational rays, and electrostatic equipotential lines. Left-hand side: spherical-section cathode and focusing electrode. Right-hand side: shaped anode and output tube. $V_0 = 20$ kV, $I_0 = 1$ A. Calculation extends 0.05 m in radius and 0.08 m along the z axis. (a) Initial run—most of the available current strikes the anode. (b) With a corrected focusing electrode, the full current enters the output tube. Note the effect of nonlinear electrical forces on the peripheral rays.

0.35 μperv, so we expect the analytic approximations are valid. The gun has an oxide cathode with a useful emission of $\geqslant 0.25$ A/cm^2. For a 1-A beam, the cathode area is $A_c = 4$ cm^2. We pick a convergence ratio of 3:1 in the extraction gap, $\rho_c/\rho_a = 3$. Table 6.1 gives the corresponding value of the Langmuir function as $\alpha(3)^2 \cong 2.5$. Equation (7.25) implies that the gun subtends an angle $\theta = 20°$, a reasonable value. Substituting for A_c and θ in Eq. (7.23), we find that the radii of curvature of the cathode and anode are $\rho_c = 3.2$ cm and $\rho_a = 1.07$ cm. The cathode has radius $r_c = \rho_c \sin \theta = 1.1$ cm, while $r_a = 0.36$ cm. The negative lens effect reduces the envelope convergence angle to $\theta' = (20°)(1 - \frac{3}{8}) = 12.5°$—the projected convergence point is 1.6 cm from the anode. Inserting values of r_a and θ' in Eq. (7.31) gives a very small value for r_{min}. Hence, we expect that emittance determines the waist radius and that the waist is close to the projected convergence point.

The parameters derived from the analytic theory were used in the ray tracing

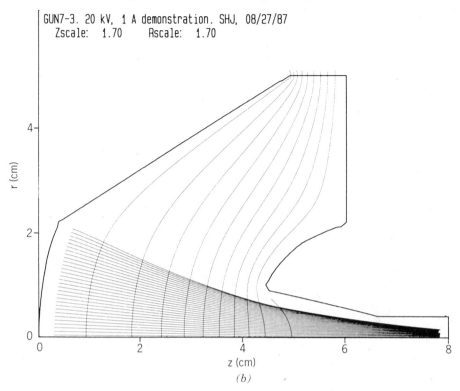

GUN7-3. 20 kV, 1 A demonstration. SHJ, 08/27/87
  Zscale: 1.70   Rscale: 1.70

Figure 7.7 (*Continued*)

program EGUN (Section 7.3) to derive the data of Figs. 7.7a and 7.7b. The cathode and anode were spherical segments. In the first run (Fig. 7.7a), the focusing electrode was a conical segment inclined at an angle of 22.5° with respect to the edge of the cathode. Note the expansion of the beam envelope—focusing was insufficient and most of the beam intercepted the anode. The high total current of 2.0 A resulted from enhanced emission at the edge of the cathode. In the next run (Fig. 7.7b), the angle of the focusing electrode was raised to 37.5°. Here, all the cathode current emerged from the anode aperture. The total current was 0.974 A, close to the predicted value. The crossing orbits in the output of the second run resulted from nonlinear electric forces. We could refine the focusing electrode geometry further to minimize this effect.

7.3. HIGH-PERVEANCE GUNS AND RAY-TRACING CODES

A high-perveance cylindrical electron gun ($P > 1 \,\mu$perv) has a large extraction aperture that strongly perturbs the electric field at both the anode and cathode. Figure 7.8 illustrates the difficulties associated with increased perveance. Figure

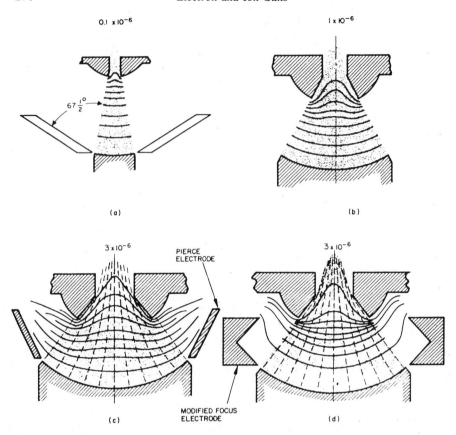

Figure 7.8   Problems of high-perveance gun design. (*a*) Low-perveance gun follows the Pierce design procedure—anode aperture has a negligible effect on particle extraction. (*b*) Moderate perveance converging gun—anode aperture has a small effect on electric fields at the cathode. (*c*) High-perveance gun—anode aperture reduces the electric field at the cathode center. (*d*) High-perveance gun—modified focus electrode to produce an almost uniform electric field on the cathode surface. (From *Ion Beams with Applications to Ion Implantation,* by R. G. Wilson and G. R. Brewer, used by permission, John Wiley and Sons.)

7.8*a* shows a low-perveance gun with small aperture diameter. The beam convergence angle is small and the electrodes are similar to those of a planar gun. In the medium perveance gun of Fig. 7.8*b,* electric field distortions are concentrated at the anode. Here, the theory of Section 7.2 gives useful predictions of the gun performance. The theory cannot describe the high-perveance gun of Fig. 7.8*c.* The large anode aperture distorts electric fields throughout the extraction gap, leading to two problems:

1. The axial electric field is lower at the center of the cathode. The resulting beam has nonuniform current density that can lead to emittance growth during acceleration and transport.

2. The transverse electric field components cause particle deflections. A large fraction of the beam in Fig. 7.8c strikes the anode.

We can compensate for the effect of the large anode with focusing electrodes (Fig. 7.8d). The extensions to the Pierce electrode reduce the electric field at the periphery of the cathode, resulting in more uniform current density. The extensions also bend the electric field lines so that all electrons pass through the aperture.

Figure 7.9 shows a design for an optimized 1.4-μperv gun. Figure 7.9a illustrates the gun geometry, equipotential lines, and selected particle orbits. Figure 7.9b plots the radial profile of extracted current density. Although all electrons exit through the anode, the output of the high-perveance gun is not ideal. The beam current density is lower at the center and nonlinear electric fields augment the emittance. These effects are sensitive to details of the electrode shapes. From the complexity of fields and particle orbits of Fig. 7.9, we can see that analytic calculations give little guidance for high-perveance gun design—numerical methods are essential. The most common design tool is the ray-tracing computer program.

We use ray-tracing programs to find steady-state characteristics of electron or ion guns. The main validity condition is that changes in the applied voltage take place slowly compared with the transit time for particles across the extraction gap. Then, we do not need detailed information on individual time-dependent orbits $r(t)$. We know that all particles emitted from the same point of the source follow identical orbits; therefore, we only need to find the traces of particle orbits $r(z)$. Orbit traces are sometimes called *rays*, a term derived from light optics. In a steady-state gun, a continuous stream of electrons populates a ray. The electrons of a ray contribute to the electric fields through their space charge. If we know the rays emerging from different regions of the source and the current density carried by each ray, we can find static electric and magnetic fields in the gun. Sometimes, transit-time effects in extractors are important. For example, laser-controlled photocathodes can generate electron pulses in the picosecond range. To describe these guns, we must use a full computer simulation program that can handle time-dependent processes.

In this section, we will concentrate on EGUN, a program developed by W. B. Herrmannsfeldt at the Stanford Linear Accelerator Center. The program describes two-dimensional structures, either (r, z) for cylindrical beams or (x, z) for extended sheet beams. We shall study electron emission from a thin rod to illustrate the ray-tracing method. Figure 7.10 shows the gun geometry and the square mesh used for the finite-difference solution of the Poisson equation. A space-charge-limited flux of electrons leaves the surface of the rod. They cross the acceleration gap and enter a drift region with a solenoidal magnetic lens. We provide the program with information on the shapes of biased metal surfaces and dielectrics to determine applied electric fields. We also specify the location and current of coils to generate focusing magnetic fields. Finally, we identify particle sources and state whether space charge or source effects limit the emission.

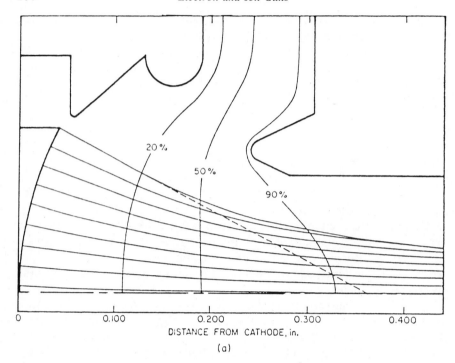

DISTANCE FROM CATHODE, in.

(a)

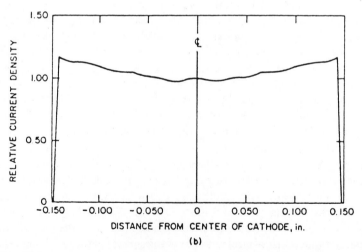

DISTANCE FROM CENTER OF CATHODE, in.

(b)

Figure 7.9 High-perveance electron gun design, $P = 1.4\,\mu$perv. Top: rays, electrode outlines, and equipotential lines. Bottom: current density distribution at cathode.

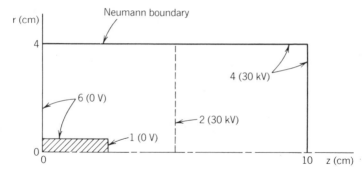

Figure 7.10 Electron emission from a rod cathode, boundaries and parameters for a numerical calculation using the EGUN code. Region 1: emitting cathode surface. Region 6: remainder of the cathode surface. Region 2: anode foil, transparent to electrons. Region 4: absorbing surface at anode potential.

The program proceeds in steps to find an approximate solution by iteration. The first step is to estimate the electric fields without space charge. Because of the possibility of complex internal boundaries, the program applies successive over relaxation [CPA, Section 4.2] to find the potential on the mesh. The program uses interpolation methods to represent curved boundaries accurately on the square mesh. The potential function gives values for the vacuum electric fields $E_r(r, z)$ and $E_z(r, z)$ in the computational region.

The second step is to calculate the orbits of several particles in the vacuum fields to define initial rays. Particle emission is uniform over the source area—the rays are weighted by a current density factor to account for variations in the normal electric field at the source and differences in the area element represented by the particle. Given the initial currents and ray trajectories, the next step is to assign beam charge to the mesh points. This follows the standard particle-in-cell method. When a ray passes through a volume element near a mesh point, the program increments the charge density at the point by an amount proportional to the current carried by the ray and the relative amount of time the particle spends in the element. The program also estimates toroidal magnetic fields generated by the beam particles.

The fourth step is to solve the Poisson equation with the effects of the biased boundaries and the beam space charge included. If the contribution of space charge is moderate, we expect that the resulting field predictions are closer to the actual fields than the initial vacuum prediction. The program next recalculates the orbits of the test particles in the corrected electric fields with beam-generated magnetic field forces included. For most gun designs, these orbits are closer to those in an actual gun than the first set. In assigning current density to the corrected rays, the program makes a local Child limit calculation with the corrected electric fields near the source surface. Subsequent interations of the program apply the following operations: (1) assignment of space charge and calculation of beam-generated magnetic fields, (2) solution of the Poisson equation, (3) recalculation of source emission using the Child law, (4) calculation of test orbits, (5) assignment of current to rays. A run usually converges with fewer than 10 repetitions.

Figure 7.11 shows output for the example of Fig. 7.10. The calculation used 27 test rays and 6 iterations. The final current was 17.2 A, accurate to about 1%. The example of the bare rod cathode solution shows the importance of Pierce electrodes to suppress edge emission and to produce good quality beams. We can compare the code results to the prediction of Eq. (6.33). The planar Child law prediction for a gap with $V_0 = 30 \, \text{kV}$ and $d = 0.025 \, \text{m}$ is

$$I = (1.94 \times 10^4)(A_{\text{eff}}) \quad \text{A/m}^2. \tag{7.32}$$

where A_{eff} is an effective emission area. If we set A_{eff} equal to the surface area of

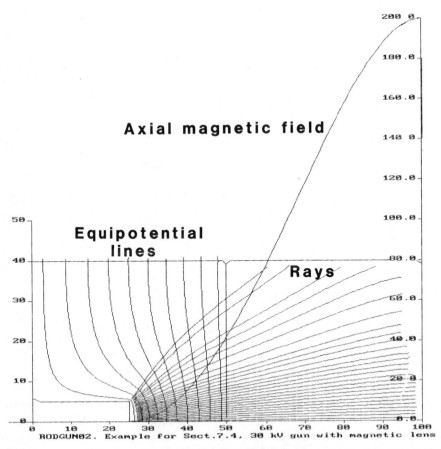

Figure 7.11 Numerical calculation of space-charge-limited electron emission from a 0.01-m-diameter rod using the EGUN code. Gap between cathode and anode, 0.025 m; drift region length, 0.05 m; peak solenoidal focusing field, 0.02 tesla. Predicted current, 17.2 A.

the rod, the predicted current is only 1.52 A. Because of enhanced edge emission, the actual current from the bare rod is considerably higher. The prediction of Eq. (6.33) is 18.7 A, close to the EGUN result.

7.4. HIGH-CURRENT ELECTRON SOURCES

High-current cathodes are important for microwave tubes, pulsed RF linacs, and induction linac injectors. Recently, there has been considerable interest in sources for high-brightness beams that can drive free electron lasers (Section 15.7). High-current-electron sources either have a large area or produce a high electron flux. We shall concentrate on sources that can supply high-current density ($> 10 \times 10^4$ A/m^2). Thermionic cathodes, with a long history of development, are the most practical sources for applications that require long lifetime and high duty cycle. We shall also discuss laser-driven photocathodes and surface plasma sources. These devices can supply very-high-current density and are useful for some pulsed-beam applications. For consistency with MKS units, we shall give all current densities in amperes per square meter. The reader should note that most cathode literature quotes current density in amperes per square centimeter.

Thermionic sources emit electrons when heated to high temperature. The electrons that escape constitute the tail of a Maxwell distribution with enough energy to overcome the surface potential barrier of the material. The magnitude of the surface potential is the material *work function* ϕ_w. The Richardson–Dushman law describes the emission current density from a thermionic source:

$$j_e = AT^2 \exp(-11600\phi_w/T). \tag{7.33}$$

In Eq. (7.33), A is a constant that depends on the material and T is the cathode temperature in degrees Kelvin. The equation shows that the current density rises rapidly with temperature. The radiation power loss from the cathode also increases with temperature—the power flux scales as T^4. The work function of large-area, high-current cathodes must be low to achieve high-current density with manageable thermal power loss. Equation (7.33) implies that small differences in the work function or temperature of a thermionic cathode lead to substantial variations of the available current density. For this reason, electron extractors with thermionic cathodes are usually operated in a space-charge-limited mode for uniform current density (Section 6.2).

We must add corrections to the emission law for a thermionic cathode exposed to an electric field. An extracting electric field reduces the surface electrostatic potential barrier. Tunneling, an effect predicted by quantum mechanics, raises the current density above the value predicted by Eq. (7.33). The corrected Richardson–Dushman equation is

$$j_e = AT^2 \exp(139E_s^{1/2}/T - 11600\phi/T). \tag{7.34}$$

Equation (7.34) is the Schottky equation. The quantity E_s is the electric field normal to the surface in kilovolts per centimeter.

The properties of different thermionic cathode materials are usually compared in terms of the *zero-field current density*, j_{e0}. The zero-field current density is the value given by Eq. (7.34) when $E_s = 0$. We can determine the value of j_{e0} at a particular temperature by measuring extracted current density at different values of extraction voltage and making a plot of $\ln((j_e))$ versus the square root of the applied vacuum electric field $E_0^{1/2}$. The resulting graph, illustrated in Fig. 7.12, is called a Schottky plot. At low values of E_0, there is significant negative space charge near the cathode—the negative electric field suppresses the current. At values of E_0 where all electrons leave the cathode, the electric field on the surface is roughly equal to the vacuum field. We can find the zero-field current by extrapolating the measurement to the $E_0 = 0$ axis (Fig. 7.12).

Commercial thermionic cathodes consist of a high-temperature metal substrate coated with a material with low work function. A coating of free barium on tungsten has a particularly low value of ϕ_s. Unfortunately, barium evaporates rapidly at high temperature. *Dispenser cathodes* solve the problem with an internal reservoir to replenish the barium layer continuously. Dispenser cathodes are fabricated by impregnating porous tungsten with chemical compounds that generate barium when heated. The barium migrates through the tungsten matrix at a high enough rate to maintain a surface layer.

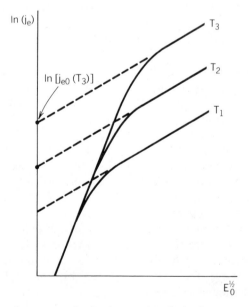

Figure 7.12 Schottky plot, current density from a thermionic electron source as a function of temperature and applied electric field.

Available dispenser cathodes generate current density in the range $20 \times 10^4 \, A/m^2$ at a maximum operating temperature of 1100°C. It is possible to fabricate and heat large cathodes—surfaces as large as $100 \, cm^2$ have produced kiloampere electron beams. The work function of the surface may be severely degraded by the presence of background gas or metal vapors. To avoid *cathode poisoning*, dispenser cathodes require a clean vacuum less than 5×10^{-7} torr. Evaporation of active material limits the lifetime of dispenser cathodes. Over the life of a device, there is a gradual decrease in emission since the barium must migrate from deeper layers of the substrate. High-current-density operation demands high cathode temperature—the penalty is rapid barium evaporation and a shortened lifetime. Dispenser cathodes have lifetimes of about 10,000 h at an operating temperature of 1100°C.

A variety of dispenser cathodes have been developed over three decades of research. The three types in common use are the B type, the M type and the scandate cathode:

1. The B-type cathode, developed in 1955, is the most widely used electron source for commercial devices. It consists of porous tungsten impregnated with a mixture of BaO and Al_2O_3. These materials react chemically to liberate free barium. The chemical CaO is also incorporated in the impregnant— observations show that it reduces the barium sublimation rate and increases emission properties. The names of B types cathodes reflect the chemical composition of the impregnant. For example, a 532 cathode contains fives parts of BaO to three parts of CaO and two parts of Al_2O_3.

2. The M-type cathode is a B-type cathode with the addition of a thin surface layer of ruthenium–osmium. The layer must be thin enough to permit passage of barium—the thickness ranges from 2000 to 10000 Å. For a given temperature, the presence of the surface layer doubles the emission current density. M-type cathodes have longer lifetime than B types since they can be run at lower temperature. On the other hand, damage to the thin surface layer by processes such as ion bombardment may degrade the emission properties.

3. A scandate cathode has Sc_2O_3 mixed with the standard impregnant materials of B-type cathode. The current density from a scandate cathode is almost identical to that from an M-type cathode. Since the scandate cathode does not rely on a thin surface film, it is not as susceptible to mechanical damage and poisoning.

Table 7.1 lists Richardson–Dushman equation coefficients for 411 cathodes. This type of cathode achieves high current density. The work function of a dispenser cathode surface depends on the production, migration, and evaporation rates of barium; therefore, it varies with temperature. Table 7.1 includes the work function temperature coefficient α_c defined by

$$\phi_w(T) = \phi_0 + \alpha_c T. \tag{7.35}$$

TABLE 7.1   Emission Constants

Type	ϕ_0 (V)	α_c (V/K)	A (A/m^{2}K^2)
411	1.67	2.82×10^{-4}	3.68×10^6
411M	1.43	3.99×10^{-4}	3.50×10^6
411 scandate	1.43	4.01×10^{-4}	3.52×10^6
LaB$_6$	2.66	—	2.90×10^5

Source: J. L. Cronin, *IEEE Proc.*, **128**, 19 (1981), and J. M. Lafferty, *J. Appl. Phys.*, **22**, 299 (1951).

where T is the temperature in degrees Kelvin. As an example, the work function of a 411M cathode at 1100°C is $\phi_w = 1.98$ V. The source-limited current density is 35.8×10^4 A/m^2.

Lanthanum hexaboride, LaB$_6$, is an alternative to dispenser cathodes—it has some advantages for pulsed-beam accelerators. The homogeneous material has adequate mechanical strength and an inherently low work function. Emission from LaB$_6$ does not depend on an active surface layer. The material is resistant to poisoning, maintaining its emission properties at pressures in the 10^{-5} torr range. Also, there is less problem with evaporation of the active material. Lanthanum hexaboride may have application in high-current pulsed accelerators that often have poor vacuum conditions and hydrocarbon insulators.

The drawback of LaB$_6$ is that it has a higher work function than dispenser cathodes and requires higher temperature. Table 7.1 lists the Richardson–Dushman coefficients. A practical temperature limit is about 1700°C—higher temperatures result in evaporation of the cathode material. At the peak operating temperature, a 15-cm^2 cathode requires about 3 kW of heater power input. This power level in vacuum presents problems of thermal power management.

Free-electron-laser (FEL) applications require beams with high normalized brightness. Values as high as 10^{10} A/(m-rad)^2 may be necessary for a short-wavelength FEL. In principle, it is feasible to achieve these brightness levels with existing thermionic cathodes. We can estimate electron source brightness capabilities if we postulate that the beam divergence results mainly from the cathode temperature. Suppose that electrons from the cathode have a Maxwell distribution of transverse velocity with a temperature T equal to that of the cathode. From Eq. (2.62), a beam accelerated to velocity v_z has an angular divergence of

$$(\langle \Delta\theta^2 \rangle)^{1/2} \cong (\langle v_x^2 \rangle)^{1/2}/v_z = (kT/m_0)^{1/2}/v_z. \tag{7.36}$$

If the cathode has radius r_c, the normalized emittance of the beam is approximately

$$\varepsilon_n \cong r_c(v_z/c)(kT/m_0)^{1/2}/v_z = r_c(kT/m_0c^2)^{1/2}. \tag{7.37}$$

From Eq. (3.34), the normalized brightness is

$$B_n = \frac{I}{\pi^2 \varepsilon_n^2} \cong \frac{(\pi r_c^2) j_c}{\pi^2 r_c^2 (kT/m_0 c^2)} = \frac{j_c}{\pi} \left[\frac{m_0 c^2}{kT} \right]. \tag{7.38}$$

The quantity j_c is the available current density. Note that factors related to the cathode geometry and beam energy have canceled—normalized brightness is an inherent property of the cathode material.

As an application example, consider a LaB_6 cathode operating at $1900°K$, so that $kT = 0.16 \, eV$. The source-limited current from Table 7.1 is $j_c \cong 10 \, A/cm^2$, leading to a predicted maximum brightness level of $B_n = 10^{11} \, A/m^2\text{-rad}^2$. The observed brightness for a large area LaB_6 cathode is lower, in the range $3 \times 10^9 \, A/m^2\text{-rad}^2$. The discrepancy could result from spatial variations of work function and surface roughness.

Photoemissive cathodes have long been used in electrooptical devices at low current density. The availability of high-intensity pulsed lasers has prompted investigations of photocathodes for high-power beams. The principle of the photocathode is simple—a high-power laser irradiates a surface coated with a low-work-function material—the energy of the laser photons is high enough to liberate electrons. The available electron flux is directly proportional to the photon flux.

Laser-driven photocathodes have some advantages compared with thermionic cathodes:

1. The cathodes operate at room temperature, simplifying the mechanical design of the electron gun.
2. Power coupling is through photons rather than direct electrical connections—a photocathode can operate on the high-voltage terminal of an electrostatic accelerator.
3. It is possible to modulate the output electron beam at high frequency by varying the photon flux.
4. The cathodes generate high-brightness beams because the photoelectrons have a small average transverse energy ($<0.2 \, eV$).
5. High values of pulsed current density are possible.

Experiments have demonstrated a current density of $200 \times 10^4 \, A/m^2$ from a small Cs_3Sb photocathode with frontal illumination by a 50-ns laser pulse.[2] One of the most interesting features of laser cathodes is the ability to initiate and to extinguish electron flow rapidly.[3] Current pulses as short as 35 ps have been obtained.

Despite their advantages, laser-driven cathodes are not a panacea for the problems of high-current sources. They have several drawbacks:

---

[2] C. H. Lea and P. Oettinger, *IEEE Trans. Nucl. Sci.*, **NS-32**, 3045 (1985).
[3] J. S. Fraser, *IEEE Trans. Nucl. Sci.*, **NS-32**, 1791 (1985).

1. The simplicity of the cathode is offset by the complexity and size of the pulsed laser system.
2. Present high-current-density experiments use frontal illumination. An effective method to backlight a large area cathode has not yet been demonstrated.
3. Photoemissive coatings are sensitive to poisoning—Cs_3Sb requires a clean vacuum in the range 10^{-10} torr.

Laser-driven photocathodes are not practical for continuous or long-pulse electron beams. Consider, for example, a photocathode with an emission efficiency of 10%. Suppose that photons of energy 3 eV are necessary to liberate electrons. In order to generate a current density of 100×10^4 A/m², the incident photon energy flux must exceed 30 MW/m². The high-power flux deposited in a thin layer may create surface plasmas or damage the cathode.

At present, the only way to generate an electron current density exceeding 100×10^4 A/m² over large areas is from a dense plasma covering the cathode surface. The effective work function of a plasma is zero. The available current density from a surface plasma source can exceed 1000×10^4 A/m². Surface plasma sources function only for short pulses—expansion of the unconfined plasma degrades the optics of the injector and ultimately leads to a short circuit. The width of the extraction gap and the average expansion velocity of the plasma determine the useful pulse length of the injector. Depending on the current density, surface plasma sources are useful for beam pulses in the range 50 ns to 1 μs.

The easiest way to create a large-area surface plasma is to expose a metal cathode surface to a strong electric field. Figure 7.13a shows the mechanism of plasma formation. Most metal surfaces are covered with microscopic protrusions. These may occur naturally during fabrication (whiskers) or they may be added to the surface (machined ridges, embedded points). Because of electric field enhancement, field emission of a small electron current occurs near the tips of the protrusions. Even though the current is small, it passes through a small cross section, causing local heating and vaporization of material. The process creates dense, cold plasmas at the emission sites. An array of sites can generate very high current density.

Resistive materials, such as graphite, are very effective for the generation of surface plasmas. Woven graphite and cut graphite strands give a rapid initiation of electron current. Insulating materials on a metal substrate can also create plasmas effectively. Here, strong electric fields cause vacuum insulator breakdown (Fig. 7.13b). A fine array of insulating strands gives a moderately uniform plasma density over a large area. Ordinary velvet cloth often gives acceptable results.

Many high-power pulsed-beam experiments have surface plasma sources, mainly because they involve no technology. The main problem with these sources is plasma closure and gap shorting. Plasma expansion velocities higher than 10 cm/μs are usually observed at high-current density. After a pulse, gas and

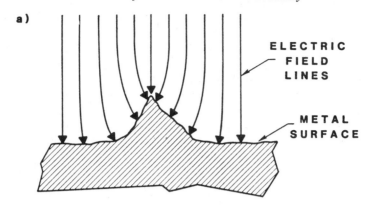

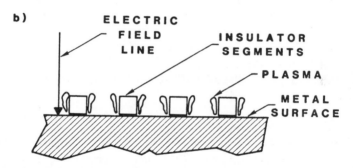

Figure 7.13 Production of surface plasmas for electron extraction. (*a*) Concentration of electric field intensity at a whisker on a cathode surface. (*b*) Plasma formation by vacuum breakdown along the surface of an exposed insulator.

plasma fill the extractor gap. The time necessary to pump out the background material limits the repetition rate. Late-time short circuits may channel energy remaining in the pulse modulator, leading to cathode damage. Uncontrolled surface plasma sources are not suitable for high-brightness beams. A typical brightness level for a high-current extractor with a velvet surface breakdown cathode is 3×10^7 A/(m-rad)^2.

7.5. EXTRACTION OF IONS AT A FREE PLASMA BOUNDARY

Although some ion species can be generated directly from solid surfaces, most high-flux ion sources rely on extraction from a plasma. In this section, we shall discuss the physical principles of ion extraction from an unconfined plasma. In particular, we shall concentrate on the boundary between a field-free plasma and the extraction gap with its high electric field. First, we must review some properties of plasmas to understand limitations on ion flux and beam optics. A plasma is a mixture of ions and electrons. The density is high enough so that long-range collective electromagnetic fields make the main contribution to particle motions. Six macroscopic quantities are usually sufficient to characterize a source plasma: electron density, ion density, average electron velocity, average ion velocity, electron temperature, and ion temperature. We shall review some common parameters for source plasma, starting with particle density.

Most plasmas are almost electrically neutral. For singly charged ions, the electron and ion densities are approximately equal:

$$n_i \cong n_e = n_0. \tag{7.39}$$

Densities of ion source plasmas are in the range $10^{16}-10^{19}$ m^{-3}. At these values, the slightest imbalance between n_e and n_i would result in large electric fields. Without an applied magnetic field, the electric fields shift mobile electrons to cancel any local charge imbalance. Sources with a high electron temperature, such as Penning discharges and metal-vapor vacuum arcs, can produce a mixture of ion charge states. Also, some sources may create mixed ion species. We must modify Eq. (7.39) when there are several types of ions. For example, in a carbon source with C^+ and C^{++}, the neutrality condition is

$$n_i^+ + 2n_i^{++} \cong n_e. \tag{7.40}$$

The plasmas of ion sources that produce current density in the range 0.01–1 A/cm^2 are usually stationary—the directed velocities $\langle v_e \rangle$ and $\langle v_i \rangle$ are approximately equal to zero. Sources have been developed for beam-driven inertial fusion to produce plasmas with a high value of directed ion velocity. These sources can achieve pulsed current densities in the range > 100 A/cm^2. Section 12.5 discusses methods to accelerate plasmas.

Plasmas are often produced by an electric discharge, resulting in electron temperatures exceeding 10,000 K. The collision rate is usually high enough in a steady-state plasma so that the electrons are in local thermodynamic equilibrium. The velocity distribution of plasma electrons is ordinarily close to a Maxwell distribution—the electron temperature is in the range $kT_e \sim 1$–10 eV. Collisions between electrons and ions are ineffective at transferring momentum to the ions. In most sources with a moderate ion residence time, the temperature of the ions is usually low, $kT_i < 1$ eV.

The random particle motions associated with nonzero electron and ion temperatures result in particle transport, even with no applied forces. In stationary plasmas, the *thermal fluxes* of ions and electrons determine the available current density. We shall calculate the ion flux through a plane in a uniform, collisionless plasma—Fig. 7.14 illustrates the model. We want to find the total number of ions that cross the plane at $z = 0$. The ions have a Maxwell velocity distribution. Following Eq. (2.59), the normalized distribution of

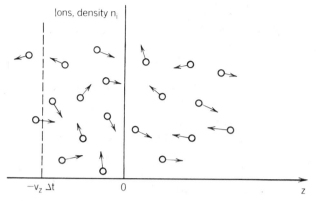

Figure 7.14 Calculation of flux through a plane for ions randomly distribution in velocity.

velocities in the z direction is

$$g(v_z) = \left[\frac{2m_i}{\pi k T_i} \right]^{1/2} \exp\left[\frac{-m_i v_z^2}{2k T_i} \right]. \tag{7.41}$$

The quantity $\Delta N(v_z)$ is the number of particles with velocity in the interval Δv_z at v_z that pass through a unit area in the plane in a time Δt. Particles with velocity v_z that cross the plane in the interval $0 < t < \Delta t$ are located between $-v_z \Delta t < z < 0$. The number of ions in a volume of length $v_z \Delta t$ and unit cross-sectional area is $n_i v_z \Delta t$. Of these ions, the fraction $g(v_z) dv_z$ is in the velocity range of interest. Combining expressions, the differential flux is

$$\frac{\Delta N(v_z)}{\Delta t} = -n_i v_z \left[\frac{2m_i}{\pi k T_i} \right]^{1/2} \exp\left[\frac{-m_i v_z^2}{2k T_i} \right] dv_z. \tag{7.42}$$

We derive the total flux of particles by integrating Eq. (7.42) over the range $-\infty \leqslant v_z \leqslant 0$. The thermal flux in the $+z$ direction is

$$(dN/dt)^+ = n_i \sqrt{k T_i / 2\pi m_i}. \tag{7.43}$$

We can write Eq. (7.43) in terms of the average *speed* of the ions, $\langle v_i \rangle = (8k T_i / \pi m_i)^{1/2}$:

$$(dN/dt)^+ = n_i \langle v_i \rangle / 4. \tag{7.44}$$

In a hydrogen plasma with $n_i = 10^{19}$ m^{-3} and $kT_i = 1$ eV, the thermal ion flux is 3.9×10^{22} ions/sec-m^2. This figure corresponds to an equivalent proton current density of 6.3×10^3 A/m^2 (0.63 A/cm^2). We can find the thermal electron flux by substituting the electron mass and temperature in Eq. (7.43). For an electron temperature of $10\,\mathrm{eV}$, the equivalent electron current density is $8.5 \times 10^5 \, \mathrm{A/m^2}$ ($85 \, \mathrm{A/cm^2}$). Because of their light mass, electrons move rapidly through a plasma.

We shall now address processes that occur on the *ion emission surface*, the boundary between the field-free plasma and region of high electric field in an extraction gap. To understand the nature of the interface, we shall start with a simplified plasma model. Suppose that the plasma electrons are cold, $kT_e = 0$. In this limit, the electrons provide complete cancellation of electric fields inside the plasma volume. The electron and ion densities are exactly equal, $n_e = n_i = n_0$. Assume further that the ions are cold but have a directed velocity in the z direction, v_0. The available ion current density is

$$j_0 = e n_0 v_0. \qquad (7.45)$$

Figure 7.15 shows the model geometry. The plasma extends infinitely in the x and y directions. A source electrode at $z = -\infty$ generates plasma electrons and ions at ground potential. A biased electrode at potential $-V_0$ and position $z = d$ extracts ions. The ion emission surface is at $z = 0$.

Inside the plasma region, the electron and ion densities are equal and there is no electric field. As a result, the electrostatic potential equals zero throughout the plasma. In the extraction gap, the electron density is zero. Since the ion current density equals j_0 at all positions, the equilibrium ion density is inversely proportional to the ion velocity. The electrostatic potential varies from 0 to V_0 over the distance d. To solve the Poisson equation to find how ϕ varies across the gap, we can apply the condition that the electric field is continuous over the emission surface; therefore, $d\phi(0)/dz = 0$.

The conditions in the gap are the same as those for space-charge-limited ion flow (Section 5.2). For a consistent steady-state solution, the space-charge-limited ion current density must equal the available thermal flux from the plasma j_0. In other words, for given values of j_0 and V_0, we must adjust d to give the

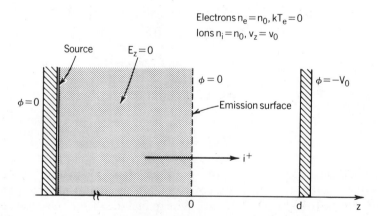

Figure 7.15 Geometry for calculating ion extraction from a plasma with cold electrons.

correct space-charge-limited flow. In the limit that $m_i v_0^2/2 \ll eV_0$, Eq. (5.48) implies that the correct spacing is

$$d = \left[\left(\frac{4\varepsilon_0}{9} \right) \left(\frac{2e}{m_i} \right)^{1/2} \frac{V_0^{3/2}}{en_0 v_0} \right]^{1/2}. \tag{7.46}$$

Equation (7.46) shows that the position of the ion emission surface relative to the extractor electrode depends on the voltage applied to the extraction gap and the properties of the plasma source.

The interdependence between the parameters of the plasma source and the extraction gap is important for the design of ion guns. Figure 7.16 shows a high-flux ion injector. A finite width beam accelerates between an anode and cathode separated by spacing d and voltage V_0 and exits through an aperture. The width of the aperture must be comparable to or smaller than d. To avoid wasted ion flux and bombardment of the cathode, we admit plasma to the extraction gap through an anode aperture opposite the cathode aperture. The figure also shows focusing

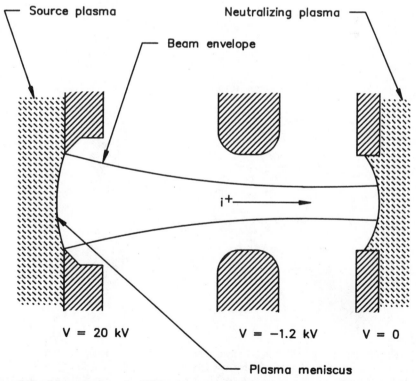

Figure 7.16 Schematic diagram of high-current ion beam extraction from a plasma. Figure shows representative voltages, a source plasma meniscus, a converging beam, and a decel gap to suppress electron backstreaming.

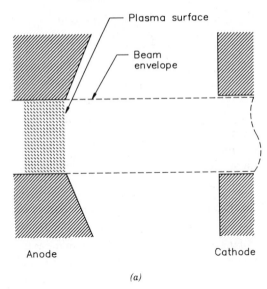

(a)

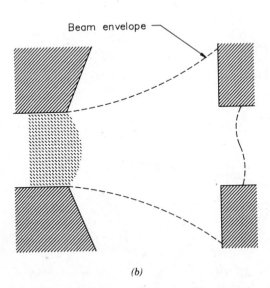

(b)

Figure 7.17 Effect of available plasma ion flux on beam optics in an extraction gap. (*a*) Pierce gun—plasma ion flux and space-charge-limited ion flux balanced to define a flat emission surface. (*b*) Excessive plasma ion flux causes a bulge of the emission surface and a diverging ion beam. (*c*) Reduction of plasma ion flux leads to a concave emission surface and a converging beam.

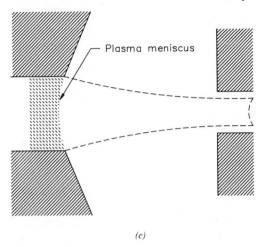

(c)

Figure 7.17 *(Continued)*

electrodes on the anode. Section 7.8 describes the function of the extra electrode at the cathode.

Neglecting the field perturbations of the cathode aperture, we know that there is an ideal solution for space-charge flow in the extractor of Fig. 7.16. When the plasma flux satisfies Eq. (7.46), the flat ion emission surface is flush with the edge of the focusing electrode (Fig. 7.17a). From the discussion of Section 7.1, the resulting beam is parallel. Fig. 7.17b shows how the solution changes if we increase the plasma flux while maintaining a constant extractor voltage. Since the available ion current density exceeds the space-charge-limited value, the ion emission surface protrudes beyond the anode aperture. Ion flow across the extraction gap increases because of the combined effects of the decreased gap spacing and increased emission area. The extracted ions no longer constitute a parallel beam—some ions may strike the cathode, resulting in gap shorting or electrode damage. In contrast to electron extractors with a thermionic source, the optics of an ion gun is coupled to the properties of the plasma source. Therefore, plasma sources for ion extraction must have high stability.

We can take advantage of the dependence of the shape of the ion emission surface on the plasma flux. As with electron guns, we would like to generate a converging beam so that we can use a small exit aperture. To do this, we need a shaped source surface. We can create a concave ion emission surface, known as a *plasma meniscus*, by reducing the plasma flux below the level for a planar emission surface. Then, the plasma surface recedes into the anode aperture (Fig. 7.17c). The effective gap width is larger at the center of the plasma so that space-charge flow is matched to the lower flux. The edge of the plasma cannot recede into the aperture—in a sense, it is tied to the surface of the anode. If the plasma on the edge were to move back, there would be a strong reduction in the electric field normal to the surface because of electrostatic shielding by the metal electrode.

The ion emission surface must remain close to the electrode face to maintain a balance between the plasma flux and the extracted flux.

We can calculate the shape of the plasma meniscus with ray tracing codes. In the iterative calculation, the plasma surface evolves to a configuration that has almost constant normal electric field over the cross section. To first order, the plasma meniscus approaches a spherical segment. The radius of curvature depends on the magnitude of the plasma ion flux compared to the planar value. Ideally, the gun should generate ion orbits similar to the electron orbits of Fig. 7.7. Plasma sources are more difficult to operate than thermionic cathodes. We saw in Section 6.2 that the current from an electron extractor is uniform because of space-charge effects, even if the flux from the thermionic source varies in space or time. The flux from ion extractors can be nonuniform because the plasma can move into the acceleration gap. As a result, the plasma source must be very stable. Also, sources for multiaperture guns (Section 7.8) must provide uniform plasma flux over large areas.

In the discussion up to this point, we neglected the effects of electron temperature. Energetic electrons can create electric fields that affect the available ion flux from the plasma. A complete self-consistent solution for ion extraction from a thermal plasma can be complex. In the following discussion, we shall estimate the available ion flux using physical insights rather than detailed mathematics. The results agree with experiments and detailed numerical studies.

Figure 7.18a shows the one-dimensional geometry of the model. Ions are extracted from a plasma emission surface at $z = 0$. Plasma electrons have uniform temperature, kT_e, while the ions are cold. We allow the possibility that the plasma properties at the emission surface differ from those in the bulk of the plasma—we denote the potential at the emission surface as $-\phi_s$, while the bulk of the plasma has $\phi = 0$. We assume that the electron and ion densities at the surface are approximately equal. The density at the surface is n_s while the bulk plasma density is n_0. The electric field throughout the plasma is small—we shall apply the condition $d\phi/dz \cong 0$ on both sides of the emission surface. The emission surface maintains a constant position only if there is a flux of incident ions. We postulate that the flux arises from a directed ion velocity v_0. In the following calculation, we want to see if there are limits on v_0 for valid solutions of the Poisson equation in the extraction gap and to find how the ions acquire a directed velocity.

Solution of the Poisson equation in the extraction gap is more difficult when electrons have a nonzero temperature. Thermal electrons can penetrate into the gap—we must include the electron contribution to the density. Suppose that both ϕ_s and kT_e/e are small compared with V_0, the gap voltage. In this case, electrons remain localized near the anode in thermal equilibrium. We can relate the electron density to the electrostatic potential through Eq. (2.134). In the region $0 \leqslant z \leqslant d$, the electron density is

$$n_e \cong n_s \exp[e(\phi - \phi_s)/kT_e]. \qquad (7.47)$$

We can also express the ion density in terms of the electrostatic potential. If the

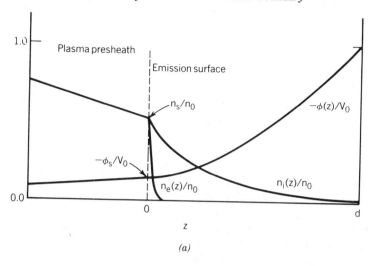

(a)

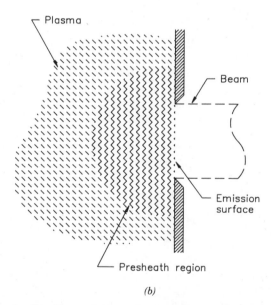

(b)

Figure 7.18 Ion extraction from a plasma with hot electrons. (*a*) Schematic diagram of spatial variation of electrostatic potential, plasma electron density, and ion density in and near a one-dimensional extraction gap. (*b*) Expanded presheath region near a finite-width extractor.

ions of mass m_i have directed velocity v_0 at $z = 0$, then

$$n_i = \frac{n_s v_0}{[v_0^2 - 2e(\phi - \phi_s)/m_i]^{1/2}}. \tag{7.48}$$

If we define the dimensionless potential, $\Phi = -e(\phi - \phi_s)/kT_e$, the Poisson equation is

$$\frac{d^2\Phi}{dz^2} = -\left(\frac{en_s}{\varepsilon_0 kT_e}\right)\left[\left(1 + \frac{2kT_e}{m_i v_0^2}\Phi\right)^{-1/2} - \exp(-\Phi)\right]. \tag{7.49}$$

The normalized potential must be a monotonically increasing function of z. Negative values of Φ (corresponding to positive values of ϕ) would violate the conditions leading to the electron density expression of Eq. (7.47). The bracketed term on the right-hand side of Eq. (7.49) governs the behavior of the solution near the emission surface. The condition $d\Phi(0)/dz \cong 0$ must hold; therefore, the second derivative of Φ is positive near $z = 0$. When this condition is not true, the curve describing Φ has a negative inflection at the emission surface, leading to negative values of Φ. Figure 7.19 shows the bracketed term of Eq. (7.49) for several values of the parameter $(2kT_e/m_i v_0^2)$. We see that

$$(2kT_e/m_i v_0^2) < 2, \tag{7.50}$$

for a valid solution. This observation gives the following condition on the

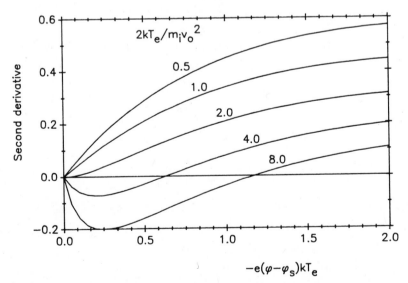

Figure 7.19 Function for the calculation of ion extraction from a plasma with hot electrons. Abscissa: $\Phi = -e(\phi - \phi_s)/kT_e$, Ordinate: $(1 + 2kT_e\Phi/m_i v_0^2)^{-1/2} - \exp(-\Phi)$.

directed ion velocity:

$$v_0^2 \geqslant kT_e/m_i. \tag{7.51}$$

Equation (7.51) is known as the *Bohm sheath criterion*. In plasma physics, a *sheath* is a narrow region of electric fields in a plasma where ion and electron densities are unequal.

In most ion-source plasmas, the ion temperature is small compared with the electron temperature,

$$kT_i \ll kT_e. \tag{7.52}$$

The condition of Eq. (7.52) implies that the ion thermal velocity is much less than the directed velocity necessary to meet the Bohm sheath criterion. Therefore, the ion flux that moves toward the emission surface must arise from processes other than thermal diffusion. The accepted theory holds that the electron and ion densities readjust slightly in a region of the plasma adjacent to the emission surface. This region is sometimes called the *presheath*. A charge imbalance leads to electric fields in the plasma—these fields are much smaller than the applied field in the extraction gap. There is a drop in the electrostatic potential ϕ_s between the bulk of the plasma and the emission surface associated with the fields of the presheath. We can find a consistent solution to the Poisson equation in the extraction gap if the potential drop is sufficient to give the ions a directed velocity that satisfies Eq. (7.51). This condition implies the following value of electrostatic potential at the emission surface

$$\phi_s = -kT_e/2. \tag{7.53}$$

Figure 7.18a illustrates variations of potential, ion density, and electron density in the presheath and emission surface sheath.

The negative value of ϕ at the emission surface means that the electron density is lower than the value in the bulk of the plasma, $n_s < n_0$. Also, the condition of small electric fields in the plasma implies that the ion density almost equals the electron density in the region $z < 0$. Equation (7.47) gives the following value for the particle densities at $z = 0$:

$$n_s = n_0 \exp(\phi_s/kT_e) = 0.6n_0. \tag{7.54}$$

Combining Eqs. (7.51) and (7.54), we can estimate the available ion current from the boundary of a free plasma with low ion temperature:

$$j_B = 0.6en_0\sqrt{kT_e/m_i}. \tag{7.55}$$

The quantity j_B is the *Bohm current density*—Eq. (7.55) gives a good estimate of the available ion current density from most source plasmas. We can balance the

Bohm current density against the Child law value in an extraction gap to find the shape of the plasma meniscus.

The model that we have developed contains a paradox if we limit attention to a purely one-dimensional plasma. The model depends on the condition that the electric field is localized near the emission surface—it falls to zero in the bulk of the plasma. On the other hand, a one-dimensional model demands that the ion flux is equal at all axial positions. Since the current density from thermal ion motion is usually lower than j_B, we must conclude that the presheath electric field that accelerates the ions extends to $z = -\infty$. We can resolve the problem is a one-dimensional theory by allowing electric fields to extend to an ion source plane at negative z. With this assumption, detailed numerical solutions predict matched current density values close to the predictions of Eq. (7.55). The geometry of real ion exractors is not one dimensional. Usually, the area of the anode occupied by apertures is a fraction of the total area. For this case, the cross-sectional area of the presheath can increase with distance from the emission surface. There is a distance where the area of the expanded presheath region is sufficient so that the Bohm current can be supplied by thermal diffusion of ions into the presheath. Figure 7.18*b* illustrates this process schematically.

7.6. PLASMA ION SOURCES

It is more difficult to produce ions than electrons. As a result, there is a wide diversity of ion source types, each with its relative merits. A review of the physics and technology of ion sources would occupy a volume. Here, we will pursue a more modest goal, listing some high-flux plasma ion sources and describing the basic principles of their operation.

We choose a particular plasma source by its relative species content (atomic species and charge states of a given species), available current density and maximum useful emission area. Also, the source plasma should have good temporal stability for ion extraction from a free surface. For large-area extractors, the plasma flux should be uniform in space so that the beam optics is identical in all apertures. To achieve low beam emittance, the temperature of ions in the plasma should be low. Two other characteristics are important for applications—the *conversion efficiency* and the *ionization efficiency*. Conversion efficiency is ratio of the total current of ions available at the extractor to the total electron current in the source. Conversion efficiency varies widely between sources—it is usually much smaller than unity. In gas-fed sources, the ionization efficiency is the probability that an atom is ionized as it crosses the source volume. The quantity determines *gas loading* by the source. Differential pumping of emitted source gas is a major technology problem in many high-flux ion beam systems.

Most sources for industrial and research applications rely on ionization of gas by a stream of energetic electrons. To begin, we shall review some constraints on the ionization process. We shall use the simplified source geometry of Fig. 7.20.

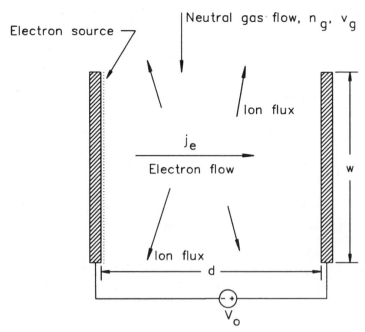

Figure 7.20 Simplified geometry of a plasma source based on ionization of a flowing gas by electrons.

Neutral gas at density n_g flows through an electron stream created by two parallel plates. The distance across the stream is w—the spacing between the plates is d. One plate, at voltage $-V_0$, emits electrons at current density j_e. Energetic electrons in the stream collide with gas atoms, stripping atomic electrons to produce ions. The region between the plates contains a low-temperature plasma. With no applied magnetic field, the plasma concentrates the electric field between the plates to a narrow cathode sheath with a width about equal to a Debye length. As a result, electrons accelerate to a kinetic energy of eV_0 before entering the bulk of the gas. Because of the low electric fields, secondary electrons produced in the plasma do not gain enough kinetic energy to create further ionization.

The differential ionization coefficient S describes the ability of primary electrons to generate ions. The product of S times the gas density equals the total number of ions (in all charge states) produced per meter of primary electron path length:

$$\text{(number of ions per electron)/meter} = Sn_g. \tag{7.56}$$

Figure 7.21 shows a plot of S as a function of electron kinetic energy for different gasses. In the range shown, most of the resulting ions are in the $+1$ charge state.

We can estimate the conversion efficiency of the source of Fig. 7.20 in terms of

S. The total number of ions generated for each primary electron is

$$\text{ions/electron} = \langle S \rangle n_g d, \tag{7.57}$$

where $\langle S \rangle$ is an average over the kinetic energy of electrons in the gas. The ratio of source ion current to the discharge electron current is

$$I_i/I_e < \langle S \rangle n_g d. \tag{7.58}$$

The inequality accounts for the fraction of ions that do not reach the extraction gap. As an example, suppose we have an argon ion source with a filament voltage of -300 V. From Fig. 7.21, we estimate that $\langle S \rangle \sim 10^{-20} \text{ m}^{-2}$. We take an electron transit length of 0.03 m and a gas pressure of $P = 10$ mtorr. The conversion from gas pressure to atomic density (at 0°C) is

$$n_g(\text{m}^{-3}) = (7.0 \times 10^{19})P(\text{mtorr}) \quad \text{[diatomic molecules]}, \tag{7.59}$$

$$n_g(\text{m}^{-3}) = (3.5 \times 10^{19})P(\text{mtorr}) \quad \text{[monatomic molecules]} \tag{7.60}$$

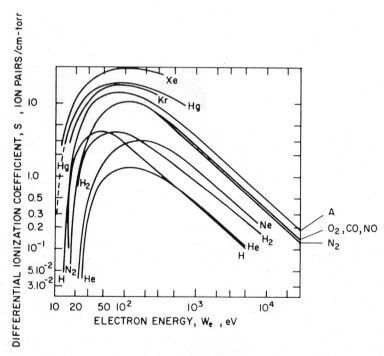

Figure 7.21 The quantity S, the number of ions produced by an electron of energy W_e per centimeter path length at 1 torr pressure. (From *Ion Beams with Applications to Ion Implantation* by R. G. Wilson and G. R. Brewer. Used by permission, John Wiley and Sons.)

The atomic density is $3.5 \times 10^{20}\,\text{m}^{-3}$. Inserting values into Eq. (7.58) gives $I_i/I_e < 0.1$.

The gas utilization efficiency ε_g equals the probability of ionization of an atom crossing the electron stream. Ideally, all entering atoms should be ionized—in practice, ε_g is much smaller than unity. The total number of ions produced per unit time in a volume $wd\Delta z$ is

$$\text{ions/sec} = (j_e w\Delta z/e)(\langle S\rangle n_g d). \qquad (7.61)$$

If v_g is the average thermal velocity of ions, the total number of atoms entering the electron stream per second is

$$\text{atoms/sec} = (d\Delta z)(n_g v_g). \qquad (7.62)$$

The gas utilization fraction is the ratio of Eq. (7.61) to Eq. (7.62):

$$\varepsilon_g = j_e\langle S\rangle w/ev_g. \qquad (7.63)$$

Argon atoms at room temperature have a thermal velocity $v_g = 244$ m/sec. For an electron current density of $j_e = 1000\,\text{A/m}^2$ (0.1 A/cm^2), Eq. (7.63) predicts a gas utilization fraction of only about $\varepsilon_g \sim 0.01$.

In designing a plasma source, we would like to achieve the highest possible values of I_i/I_e and ε_g. Following Eq. (7.58), there are two approaches to raise I_i/I_e: increase the gas density (n_g) or extend the electron path length (d). We avoid the first option because it results in higher gas loading. There are several ways to increase d. For example, we can apply a magnetic field in the ionization region that forces the electron to follow longer paths. To raise ε_g, we must have higher primary electron current density (j_e). For a given source current density, we can use reflex orbits for higher effective j_e. Another way to increase the electron current is to use secondary electrons to create additional ionizations. We can increase the kinetic energy of secondaries through plasma instabilities or by an electric field in the plasma volume.

In the rest of this section, we shall look briefly at some high-flux plasma sources. We shall start with gas sources—the first example is the large-area source used to create neutral beams for fusion research. These sources are conceptually simple—the ionization process is similar to that of Fig. 7.20. We shall then move to more complex gas sources and discuss how they reduce the electron source current and improve gas utilization. We then review the properties of a source that does not rely on gas injection, the metal-vapor vacuum arc. We conclude with a discussion of sources that can provide current densities exceeding $100 \times 10^4\,\text{A/m}^2$ for high-current pulsed beams.

A. Large-Area Plasma Sources

Large-area plasma sources supply ions to multi-aperture electrostatic accelerators for fusion research. The energy of the output beams is about 50 to 100 keV.

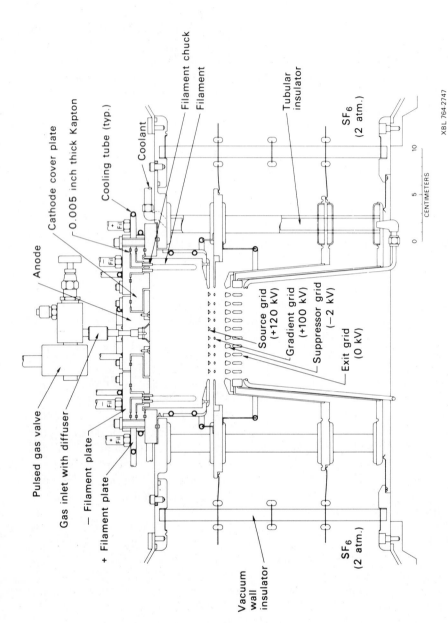

Pulsed gas valve

Gas inlet with diffuser

− Filament plate

+ Filament plate

Anode

Cathode cover plate

0.005 inch thick Kapton

Cooling tube (typ.)

Coolant

Filament chuck

Filament

Tubular insulator

SF₆ (2 atm.)

Source grid (+120 kV)

Gradient grid (+100 kV)

Suppressor grid (−2 kV)

Exit grid (0 kV)

Vacuum wall insulator

SF₆ (2 atm.)

CENTIMETERS

0   5   10

XBL 764-2747

Figure 7.22 Large area ion source. (Courtesy W. Cooper, Lawrence Berkeley Laboratory.)

304

The source of Fig. 7.22 generates current densities of H^+ or D^+ in the range 0.5 $\times 10^4$ A/m^2 over areas exceeding 0.02 m^2. As in Fig. 7.20, free-electron flow through a gas creates ions—the gas pressure is in the range 10–20 mtorr. The electron sources are multiple thermionic filaments biased to about -300 V. Some of the ions created in the large-volume plasma drift through apertures in the front wall to the extraction gap.

The large-area source has low conversion efficiency—the electron current to heat the filaments and to drive the discharge is in the kiloampere range. The sources also emit a substantial amount of gas. Nonetheless, the source has decided advantages for multiaperture ion beam generation. The plasma is very stable because the electron density is low and there are no confining magnetic fields. The ion flux is constant in time and has excellent spatial uniformity. The temperature of ions from the source is low because of plasma stability and the short ion residence time. Good beam aiming and low ion temperature are essential for the application—divergence angles must be less than 10 mrad.

B. Magnetic Bucket Source

Application of *magnetic bucket* plasma confinement improves the characteristics of large-area plasma sources. Figure 7.23 shows the geometry of a bucket source. It is similar to the large-area source, except that an array of permanent magnets covers the wall of the plasma generation chamber. The wall acts as the anode for the electron discharge. The magnets produce multipole fields of a few kilogauss localized near the wall that reflect a portion of the incident electrons. In plasma physics, this process is called *magnetic cusp confinement*. Even though confinement is imperfect, the magnetic buckets improve the performance of the ion source. For a given source current, partial electron reflection leads to a higher density of energetic electrons in the plasma.

The bucket source can operate at lower gas pressure than a simple large-area source. Therefore, a bucket source has reduced gas loading and lower primary electron current. There is little sacrifice in the quality of the plasma to gain these improvements—the cusp field geometry gives stable plasma confinement. In fact, confinement may improve plasma stability because the primary electrons have a more isotropic velocity distribution. Since ionization takes place mainly in a field-free region, magnetic fields do not contribute to the emittance of the extracted ion beam.

C. Penning Ionization Gauge Source

The Penning Ionization Gauge (PIG) source uses a self-sustained discharge that requires no electron source. The geometry of a large-area positive ion source is shown in Fig. 7.24. The plasma region has an applied axial magnetic field and a radial electric field generated by a cylindrical anode. The fields trap electrons in the source volume. The magnetic field provides radial confinement and the longitudinal component of the electric field provides axial confinement.

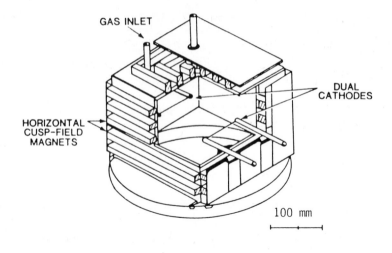

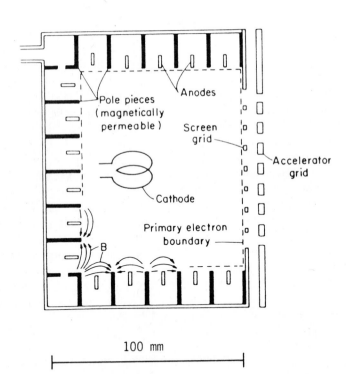

Figure 7.23 Magnetic bucket ion source. (From *The Physics and Technology of Ion Sources*, edited by I. G. Brown. Used by permission, John Wiley and Sons.)

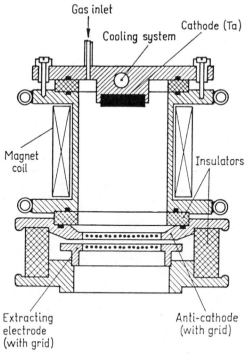

Figure 7.24 Penning-type ion source. (From *Atom and Ion Sources*, by L. Valyi. Used by permission, John Wiley and Sons.)

Electric fields can exist in the plasma because the magnetic field inhibits radial electron motion. A free electron generated in the plasma region follows a complex drift orbit in the crossed fields. Certain combinations of gas pressure, magnetic field, and anode voltage lead to a diffuse discharge. The discharge is self-sustained if the electrons in the plasma volume gain enough energy to cause further ionization and if the lifetime of free electrons is longer than the mean time to ionize a gas atom. When the second condition holds, each electron creates at least one additional electron before being lost. Ultimately, electrons migrate to the anode through the combined effects of collisions with atoms and the high-frequency electric fields of instabilities. Ions are extracted axially by applying a small positive voltage to the upstream source electrode.

The PIG source has good energy efficiency. Since it requires no thermionic sources, it is simple to operate for long periods of time. The source has disadvantages for the generation of low-emittance ion beams. The discharge relies on complex collective processes that may change abruptly with variations in operating parameters. PIG discharges are invariably unstable. In some regimes, plasma microinstabilities lead to enhanced ion temperature. In other regimes, macroinstabilities cause large spatial and temporal variations of the available plasma flux. The main problem is that ion generation occurs in a strong

a)

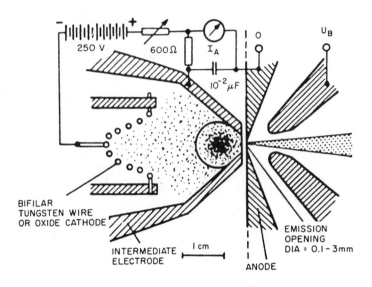

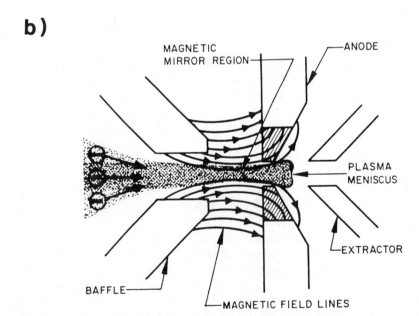

Figure 7.25 Duoplasmatron ion source. (*a*) Components and driving circuit. (*b*) Expanded view of orifice region. (From *Ion Beams with Applications to Ion Implantation* by R. G. Wilson and G. R. Brewer. Used by permission, John Wiley and Sons.)

applied axial magnetic field. Section 9.2 shows that this leads to enhanced beam emittance.

D. Duoplasmatron

The duoplasmatron is widely used for moderate current beams in research and industrial applications. The device, illustrated in Fig. 7.25, makes effective use of the primary electron current and reduces gas loading by funneling both the gas and the primary electron stream through a narrow orifice. The gas pressure in the interaction region can be high without excessive gas loading. Plasma generation processes in the duoplasmatron are complex—development of the device has been largely empirical. The basic principles of operation are:

1. A negatively biased cathode filament generates electrons that accelerate toward the anode electrode through a low-density plasma.

2. An intermediate focusing electrode with a narrow aperture (~ 5 mm diameter) stands between the cathode and anode. If the potential of the electrode has a value between that of cathode and anode, a portion of the electron current is forced to flow through the aperture to the anode.

3. An axial magnetic field reduces electron collection by the focusing electrode. The electrode and portions of the anode are composed of soft iron to direct the magnetic field to the orifice region.

4. In applications with moderate beam current, a voltage in the range ~ 50 kV extracts ions directly from the intense plasma bubble at the orifice.

The main advantage of the duoplasmatron is that it has high ionization efficiency—sometimes, ε_g may be as high as 90%. The major technology problem for high-current applications is cooling the anode insert, which is subject to intense electron bombardment.

A variant of the duoplasmatron has been applied as a source for large-area multiaperture extractors. The device is distinguished, in part, by having the least poetic name in charged particle beam technology, the *duopigatron*. The duopigatron is a duoplasmatron combined with a PIG source (Fig. 7.26). The duoplasmatron section injects seed electrons, plasma, and gas into an axial crossed-field discharge. Drifting electrons in the PIG region complete the ionization of the gas. The plasma drifts through an expansion region to a multiaperture extractor. Although the duopigatron has lower gas loading than the large-area source, the available ion beam current is about an order of magnitude lower. The emittance of extracted beams is higher because of magnetic fields and plasma turbulence in the source.

E. Metal-Vapor Vacuum Arc

A metal-vapor vacuum arc is an intense discharge between electrodes at high vacuum. Evaporation of the cathode supplies material to form the conducting

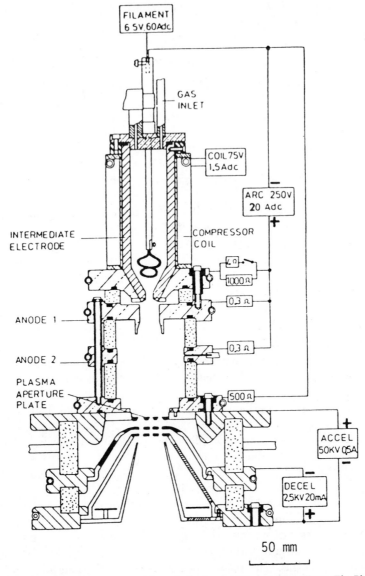

Figure 7.26 Duopigatron ion source with accel–decel extractor electrodes. (From *The Physics and Technology of Ion Sources*, edited by I. G. Brown. Used by permission, John Wiley and Sons.)

plasma between the electrodes. The importance of vacuum arc phenomena in high-power circuit breakers has lead to studies for many decades. Only recently have vacuum arcs been applied to high-flux ion extraction. Vacuum arcs have several useful properties:

1. The plasma that expands from the arc region has high available ion flux.

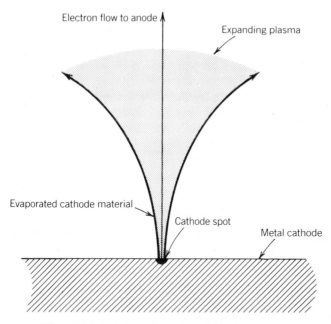

Figure 7.27 Principle of the vacuum arc ion source.

2. The arcs can generate a wide variety of previously unobtainable metal and semiconductor ions with a high degree of purity.
3. The sourcs produce few neutrals, reducing problems of gas loading.
4. Vacuum arc sources are well suited to pulsed extractors—the plasma flux reaches equilibrium level a few microseconds after initiation.
5. For some electrode materials, the sources generate a high fraction of multiply stripped ions.
6. They are simple and reliable.

Figure 2.27 illustrates the principle of the metal-vapor vacuum arc. A discharge flows in a plasma between two electrodes. Electrons are created by thermionic emission from a small region ($\sim 1 \, \mu$m) at very high temperature on the negative electrode called the *cathode spot*. Evaporation of metal at the spot supplies material. The intense electron flow converts the expanding neutrals to a dense, highly ionized plasma. At the cathode spot, the electron current density may exceed $10^{13} \, \text{A/m}^2$, maintaining surface temperatures over 5000°C. Plasma from the spot conducts the high-current density of electrons away from the cathode. Vacuum arcs carry current in the range 100–300 A with voltage between electrodes of ~ 20 V.

Because of the intensity of the discharge near the cathode spot, vacuum arcs may produce high fractions of multiply stripped ions. For example, beams with

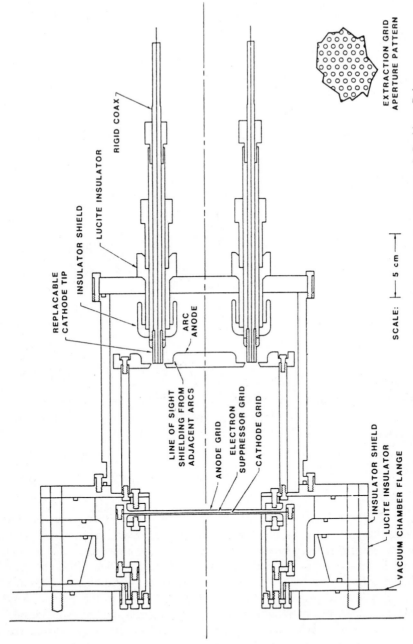

Figure 7.28 Multiple carbon-arc ion source and multiaperture extractor. (Courtesy C. Burkhart, Pulse Sciences Incorporated.)

REPLACABLE CATHODE TIP

INSULATOR SHIELD

LUCITE INSULATOR

RIGID COAX

ARC ANODE

LINE OF SIGHT SHIELDING FROM ADJACENT ARCS

ANODE GRID

ELECTRON SUPPRESSOR GRID

CATHODE GRID

INSULATOR SHIELD

LUCITE INSULATOR

VACUUM CHAMBER FLANGE

EXTRACTION GRID APERTURE PATTERN

SCALE: |—— 5 cm ——|

80% of the current carried by doubly charged ions have been extracted from the plasma generated by magnesium and titanium cathodes. The main problem of vacuum arc sources is that there are substantial ($\sim 20\%$) variations of available flux with time. These variations occur because the plasma near the emission point is unstable—the cathode spot moves and may periodically extinguish. The use of multiple arc sources or grid-controlled extractors (Section 7.7) can reduce the problem.

Figure 7.28 shows a multiple arc source for ion beam generation. The source generates 0.5×10^4 A/m^2 of C$^+$ ions over an area of 4×10^{-4} m^2. The cathode connects through ballast resistors to a pulse-forming network with about 1-kV standing voltage. A pulsed plasma injected into the gap initiates the arc. In Fig. 7.28, a low-energy spark on an insulator generates the trigger plasma. Cathode cooling limits pulse length and repetition rate—a typical extraction pulse is 1 msec.

F. Intense Pulsed Ion Beam Sources

Accelerators of intense pulsed light ion beams have potential application as drivers for inertial fusion reactors. Source requirements are extreme—typical values are 1000×10^4 A/m^2 of pure Li$^+$ for a pulse length 0.1–1 μsec. Figure 7.29 illustrates one approach to achieve intense plasma pulses. A metal-vapor vacuum arc creates a spatially extended plasma over a long time scale ($\geqslant 10\ \mu$sec). During the short extraction time, a pulsed magnetic field drives the plasma into the extraction gap. With axial and radial plasma compression, high current densities are possible.

We shall use a result from Section 12.5, which describes the theory of the magnetic acceleration of plasmas. A pulsed magnetic field of magnitude B_0 drives a highly conducting plasma at a velocity of

$$v_{\mathrm{p}} \cong [B_0^2/2\mu_0 n_0 m_{\mathrm{i}})^{1/2}. \tag{7.64}$$

In Eq. (7.64), n_0 is the plasma density before acceleration and m_{i} is the ion mass. The ion current density of a magnetically accelerated plasma with singly charged ions is roughly

$$j_{\mathrm{i}} \sim eB_0 [n_0/2\mu_0 m_{\mathrm{i}}]^{1/2}. \tag{7.65}$$

Moderate values of magnetic field and plasma density give high values of current density. As an example, suppose we have Li$^+$ ions stored at a density of 10^{20} m^{-3} and accelerated by a pulsed field of $B_0 = 0.2$ tesla. The velocity of the plasma front is about 3×10^5 m/sec, corresponding to a directed energy of 3.3 keV. If the initial length of the lithium plasma is 0.3 m, the rise time of the magnetic field should be $\leqslant 1\ \mu$sec. The current density predicted by Eq. (7.65) is over 400 A/cm^2.

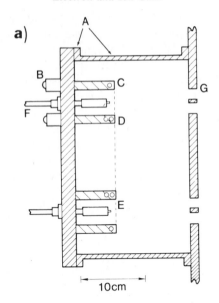

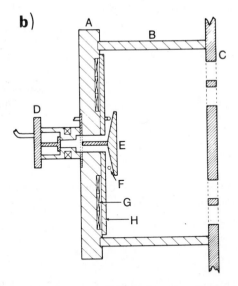

Figure 7.29 Pulsed ion sources using magnetic field acceleration of a plasma. (*a*) Source for a multi-kA proton gun. *A*: vacuum chamber; *B*: feedthroughs for pulsed magnet coils; *C* and *D*: coils to generate radial magnetic field: *E*: hydrocarbon spark source for protons; *F*: high-voltage feedthroughs; *G*: magnetically insulated extraction gap. (*b*) Source for high-current nitrogen and argon beams. *A*: insulating flange; *B*: vacuum chamber; *C*: magnetically insulated extraction gap; *D*: fast pulsed gas valve; *E*: radial gas nozzle; *F*: preionizer electrode; *G*: pulsed pancake magnet coil; *H*: ceramic faceplate. (*c*) Magnetic field lines from a pulsed pancake coil adjacent to an accelerated plasma. (Field calculation courtesy of J. Freeman, Sandia National Laboratories.)

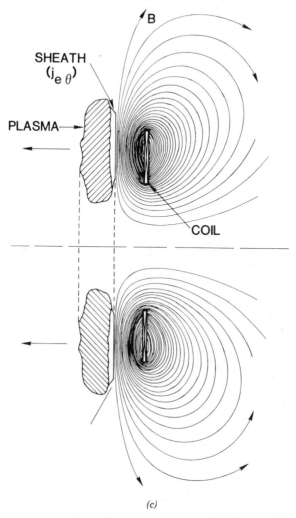

(c)

Figure 7.29 (*Continued*)

7.7. CHARGED-PARTICLE EXTRACTION FROM GRID-CONTROLLED PLASMAS

Section 7.5 described ion extraction from a free plasma surface. The term "free" implies that there are no confining forces at the plasma surface. A drawback of this method is that the optics of the output beam depends sensitively on the plasma properties. Many potentially useful ion sources, such as metal-vapor vacuum arcs, do not generate a constant ion flux. Free surface extraction presents another problem—it is not compatible with rapidly pulsed ion beams (100 nsec–10 μsec). Most plasma sources take a longer time to reach equilibrium operation.

Unless the extraction voltage is present during source initiation, a precursor plasma fills the extraction gap. The presence of a *plasma prefill* often causes breakdowns at high voltage.

To solve these problems, we can use biased electrostatic grids to confine an extraction plasma. For ion generation, plasma control grids uncouple the beam optics from variations of the plasma source and prevent plasma prefill. In this section, we shall discuss electrostatic confinement of plasmas. The process also has application to plasma cathodes as well as ion sources.

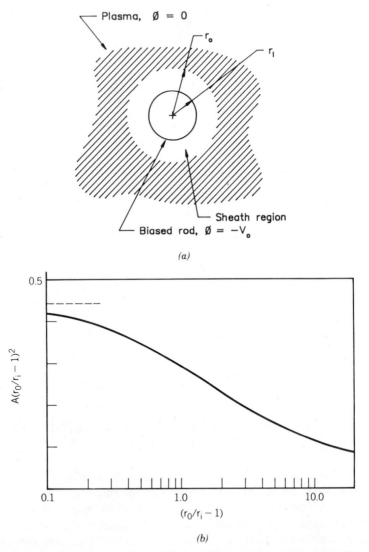

Figure 7.30 Grounded plasma near a rod with a negative bias voltage. (*a*) Geometry. (*b*) Numerical solutions for the space-charge-limited flow function for converging cylindrical flow.

To understand electrostatic plasma confinement, consider the geometry illustrated in Fig. 7.30a. A cylindrical rod of radius r_i is immersed in a uniform plasma of infinite extent. Contact with surrounding electrodes fixes the plasma potential near $\phi = 0$. We bias the rod to a negative potential, $\phi = -V_0$. The electron temperature of the plasma is small, $kT_e \ll eV_0$. We can find a value for the Bohm current density, j_B (Section 7.6) from the electron temperature and ion density.

The biased rod collects ions. Following Section 7.6, we can divide the volume outside the rod into two regions: the uniform plasma with zero electric fields and a sheath with strong electric fields. We denote the sheath radius as r_0. We can calculate r_0 by the following procedure. First, for a given ion species, we solve the problem of self-consistent space-charge flow for inwardly directed ions in a cylindrical geometry. We apply the conditions that the ions have small energy at r_0, the plasma boundary, that the radial derivative of electrostatic potential is almost zero at this position, and that the total potential drop across the sheath is $-V_0$. We next adjust the radius of the plasma boundary until the space-charge-limited current density at r_0 equals j_B. From Section 7.5, we know that the width of the sheath increases with bias voltage for given plasma properties.

Since we have already studied many examples of one-dimensional space-charge flow, we shall briefly summarize results for nonrelativistic flow in a cylindrical geometry. The Poisson equation has the form

$$\frac{d^2\Phi}{dR^2} = -\frac{1}{R}\frac{d\Phi}{dR} + \frac{A}{\sqrt{\Phi}},\qquad (7.66)$$

where

$$\Phi = -\phi/V_0, \qquad R = r/r_i,$$

and

$$A = j_B r_i r_0/\varepsilon_0 V_0^{3/2}(2e/m_i)^{1/2}.$$

The quantity j_B is the available plasma flux, m_i is the mass of the singly charged ion, r_i is the radius of the enclosed ion collector at $\phi = -V_0$, and r_0 is the radius of the ion source at $\phi = 0$. The solution of Eq. (7.66) constrains the value of A. The boundary conditions for the solution are

$$\Phi(1) = 1,\qquad (7.67)$$

$$\Phi(r_0/r_i) = 0,$$

and

$$d\Phi(r_0/r_i)/dR = 0.$$

Figure 7.30b summarizes numerical solutions of Eq. (7.66). The figure plots $A(r_0/r_i - 1)^2$ versus $(r_0/r_i - 1)$. We shall use the results after discussing electrostatic plasma confinement by a biased wire array.

Figure 7.31 shows a cross-section view of a negatively biased metal mesh immersed in a flowing plasma. We can approximate the mesh as an array of circular wires separated by distance δ. Figure 7.31a shows a case where the voltage applied to the wires is low. A single wire extracts ions from the plasma over a sheath width r_o, where $r_o < \delta/2$. Here, the sheaths are separated and each wire acts independently. In the limit of narrow sheaths, there are regions of field-free plasma between the wires. Although the wires collect some of the

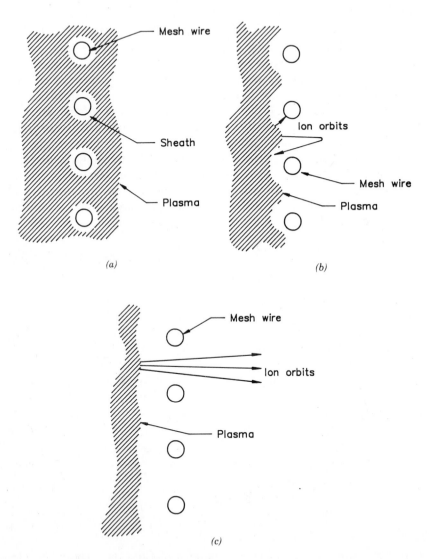

Figure 7.31 Inhibition of plasma flow by a wire mesh with a negative bias voltage. (a) At low bias voltage, separate ion extraction sheaths form at each wire, allowing plasma flow. (b) Moderate bias voltage gives coalesced sheaths and inhibition of plasma electron flow. (c) High bias voltages gives an approximately planar sheath surface for ion extraction.

plasma ions, the remainder drifts to the opposite side of the wire array. The result is that the biased mesh of Fig. 7.31*a* does not confine the plasma.

If we increase the magnitude of the mesh voltage, the sheaths broaden and may ultimately coalesce (Fig. 7.31*b*). In this case, plasma cannot penetrate through the mesh. The potential at all points between the wires is negative; therefore, low-energy plasma electrons cannot cross the mesh plane. The plasma ions are either collected directly by the wires or accelerated through the mesh. The bare ion space charge on the other side creates a positive space-charge potential, resulting in ion reflection (Fig. 7.31*b*). Ions reflected at the *virtual anode* either strike a wire or return to the plasma. By stopping the plasma electrons, the biased wire array prevents any plasma particles from penetrating through the mesh—the wire array provides electrostatic confinement.

A rough criterion for plasma confinement by the mesh is that

$$r_o > \delta/2. \tag{7.68}$$

An example will show how we can apply the cylindrical flow solutions to derive the confinement condition. We take a 100×100 (to the inch) mesh woven of 1-mil wires. For this geometry, $r_i = 1.25 \times 10^{-5}$ m and $r_o = 1.25 \times 10^{-4}$ m. We want to confine a singly ionized carbon plasma with $kT_e = 10$ eV and $n_i = 5 \times 10^{17}$ m^{-3}. From Eq. (7.55), the available ion current density from the plasma is 0.043×10^4 A/m^2. Inspection of Fig. 7.30*b* for $r_o/r_i = 10$ implies that $A \cong 0.12/81 = 1.5 \times 10^{-3}$. Inserting the given parameters into Eq. (7.68) gives the condition $V_0 > 30$ V for plasma confinement. If the wire voltage is much higher, the sheaths are large and the emission surface approaches a plane (Fig. 7.31*c*)

In the preceding analysis, we neglected the effects of plasma electron temperature. We can make a simple estimate of temperature effect in the limit of an ideal mesh—the wire array is infinitely fine with 100% transparency. Here, the sheath edge is planar. If the electrons have a Maxwell distribution with temperature kT_e, the electron current density incident on the sheath edge is [Eq. (2.61)]

$$j_e \cong 0.4 e n_0 (kT_e/m_e)^{1/2}. \tag{7.69}$$

Because of their low mass, the flux of electrons from the plasma is much larger than the ion flux. Confinement of the plasma depends on reflection of the ions from a virtual anode. The virtual anode cannot form if electrons penetrate the mesh. Since the electrons have a Maxwell distribution, we know that a mesh with bias voltage $-V_0 \sim kT_e/e$ does not reflect all incident electrons. The marginal voltage for plasma confinement allows one high-energy electron to cross the potential barrier for every incident ion. For a mesh voltage lower than this value, all incident ions can propagate past the mesh.

If the mesh has voltage $-V_0$, the fraction of plasma electrons that cross the potential barrier is $f_e = \exp(-eV_0/kT_e)$. In order to have plasma containment, we must reduce the electron flux by a factor greater than $(0.6/0.4)(m_e/m_i)^{1/2}$, or

$$V_0 > (kT_e/2e)\ln(0.667m_i/m_e). \tag{7.70}$$

For a carbon plasma with electron temperature $kT_e = 10\,\text{eV}$, Eq. (7.70) implies a mesh voltage $V_0 > 48\,\text{V}$. To find the required stopping voltage for a given plasma and mesh geometry, we take the highest value of Eqs. (7.68) and (7.70).

Figure 7.32 shows a geometry for the application of electrostatic confinement to ion extraction. The system consists of a plasma source, a plasma grounding

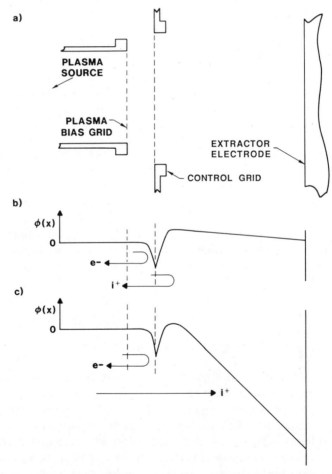

Figure 7.32 Extraction of ions using electrostatic control of a plasma by a negatively biased anode mesh. (*a*) Geometry of a planar injector. (*b*) Particle flows and the spatial variation of electrostatic potential without an extraction voltage. Figure illustrates the formation of a virtual anode. (*c*) Particle flows and the spatial variation of electrostatic potential with an applied extraction voltage.

grid, and a biased control grid. The control grid defines the surface of ion emission and acts as the anode of the extraction gap. Figure 7.32*a* illustrates particle flow and electrostatic potentials when there is no voltage applied to the extraction gap. The source electrodes and grounding grid clamp the potential of the expanding plasma to $\phi = 0$. The control grid has a negative voltage high enough to stop almost all plasma electrons; therefore, there is no plasma prefill. Because of the positive space charge in the extraction gap near the control grid, plasma ions are reflected. The control voltage is typically -100 V while the extraction voltage is many kilovolts—the distance ions travel from the control mesh is small compared with the extraction gap spacing, *d*. Figure 7.32*b* shows particle flow and electrostatic potential variations with a voltage applied to the extraction cathode. The voltage reduces the positive space-charge potential in the gap—a portion of the ion flux accelerated through the control mesh continues across the gap. Following Section 6.2, the transported current density is close to the Child law prediction [Eq. (5.48)].

The ion current density from a grid-controlled extractor is limited by space-charge forces rather than by the source properties. Operation of the device is analogous to thermionic cathodes (Section 6.2). Ions with a directed energy continually emerge from the control grid surface. Without an extraction voltage, the ions return to the grid. This behavior contrasts with emission from a free plasma surface where both electrons and ions can enter the gap. In summary, electrostatic confinement by a control grid prevents plasma prefill and uncouples the current density of an extracted ion beam from variations of the plasma source.

Controlled plasmas also have application to high-current pulsed electron beams. At low duty cycle, plasma sources are easier to construct and are more energy efficient than thermionic sources. Unfortunately, there is a fundamental problem with electron extraction from a free plasma boundary. The electron thermal flux is much higher than the ion flux. If the extracted current density is less than the thermal flux, the plasma expands into the gap. Extraction of a current density equal to the thermal electron flux causes a velocity-space anisotropy in the plasma. The nonuniform electron velocity distribution results in a two-stream instability (Section 14.1). The instability introduces strong variations in the electron flux and enhanced emittance in the extracted beam. The two stream instability is not important for ion extraction—the rapid response of the plasma electrons cancels space-charge bunching of the ion flow.

Although electrostatic confinement is effective for ion extraction, it cannot be applied to electrons. Confinement would require a positively biased control grid to stop ions. The grid would draw a large electron current, resulting in a plasma instability. Figure 7.33 illustrates an alternate approach to the generation of high-current pulsed electron beams. Here, a negatively biased control grid acts as a *plasma switch*. The grid restrains the plasma until electrons are required in the extraction gap. To initiate electron flow, we drop the control grid voltage to zero, allowing free expansion of the plasma into a low-voltage primary extraction gap. The choice of gap voltage and width ensures that the

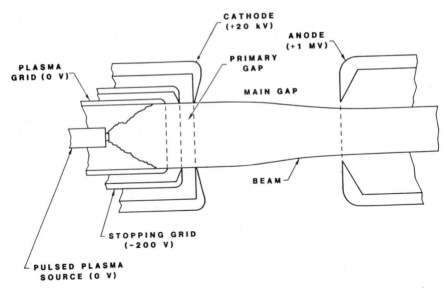

Figure 7.33 Multigap electron extractor for electron beam formation from a plasma. Figure shows representative voltages for a relativistic electron beam.

space-charge-limited flux is small compared with the electron thermal flux. As time passes, the plasma expands into the primary gap at the velocity of the ion directed motion; therefore, the space-charge-limited current in the primary gap grows. The extracted electrons travel into a high-voltage extraction gap. If the output current density of the primary gap exceeds the space-charge limit in the extraction gap, then the output beam current is constant. This method applies only to pulsed beams, since the primary gap ultimately fills with plasma. For plasmas from metal-vapor vacuum arcs, practical extraction pulselengths are in the range $\leqslant 100$ nsec.

7.8. ION EXTRACTORS

In this section, we shall amplify the discussion of Section 7.5 by looking at some additional features of ion extractors. In particular, we shall discuss the use of multiple extraction apertures, electron traps, and multiple acceleration gaps for the generation of high-current low-energy ion beams.

Electrostatic ion accelerators that create beams in the energy range 20–200 keV have applications in fusion research and space thrusters. A 50-A, 50-kV fusion accelerator has a perveance of over 4μperv, far beyond the capability of a single extraction gap—we must use multiple extraction structures to create an array of beams that combine after leaving the gun. Figure 7.34 illustrates a common extractor geometry. Ion flux from a single plasma source passes through an anode with multiple slots. Medium-perveance

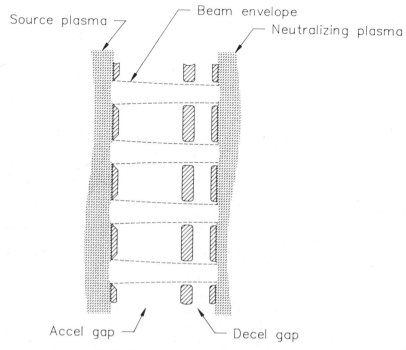

Figure 7.34 Multiple aperture ion extractor with decel electrode.

sheet beams emerge from each aperture. The individual beams must be spaced far enough apart so that they are electrically isolated by induced charges in the electrodes.

The high-perveance output beams from multiaperture extractors cannot propagate in vacuum because of longitudinal space-charge limits (Section 5.2). Fortunately, the transport problem is solved automatically in high-perveance ion accelerators with plasma sources based on gas injection. Neutral gas escaping from the source fills the region downstream from the extractor. Beam ions collide with gas atoms, creating a weakly ionized, quiescent plasma. Neutralization by the plasma almost cancels the space-charge fields of the beam—the emerging ions follow ballistic orbits for long distances in the transport region.

The main problem of plasma transport is that the extraction gap can pull electrons from the plasma and accelerate them back to the ion source. Depending on the downstream plasma density, electron acceleration may load the accelerator and damage the ion source. To use plasma transport, we must isolate the plasma from the acceleration gap. Usually, an *accel–decel* geometry provides isolation. The extractor of Fig. 7.34 has an additional electrode at a negative potential interposed between the anode and grounded output electrodes. The decel electrode creates an electrostatic potential boundary that prevents plasma electrons from entering the acceleration gap. The region of negative axial electric field is called the *electron trap*.

The voltage applied to the decel electrode must be high enough to ensure that the electrostatic potential is negative over the aperture cross section. Furthermore, the voltage must prevent plasma penetration. We derived a criterion for electrostatic plasma confinement in Section 7.7—the electrostatic sheath width should be larger than the halfwidth of the aperture or slot. As an example, suppose a slot of a multiaperture extractor produces a 30-keV-proton beam at a current density of $0.5 \times 10^4 \, A/m^2$—the beam density is $1.3 \times 10^{16} \, m^{-3}$. We take a neutralizing hydrogen plasma with density of $5 \times 10^{16} \, m^{-3}$ and electron temperature of 2 eV. The plasma Bohm current density is about $20 \, A/m^2$. We apply the cylindrical results of Section 6.3 to estimate the sheath width. For an aperture halfwidth of 3.5 mm, we take $r_i = 3.5$ mm and $r_o = 7.0$ mm. Solution of Eq. (6.26) gives a value of $A \cong 0.1$. Inserting parameters into Eq. (6.27), we find a minimum trap voltage of -1 kV.

Figure 7.35 shows results from the ray tracing program WOLF for a single gap extractor with an electron trap to generate a 20-keV-deuteron beam at $0.65 \times 10^4 \, A/m^2$. The program differs from the electron gun program EGUN discussed in Section 7.3—it carries out an iterative calculation to find the shapes of emission surfaces on the source and neutralizing plasmas. If n_i and T_e are uniform over the plasma volume, then the plasma boundary is defined by the condition that the emission current density normal to the surface is uniform. The program moves elements of the plasma surfaces to seek such a solution. The concave source plasma of Fig. 7.35 gives a converging extracted beam. The choice of electrode shapes ensures that the beam envelope is parallel to the axis at the point where the beam enters the neutralizing plasma. The program predicts a root-mean-square angular divergence for the output beam of 1.93°. The theoretical minimum divergence for the assumed ion temperature is about 0.5°—nonlinear transverse electric fields make a significant contribution to beam emittance.

Additional electrodes are often incorporated in high-voltage ion guns to increase the allowed current. If we divide the accelerating voltage between two or more gaps, we can achieve higher current density for a given value of peak electric field. We saw in Section 5.2 that we can write the space-charge-limited ion current density for a planar gap with voltage V_0 and width d in the form

$$j = \frac{4\varepsilon_0}{9} \left[\frac{2e}{m_i} \right]^{1/2} \frac{(E_{max}/1.33)^2}{V_0^{1/2}} \qquad (7.71)$$

The quantity E_{max} is electric field in volts per meter on the negative electrode. For a given electric field, the current decreases at high voltage.

The fields applied to electrodes of an ion extractor are limited by breakdown. Depending on electrode conditioning, the maximum safe field stress on the negative electrodes is about 5–10 MV/m. A dual-gap electrostatic accelerator extracts ions in the low-voltage gap then accelerates them to full voltage in a

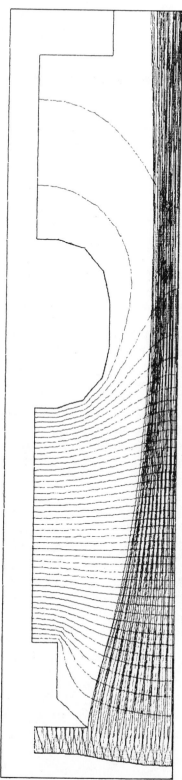

Figure 7.35 Design of a 10-A, 20-kV deuteron gun with accel and decel electrodes using the WOLF code. One half of one extraction slot is illustrated. Self-consistent determination of plasma emission surface is at the left hand side. (Courtesy W. Cooper. Lawrence Berkeley Laboratory.)

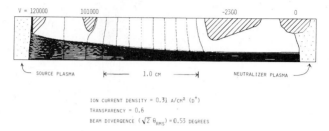

Figure 7.36 Numerical calculation of dual-gap, accel–decel ion extractor using the WOLF code. Figure shows electrodes, computational rays, and lines of constant electrostatic potential. (Courtesy W. Copper, Lawrence Berkeley Laboratory.)

second gap. The current density in the first gap follows Eq. (7.71), while the second gap limits are set by Eq. (6.6). The longitudinal space-charge limit is higher in the second gap because the ions enter at high velocity. Figure 7.36 shows a dual-gap gun for a 120-keV sheet beam of protons. The area-averaged current density is 0.3×10^4 A/m² from a structure with 60% transparency. The electron trap voltage is -2.3 kV—note the self-consistent plasma emission surfaces for both the source and neutralizing plasmas. Because of the high output energy, the exit beam divergence angle is only 0.53°.

For electrostatic acceleration of ion beams to MeV energies, the limits on current are set mainly by transverse beam confinement. The focusing force of transverse electric fields in cylindrical or planar structures drops rapidly with ion energy. The transverse field magnitude must increase as the beam accelerates to maintain a matched equilibrium—the axial field magnitude also rises. From the discussion of the Pierce column in Section 7.1, we know that the axial electric field must scale roughly as $z^{4/3}$. The voltage limit on an electrostatic

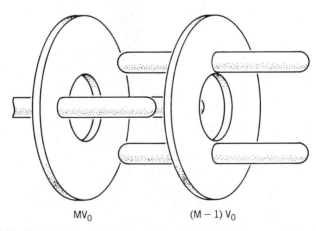

MV_0 $(M-1) V_0$

Figure 7.37 Electrostatic acceleration column with supplemental quadrupole focusing.

extraction column is set by electrical breakdown. For example, if we limit the electric field to $E_{max} = 5\,\text{MV/m}$, Eq. (7.71) implies that the maximum current density for a 1 MeV C^+ beam is $0.022 \times 10^4\,\text{A/m}^2$. We can achieve higher current if we relax the condition of cylindrical or planar symmetry for the applied fields. Figure 7.37 shows an acceleration column with shaped electrodes to generate quadrupole field components. The electrostatic quadrupole fields supplement the standard electrostatic focusing. The quadrupole fields increase with distance down the column—they provide the bulk of the focusing force at the exit. The gradual transition to quadrupole focusing reduces the problem of beam matching to downstream quadrupole arrays.

8

High-Power Pulsed Electron and Ion Diodes

Progress in any field of technology results largely from efforts to extend limits of performance. The book *Principles of Charged Particle Acceleration* showed how the goal of achieving the highest possible values of charged-particle kinetic energy was largely responsible for the evolution of accelerators from the earliest cyclotrons to the present generation of synchrotrons and beam colliders. The challenge in collective-beam physics is to attain increasingly higher beam current and power. While high-energy physics was the catalyst for the development of conventional accelerations, inertial fusion is a main motivation for research on high-power beams. The requirements for beams to drive inertial fusion targets are daunting. Beam power of over 100 million MW must strike a target a few millimeters in diameter in about a 10 nsec pulse. Present approaches to ion-beam-driven fusion aim for currents from 10 kA to 50 MA, depending on the kinetic energy of the ions.

Although a successful fusion driver has not yet been developed, the field has generated unique approaches to collective-beam physics. In this chapter, we shall study diodes that generate intense electron and ion beams. Here, the term diode implies a two-electrode acceleration gap. These devices have generated beam power density exceeding 10^{16} W/m^2.

Sections 8.1, 8.4, and 8.5 review background topics relevant to high-current injectors. Section 8.1 describes the motion of electrons in perpendicular electric and magnetic fields. A strong magnetic field can inhibit electron flow across a high-voltage gap. Section 8.4 discusses magnetic insulation of the high-power transmission lines that conduct pulsed power to diodes. Here, the magnetic field generated by the flow of current to the load is strong enough to prevent electron

328

motion across the vacuum gap between the inner and outer conductors. The lines operate at high electric fields and carry a large electromagnetic energy flux. Section 8.5 covers plasma erosion, the depletion of plasma ions and electrons from a gap after the application of a voltage. Plasma prefills can supply ions in high-current diodes. Plasma-filled gaps also have application to pulsed-power switching.

Section 8.2 describes pinched-electron-beam diodes. These devices create electron beams with current exceeding $100 \, \text{kA}$ and kinetic energy of $\geqslant 1 \, \text{MeV}$. The magnetic field generated by the electron flow is so high that the electrons pinch to a small spot. The resulting local current density may exceed $10^6 \times 10^4 \, \text{A/m}^2$. Despite the complex geometry of the beam, we shall find that it is possible to develop a self-consistent model to describe the electron flow. Section 8.3 reviews the operation of electron diodes with a strong applied axial magnetic field to prevent beam pinching. These diodes have application to pulsed-power electron accelerators and generators for intense microwave radiation.

Sections 8.6 through 8.9 outline methods to generate high-current ion beams. Fusion applications demand ion current density above $1000 \times 10^4 \, \text{A/m}^2$. There are two problems that must be solved to generate such high ion fluxes. First, we must prevent electron breakdown of the ion diode at high values of applied electric fields. Second, we must circumvent space-charge limits on ion current density from an extractor. We shall review the principles of two ion beam sources, the reflex triode and the magnetically insulated diode. Both devices achieve ion current density more than 100 times higher than the Child law prediction.

8.1. MOTION OF ELECTRONS IN CROSSED ELECTRIC AND MAGNETIC FIELDS

Several microwave sources, such as magnetrons and gyrotrons, depend on the motion of electrons in crossed electric and magnetic fields. The term *crossed field* means that the electric and magnetic fields are perpendicular. In this section, we shall discuss electron motion in crossed fields and emphasize the idea of *magnetic insulation*, an important process for high-power electron and ion diodes. Magnetic insulation prevents the motion of electrons across a gap with a high applied voltage.

Initially, we limit attention to single-electron motion, neglecting the contribution of free electrons to the fields. Figure 8.1 illustrates a one-dimensional magnetically insulated gap. Electrons emerge from a plane cathode at $z = 0$. An anode at $z = d$ has a bias voltage V_0. The electric field is

$$\mathbf{E} = -E_0 \mathbf{z} = -(V_0/d)\hat{\mathbf{z}}. \tag{8.1}$$

A uniform magnetic field is applied in the y direction:

$$\mathbf{B} = B_0 \hat{\mathbf{y}}. \tag{8.2}$$

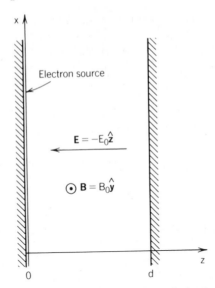

Figure 8.1   Geometry of a planar magnetically insulated gap.

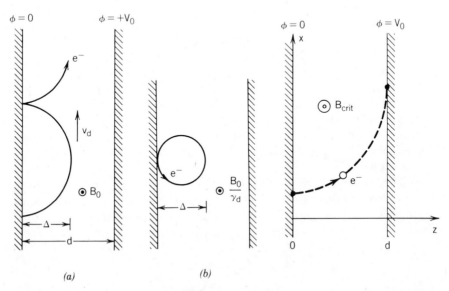

(a)   *(b)*

Figure 8.2   Drift orbits of electrons with zero total energy in a planar magnetically insulated gap. (*a*) Scalloped drift orbit viewed in the stationary frame. (*b*) Drift orbit viewed in a frame of reference moving at the electron drift velocity.

Figure 8.2a shows the trajectory of an electron created on the cathode with zero kinetic energy. The electric field pulls the particle toward the anode. As the electron gains velocity in the z direction, the magnetic field bends its orbit in the x direction. Ultimately, the electron returns to the cathode with zero kinetic energy. If the electron is not absorbed, it follows a scalloped orbit. Note in Fig. 8.2a that motion in the z direction is periodic, while the electron has a cumulative displacement along the x direction. This type of motion is called a particle *drift* (Section 10.7). We denote the average velocity along x as v_d, and the maximum distance the electron moves away from the cathode as Δ. If $\Delta < d$, electrons cannot cross to the anode and the gap is magnetically insulated.

We can find v_d and Δ in terms of B_0 and E_0 by making a relativistic transformation to a frame of reference that moves at velocity v_d in the positive x direction. We want to find electric and magnetic fields as they appear in the moving frame for given quantities in the stationary frame. Following Section 5.5, the field relationships are

$$E'_z = \gamma_d(E_z + v_d B_y), \tag{8.3}$$

$$B'_y = \gamma_d(B_y + v_d E_z/c^2). \tag{8.4}$$

The prime symbol marks quantities measured in the moving frame. The quantity γ_d is a function of the drift velocity,

$$\gamma_d = \left[\frac{1}{1 - (v_d/c)^2} \right]^{1/2}. \tag{8.5}$$

We can see in Eq. (8.3) that the electric field vanishes in the moving frame if the transformation velocity equals

$$v_d = E_0/B_0. \tag{8.6}$$

Then, the electron motion is a simple gyration in the transformed magnetic field (Fig. 8.2b). Since motion in the moving frame is purely oscillatory, the velocity of Eq. (8.6) equals the average drift velocity of the centroid of the electron orbit. This velocity is called the $\mathbf{E} \times \mathbf{B}$ ("$\mathbf{E}$ cross $\mathbf{B}$") drift velocity—its direction is parallel to the cross product of field vectors. Note that if

$$E_0 \geq cB_0, \tag{8.7}$$

then the velocity exceeds the speed of light and the transformation is invalid. When the condition of Eq. (8.7) holds, there is no frame in which electron motion is oscillatory—the electron moves without limit in the negative y direction and never returns to the cathode.

When $E_0 < cB_0$, we can calculate the quantity Δ by applying the following facts:

1. Equation (8.4) implies that the magnitude of the magnetic field in the moving frame is lower by a factor of γ, $B'_y = B_0/\gamma$.
2. The electron follows a circular gyration orbit in the constant magnetic field.
3. At the point of contact with the cathode, the electron has zero total velocity in the stationary frame; therefore, it has velocity $\mathbf{v} = -v_d\mathbf{x}$ in the moving frame.

The radius of the circular electron orbit in the moving frame is

$$r_g = \frac{\gamma m_0 |\mathbf{v}|}{eB_0/\gamma} = \frac{\gamma^2 m_0 v_d}{eB_0}. \tag{8.8}$$

The maximum excursion from the cathode is twice the gyroradius,

$$\Delta = 2r_g. \tag{8.9}$$

The maximum electron displacement from the cathode is the same in the stationary frame since Lorentz transformations do not change transverse dimensions.

The criterion for magnetic insulation is that $\Delta \leqslant d$. The *critical magnetic field*, B_{crit}, is the field magnitude that gives $d = \Delta$. For a given electric field, a magnetic field larger than B_{crit} prevents electrons from crossing to the anode. We can combine Eqs. (8.1), (8.5), (8.6), (8.8), and (8.9) to give an expression for B_{crit}:

$$B_{\text{crit}} = B^*\left[1 + \frac{eV_0}{2m_ec^2}\right]^{1/2}, \tag{8.10}$$

where

$$B^* = \frac{1}{d}\left[\frac{2V_0m_0}{e}\right]^{1/2}. \tag{8.11}$$

We have written Eq. (8.10) in terms of a nonrelativistic factor, B^*, and a relativistic correction term. The relativistic term is important when $V_0 \geqslant m_0c^2/e = 0.511\,\text{MV}$. Figure 8.3 shows $B_{\text{crit}}d$ as a function of V_0. We see that moderate magnetic fields can insulate high-voltage gaps. For example, the critical field for a gap of width 0.02 m with applied voltage 1 MV is $B_{\text{crit}} = 0.24$ tesla.

We can derive Eq. (8.10) by a different method that uses conservation of energy and canonical momentum. In an equilibrium gap, the total energy of electrons is a conserved quantity. The canonical angular momentum in the x direction, P_x, is also a constant of the motion since all forces are uniform along x. For a vector potential $A_x(z)$, we can write the canonical momentum as

$$P_x = \gamma(z)m_0v_x(z) - eA_x(z). \tag{8.12}$$

In Eq. (8.12), $\gamma(z)$ refers to the relativistic energy factor of an electron observed in

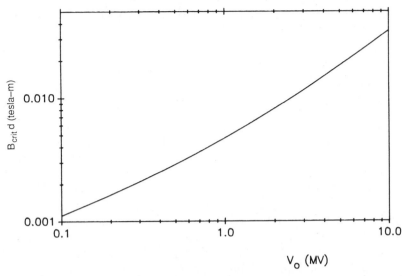

Figure 8.3   Plot of $B_{crit}d$ for a planar magnetically insulated gap as a function of the applied voltage, V_0.

the stationary frame. The vector potential is related to the magnetic field by

$$B_y = \frac{\partial A_x}{\partial z}. \tag{8.13}$$

Taking A_x equal to zero at cathode, integration of Eq. (8.13) gives

$$A_x(d) = \int_0^d B_y(z)\, dz. \tag{8.14}$$

If the field is uniform in z, Eq. (8.14) becomes

$$A_x(d) = B_0 d. \tag{8.15}$$

We can find the critical magnetic field by comparing the properties of fields and electron orbits at the cathode and anode. At the cathode, the vector potential is zero [$A_x(0) = 0$] and electrons have no kinetic energy ($v_z = 0, v_x = 0$). Therefore, electrons leaving the cathode have zero canonical momentum, $P_x = 0$. An applied magnetic field equal to B_{crit} gives electron orbits that just reach the anode. Conservation of total energy implies the electrons have a relativistic γ factor of

$$\gamma(d) = 1 + eV_0/m_0 c^2, \tag{8.16}$$

at the anode. We note that $v_z(d) = 0$ and $v_x(d)$ has a maximum value at the anode. Setting $P_x = 0$, Eq. (8.12) becomes

$$\gamma(d)v_x(d)/c = eA_x(d)/m_0 = eB_0d/m_0. \tag{8.17}$$

Noting that

$$\gamma(d)v_x(d)/c = [\gamma(d)^2 - 1]^{1/2}, \tag{8.18}$$

we can combine Eqs. (8.17) and (8.18) to yield the same relation as Eq. (8.10).

Equations (8.16) and (8.17) are valid even when trapped electrons modify the electric and magnetic fields. Electron space charge reduces E_z near the cathode, concentrating the field near the anode. Although the electric field varies in the y direction, the total voltage across the gap must equal V_0. Therefore, the equation of conservation of energy applied to electrons that reach the anode does not change. The drifting electrons carry a current in the x direction. We shall see that the electron density is high in relativistic gaps ($eV_0 \geqslant m_0c^2$). The corresponding high current can significantly change the distribution of magnetic field in the gap.

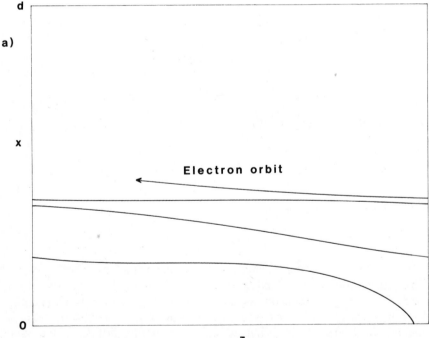

Figure 8.4 Numerical calculations of electron orbits in a magnetically insulated gap with a constant magnetic field and a slowly rising voltage. The voltage rises linearly by a factor of 99. The initial voltage corresponds to $B = 0.1B_{crit}$, while the final voltage gives $B = B_{crit}$. (a) Voltage risetime: $15/\omega_{ge}$. (b) Voltage risetime: $50/\omega_{ge}$. (c) Voltage risetime: $200/\omega_{ge}$.

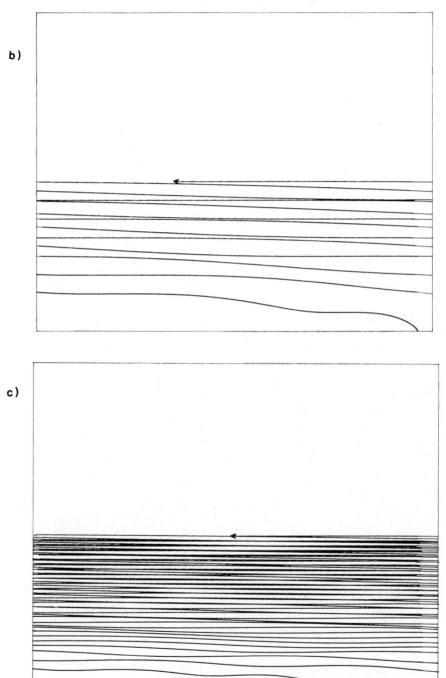

Figure 8.4 (*Continued*)

The trapped electrons are diamagnetic—they reduce the magnetic field near the cathode. If the cathode and anode are perfect conductors and the electron density enters rapidly, the total number of magnetic field lines in the gap is constant. As a result, the integral of Eq. (8.14) equals $B_0 d$, even though B_y varies across the gap.

Electrons follow the cycloid orbits of Fig. 8.2a when (1) they enter the gap at the cathode with zero kinetic energy and (2) the gap voltage has the constant value V_0. Other types of orbits are possible if we relax these conditions. For example, we could describe the motion of electrons emitted from the cathode during a slow rise of the gap voltage to a final value V_0. This type of motion occurs during the risetime of a pulsed voltage waveform applied to the gap. Here, the term "slow" means that the risetime is much longer that the period for electron gyration in the magnetic field, $2\pi/\omega_{ge}$. In this limit, we find that the electrons follow laminar orbits in the $\mathbf{E} \times \mathbf{B}$ direction rather than cycloid drift orbits.

Drift orbit theory predicts the nature of cross field electron motion with slow changes of electric field. In a rising field, the polarization drift carries electrons away from the cathode while they move sideways at the $\mathbf{E} \times \mathbf{B}$ drift velocity. Figure 8.4 shows numerical calculations of a nonrelativistic electron orbit—the electron leaves the cathode at $t = 0$. The magnetic field B_0 is uniform and constant—the gap voltage rises linearly to a voltage that gives $B_0 = B_{crit}$. The sequence of figures shows increasing voltage risetimes, $15/\omega_{ge}$ in Fig. 8.4a, $50/\omega_{ge}$ in Fig. 8.4b, $200/\omega_{ge}$ in Fig. 8.4c. With a long risetime, the electron orbit approaches a straight line with a small component of velocity in the y direction. Note that the electron travels only halfway across the gap—with the same values of magnetic field and voltage cycloid orbits would reach the anode. The laminar orbits have the same average position in z and velocity in x, but they do not have an oscillatory component of motion.

We can find the distance that a laminar electron orbit moves from the cathode during the voltage rise by applying conservation of canonical angular momentum. In the nonrelativistic limit, the final orbit satisfies the following three equations:

$$m_e v_x(z) - (eB_0 d)(z/d) = 0, \qquad (8.19)$$

$$v_x(z) = V_0/B_0 d, \qquad (8.20)$$

$$v_z(z) \cong 0. \qquad (8.21)$$

Combining Eqs. (8.19) and (8.20) and taking $B_0 = B^*$ of Eq. (8.11), we find that the final position is $z = d/2$. The kinetic energy of the electron at peak voltage is

$$m_e v_x^2/2 \cong eV_0 d/4.$$

This energy equals half the change in energy for an electron that moves from the cathode to the middle of the gap at full voltage. The discrepancy results from the fact the average electric field during the time an electron moves across the gap equals half the peak field.

8.2. PINCHED ELECTRON BEAM DIODES

Experiments have shown that two-electrode vacuum gaps driven by a pulsed-power generator can create tightly pinched electron beams. Figure 8.5a illustrates an electron beam diode that consists of a planar cathode of radius R_c separated from an anode by a vacuum gap of width d. The cathode emits a high current density of electrons by the surface plasma mechanism discussed in Section 7.4. With beam current $\sim 100\,\text{kA}$ and voltage $\sim 1\,\text{MV}$, almost all electrons emitted from the cathode compress to a tight focus on the axis at the anode. Current densities exceeding $1\,\text{MA/cm}^2$ have been observed. The electron flow is called a *superpinched electron beam*. The beams have application to the simulation of high-power-density processes. In contrast to the conventional electron guns of Chapter 7, superpinched beams have perveance values that range from 100 to 1000 μperv.

Pinching results from the strong azimuthal magnetic field generated by the high-current electron beam. A pinch occurs when the magnetic field at the edge of the cathode is strong enough to bend electron orbits so that they cannot cross directly to the anode—the magnetic field insulates the edge of the cathode. We can estimate the conditions for the existence of a pinch. When the beam carries low current, the magnetic field is low and electrons move directly across the gap. At low current, the one-dimensional Child law gives a good estimate of j_e. We can write the space-charge-limited current in terms of the relativistic energy factor for electron arriving at the anode,

$$\gamma_0 = 1 + eV_0/m_e c^2. \tag{8.22}$$

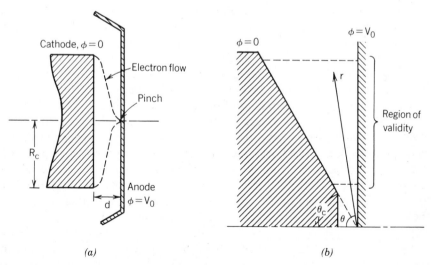

(a)   (b)

Figure 8.5 Pinched electron beam diode. (a) Geometry of a cylindrical diode. (b) Idealized conical diode geometry used for the parapotential flow model.

We find that

$$I \cong \frac{4\varepsilon_0}{m_e} \left[\frac{2e}{m_e} \right]^{1/2} \frac{(\gamma_0 - 1)^{3/2}(m_e c^2)^{3/2}}{d^2 e^{3/2}} (\pi R_c^2) \Theta(\gamma_0). \qquad (8.23)$$

The quantity $\Theta(\gamma_0)$ is a correction factor equal to the ratio of the relativistic space-charge-limited current density to the nonrelativistic Child law current—we can derive $\Theta(\gamma_0)$ from Fig. 6.9—in the energy range of interest, it is comparable to or less than unity.

Electrons move radially inward when the field at the cathode edge satisfies the magnetic insulation condition. We can substitute Eq. (8.22) into Eq. (8.10) to derive a condition for magnetic cutoff at the outer radius of the cathode:

$$I_c = \left[\frac{2\pi m_e c}{e\mu_0} \right] \frac{R_c}{d} [\gamma_0^2 - 1]^{1/2}. \qquad (8.24)$$

The electrons pinch when the space-charge-limited current of Eq. (8.23) approaches the cutoff current of Eq. (8.24). Setting the currents equal gives the following criterion for a pinched beam:

$$\frac{\sqrt{8}}{9} \left(\frac{R_c}{d} \right) \frac{(\gamma_0 - 1)^{3/2} \Theta(\gamma_0)}{(\gamma_0^2 - 1)^{1/2}} > 1. \qquad (8.25)$$

Equation (8.25) shows that we need a high value of the diode aspect ratio (R_c/d) and voltage (γ_0) to achieve a pinch. For a voltage $V_0 = 1$ MV, Eq. (8.25) implies that R_c/d must be greater than 3.2.

A primary theoretical challenge of pinched beam diodes is to explain how electrons move to the axis. Although the diodes are geometrically simple, the electromagnetic field distributions and the motion of electrons are complex. Particle simulations are essential for detailed results. Nonetheless, an analytic theory, the *parapotential model*, describes many of the features of the pinched electron beam diode and other crossed-field devices. Although the model simplifies the device physics, it gives valuable insights into the nature of pinched flow and generates good estimates of the net diode current.

The parapotential model seeks a self-consistent equilibrium solution for electron flow with the simplest possible particle orbits. We saw in Section 8.2 that electron orbits in a crossed field have an oscillation superimposed on an $\mathbf{E} \times \mathbf{B}$ drift. It is difficult to calculate the space-charge density when electrons follow scalloped orbits. Section 8.2 also showed that under some conditions electrons follow straight-line orbits. The flow of electrons is laminar if their velocity in the direction normal to the fields satisfies the condition

$$\mathbf{E} = -\mathbf{v} \times \mathbf{B}. \qquad (8.26)$$

We shall seek conditions that lead to a laminar flow equilibrium. In the pinched beam diode, electrons move radially inward through regions of varying **E** and **B**. For laminar flow, Eq. (8.26) must hold at all radii. The equation also implies that the electrons move perpendicular to electric field lines; therefore, the orbits lie on lines of constant electrostatic potential. This fact motivates the term "parapotential"—the prefix "para" means "in the same direction." Self-consistent crossed-field equilibria with laminar electron orbits are called *Brillouin flow solutions*. Section 10.3 describes another Brillouin flow equilibrium for electrons confined in a cylindrical beam.

The parapotential model uses the geometry of Fig. 8.5b. Electrons follow laminar orbits between a biased anode plate at $V = + V_0$ and a grounded conical cathode. The electric field points mainly in the axial direction, while the magnetic field is azimuthal. The electrons move along predominantly radial equipotential lines that converge at the apex of the cone. The apex is a singular point—the applied electric field and electron density diverge to infinity. The model does not give a realistic representation close to the axis. Another problem not addressed by the parapotential model is how electrons reach the appropriate equipotential lines at large radius. The purpose of the model is to investigate how electrons move from the edge of the diode toward the center. We must use other methods, such as computer simulations, to study processes at the diode periphery and the pinch.

As in the Child law derivation of Section 5.2, we assume that the distribution of electrons is singular. The total energy of all electrons referenced to the potential of the cathode is zero. Equation (8.26) determines the total electron velocity. Since all electrons at the same position have the same velocity, the fluid equations of particle and momentum conservation are sufficient to describe the equilibrium. Electric fields arise from the voltage applied between electrodes and the space charge of the electrons. The azimuthal magnetic field results mainly from the flow of electrons toward the apex. We also include the possibility that current flows along the cathode surface to create a component of applied magnetic field. We will use the spherical coordinate system of Fig. 8.5b for the analysis. Cylindrical symmetry implies that all quantities are independent of the azimuthal coordinate ϕ. The coordinate r is the total distance from apex, while θ is the angle relative to the z axis. The equation $\theta = \theta_c$ defines the cathode surface, while the anode surface corresponds to $\theta = \pi/2$. Finally, the variable R denotes the radial distance from the axis:

$$R = r \sin \theta. \tag{8.27}$$

We assume that the electron streamlines and equipotential lines lie on conical surfaces that radiate from the apex. An equivalent statement is that the electrostatic potential depends only on the coordinate θ:

$$V(r, \theta, \phi) = V(\theta). \tag{8.28}$$

We can justify Eq. (8.28) by showing that it is consistent with Eq. (8.26). Equation

(8.28) implies that the electric field has a component only in the θ direction, given by

$$E_\theta = -(1/r)\, dV/d\theta. \tag{8.29}$$

The condition that equipotential lines lie on conical surfaces implies that $dV/d\theta$ is constant along a line of constant θ. Therefore, the electric field varies as

$$E_\theta \sim 1/r \sim (\sin\theta)/R. \tag{8.30}$$

The toroidal magnetic field on an equipotential line with angle θ equals

$$B_\phi = \mu_0 I(\theta)/2\pi R, \tag{8.31}$$

where $I(\theta)$ is the net enclosed axial current. The current includes contributions from electrons that flow on equipotential lines between the given surface and the axis and the bias current that may flow along the cathode. In steady state, the electron current flows inward along radial streamlines, crosses to the anode at the apex, flows outward, and returns to appropriate equipotential lines at large radius. By the law of continuity of current, the total axial current inside an equipotential line equals the sum of the total currents that flow along enclosed equipotential lines; the quantity depends only on θ. The implication is that

$$B_\phi \sim 1/R. \tag{8.32}$$

Comparison of the field variations of Eqs. (8.30) and (8.32) with Eq. (8.26) shows that the electron velocity in the $-r$ direction depends only on θ. As the electrons flow toward the apex at constant velocity, the volume they occupy decreases as $1/r^2$. Therefore, the space-charge density of electrons has the form

$$\rho(r,\theta) = g(\theta)/r^2. \tag{8.33}$$

We now have sufficient information to construct equations that describe the ideal diode. With no variations in r and ϕ, the Poisson equation for electrostatic potential V is

$$\frac{1}{r^2 \sin\theta}\frac{d}{d\theta}\left(\sin\theta\frac{dV}{d\theta}\right) = -\frac{g(\theta)}{\varepsilon_0 r^2}. \tag{8.34}$$

We can write Eq. (8.34) as

$$\frac{d^2 V}{d\theta^2} + \frac{\cos\theta}{\sin\theta}\frac{dV}{d\theta} = -\frac{g(\theta)}{\varepsilon_0}. \tag{8.35}$$

We shall express all results in terms of the relativistic energy factor:

$$\gamma(\theta) = 1 + eV(\theta)/m_ec^2. \tag{8.36}$$

The electron velocity along an equipotential line, $v_r(\theta)$, is related to γ by

$$v_r(\theta) = -\beta(\theta)c = -c[1 - 1/\gamma(\theta)^2]^{1/2}. \tag{8.37}$$

The radial current density, $j_r(\theta)$, is the product of the radial velocity and the charge density:

$$j_r(\theta) = -g(\theta)\beta(\theta)c/r^2. \tag{8.38}$$

We can express the total current enclosed within an equipotential line at θ in terms of the radial current density:

$$I(\theta) = \int_{\theta_c}^{\theta} (2\pi r^2 \sin\theta'\, d\theta')j_r(\theta') + I_b \tag{8.39}$$

$$= \int_{\theta_c}^{\theta} -2\pi g(\theta')\beta(\theta')c \sin\theta'\, d\theta' + I_b.$$

The quantity I_b is the cathode bias current. To complete the set of equilibrium equations, we combine Eqs. (8.26), (8.27), (8.29), (8.31) and (8.37) to give:

$$\frac{dV}{d\theta} = \frac{\mu_0 c\beta(\theta)I(\theta)}{2\pi \sin\theta}. \tag{8.40}$$

We rewrite Eq. (8.40) in terms of γ:

$$I(\theta) = \left[\frac{2\pi m_e c}{e\mu_0}\right]\frac{\sin\theta}{\beta(\theta)}\frac{d\gamma(\theta)}{d\theta}. \tag{8.41}$$

As in the Child law derivation, our strategy is to combine all relationships into a single equation for the electrostatic potential. Solution of the equation with the proper boundary conditions leads to the self-consistent potential. The first step is to take the derivative of Eq. (8.39) and substitute for $g(\theta)$ in Eq. (8.35). Next, we eliminate $dI/d\theta$ from the resulting equation by taking the derivative of Eq. (8.40). This gives an equation that involves only the dependent variable. The final step is to rewrite the equation in terms of $\gamma(\theta)$ using Eq. (8.36):

$$\frac{d^2\gamma}{d\theta^2} + \frac{\cos\theta}{\sin\theta}\frac{d\gamma}{d\theta} = \frac{\gamma}{\gamma^2 - 1}\left(\frac{d\gamma}{d\theta}\right)^2. \tag{8.42}$$

Saturated parapotential flow results when electrons occupy all equipotential lines between the cathode and anode. For this case the boundary conditions for Eq. (8.42) are

$$\gamma(\theta_c) = 1 \qquad \text{(cathode)}, \tag{8.43}$$

$$\gamma(\pi/2) = \gamma_0 = 1 + eV_0/m_e c^2 \qquad \text{(anode)}. \tag{8.44}$$

Given the function $\gamma(\theta)$, we can find the current carried on equipotential lines and the total current flow in the diode from Eq. (8.41).

We can simplify Eq. (8.42) with the substitution $\chi = \ln [\tan (\theta/2)]$:

$$\frac{d^2\gamma}{d\chi^2} = \frac{\gamma}{\gamma^2 - 1} \left(\frac{d\gamma}{d\chi} \right)^2. \tag{8.45}$$

The solution of Eq. (8.45) that satisfies Eqs. (8.43) and (8.44) is

$$\chi = C_1 \ln [\gamma + (\gamma^2 - 1)^{1/2}] + C_2, \tag{8.46}$$

where

$$C_1 = -\ln [\tan (\theta_c/2)]/\ln [\gamma_0 + (\gamma_0^2 - 1)^{1/2}] \tag{8.47}$$

and

$$C_2 = \ln [\tan (\theta_c/2)]. \tag{8.48}$$

We write the current contained within angle θ in terms of the solution for $\gamma(\chi)$ using the chain rule of derivatives, $d\gamma/d\theta = (d\gamma/d\chi)(d\chi/d\theta)$:

$$I(\theta) = [2\pi m_e c/e\mu_0]\gamma/C_1. \tag{8.49}$$

In the limit $R_c/d \gg 1$, a series expansion of the transcendental functions in the constant C_1 leads to the result

$$-\ln [\tan (\theta_c/2)] \cong d/R_c. \tag{8.50}$$

Combining Eqs. (8.47), (8.49), and (8.50), the net current flowing along equipotential lines inside the line with energy factor γ is approximately

$$I(\theta) = \left[\frac{2\pi m_e c}{e\mu_0} \right] \frac{R_c}{d} \gamma \ln [\gamma_0 + (\gamma_0 - 1)^{1/2}]. \tag{8.51}$$

For the total current, we take $\gamma = \gamma_0$ in Eq. (8.51). The expression for the total saturated parapotential current in a pinched electron beam diode is

$$I_0 = \left[\frac{R_c}{d} \right] \left[\frac{2\pi m_e c}{e\mu_0} \right] \gamma_0 \ln [\gamma_0 + (\gamma_0^2 - 1)^{1/2}]. \tag{8.52}$$

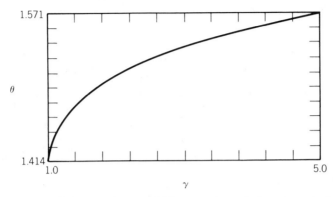

Figure 8.6 Variation of electrostatic potential between the cathode and anode for saturated parapotential flow. $\gamma = 1 + e\phi/m_e c^2$ versus θ.

The second quantity in brackets on the right-hand side of Eq. (8.52) equals 8.5 kA. Figure 8.6 shows a plot of the variation of potential between the cathode and anode for saturated flow. The charge density of the electrons cancels the electric field near the cathode while enhancing the field at the anode.

Equation (8.51) implies that the current inside the equipotential line with $\gamma = 1$ does not equal zero. A valid Brillouin flow solution requires that current flows along the cathode surface. The cathode current is the boundary current of Eq. (8.39). Setting $\gamma = 1$ in Eq. (8.51) gives

$$I_b = I(\theta_c) = I_0/\gamma_0. \tag{8.53}$$

The magnetic field generated by the boundary current is analogous to the external magnetic field required for the nonrelativistic Brillouin flow solution of Section 10.4. At high values of γ_0, electron flow generates most of the confining magnetic field.

We can remove the problem of the singularity at the axis by recognizing that in an actual diode electrons have axial oscillations superimposed on the radial drift. Depending on the oscillation amplitude, electrons strike the anode at varying distances from the axis. In a sense, the beam emittance limits the focus—electrons in perfect parapotential flow could be focused to a point. We must also explain how electrons reach the appropriate equipotential lines at large radius. To answer this question, we note that electron diodes are usually located at the end of a magnetically insulated transmission line with high electric field (Section 8.4). The electrons moving down the line are exposed to a rising electric field as they approach the edge of the diode. The polarization drift spreads the electrons over a good fraction of the diode gap width. The injected electrons have orbits that are intermediate between laminar and scalloped. The power density in the pinch is high if the electric field rises smoothly along the length of the vacuum electrodes, leading to laminar orbits in the diode.

Most diodes are cylindrical rather than conical, with a uniform gap width d. Despite this fact, conical equipotential surfaces still provide a good description if electrons entering at the edge of the diode dominate the flow. The electric field generated by these electrons suppresses emission from inner regions of the cathode. The compression of electron space charge near the axis and the increased magnetic field require that the axial electric field increases near the axis in an equilibrium. The result is compression of the electrons toward the anode as they approach the axis. The parapotential model prediction for the net current from a cylindrical diode is in good agreement with experiments if the effective cathode angle is taken as

$$\theta_c \cong R_c/d. \tag{8.54}$$

A point of discussion for the validity of the parapotential model is the nature of the bias current I_b. The most likely source is ion flow from the anode to the cathode. Experimental studies of pinched-beam diodes have shown that ions generated on the anode by electron bombardment play an important role. They add to the net diode current and partially neutralize electron space charge to give tight pinches.

Sometimes, ion flow may account for a large fraction of the net diode current. As a result, the pinched-beam diode has been used for the generation of intense ion beams. In an ion diode, the strong beam-generated magnetic fields prevent direct electron flow; the fraction of the current carried by ions can be much higher than the prediction of the one-dimensional bipolar flow model [Eq. (6.39)]. In turn, the trapped electrons in the acceleration gap counter the ion space charge, giving ion current densities that exceed the Child law value [Eq. (5.48)].

Figure 8.7 illustrates the motion of electrons and ions in a pinched-beam diode. The figure shows particle orbits from a two-dimensional simulation of a 600-kV diode. Because of their high mass, ions travel almost directly across the gap, while the electrons follow complex drift orbits. The simulation confirms the general features of the parapotential model. The equipotential surfaces are conical and the electron space charge compresses electric fields toward the anode. The electron orbit shown consists of a drift along an equipotential line with a superimposed axial oscillation.

We can estimate the fraction of diode current carried by ions from the constraint of global charge balance. The emission of electrons and ions is space-charge-limited; therefore, the normal electric fields at the cathode and anode surfaces equal zero. By Gauss's law, the total number of ions in the diode at any time, N_i, must equal the total number of electrons, N_e. We define the quantity t_i as the average residence time for an ion in the gap. We can estimate t_i from the average electric field, V_0/d:

$$t_i \cong 2d/\sqrt{2eV_0/m_i} = d\sqrt{2}\sqrt{m_i/m_e}/c\sqrt{\gamma_0 - 1}. \tag{8.55}$$

The electron residence time, t_e, depends on the drift orbit. We can estimate the

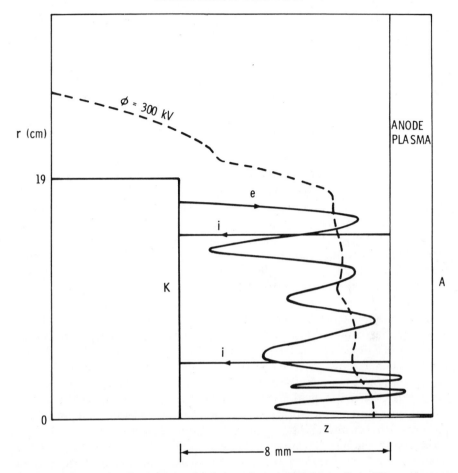

Figure 8.7 Particle-in-cell computer simulation of a pinched beam diode. Figure shows a representative electron orbit, ion orbits, and a line of constant electrostatic potential; 500-kV applied voltage, $I_e = 285 \, kA$, and $I_i = 300 \, kA$. (Courtesy J. Poukey, Sandia National Laboratories.)

minimum electron residence time by taking a straight-line orbit (as in parapotential theory) and recognizing that there is an upper limit on the radial drift velocity:

$$v_r \leqslant \beta_0 c = c(\gamma_0^2 - 1)^{1/2}/\gamma_0. \tag{8.56}$$

If the electrons enter at the edge of the diode, the electron residence time is

$$t_e \geqslant R_c \gamma_0 / c (\gamma_0^2 - 1)^{1/2}. \tag{8.57}$$

The ion current equals the total number of ions in the gap at any time divided by the average residence time:

$$I_i = N_i/t_i. \tag{8.58}$$

Since $N_e = N_i$, the ratio of ion to electron current is inversely proportional to the residence time ratio:

$$\frac{I_i}{I_e} \simeq \frac{t_e}{t_i} = \frac{1}{\sqrt{2}}\left(\frac{m_e}{m_i}\right)^{1/2}\left(\frac{\gamma_0}{\sqrt{\gamma_0 + 1}}\right)^{1/2}\left(\frac{R_c}{d}\right). \tag{8.59}$$

We can inspect the right-hand side of Eq. (8.59) to find parameters that affect the ion current fraction. Scaling with the mass ratio is the same as that for one-dimensional bipolar flow. For the same charge density and diode voltage, the ion current density is lower because the ions move slower. The second term represents relativistic effects on the flow. The ion velocity increases as the square root of the diode voltage, while the electron velocity asymptotically approaches the speed of light for $\gamma_0 > 1$. The third term represents the difference in electron and ion orbits; the electrons travel a longer distance.

The following example illustrates the utility of a pinched-beam diode for ion generation. Suppose we want to create a proton beam using a pulsed-power generator that can drive 300 kA at 1 MV ($\gamma_0 = 2.96$). The first step is to find the geometry of a diode matched to the generator. If the gap spacing is $d = 0.015$ m, then Eq. (8.52) shows that the cathode radius should be $R_c = 0.19$ m for $I_e = 300$ kA. Substitution into Eq. (8.25) shows that the diode parameters satisfy the pinch criterion. Equation (8.59) implies that $I_i/I_e \cong 0.31$. This figure is much higher than the ratio of 0.023 predicted for one-dimensional bipolar flow. The ion flow constitutes 24% of the total diode current, or $I_i = 72$ kA. The prediction of the one-dimensional, single-species Child law [Eq. (5.48)] for a diode area $\pi R_c^2 = 0.113$ m^2 is $I_i = 27$ kA. The ion flow enhancement factor arising from electron space charge in the pinched-beam diode is 2.7, exceeding the factor of 1.86 for bipolar flow. Although pinched-beam diodes generate high net ion current, the resulting beams usually have poor emittance because of the strong magnetic fields in the diode and instabilities of the anode plasma.

8.3. ELECTRON DIODES WITH STRONG APPLIED MAGNETIC FIELDS

Although pinched electron beams have some specialized applications, most of the time we need beams with better directionality. Applications for extended high-current beams include high-power microwave devices and large-area gas lasers. We can circumvent electron beam pinching by applying a magnetic field in the direction of electron flow. The net magnetic field forces electrons to follow spiral orbits directly across the acceleration gap.

A rough criterion for pinch suppression is that the axial field, B_0, should be comparable to or greater than the beam-generated field on the envelope, B_s. In a cylindrical gun, the condition is

$$B_0 \geqslant B_s = \mu_0 I / 2\pi R.$$

The quantity I_0 is the total beam current and R is the envelope radius. For the sheet beam geometry typical of beam-controlled gas lasers, the maximum beam-generated field is

$$B_s = \mu_0 J / 2, \tag{8.60}$$

where J is the beam current per meter. As an example, imagine a laser with 500 kA of current spread over a 1-m length. The axial field applied over the large gun area must exceed 0.3 tesla.

An axial magnetic field of infinite magnitude is necessary to achieve laminar electron flow in a high-current diode. For a finite-field amplitude, the inclination of net field lines results in angular deflections of electron orbits. Figure 8.8 illustrates the cause of angular divergence. The net field lines at the edge of the cathode incline at an angle $\tan^{-1}(B_s/B_0)$ relative to the axis, while the electric field line is axial. Accelerated electrons follow spiral orbits around a magnetic field line. They emerge at angles that depend on the number of rotations they make around the field line. We could apply the numerical methods of Section 2.3 for accurate predictions of single-particle electron orbits in an inclined magnetic field. Here, we shall limit the development to a rough estimate of orbit characteristics.

We assume that the magnetic field is strong—electrons follow a field line and complete more than one rotation during acceleration across the gap. The

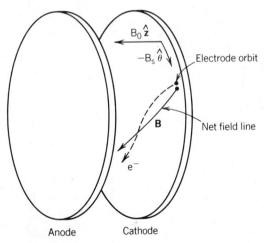

Figure 8.8 Motion of electrons near the edge of a high-current electron diode with a beam-generated toroidal field, B_s, and an applied solenoidal field, B_0.

electrons leave the cathode parallel to z and enter a region of magnetic field inclined at an angle θ_f. For a cylindrical diode, the maximum value of θ_f occurs at the edge:

$$\theta_f \leqslant \tan^{-1}(\mu_0 I / 2\pi R B_0). \tag{8.61}$$

The electrons follow a spiral orbit around the field line. The inclination angles of orbits about the axis vary between 0 and $2\theta_f$. We expect that electrons emerging at the anode have a spread in angle between the limits $\pm 2\theta_f$. Equation (8.61) shows that the applied field must be strong, $B_0 \gg B_s$, to generate low-divergence beams at high beam current. Technology limits the magnitude of the magnetic field in electron diodes. For example, in an electron-beam-driven laser, the magnet coils are subject to substantial forces and conventional magnets require high power input to produce fields over the large volume.

Many pulsed microwave devices require high-current-density beams that would melt anode foils or meshes. With an applied axial magnetic field, it is possible to generate high-current beams that do not pass through the anode. Devices based on this principle are called *foilless electron diodes*. Figure 8.9 shows a typical geometry. An annular cathode immersed in an axial field emits electrons. The axial component of electric field between the cathode and a ring anode accelerates the electrons while the axial magnetic field confines them so that they miss the anode and continue into a transport region. The diode of Fig. 8.9 is useful only for the generation of annular beams since the axial electric field

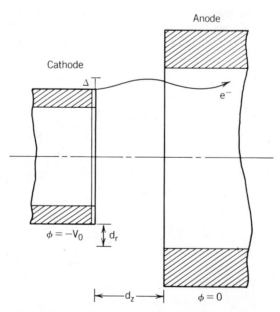

Figure 8.9 Schematic diagram of a foilless diode with definitions of geometric quantities.

magnitude is small on the axis. For uniform emission, the thickness of the annulus cannot exceed the gap width between the cathode and anode.

Transverse electric forces act on electrons passing through a foilless diode. The applied magnetic field must be strong enough to keep electrons from striking the anode ring. We can use results derived in Section 8.1 to estimate the required magnetic field. To simplify the model, we shall neglect the beam-generated magnetic field. Let E_r represent the average radial electric field near the anode where electrons have kinetic energy $eV_0 = (\gamma - 1)m_e c^2$. In this region, electrons follow scalloped transverse orbits. Although the drift motion near the anode is nonrelativistic ($v_d \ll c$), the electrons have an effective mass of γm_e. Adopting Eq. (8.8), we find a maximum radial excursion of

$$\Delta = 2\gamma m_e E_r / e B_0^2. \tag{8.62}$$

To estimate the minimum value of B_0, we take $\Delta = d_r$ and $E_r \sim V_0/d_r$:

$$B_0 \geqslant (2\gamma m_e V_0 / e d_r^2)^{1/2} = m_e c / e d_r [2\gamma(\gamma - 1)]^{1/2}. \tag{8.63}$$

As an example, suppose a diode has $V_0 = 500\,\text{kV}$ and $d_r = 0.01$. The insulating magnetic field must exceed 0.34 tesla.

Transverse electric field components in a foilless diode contribute to the beam angular divergence at the anode. The transverse kinetic energy varies from zero to a maximum value of roughly $eE_r\Delta$. Substituting from Eq. (8.62), we find that the maximum transverse energy is roughly $(E_r/cB_0)^2 \gamma m_e c^2$. In practical diodes, $E_r \ll cB_0$. Therefore, the transverse energy is approximately equal to $\gamma m_e v_r^2$. For highly relativistic electrons, the divergence angle at the anode is about

$$\Delta\theta \cong v_r/c \cong E_0/B_0. \tag{8.64}$$

Equation (8.64) shows that a strong applied magnetic field is necessary for a low-emittance beam.

Several analytic models have been developed to predict the net current from a foilless diode. Solutions are complex, even with an infinite magnetic field. Few results can be expressed in a simple form. In designing a foilless diode, it is generally more efficient to estimate the total current from scaling laws, and then to make accurate calculations with ray tracing programs or simulations. Applying Eq. (5.48), the space-charge-limited current from the foilless diode of Fig. 8.9 is about

$$I_0 \sim \left[\frac{4\varepsilon_0}{9}\right]\left[\frac{2e}{m_e}\right]^{1/2} V_0^{3/2} \left[\frac{2\pi R}{d_z}\right]. \tag{8.65}$$

In Eq. (8.65), we assume that the radial width for active emission from the cathode is comparable to d_z.

Figure 8.10 shows a computer simulation of electron flow across a foilless

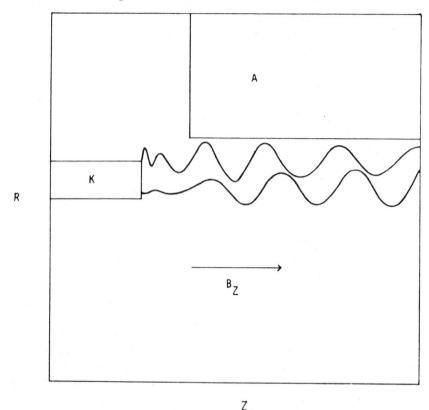

Figure 8.10 Particle-in-cell computer simulation of a foilless diode. Figure shows electrode geometry and superimposed electron orbits. $V_0 = 4\,\mathrm{MV}$, $B_0 = 1.7$ tesla, $I = 66\,\mathrm{kA}$. (Courtesy J. Poukey, Sandia National Laboratories.)

diode for injection into a high-current pulsed accelerator. Note that the radial scale is highly expanded. The diode has a small cathode radius, $R = 0.01\,m$, and an extended gap length, $d_z = 0.02$ m. The diode voltage is $V_0 = 2\,\mathrm{MV}$ and the applied field is $B_0 = 2$ tesla. The code predicts a net current of 38 kA. The diode parameters were chosen to give a low-divergence beam. Figure 8.10 shows the r–z projections of the beam envelope on the inside and outside of the annulus. The envelope oscillations result from transverse electric fields and the beam-generated magnetic field ($B_s \sim 0.76$ tesla). Also, we shall see in Section 10.2 that high-current beams are not in a radial force balance when they leave an immersed cathode. As the beam propagates downstream, nonlinear forces cause phase mixing and damping of envelope oscillations. Force mismatches in the diode ultimately contribute to increased beam emittance.

    Figure 8.11 shows a second foilless high-current electron-gun design accomplished with a ray tracing program. The magnetron gun uses a combination of

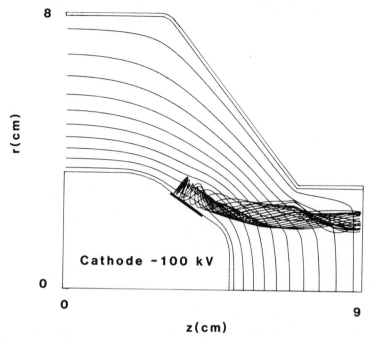

Figure 8.11   Numerical calculation of emission from a magnetron gun using the EGUN code. A converging magnetic field increases from 0.087 tesla to 0.1 tesla over the 0.03 m distance from the cathode emission point to the transport tube—the field is uniform in the transport tube. $V_0 = 100\,\text{kV}$, $I = 340\,\text{A}$.

axial and radial electric fields with a converging magnetic field to create a rotating annular electron beam in the downstream solenoidal field. The injector geometry is often used in high-power gyrotron microwave sources.

8.4.   MAGNETIC INSULATION OF HIGH-POWER TRANSMISSION LINES

In this section, we shall review the subject of magnetically insulated power flow in a vacuum transmission line. The process is essential for the efficient transfer of energy from pulsed power modulators to high-current electron or ion diodes. The transmission lines that connect the power source to the diode must operate at electric field levels well beyond the breakdown limit. Sometimes, the magnetic fields from the flow of current of the load can inhibit a line short circuit from electron leakage. A strong magnetic field can insulate the line—breakdown electrons follow drift orbits toward the load instead of crossing the vacuum gap.

The importance of magnetic insulation is apparent if we consider power flow in high-current electron and ion diodes. Existing pulsed-power generators can supply power at levels exceeding 10^{12} W (1 TW) for pulse lengths in the range

50–100 nsec. To achieve intense focused beams from diodes, it is necessary to deliver electromagnetic energy at high-power density. The energy flux of a pulse with electric field amplitude E_0 in a vacuum transmission line is

$$S\,(\text{W/m}^2) = E_0^2/(\mu_0/\varepsilon_0)^{1/2}. \tag{8.66}$$

The quantity $(\mu_0/\varepsilon_0)^{1/2}$ is the *free space impedance*, equal to 377 Ω. Breakdowns limit the peak energy flux—at a field of 20 MV/m, we find that $S = 10^{12}$ W/m^2. The figure implies that the transmission line area for a 1-TW diode must exceed 1 m^2. The inductance of such a large-area line is too high to achieve fast power flow risetimes at the diode ($\leqslant 10$ nsec).

Magnetically insulated lines transport high-power density partly because they operate with electric fields well above the breakdown level. Figure 8.12 shows power flow with magnetic insulation in a coaxial cylindrical transmission line with negative voltage on the center conductor. Electrons generated along the length of the line move away from the power generator toward the load under the combined action of electric and magnetic fields. The magnetic field results from the sum of current carried by drifting electrons and current flow along the center conductor. If a pinched-beam diode terminates the line, the drifting electrons can contribute to the pinched flow in the diode.

We can use the parapotential model to gain insight into the nature of magnetically insulated electron flow. With minor modifications, the theory of Section 8.2 applies to geometries besides concentric cones. Suppose we have a parallel-plate transmission line where the plates are separated by a gap x_0—the

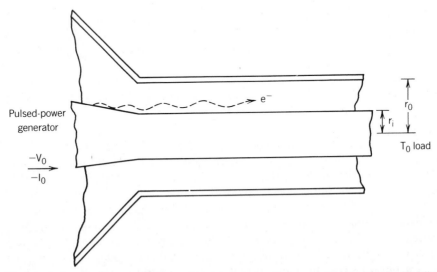

Figure 8.12 Schematic view of a magnetically insulated vacuum transmission line. Dashed line shows the orbit of an electron created in the rising electric field region. The electron drifts downstream in the combined electric and magnetic fields.

electron distribution has uniform properties in y and z. Electrons drift in the z direction along equipotential surfaces; the electrostatic potential varies only with x. It is straightforward to show that Eq. (8.45) still determines the potential with the substitution $\chi = x/x_0$. The saturated parapotential current for a parallel plate transmission line of width y_0 is

$$I_0 = \left[\frac{y_0}{2\pi x_0} \right] \left[\frac{2\pi m_e c}{e\mu_0} \right] \gamma_0 \ln [\gamma_0 + (\gamma_0^2 - 1)^{1/2}]. \tag{8.67}$$

Equation (8.67) holds if electron flow extends to the anode plate. The quantity I_0 is the sum of the current of drifting electrons and the current in the cathode plate. Note that Eq. (8.67) is identical to Eq. (8.52) except for the first term, a geometric factor.

Most pulsed-power generators have coaxial cylindrical output lines. The saturated axial parapotential current between the coaxial cylinders of Fig. 8.12 is

$$I_0 = \left[\frac{1}{\ln(r_o/r_i)} \right] \left[\frac{2\pi m_e c}{e\mu_0} \right] \gamma_0 \ln [\gamma_0 + (\gamma_0^2 - 1)^{1/2}]. \tag{8.68}$$

The quantities r_i and r_o are the inner and outer radii of the cylinders. We can use Eq. (8.68) to estimate the impedance of a magnetically insulated line that carries a step-function voltage pulse. The impedance of a transmission line equals the magnitude of the voltage between the inner and outer conductors divided by the total current flowing through the line, or

$$Z_m = (V_0/I_0) = (m_e c^2/e)(\gamma_0 - 1)/I_0. \tag{8.69}$$

Inserting the saturated parapotential prediction for I_0 gives an expression for the line impedance with free electron flow:

$$Z_m = Z_0 \left[\frac{\gamma_0 - 1}{\gamma_0 \ln [\gamma_0 + (\gamma_0^2 - 1)^{1/2}]} \right]. \tag{8.70}$$

The quantity Z_0 is the impedance for a conventional coaxial line:

$$Z_0 = \left[\frac{\mu_0}{\varepsilon_0} \right]^{1/2} \frac{\ln(r_o/r_i)}{2\pi}. \tag{8.71}$$

Equation (8.70) provides a guideline to match a power transfer section to a diode with known properties. The bracketed quantity in the equation is a correction factor to account for free drifting electrons. Figure 8.13 shows a plot of the correction factor as a function of voltage—note that it is always less than unity. For the same voltage, a magnetically insulated line carries higher energy density than a conventional line. In summary, there are two properties of a

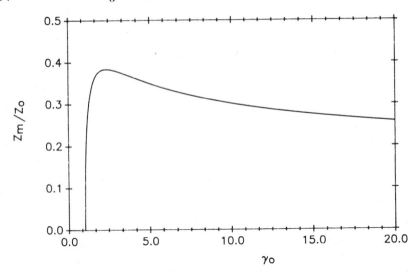

Figure 8.13 Ratio of the impedance of a cylindrical magnetically insulated transmission line (Z_m) to the vacuum impedance (Z_0) for saturated parapotential flow as a function of $\gamma_0 = 1 + eV_0/m_ec^2$.

magnetically insulated transmission line that allow high power density: high electric field magnitude and low impedance.

The details of the inception and termination of magnetically insulated flow at the ends of a line are complex. Again, computer simulations are the best approach for detailed predictions. We can gain some useful qualitative insights. For example, consider the initiation of parapotential flow in a line. Figure 8.14 illustrates an extended line attached to a pulsed power generator. The line has a high-vacuum environment while the generator is insulated with oil or purified water. A vacuum interface separates the two regions. The interface is a significant impediment to power flow in the system. The breakdown field on the vacuum side

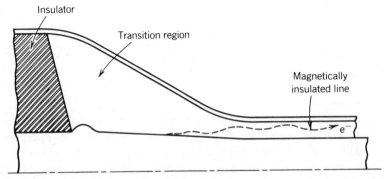

Figure 8.14 Vacuum interface region of a pulsed power generator showing creation of drifting electrons in the region of spatially varying electric field.

[CPA, Section 9.3] is only about $\sim 10\,\text{MV/m}$. The interface provides an impedance mismatch between the dielectric and vacuum transmission lines. The common strategy to limit its effect is to make the transition region as short as possible. Then we can treat the interface region as a series inductor that limits the rise time for downstream power flow. The electric field is low near the interface and increases along the transition to the magnetically insulated line. At some point, electron emission begins (Fig. 8.14). The electrons experience a rising electric field as they drift downstream. The combination of the rising electric field and emission of more electrons pushes the drifting electrons toward the anode. As in the pinched-electron-beam diode, it is unlikely that pure parapotential flow results. Electron orbits in a real line probably resemble those of the simulation of Fig. 8.7. If the line voltage for a given net current rises above the level for magnetic cutoff, electron losses occur at the line entrance. Therefore, we expect that the voltage is regulated in the downstream line so that electron flow is close to the saturated parapotential value.

The connection to a load at the end of the magnetically insulated line usually involves a geometric discontinuity. For example, in the transition to a pinched-electron-beam diode, the gap width decreases at the edge of the diode. For a smaller gap width, Eq. (8.70) shows that the impedance at constant voltage decreases; therefore, a steady-state solution requires higher current. This transition presents little problem. For example, extra electrons can enter at the edge of a pinched beam diode to satisfy conditions for parapotential flow in the compressed geometry. Sometimes the gap width increases at the end of a magnetically insulated line—the matched impedance is higher. Drifting electrons must be removed to maintain continuity of current at constant voltage. In steady-state fields, it is energetically impossible for electrons to return to the cathode. For such a discontinuity, the drifting electrons drive time-varying electric fields. The instability carries electrons to the anode side, resulting in losses. The mechanism of electron bunching in a magnetically insulated line and the coupling of longitudinal energy to a microwave instability is the basis for the operation of the magnetron (Section 15.6).

Magnetically insulated transmission lines cannot transport power over an indefinite length. The electrical transit time of the line must be shorter than the voltage pulse length. If we apply an input voltage pulse with a rise time shorter than the electrical transit time of the line, the downstream section of the line must act as the load for the magnetically insulated power flow. By Eq. (8.70), the vacuum line impedance is always greater than the impedance for magnetically insulated flow, $Z_0 > Z_m$. The implication is that a steady-state solution with no particle loss is impossible with a short voltage rise time. For effective magnetic insulation, the power must rise over many electromagnetic transits of the line. Then, the current that flows through the line to the low-impedance load can rise to levels above V_0/Z_0. If the voltage rises in time Δt, we expect to observe electron losses if the length of the magnetically insulated line is

$$L > c\Delta t/2. \tag{8.72}$$

As an example, for $\Delta t = 10\,\mathrm{nsec}$, the line length must be less than $1.5\,\mathrm{m}$ for complete insulation.

8.5. PLASMA EROSION

The term *plasma erosion* signifies the removal of a plasma from a gap by a pulsed electric field. In response to the field, electrons and ions flow to opposite electrodes. With no replenishment of particles, the field ultimately clears the gap. Although plasma erosion may occur in a conventional ion extractor with a pulsed acceleration field, the main application of the process is to pulsed-power switching and intense ion beam diodes. Ion diodes may operate at very high current densities exceeding $1\,\mathrm{kA/cm^2}$. One approach to supply ions is to fill the diode gap with a high-density plasma before the voltage pulse. The stored ions can support high flux for a short pulse. A further advantage of plasma prefill is that the ion diode exhibits a rising impedance during the voltage pulse—the consequent ramped voltage waveform may be useful for power multiplication by longitudinal ion bunching.

The rising gap impedance associated with plasma erosion is important for applications to pulsed-power switching. Initially, large currents flow with small voltage drop when plasma fills the gap. As plasma leaves, particles must cross a widening vacuum region to reach the electrodes. Then the Child law limits the current density in the vacuum region. At later times, increased voltage is required to drive particle flow—the impedance rises. Devices based on plasma erosion have potential applications as high-power opening switches.

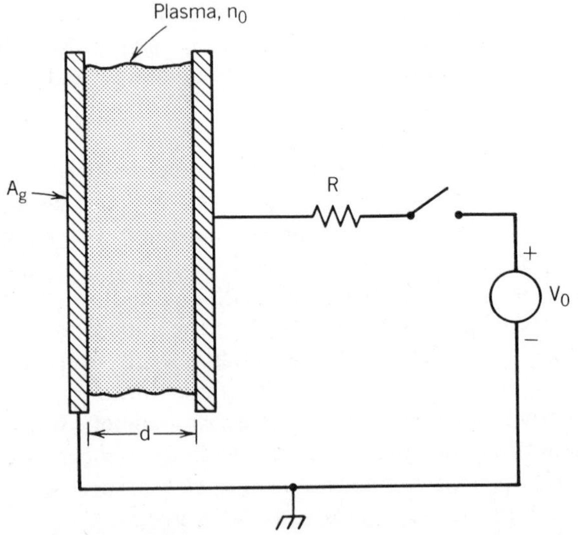

Figure 8.15 Planar plasma-filled gap driven by a simple pulsed power circuit.

The theory of plasma erosion can be complex in high-power systems with three-dimensional geometric variations and strong magnetic fields generated by the current flow. To develop an analytic model, we shall limit the discussion to moderate power density and adopt several simplifying assumptions. Figure 8.15 shows the geometry of a planar gap. We neglect magnetic fields and take the gap width d as small compared with the system width. As a result, we can treat particle orbits as one-dimensional. A uniform plasma with equal electron and ion densities, n_0, initially fills the gap. The plasma electrons are cold, so that the plasma excludes electric fields. An external circuit consisting of an ideal voltage source, a series resistor R, and a switch (Fig. 8.15) controls plasma erosion. The output voltage from the source and switch has the time variation

$$V(t) = V_0 H(t), \tag{8.73}$$

where $H(t) = 0$ for $t < 0$ and $H(t) = 1$ for $t > 0$.

To begin, we will develop a model that describes a plasma-filled ion extractor. The only plasma particles that enter the problem are those in the gap at $t = 0$— we assume that the electrodes do not emit ions or electrons. Figure 8.16 illustrates the sequence of events that follows application of a voltage at $t = 0$. The rising electric field immediately pulls mobile plasma electrons to the anode. Because of electron loss, the electrostatic potential of the plasma rises until it is near the anode potential. The potential is almost uniform throughout the highly conductive plasma—most of the gap voltage drop occurs across a narrow sheath at the negative electrode (Fig. 8.16a). The rising plasma potential impedes further electron loss—the electron loss rate approaches the ion loss rate to maintain average charge balance. The electric fields in the narrow sheath at the cathode accelerate ions from the plasma.

Just after switching ($t = 0^+$), the source voltage appears across the series resistor because current flows easily across the gap. The initial circuit current is

$$I_0 = I(t = 0^+) = V_0/R. \tag{8.74}$$

One electron leaves the gap for each ion accelerated across the sheath. Therefore, ion depletion accounts for half the circuit current while electron depletion carries the rest. Therefore, the initial ion current across the cathode sheath is $i = I_0/2R$. As ions leave the region near the cathode, a vacuum region of increasing width forms (Fig. 8.16b). We denote the width of the vacuum region as λ. The gap voltage $v(t)$ rises to maintain ion flow across the growing vacuum sheath (Fig. 8.16c). The process continues until all plasma ions leave the gap.

We can construct a simple model to describe plasma erosion in the limit that the ion transit time over the sheath width λ is much shorter than the time for a substantial change in λ. This condition is valid for low-mass ions. Then, we can approximate the ion flux across the cathode sheath by the equilibrium value

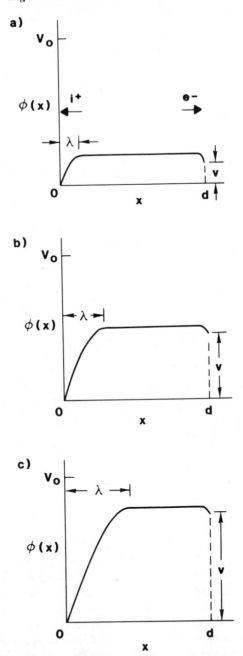

Figure 8.16 Sequence of events during the erosion of plasma from a planar gap driven by the circuit of Fig. 8.15. Positive electrode on the right-hand side. Immediately after voltage is applied (*a*), electrodes are swept from the gap leaving a positive plasma potential. Ions exit at the left-hand side, resulting in an electrostatic sheath at the cathode. The ion extraction sheath widens later in time (*b* and *c*).

predicted by the Child law. Following Section 5.2, the ion current density is

$$j_i \cong \left[\frac{4\varepsilon_0}{9} \right] \left[\frac{2e}{m_i} \right]^{1/2} \frac{v^{3/2}}{\lambda^2}. \tag{8.75}$$

The total circuit current over an erosion gap of area A_g is

$$i = 2A_g j_i. \tag{8.76}$$

The factor of two arises because there is an equal depletion of electrons. The circuit current is related to gap voltage and sheath width by

$$i(t) = 2A_g \left[\frac{4\varepsilon_0}{9} \right] \left[\frac{2e}{m_i} \right]^{1/2} \frac{v(t)^{3/2}}{\lambda(t)^2}. \tag{8.77}$$

For the series resistor circuit, the gap voltage and circuit current satisfy the equation

$$v(t) = V_0 - i(t)R. \tag{8.78}$$

To find a solution for $v(t)$, $i(t)$, and $\lambda(t)$, we must supplement Eqs. (8.77) and (8.78). We introduce a third equation to represent the conservation of charge. If the width of the vacuum sheath increases by an amount $\Delta\lambda$ in a time Δt, then the total ion and electron charge removed from the gap is

$$\Delta Q = 2(en_0)(A_g\Delta\lambda). \tag{8.79}$$

Dividing both sides of Eq. (8.79) by Δt, we find

$$i(t) = (2en_0A_g)d\lambda/dt. \tag{8.80}$$

We can rewrite Eqs. (8.77), (8.78), and (8.80) in a dimensionless form to show that the solutions are governed by a single parameter. We define the dimensionless gap voltage, current, and vacuum sheath width as

$$V = v/V_0, \tag{8.81}$$

$$I = i/I_0, \tag{8.82}$$

$$\Lambda = \lambda/d. \tag{8.83}$$

We take the characteristic time scale equal to the total initial ion charge in the gap divided by the maximum circuit current, I_0. The dimensionless time is

$$\tau = t/(en_0dA_g/I_0). \tag{8.84}$$

The dimensionless equations are

$$I = \frac{V^{3/2}}{(I_0/I_f)\Lambda^2},$$ (8.85)

$$V = 1 - I,$$ (8.86)

and

$$I = 2(d\Lambda/d\tau).$$ (8.87)

The quantity I_f is the current that flows in the gap just before depletion of the plasma. This current equals twice the Child limited ion current across a vacuum gap of width d:

$$I_f = 2A_g \left[\frac{4\varepsilon_0}{9} \right] \left[\frac{2e}{m_i} \right]^{1/2} \frac{V_0^{3/2}}{d^2}.$$ (8.88)

The ratio I_0/I_f determines the nature of plasma erosion. If the gap has a dense plasma fill and has a large impedance change during plasma depletion, the final current is much smaller than the initial current, or

$$(I_0/I_f) \gg 1.$$ (8.89)

We combine Eqs. (8.85), (8.86), and (8.87) into a single equation for the width of the vacuum region at the cathode:

$$\frac{d\Lambda}{d\tau} = \frac{[1 - 2(d\Lambda/d\tau)]^{3/2}}{(I_0/I_f)\Lambda^2}.$$ (8.90)

The initial conditions on current and voltage are $I(0) \cong 1$ and $V(0) \cong 0$. Equations (8.85) and (8.87) imply that

$$\Lambda(0) = 0,$$ (8.91)

$$d\Lambda/d\tau = 1/2.$$ (8.92)

The calculation ends when all particles leave the gap, or

$$\Lambda(\tau_{max}) = 1.$$ (8.93)

Although Eq. (8.90) is nonlinear, we can easily solve it numerically with a two-step integration procedure. We must find the quantity $d\Lambda/d\tau$ by iteration at the full and half time steps. Given $\Lambda(\tau)$, Eqs. (8.86) and (8.87) give the normalized gap voltage and current.

Figure 8.17a gives a plot of $V(\tau)$ for a choice of $(I_0/I_f) = 500$. Initially, the

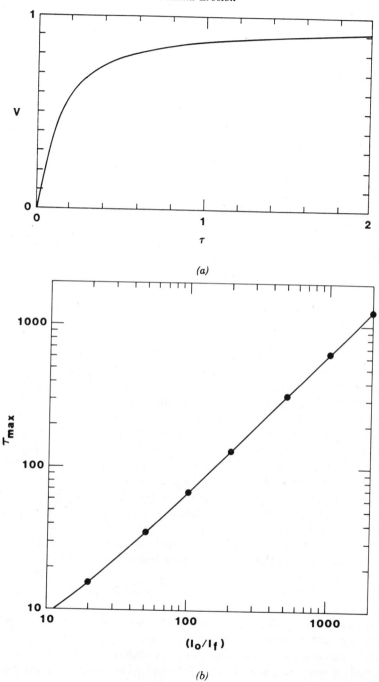

(a)

(b)

Figure 8.17 Solution of the plasma erosion equations for a planar gap. (a) Normalized voltage versus time for $I_0/I_f = 500$. (b) Clearing time for complete removal of plasma as a function of I_0/I_f. (c) Opening time at which voltage reaches 80% of its peak value as a function of I_0/I_f.

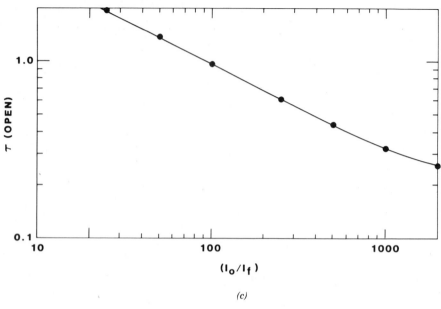

(c)

Figure 8.17   (*Continued*)

voltage rises rapidly as the sheath edge moves away from the cathode. The high voltage is necessary to accelerate ions over the vacuum region. Subsequently, there is a long period of slow plasma depletion at low current. In the example of Fig. 8.17a, the plasma does not clear until $\tau = \tau_{max} = 330$. Figure 8.17b shows the normalized plasma clearing time, τ_{max}, as a function of (I_0/I_f). The variation is almost linear—the removal of particles from the gap takes longer at high plasma density and large gap spacing. Figure 8.17c shows the opening time, the interval for the voltage to rise to $0.8V_0$. The opening time is much shorter than the clearing time. The opening time does not depend strongly on the gap geometry or the plasma properties—it drops with increased (I_0/I_f). The electric field at the cathode is a quantity of practical interest. If the field is too high, the cathode could emit electrons and the gap may not make a transition to high impedance. Equation (5.52) implies that the cathode electric field is

$$E = 4v(t)/3\lambda(t) = (4V_0/3d)[1 - 2(d\Lambda/d\tau)]/\Lambda. \tag{8.94}$$

A specific example illustrates some implications of the results. Suppose a plasma gap and driving circuit have the following properties: $V_0 = 100\,\text{kV}$, $I_0 = 2\,\text{kA}$, and $d = 0.05\,\text{m}$. The gap is circular with diameter $0.15\,\text{m}$—it carries an initial current density of $11 \times 10^4\,\text{A/m}^2$. The parameters imply that $(I_0/I_f) = 83$. Figure 8.18 shows variations of $v(t)$, $i(t)$, and the cathode electric field for the example. From Fig. 8.17c we find an opening time of only 20 nsec, while the clearing time is over 1 μsec. The example shows that plasma erosion can

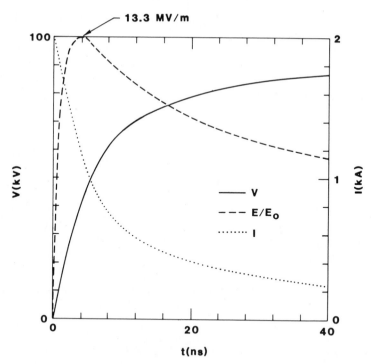

Figure 8.18 Solution of the plasma erosion equations for a planar gap. Dimensional quantities for $I_0/I_f = 83$. Voltage, normalized electric field, and current as functions of time.

give fast switching times for isolated pulses, but it is not useful for high-frequency applications. Analysis of the time-dependent voltage and current shows that the energy deposited in the negative electrode over the clearing time is small, less than $50 \, mJ/cm^2$. The electric field on the negative electrode reaches a peak value of $13 \, MV/m$ about 4 nsec after application of the voltage.

We can apply a similar approach to study the properties of another device, the plasma-erosion-opening switch. These switches are sometimes connected in parallel with high-current ion diodes to sharpen the voltage rise time at the end of a magnetically insulated transmission line. The main difference from the previous calculation is that we take the cathode as an unlimited source of electrons. We shall again neglect the effects of magnetic fields. This assumption is questionable since plasma-erosion-opening switches operate at very high current density. Although the magnetic field pressure on a plasma probably plays a major in the behavior of the switch, it is nonetheless useful to carry out a one-dimensional analysis as a baseline. Initially, electrons emitted from the cathode flow through a dense plasma so that the gap impedance is almost zero. Again, the mobility of the electrons maintains the plasma near the potential of the positive electrode; therefore, voltage across the switch appears in a sheath adjacent to the cathode. Acceleration across the sheath results in depletion of ions in the gap.

Equation (8.86) still holds if the gap connects to a stepped voltage source of magnitude V_0 through a series resistor R. Furthermore, we retain Eq. (8.87), which states that ion charge is conserved. The main difference is that the current that results from plasma depletion is much smaller than the current carried by electrons emitted from the cathode. For a carbon plasma, the space-charge-limited ion current across the vacuum sheath is only 0.7% of the emitted electron flow. Neglecting the plasma contribution, the total circuit current carried by electrons in the nonrelativistic limit is

$$i(t) = 1.86 A_g \left[\frac{4\varepsilon_0}{9} \right] \left[\frac{2e}{m_e} \right]^{1/2} \frac{v(t)^{3/2}}{\lambda(t)^2}. \tag{8.95}$$

The quantities in Eq. (8.95) have the same definitions as in Eq. (8.77). The two differences between the equations are that (1) there is a multiplying factor of 1.86 to account for bipolar flow in the vacuum region and (2) the ion mass has been replaced by the mass of the electron. We can apply the same set of dimensionless equations [Eqs. (8.85), (8.86), and (8.87)] to the plasma-erosion-opening switch if we define I_f as the space-charge-limited current of emitted electrons across the full vacuum gap under bipolar flow conditions:

$$I_f = 1.86 A_g \left[\frac{4\varepsilon_0}{9} \right] \left[\frac{2e}{m_e} \right]^{1/2} \frac{V_0^{3/2}}{d^2}. \tag{8.96}$$

As an example, suppose we have $(I_0/I_f) = 10$. For gap parameters and circuit parameters of $d = 0.05\,\text{m}$, $A_g = 10^{-2}\,\text{m}^2$, $V_0 = 250\,\text{kV}$, we find that $I_f = 1.8 \times 10^5\,\text{A}$ and $I_0 = 1.8\,\text{MA}$. For a plasma density of $n_0 = 10^{20}\,\text{m}^{-3}$ the scale time is about $(e n_0 A_g d/I_0) = 4$ nsec. The predicted opening time is about three times this scale value, or 12 nsec. This time is in the range observed in experiments.

8.6. REFLEX TRIODE

The fundamental problem for the generation of high-current ion beams is energy loss to electron flow. In a one-dimensional diode, the Child law [Eq. (5.48)] implies that the electric field to generate an ion beam is much higher than the field to generate an equal current density of electrons. We need strong electric fields to produce intense ion beams—for a current density of $100 \times 10^4\,\text{A/m}^2$ in a 1-MV proton extractor, we must apply an electric field of $140\,\text{MV/m}$. At such a high field, all exposed negative surfaces in the diode act as electron sources because of surface plasma formation (Section 7.4).

With unlimited sources of both ions and electrons, the flow in a one-dimensional gap at moderate voltage is bipolar and follows the expressions of Section 6.4. The ion flow is smaller than the counterstreaming electron flow by a factor

$$j_i/j_e \cong \alpha = \sqrt{m_e/m_i}. \tag{8.97}$$

For protons the fluxes are $j_i/j_e \cong 0.0233$. The energy efficiency of a simple bipolar diode is only

$$j_i/(j_i + j_e) = \alpha/(1 + \alpha), \tag{8.98}$$

High extraction voltages where electron flow is relativistic improve the efficiency. Nonetheless, at all voltages electron leakage current is too high for practical devices.

The *reflex triode* solves some of the problems of space-charge flow at high electric field. Experiments with this device have demonstrated the generation of neutralized ion beams at high-current density with energy efficiency significantly better than the prediction of Eq. (8.98). In this section, we will study the basic reflex triode. We assume a one-dimensional system with monoenergetic circulating electrons. In the next section, we shall construct a theory with more realistic electron energy spectra. We shall see that some electron distributions lead to ion flow that is greatly enhanced beyond the Child limit.

Figure 8.19 shows the geometry of the reflex triode. Two cathodes are placed symmetrically a distance d from a central anode. The cathode surfaces are grids to allow extraction of ions. The anode is also a grid with a high geometric transparency factor T. After application of a strong pulsed voltage, the anode can supply space-charge-limited ion flux in both directions. In many experiments, the anode is composed of insulating wires or a plastic foil. The applied electric field in

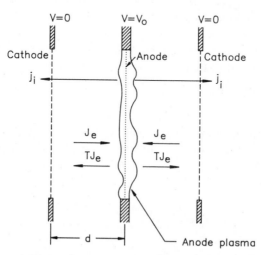

Figure 8.19 Geometry of the symmetric reflex triode. The figure shows flows of electrons and ions for an ideal one-dimensional system.

the triode induces breakdowns along the insulators, generating a dense plasma. The electric fields also create surface plasmas on the cathodes. We shall use the following assumptions to analyze equilibrium particle flow in the device:

1. The anode and cathodes supply unlimited fluxes of ions and electrons. As a result, the electric field equals zero on all electrode surfaces.
2. Particles move only in the z direction. In most experiments, an applied axial magnetic field limits transverse motion of the electrons.
3. An electron that strikes the anode grid is absorbed. Section 8.7 discusses the consequences of electron scattering and partial energy loss in the anode.
4. Particle motion is nonrelativistic. We shall discuss relativistic corrections after developing the basic model.
5. The system is in equilibrium—the electron and ion fluxes are steady in the triode gaps.

Expressions for electrostatic potential and particle densities in the triode of Fig. 8.19 are symmetric on both sides of the anode. The figure illustrates the flow of particles. Suppose a space-charge-limited ion current j_i leaves both sides of the anode. We define the quantity J_e as the current density of electrons directed toward the anode from either cathode. The fraction of incident electron current density that passes through the anode is TJ_e. By conservation of energy, the transmitted electrons travel to the opposing cathode where they are absorbed. It is easy to find the self-consistent space-charge solution in the triode if we recognize the following facts:

1. The solution of the one-dimensional Poisson equation depends on the magnitude of the total electron density, not the direction of electron motion. The incoming and outgoing electrons contribute to a net negative space-charge density.
2. If $\phi(z)$ is the electrostatic potential, the magnitude of the electron velocity for both incoming and outgoing electrons is proportional to $1/\phi(z)^{1/2}$.
3. The condition of zero electric field at the cathode constrains the net electron density.

Disregarding the direction of electron motion, we recognize that the space-charge solution in a gap of the reflex triode is identical to that for bipolar flow. Applying the results of Section 6.4, we conclude that the ion current density equals

$$j_i = \frac{(1.86)4\varepsilon_0}{9} \left[\frac{2e}{m_i}\right]^{1/2} \frac{V_0^{3/2}}{d^2}. \tag{8.99}$$

The condition $E_z = 0$ on all electrodes implies that the total number of electrons in a gap equals the total number of ions. An equivalent statement is that the total electron flux (ignoring direction) equals the total ion flux multiplied by $(m_i/m_e)^{1/2}$, or

$$J_e(1 + T) = j_i/\alpha. \tag{8.100}$$

The reflex triode has improved efficiency for ion generation because only a small fraction of the electrons that leave the cathode strike the anode. The current density of electrons absorbed at the anode is

$$j_e = (1 - T)J_e. \tag{8.101}$$

An accelerated ion that passes into the field-free region outside the triode can easily capture an electron from the cathode. The net result is that neutralized ion beams leave both sides of the device. Section 11.1 discusses the mechanisms of electron capture for ion beam neutralization.

A good figure of merit for a triode is the ratio of the net ion current exiting the device to the electron current lost on the anode. We can combine Eqs. (8.100) and Eq. (8.101) to give the ratio

$$\frac{j_i}{j_e} = \left[\frac{m_e}{m_i}\right]^{1/2} \frac{1 + T}{1 - T}. \tag{8.102}$$

In the limit of a solid anode, $T \to 0$, Eq. (8.102) reduces to the standard ratio for bipolar flow [Eq. (6.39)]. In contrast, at high transparency, $T \to 1$, the electron loss current is much lower. For example, the ratio for proton generation with $T = 0.95$ is $j_i/j_e = 0.91$. The efficiency for ion production is almost 50% compared with 2.3% for a solid anode.

When a reflex triode has an applied voltage in the megavolt range, the net electron and ion fluxes are no longer related by Eq. (8.100). To describe the triode, we must derive relativistic expressions for bipolar flow following the methods of Sections 6.4 and 6.5. The calculation is easy in the limit of ultrarelativistic electrons $(eV_0 \gg m_e c^2)$. We assume that the electron density is almost constant over the width of the gap, independent of ϕ. The resulting ion current density is

$$j_i = \frac{\pi^2 \varepsilon_0}{4} \left[\frac{2e}{m_i}\right] \frac{V_0^{3/2}}{d^2}. \tag{8.103}$$

Relativistic electrons enhance the ion current density by a factor of 5.55 compared with 1.86 for nonrelativistic voltages. The net electron flux is related to the ion flux by

$$J_e(1 + T) = \left[\frac{2m_i c^2}{eV_0}\right]^{1/2} j_i. \tag{8.104}$$

The figure of merit for an ultrarelativistic reflex triode is

$$\frac{j_i}{j_e} = \left[\frac{eV_0}{2m_i c^2}\right]^{1/2} \frac{1+T}{1-T}. \tag{8.105}$$

As an example, if $V_0 = 2\,\text{MeV}$ and $T = 0.95$, Eq. (8.105) predicts a current density ratio of $j_i/j_e \geqslant 1.27$.

Figure 8.20 shows an interesting variant of the reflex triode that provides a good demonstration of collective space-charge effects. The device generates bidirectional ion beams with only one real cathode. This cathode, at ground potential, is separated by a gap of width d_1 from an anode pulsed to voltage $+V_0$. Again, the cathode and anode can supply unlimited fluxes of electrons and ions. The downstream propagation region is surrounded by walls at ground potential. Figure 8.20 shows a metal transport tube and a distant target at ground potential. Although the system resembles an electron gun, experiments have shown that it generates neutralized ion beams of energy eV_0 in both the upstream and downstream directions.

A space-charge-limited electron current density, J_e, leaves the real cathode. We take monoenergetic electrons with only axial velocity. A current density TJ_e passes through the anode into the drift region. Without ions, we know that the negative space charge reduces the potential in the drift region. At a distance d_2 from the anode, the potential reaches $\phi = 0$. The virtual cathode reflects the electrons. If $T = 1$, the space charge downstream from the anode is a mirror

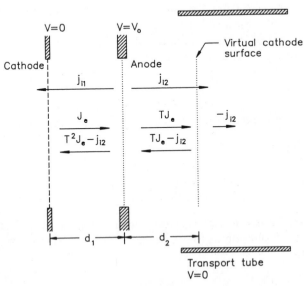

Figure 8.20 One-dimensional reflex triode with extraction through a virtual cathode on the right-hand side.

image of the solution on the upstream side; therefore, $d_2 = d_1$. If $T < 1$, the electron flux on the virtual cathode side is lower, so that $d_2 > d_1$.

Accounting for particles is more difficult when we add ions—Fig. 8.20 clarifies particle flows in the asymmetric reflex triode. We shall postulate that a virtual cathode surface exists and then show that we can derive consistent current densities in the two regions of the triode. Again, the net electron density is proportional to $\phi^{-1/2}$—the condition E_z equals zero on the physical electrodes and at the virtual cathode determines the magnitude of the electron density. The ion current densities on both sides equal the bipolar flow expression [Eq. (8.99)] with the appropriate gap spacing, d_1 or d_2. The ratio of ion current density on the virtual cathode side to that on the real cathode side is

$$j_{i2}/j_{i1} = (d_1/d_2)^2. \tag{8.106}$$

A current density of electrons denoted by J_e leaves the real cathode—a fraction TJ_e continues through the anode. Since the virtual cathode has no connection to an external power source, the total current arriving at the surface must equal zero for a steady-state solution. A small fraction of the electron flux arriving at the virtual cathode continues in the forward direction to neutralize the extracted ions. The current density of lost electrons equals j_{i2}. The bulk of electrons arriving at the virtual cathode are reflected. The return electron current density on the right-hand side equals $TJ_e - j_{i2}$. A current density $T(TJ_e - j_{i2})$ passes back through the anode mesh and returns to the real cathode.

We can find the geometric parameters and particle fluxes by the same method that we used for the symmetric reflex triode. On the real cathode side, the condition of equal numbers of electrons and ions gives the following relationship:

$$J_e + J_e T^2 - j_{i2} T = (m_i/m_e)^{1/2} j_{i1}. \tag{8.107}$$

On the virtual cathode side, Eq. (8.107) becomes

$$2J_e T - j_{i2} = (m_i/m_e)^{1/2} j_{i2}. \tag{8.108}$$

We can combine Eqs. (8.97), (8.107), and (8.108) to give the ratio of ion current densities on each side:

$$\frac{j_{i2}}{j_{i1}} = \frac{2T}{(1+\alpha)(1+T^2) - 2\alpha T^2}. \tag{8.109}$$

Equation (8.109) leads to an expression for the ratio of the gap widths:

$$\frac{d_2}{d_1} = \left[\frac{(1+\alpha)(1+T^2) - 2\alpha T^2}{2T}\right]^{1/2}. \tag{8.110}$$

Finally, we can calculate a figure of merit for the device. We define the quality factor as the ratio of the ion current through the virtual cathode, j_{i2}, to the total electron current from the real cathode to the anode, j_e. The quantity j_e equals

$$j_e = J_e(1 - T) + (J_e T - j_{i2})(1 - T) = J_e(1 - T^2) - j_{i2}(1 - T). \quad (8.111)$$

Combining Eqs. (8.108) and (8.111), the figure of merit is

$$\frac{j_{i2}}{j_e} = \frac{2T}{(1 - T)^2 + (1 - T^2)/\alpha}. \quad (8.112)$$

As an example, suppose we have a proton extractor where $\alpha = 1/(1843)^{1/2} = 0.0233$. The anode transparency factor is $T = 0.95$. Eqation (8.112) predicts that the ion current passing through the virtual cathode equals 45% of the total electron current lost to the anode. The ion fluxes on each side of the anode have equal magnitudes to within 0.3%.

8.7. LOW-IMPEDANCE REFLEX TRIODE

The predicted behavior of the reflex triode changes dramatically if we relax the condition of purely one-dimensional motion. If we permit transverse velocity components, the total electron energy projected in the longitudinal direction is no longer a conserved quantity. If transverse fields or collisions deflect the reflexing electrons, the triode generates ion current density far beyond the Child limit.

Ion motion in a reflex triode is always simple. Ions follow almost straight-line orbits and leave the device immediately after crossing the extraction gap. The key to enhanced current density in the reflex triode is the electron motion, which can be complex. Several processes can modify the orbits of the reflexing electrons:

1. It we use a thin foil anode instead of wires, electrons suffer small angle scattering (Section 10.4) each time they pass through. Although a scattered electron retains almost the same net kinetic energy, the collision reduces the energy projected in the axial direction.

2. Transverse electric field components at the edge of the triode may increase the transverse energy of electrons.

3. The toroidal magnetic field resulting from ion and electron currents deflects electrons.

Figure 8.21 illustrates the effect of foil scattering on electron orbits in a symmetric reflex triode. Scattered electrons have reduced kinetic energy projected in the z direction. Conservation of energy prohibits them from returning to the cathode—they pass back through the anode foil. With no transverse forces, electrons reenter the anode at the same angle as they left. Electrons may make several additional transits through the anode. Each time, random scattering

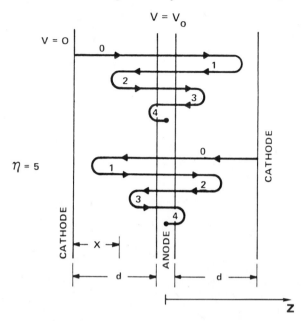

Figure 8.21 Schematic view of the loss of axial electron energy in a reflex triode by scattering in a foil anode. (Courtesy J. Creedon, Physics International Company.)

processes modify their angles. After many passes, the electrons scatter to a grazing incidence angle and are absorbed.

To model the reflex triode with scattering, we must recognize some differences from the model of Section 8.6:

1. Individual electrons follow a complex decay history.
2. The anode collects all electrons that leave the cathode.
3. At any time, the extraction gaps contain electrons with a broad distribution of energies.

Again, we seek a solution with steady-state electron and ion flow. For simplicity, we assume that transverse components of electron velocity result only from scattering in the foil. We take electrodes of infinite width so that there are only z-directed electric fields. We neglect the effects of beam-generated magnetic fields. Finally, we limit the treatment to nonrelativistic electron dynamics.

Finding a self-consistent equilibrium is not difficult if we recognize that the computation can be divided into two independent parts: (1) calculation of the spectrum of electron energy and (2) use of the spectrum to find the electron density as a function of electrostatic potential. An expression for the electron density leads to a self-consistent solution of the Poisson equation. To begin, consider how we can find the electron spectrum. Inside the foil, electrons move

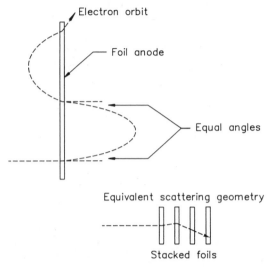

Figure 8.22 Calculation of electron scattering in a one-dimensional reflex triode. Top: Schematic view of electron orbits in a reflex triode. Bottom: Equivalent system.

independently. The high density of conduction electrons in the metal cancels long-range electric fields generated by the reflexing electrons. Furthermore, electrons return to the anode at the same angle as they exited, independent of the nature of the space-charge solution in the vacuum gap. In other words, the transverse and axial kinetic energy components do not change in the gap. Figure 8.22 shows a physically equivalent geometry to calculate scattering— anode foils are stacked in sequence with no space between. We can use probability theory or Monte Carlo calculations to find the electron distribution at each foil boundary. The output from the calculation is an electron spectrum for each pass through the anode foil. We can sum over the set of spectra to find the net electron axial energy distribution at the anode foil surface.

We can use the net spectrum to find the relative density function. Then, we can normalize the density by applying the condition that $E_z = 0$ on all electrodes. Section 15.1 discusses the numerical calculation of density for an arbitrary longitudinal energy spectrum. Here, we shall follow the simplified model of Creedon[1] that illustrates the essential physical processes. Figure 8.21 shows an idealized electron history. The quantity η equals the number of times an electron passes through the anode foil, including the final pass where it is absorbed. Figure 8.21 shows an orbit where $\eta = 5$. The quantity W_z is the total energy of an electron in the axial direction:

$$W_z = \frac{m_0 v_z^2}{2} + e[V_0 - \phi(z)]. \tag{8.113}$$

[1]J. Creedon, I. D. Smith, and D. S. Prono, *Phys. Rev. Lett.*, **35**, 91 (1975).

Without scattering, the total axial energy is a constant, $W_z = eV_0$. Foil scattering reduces W_z—the quantity drops to zero when the anode absorbs the electron.

To complete the definition of quantities, we take the flux function f as the total flux of scattered electrons leaving the anode surface in either direction. Referring to Fig. 8.21, f equals the sum of the partial fluxes from the orbits marked 1, 2, 3, and 4. We take f_e as the flux of incoming electrons (orbit 0) from either side. Since electrons are not absorbed until the final pass, conservation of charge implies that the incoming and outgoing electron flows at the foil are equal. As a result the total flux of scattered electrons in the gap is related to the incoming flux by

$$f = (\eta - 1)f_e. \tag{8.114}$$

We express the total flux in terms of the axial energy distribution by defining the differential flux spectrum, df/dW_z. The differential flux equals the number of electrons in the energy interval dW_z leaving one anode surface per second per square meter. In the limit that $\eta \gg 1$, we can approximate df/dW_z as a continuous function with the property

$$f = \int_0^{eV_0} dW_z \frac{df}{dW_z}. \tag{8.115}$$

We can find the electron density in the gap as a function of the differential flux spectrum if the electrostatic potential $\phi(z)$ decreases monotonically from the anode to the cathode. The condition of equilibrium flow implies that the incremental flux associated with each energy group, df, is independent of position. The differential density associated with a particular energy group of electrons that travel toward the anode simply is

$$dn_e(z) = df/v_e(z), \tag{8.116}$$

where

$$v_e(z) = \left[\frac{2[W_z - eV_0 + e\phi(z)]}{m_0} \right]^{1/2}. \tag{8.117}$$

The total electron density at a position z is the sum over all energy groups. We must remember two facts to perform the sum:

1. Contributions to the density at position z are included only for those electrons that have high enough kinetic energy to reach the position, $W_z \geqslant e[V_0 - \phi(z)]$.

2. At any point, the scattered particles have components traveling in both directions, while the unscattered electrons only travel inward.

Given the flux spectrum, the total electron density is

$$n_e(\phi) = f_e\left[\frac{2e\phi}{m_0}\right]^{-1/2} + 2\int_{e(V_0-\phi)}^{eV_0} dW_z \frac{df}{dW_z}\left[\frac{2[W_z - eV_0 + e\phi(z)]}{m_0}\right]^{-1/2}.$$

(8.118)

The first term in Eq. (8.118) represents the contribution from unscattered electrons—it varies as $1/v_e(z)$. The second term represents scattered electrons—note the factor of 2 to account for both velocity components. The density is a function of the electrostatic potential. We can substitute the expression into the Poisson equation to find a self-consistent space-charge solution following the method of Section 5.2.

The nature of electron scattering influences the calculation through the spectral function, df/dW_z. As an example, suppose that electrons lose equal decrements of axial energy on each pass through the foil. The continuous approximation to the spectral function is a constant given by

$$\frac{df}{dW_z} = \frac{f_e}{eV_0/(\eta - 1)}.$$

(8.119)

Substituting Eq. (8.119) into Eq. (8.118) and performing the integral gives

$$n_e(\phi) = \frac{f_e}{\sqrt{2e/m_0}}\left[\frac{1}{\sqrt{\phi}} + \frac{4(\eta - 1)\sqrt{\phi}}{V_0}\right].$$

(8.120)

Ions in the reflex triode are created at the anode and immediately leave through the cathode. Following Section 5.2, the ion density is given by

$$n_i(\phi) = f_i\left[\frac{2e(V_0 - \phi)}{m_i}\right]^{-1/2}.$$

(8.121)

where m_i is the ion rest mass and f_i is the ion flux through the cathode.

The ion current density leaving one side of the reflex triode is $j_i = en_i(2eV_0/m_i)^{1/2}$. The current density of unscattered electrons leaving one cathode, j_e, equals ef_e. Since the anode ultimately captures all electrons, j_e is also the electron-loss current density. Combining Eqs. (8.120) and (8.121) gives the following form for the Poisson equation on one side of the reflex triode:

$$\frac{d^2\phi}{dz^2} = \left[\frac{j_e d^2}{\varepsilon_0\sqrt{2e/m_0}V_0^{3/2}}\right]\left[\phi^{-1/2} + \left(\frac{f_i}{f_e}\right)\left(\frac{m_i}{m_e}\right)^{1/2}(V_0 - \phi)^{-1/2} + 4(\eta - 1)\phi^{1/2}\right].$$

(8.122)

Multiplying both sides of Eq. (8.122) by $2(d\phi/dz)$, we can find the first integral.

Combining the result with the condition that $d\phi/dz = 0$ on all electrodes implies the following relationship:

$$\frac{j_i}{j_e} = \left[\frac{m_e}{m_i}\right]^{1/2}\left[1 + \frac{4(\eta - 1)}{3}\right].$$

(8.123)

The ratio of ion current density to electron-loss current density in Eq. (8.123) is a figure of merit for the triode. We must perform the second integral numerically. Following the method of Section 6.4, we can write the solution in terms of a definite integral. The normalized electron loss current density is

$$j_e/j_e(CL) = 9G^2/16,$$

(8.124)

where the normalizing factor, $j_e(CL)$, is the Child law value of electron current density in a gap of width d with applied voltage V_0 [Eq. (5.48)]. The factor G equals

$$G = \int_0^1 \frac{d\Phi}{(\phi^{1/2} + [(4\eta - 1)/3][(1 - \phi)^{1/2} - 1] + 4(\eta - 1)\phi^{3/2})^{1/2}}.$$

(8.125)

Equations (8.123) and (8.124) have some interesting physical implications. For a given spectral function, the free parameter in the equations is the number of passes through the foil, η. The anode foil thickness, the foil material, and the applied voltage combine to determine η. A thin foil or a high applied voltage gives a value of η much larger than unity. Figure 8.23 shows a plot of $j_e/j_e(CL)$, $(j_i/j_e)(m_i/m_e)^{1/2}$, and (j_i/j_e) for protons. Note that when $\eta = 1$, the value of $j_e/j_e(CL)$ equals 1.86 and j_i/j_e equals 0.0233. These results are identical to those derived for simple bipolar flow (Section 6.4) since the condition $\eta = 1$ corresponds to a solid anode that absorbs electrons on their first transit.

As the number of electron transits increases, the electron-loss current drops and the fraction of triode current carried by the ions increases proportional to η. The most interesting result is that both the electron and ion current densities approach values far beyond the Child limit when η approaches 11. The model predicts that particle fluxes are unlimited when $\eta \geq 11$. In an actual system, the result implies that properties of the driving circuit rather than space-charge effects limit the current. We can find the cause of the current density divergence by inspecting the expression for the electron density, Eq. (8.120). The first term represents unscattered electrons. Since the density peaks at the cathode, the negative charge has little effect on ion flow. In contrast, the density of scattered electrons, represented by the second term, is maximum at the anode. The parameter η determines the proportion of scattered electrons compared with entering electrons. At a critical value, η_{crit}, the electron density at the anode is high enough to neutralize the space charge of extracted ions. The resulting

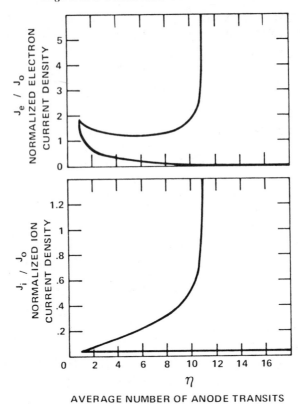

Figure 8.23 Predictions for space-charge-limited flow in a one-dimensional reflex triode with a uniform electron axial energy spectrum. (*a*) Electron current density normalized to the Child law prediction for single-species flow in a simple gap. Lower line shows the prediction for a monoenergetic electron distribution. (*b*) Ion current density normalized to Child law prediction. Lower line shows the prediction for a monoenergetic electron distribution. (Courtesy J. Creedon, Physics International Company.)

enhanced ion flow allows more electrons to enter so that the current density grows without limit. The value of η_{crit} depends on df/dW_z—it is typically in the range 5–15 for electron spectra resulting from scattering.

Suppose we apply a high pulsed voltage to a reflex triode from a generator with a nonzero output impedance. Initially, the value of η is higher than η_{crit} because of the high voltage. The triode current rises rapidly, loading the generator. Ultimately, the voltage drops to a level slightly below η_{crit} and the system approaches an equilibrium. Because η_{crit} depends on voltage, a reflex triode with anode foil acts like a high-power Zener diode. It operates at constant voltage, almost independent of the current.

Experiments have shown ion current densities exceeding 1 kA/cm² from reflex triodes with anode foils, more than two orders of magnitude higher than the

Child law prediction. The price of the enhanced ion flux is lower energy efficiency. The maximum ratio of ion to electron loss current density is

$$\frac{j_i}{j_e} = \left[\frac{m_e}{m_i}\right]^{1/2}\left[1 + \frac{4(\eta_{crit} - 1)}{3}\right]. \qquad (8.126)$$

For the energy spectrum of Eq. (8.119), Eq. (8.126) implies that $j_i/j_e = 0.365$. The corresponding energy efficiency of device is only 27%.

8.8. MAGNETICALLY INSULATED ION DIODE

The generation of high-intensity ion beams depends on the suppression of electron flow in regions of strong electric fields. The magnetically insulated diode uses a magnetic field perpendicular to the accelerating electric field to stop electrons while allowing ions to pass with a small deflection. The magnetically insulated diode has two main advantages compared with the reflex triode. First, electron losses in the device can be small, leading to high energy conversion efficiency. Second the diode does not use thin foils for beam generation or extraction.

Figure 8.24 shows an idealized planar magnetically insulated diode. Ions flow between infinite parallel plates with separation d and voltage difference V_0. The applied electric field is in the axial direction, $E_z = -V_0/d$; the magnetic field is normal to the ion flow, $B_y = B_0$. Ions enter the acceleration gap from an anode plasma. We assume that the negative electrode can supply an unlimited number of electrons. Without a magnetic field, electrons stream across the gap

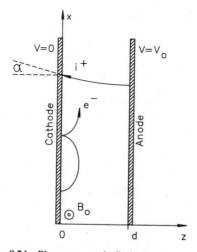

Figure 8.24 Planar magnetically insulated ion diode.

and the ratio of ion to electron current equals $(m_e/m_i)^{1/2}$. Applied magnetic fields stronger than the critical value, B_{crit}, insulate the gap by confining electrons to the cathode region. From the discussion of Section 8.1, the electron distribution in an insulated diode extends over a layer of thickness.

$$\Delta \cong d \left[\left(\frac{B_0}{B^*} \right)^2 - \left(\frac{eV_0}{2m_e c^2} \right)^2 \right]^{-1}. \tag{8.127}$$

Equation (8.11) defines the quantity B^*.

In principle, we can suppress electron flow completely in a magnetically insulated diode. In practice, there are always electron losses. The fraction of current carried by electrons is small in diodes operating at moderate currents ($< 10\,kA$), usually less than 10%. Even though the loss current is small, the space charge of electrons trapped in the gap can significantly enhance the magnitude of the ion flux. We shall see in Section 8.9 that the ion current density in magnetically insulated diodes may exceed the Child limit by more than an order of magnitude.

To begin, we shall discuss magnetic bending of ion orbits in an insulated gap. Although ion deflections are much less than those of electrons, small angular errors may increase the emittance of the extracted beam. We shall use nonrelativistic equations to find the deflection angle, $\Delta \alpha$. The equation for transverse ion motion is

$$m_i(dv_x/dt) = - ev_z B_y(z). \tag{8.128}$$

For small $\Delta \alpha$, we can rewrite Eq. (8.128) in terms of an axial derivative,

$$dv_x/dz \cong - eB_y(z)/m_i. \tag{8.129}$$

Equation (8.129) has the solution

$$v_x(d) = (e/m_i) \int_0^d B_y(z)dz. \tag{8.130}$$

Following the discussion of Section 8.1, if B_0 is the applied magnetic field, then the integral in Eq. (8.130) equals $B_0 d$ for short voltage pulses. Equation (8.130) gives an expression for the exit angle of ions emitted normal to the anode:

$$\Delta \alpha \cong \frac{v_x(d)}{v_z(d)} = \frac{eB_0 d/m_i}{(2eV_0/m_i)^{1/2}}. \tag{8.131}$$

An alternative form of Eq. (8.131) shows scaling with applied magnetic field:

$$\Delta \alpha = \left[\frac{B_0}{B^*} \right] \left[\frac{m_e}{m_i} \right]^{1/2}. \tag{8.132}$$

The first term on the right-hand side of Eq. (8.132) equals the ratio of the applied magnetic field to the nonrelativistic insulating field. This term must exceed unity, while the second term in Eq. (8.132) is much smaller than one. As an example, consider protons in a diode operating at about twice the critical insulating field. Equation (8.132) gives an angular deflection of 0.047 radians (2.7°).

To find the space-charge-limited flux of ions across the magnetic diode of Fig. 8.24, we must include the effects of the space charge of electrons with curved orbits. The solution is more difficult than the bipolar flow computation (Section 6.4). Although we shall not follow all details of the model, it is useful to review the origin and meaning of the governing equations. We simplify the physics with some limiting conditions. The diode is one dimensional, with infinite extent in the x and y directions. We neglect magnetic fields arising from axial ion current but include the transverse magnetic field created by electron drift. We take ion orbits as straight lines across the diode, ignoring the small magnetic deflection. The electrons leave the cathode with zero kinetic energy; therefore, all electrons have total energy and canonical angular momentum equal to zero. Although the model holds for moderate ion current, it does not include several important processes that occur in high-current (> 0.1 MA) ion diodes. In these devices, the magnetic field associated with ion flow can distort electron drifts, resulting in electron losses and enhancement of the ion current density.

In equilibrium, the ion flux is independent of axial position. A similar constraint applies to the axial electron flux in accessible regions. Ions move only in the axial direction, while electrons have velocity components v_x and v_z. The expression of conservation of total ion energy takes the form

$$m_i v_{iz}(z)^2/2 + e\phi(z) = eV_0, \tag{8.133}$$

where $\phi(z)$ is the electrostatic potential. For electrons, the relativistic equation of energy conservation is

$$[\gamma(z) - 1]m_e c^2 = e\phi(z), \tag{8.134}$$

where

$$\gamma(z) = \frac{1}{[1 - (v_{ex}^2 + v_{ez}^2)/c^2]^{1/2}}. \tag{8.135}$$

The electrostatic potential is related to the density of ions and electrons through the Poisson equation:

$$\frac{\partial^2 \phi(z)}{\partial z^2} = \frac{e}{\varepsilon_0}[n_e(z) - n_i(z)]. \tag{8.136}$$

The canonical momentun P_x is a constant of electron motion, since forces are uniform in the x direction. The angular momentum of all electrons equals zero if we take the vector potential A_x equal to zero at the cathode and assume that the electrons exit normal to the surface $[v_x(0) = 0]$. The condition $P_x = 0$ means that

$$\gamma(z)m_e v_{ex}(z) = eA_x(z). \tag{8.137}$$

The vector potential is related to the diode magnetic field by

$$B_y(z) = \frac{\partial A_x(z)}{\partial z}. \tag{8.138}$$

We can relate the magnetic field to the electron current density through the Maxwell equation:

$$\frac{\partial B_y(z)}{\partial z} = -\mu_0 j_{ex}(z) = -\mu_0 e n_e(z) v_{ex}(z). \tag{8.139}$$

The combination of Eqs. (8.138) and (8.139) gives a differential equation for the vector potential:

$$\frac{\partial^2 A_x(z)}{\partial z^2} = -\mu_0 n_e(z) v_{ex}(z). \tag{8.140}$$

To close the set of equations, we need to connect the ion and electron densities to the field quantities. The constant flux condition implies the following condition for the ion density:

$$n_i(z) = j_i/e v_{iz}(z). \tag{8.141}$$

The quantity $v_{iz}(z)$ depends on the electrostatic potential through Eq. (8.133). The corresponding equation for the electron density has the form

$$n_e(z) = j_e/e v_{ez}(z). \tag{1.142}$$

The relation between the electron density and the potential functions is more complex. The electron density depends on both ϕ and A_x through Eqs. (8.134), (8.135), (8.137), (8.140), and (8.142). In addition, we must include the fact that the magnetic field limits electron excursions to a fraction of the gap width.

The boundary conditions for electrostatic potential at the cathode and anode are

$$\phi(0) = 0, \qquad \phi(d) = V_0. \tag{8.143}$$

The condition that electron and ion flows are space-charge-limited implies that the derivatives of the potential must be zero on both surfaces:

$$\frac{d\phi(0)}{dz} = 0, \qquad \frac{d\phi(d)}{dz} = 0. \tag{8.144}$$

We set the vector potential equal to zero on the cathode:

$$A_x(0) = 0. \tag{8.145}$$

Also, we take the integral of magnetic flux integral as

$$\int_0^d B_y(z)\,dz = B_0 d. \tag{8.146}$$

Comparison of Eqs. (8.138) and (8.146) gives a second boundary condition for the vector potential:

$$A_x(d) = B_0 d. \tag{8.147}$$

The final condition is that the electrostatic and vector potentials are continuous across the transition at the electron turning point.

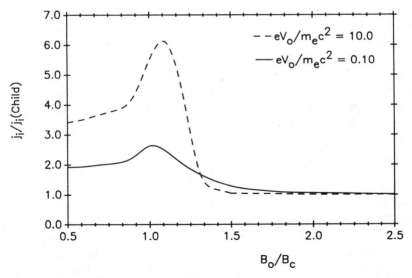

Figure 8.25 Predictions of steady-state ion flow in a planar, magnetically insulated gap based on numerical solutions of fluid equations. (Adapted from K. Bergeron, *Appl. Phys. Lett.*, **28**, 306 (1976).)

Given values of v_0, B_0, and d, we can solve the equations to find self-consistent values of j_i and j_e. Here, we shall merely quote the numerical results of Bergeron[2] Figure 8.25 shows normalized ion current in a magnetically insulated diode as a function of the ratio of the applied magnetic field to the critical insulating field. The calculation extends over both insulated and noninsulated regimes for the diode. The quantity j_{ic} is the single species Child limit in a gap with voltage V_0 and spacing d [Eq. (5.48)].

The solid line in Fig. 8.25 shows results for a low applied voltage, $eV_0 = (0.1)m_e c^2$, where electron motion is nonrelativistic. With no applied magnetic field ($B_0/B_{crit} = 0$), electrons and ions flow freely across the gap. The normalized current densities approach the bipolar flow limited, $(j_e/j_{eC}) \cong (j_i/j_{iC}) \cong 1.86$. High values of applied magnetic field ($B_0/B_{crit} \gg 1$) confine the electrons close to the cathode. In this case, the electron space charge has little effect on ion flow, so that $(j_i/j_{iC} \cong 1)$. The most interesting features appear just above the critical insulating field ($B_0/B_{crit} \geqslant 1$). The ion current density exceeds the bipolar flow limit. The enhancement factor results from the curvature of electron orbits in the magnetic field—the electrons spend more time near the anode. The resulting negative space charge partially neutralizes the ion space charge.

The ion flux is higher when electron motion is relativistic. At high voltage, the electrons move at a speed close to that of light. Because of the relativistic saturation of velocity, electrons spend a longer fraction of time near the anode. The dashed line of Fig. 8.25 shows results for a relativistic diode [$eV_0 = (10.0)m_e c^2$]. The ion flux exceeds the single species Child law prediction by a factor of 6.3

Ion flow enhancement in magnetically insulated diodes is a critical issue for proposed light-ion inertial fusion drivers. High focused power density ($\geqslant 10^{14}$ W/cm^2) requires high-current-density ion diodes. The enhancement factors for magnetically insulated diodes predicted by equilibrium models have not played a role in experiments. Real diodes are subject to plasma closure, geometric asymmetries, and electron flow instabilities. To achieve efficient conversion of power to ion flux, the diodes must operate with applied field in the range $B_0 > 1.5B_{crit}$. In this regime, the enhancement predicted by equilibrium models is small. In order to account for the high ion fluxes observed in experiments, we must include nonequilibrium processes that introduce a spread in the electron canonical momentum (Section 8.9).

Up to now, we have discussed magnetic diodes with infinite transverse extent. A real diode has a finite length in the direction of electron drift—this can lead to electron loss. Figure 8.26 illustrates the loss mechanism. Suppose a parallel-plate diode extends from $x = 0$ to $x = L$ along the drift direction. Conditions for the equilibrium model are not valid at the downstream edge of the diode, $x = L$. Here, the electric field amplitude decreases along the direction

[2]K. Bergeron, *Appl. Phys. Lett.*, **28**, 306 (1976).

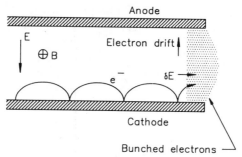

Figure 8.26 Mechanism for electron loss in a magnetically insulated gap with finite length along the drift direction. The axial electric field of bunched electrons results in drift motion across the gap.

of electric drift. Therefore, the magnitude of the $\mathbf{E} \times \mathbf{B}$ velocity drops for $x > L$. The electron flux that approaches the edge exceeds the flux that can leave. The continuity equation [Eq. (2.108)] implies that there must be a time variation of electron space charge near the edge. Bunching of the drifting electrons creates electric field components in the x direction. The field E_x causes a component of drift velocity across the gap. A group of electrons crosses to the anode, followed by growth of another electron clump at the edge. Experiments with magnetically insulated diodes with finite parallel-plate geometry show large electron losses at the downstream edge and emission of strong microwave radiation. The accumulated electron space charge at the edge also results in enhancement of the ion current density by factors of 10–100.

To design an efficient ion diode, we must reduce electron losses by ensuring that there is geometric continuity along the drift direction. Figure 8.27 illustrates three diodes with uniform geometry in the drift direction. The device of Fig. 8.27a consists of two coaxial cylinders. The electric field is radial and the applied magnetic field is axial. If the applied field is much stronger than the toroidal field created by power flow to the diode, the electrons drift in the azimuthal direction. Experiments on cylindrical diodes at moderate current show low electron loss and an ion current density close to the single-species Child law prediction. At high current, the summation of the axial magnetic field and the toroidal field from diode power flow produce helical magnetic field lines. The electron drift velocity has an axial component away from the power feed. The result is that electrons move to the end of the coaxial diode. The charge accumulation leads to electron losses and enhanced local ion flow.

The coaxial diode produces a diverging or converging ion flow. For most applications, we want a directed ion beam traveling in the axial direction. Figure 8.27b shows a diode that can generate such a beam. The diode gap has an axial electric field insulated by a radial magnetic field. Trapped electrons drift in the azimuthal direction. The diode produces an annular ion beam—the axis cannot be used for acceleration since the radial magnetic field equals zero at $r = 0$ for any arrangement of coils with cylindrical symmetry.

The radial field diode can produce low-emittance ion beams if the anode surface lies on a magnetic field boundary (Fig. 8.27b). From such a configuration there is no axial magnetic flux included within any position on the anode. All ions that leave normal to the anode surface have canonical angular momentum equal to zero. Although the downstream magnetic field displaces ions in the azimuthal direction, they emerge into the field-free propagation region with no azimuthal velocity.

Figure 8.27c shows another diode with symmetry along the electron drift direction that creates ions with zero canonical angular momentum. The *barrel diode* produces a beam that focuses in two dimensions. In light-ion fusion experiments, the diode has produced focused proton beams with power density exceeding 5 TW/cm². The barrel diode has pulsed magnet coils inside the anode to generate the field pattern of Fig. 8.27c. The anode is a metal cylinder that excludes the pulsed magnetic field. Therefore, the net magnetic flux inside the anode is close to zero. Focusing in the z direction is achieved by shaping the anode. The net magnetic fields in a multi-MA barrel diode are much different from the applied fields. With field components from diode current flow included, electrons drift in both the azimuthal and axial directions, converging to the diode midplane. The ion flow is enhanced strongly near the midplane.

Magnetic diodes have a unique and useful property—high-perveance ion beams can be extracted through a large-diameter cathode aperture with no foil or mesh. The ions exit through a surface defined by space charge. The surface, called the *extraction virtual cathode*, has no counterpart in conventional ion optics. The virtual cathode in a magnetically insulated ion diode is different

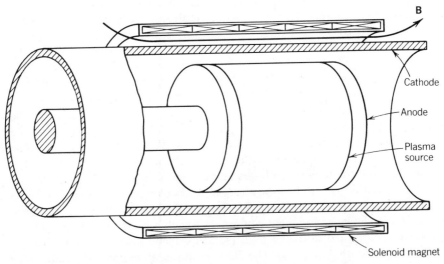

Figure 8.27(a, b) Magnetically insulated diodes with uniformity along the electron drift direction. (a) Coaxial cylinders with an axial magnetic field. (b) Radial magnetic field ion gun.

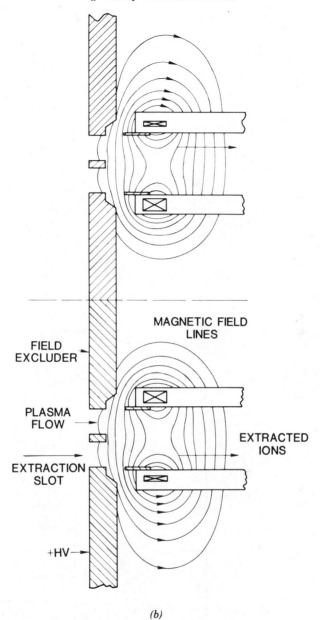

FIELD
EXCLUDER

MAGNETIC FIELD
LINES

PLASMA
FLOW

EXTRACTED
IONS

EXTRACTION
SLOT

+HV

(b)

Figure 8.27   (*Continued*)

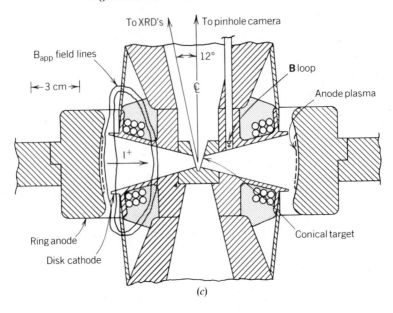

(c)

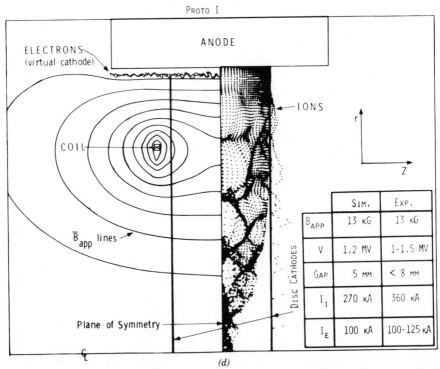

(d)

Figure 8.27(c, d)   (c) Barrel diode for focusing experiments. (d) Particle-in-cell computer simulation of ion trajectories and electron neutralization in a barrel diode. (Simulation courtesy of J. Quintenz, Sandia National Laboratories.)

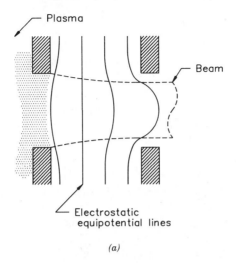

(a)

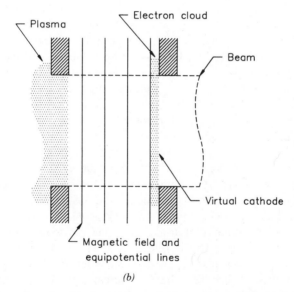

(b)

Figure 8.28 Formation of a virtual extraction cathode in a magnetically insulated ion diode by electron flow along magnetic field lines.

from the phenomenon that we discussed in Section 5.2. Figure 8.28 illustrates how electron space charge defines an extraction surface. Electrons from the cathode electrode move freely along magnetic field lines. Electron flow acts to cancel electric fields along the magnetic field lines that arise from the applied voltage and ion space charge. Section 11.2 describes the process of electron neutralization along field lines in detail. The result is that magnetic field lines define electrostatic equipotential surfaces. In Fig. 8.28, a cloud of electrons from the cathode potential covers the cathode aperture. The resulting electric fields are close to those that would result with an aperture covered with a conducting foil. There is no perveance limit for ion beams from magnetic diodes with extraction apertures. Experiments have shown that ion beams can be extracted through apertures with diameter more than an order of magnetic larger than the acceleration gap width.

8.9. ION FLOW ENHANCEMENT IN MAGNETICALLY INSULATED DIODES

The magnetic field in a high-current ion diode may serve two functions. Besides insulating the diode, the field can also trap electrons. The accumulated negative space charge leads to ion flux beyond the single-species Child limit. We have already seen that electron clumping causes local ion flow enhancement in a magnetic diode—an example is the boundary of a finite-length parallel-plate diode. In this device, electron loss accompanies the increased ion current. Experiments have shown that the ion current density in diodes with uniformity along the electron drift direction rises over long voltage pulses ($\sim 1\,\mu$sec). Enhancement factors exceeding 100 have been achieved in coaxial and radial field diodes (Section 8.8), yielding ion current density comparable to that obtained from low-impedance reflex triodes. In magnetic diodes, high ion current is not necessarily accompanied by high electron-loss current. In this section, we shall review computer simulations results to understand the origin of ion flow enhancement in magnetic diodes.

We know that equilibrium theories do not predict strong ion flow enhancement; therefore, we must add time variations to explain the results. Nonetheless, we recognize that time-dependent processes have little effect on the motion of individual ions. At magnetic fields of interest, ions move freely across the gap with a density proportional to $(V_0 - \phi)^{-1/2}$—the ion distribution is always a delta function in total axial energy. The key to ion flow enhancement is variation of the electron distribution with time. The ion flux is higher if the electron density increases near the anode. Such a density shift occurs if we relax the condition that all electrons have identical total energy and canonical momentum in the drift direction.

To begin, we shall introduce methods to categorize electron distributions in magnetically insulated diodes. The familiar phase-space plot at an instant of time is not informative because electrons in the gap have large orbit excursions—the velocity and position of an electron changes much more rapidly than the time scale for the growth of ion current density. It is more useful to plot the electron distribution in terms of the constants of motion: W, the total energy, and P_x, the canonical momentum in the drift direction. The plot displays allowed parameters for electron containment and the effect of electron diffusion in response to time-varying forces. The W-P_x coordinate of an electron subject to field perturbations moves slowly through the allowed region.

We can easily calculate the region of allowed orbits for magnetically confined electrons in W-P_x space if we adopt two limiting conditions:

1. Electron motion in the diode is nonrelativistic ($eV_0 \ll m_e c^2$).
2. Particle contributions to the electric and magnetic fields are small.

The second condition means that the electric and magnetic fields are uniform: $B_y = B_0$ and $E_z = -V_0/d$ (Fig. 8.24). Electron orbits in crossed fields can be laminar, scalloped, or a mixture of the two. To begin, consider laminar electron flow (Section 8.1). The orbits have only a component of velocity in the x direction—the magnetic force exactly balances the electric force. All electrons have the same velocity:

$$v_x = V_0/B_0 d. \tag{8.148}$$

The total electron energy is the sum of kinetic plus potential contributions:

$$W = eV_0 \left[\frac{1}{4} \left(\frac{B^*}{B_0} \right)^2 - \left(\frac{z}{d} \right) \right]. \tag{8.149}$$

The first term on the right-hand side of Eq. (8.149) equals $m_e v_x^2/2$—Eq. (8.11) defines the quantity B^*. The second term is the potential energy in the electrostatic field. Similarly, we can write the canonical momentum along x as

$$P_x = (eB_0 d) \left[\frac{1}{2} \left(\frac{B^*}{B_0} \right)^2 - \left(\frac{z}{d} \right) \right]. \tag{8.150}$$

Equation (8.149) and (8.150) are the parametric equations of a straight line in $W - P_x$ space. Figure 8.29a plots the solution for a choice $B_0/B^* = 2$. Electrons adjacent to the cathode ($x = 0$) have a positive total energy because they move at the $\mathbf{E} \times \mathbf{B}$ velocity. If we include particle-generated fields, the axial electric field approaches zero near the cathode. Therefore, laminar flow electrons at $x = 0$ have zero energy and canonical momentum. With particle field corrections, the laminar distribution line interesects the origin in $W - P_x$ space (dashed line in Fig. 8.29a). Similarly, the condition $E_z = 0$ holds at the anode for space-charge-limited emission of ions. Therefore, the laminar electron distribution line passes through the point $W = -eV_0$, $P_x = -eB_0 d$.

We can visualize the generalization to nonlaminar orbits readily in a frame of reference that moves at $v_x = V_0/B_0 d$. In the nonrelativistic limit, the fields in the transformed frame are $E'_z = 0$ and $B'_y \cong B_0$. In the drift frame, a laminar electron orbit is a point at a position $z' = z$. We can define a class of orbits with increasing kinetic energy about each point in the drift frame. The particles gyrate about z'

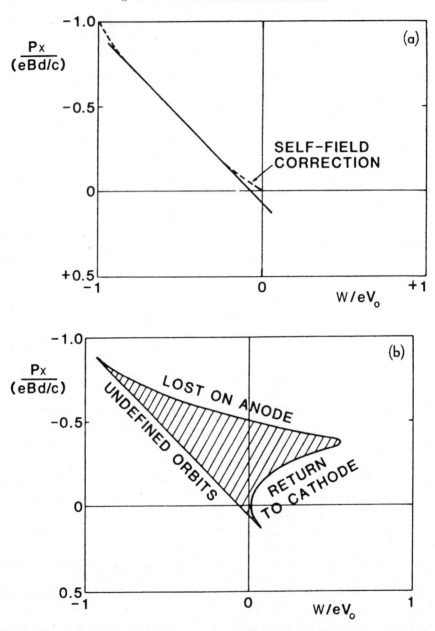

Figure 8.29 Magnetically insulated ion diode with electron diffusion. Allowed regions for electron orbit vectors in $W - P_x$ space. (*a*) Laminar electrons formed during a slowly rising voltage. Dashed line shows corrections for effects of space-charge electric fields. (*b*) Extended allowed region with nonlaminar orbits.

with a gyroradius, r'_g. The gyroradius depends on the excess kinetic energy an electron has compared with an electron with a laminar orbit. The difference in kinetic energy is related to the gyroradius by

$$\Delta W = eV_0(B_0/B^*)^2(r'_g/d)^2. \tag{8.151}$$

Geometric constraints on the gyroradius limit the allowed kinetic energy difference. Trapped electrons cannot strike the cathode; therefore,

$$r'_g < z'. \tag{8.152}$$

Similarly, the electrons cannot strike the anode:

$$r'_g < (d - z'). \tag{8.153}$$

The maximum value of r'_g depends on whether Eq. (8.152) or (8.153) is smaller. Electrons with extra kinetic energy follow cycloid orbits in the accelerator frame. All electrons that share the same center of gyration in the drift frame, z', have the same value of canonical momentum in the laboratory frame. We can prove this fact by noting that all the particle orbits centered on z' have velocity $v_x = V_0/B_0 d$ in the accelerator frame when they cross the position $z = z'$.

The allowed space for trapped electrons extends along the W axis at each position of P_x. The shape resembles that of Fig. 8.29b. The conditions of Eqs. (8.152) and (8.153) define the boundaries. Particle contributions to the electric and magnetic fields deform the shape so that it passes through the points $(0, 0)$ and $(-eV_0, -eB_0 d)$. The change in the shape of the allowed region with relativistic corrections is small.

In W-P_x space, the equilibrium distribution of the model of Section 8.8 is a point at the origin—all electrons created on the cathode have $W = 0$ and $P_x = 0$. When diffusion processes act on the electrons, the total energy and canonical momentum can change over time scales much longer than the orbit oscillation period. We expect that electrons migrate toward the anode, resulting in enhanced ion flow. A self-consistent description of the process is complex. Here, we shall review the results of a computer simulation, using the W-P_x diagram as a guide. The one-dimensional program treats self-consistent electron and ion orbits in the presence of applied and particle-generated electric and magnetic fields. The field calculation uses the quasi-static approximation, excluding displacement currents and inductive electric fields. The requirement of vanishing electron fields on the boundaries determines the emission of ions and electron.

Several processes, such as geometric imperfections in the electrodes, can cause electron diffusion. The inclusion of these processes, which involve field variations along the x direction, is beyond the capability of a one-dimensional program. Therefore, we limit the investigation to a specific question: how is the self-consistent ion flow in a magnetically insulated diode affected by electron diffusion? We introduce diffusion by adding a phenomenological transverse

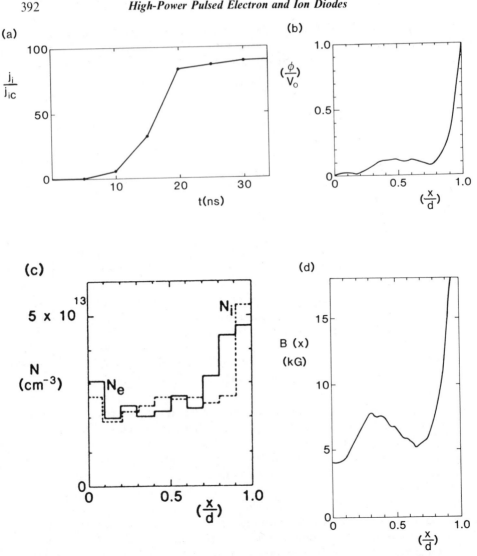

Figure 8.30 Particle-in-cell computer simulation of a magnetically insulated ion diode with electron diffusion. Electrons migrate across an initially empty gap in approximately 5 nsec. $V_0 = 1$ MV, $B_0 = 2$ tesla, $B_{crit} = 0.96$ tesla, perturbation electric field amplitude ~ 100 kV. (*a*) Ion current density normalized to the Child law prediction. (*b*) Spatial variation of normalized potential at 30 nsec. (*c*) Spatial variation of particle densities at 30 nsec. (*d*) Spatial variation of magnetic field at 30 nsec. (Courtesy J. Poukey, Sandia National Laboratories.)

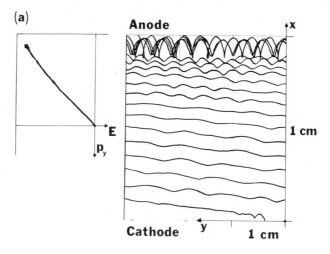

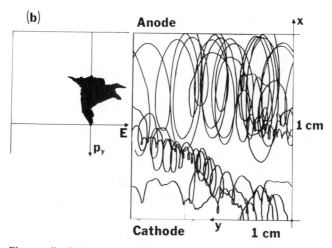

Figure 8.31 Electron distributions in $W - P_x$ and test electron orbits for the simulation of Fig. 8.30. (a) Distribution at $t = 5$ nsec and an orbit trace of an electron that reaches the anode at 5 nsec. (b) Distribution at $t = 20$ nsec and an orbit trace of an electron that reaches the anode at 20 nsec. (Courtesy J. Poukey, Sandia National Laboratories.)

electric field variation of the form

$$E_x = E_{x0}[1 - \cos(\omega_0 t)][x(x - d)/d^2]. \tag{8.154}$$

The spatial factor ensures that the transverse electric field equals zero at the electrodes. The results are insensitive to the exact choice of perturbation frequency, ω_0. The field amplitude, E_{x0}, is a free parameter. For practical diodes, we know that E_{x0} is much smaller than the applied field, $E_{x0} \ll V_0/d$. We adjust the perturbation so that the time variation of ion flux matches the variation observed in experiments. The simulation then gives an idea of the amplitude of field perturbations in the experiment.

Figure 8.30 illustrates some simulation results. To start the run, electrons and ions are injected into a diode with $B_0/B_{crit} \gg 1$, the full applied voltage, and no perturbation field. After several time steps, the self-consistent ion flow closely approaches the space-charge-limited value predicted by the theory of Section 8.8. At a time taken as $t = 0$, the perturbation field is turned on. Figure 8.30a shows the subsequent variation of ion current density (normalized to the space-charge limit). The normalized current density increases a factor of 80 over an interval close to the electron diffusion time across the gap. Figure 8.30b shows the spatial variation of electrostatic potential in the final state. The axial electric field is concentrated close to the anode, accounting for the enhanced ion flow. A plot of particle densities in the final state (Fig. 8.30c) shows that both the electron and ion density are high near the anode. The augmented electron density results, in part, from electron diamagnetism which compresses the magnetic flux against the anode (Fig. 8.30d). In the final state, electron loss accounts for only 25% of the total current flow in the diode.

Figure 8.31a plots a folded orbit for a test electron crossing the gap during the early phase of current density growth ($t = 5$ nsec). The figure also shows a W-P_x diagram of the electron distribution at the same time. During the time initial electrons move across the gap, the electron distribution is almost laminar. The ion current density increases when the electrons reach the vicinity of the anode. In turn, the enhanced ion flow allows more electrons to enter the gap. In the advanced stage of enhanced ion flow with strong diamagnetic effects, the electrons make a transition to scalloped orbits. Figure 8.31b shows a folded test electron orbit and W-P_x plot at $t = 20$ nsec. The distribution is highly nonlaminar. The bounded region of Fig. 8.31b has many features in common with the boundary of allowed orbits shown in Fig. 8.29.

9

Paraxial Beam Transport with Space Charge

In this chapter we begin the study of beam transport. We shall develop equations to describe beam focusing in long accelerators or beam-transport lines. The term focusing means the application of transverse forces to maintain a small beam radius about a main axis. We shall concentrate on paraxial beams. Here, particle orbits have small angles with respect to the main axis—the components of transverse velocity are much smaller than the axial velocity. Furthermore, the space-charge potential energy in the beam volume is much less than the average particle kinetic energy. The paraxial model gives a good description of beams in high-energy accelerators and in low-current devices such as electron microscopes. The approximation is not valid for the initial stages of high-current accelerators and for high-perveance devices like microwave tubes. We shall discuss transport of beams with strong space-charge effects in following chapters.

Section 9.1 derives the envelope equation for sheet beams, while Section 9.2 covers cylindrical beams. Envelope equations are useful for first-order designs of transport systems. The idea is to seek conditions for global force balance rather than to pursue detailed solutions of self-consistent distributions. Envelope equations specify transverse force balance at the root-mean-squared beam dimension. They predict the variation of beam width along z in response to external forces and those generated by the beam. Section 9.3 presents another example of envelope equations for application to a periodic arrays of quadrupole lenses. We use the KV distribution to derive self-consistent equations. The resulting KV equations describe beam propagation in a system where the applied forces are independent in the x and y directions but the space-charge forces couple particle motion in x and y.

Section 9.4 reviews a practical application of envelope equations, calculation of the maximum allowed current in a transport system. We shall concentrate on periodic lens arrays and seek a condition of transverse force balance averaged over the length of a focusing cell. In periodic systems, there is a well-defined current limit. The average applied focusing force must be low enough so that the vacuum phase advance of individual orbits is in the range $\mu_0 < 180°$. This defines an upper limit on the sum of emittance and space-charge forces and hence a limit on beam current. Section 9.5 describes a method to circumvent constraints on ion beam flux by dividing the current between several small beams. This method is technologically feasible in an array of parallel electrostatic quadrupole channels.

Section 9.6 discusses limits on beam current that result from axial space-charge electric fields. These constraints are important for accelerators where the beam must remain synchronized to an accelerating wave. Even if the electrostatic potential energy in the beam volume is small, the electric fields may be strong enough to push particles out of the bucket of an RF accelerator. Similar processes occur for high-current ion beams in induction linear accelerators.

9.1. ENVELOPE EQUATION FOR SHEET BEAMS

Often in the design of accelerators and beam-transport systems, we do not need to describe transverse beam distributions in detail. It is usually sufficient to know properties of the beam averaged over the cross section. We seek an estimate of the beam radius and envelope angle to make sure that the beam fits through the bore of acceleration gaps and lenses. We use envelope equations to make such estimates. The equations apply conditions of force balance on the beam periphery to predict the envelope trace, the width of the beam as a function of axial position.

In this section, we shall derive the envelope equation for a sheet beam (Section 5.1). Section 9.2 extends the model to cylindrical beams. Envelope equations apply to paraxial beams. The term *paraxial* means that the inclination angles of particle orbits are small, $x' \ll 1$. As a result, changes in the beam width take place over distances much longer than the transverse beam dimension. This condition leads to major simplification of the theory—we can use special expressions for applied transverse forces and we can approximate beam-generated forces with expressions for an infinite-length beam. We treat fields in the static limit. The paraxial condition also means that the axial kinetic energy component is much larger than the transverse kinetic energy. The implication is that all beam particles have about the same axial kinetic energy at position z, independent of their transverse motions. Envelope equations are most useful for steady-state beams. They also describe pulsed beams if the beam length is much longer than the width. The equations are not suited to short beam pulses, such as the bunches in a RF accelerator. In this device, axial and transverse particle motions are coupled, and fields must be described by the full set of Maxwell equations.

We consider a sheet beam that travels in the z direction and has infinite extent

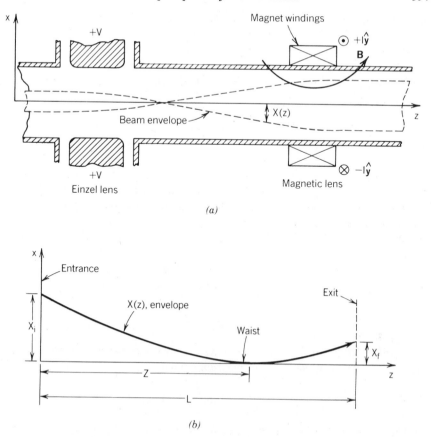

Figure 9.1 Basis of the envelope equation for a sheet beam. (*a*) Focusing by electrostatic and magnetic lenses that extend in the *y* direction. (*b*) Propagation of a high-current laminar sheet beam in a field-free drift region.

in the *y* direction (Fig. 9.1). The beam is symmetric about the line $x = 0$. The quantity $X(z)$ is the envelope half-width at position *z*. To begin, we shall make an inventory of forces on the periphery of the beam. Figure 9.1 illustrates the types of electric fields and magnetic fields that can be used to focus extended sheet beams—the fields are uniform in the *y* direction. Applied electric fields can both accelerate and focus particles. In the paraxial limit for a sheet beam, $E_y = 0$ and $E_x \ll E_z$. The transverse component of the static electric field is related to the axial field by [CPA, Section 6.2]

$$E_x(x, z) \cong - x \left[\partial E_z(0, z)/\partial z \right]. \tag{9.1}$$

We can show that the field expression of Eq. (9.1) is consistent with $\nabla \cdot \mathbf{E} = 0$ if E_z is almost uniform over the beam width. The axial electric field changes the kinetic energy of beam particles:

$$\partial(\gamma m_0 c^2)/\partial z \cong q E_z(0, z). \tag{9.2}$$

Combining Eqs. (9.1) and (9.2), we can write the transverse applied electric force on the envelope as

$$F_x \cong - X(m_0 c^2)\gamma'' \qquad \text{[applied electric field]}, \tag{9.3}$$

where $\gamma'' = \partial^2 \gamma/\partial z^2$.

Currents that flow in the y direction generate magnetic forces that focus the sheet beam. For such currents, the only nonzero component of the vector potential is A_y. Since there are no y-directed forces, the canonical momentum of particles in y is a conserved quantity:

$$P_y = P_0 = \gamma m_0 v_y + q A_y. \tag{9.4}$$

The vector potential is related to the axial magnetic field by

$$\partial A_y/\partial x = B_z. \tag{9.5}$$

In the paraxial limit, B_z is almost constant over the beam cross section, so that $A_y \cong B_z x$. We can write the canonical momentum as

$$P_y = P_0 \cong \gamma m_0 v_y + q B_z x. \tag{9.6}$$

The magnetic field is usually localized to discrete lens regions. If there is no magnetic field at the source and particles leave perpendicular to the surface ($v_y = 0$), then all particles have zero canonical momentum,

$$P_0 = 0. \tag{9.7}$$

In the following discussion, we will apply the condition of Eq. (9.7)—the particle source is not immersed in the magnetic field. Using Eq. (9.5), the velocity in the y direction is related to the on-axis axial magnetic field by

$$v_y(z) \cong - [q B_z(0, z)/\gamma m_0] X. \tag{9.8}$$

The magnetic force in the x direction equals $q v_y B_z$. At the envelope, the applied transverse magnetic force is

$$F_x \cong - [q^2 B_z^2(0, z)/\gamma m_0] X \qquad \text{[applied magnetic field]}. \tag{9.9}$$

Electric and magnetic forces also arise from beam-generated fields. The electric force acting on the envelope of a sheet beam carrying a current per unit length (along y) of J A/m is [Eq. (5.16)]:

$$F_x = (qJ/2\varepsilon_0\beta c)X \qquad \text{[beam electric field]}. \qquad (9.10)$$

For a paraxial beam, the beam-generated magnetic force equals the electric force multiple by $-\beta^2$. The total beam-generated force is

$$F_x = [\gamma m_0(\beta c)^2]K_x X \qquad \text{[beam force]}. \qquad (9.11)$$

The quantity K_x is the generalized perveance for a sheet beam,

$$K_x = \frac{qJ}{2\varepsilon_0 m_0(\gamma\beta c)^3} \qquad (9.12)$$

Note that K_x has dimensions of m^{-1}. To complete the list, we can also include the transverse force associated with emittance derived in Section 3.5.

The beam envelope follows an equation of motion of the form

$$\frac{d[\gamma m_0(dX/dt)]}{dt} = \sum F_x. \qquad (9.13)$$

The right-hand side of the equation is the summation of all transverse forces. We shall convert Eq. (9.13) to a trace equation that gives the beam width as a function of axial position, $X(z)$. Since all particles at a location z have the same axial velocity, we can change time derivatives to axial derivatives using the chain rule,

$$\frac{dX}{dt} = \frac{dX}{dz}\frac{dz}{dt} = (\beta c)X'. \qquad (9.14)$$

Noting that γ varies with z in the presence of an accelerating electric force, the left-hand side of Eq. (9.13) becomes

$$\frac{d[\gamma m_0(dX/dt)]}{dt} = \frac{d[\gamma m_0\beta cX']}{dt} \qquad (9.15)$$
$$= m_0\beta^2 c^2[\gamma\beta X'' + \gamma\beta'X' + \gamma'\beta X'].$$

By manipulating Eqs. (9.13) and (9.14), we can show that

$$\gamma\beta' + \beta\gamma' = \gamma'/\beta. \qquad (9.16)$$

Addition of the individual force terms to the right-hand side of Eq. (9.13) gives the following second-order differential equation for $X(z)$:

$$X'' = -(\gamma'X'/\gamma\beta^2) - (\gamma''/\gamma\beta^2)X - (qB_z/\gamma m_0\beta c)^2 X + K_x + \varepsilon_x^2/X^3. \qquad (9.17)$$

The first term on the right-hand side of Eq. (9.17) contributes to a decrease in the envelope angle when the beam accelerates ($\gamma' > 0$). The second term represents electrostatic focusing from transverse field components in acceleration gaps (Fig. 9.1) or einzel lenses [CPA, Chapter 6]. The second term corresponds to focusing by the magnetic lens of Fig. 9.1. The fifth term gives defocusing by beam-generated forces, while the last term represents the effect of emittance.

As an example of the application of Eq. (9.17), consider propagation of a laminar sheet beam in a vacuum drift space with no applied forces. We want to find the maximum allowed beam current for a given drift length L, entrance half-width X_i, and exit half-width X_f. From the discussion of Section 5.4, we seek a solution where the beam converges as it enters and passes through a waist between the entrance and exit. The solution for a sheet beam is easier than a cylindrical beam—the beam-generated force is independent of the envelope width. As a result, the width of a zero emittance beam can approach zero at the waist. The reduced envelope equation is

$$X'' = K_x. \tag{9.18}$$

If we set the envelope width equal to zero at the waist, the solution of Eq. (9.18) is

$$X(z) = K_x z^2/2. \tag{9.19}$$

The quantity z is the distance from the waist. If Z_i is the distance from the entrance to the waist, we can show that

$$Z = \frac{L(X_i/X_f)^{1/2}}{[1 + (X_i/X_f)^{1/2}]}. \tag{9.20}$$

When Eq. (9.20) holds, the matched value of generalized perveance is

$$K_x = 2X_i/Z^2. \tag{9.21}$$

Suppose we have an electron beam for an intense pulsed microwave source with kinetic energy equal to 1 MeV. Take $L = 0.3$ m, $X_i = 0.01$ m, and $X_f = 0.01$ m. The total length of the beam along the y direction is 0.2 m—the one-dimensional approximation is well satisfied. Substitution into Eqs. (9.20) and (9.21) implies that $K_x = 0.89$. Solving Eq. (9.12) with $\gamma = 2.96$ and $\beta = 0.941$ gives $J = 52.3$ kA/m. The maximum beam current consistent with the constraints is 10.5 kA.

9.2. PARAXIAL RAY EQUATION

It is easy to carry out the derivation of Section 9.1 in polar coordinates for application to cylindrical beams. The result, the *paraxial ray equation*, is one of the most widely used relationships is charged-particle-beam optics. Although the equation is limited to cylindrical beams and optical elements with azimuthal symmetry, it describes a broad class of practical devices. The paraxial ray equation is useful for the study of low-current beams in cathode ray tubes and electron microscopes. It is also valuable for the design of high-current microwave tubes and induction linacs.

In a cylindrical system, symmetry permits only certain components of electric and magnetic field:

1. axial and radial components of the applied electric field,
2. radial electric field resulting from space charge,
3. axial and radial magnetic field components generated by axicentered circular coils, and
4. beam-generated toroidal magnetic field.

In the paraxial limit, we can relate the radial components of *applied* fields to the axial field by

$$E_r(r, z) \cong -(r/2) \ \partial E_z(0, z)/\partial z, \tag{9.22}$$

$$B_r(r, z) \cong -(r/2) \ \partial B_z(0, z)/\partial z. \tag{9.23}$$

The particles of cylindrical paraxial beams have small transverse velocity, v_r, $v_\theta \ll v_z$. Furthermore, particles have almost the same kinetic energy, $(\gamma-1)m_0c^2$, and axial velocity, $v_z/c = \beta$. At all positions in an unneutralized beam, the beam-generated magnetic force equals the beam electric force multiplied by $-\beta^2$. Particles gain azimuthal velocity when they move through the radial magnetic fields of a solenoidal lens. For forces with cylindrical symmetry, the canonical angular momentum is a constant of particle motion:

$$\gamma m_0 r v_\theta + q r A_\theta = P_\theta = \text{constant.} \tag{9.24}$$

The quantity A_θ is the vector potential generated by azimuthal currents.

Following the methods of Section 9.1, we derive the following equation for the axial variation of the envelope of a cylindrical beam:

$$R'' = \underset{(1)}{-\frac{\gamma'R'}{\beta^2\gamma}} - \underset{(2)}{\left[\frac{\gamma''}{2\beta^2\gamma}\right]R} - \underset{(3)}{\left[\frac{qB_z(0,z)}{2\beta\gamma m_0 c}\right]^2 R} + \underset{(4)}{\frac{\varepsilon^2}{R^3}} + \underset{(5)}{\left[\frac{q\psi_0}{2\pi\beta\gamma m_0 c}\right]^2 \frac{1}{R^3}} + \underset{(6)}{\frac{K}{R}}.$$

$$\tag{9.25}$$

The quantity R is the envelope radius of the beam—the prime symbol denotes a derivative taken with respect to z.

The first three terms on the right-hand side of Eq. (9.25) represent focusing processes. Term 1 arises from acceleration, or a change of γ with position along the axis. The term reduces the envelope angle, R', when particles accelerate. Term 2 represents electrostatic focusing from radial components of applied electric fields. The effect is important in low-energy-electron optical systems or in acceleration columns. Usually, we can neglect the term when beams have high kinetic energy. Term 3 describes magnetic focusing from applied solenoidal fields. Solenoidal lenses are common in electron-beam-transport systems at low or moderate energy (< 50 MeV). Solenoidal fields are seldom used for ion transport since they are ineffective for focusing heavy particles.

Terms 4, 5, and 6 describe defocusing processes. Term 4 is the emittance force of Section 3.5. Terms is important when the particle source is immersed in an axial magnetic field—eytracted particles have nonzero canonical angular momentum. The quantity ψ_0 is the total magnetic flux enclosed within the beam envelope at the source:

$$\psi_0 = \int_0^{R_s} 2\pi R \, dR \, B_z(R, Z_s). \tag{9.26}$$

In Eq. (9.26), R_s is the radius of the beam at a source at position $z = Z_s$. This term has the same $1/R^3$ variation as emittance. We know from the theory of motion in a central force that particles with nonzero P_θ cannot pass through the axis. Finally, term 6 gives the combined action of beam-generated electric and magnetic fields (Section 5.1). The quantity K is the generalized perveance of Eq. (5.88).

We often apply an alternative form of the paraxial ray equation for low-energy-electron or ion beams. For low energies it is convenient to use the on-axis absolute potential, $\phi(0, z)$, instead of the relativistic γ. The absolute potential equals zero at the particle source—it is related to γ by

$$\phi(0, z) = (\gamma - 1)m_0 c^2. \tag{9.27}$$

Substituting Eq. (9.27) into Eq. (9.25), we find

$$R'' = -\frac{\phi' R'}{2\phi} - \left[\frac{\phi''}{4\phi}\right] R - \left[\frac{q B_z(0, z)^2}{8 m_0 \phi}\right] R + \frac{\varepsilon^2}{R^3} + \left[\frac{q \psi_0^2}{8\pi^2 m_0 \phi}\right] \frac{1}{R^3} + \frac{K}{R} \tag{9.28}$$

To illustrate the application of Eq. (9.25), we shall review the basis of a computer program to predict beam dynamics in a linear induction accelerator. These machines generate pulsed beams with current in the range 1–10 kA and energy of 5–50 MeV. Despite the high current, the energetic electrons have paraxial orbits. Through careful design, the axial magnetic field equals zero on the cathodes of most induction accelerator injectors; accordingly, we take $\psi_0 = 0$.

Solenoidal lenses or extended axial magnetic fields provide focusing in all existing induction accelerators. To describe magnetic focusing, we only need to know the axial variation of $B_z(0, z)$. We can approximate any solenoidal magnetic focusing system as a set of windings with different axial position and radius. The net magnetic field along the axis is the sum of contributions from individual coils. The following formula gives $B_z(0, z)$ for a solenoid with a uniform density of windings. The coil has radius r_i, midpoint position z_i, length L_i, and a net current of I_i amp-turns:

$$B_{z_i}(0, z) = \left[\frac{\mu_0 I_i}{2L_i}\right]\left\{\frac{z_i + L_i/2 - z}{[(z_i + L_i/2)^2 - 2(z_i + L_i/2)z + r_i^2 + z^2]^{1/2}}\right.$$

$$\left. - \frac{z_i - L_i/2 - z}{[(z_i - L_i/2)^2 - 2(z_i - L_i/2)z + r_i^2 + z^2]^{1/2}}\right\}. \tag{9.29}$$

To specify the focusing system, we need only supply a list of solenoid parameters (r_i, z_i, L_i, I_i). The computer program performs a sum to calculate the net field at all positions.

Equation (9.25) is not valid within the beam injector—here, axial fields vary over a scale length comparable to the beam radius. Instead, we must use a ray tracing code (Section 7.3) to supply initial conditions for downstream transport calculations with the paraxial ray equation. The required injection parameters are γ_0, ε_0, K_0, R_0, and R_0'. Particle acceleration in induction linacs takes place in narrow gaps located at spatial positions z_j. Usually, the gaps have the same value of applied voltage V_0. Electrostatic focusing in the gaps is negligible compared with magnetic focusing. Rather than integrate Eq. (9.25) through the gaps, it is more convenient to apply the following step conditions based on conservation of energy and momentum:

$$\gamma_j = \gamma_{j-1} + eV_0/m_0c^2 = \gamma_0 + jeV_0/m_0c^2, \tag{9.30}$$

$$\beta_j = (1 - 1/\gamma_j^2)^{1/2}, \tag{9.31}$$

$$R_j = R_{j-1}, \tag{9.32}$$

$$R_j' = R_{j-1}'(\beta_{j-1}\gamma_{j-1}/\beta_j\gamma_j), \tag{9.33}$$

$$\varepsilon_j = \varepsilon_{j-1}(\beta_{j-1}\gamma_{j-1}/\beta_j\gamma_j), \tag{9.34}$$

$$K_j = K_{j-1}(\beta_{j-1}\gamma_{j-1}/\beta_j\gamma_j)^3. \tag{9.35}$$

Figure 9.2 shows results of a numerical calculation to find the effect of discontinuities in solenoidal field windings at acceleration gaps.

We can add an approximate term to the paraxial ray equation that is useful to represent the effects of periodic focusing systems. Figure 9.3 shows two such systems, the einzel lens array (Fig. 9.3a) and the periodic permanent magnet (PPM) array (Fig. 9.3b). We shall ignore details of motion through individual

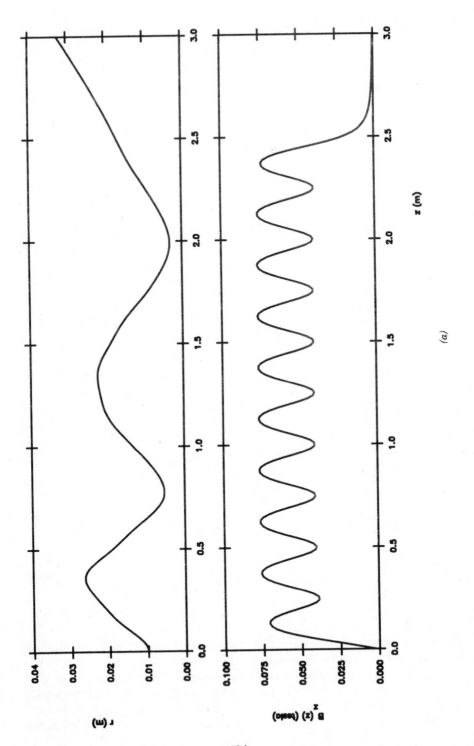

(a)

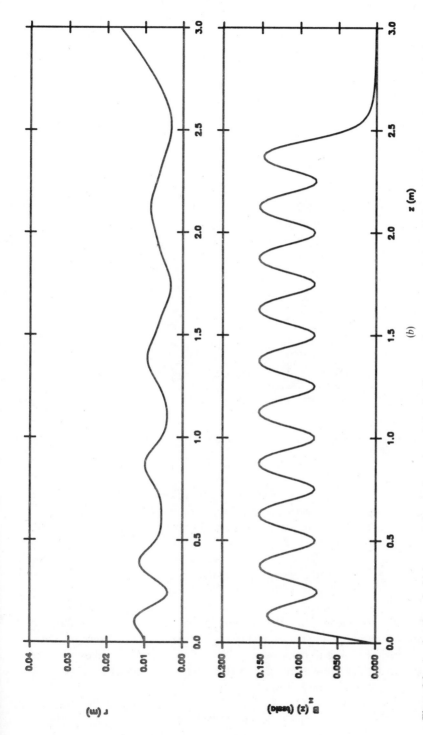

(b)

Figure 9.2 Acceleration of a high-current electron beam in an induction linear accelerator—numerical results from PARAX5 code. Top graphs in figures show beam radius as a function of position in accelerator and in a downstream drift region. Bottom graphs show axial variation of solenoidal magnetic field. Injected beam parameters: 2-kA current; 10-MeV kinetic energy; 0.01-m radius; zero envelope angle; 8.8×10^{-5} π-m rad emmitance. Accelerator parameters: ten acceleration gaps, 250 kV each, at 0.25 m, 0.50 m, 0.75 m, ... : ten solenoidal coils, 7.5-cm radius, 0.15-cm length, spaced between acceleration gaps; plus one bucking coil behind cathode. (*a*) Total current per solenoid: 1.19×10^5 A-turns. (*b*) Total current per solenoid: 2.38×10^5 A-turns.

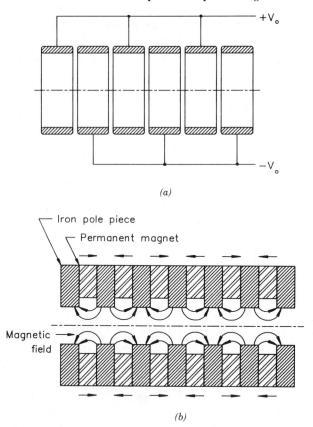

(a)

(b)

Figure 9.3 Periodic focusing systems with cylindrical symmetry. (a) Einzel lens array. (b) Periodic-permanent-magnet (PPM) array.

lenses. Instead, we simplify the calculation by invoking properties of particle orbits in periodic systems with linear applied forces. The displacement of a particle orbit at the boundary of the nth lens in an array obeys the equation [CPA, Chapter 8]

$$r_n = r_0 \cos(n\mu_0 + \phi). \tag{9.36}$$

The quantity ϕ is a phase factor, while μ_0 is the *vacuum phase advance* per lens.

To illustrate the meaning of the vacuum phase advance, Fig. 9.4 shows a numerical calculation of a particle orbit in a PPM array for the choice $\mu_0 = 90°$. The orbit consists of long-term harmonic motion with superimposed oscillations on the scale length of individual lenses. If the length of a focusing cell is L, the long-term harmonic motion follows the equation

$$r(z) \cong r_0 \cos[(\mu_0/L)z + \phi]. \tag{9.37}$$

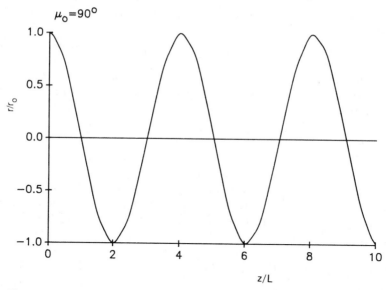

Figure 9.4 Numerical calculation of a particle orbit in a PPM array with $\mu_0 = 90°$.

The orbit of Eq. (9.37) is the solution of the equation

$$r'' = -(\mu_0/L)^2 r. \tag{9.38}$$

If we ignore small-scale oscillations and assume that only periodic forces act on the beam, Eq. (9.38) leads to the following form for the paraxial ray equation:

$$R'' = -\left[\frac{\mu_0}{L}\right]^2 R + \frac{\varepsilon^2}{R^3} + \frac{K}{R}. \tag{9.39}$$

We shall use Eq. (9.39) in Section 9.4 to estimate beam transport limits in the limit $\mu_0 \ll 1$.

9.3. ENVELOPE EQUATION IN A QUADRUPOLE LENS ARRAY

Quadrupole lens arrays are the most widely used focusing systems for high-energy particle accelerators. In this section, we shall develop equations that describe the propagation of beams with strong self-fields through quadrupole lenses. The equations are often applied to beam dynamics in storage rings. The theory of quadrupole transport with beam-generated fields is also useful to predict limits on beam current in high-current RF linacs and heavy-ion induction accelerators for inertial fusion. To construct a simple theory, we assume that the beam has a KV distribution (Section 6.7) in the transverse direction.

The transverse forces in a quadrupole lens array vary periodically along z. In one transverse direction, the force alternately focuses and defocuses. When the force in the y direction focuses, the force in the x direction defocuses. A beam in such a system does not have cylindrical symmetry. The calculation of electric and magnetic fields for three-dimensional variations of beam density is complex. We shall reduce the problem to two dimensions by taking the scale length for axial variations of density to be much longer than the beam width. Again, we use field expressions for an infinite-length beam. The remaining two-dimensional calculation is simple if the beam has a KV distribution in transverse position and angle. Section 6.7 showed that such a distribution corresponds to a beam with elliptical cross section and uniform charge density. The resulting electric fields *within the beam* are linear and separable in x and y. We must recognize that theories based on the KV distribution have limitations—for instance, they cannot represent nonlinear space-charge forces that often lead to emittance growth.

We shall combine results from Sections 6.7 and 4.2 to construct a self-consistent theory. Particle motions in the x and y directions are decoupled in an ideal quadrupole array. Matched distributions fill boundary ellipses in x-x' or y-y' space. As a beam propagates, the areas of the x and y ellipses remain constant, although transverse forces may modify their shape and orientation. The derivations of Section 4.2 were carried out for a general linear transverse force—here, we take a force that results from the combination of applied fields and beam-generated fields.

In the x direction, the distribution boundary depends on the transport functions $\alpha_x(z)$, $\beta_x(z)$, and $\gamma_x(z)$ through the equation

$$\gamma_x x^2 + 2\alpha_x x x' + \beta_x x'^2 = \varepsilon_x. \tag{9.40}$$

The quantity ε_x is the beam emittance in the x direction. The transport functions depend on the properties of the distribution at injection and the variation of the total x-directed force. We shall denote the boundary of the distribution in x (the envelope width) as X. The envelope width is related to the transport parameters by

$$X = \sqrt{\beta_x \varepsilon_x}. \tag{9.41}$$

Similarly, in the y direction, the distribution boundary lies on the curve:

$$\gamma_y y^2 + 2\alpha_y y y' + \beta_y y'^2 = \varepsilon_y. \tag{9.42}$$

The envelope width in y is

$$Y = \sqrt{\beta_y \varepsilon_y}. \tag{9.43}$$

The treatment of the KV distribution in Section 6.7 was limited to upright elliptical distributions. We shall extend the model to skewed ellipses, since we expect that the beam envelope converges and diverges in a quadrupole array. The KV distribution is a hollow shell in trace space with particle orbit coordinates distributed uniformly over a hyperellipsoid. If we assume that the emittances are equal in the x and y directions ($\varepsilon = \varepsilon_x = \varepsilon_y$), the KV distribution function is

$$f_{KV}(x, x', y, y') = A\delta[\gamma_x x^2 + 2\alpha_x xx' + \beta_x x'^2 + \gamma_y y^2 + 2\alpha_y yy' + \beta_y y'^2 - 2\varepsilon].$$
(9.44)

We calculate the configuration-space density of the distribution by taking integrals over x' and y'. The method is discussed in [I. M. Kapchinskij and V. V. Vladimirskij, *Proc. Int. Conf. on High Energy Accelerators* (CERN, Geneva, 1959), p. 274.]; it is similar to the treatment of Section 6.7.

The resulting density is similar to that for an upright ellipse. At each axial position, the beam has an elliptical cross section with envelope half-widths of $X(z)$ and $Y(z)$. The charge density $\rho(z)$ is uniform over the cross section. It is related to the total beam current I by

$$\rho(z) = I/\pi X(z) Y(z)v_z.$$
(9.45)

We can connect the distribution normalization constant A to the charge density by

$$A = \rho(z)\sqrt{\beta_x(z)} \sqrt{\beta_y(z)}/e\pi.$$
(9.46)

where β_x and β_y are related to X and Y through Eqs. (9.41) and (9.43).

Expressions for the transverse electric fields of a uniform-charge-density beam with elliptical cross section are simple. The electrostatic potential inside a long beam is

$$\phi(x, y, z) = -\frac{I}{4\pi\varepsilon_0 XY v_z}\left[x^2 + y^2 - \left(\frac{X - Y}{X + Y}\right)(x^2 - y^2)\right].$$
(9.47)

We can find the transverse electric field in the x direction by taking a partial derivative of Eq. (9.47), $E_x = -\partial\phi/\partial x$. The beam-generated electrical force is qE_x. From the discussion of Section 5.1, we know that the magnetic force equals the electric force multiplied by $-(v_z/c)^2$. Adding the beam-generated force and the applied force of the quadrupole lenses leads to the following equation of motion in the x direction:

$$\gamma m_0 \frac{d^2x}{dt^2} = \left[\frac{I}{2\pi\varepsilon_0\gamma^2 v_z} \frac{2}{X(z)[X(z) + Y(z)]}\right] - (\gamma m_0 v_z)[\kappa_x g_x(v_z t)]x.$$
(9.48)

The factor $1/\gamma^2$ in the first term on the right-hand side of Eq. (9.48) accounts for the combined electric and magnetic forces generated by the beam. The function $g_x(v_z t)$ gives the periodic variation of transverse applied force as the particle moves through the focusing system. It varies between -1 and $+1$. The quantity κ in Eq. (9.48) is the standard strength parameter for quadrupole lenses [CPA, Section 6.10]. For magnetic lenses, κ equals

$$\kappa = qB_0/\gamma m_0 a \beta_z c. \tag{9.49}$$

The quantity a is the distance from the axis to the closest point on a pole face and B_0 is the magnetic field at the point. For an electrostatic quadrupole with pole-tip field E_0, the strength parameter is

$$\kappa = qE_0/\gamma m_0 a(\beta_z c)^2. \tag{9.50}$$

We can apply the methods of Section 9.1 to derive envelope equations using the information in Eq. (9.48). We evaluate applied and beam-generated forces at the beam boundary, convert time derivatives to axial derivatives, and include an emittance force term to represent the effect of particle velocity spread. The result is the set of KV equations

$$\frac{d^2 X}{dz^2} = -\kappa_x g_x(z) X + \frac{2K}{X+Y} + \frac{\varepsilon_x^2}{X^3} = 0 \tag{9.51}$$

and

$$\frac{d^2 Y}{dz^2} = -\kappa_y g_y(z) Y + \frac{2K}{X+Y} + \frac{\varepsilon_y^2}{Y^3} = 0. \tag{9.52}$$

The quantity K is the generalized perveance [Eq. (5.88)]. Equations (9.51) and (9.52) have wide utility in accelerator theory. Their solution gives self-consistent predictions of the axial variation of envelope widths. Note that envelope oscillations in the x and y directions are coupled through the beam-generated forces.

We can easily solve Eqs. (9.51) and (9.52) numerically for choices of K, κ_x, κ_y, $g_x(z)$, and $g_y(z)$ combined with initial conditions, $X(0)$, $X'(0)$, $Y(0)$, $Y'(0)$. The KV equations have an important application to the design of storage rings. In these devices, it is imperative to avoid orbital resonance instabilities [CPA Sections 7.3 and 8.7] that cause particle loss. A damaging instability occurs when particle orbits have an integer number of betatron oscillations in one revolution around the ring. With no beam-generated forces, we can write the condition for an integer resonance as

$$\mu_0 = 2\pi M/N. \tag{9.53}$$

Following the discussion of Section 1.3, the quantity μ_0 is the vacuum phase advance for an orbit in a quadrupole focusing cell. In Eq. (9.53), N is the number of focusing cells around the ring and M is an integer equal to the number of betatron oscillations. Normally, an accelerator operator adjusts the strength of focusing quadrupoles to avoid the condition of Eq. (9.53). The machine setting for a choice of μ_0 is often called the *vacuum tune*.

An inspection of Eq. (9.48) shows that beam-generated forces change the wavelength for single-particle betatron oscillations. This change may be significant in storage rings, since the goal is to contain the maximum number of particles. All forces in Eq. (9.48) are linear; we know that the displacement of a particle at boundaries between the quadrupole cells follows an equation of the form

$$X_n = A\cos(n\mu + \phi). \tag{9.54}$$

The quantity μ is the phase advance per cell with combined applied and beam forces. The beam-generated forces point away from the axis. Since they counteract the applied focusing forces, the condition $\mu \leqslant \mu_0$ always holds. The quantity $\mu_0 - \mu$ is called the *space-charge tune shift* or the *space-charge tune depression*.

Suppose we have a storage ring where the vacuum phase advance does not satisfy Eq. (9.53). Initially, the current is low so that $\mu = \mu_0$ and particle orbits are stable. As more current enters the ring, the phase advance changes from the vacuum value. With enough current, the shifted phase advance may meet the condition for an instability:

$$\mu = 2\pi M/N. \tag{9.55}$$

To avoid orbital resonances, the machine tune must change as the trapped current in the storage ring rises.

We can clarify the application of the KV equations to a storage ring by reviewing possible steps of a study of beam dynamics at a particular value of current. First, we solve the nonlinear envelope equations numerically for a choice of beam energy (γ), current (I), emittance ($\varepsilon_x, \varepsilon_y$), focusing system periodicity [$g_x(z)$, $g_y(z)$], and lens strength (κ_x, κ_y). If the quadrupole lenses around the ring are not uniform, we can treat κ_x and κ_y as functions of z. The calculation predicts the beam envelope, $X(z)$ and $Y(z)$, or the beta functions, $\beta_x(z)$ and $\beta_y(z)$. Usually, we want certain values for the β functions at different locations in the storage ring. For example, the functions should be small at the interaction region of colliding beams. We adjust the forms of κ_x and κ_y and repeat the envelope calculation until the β functions have the desired variation. We seek a *stationary solution* where the beta functions return to their initial values after a transit around the ring. For known $X(z)$ and $Y(z)$, we can solve Eq. (9.48) and a similar equation in the y direction to find the single-particle orbits and the phase advances $\mu_x(z)$ and $\mu_y(z)$. We sum the phase advances over the focusing cells of the ring to find the number of betatron wavelengths per revolution. If the results show a resonance, we must repeat the calculation for a different machine tune.

9.4. LIMITING CURRENT FOR PARAXIAL BEAMS

In this section, we shall derive expressions for the maximum current that can be transported in accelerators. We shall emphasize paraxial beams in periodic focusing systems. Here, the requirement for transverse force balance on the beam envelope sets an upper limit on the magnitude of beam-generated forces. In periodic systems, the applied focusing forces cannot be so large that they cause single-particle orbit instabilities.

To illustrate the method, we begin with a simple calculation of transverse equilibrium in a continuous focusing force. We take a cylindrical, paraxial electron beam in a uniform solenoid field. The beam has high kinetic energy— longitudinal space-charge effects (Section 5.3) are not significant. All electrons have about the same energy—the relativistic factors β and γ represent average properties. We find a condition for radial force balance on the beam envelope by setting $R'' = 0$ in Eq. (9.25):

$$K = (qB_0/2\beta\gamma m_0 c)^2 R^2 + \varepsilon^2/R^2. \tag{9.56}$$

For given values of the emittance and the focusing field magnitude, Eq. (9.56) gives a value for the generalized perveance. Knowing K, we can find the matched beam current I.

We can write Eq. (9.56) in an alternative form by introducing the focusing system acceptance, α. Following Section 3.7, the acceptance equals the allowed beam emittance for a given envelope radius when there are no beam-generated forces. Setting $K = 0$ and $\varepsilon = \alpha$ in Eq. (9.56) gives

$$\alpha^2 = (qB_0/2\beta\gamma m_0 c)^2 R^4. \tag{9.57}$$

Substituting Eq. (9.57) into Eq. (9.56) and expanding the expression for the generalized perveance gives an equation for the matched beam current:

$$I = \left[\frac{\pi\varepsilon_0 ec}{2m_0}\right](\beta\gamma)(B_0 R)^2 \left[1 - \frac{\varepsilon^2}{\alpha^2}\right]. \tag{9.58}$$

The last term on the right-hand side of Eq. (9.58) is less than or equal to unity. If $\varepsilon = \alpha$, beam defocusing arises from emittance only. Then, the current must equal zero if the beam is in equilibrium. If there is no emittance, the beam-generated forces exactly balance the focusing force of the axial magnetic field. Here, particle flow is laminar and the allowed current has a maximum value.

Equation (9.58) shows why solenoidal focusing is ineffective for ions. Ions have much higher mass than electrons and the β factor is smaller for equal kinetic energies. In practical units, Eq. (9.58) gives the following current limit for electrons:

$$I \leqslant (7.33 \times 10^8)(\beta\gamma)(B_0 R)^2 [1 - (\varepsilon^2/\alpha^2)]. \tag{9.59}$$

The quantities in Eq. (9.59) have the units: I (amperes), B_0 (tesla), R (meters). The limit on power (in watts) for a beam in a solenoidal field is

$$P \leqslant (3.75 \times 10^{14})[\beta\gamma(\gamma - 1)](B_0 R)^2[1 - (\varepsilon^2/\alpha^2)]. \qquad (9.60)$$

The allowed beam power rises rapidly with increasing γ. As an example, suppose we inject a 2.5-MeV electron beam into an induction accelerator with solenoidal focusing. We want to find the minimum magnetic field to contain a 10-kA beam with radius $R = 0.02$m. For a laminar beam, Eq. (9.59) implies that $B_0 \times 7.6 \times 10^{-2}$ tesla. The field must be higher if the beam has significant emittance. It is important to note that when the focusing force is continuous, transverse equilibrium sets no limit on the beam current. Technology rather than beam physics determines the maximum value of B_0. In contrast, periodic systems have definite limits set by the properties of particle orbits.

To initiate the study of periodic focusing systems, we shall consider an array of solenoidal lenses (Fig. 9.5). The field direction reverses in alternate lenses. If the length of the coils is comparable to their radius, the on-axis magnetic field has variation

$$B_z(0, z) \cong B_0 \sin (\pi z/l). \qquad (9.61)$$

The quantity B_0 is the peak axial magnetic field and l is the length of a single coil. The variation of Eq. (9.61) also represents the fields of periodic-permanent-magnet arrays (Section 10.9). Even though the field polarity reverses, we identify a

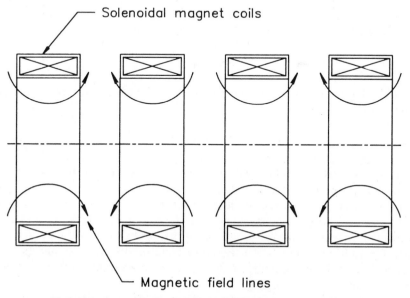

Figure 9.5 Array of solenoidal lenses with alternating field polarity.

focusing cell as a single magnetic lens. This is because the focusing force of a solenoidal lens is independent of the field direction—the radial forces of all lenses are identical.

To estimate the limiting current for the solenoid lens array, we insert Eq. (9.61) into the paraxial ray equation and take an axial average of the focusing force. The averaging process is valid when envelope oscillations are small—then, we can replace the periodic focusing force with an equivalent continuous force. The result gives a good approximation if the beam envelope oscillations are much smaller than R. The limiting current is

$$I \leqslant (7.33 \times 10^8)(\beta\gamma)(B_0 R)^2 \langle \sin^2(\pi z/l) \rangle [1 - (\varepsilon^2/\alpha^2)]$$
$$= (3.67 \times 10^8)(\beta\gamma)(B_0 R)^2 [1 - (\varepsilon^2/\alpha^2)]. \tag{9.62}$$

Equation (9.62) is a necessary condition for successful beam transport, but it is not sufficient. In the periodic system, we must be certain that the parameters in the equation do not correspond to unstable particle orbits—the vacuum phase advance must be in the range $\mu_0 \leqslant \pi$. Comparison to Eq. (9.38) shows that the vacuum phase advance in a solenoidal lens array is given by

$$(\mu_0/l)^2 R \cong (qB_0/2\beta\gamma m_0 c)^2 R \tag{9.63}$$

or

$$\mu_0 \cong eB_0 l/2\beta\gamma m_0 c \leqslant \pi. \tag{9.64}$$

Depending on the system parameters, either Eq. (9.62) or Equation (9.64) may provide the most stringent limit. For example, if the field magnitude is the main limiting factor and the phase advance is always small, Eq. (9.62) implies that electron beam current scales as

$$I \sim (\beta\gamma)(B_0 R)^2. \tag{9.65}$$

On the other hand, if phase advance is the main limiting factor, then Eq. (9.64) implies that the current scales as

$$I \sim (\beta c)^3 R^2. \tag{9.66}$$

The lesson is that we can find widely different scaling laws for the same process, depending on which factors we hold constant.

As an example of current limits in a PPM array, suppose we have a low-emittance, 20-keV electron beam ($\gamma = 1.04$, $\beta = 0.272$). The characteristics of the permanent magnets limit the properties of the focal system. We take $B_0 = 0.05$ tesla, $R = 0.005$ m, and $l = 0.02$ m. Equation (9.62) predicts a maximum current of $I \leqslant 26$ A for a laminar beam; Eq. (9.64) predicts an acceptable value of vacuum phase advance, $\mu_0 \cong 1.04$ (59°).

We can derive an alternative form of Eq. (9.62) for periodic systems—the modified equation shows the dependence on the vacuum phase advance explicitly. The expression involves the net particle orbit phase advance, μ. Let F_f be the applied focusing force at the beam envelope averaged over a cell. The quantity F_d is the defocusing force from beam-generated electric and magnetic fields. According to Eq. (9.38), the vacuum phase advance for a single-particle orbit is related to the magnitude of the applied focusing force by

$$\mu_0^2 \sim F_f. \tag{9.67}$$

Similarly, the single-particle phase advance in the presence of beam-generated forces is proportional to the total focusing force:

$$\mu^2 \sim (F_f - F_d). \tag{9.68}$$

Taking the ratio of Eqs. (9.67) and (9.68) gives

$$\frac{\mu}{\mu_0} = \left[1 - \frac{F_d}{F_f}\right]^{1/2}. \tag{9.69}$$

Equation (9.69) shows that if the beam-generated force equals the focusing force, $F_d = F_f$, then $\mu = 0$. Here, particles follow laminar orbits. When $\mu = 0$, the beam must have zero emittance for radial force balance on the envelope. Conversely, emittance dominates the equilibrium when the beam-generated forces are zero, or $\mu = \mu_0$.

The condition for radial force balance on the beam envelope is

$$F_f - F_d - \varepsilon^2/R^3 = 0. \tag{9.70}$$

From the discussion at the beginning of this section, the acceptance equals the allowed emittance when there are no beam-generated forces. The acceptance satisfies the equation

$$F_f - \alpha^2/R^3 = 0. \tag{9.71}$$

Combining Eqs. (9.70) and (9.71) and comparing the result to Eq. (9.69), we find that:

$$(\mu/\mu_0) = (\varepsilon/\alpha). \tag{9.72}$$

We apply Eq. (9.72) to write a general expression for radial force balance in a periodic focusing system:

$$K \leqslant (R/L)^2(\mu_0^2 - \mu^2). \tag{9.73}$$

The quantity L is the length of a focusing cell and R is the maximum allowed envelope radius. Note that Eq. (9.37) holds when the focusing force acts almost continuously ($\mu_0 < 1$).

We can derive a similar equation for current limits in a quadrupole array. We can simplify the KV equations [eqs. (9.51) and (9.52)] for a matched beam with the following assumptions:

1. Envelope oscillations are small compared with the beam radius.
2. The quadrupole lenses have equal focusing properties in the x and y directions, $\kappa = \kappa_x = \kappa_y$, $g_x(z) = -g_y(z)$.
3. The beam emittances are equal in the x and y directions, $\varepsilon = \varepsilon_x = \varepsilon_y$.

Taking $X \cong Y$, the condition for transverse force balance on the envelope is

$$-X(\mu_0/L)^2 + K/X + \varepsilon^2/X^3 = 0. \tag{9.74}$$

We can write Eq. (9.74) in the form

$$K = \left[\frac{\mu_0 X}{L}\right]^2 \left[1 - \left(\frac{\mu}{\mu_0}\right)^2\right]. \tag{9.75}$$

The vacuum phase advance in a quadrupole lens array depends on the geometry of the lenses and the applied field strength. Usually, the lens length is much larger than the bore width. Then, we can neglect fringing fields and take $g_x(z)$ as a step function. To begin, we shall concentrate on an FD focusing cell consisting of two identical lenses with an angular offset of $90°$. We can calculate the vacuum phase advance using transfer matrix theory [CPA, Chapter 8]. The result for the x direction is

$$\cos \mu_0 = \cos(\sqrt{\kappa}l)\cosh(\sqrt{\kappa}l), \tag{9.76}$$

where κ is the quadrupole strength parameter of Eq. (9.49) or (9.50). The quantity l is the length of a single lens. For the FD geometry, the lens length equals half the focusing cell length, $l = L/2$. When $\mu_0 < 1$ and $\sqrt{\kappa}l < 1$, we can expand the trigonometric and hyperbolic functions of Eq. (9.76) in a power series to give

$$1 - \mu_0^2/2 + \cdots \cong [1 - \kappa l^2/2 + \kappa^2 l^4/24 - \cdots][1 + \kappa l^2/2 + \kappa^2 l^4/24 + \cdots]. \tag{9.77}$$

Solving Eq. (9.77), we find the vacuum phase advance:

$$\mu_0 \cong \kappa l^2/\sqrt{3}. \tag{9.78}$$

Combining Eqs. (9.75) and (9.78) leads to the following expression for the limiting current for ions in a quadrupole transport channel with magnetic lenses:

$$I_{FD} = \left[\frac{\pi\varepsilon_0 ec}{6m_p}\right]\left[\frac{Z}{A}(lB_0^2)\left(\frac{X}{a}\right)^2(\gamma\beta)\right]\left[1 - \left(\frac{\mu}{\mu_0}\right)^2\right] \tag{9.79}$$

The first term on the right-hand side of Eq. (9.79) equals 1.33×10^5. In the second term, Z is the ionization state of the ions, A is the ratio of the ion mass to that of the proton, l is lens length, and B_0 is the pole tip magnetic field.

As an application of Eq. (9.79), consider a beam in a high-current ion induction linac. A 5-MeV C^+ beam ($Z = 1$, $A = 12$, $\gamma \cong 1$, $\beta = 0.03$) fills half the available bore ($X/a = 0.5$) of an array of magnetic quadrupole lenses. The cell length is $L = 0.3$ m and the magnetic field is $B_0 = 1$ tesla. Inserting parameters into Eq. (9.79) gives a beam current of 5.6 A and power of 2.8×10^6 W. For conservative transport, we take $\mu_0 \leqslant 1$. Equation (9.78) implies that $\sqrt{\kappa l} \leqslant 1.3$. Substitution in Eq. (9.49) shows that the quadrupole bore radius should be in the range $a \geqslant 0.02$ cm.

The limiting current expression for a FODO channel is similar to Eq. (9.79). Suppose that the channel consists of identical F and D lenses of lengh l separated by a drift distance D. We can write the current limit as the product of the FD channel expression and a geometric correction factor:

$$I_{\text{FODO}} = I_{\text{FD}} g(D/l) \tag{9.80}$$

For small phase advance, the correction factor is[1]

$$g\left(\frac{D}{l}\right) \cong \frac{[1 + 4(D/l) + 3(D/l)^2]^{1/2}[1 + 3(D/l)]^{1/2}}{(1 + D/l)^{3/2}}. \tag{9.81}$$

Figure 9.6 shows a plot of $g(D/l)$.

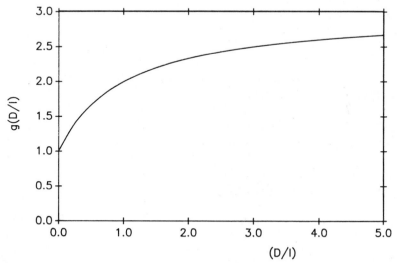

Figure 9.6 Correction factor for limiting current in quadrupole lens arrays, *FODO* versus *FD* configurations. $g(D/l) = I_{FODO}/I_{FD}$, lens length is l, drift distance between lenses is D. (Adapted from M. Reiser, *Part. Accel.* **8**, 167 (1978).)

[1]M. Reiser, *Part. Accel.* **8**, 167 (1978).

To estimate current limits in quadrupole channels, we must have criteria to choose μ_0 and μ in Eq. (9.79). The vacuum phase advance cannot exceed 180°. If we increase the lens strength so that $\mu_0 > \pi$, a beam with strong space-charge fields is subject to a destructive envelope instability (Section 13.1). For paraxial beams with low emittance, beam simulations and experiments have shown that periodic focusing forces can excite coherent oscillations in the beam when $\mu_0 \geqslant 90°$.[2] These oscillations result in the growth of transverse emittance. Most studies show that beams are stable when $\mu_0 < 90°$. When the vacuum phase advance is low, experiments[3] show that uniform-density beams propagate with no emittance growth, even when the net phase advance approaches zero, $\mu \to 0$.

Space-charge-dominated beams with nonuniform density suffer emittance growth in a linear focusing system. To see the origin of this effect, imagine that we inject a beam with a nonuniform density into a focusing channel with linear applied forces. A laminar beam cannot achieve radial force balance over the entire cross section. For the conditions givens, the emittance force is zero, the focusing force is linear, but the beam-generated force is nonlinear. Because of local force imbalance, parts of the beam accelerate in the transverse direction. The transverse velocity components of particles mix after several betatron oscillations. The additional transverse pressure force gives detailed force balance over the beam width at the expense of increased emittance. Therefore, we expect that a laminar equilibrium is impossible for a nonuniform beam.

Computer simulations and analytic theory lead to an equation for emittance growth in a nonuniform, space-charge-dominated beam.[4] The following form holds for nonrelativistic ion beams:

$$\frac{\varepsilon_f}{\varepsilon_i} = \left[1 + \frac{U - U_0}{2U_0}\left(\frac{\mu_0^2}{\mu^2} - 1 \right) \right]. \qquad (9.82)$$

In Eq. (9.82), ε_i is the RMS emittance at injection, μ_0 is the single-particle vacuum phase advance, μ is the approximate phase advance with space-charge effects, and ε_f is final emittance. Two new quantities appear in Eq. (9.82), U_0 and U. The quantity U is the electrostatic energy per unit length of beam-generated electric fields. The field energy equals the integral of $\varepsilon_0 E^2/2$ over the beam cross section. The quantity U_0 is the electrostatic field energy of a uniform-density beam with the same current and envelope radius. We see in Eq. (9.82) that there is no emittance growth if the injected beam has uniform density because $U = U_0$. Usually, the beams emerging from injectors and preaccelerators are nonuniform. As an example, the field energy factor for a Gaussian beam is $(U - U_0)/U_0 = 0.31$.

[2] I. Hoffman, L. J. Laslett, L. Smith, and I. Haber, *Particle Accel.*, **13**, 145 (1983).
[3] M. G. Tiefenback, "Space Charge Limits on the Transport of Ion Beams in a Long Alternating Gradient System" (Lawrence Berkeley Laboratory, LBL-22465, 1986).
[4] T. P. Wangler, K. R. Crandall, R. S. Mills, and M. Reiser, *IEEE Trans. Nucl. Sci.* **NS-32**, 2196 (1985).

9.5. MULTIPLE-BEAM ION TRANSPORT

One strategy to increase the limiting current in a high-flux ion accelerator is to divide a beam into many segments, each with its own focusing system. If the segments are isolated from one another, both the longitudinal and transverse electric fields are smaller. Then, the accelerator can contain higher net current at the expense of more complex transport hardware. In the early stages of an ion accelerator where longitudinal limits (Section 5.3) are important, multiple-beam systems have a clear advantage. Equation (5.68) shows that the longitudinal current limit at a given energy depends on the beam current and the ratio of wall to beam radius. If the space-charge potential is the main problem, then a system that divides a beam in N smaller beams can transport N times higher current.

Multiple-beam systems are important for neutral-particle injectors for magnetic fusion and electrostatic ion propulsion devices. Figure 9.7a shows a common geometry—ions are extracted from multiple slots and accelerate through one or more gaps. If individual sheet beams are spaced far enough apart, induced charges in the electrodes prevent the electric field of one beam from interfering with another. The individual beams have low perveance, allowing the application of conventional ion extraction techniques. Multiple-beam transport is not very useful for relativistic electrons. Although the geometry reduces the electric fields of such beams, the magnetic field in acceleration gaps is usually unaffected. Therefore, the net current of multiple electron beams must be less than the Alfven limit (Section 12.7).

Recently, there has been considerable interest in multiple-beam transport of high-energy ions for application to ion implantation and inertial fusion; the idea is to use parallel channels of electrostatic quadrupole lenses (Fig. 9.7b). In comparision with magnetic lenses, electrostatic quadrupole focusing has two advantages for high-current ion beam transport

1. Electric fields deflect nonrelativistic ions more effectively than magnetic fields.
2. Miniature magnetic quadrupole lenses are difficult to fabricate and to operate because of cooling problems.

We shall use material from several previous sections to quantify the advantages of multiple-ion-beam transport. The derivations also illustrate some techniques and precautions for scaling studies.

We shall calculate the current limited by transverse force balance in both single-beam and multiple-beam electrostatic quadrupole arrays. The single beam carries current I—the multiple-beam system has N channels, each with current i. We want to find the relationship between Ni and I. We are interested only in transverse beam confinement, so we assume that the kinetic energy of ions is always well above the electrostatic potential energy. To make a fair comparison, we must define common properties of the systems:

1. The beams consist of ions with identical rest mass, m_i, and kinetic energy.
2. The electrostatic quadrupoles have the same value of pole tip electric field, E_0.
3. Space-charge fields are more important than emittance in the transverse force balance—we shall take $\mu \cong 0$ for the single and multiple beams.
4. The multiple beams and the focusing electrodes occupy the same transverse area as the single-beam transport system.

We shall start with the properties of the single beam. The maximum available radius is R_B and R is the radius of the single beam. The fill factor, F, is the fraction of the total cross section available for beam transport. Quadrupole electrodes, insulators, and voltage leads occupy the rest of the area. The radius of the available transport region equals A, the distance from the axis to the tip of a quadrupole electrode. By the definition of the fill fraction,

$$A = R_B/(F)^{1/2}. \tag{9.83}$$

We must include a safety factor so that peripheral ions in the beam do not strike the electrodes. We define another quantity, ζ, as the ratio of the beam radius to the radius of the available transport area,

$$R = \zeta A. \tag{9.84}$$

We assume that F and ζ are the same for both the large and small beams. Since both systems have the same cross-section area, the radius of one of the multiple beams, r, is

$$r = R/N^{1/2}. \tag{9.85}$$

The quantity L is the length of a quadrupole focusing cell for the single-beam system, while l is the cell length for multiple beams. From Section 9.4, we know that the generalized perveance of the laminar single beam is related to the focusing cell length by

$$K = \left(\frac{\mu_0 R}{L}\right)^2. \tag{9.86}$$

The quantity μ_0 is the vacuum phase advance per focusing cell. The generalized

---

Figure 9.7 High current transport of multiple ion beams. (*a*) Side-view of a multiaperture electrostatic extractor. The electrode openings have cylindrical symmetry about each small beam. (*b*) Endview of a transverse array of multiple electrostatic quadrupole lenses. The figure shows electrode polarities and beam shapes at the center of a lens set.

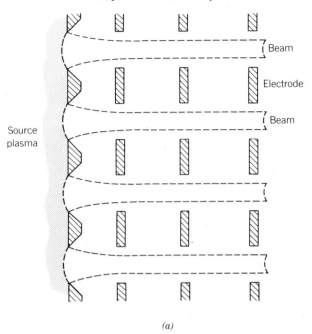

(a)

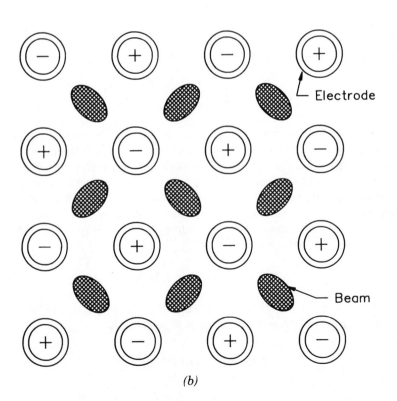

(b)

perveance for one of the multiple beams, k, satisfies the equation

$$k = \left(\frac{\mu_0 r}{l}\right)^2.$$ (9.87)

We assume that both focusing systems have the same value of μ_0.

From the discussion of Section 9.4, we can write the vacuum phase advance for the single beam in terms of the properties of the electrostatic quadrupole lens by

$$\mu_0 = \kappa_e L^2/4(3)^{1/2}.$$ (9.88)

The quantity κ_e is the electrostatic lens strength constant [Eq. (9.50)]:

$$\kappa_e = eE_0/\gamma m_0(\beta c)^2(R/\xi).$$ (9.89)

Combining Eqs. (9.86), (9.88), and (9.89) gives the following scaling expression for the generalized perveance:

$$K = \left(\frac{eE_0\zeta}{\gamma m_0\beta^2 c^2}\right)^2 \frac{L^2}{48}.$$ (9.90)

Following the same reasoning, we find the generalized perveance for one of the multiple beams:

$$k = \left(\frac{eE_0\zeta}{\gamma m_0\beta^2 c^2}\right)^2 \frac{l^2}{48}.$$ (9.91)

Equation (9.90) and (9.91) are relationships between generalized perveance and the lengths of the focusing cells. The cell lengths are not independent of other beam properties—they must be short enough so that envelope oscillations are small. To find constraints on L and l, we assume that the fraction change in the envelope width is the same for both the single beam and the multiple beams. For a rough estimate of the fractional envelope variation, we represent the beams as cylinders of radius R and r that expand in free space over lengths $L/2$ and $l/2$. In the expansions, we require that

$$(R + \Delta R)/R = (r + \Delta r)/r.$$ (9.92)

Section 5.4 gives the distance for expansion of a cylindrical beam with generalized perveance K from a radius of R to $R + \Delta R$:

$$L = RF[(R + \Delta R)/R]/(2K)^{1/2}.$$ (9.93)

Table 5.1 lists the function $F[(R + \Delta R)/R]$. For one of the multiple beams, the expansion distance is

$$l = rF[(r + \Delta r)/r]/(2k)^{1/2}. \tag{9.94}$$

The function F has the same value in Eqs. (9.93) and (9.94). Combining Eqs. (9.90), (9.91), (9.93), and (9.94), we find the following relationship for beams in the two focusing systems:

$$K/k = I/i = R/r. \tag{9.95}$$

Substitution from Eq. (9.85) gives the desired scaling relationship:

$$i = I/(N)^{1/2}. \tag{9.96}$$

The total current in the multiple-beam system equals

$$Ni = I(N)^{1/2}. \tag{9.97}$$

Equation (9.87) implies that, in principle, a system of parallel electrostatic quadrupole channels can carry higher current than a single channel. The total current scales as the square root of N—a 16-beam system can carry four times as much current as a single beam. The inherent assumptions in the derivation are that the beams are laminar and that the quadrupole lenses have equal electric fields. Also, we should note that Eq. (9.87) depends on the condition that the electrodes of the multiple quadrupole array occupy the same fraction of the cross section as those in the single quadrupole channel. The advantage of multiple-beam transport is reduced if extra voltage leads and supports consume space.

9.6. LONGITUDINAL SPACE-CHARGE LIMITS IN RF ACCELERATORS AND INDUCTION LINACS

Beam-generated axial electric fields can limit the beam current in RF accelerators and induction linacs. In these devices, the beam must be confined in the longitudinal direction. Beam-generated axial forces can be important, even if the electrostatic potential energy of a beam is much smaller than the kinetic energy. We shall focus our attention on nonrelativistic ion beams—the effects of axial fields are small for relativistic particles.

Ions in RF accelerators must remain in specific phase regions of the accelerating wave [CPA, Chapter 13]. The electric field of a traveling wave can provide stable axial confinement for ions that are localized along z and have a small spread in kinetic energy. The wave creates a potential well for ion confinement called an *RF bucket*. Ions that escape from the bucket quickly lose their synchronization with the wave and are no longer accelerated. Space-charge electric fields can drive ions out of an RF bucket. This process set limits on the current in the accelerator.

We can represent the accelerating electric field in any RF accelerator as a traveling wave with the form

$$E_{zw}(z,\ t) = E_0(z) \sin \{\omega[t - \int dz/v_s(z)]\}. \qquad (9.98)$$

For simplicity, we will take E_0 as constant in the z direction. The quantity ω equals $2\pi f$, where f is the RF frequency. The frequency is constant over the length of the accelerator. The quantity v_s is the synchronous velocity, the velocity of an ideal particle at the center of a bunch. The synchronous particle remains at a position of constant phase in the wave. In the following, the quantity ϕ_s is the synchronous phase. The design of the accelerator ensures that the wave phase velocity increases with distance to match the growing velocity of the synchronous particle.

We can easily describe ion motion in a frame of reference that moves at velocity $v_s(z)$. In this frame, the accelerating wave appears to be at rest. We let z' represent distance measured from the zero crossing (positive to negative) of the electric field in the wave rest frame. In terms of z', the force exerted on ions by the

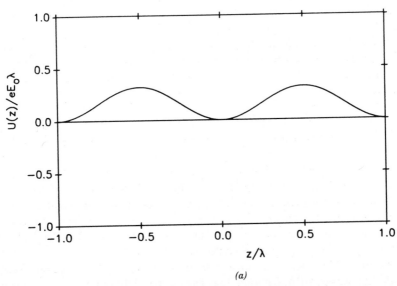

(a)

Figure 9.8 Longitudinal confinement of ions in an RF linac. Synchronous particle phase, $\phi_s = 60°$. (a) Spatial variation of the electrostatic potential energy of ions viewed in the rest frame of a traveling wave. (b) Spatial variation of the the axial force on an ion, viewed in the rest frame of an accelerating traveling wave. (c) Spatial variation of the total potential energy of ions viewed in the rest frame of an accelerating traveling wave.

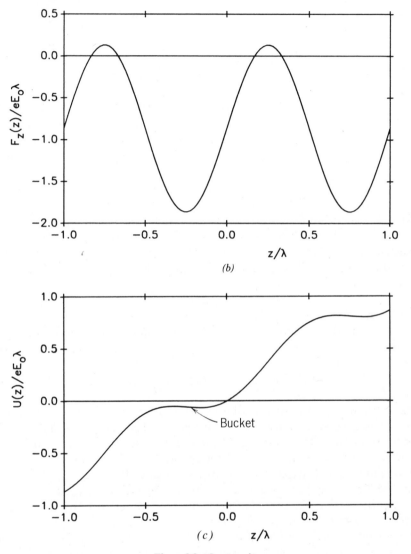

Figure 9.8 (*Continued*)

traveling wave is

$$F_w(z') = eE_{zw}(z') = -eE_0 \sin(2\pi z'/\lambda). \tag{9.99}$$

The quantity λ in Eq. (9.99) is the wavelength of the traveling wave at position z:

$$\lambda = 2\pi v_s/\omega. \tag{9.100}$$

We shall neglect changes in λ that occur as the wave accelerates. Integration of

Eq. (9.99) gives the rest frame electrostatic potential energy associated with the wave field:

$$\varepsilon\phi_w(z') = \frac{eE_0\lambda}{2\pi}\left[1 - \cos\left(\frac{2\pi z'}{\lambda}\right)\right].$$ (9.101)

Figure 9.8a shows the variation of potential.

To calculate the axial force balance for trapped ions, we must remember that the wave rest frame is not inertial. We can represent wave acceleration by introducing a fictitious retarding force in the rest frame:

$$F_i(z') = -d(m_i v_s)/dt.$$ (9.102)

To simplify the model, we assume that the wave acceleration is uniform throughout the accelerator. For the synchronous particle, the retarding force exactly balances acceleration by the electric field—the particle remains at a constant value of z'. The wave force on the synchronous particle is

$$F_{ws} = eE_0 \sin\phi_s = -F_i.$$ (9.103)

Combining Eqs. (9.99) and (9.103), the total applied force in the beam rest frame is

$$F(z') = F_w(z) + F_i = -eE_0[\sin(2\pi z'/\lambda) - \sin(\phi_s)].$$ (9.104)

Figure 9.8b shows the variation of total force with z' for $\phi_s = 60°$. Integration of Eq. (9.104) gives an expression for U_c, the potential energy of ions in the applied forces (Fig. 9.8c). The potential defines a well in the region around the synchronous particle. In this region, the applied forces can counteract the effects of longitudinal velocity spread and space charge to trap a group of ions. The trapped ions constitute a *microbunch*—they remain synchronized to the wave during acceleration. The output of an RF linac consists of a train of microbunches separated in time by an interval $2\pi/\omega$.

The depth of the confining potential well, ΔU_c, is important for the estimation of space-charge limits. The depth equals the spatial integral of the force from the synchronous particle position to the top of the well. The range of integration is

$$\lambda\phi_s/2\pi \leqslant z' \leqslant \lambda(\pi - \phi_s)/2\pi.$$ (9.105)

The resulting well depth is

$$\Delta U_c = \frac{eE_0\lambda}{\pi}\Psi(\phi_s),$$ (9.106)

where

$$\Psi(\phi_s) = [\cos(\phi_s) - \cos(\pi - \phi_s) - \sin(\phi_s)(\pi/2 - \phi_s)]/2.$$ (9.107)

Figure 9.9 shows the function $\Psi(\phi_s)$. It drops rapidly as ϕ_s varies from 0 to $\pi/2$. Although a choice of $\phi_s = 0$ gives good particle confinement, there is no average acceleration of the microbunch [Eq. (9.103)]. Acceleration is strongest at $\phi_s = 90°$, but there is no containment—the number of confined particles in a microbunch approaches zero.

Given a value for ΔU_c, we can estimate the properties of ions in a microbunch. For example, the spread in rest frame kinetic energy at the synchronous particle position must be less than ΔU_c. We can transform the kinetic energy spread to the accelerator frame—the result is the familiar longitudinal acceptance diagram [CPA, Chapter 13]. In this section, we will concentrate on space-charge limits, assuming that the beam has small axial emittance. The beam-generated electric force pushes particles out of the bucket if the peak space-charge electrostatic potential energy, ΔU_e, exceeds the well depth. The condition

$$\Delta U_e = \Delta U_c. \tag{9.108}$$

defines the space-charge limit for ion transport.

We can apply the discussion of Section 6.7 to calculate ΔU_e for an ellipsoidal microbunch in free space. The calculation for arbitrary bunch shapes in the presence of surrounding structures can be complex. We limit consideration to a cylindrical microbunch of ions with radius r_0 and length l in a conducting pipe of radius r_w. The beam density drops to zero at the ends—the bunch length is much

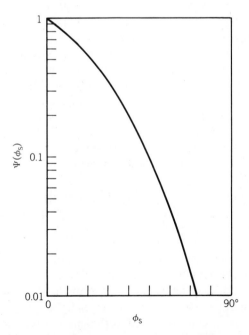

Figure 9.9 RF bucket depth function, $\Psi(\phi_s)$, as a function of the synchronous phase, ϕ_s.

longer than r_w. The density of the nonrelativistic beam is about the same when observed in the rest and accelerator frames.

To find the difference in electrostatic potential energy over the beam length, we calculate the rest frame electrostatic potential, ϕ, at the center and end of the microbunch:

$$\Delta U_e = e[\phi(0) - \phi(l/2)]. \tag{9.109}$$

In Section 5.3, we found that the electrostatic potential of a long cylinder of charge is

$$e\phi(0) = [eI_0/4\varepsilon_0 \beta c][1 + 2\ln(r_w/r_b)]. \tag{9.110}$$

The quantity I_0 is the peak current of the microbunch in the accelerator frame. For an ellipsoidal beam, the electrostatic potential approaches zero at the ends, $\phi(l/2) \sim 0$. The combination of Eqs. (9.106), (9.109), and (9.110) gives a limit for the peak current:

$$I_0 \leqslant \frac{4F(\phi_s)E_0 \lambda \varepsilon_0 \beta c}{\pi[1 + 2\ln(r_w/r_b)]}. \tag{9.111}$$

Equation (9.111) applies to a beam with no longitudinal energy spread. The allowed current is lower for beams with non-zero axial emittance.

As example, suppose we have a proton beam near the entrance of a 200-MHz RF linac. We take the following characteristic parameters: $E_0 = 2.5\,\text{MV/m}$, $r_w/r_b = 4$, $T_i = 2\,\text{MeV}$, $\phi_s = 70°$, and $\beta = 6.5 \times 10^{-2}$. For synchronous acceleration, the RF cavities create a traveling wave with phase velocity βc and wavelength $\lambda = \beta c/(2 \times 10^8) = 0.0972\,\text{m}$. Figure 9.9 shows that the bucket depth function is $F(70°) = 1.5 \times 10^{-2}$. Substitution in Eq. (9.111) gives a peak microbunch current of $I_0 = 0.21$ A. High-power proton linacs often require larger values of I_0. One strategy to raise the trapped charge is to run the initial stages of the accelerator at reduced ϕ_s, increasing the depth of the bucket. The relative importance of space-charge forces decreases as the ions accelerate. Therefore, the synchronous phase can be raised in downstream cavities to raise the average accelerator gradient.

The effects of beam-generated axial fields decrease with acceleration because the axial length of the bucket grows. At constant RF frequency, the wavelength of the traveling wave is proportional to βc, the beam velocity. If the accelerating gradient, E_0, remains constant, Eq. (9.106) shows that the bucket depth grows with β. Relativistic effects also help axial confinement. To carry out a relativistically correct calculation, we must be careful when comparing quantities in the accelerator and wave rest frame. For example, the space-charge density observed in the rest frame is a factor of γ lower than the density in the accelerator frame because of Lorentz contraction. Also, the length of the bucket is a factor of γ higher in the rest frame. The amplitude of the accelerating field, E_0, is the same in both frames. The result is that the peak microbunch current, I_0, is higher by a factor of γ^2 for relativistic beams. Longitudinal space-charge effects are negligible in RF linacs for electrons.

We can apply a similar approach to the longitudinal confinement of nonrelativistic ions in induction linacs. In this accelerator, a beam passes through a set of acceleration gaps with pulsed voltage. The pulse length may range from 0.05 to 5 μsec. The synchronization condition is simple—the beam must cross each gap during the applied voltage pulse and the beam pulse length must be less than or equal to the voltage pulse length.

If the voltage waveforms in the acceleration gaps are square pulses, the gaps provide no longitudinal confinement for ions. The time for a beam to pass through a gap increases because of the effects of longitudinal velocity spread and space-charge forces. We can contain a beam bunch with shaped voltage pulses, such as the ramp waveform of Fig. 9.10a. An ion that crosses the gap late in the pulse experiences an increased accelerating voltage. The ion emerges with enhanced velocity and overtakes the other ions in the bunch. Conversely, ions that arrive early are slowed.

We can describe confinement of a nonrelativistic beam by shaped voltage pulses if we adopt some simplifying assumptions. The voltage waveforms in all gaps follow the linear ramp of Fig. 9.10a. We take the gap voltage as a given function of time, neglecting loading of the waveform by the beam current. Since we are concerned mainly with longitudinal confinement, we shall neglect average acceleration of the ion bunch. We remove the constant part of the gap voltage, giving a voltage waveform

$$\Delta V(t) = \Delta V t / \Delta t_{\mathrm{p}}, \tag{9.112}$$

for $-\Delta t_{\mathrm{p}}/2 \leqslant t \leqslant \Delta t_{\mathrm{p}}/2$. The length of the beam bunch is

$$L_{\mathrm{b}} = \beta c \Delta t_{\mathrm{p}}. \tag{9.113}$$

We take L_{g} as the distance between acceleration gaps while d is the width of the gaps. Finally, we define ζ as the position of a particle in the bunch with reference to the synchronous particle. The dashed line in Fig. 9.10a shows the confining component of the accelerating voltage as a function of ζ.

Ions oscillate in ζ as the beam propagates. Assume that the beam passes through many gaps in the period of a longitudinal oscillation. With this condition, we can represent the gap electric fields by a continuous, time-averaged function. The time-averaged axial electric force is

$$eE_z(\zeta) = (e\Delta V/d)(d/L_{\mathrm{g}})(\zeta/L_{\mathrm{b}}), \tag{9.114}$$

in the region $-L_{\mathrm{b}}/2 \leqslant \zeta \leqslant L_{\mathrm{b}}/2$. Integrating Eq. (9.114) gives the confining potential

$$U_{\mathrm{c}}(\zeta) = (e\Delta V)[\zeta^2/(2L_{\mathrm{b}}L_{\mathrm{g}})]. \tag{9.115}$$

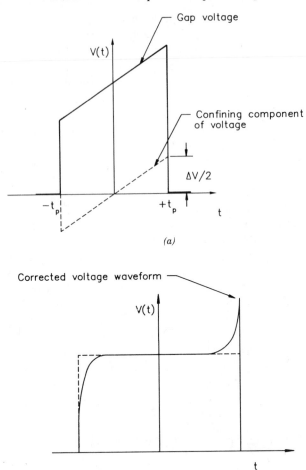

Figure 9.10   Voltage waveforms in induction linear accelerators for ions. (*a*) Definition of parameters of a ramped accelerating voltage waveform. The dashed line shows the time-varying part used for the drifting beam analysis. (*b*) Possible waveform to confine ion beams with a nonlinear distribution of axial space charge force.

The depth of the confining potential well is

$$\Delta U_c = e\Delta V(L_b/8L_g). \tag{9.116}$$

For a beam with no longitudinal velocity spread, we derive the space-charge limit by setting the energy associated with the space-charge electrostatic potential equal to the depth of the confining potential. We shall neglect acceleration of the beam. If we again take a long ellipsoid of charge in a conducting pipe, Eq. (9.110) gives an estimate for ΔU_e in terms of the peak current in beam bunch, I_0. The current limit for longitudinal space-charge effects is

$$I_0 \leqslant \frac{\Delta V \varepsilon_0 \beta c}{2[1 + 2 \ln (r_w/r_b)]} \left(\frac{L_b}{L_g} \right). \qquad (9.117)$$

As application example, consider the following parameters for the high-energy portion of an inertial fusion induction accelerator. The beam consists of a 100-nsec pulse of Xe^{131} ions with a kinetic energy of 5 GeV and peak current $I_0 = 1$ kA. The distance between acceleration gaps is $L_g = 0.7$ m and $r_w/r_b = 4$. The ions have a velocity of $\beta c = 0.28c$, giving a bunch length of $L_b = 8.3$ m. The ion bunch extends over many acceleration gaps. Inserting quantities into Eq. (9.117) shows that we must superimpose a large voltage ramp with $\Delta V = 0.88$ MV on the accelerating voltage in each gap to maintain a constant beam length. The voltage waveform in an actual machine may differ considerably from the simple ramp of Fig. 9.10a. The axial space-charge forces of a long beam bunch in a conducting pipe vary nonlinearly with axial position. The confining forces must be concentrated near the ends of the bunch for axial force balance at all positions. Figure 9.10b illustrates a waveform to counteract nonlinear space-charge forces.

10

High-Current Electron Beam Transport under Vacuum

In this chapter we shall study the vacuum transport of electron beams when the effects of beam-generated fields are strong. Much of the original theory was developed for application to microwave devices such as high-power traveling-wave tubes. These tubes use low-energy, high-perveance electron beams in the range of 1–10 A. Recent applications include the transport of relativistic beams in the range of $> 1\,\text{kA}$ from pulsed electron diodes and linear induction accelerators.

Focusing by solenoidal magnetic fields is the best method for high-current electron beams at low to moderate energy. While the forces in quadrupole arrays are alternately outward and inward, the forces from solenoid lenses continually focus a beam. Quadrupole lenses are not useful at low energy because high-perveance beams expand too much in the defocusing lenses.

Four sections of this chapter review background material that is useful for a wide range of electron beam applications. Section 10.1 describes the motion of single electrons entering and exiting a solenoidal magnetic field. Section 10.7 discusses the transverse drift motion of single particles and beams in solenoidal fields. In a magnetic field, we shall find that transverse perturbation forces do not cause an acceleration in the direction of the force. Instead, the beam follows a constant-velocity drift orbit perpendicular to the force. Drift motions can lead to beam loss. Section 10.4 reviews the interaction of electrons with solid matter. Electrons lose kinetic energy from collisions with atomic electrons. They also suffer angular deflections by collisions with electrons and nuclei in the medium. We shall review equations for energy loss and scattering that have several applications in accelerator technology. We shall apply the results to study the feasibility of electron beam guiding by material structures. Section 10.6 derives equations for the induced charge and return current in metal vacuum-chamber walls surrounding a pulsed beam. Although the wall forces can help to steer a beam, they can also drive instabilities. We shall study resistive wall instabilities in Sections 13.6 and 14.4.

Sections 10.2, 10.3, and 10.5 concentrate on methods to focus high-current beams. Section 10.2 covers focusing by a uniform solenoid field when the electron source is located inside the field region. Section 10.3 describes solenoid focusing with an external cathode. The electrons enter the field through a transition region. For this type of injection, we can achieve a matched laminar beam equilibrium with no envelope oscillations. Section 10.5 reviews the use of metal foils or meshes to focus high-current electron beams. The induced charge on the meshes partially cancels the beam-generated electric field. As a result, relativistic beams can propagate in a self-pinched equilibrium. The method is feasible because of the low stopping power of electrons in matter. A mesh focusing transport system can contain very intense electron beams with current approaching 1 MA.

Sections 10.8 and 10.9 discuss methods to steer electron beams. This topic is important for high-power circular accelerators such as recirculating induction accelerators and high-current betatrons. Section 10.8 describes the transverse motion of beams in a toroidal magnetic field in response to centrifugal force. We shall see that a uniform axial field provides no centering force for the beam— small transverse force errors result in a drift to the walls. Section 10.9 reviews focusing and steering of intense electron beams by a periodic magnetic cusp field. This field pattern results from arrays of permanent magnets or conventional solenoid lenses with alternating polarity. The field geometry provides a centering force that maintains a beam on the system axis.

10.1. MOTION OF ELECTRONS THROUGH A MAGNETIC CUSP

Solenoid magnets are often applied to high-current electron beam transport. Sometimes, the beam remains within the magnetic field throughout the acceleration and transport processes. More commonly, the electron source is outside the magnet. The beam enters the solenoidal field for transport in the accelerator and leaves the field before traveling to its final destination. For this arrangement, it is essential to understand the motion of electrons in the transition region between a solenoidal field and free space. This section summarizes the motion of single electrons at a field boundary without beam-generated forces. We shall include the effects of beam fields in Sections 10.2 and 10.3.

Figure 10.1a shows an infinite-length solenoid. An azimuthal current sheet creates an axial magnetic field. The current sheet usually consists of a helical winding of wire with many turns. The magnetic field inside the winding is uniform in space, $B_z(r, z) = B_0 z$. If the coil has N turns per meter and carries current I, then

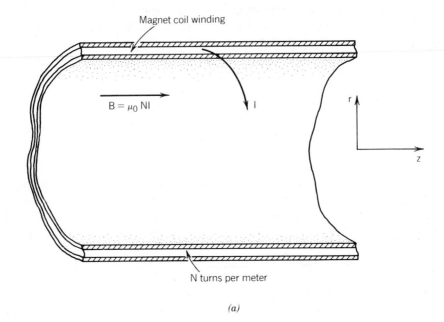

(a)

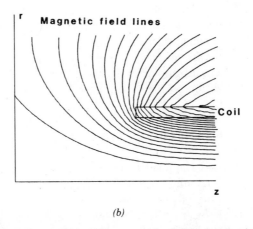

(b)

Figure 10.1   Solenoidal magnetic fields. (*a*) Geometry of an infinite-length solenoid. (*b*) Field lines at the end of a uniform solenoid winding using the POISSON code. (*c*) Field lines at the end of a solenoid with an iron flux return structure. (*d*) Field lines between two uniform solenoid windings with opposite polarity—a symmetric magnetic cusp.

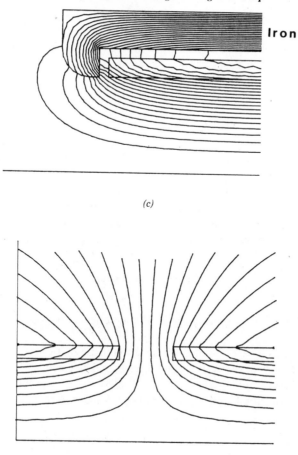

(c)

(d)

Figure 10.1 *(Continued)*

the field magnitude is

$$B_0 = \mu_0 NI. \tag{10.1}$$

We shall investigate electron motion at the end of a finite length solenoid. Figure 10.1*b* shows numerically calculated field lines. The lines emerge from the coil and spread radially. The axial field magnitude drops over a distance comparable to the coil radius. To reduce the width of the transition region, we can add an iron pole at the end of the coil (Fig. 10.1*c*). Finally, Fig. 10.1*d* shows the transition region between two axi-centered solenoids with opposite field polarity. This geometry is a *magnetic cusp*. The opposing currents of the coils create strong radial fields in a narrow intervening region.

The radial magnetic field at the end of a solenoid deflects entering electrons in the azimuthal direction. The deflection results from the $v_z \times B_r$ force. We can apply conservation principles to find some properties of the electron orbits. Since forces arise only from a static magnetic field, the total energy of an electron is constant:

$$\gamma = \gamma_0 = \text{constant.} \qquad (10.2)$$

The magnetic force has azimuthal symmetry; therefore, the canonical angular momentum, P_θ, is constant:

$$P_\theta = \gamma m_0 r v_\theta - e r A_\theta = \text{constant.} \qquad (10.3)$$

In Eq. (10.3), A_θ is the azimuthal component of the vector potential, related to the axial component of magnetic field by

$$A_\theta(r) = \frac{1}{r} \int_0^r r' B_z(r') dr'. \qquad (10.4)$$

The integral in Eq. (10.4) equals the axial magnetic flux included within a radius r divided by 2π.

We shall consider motion of electrons into the magnetic field region $z > 0$ from one of the planes marked A, B, or C in Fig. 10.2. The electrons enter with kinetic energy $(\gamma - 1)m_0 c^2$ and initial velocity $\mathbf{v} = \beta c \mathbf{z}$. Analysis of the motion is simplified if we assume that the length of the magnetic transition region, Δz, is short. We

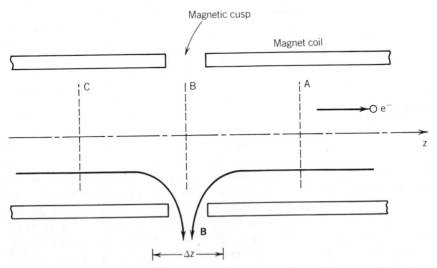

Figure 10.2 Axial injection of electrons near a symmetric magnetic cusp. Planes marked A, B, and C are possible locations for an electron gun.

adopt the condition that Δz is much smaller than the transit distance an electron travels during a gyration in the magnetic field:

$$\Delta z \ll \beta c / \omega_g. \tag{10.5}$$

The quantity ω_g is the electron gyrofrequency, $\omega_g = eB_0 / \gamma m_0$. When the condition of Eq. (10.5) holds, the change in the radial position of an electron moving through the transition is small.

The electron source at plane A is an *immersed source* that generates electrons within the solenoidal field. There is no transverse force that acts on electrons that leave the source in the axial direction; therefore, they follow straight-line orbits. A plot of the projected electron orbit in the r-θ plane is a single point (Fig. 10.3).

The plane marked B is a symmetry axis between opposing solenoidal fields of equal magnitude. The included axial flux equals zero at all radial positions in the plane, $A_\theta(r, 0) = 0$. An electron source at plane B is a *nonimmersed* injector. Electrons leaving the plane in the axial direction have $P_\theta = 0$, since both v_θ and A_θ equal zero. Similar conditions apply at a source located a long distance outside a single solenoid. After the electrons cross the transition region into the solenoidal field, the vector potential at the injection radius is approximately $A_\theta = - rB_0/2$. Conservation of P_θ implies that electrons have an azimuthal velocity

$$v_\theta = - erB_0/2\gamma m_0, \tag{10.6}$$

in the solenoid. Electron orbits projected in the r-θ plane are circles, with

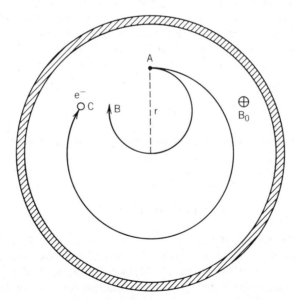

Figure 10.3 Orbit projections in the r-θ plane for electrons injected near a symmetric magnetic cusp. The letters A, B, and C refer to the locations of injectors in Figure 10.2.

gyroradius

$$r_g = \gamma m_0 v_\theta / e B_0. \tag{10.7}$$

Substitution of Eq. (10.6) into Eq. (10.7) shows that r_g equals one-half of the injection radius, r. Figure 10.3 shows the orbit projection in the r-θ plane— electrons pass through the axis. When viewed in the r-z plane, an electron orbit crosses the axis periodically at points separated by roughly $\pi \beta c / \omega_g$.

Electrons emerging from a source at plane C have nonzero canonical angular momentum. If the field magnitude in the upstream solenoid is $-\alpha B_0$, then

$$P_\theta = \alpha e r^2 B_0 / 2. \tag{10.8}$$

After crossing the cusp transition, the azimuthal velocity of electrons is

$$v_\theta = (e r B_0 / 2 \gamma m_0)(1 + \alpha). \tag{10.9}$$

For the special case of a symmetric cusp ($\alpha = 1$), the value of v_θ is twice that of Eq. (10.6). Equation (10.9) implies that the electron gyroradius in the down-stream field equals the injection radius, $r_g = r$. The orbit projected in the r-θ plane is a circle centered on the axis (Fig. 10.3). Symmetric cusps are commonly used to generate rotating electron beams for experiments on plasma confinement, collective ion acceleration, and microwave generation.

Electrons that pass through a magnetic transition gain azimuthal velocity at the expense of axial velocity. This process sets limits on propagation. Electrons may be reflected at a cusp if they enter a strong field at large radius. We can estimate conditions for electron reflection by applying conservation of energy. We shall address the special cases of a half-cusp with nonimmersed injector (plane B) and a symmetric cusp (plane C with $\alpha = 1$). The total velocity of electrons is constant:

$$v_z^2 = (\beta c)^2 - v_\theta^2 - v_r^2. \tag{10.10}$$

For the nonimmersed injector, Eq. (10.6) shows that the axial velocity at the cusp exit is

$$v_z^2 = (\beta c)^2 - r^2 (e B_0 / 2 \gamma m_0)^2 \tag{10.11}$$
$$= (\beta c)^2 [1 - (r / 2 r_g)^2].$$

An electron crosses the magnetic transition only if $r_g > r/2$. As an example, suppose we want to inject a 20-keV, 0.01-m-diameter electron beam into a solenoid. If we set $r_g = 0.0025$ m, Eq. (10.11) implies that the field magnitude must be less than 2 tesla for beam transport. The condition for transmission through a symmetric cusp is more stringent, $r_g > r$. For the beam parameters given, the condition means that $B_0 < 1$ tesla.

10.2. PROPAGATION OF BEAMS FROM AN IMMERSED CATHODE

Cathodes immersed in an axial magnetic field are found in many devices that use high-current electron beams. Usually, the injector cathode, injector anode, beam-transport section, and target are all located in a strong axial magnetic field, $B_0\mathbf{z}$. The field counteracts defocusing beam-generated forces—electrons are tied to the field lines. In a strong solenoidal field, transport of electron beams is effective over a wide range of beam current. An immersed cathode is the only practical option for magnetic focusing of very-high-current beams. It is difficult to design injectors with a nonimmersed cathode for current higher than 10 kA.

A complete derivation of a self-consistent equilibrium for a relativistic electron beam in an applied magnetic field is complex. Often, the results of such a model are strongly dependent on assumptions about the beam distribution. In this section, we shall construct approximate models that illustrate many of the physical principles of electron beam transport in an axial magnetic field. In particular, we will concentrate on two special beam geometries:

1. A paraxial, cylindrical beam in a region far from the injector.
2. A large-area sheet beam in a plasma discharge.

The first geometry is a good model for beams in many microwave devices. The second geometry represents the control beam in a transverse discharge gas laser.

Figure 10.4 illustrates an injector for a circular beam. The device generates a uniform current density of axially directed electrons from the axis to a radius r_0. The emerging beam has density n_0. An electron beam leaving the injector does not have a balance of radial forces. When the electrons have no azimuthal velocity, v_θ, the axial magnetic field exerts no radial force on the beam. The beam

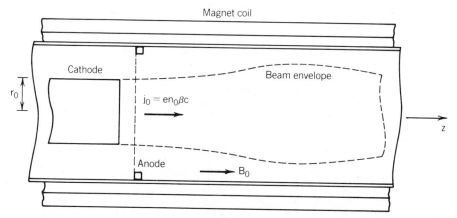

Figure 10.4 Geometry of a cylindrical electron injector immersed in a magnetic field.

expands radially because of the unbalanced space-charge force. The axial magnetic field converts the radial velocity of electrons to azimuthal velocity, resulting in a focusing force. Ultimately, the beam returns to its original radius, and expansion begins again. We can describe the envelope oscillations of a high-current electron beam in a magnetic field by a simple model if we adopt the following assumptions:

1. The beam current is well below the longitudinal space-charge limit. All electrons have about the same kinetic energy, $(\gamma-1)m_e c^2$.

2. Electron orbits beam are paraxial; therefore, all particles have about the same axial velocity, βc.

3. We ignore details of the beam equilibrium over the cross section and consider only electron trajectories on the beam envelope. We neglect contributions of beam emittance to the envelope oscillations.

4. Since electric fields near the injector vary in two dimensions, we shall limit attention to electron motion in a downstream region. Here, we can use electric field expressions for an infinite-length beam. Neglecting processes near the injector is a good approximation if the wavelength for envelope oscillations is much longer than r_0.

5. The change of particle energy associated with envelope oscillations is much smaller than $m_0 c^2$. As a result, we can use nonrelativistic equations for transverse electron motion with an adjusted mass, γm_0.

Again, the canonical angular momentum of electrons is constant in the cylindrical system. We can express the conservation principle as

$$\gamma m_0 r(z)v_\theta - er(z)^2 B_0/2 = -er_0^2 B_0/2. \tag{10.12}$$

The quantity r_0 is the envelope radius at the injection point, while $r(z)$ is the trace of the oscillating envelope radius. We can rewrite Eq. (10.12) as

$$\frac{v_\theta}{r} = \frac{eB_0}{2\gamma m_0}\left[1 - \left(\frac{r_0}{r}\right)^2\right]. \tag{10.13}$$

When the envelope oscillation wavelength is much larger than r, the combined radial electric and magnetic forces of the beam are given by

$$F_r = \frac{e^2 n_0 r_0^2}{2\varepsilon_0 \gamma^2 r}. \tag{10.14}$$

The time-dependent envelope motion follows the equation

$$\gamma m_0 \frac{d^2 r}{dt^2} = \frac{e^2 n_0 r_0^2}{2\gamma^2 \varepsilon_0 r} + \gamma m_0 r\left(\frac{v_\theta}{r}\right)^2 - er\left(\frac{v_\theta}{r}\right)B_0. \tag{10.15}$$

We can convert Eq. (10.15) to a trace equation using the methods of Section 9.1. To make comparisons to the Brillouin flow solutions of Section 10.3, we will write the trace equation in terms of characteristic frequencies instead of the generalized perveance. One fundamental scaling parameter is the magnetic gyrofrequency of beam electrons:

$$\omega_{gb} = eB_0/\gamma m_0. \tag{10.16}$$

We also use the beam plasma frequency:

$$\omega_{pb} = (e^2 n_0/\gamma m_0 \varepsilon_0)^{1/2}. \tag{10.17}$$

We shall see in Section 12.3 that ω_{pb} is the characteristic space-charge oscillation frequency for electrons. Since the current problem involves an interaction between magnetic and space-charge forces, we expect that the solutions for envelope oscillations are governed by the ratio ω_{pb}/ω_{gb}.

We define the dimensionless variables

$$R = r/r_0 \tag{10.18}$$

and

$$Z = z/(\beta c/\omega_{gb}). \tag{10.19}$$

Using Eqs. (10.16), (10.17), (10.18), and (10.19), Eq. (10.15) becomes

$$\frac{d^2 R}{dZ^2} = \left[\frac{\omega_{pb}}{\gamma \omega_{gb}} \right]^2 \frac{1}{2R} - \frac{R}{4}\left[1 - \frac{1}{R^4} \right]. \tag{10.20}$$

It is easy to solve this nonlinear equation by numerical methods. The starting conditions at $Z = 0$ are $R(0) = 1$ and $dR(0)/dZ = 0$. Figure 10.5 plots solutions for several values of $\omega_{pb}/\gamma\omega_{gb}$. The amplitude of envelope oscillations grows if the beam density increases or if the magnetic field magnitude drops.

Figure 10.6 plots the amplitude of envelope oscillations as a function of $\omega_{gb}/\gamma\omega_{pb}$. As an example, suppose that we want to transport a 50-keV, 50-A electron beam with radius 0.025 m. Depending on the radius of the transport tube, the maximum beam-generated electrostatic potential is only about 5 keV. The beam density is $n_0 = 1.3 \times 10^{15}\,\mathrm{m}^{-3}$—the relativistic plasma frequency is $\omega_{pb} = 1.9 \times 10^9\,\mathrm{sec}^{-1}$. For envelope oscillations less than 10% of the beam radius, we find from Fig. 10.6 that $\omega_{pb}/\gamma\omega_{gb} < 0.29$. The magnetic field should exceed $B_0 = 0.037$ tesla.

Next, we shall discuss a sheet beam gun to control a large-area gas laser discharge (Fig. 10.7). We want to find the envelope trace for a beam traveling through the discharge plasma in an axial magnetic field. In an electron-beam-controlled laser, the discharge is stable with current density proportional to the

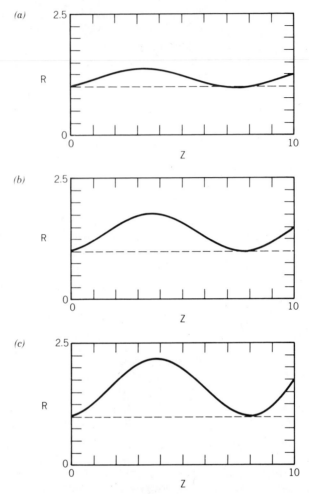

Figure 10.5 Numerical solutions of the envelope equation for an electron beam from an immersed injector in a uniform axial magnetic field. Normalized radius, $R = r/r_0$, as a function of normalized axial position, $Z = z/(\beta c/\omega_{gb})$: (a) $\omega_{pb}/\gamma\omega_{gb} = 0.50$; (b) $\omega_{pb}/\gamma\omega_{gb} = 0.75$; (c) $\omega_{pb}/\gamma\omega_{gb} = 1.00$.

beam current density—we denote the proportionality constant as M. In a practical system, M is much larger than unity. Again, we shall adopt some assumptions to construct a simple envelope equation:

1. The discharge plasma neutralizes all electric fields.
2. A space-charge-limited diode with voltage V_0 and spacing d generates the electron beam. The current density at injection is uniform over width $\pm x_0$ and all electron orbits have zero inclination angle, $x' = 0$.
3. We neglect electron energy loss to background gas or plasma.

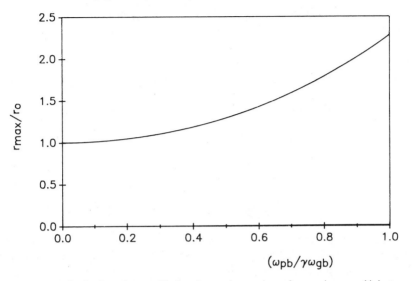

Figure 10.6 Amplitude of envelope oscillations for an electron beam from an immersed injector as a function of $\omega_{pb}/\gamma\omega_{gb}$.

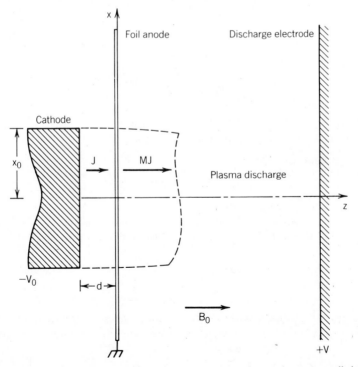

Figure 10.7 Schematic diagram of a sheet beam injector for an electron-beam-controlled gas laser.

4. The transverse distribution of plasma current density is proportional to that of the beam.

The last assumption is questionable—plasma electrons are strongly tied to field lines. At best, the model gives an estimate of the envelope shape for small changes in width.

Again, we employ a nonrelativistic transverse equation with an adjusted mass. We must modify the sheet beam envelope equation [Eq. (9.17)] to include the nonzero canonical momentum, P_y, of electrons created in a magnetic field and current amplification in the discharge. The result is

$$\frac{d^2 X}{dz^2} = \frac{-e(MJ)}{2\varepsilon_0 \gamma m_0 \beta c^3} + \left[\frac{eB_0}{\gamma m_0 \beta c}\right]^2 (x_0 - X). \tag{10.21}$$

The forces acting on the beam arise from the amplified beam-generated magnetic field and the applied axial magnetic field, B_0. The term with the initial envelope width, x_0, represents the effect of nonzero P_y. The axial magnetic field exerts a force only when the envelope width shifts from x_0. The quantity J is the current per meter of the sheet beam. In the nonrelativistic limit, the linear current density is related to the gun voltage V_0 and spacing d by

$$J = (2x_0)\frac{4\varepsilon_0}{9}\left[\frac{2e}{m_0}\right]^{1/2}\frac{V_0^{3/2}}{d^2}. \tag{10.22}$$

The relativistic velocity factor in Eq. (10.21) is

$$\beta c \cong (2eV_0/m_0)^{1/2}. \tag{10.23}$$

Substituting from Eqs. (10.22) and (10.23) and inserting the dimensionless variables, $X = x/d$ and $Z = z/d$, we can write Eq. (10.21) in the form:

$$\frac{d^2 X}{dZ^2} = -A_1 + A_2(X_0 - X). \tag{10.24}$$

The first term on the right-hand side represents pinching in the combined magnetic field of the beam and discharge; the dimensionless constant A_1 is

$$A_1 = \frac{4M}{9}\frac{x_0}{d}\frac{eV_0}{m_0 c^2}. \tag{10.25}$$

The second term represents the effect of the applied magnetic field when the envelope width shifts from the injection value. The dimensionless constant A_2 is

$$A_2 = \frac{B_0^2}{(2V_0 m_0/ed^2)}. \tag{10.26}$$

The width of electron beams to control laser discharges should remain constant. We can use Eq. (10.24) to find the required magnetic field magnitude in the discharge region. For example, suppose we have a 100-keV electron beam emerging from a gun with gap width $d = 0.05$ m. We take a beam half-width of $x_0 = 0.3$ m and a discharge current amplification factor of $M = 10$. The parameter A_1 equals 5.22. We find that a value of $A_2 = 20$ gives envelope oscillations that are $\pm 5\%$ of the original beam width. The corresponding applied field is $B_0 = 0.095$ tesla.

10.3. BRILLOUIN EQUILIBRIUM OF A CYLINDRICAL ELECTRON BEAM

In Section 10.2, we found that a cylindrical electron beam generated within a solenoidal field cannot propagate in a matched equilibrium. Here, the term *matched* means that there are no oscillations of the envelope radius. The matched condition is desirable when a beam must propagate through a narrow bore, as in a traveling-wave tube. A matched beam can be confined within a given radius by a minimum value of applied magnetic field.

We will show in this section that an ideal matched equilibrium is possible for a laminar beam created outside the magnetic field volume. The solution, valid for nonrelativistic electrons, is called the *Brillouin equilibrium* [L. Brillouin, *Phys. Rev.* **67**, 260 (1945)]. Figure 10.8 illustrates the geometry of the calculation. The beam, the surrounding structures, and the electromagnetic fields have cylindrical symmetry. A gun located in a plane where $B_z(r, z) = 0$ generates a steady-state, circular electron beam. Figure 10.8 shows how the zero-flux condition can be achieved using a bucking coil behind the cathode. The beam emerging from an ideal gun has zero emittance and radially uniform current density. At the anode, all electrons have the same kinetic energy, eV_0, and axial velocity.

At the gun, the beam has a simple distribution function—it is singular in total

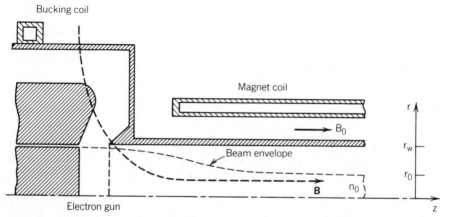

Figure 10.8 Injector to create a magnetically confined electron beam in a Brillouin equilibrium.

energy and canonical angular momentum:

$$f(T, P_\theta) \sim \delta(T - eV_0)\delta(P_\theta). \qquad (10.27)$$

The total energy T is the sum of the nonrelativistic kinetic energy and the potential energy in the electrostatic fields. Equation (10.27) is a valid equilibrium distribution since both T and P_θ are constants of the particle motion. The total energy is constant because electric fields are static; the canonical angular momentum is constant because the forces have azimuthal symmetry.

Electrons leaving the gun follow complex orbits through the transition region at the edge of the magnet before entering the uniform solenoidal field. Nonetheless, the beam distribution retains the form of Eq. (10.27). We shall concentrate on the beam characteristics at a point far from the transition region. The azimuthal and axial derivatives of all field and beam quantities equal zero. The problem reduces to the calculation of a self-consistent radial equilibrium with a given distribution function. We shall seek an equilibrium solution where the beam density, n_0, is uniform from the axis to the envelope radius, r_0.

Electrons gain a component of azimuthal velocity when they pass through the radial magnetic field in the transition region. In the uniform field region, the condition $P_\theta = 0$ gives an expression for the azimuthal velocity of the nonrelativistic electrons as a function of radius:

$$v_\theta = (eB_0/2m_e)r. \qquad (10.28)$$

The rotational motion of the electrons generates an axial magnetic field. For the time being, we assume that the beam-generated axial field is much smaller than the applied field.

Equation (10.28) predicts that the azimuthal velocity is linearly proportional to radial position. This condition implies that all electrons rotate about the axis at the same angular velocity:

$$\frac{d\theta}{dt} = \frac{v_\theta}{r} = \frac{eB_0}{2m_e} = \frac{\omega_{go}}{2}. \qquad (10.29)$$

The quantity ω_{go} is the gyrofrequency for nonrelativistic electrons in the magnetic field B_0:

$$\omega_{go} = eB_0/m_e. \qquad (10.30)$$

Since the angular velocity is independent of radius, the entire beam rotates at the same rate. Cylindrical equilibria with this property are called *rigid rotor equilibria*. The angular rotation frequency of the beam is the *Larmor frequency*,

$$\omega_L = \omega_{go}/2. \qquad (10.31)$$

The focusing force from the azimuthal magnetic field of a cylindrical electron beam is about a factor β^2 less than the defocusing space-charge force [Eq. (5.25)]. The contribution of magnetic force is small in a nonrelativistic beam. Neglecting the beam-generated magnetic force, the following equation describes the radial motion of individual electrons with $P_\theta = 0$:

$$m_e \frac{d^2r}{dt^2} = \frac{e^2 n_0 r}{2\varepsilon_0} + \frac{m_e}{r}\left[\frac{erB_0}{2m_e}\right]^2 - \frac{(eB_0)^2 r}{2m_e}. \tag{10.32}$$

The first term on right-hand side of Eq. (10.32) represents the space-charge force [Eq. (5.26)], the second term corresponds to the centrifugal force $(m_e v_\theta^2/r)$, and the last term is the focusing magnetic force $(v_\theta \times B_z)$. Setting the right-hand side equal to zero gives the condition for a matched beam. Note that all terms are linearly proportional to r; therefore, force balance applies at all radii when the beam has no emittance. The matched condition is

$$e^2 n_0/2\varepsilon_0 m_e = (eB_0/m_e)^2/4. \tag{10.33}$$

We can write Eq. (10.33) in an alternative form that incorporates the beam plasma frequency [Eq. (10.17)]. Substituting Eq. (10.17) with $\gamma = 1$ gives the Brillouin condition:

$$\omega_{pbo}/\omega_{go} = 1/\sqrt{2}. \tag{10.34}$$

In the rigid-rotor equilibrium, electrons follow circular transverse orbits under the combined actions of the space-charge electric field and the applied magnetic field. The electrons move normal to both the electric and magnetic fields. Charged particle equilibria in crossed field geometries with laminar orbits aligned in the $\mathbf{E} \times \mathbf{B}$ direction are called *Brillouin flow solutions*. We already studied examples in different geometries in Sections 8.2 and 8.9. When the electrons have a transverse velocity spread, radial oscillations are superimposed on the azimuthal drifts.

To complete the model, we must investigate the radial variation of axial velocity and verify conservation of total energy. The condition of constant energy is

$$(m_e/2)(v_z^2 + v_\theta^2) - e\phi(r) = eV_0. \tag{10.35}$$

The quantity $\phi(r)$ is the absolute electrostatic potential in the beam volume. With the assumption of uniform density, the radial variation of potential is [Eq. (5.28)]

$$\phi(r) = \phi(0) + en_0 r^2/4\varepsilon_0. \tag{10.36}$$

Equation (5.67) shows that the potential at the center of a circular beam of radius

r_0 propagating in pipe of radius r_w is

$$\phi(0) = -\frac{en_0r_0^2}{4\pi\varepsilon_0}\left[1 + 2\ln\left(\frac{r_w}{r_0}\right)\right].$$ (10.37)

Combining the condition $P_\theta = 0$ with Eq. (10.33), we find the following expression for the azimuthal component of kinetic energy:

$$m_e v_\theta^2/2 = (e^2 n_0/4\varepsilon_0)r^2.$$ (10.38)

Substituting Eqs. (10.37) and (10.38) into the energy balance equation, Eq. (10.35), gives

$$\frac{m_e v_z^2}{2} = eV_0 - \frac{e^2 n_0 r_0^2}{4\varepsilon_0}\left[1 + 2\ln\left(\frac{r_w}{r_0}\right)\right].$$ (10.39)

Note that terms with a radial dependence have canceled and do not appear in Eq. (10.39). As a result, all the electrons in the beam travel at the same axial velocity, independent of radius. Although electrons near the axis sacrifice more energy to enter the region of high electrostatic potential, they do not acquire as much rotational energy.

We can use Eq. (10.39) to find limits on current in a Brillouin beam. The scaling laws are similar to those of Sections 5.3 and 9.4. The main difference is that the present model gives an exact prediction that accounts for variations of axial velocity. For a given beam density, n_0, radius, r_0, and uniform axial velocity, v_z, the beam current is

$$I = (\pi r_0^2)en_0 v_z.$$ (10.40)

Substituting for v_z from Eq. (10.39), Eq. (10.40) becomes

$$I = (\pi r_0^2 en_0)\left[\frac{2eV_0}{m_e} - \frac{e(\pi r_0^2 en_0)}{2\pi\varepsilon_0 m_e}\left\{1 + 2\ln\left(\frac{r_w}{r_0}\right)\right\}\right]^{1/2}.$$ (10.41)

Equation (10.41) has the form

$$I = \lambda(A - B\lambda)^{1/2},$$ (10.42)

where

$$\lambda = \pi r_0^2 en_0, \qquad A = 2eV_0/m_e$$

and

$$B = (e/2\pi\varepsilon_0 m_e)[1 + 2\ln(r_w/r_0)].$$

The quantity λ is the line density of beam charge, which we take as a variable. Setting $dI/d\lambda = 0$ gives a value of λ for the maximum current:

$$\lambda_m = 2A/3B. \tag{10.43}$$

The combination of Eqs. (10.41), (10.42), and (10.43) gives a relationship for the maximum current that can be carried by a laminar electron beam in a uniform solenoidal field:

$$I_m = \left[\frac{8\pi}{3}\right]\left[\frac{2e}{3m_e}\right]^{1/2} \frac{\varepsilon_0 V_0^{3/2}}{[1 + 2\ln(r_w/r_0)]} \tag{10.44}$$

or

$$I_m = (2.54 \times 10^{-5})V_0^{3/2}/[1 + 2\ln(r_w/r_0)].$$

At a low value of λ, the beam current is directly proportional to the line charge density of the beam. When λ exceeds λ_m, the reduction in axial velocity counterbalances the increased line charge, reducing the net current. For $\lambda = \lambda_m$, the longitudinal energy equals,

$$m_e v_z^2/2 = 2eV_0/3. \tag{10.45}$$

The following example illustrates application of the results. Suppose we want to transport a 10-A electron beam with kinetic energy of 30 keV. The beam has radius $r_0 = 0.01$ m and propagates in a tube of radius $r_w = 0.03$ m. For the given parameters, Eq. (10.44) shows that I_m equals 41.3 A. We find the line charge density by solving Eq. (10.42) with $A = 1.055 \times 10^{16}$, $B = 1.011 \times 10^{22}$, and $I = 10$. A root solver gives the value of λ as 1.02525×10^{-7} (coulombs/m); the associated value of beam density is $n_0 = 2.04 \times 10^{15}\,\text{m}^{-3}$. Substituting the density into Eq. (10.42) shows that the focusing field should be $B_0 = 0.0205$.

All models in collective-beam physics involve approximations. Deriving a result is only a portion of any analysis; we must also find limits of validity for the model. Here, we shall find when it is valid to neglect the beam-generated components of axial magnetic field. To make an estimate, we calculate the axial field generate by a beam in an ideal Brillouin equilibrium and require that its magnitude be much smaller than B_0.

Equation (10.28) gives an expression for the total azimuthal current per length of a Brillouin beam:

$$J_b = \int_0^{r_0} en_0 v_\theta(r')dr' = \frac{e^2 n_0 B_0 r_0^2}{4m_e}. \tag{10.46}$$

The magnitude of the beam-generated magnetic field is $\mu_0 J_b$. The condition for a small beam contribution to the axial magnetic field is

$$\mu_0 J_b / B_0 \ll 1$$

or

$$(\omega_L r_0 c)^2 \ll 1. \qquad (10.47)$$

Equation (10.47) follows from the Brillouin condition. The equation implies that the beam-generated axial magnetic field is negligible if the azimuthal velocity of particles on the beam envelope is much less than the speed of light. This condition holds for most applications. An exception is the rotating relativistic electron ring.[1] This configuration has been studied for potential applications to collective acceleration and fusion plasma confinement. Electron rings generate strong axial magnetic fields—with sufficient current, the magnitude of the beam-generated fields can exceed the applied field, creating a region of closed magnetic field lines.

The derivation we performed applies only to nonrelativistic electrons. The condition of constant axial velocity results from the nonrelativistic form of the energy equation [Eq. (10.35)]. The general derivation of a self-consistent equilibrium for a relativistic electron beam in a solenoidal field is much more difficult than the nonrelativistic calculation. For the general case, we must include beam-generated toroidal and axial fields and the variation of γ and β_z with radius. The calculation is simpler in the paraxial approximation. Here, the beam current is well below the space-charge limit, or

$$e\phi(0) \ll (\gamma - 1)m_e c^2. \qquad (10.48)$$

In the paraxial limit, all electrons have about the same total energy and longitudinal velocity, represented by average values of γ and β. Furthermore, the azimuthal velocity is low ($v_\theta \ll \beta c$) so that beam-generated axial fields are small. For a relativistic paraxial beam, we must make two changes to the radial force balance equation: (1) replace the rest mass, m_e, with the relativistic mass, γm_e, and (2) include the focusing force of the beam-generated toroidal magnetic field.

The relativistic radial force balance is similar to Eq. (10.32). We can write the modified Brillouin condition in terms of the relativistic beam plasma frequency:

$$\omega_{pb}^2 = e^2 n_0 / \varepsilon_0 \gamma m_e, \qquad (10.49)$$

and the relativistic gyrofrequency,

$$\omega_g = eB_0 / \gamma m_e. \qquad (10.50)$$

The result is

$$\omega_{pb}/\omega_g \cong \gamma/\sqrt{2}. \qquad (10.51)$$

[1] N. C. Christofilos, *Phys. Fluids*, **9**, 1425 (1966).

We can show that Eq. (10.51) is identical to Eq. (9.58) in the limit of a laminar beam.

10.4. INTERACTION OF ELECTRONS WITH MATTER

Electron beams that pass through a dense medium lose energy through collisions with electrons and nuclei. A fraction of the energy heats electrons in the medium, while the remainder is converted to X-ray or γ-ray radiation. Collisions also deflect the light electrons. An understanding of collisional processes in a dense background is essential for many electron beam applications:

1. High current electron diodes often have thin-foil anodes. Angular scattering in the foil limits the brightness of extracted beams.

2. In devices such as electron irradiators, the beam travels through a foil from the accelerator to a target at atmospheric pressure. Foil heating limits the time-averaged intensity of the beams.

3. Many high-energy electron diagnostic devices, such as solid-state detectors, rely on energy loss processes.

4. Angular scattering limits the propagation lengths for beams in diffuse media. Section 12.9 discusses the subject in detail.

5. Applications such as food irradiation and electron beam welding depend on energy transfer processes in materials.

6. Thin structures, such as grids or wires, can guide intense electron beams (Section 10.5). Scattering and energy-loss limit the length of such systems.

In previous sections, we concentrated on beams of charged particles that interact through long-range collective forces rather than short-range collisions. In contrast, collisions are the most important processes in dense media. Fortunately, we can usually neglect the collective forces of beams inside a metal. The free electrons in a metal cancel electric and magnetic fields generated by the beam. As a result, the orbit of each particle in an intense beam is independent of other beam particles. We can use statistical theories to predict properties of beams emerging from a foil.

We express the energy loss rate for electrons passing through a medium in terms of the *stopping power*, dT_e/dz. The stopping power equals the energy lost by a high-energy electron per unit distance of path length in a material. Here we shall use units of joules per meter—a wide variety of units may be encountered in the literature. One important energy-loss mechanism is collisions with low-energy electrons in the medium. In this interaction the electric field of the high-energy electron accelerates electrons of the medium as it passes. The electric field interactions are called *Coulomb collisions*. The collisional stopping power can be predicted by calculating the average energy transferred to randomly distributed background electrons.

The Bethe formula[2] gives the collisional stopping power for relativistic electrons:

$$\left(\frac{dT_e}{dz}\right)_c = -\left[\frac{2\pi N Z r_e^2 m_e c^2}{\beta^2}\right]$$
$$\times \left[\ln\left(\frac{m_e \gamma^2 \beta^2 c^2 T_e}{2I^2}\right) - \ln(2)\left(\frac{2}{\gamma} - 1 + \beta^2\right) + \frac{1}{\gamma^2} + \frac{(1 - 1/\gamma)^2}{8}\right].$$

$$(10.52)$$

Equation (10.52) gives the energy per meter extracted from an energetic electron with kinetic energy $T_e = (\gamma - 1)m_e c^2$ and velocity βc. The quantity N is the atomic density of the medium in atoms per cubic meter, while z is the atomic number of the medium. Note that the product NZ is proportional to the number density of electrons in the medium. The quantity I is the average ionization potential of the material—Table 10.1 gives representative values of I. Finally, r_e is the classical radius of the electron:

$$r_e = e^2/4\pi\varepsilon_0 m_e c^2 = 2.81777 \times 10^{-15}\,\text{m}. \qquad (10.53)$$

The quantity NZ is also proportional to the mass density of the medium, ρ (kg/m³). Therefore, $(dT_e/dz)_c$ is roughly proportional to ρ. Many tables quote values of the stopping power with the density factored out,

$$[dT_e/d\zeta]_c = [dT_e/dz]_c/\rho. \qquad (10.54)$$

TABLE 10.1   Mean Ionization Potential for High-Energy Charged Particles in Matter[a]

Substance	Z	I (eV)	I/Z
H$_2$	1	19	19
He	2	44	22
Be	4	64	16
Air	7.2	94	13.1
Al	13	166	12.7
Ar	18	230	12.8
Cu	29	371	12.8
Ag	47	586	12.5
Xe	54	660	12.2
Au	79	1017	12.8
Pb	82	1070	13.1

[a]Adapted from R. M. Sternheimer, *Methods of Experimental Physics*, Ed. L. Marton, Vol. 5, Part A, Academic Press, New York, 1961.

[2] H. A. Bethe, *Handbuch der Physik*, Springer, Berlin, 1933, p. 24.

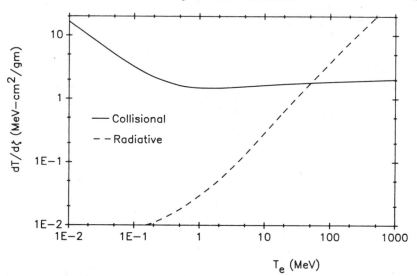

Figure 10.9 Collisional and radiative normalized stopping powers for electrons in solid aluminum as a function of kinetic energy.

The quantity $[dT_e/d\zeta]_c$ is the *normalized collisional stopping power*. We can estimate the stopping power for a variety of materials by multiplying the normalized stopping power for a known material by the appropriate value of ρ.

Figure 10.9 shows the normalized collisional stopping power for aluminum as a function of electron kinetic energy. The quantity $[dT_e/d\zeta]_c$ is large at low values of T_e because the incident electrons move slowly. They take a relatively long time to pass a background electron—the momentum transfer to the background is high. On the other hand, $[dT_e/d\zeta]_c$ is almost independent of T_e for relativistic electrons, which all move at about the same velocity.

Energetic electrons also lose energy by the emission of radiation. Electrons radiate when they are strongly deflected passing the nuclei of atoms. The radiation accompanying electron collisions is called *bremsstrahlung radiation*. The radiation stopping power quantifies the production of radiation:

$$\left(\frac{dT_e}{dz}\right)_r \cong -\left[\frac{NZ(Z+1)T_e r_e^2}{137}\right]\left[4\ln\left(\frac{2T_e}{m_e c^2}\right)-\frac{4}{3}\right]. \tag{10.55}$$

The quantities in Eq. (10.55) have the same meaning as those in Eq. (10.52).

Note that energy loss to radiation scales approximately as Z^2, while collisional loss is proportional to Z. Furthermore, bremsstrahlung losses are proportional to T_e, while the collision loss for relativistic electrons is almost independent of T_e. Ignoring small variations in the logarithmic terms, the ratio of radiation and

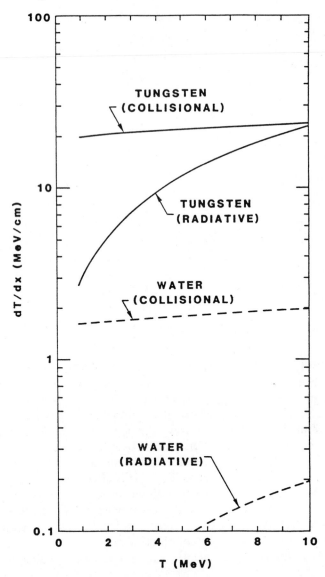

Figure 10.10 Collisional and radiative stopping powers for electrons in water and tungsten as a function of kinetic energy over a range of interest for food-processing applications.

collisional stopping powers is

$$\frac{(dT_e/dx)_r}{(dT_e/dx)_c} \cong \frac{ZT_e(\text{eV})}{7 \times 10^8}. \tag{10.56}$$

Note that T_e is expressed in eV. Equation (10.56) implies that bremsstrahlung may be the dominant energy-loss mechanism for high-energy electron beams in a medium with high atomic number. In applications where electrons create γ-rays, it is best to use a high-energy beam and a heavy target material, such as tungsten or uranium. On the other hand, if we want to dump an electron beam with minimum radiation, the best target choice is a low-Z material such as graphite. Figure 10.10 shows stopping powers for collisions and radiation over the range of interest for commercial irradiation applications. The figure includes results for low-Z and high-Z targets. For high-energy electrons, other energy-loss processes are possible. Beams in the energy range above 10 MeV can excite nuclear reactions. Although the probability of a nuclear reaction is usually small, the interactions may constrain beam parameters for applications such as food irradiation.

The range of a charged particle in matter is the total distance traveled before the particle comes to rest. The range is related to the total stopping power by

$$R = \int_{T_0}^{0} \frac{-dT_e}{(dT_e/dz)_c + (dT_e/dz)_r}. \tag{10.57}$$

The quantity T_0 is the initial electron kinetic energy. Although the energy-loss range is a useful quantity for high-energy ion beams, we must take care in using values of electron range. Electrons suffer substantial angular scattering as they slow down in materials. The quantity R in Eq. (10.57) is an integrated distance along a path length that differs markedly from a straight line. We cannot assume that an electron beam deposits its full energy in a target with thickness equal to the range. Often, a high fraction of electrons incident on a target scatters strongly enough to leave through the entrance surface. This process is called *backscattering*.

As an example of application of the stopping power results, consider heating of a thin foil by a high-power electron beam. Foils under vacuum have application as septum electrodes for particle extraction from a circular accelerator. Thin foils are also used for beam diagnostics and for the transport of intense electron beams (Section 10.5). There are two modes of operation for accelerators that require separate thermal analyses:

1. Pulsed accelerators with low duty cycle generate intense electron beams with high current density.
2. Continuous or high repetition-rate accelerators create beams with moderate peak current but high average power.

Pulsed-power diodes and induction linacs operate in the first mode. To analyze foil heating, we assume that the beam pulse is much shorter than the thermal conduction time through the foil and that thermal radiation from the foil is small during the beam pulse. With these conditions, the temperature rise in a homogeneous foil is

$$\Delta T = j_e \Delta t \left[\frac{dT_e}{dz} \right]_c \bigg/ \rho C_p e. \tag{10.58}$$

In Eq. (10.58), C_p is the specific heat of the material in joules per kilogram averaged over the expected temperature change. The quantity ρ is the material density in kilograms per cubic meter, j_e is beam current density in amperes per square meter, Δt is beam pulse length in seconds, and $[dT_e/dz]_c$ is the collisional stopping power in joules per meter. We use only the collisional part of the stopping power since most of the radiation generated by bremsstrahlung escapes from the foil. Since $[dT_e/dz]_c$ is proportional to the material density, Eq. (10.58) implies that the temperature rise in materials is inversely proportional to the specific heat. Low-Z elements have smaller temperature rise since they have higher values of C_p. For example, the temperature rise in a beryllium foil is only one-fourth that of a copper foil for the same beam intensity.

To illustrate the use of Eq. (10.58), suppose a 70-nsec electron beam pulse passes through a titanium foil. The quantities in the equation have the values $\rho = 4.54 \times 10^3 \, \text{kg/m}^3$, $C_p \cong 0.52 \times 10^3 \, \text{J/k}$, and $[dT_e/dz]_c = 800 \, \text{MeV/m}$. Titanium has a melting point of 1675°C—we limit the temperature rise to 1400°C. Substituting in Eq. (10.58), we find an allowed current density of $j_e = 5900 \times 10^4 \, \text{A/m}^2$, a value well above those encountered in high-power induction linacs.

A foil exposed to a continuous beam reaches a thermal radiation equilibrium. Here, the material temperature reflects a balance between input power from the beam and thermal radiation to surrounding surfaces. To simplify the analysis, we assume that surfaces in a line-of-sight from the foil are at room temperature, approximately 300 degrees Kelvin. The following equation describes energy balance over a unit area of a foil of thickness δ exposed to a relativistic beam of current density j_e:

$$j_e \left(\frac{dT_e}{dz} \right)_c \delta \bigg/ e < 2\varepsilon\sigma(T - 300)^4. \tag{10.59}$$

The quantity σ is the Stefan–Boltzmann constant

$$\sigma = 5.6697 \times 10^{-8} \, \text{J/s} - \text{m}^2 - \text{K}^4. \tag{10.60}$$

and ε is the surface emissivity. For titanium, we take $\varepsilon = 0.6$ and $T - 300 = 1400 \, \text{K}$. For a foil thickness of $\delta = 1.27 \times 10^{-5} \, \text{m}$, Eq. (10.59) implies that the time-averaged current density should be less than $j_e = 0.0026 \times 10^4 \, \text{A/m}^2$. Suppose we have a high-repetition-rate induction linac with a 70-nsec pulse length operating at 100 Hz. The peak instantaneous current density incident on the foil should be less than about 370×10^4 A/m^2.

Because of their low mass, electrons scatter in angle as they pass through matter. Although large angle deflections can occur, electron scattering results mainly from the cumulative effect of multiple small-angle collisions. An electron in matter follows a random walk in angle. An initially parallel beam emerges from a foil with a spread in inclination angle. In the limit of small angular spread, the inclination angles at the foil exit have a Gaussian probability distribution:

$$P(\theta)d\theta \cong (2\theta/\langle\theta^2\rangle)\exp(-\theta^2/\langle\theta^2\rangle)d\theta. \tag{10.61}$$

The quantity $\langle\theta^2\rangle$ is the mean squared inclination angle for electrons. The theory of Coulomb scattering combined with quantum mechanical corrections[3] gives the following expression for the change in mean squared angle with propagation distance in the material:

$$\frac{d\langle\theta^2\rangle}{dz} \cong \left[\frac{16\pi NZ(Z+1)r_e^2}{\gamma^2\beta^4}\right]\ln\left(\frac{204}{Z^{0.33}}\right). \tag{10.62}$$

Again, the quantities Z and N are properties of the medium, while γ and β characterize the incident electron beam. Note that the scattering angle decreases rapidly with increasing electron γ factor. The effect of scattering is small for particles with high relativistic mass.

As an example of the application of Eq. (10.62), suppose we extract a relativistic electron beam through a thin anode foil. A 2.5-MeV beam passes through an aluminum sheet with thickness $12.7\,\mu m$. From Eqs. (10.52) and (10.55), the electrons lose only 0.2% of their energy in the foil. In contrast, the angular scattering is significant. Equation (10.62) predicts a root-mean-squared angle of

$$(\langle\theta^2\rangle)^{1/2} = 0.15 \text{ rad } (8.5°).$$

The example illustrates a general rule for electron interactions in the energy range of 0.5–20 MeV. Scattering makes the main contribution to the decay of a beam passing through a medium. For high-energy electrons, the $1/\gamma^2$ factor reduces the importance of scattering. High-energy electrons, like ions, slow down in a straight-line trajectory.

10.5. FOIL FOCUSING OF RELATIVISTIC ELECTRON BEAMS

It is possible to transport high-current relativistic electron beams in accelerators with no applied focusing force. This possibility arises because the beam-generated magnetic force of a high-γ beam almost equals the space-charge electric force. If we can reduce the average electric force by a small amount, the magnetic force can

---

[3] See, for instance, J. D. Jackson, *Classical Electrodynamics*, 2nd ed., Wiley, New York, 1975, Section 13.8.

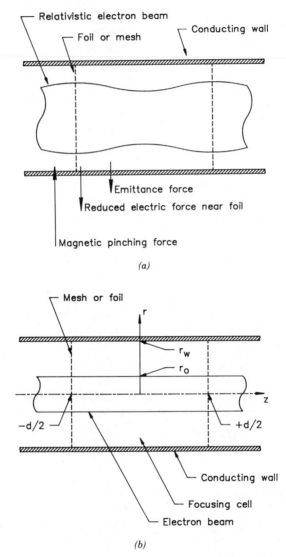

Figure 10.11 Mesh or foil focusing for relativistic electron beams. (*a*) A cylindrical foil-focusing system showing the balance of radial forces. (*b*) Geometry to calculate reduction factors for a uniform-density beam.

contain the beam in a self-focused equilibrium. Figure 10.11*a* illustrates a method to lower the average radial electric field. Conducting meshes and foils divide a transport tube into cells. Positive induced charge in the foils partially cancels the radial electric field. The foils have no effect on the magnetic field since they cannot carry axial current. Although we usually avoid placing objects in the path of a beam, foil focusing is practical in some circumstances because electrons have low stopping power [Eqs. (10.52) and (10.55)]. Foil focusing has advantages over solenoid focusing in addition to the elimination of coils and power supplies. For example, a foil array provides active centering of beams in transport tubes and may have application to curved recirculating accelerators.

We will analyze the geometry of Fig. 10.11b. A cylindrical beam of average radius r_0 propagates along the axis of a cylindrical metal tube of radius r_w. The foils divide the tube into sections of length d. If the rise time of beam current is long compared with the period of the lowest-frequency resonant mode in the cells, then we can estimate net electric fields in the static limit. For example, the period of the TM_{010} mode in a cell of radius $r_w = 0.08$ m is 0.7 nsec—the electrostatic conditions are appropriate for a beam pulse with a 10 nsec rise time.

When there are many closely spaced foils, the approximate condition for radial equilibrium of a beam is that the magnetic focusing force balances the forces of emittance and the reduced radial electric field averaged over a cell. We assume that oscillations of the beam envelope, Δr, are much smaller than r_0. To calculate electric fields, we approximate the beam as a uniform charge cylinder with charge density ρ_0. The beam current is $I_0 = \pi r_0^2 \rho_0 \beta c$. The charge density has little axial variation if the beam electrons are highly relativistic. This condition means that β is almost constant, so that the beam density is insensitive to variations of the electrostatic potential. Finally, we assume that all walls of a cell are grounded and that there are no applied focusing or accelerating forces.

The limiting conditions reduce the problem to the solution of the Poisson equation in cylindrical geometry with a given beam charge density:

$$\frac{1}{r}\frac{\partial}{\partial r}r\left[\frac{\partial\phi(r,z)}{\partial r}\right]+\left[\frac{\partial^2\phi(r,z)}{\partial z^2}\right]=-\frac{\rho(r,z)}{\varepsilon_0}. \tag{10.63}$$

We can use the solution for $\phi(r,z)$ to investigate longitudinal current limits and to derive the radial electric field:

$$E_r(r,z)=-\partial\phi/\partial r. \tag{10.64}$$

The boundary conditions for Eq. (10.63) are

$$\phi(r,-d/2)=0, \qquad \phi(r,+d/2)=0, \qquad \phi(r_w,z)=0. \tag{10.65}$$

We use the space-charge function

$$\rho(r,z)=\rho_0 \qquad (0\leqslant r\leqslant r_0), \tag{10.66}$$
$$\rho(r,z)=0 \qquad (r_0<r\leqslant r_w).$$

The space-charge density varies only in r. Applying this condition and noting the symmetry of the transport geometry, we can express the electrostatic potential as the product of a radial function times an axial function. For the cylindrical geometry, a Fourier–Bessel series expansion represents a solution to the Poisson equation inside the cell:

$$\phi(r, z) = \sum_{m = 1,2,\dots} \sum_{n = 1,2,\dots} \phi_{mn} J_0(x_m r/r_w) \cos\left[(2n - 1)\pi z/d\right]. \qquad (10.67)$$

The constants x_m give null values of the zero-order Bessel function at $r = r_w$:

$$x_1 = 2.405, \qquad x_2 = 5.520, \qquad x_3 = 8.654, \qquad x_4 = 11.792,$$

$$x_5 = 14.931, \qquad x_6 = 18.071, \qquad \dots. \qquad (10.68)$$

Note that the form of Eq. (10.67) ensures that ϕ satisfies the boundary conditions at $r = r_0$ and $z = \pm d/2$. Also, ϕ has a maximum value on the axis and has symmetry about the cell midplane, $z = 0$. The properties of orthogonal series expansions guarantee that Eq. (10.67) represents any well-behaved function $\phi(r, z)$ with a proper choice of the coefficients ϕ_{mn}.

Similarly, we resolve the space-charge function into the product of axial and radial parts:

$$\rho(r, z) = \rho_r(r) \cdot \rho_z(z). \qquad (10.69)$$

When Eq. (10.69) is true, we can represent $\rho(r, z)$ with the series expansion:

$$\rho(r, z) = \sum_{m = 1,2,\dots} \sum_{n = 1,2,\dots} \rho_{mn} J_0(x_m r/r_w) \cos\left[(2n - 1)\pi z/d\right]. \qquad (10.70)$$

For a given space-charge function, we can find the coefficients ρ_{mn}. The cylindrical Poisson equation takes a simple form if we write it in terms of the coefficients of Fourier–Bessel series. The Laplacian operator has the following effect on trigonometric and zero-order Bessel functions:

$$\frac{1}{r} \frac{\partial}{\partial r} r \left[\frac{\partial J_0(x_m r/r_w)}{\partial r} \right] = -\left[\frac{x_m}{r_w} \right]^2 J_0(x_m r/r_w) \qquad (10.71)$$

and

$$\frac{\partial^2 \cos\left[(2n - 1)\pi z/d\right]}{\partial z^2} = -\left[\frac{(2n - 1)\pi}{d} \right]^2 \cos\left[(2n - 1)\pi z/d\right]. \qquad (10.72)$$

Substituting Eqs. (10.71) and (10.72), the Poisson equation has the form

$$\phi_{mn} = \frac{\rho_{mn}}{\left[(x_m/r_w)^2 + \left[(2n - 1)\pi/d \right]^2 \right]}. \qquad (10.73)$$

To solve the problem, we find the coefficients ρ_{mn}, substitute in Eq. (10.73) to find ϕ_{mn}, and then use Eq. (10.67) to calculate $\phi(r, z)$.

We shall carry out the calculation for a uniform cylinder of charge. Following Eq. (10.69), we resolve the space-charge function into axial and radial parts. We can write the coefficients of the space-charge function as a product of radial and axial parts, $\rho_{mn} = \rho_{rm} \cdot \rho_{zn}$. In the interval $-d/2 \leqslant z \leqslant +d/2$, the axial function has constant value $\rho_z(z) = \rho_0$. Fourier series expansions represent periodic functions; therefore, we take the axial variation as a section of a periodic square-wave function. A Fourier analysis of a square wave gives the following values for axial coefficients:

$$\rho_{zn} = \frac{2\rho_0}{d} \int_{-d/2}^{d/2} dz\, \rho_z(z) \cos\left[(2n-1)\pi z/d\right]$$

$$= [4\rho_0/(2n-1)\pi] \sin\left[(2n-1)\pi/2\right]$$

$$= [4\rho_0/(2n-1)\pi](-1)^n, \tag{10.74}$$

for $n = 1, 2, 3, \ldots$.

A Bessel function analysis of the radial function

$$\rho_r(r) = 1 \qquad (0 \leqslant r \leqslant r_0), \tag{10.75}$$

$$\rho_r(r) = 0 \qquad (r_0 \geqslant r \geqslant r_w)$$

gives the terms

$$\rho_{rm} = \frac{\int_0^{r_0} r J_0(x_m r/r_w)\,dr}{\int_0^{r_w} r J_0^2(x_m r/r_w)\,dr} = 2r_0 \frac{J_1(x_m r_0/r_w)}{x_m r_w J_1^2(x_m)}. \tag{10.76}$$

Substituting Eqs. (10.76) and (10.74) into Eq. (10.73), the coefficients of the electrostatic potential expansion are

$$\phi_{mn} = \frac{8\rho_0(-1)^{n-1} r_0 J_1(x_m r_0/r_w)/\pi(2n-1)x_m r_w J_1^2(x_m)}{[(x_m/r_w)^2 + [(2n-1)\pi/d]^2]}. \tag{10.77}$$

The two main quantities of interest are the maximum value of potential, $\phi(0,0)$, and the radial electric field at the beam envelope averaged over a cell. The peak potential equals

$$\phi(0,0) = \sum_{m=1,2,\ldots}^{\infty} \sum_{n=1,2,\ldots}^{\infty} \phi_{mn}. \tag{10.78}$$

By comparing $e\phi(0,0)$ to the beam kinetic energy, we can define longitudinal space-charge limits on beam current.

We find the radial electric field by taking the radial derivative of Eq. (10.67):

$$E_r(r,z) = \sum_m^\infty \sum_n^\infty \phi_{mn} \left[\frac{x_m}{r_w} \right] J_1 \left[\frac{x_m r}{r_w} \right] \cos \left[\frac{(2n-1)\pi z}{d} \right]. \tag{10.79}$$

An integral over z from $-d/2$ to $d/2$ gives the average of electric field times d. The average radial electric field on the envelope is

$$\overline{E_r}(r_0) = \sum_m^\infty \sum_n^\infty \frac{2(-1)^{n-1}\phi_{mn}x_m J_1(x_m r_0/r_w)}{(2n-1)\pi r_w}. \tag{10.80}$$

To show the effect of the foil, we plot the ratio of $\phi(0,0)$ to the maximum potential of an infinite-length cylindrical beam in Fig. 10.12a. From Eq. (5.67), the potential at the center of a long beam with radius r_0 and space-charge density ρ_0 is

$$\phi_\infty(0) = \frac{\rho_0 r_0^2}{4\varepsilon_0} \left[1 + 2\ln\left(\frac{r_w}{r_0}\right) \right]. \tag{10.81}$$

The ratio of potentials is the *longitudinal reduction factor*, denoted as $F_L = \phi(0,0)/\phi_\infty(0)$. We can write the condition for the space-charge-limited current as

$$\frac{e I_0 F_L}{4\pi\varepsilon_0 \beta c} \left[1 + 2\ln\left(\frac{r_w}{r_0}\right) \right] < (\gamma - 1)m_e c^2. \tag{10.82}$$

Figure 10.12a shows F_L as a function of r_0/d. The curves correspond to different values of the normalized wall radius, r_w/r_0. Note that for $d \gg r_0$, the potential approaches the value for a long cylindrical beam ($F_L = 1$). In the opposite limit, $d \ll r_w$, the maximum potential approaches the value for a uniform charge sheet between plates separated by distance d [Eq. (5.12)]:

$$\phi(0,0) \cong \rho_0 d^2 / 8\varepsilon_0. \tag{10.83}$$

To combat longitudinal space-charge effects for high-current transport, we must use closely spaced foils. As an example, suppose we want to transport a 1-MeV, 25-kA beam of radius 0.02 m through a pipe with radius 0.08 m. Without foils, the electrostatic potential at the center of the beam is $\phi = -3$ MV. Reference to Fig. 10.12a shows that if the beam propagates through an array of foils spaced 0.02 m apart, the magnitude of the beam-generated electrostatic potential is only about 260 kV.

At moderate current, the main action of foils is to provide transverse beam

---

Figure 10.12 Reduction factors for a uniform density, paraxial beam. r_0: beam envelope radius; r_w: wall radius; d: cell axial length. (a) Longitudinal reduction factor, F_L. (b) Radial reduction factor, F_R.

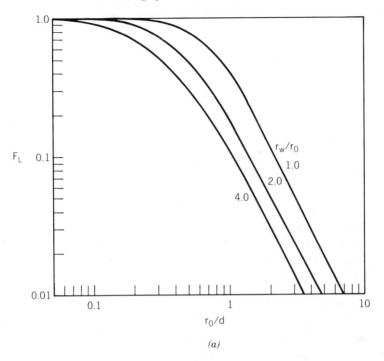

(a)

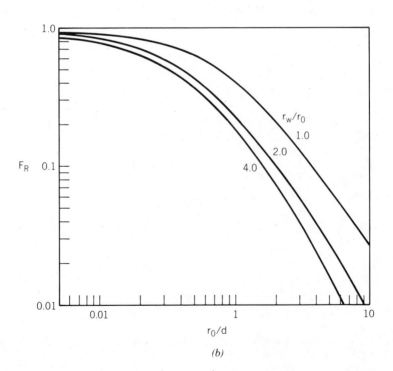

(b)

focusing. Here, we neglect longitudinal field effects and assume that the electrons have almost constant energy. Figure 10.12*b* plots the axially averaged radial electric field at the beam envelope normalized to the radial field on the boundary of an infinite-length beam. We define the radial reduction factor, F_R, as:

$$F_R = \frac{\langle E_r(r_0, z)\rangle}{\rho_0 r_0 / 2\varepsilon_0}. \tag{10.84}$$

A laminar beam is in equilibrium when the reduced average electric force balances the magnetic force. The condition for a laminar equilibrium is

$$F_R = \beta^2. \tag{10.85}$$

The quantity βc is the average axial velocity of the beam. In a beam with nonzero emittance, the inequality $F_R < \beta^2$ is a necessary condition for an equilibrium. The electric and magnetic forces in highly relativistic beams are almost balanced without foils—the value of F_R for a laminar equilibrium is close to unity. In such beams it is more convenient to use the quantity $1 - F_R$. Figure 10.13 plots $1 - F_R$ versus d/r_0. The scale on the right-hand side of the graph shows the value of beam γ for a laminar equilibrium. For example, a zero-emittance electron beam with kinetic energy of 4 MeV and radius $r_0 = 0.02$ m can propagate in a 0.04-cm-radius pipe with foils spaced 1.7 m apart. The parameters show that a few widely spaced transverse foils can guarantee radial equilibrium even at moderate energy.

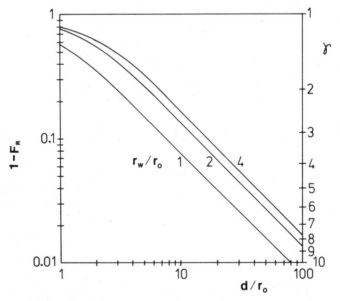

Figure 10.13 Radial reduction factor for energetic beams in long cells: $1 - F_R$ as a function of d/r_0 for different choices of wall radius.

The condition for a laminar equilibrium [Eq. (10.85)] does not depend on I_0, the beam current. Nonetheless, we must consider the value of I_0 to determine validity conditions for the model. We require that the variation of the beam envelope in a cell is small, $\Delta r \ll r_0$. A matched beam has a waist midway between the foils. The radial expansion of a paraxial cylindrical beam traveling from the waist to a foil approximately equals

$$\frac{\Delta r}{r_0} \cong \left[\frac{eI_0}{2\pi\varepsilon_0 m_0(\gamma\beta c)^3} \right]\left[\frac{d^2}{8r_0^2} \right], \tag{10.86}$$

if $\Delta r \ll r_0$. Suppose we pick a maximum value of $\Delta r/r_0$. Then, Eq. (10.86) implies that

$$\frac{d}{r_0} \leqslant (2\gamma\beta)\left[\frac{I_A\Delta r}{I_0 r_0} \right]^{1/2}, \tag{10.87}$$

where

$$I_A = (4\pi\varepsilon_0 m_e c^3/e)\beta\gamma. \tag{10.88}$$

The quantity I_A is the Alfven current—we shall study its physical interpretation in Section 12.7. Figure 10.14 plots implications of Eq. (10.87) for a choice $\Delta r/r_0 = 0.1$. The graph shows combinations of beam current and transport geometry consistent with small beam expansion as a function of γ. Even at high current and moderate γ, high values of d/r_0 are acceptable. As an example,

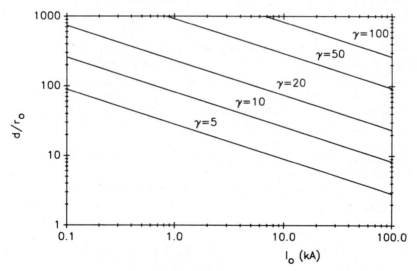

Figure 10.14 Condition for small envelope oscillations in a foil focusing system ($\Delta r/r_0 \leqslant 0.1$). Upper limit on d/r_0 as a function of beam energy and current.

consider transport of a low-emittance electron beam with 1.5-MeV kinetic energy. Take $r_0 = 0.02$ m and $r_w = 0.08$ m. The value of F_R should be 0.875 for a laminar flow equilibrium. Reference to Fig. 10.13 shows that $d/r_0 \cong 20$. Finally, Eq. (10.87) implies that the current corresponding to a $\pm 10\%$ radial oscillation is 0.9 kA.

We can derive equilibrium conditions for beams with nonzero emittance by modifying the force-balance arguments of Chapter 9. We can combine the electric and magnetic forces in a form suitable for an envelope equation to give

$$R'' = \frac{2eI_0}{\beta^2 I_A R}[F_R - \beta^2] + \frac{\varepsilon^2}{R^3}. \qquad (10.89)$$

The variable R represents an envelope radius that may vary along z, $R = r_0(z)$. Setting the right-hand side of Eq. (10.89) equal to zero gives an expression for the emittance of a matched beam. The angular divergence of a beam confined by a foil-focusing system, $\Delta\theta = \varepsilon/R$, equals

$$\Delta\theta = \left[\frac{2I_0}{\beta^2 I_A}(\beta - F_R) \right]^{1/2}. \qquad (10.90)$$

As an example of a mesh-focusing system, suppose we have a beam with $I_0 = 10$ kA, $r_0 = 0.05$ m, and $\gamma = 10$ (4.6 MeV). The Alfven current is $I_A = 174$ kA. The distance between foils is $d = 2.5$ m and the wall radius is $r_w = 0.1$ m. Figure 10.13 shows that the electric field reduction factor is $F_R = 0.973$. Since β^2 equals 0.995, the net force points radially inward. The beam current is well below the Alfven level, $I/I_A = 0.06$, so that electron motion is paraxial. Figure 10.14 shows that envelope oscillations have amplitude less than 10% of the beam radius. Substitution in Eq. (10.90) shows that the electron divergence angle for a matched beam is $\Delta\theta = 54$ mrad and $\varepsilon = 2.7 \times 10^{-3}$ π-m-rad. The result shows that foil transport can contain high-emittance beams.

An alternative formulation[4] is more useful for the calculation of equilibrium of high-energy electron beams in foil-focusing systems. Here, widely spaced foils cancel radial electric fields locally. Near the foils, the magnetic field exerts the dominant force—each foil acts like a toroidal magnetic field lens [CPA]. The beam-generated electric and magnetic forces are almost perfectly balanced in the region between foils. Here, the difference between electric and magnetic forces is small compared with the emittance force.

We can evaluate the focusing action of a single foil by solving the Poisson equation locally to derive the distribution of image charge. If $E_{rf}(r, z)$ is the radial electric field generated by the image charges, then an electron passing through the foil receives an impulse

$$\Delta p_r(r) \cong \int_{-\infty}^{+\infty} -dt\, e E_{rf}(r, z) \cong \int_{-\infty}^{+\infty} -dz(e/\beta c) E_{rf}(r, z). \qquad (10.91)$$

[4] R. Adler, *Part. Accel.*, **12**, 263 (1980).

We can rewrite Eq. (10.91) as

$$\Delta p_r(r) = - G_A(r)(I/I_A)(\gamma m_e \beta c), \tag{10.92}$$

where $G_A(r)$ is the geometry-dependent function of radius plotted in Fig. 10.15. The figure shows that the focusing force is nonlinear. The function G_A has a value of about 2.0 at the beam envelope. Suppose a beam passes through a series of foils separated by a distance d and that the change in radial position of envelope electrons is much smaller than the beam radius. We can write a simple expression for the spatially averaged force from the foils when the envelope oscillation wavelength is long compared to d:

$$\langle F_r(r) \rangle = \frac{dp_r}{dt} \cong - \frac{G_A I(\gamma m_e \beta^2 c^2)}{I_A d}. \tag{10.93}$$

We can balance the expression of Eq. (10.93) against the emittance force [Eq. (3.39)] to find conditions for a transverse equilibrium.

The wall of the transport tube of a foil-focusing system carries image charge and image current when a pulsed beam passes through. We shall derive expressions for the forces exerted by the images in the next section. In an empty tube, the net image force points outward. Transverse foils reduce the image charge force without affecting the return current force. As a result, the wall provides a restoring force that acts to maintain the beam position on the axis of

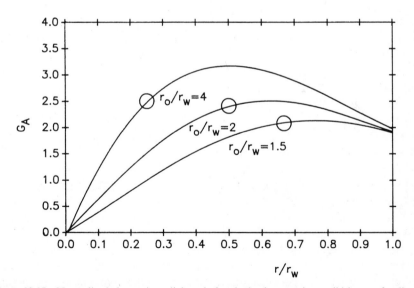

Figure 10.15 Normalized change in radial angle for single electrons in a solid beam of radius r_0 passing through a foil across a cylindrical tube of radius r_w. The function G_A is plotted as a function of r/r_w for choices of beam radius. Circles show the assumed beam radius. [Adapted from R. Adler, *Part. Accel.*, **12**, 263 (1980).]

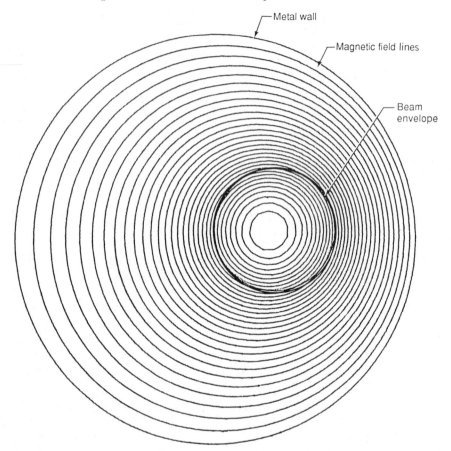

Figure 10.16　Compression of toroidal magnetic field lines generated by a displaced beam in a metal pipe. The beam has uniform current density.

the tube. Figure 10.16 illustrates the origin of the magnetic restoring force. The figure shows toroidal magnetic field lines for a uniform-current-density beam displaced from the axis. Note the compression of field lines in the direction of displacement. Particles on the side of the beam closest to the wall experience a stronger magnetic field that forces them back toward the axis.

A major issue for foil transport is the increase in beam emittance from angular scattering in the foils. This effect is easy to calculate in transport systems where the electrons have constant energy. The effect of scattering is less severe when the electrons are accelerated. Scattering in each foil decreases as the electrons gain energy, and the spacing between foils can be increased. There is no simple method to estimate emittance growth for an accelerated beam. A code to calculate emittance growth must satisfy several conditions simultaneously. The code must decide whether transverse or longitudinal limits are important, adjust emittance

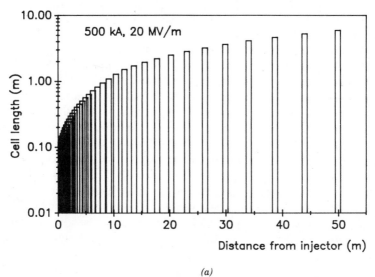

(a)

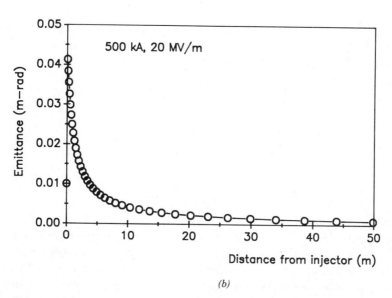

(b)

Figure 10.17 Numerical study for a high-current accelerator using a foil-focusing system. Beam current, 500 kA; injection energy, 20 MeV; output energy, 1 GeV; average acceleration gradient, 20 MV/m; aluminum foil thickness, 10 μm; beam radius, 0.05 m; wall radius, 0.07 m. (*a*) Lengths of 38 focusing cells as a function of position in the accelerator. (*b*) Beam RMS emittance as a function of position in the accelerator.

and foil scattering to account for acceleration, and ensure that envelope oscillations are small. Application of a self-consistent code gives the following sample parameters for beam transport in an induction linac. We take a 5-kA beam injected at 2.5 MeV and accelerated to a final energy of 50 MeV. The beam has constant radius $r_0 = 0.01$ m, an input emittance of 3×10^{-4} π-m-rad, and a normalized emittance of 0.0174 π-m-rad. The initial divergence angle is $1.7°$. The accelerator has an average gradient of 1 MV/m and the focusing foils are 6 μm titanium. The code predicts that the focusing system has 10 cells. The output emittance is 4×10^{-5} π-m-rad. The normalized emittance is 4.2×10^{-3} π-m-rad, enhanced a factor of 2.5 by foil scattering.

The most interesting application of a foil system is the transport of very-high-current beams in the megaampere range. Here, the main action of the foils is reduction of the beam space-charge potential. Accordingly, the spacing between foils is small. For this geometry, the high-current beams must have high emittance for a transverse force balance. Injected low-emittance beams exhibit large envelope oscillations and are subject to filamentation instabilities—these processes increase the emittance until a force balance is attained. Figure 10.17 shows a computer calculation for acceleration of a 500-kA beam to 1 GeV in a high-gradient accelerator. The beam radius is $r_0 = 0.05$ m and the transport tube radius is $r_w = 0.07$ m. The 50-m-long accelerator has an average gradient of about 20 MV/m. The beam enters with a kinetic energy of 15 MeV and a high divergence angle of about $45°$. The cell length is 0.09 m at the entrance. The system has 38 cells that increase in length. Figure 10.17a shows a plot of the cell length versus axial position, while Fig. 10.17b plots the emittance. The effect of foil scattering is negligible for the high-divergence beam in the high-gradient accelerator. The final beam emittance is 8×10^{-4} π-m-rad, giving a divergence angle of $0.9°$ for the self-pinched beam.

10.6. WALL CHARGE AND RETURN CURRENT FOR A BEAM IN A PIPE

In most accelerators and transport systems, beams propagate through metal pipes. Usually, the pipe constitutes a vacuum chamber. Sometimes the chamber geometry is complex, as in an induction linac. No matter how complex the vacuum chamber geometry may be, electric and magnetic fields near the pipe must have the properties illustrated in Fig. 10.18a. There is no beam-generated electric field outside the metal boundary. Therefore, the pipe supports an induced charge (per length) equal and opposite to that of the beam. Also, if the beam has a

---

Figure 10.18 Induced charge and return current for a beam in a cylindrical pipe. (a) An isolated high-vacuum transport tube. (b) Cross section of a pipe with a displaced beam showing nonuniform distribution of induced charge. (c) Geometry to calculate fields generated by induced charge and return current using the method of images. (d) Redistribution of return current for a resistive transport tube wall with surrounding structures.

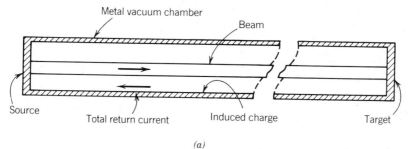

Metal vacuum chamber

Beam

Source

Total return current

Induced charge

Target

(a)

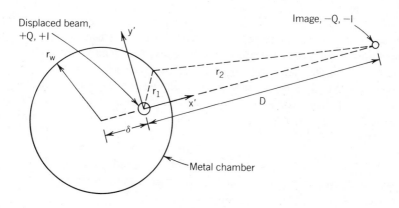

Conducting wall

Displaced electron beam

δ

Induced charge

(b)

Displaced beam, +Q, +I

Image, −Q, −I

r_w

r_2

r_1

y'

x'

D

δ

Metal chamber

(c)

Figure 10.18 *a–c*

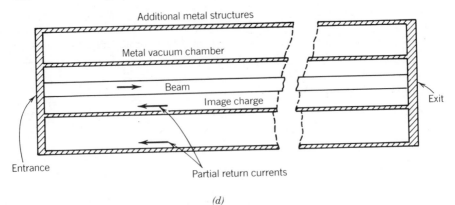

(d)

Figure 10.18 *(Continued)*

pulse length much shorter than the magnetic field penetration time in the metal, there is no beam-generated magnetic field outside the pipe. Therefore, the pipe carries a *return current* with equal magnitude and opposite direction to the beam current. Sometimes, the return current in the pipe equals the beam current regardless of the wall thickness or beam pulse length. This condition holds if the pipe is the only connection between the charged particle source and the target (Fig. 10.18a). Here, all current must return to the source along the wall to prevent charge accumulation and a high electrostatic voltage on the target. For example, suppose we have a 1-kA, 60-nsec electron beam typical of induction linacs. If the beam charge remains on a spherical target of radius 0.3 m, the target potential would equal -2 MV.

The induced wall charge attracts the particles of the beam. Conversely, the return current repels the beam particles. For a beam on the axis of a cylindrical pipe, the induced-charge and return-current distributions have azimuthal symmetry. Therefore, there is no net force on the beam from the wall. If the beam moves from the axis, the induced-charge distribution shifts to maintain the condition that the pipe is an equipotential surface. The asymmetric charge distribution exerts a transverse force on the beam. The wall force affects the equilibrium position of the beam in the pipe and may lead to instabilities.

In this section, we shall derive expressions for the electric and magnetic forces between a beam and a surrounding cylindrical pipe. We apply the expressions in Sections 10.8 and 10.9 to describe electron beam transport in accelerators with magnetic-focusing systems. We adopt some approximations to simplify the derivation:

1. The pipe has uniform radius r_w and infinite length in z. The results hold approximately for pipes that bend or change radius over distances long compared with r_w.

2. The beam is a straight cylinder with a radius much smaller than r_w.

By the second condition, we neglect details of the beam distribution in the transverse direction. To calculate wall charge, we represent the beam as a line charge and line current with zero width.

The beam has line charge Q (coulombs per meter) and current I. For paraxial particle orbits with average axial velocity βc, the two quantities are related by

$$I = Q(\beta c). \tag{10.94}$$

The wall has a total induced charge $-Q$ and carries a return current $-I$. To begin, we calculate the electric fields of a narrow beam displaced a distance δ from the axis of a pipe. Figure 10.18b shows the asymmetric distribution of wall charge. By analogy with the method of images to find the fields of a line charge adjacent to a planar conducting wall, we postulate that the fields inside the pipe from the induced-charge distribution on the circular wall are identical to those of a discrete line charge. We must find a location for the line image so that the superposition of its electric field and the field of the beam define an equipotential surface at the pipe wall.

The analysis is most convenient in a polar coordinate system centered on the beam (Fig. 10.18c). The electric field of the beam without the wall is

$$E_r = Q/2\pi\varepsilon_0 r'. \tag{10.95}$$

The electrostatic potential from the beam is

$$\phi = (Q/2\pi\varepsilon_0)\ln(r') + K. \tag{10.96}$$

The image line charge must be outside the conducting wall. The distribution of induced charge on the wall is symmetric about a line connecting the beam center to the axis of the pipe. Therefore, the wall force vector and the image line charge must also lie on this line. Suppose that the image charge is a distance D from the beam. Following Eq. (10.96), the total electrostatic potential at the wall from the beam and the image is

$$\phi = [Q\ln(r_1) - Q\ln(r_2)]/(2\pi\varepsilon_0) + K \tag{10.97}$$
$$= (Q/2\pi\varepsilon_0)\ln(r_1/r_2) + K.$$

The quantity r_1 is the distance from the beam to the wall and r_2 is the distance from the image. In terms of the beam-centered coordinate system,

$$r_1 = x'^2 + y'^2, \tag{10.98}$$
$$r_2 = (D - x')^2 + y'^2. \tag{10.99}$$

Inspection of Eq. (10.97) shows that equipotential surfaces lies on curves of constant r_1/r_2, or

$$r_2 = \alpha r_1. \tag{10.100}$$

We combine Eqs. (10.98), (10.99), and (10.100) to give

$$(D - x')^2 + y'^2 = \alpha x'^2 + \alpha y'^2. \tag{10.101}$$

The equation that defines the cylindrical wall in the beam-centered coordinate system is

$$(x' + \delta)^2 + y'^2 = r_w^2. \tag{10.102}$$

We can show that Eqs. (10.101) and (10.102) hold at all points on the conducting wall if

$$\alpha = 1 + D/\delta, \tag{10.103}$$

$$D = \delta[(r_w/\delta)^2 - 1]. \tag{10.104}$$

The wall force on the beam points radially outward along a line connecting the beam to the pipe axis. Equation (10.104) shows that the image charge moves to infinity as δ approaches 0; therefore, the image electric force vanishes. Equation (10.104) follows from the assumption of a filamentary beam—it holds only if the beam is not close to the wall. We adopt the condition that $\delta \ll r_w$. In this limit, the following expression gives the electric force from the wall on each particle of a beam in coordinates referenced to the pipe axis:

$$\mathbf{F}_{ew} \cong (qQ/2\pi\varepsilon_0 r_w^2)\delta\mathbf{r} \tag{10.105}$$

$$\cong (qI/2\pi\varepsilon_0 \beta c r_w^2)\delta\mathbf{r}.$$

The quantity $\mathbf{r}$ is a unit radial vector that points from the pipe axis to the beam center. Note that the image force varies linearly with beam displacement. Also, Eq. (10.105) represents a collective effect—the force on each particle is proportional to the total beam current.

We can apply a similar analysis to find the magnetic fields of wall return current in terms of an image line current. In a beam-centered coordinate system, the magnetic field of the beam is

$$B_\theta = \mu_0 I/2\pi r'. \tag{10.106}$$

Equation (10.106) gives a vector potential

$$A_z = (\mu_0 I/2\pi)\ln(r') + K. \tag{10.107}$$

The wall is a surface of constant A_z, since magnetic field lines must be parallel to a perfectly conducting surface. We can show that the net vector potential is a constant for a filamentary image current, $-I$, at the same position as the image charge [Eq. (10.104)]. The force from the current points toward the pipe axis. The magnetic force from the return current equals the electric force of the image charge multiplied by $-\beta^2$. For $\delta \ll r_w$, the sum of electric and magnetic wall forces on a displaced filamentary beam is

$$\mathbf{F}_w \cong \left[\frac{qI}{2\pi\varepsilon_0 r_w^2 \beta\gamma^2 c} \right] \delta\mathbf{r}. \tag{10.108}$$

Wall resistivity may affect the distribution of the return current for long beam pulses. To understand the effect, consider the schematic geometry of a beam-transport system is Fig. 10.18*d*. There are other connections between the system exit and entrance besides the vacuum-chamber pipe. For simplicity, we represent these connections by a large-radius tube around the vacuum chamber. Depending on the accelerator geometry, the outer cylinder may not be collinear with the vacuum chamber. Immediately following the beginning of a beam pulse, all return current flows through the vacuum-chamber wall, the path of lowest inductance. As time passes, resistivity of the chamber wall causes diffusion of the magnetic field into the volume between the two cylinders. At late time, the return current flows along the path of lowest resistance. We shall derive an expression for the diffusion time in Section 12.6. Field diffusion may affect the equilibrium position of a long-pulse relativistic electron beam. At early time, the image charge and return-current image are collinear—electric and magnetic wall forces almost cancel. With field diffusion, the return-current image may be displaced from the line-charge image. In such a situation unbalanced wall forces may shift the beam position.

To conclude, we shall discuss the nature of the *elastic beam approximation*. In this and several following sections, we study average displacements of a beam without addressing details of the transverse equilibrium. Often, we can represent a charged particle beam as an elastic body that maintains its shape and size when it is displaced in the transverse direction. The model holds if all applied and beam-generated forces are linear about a symmetry axis. We also include, as special linear forces, transverse displacement forces that are uniform in position. We assume that without displacement forces the beam has a matched equilibrium in the linear focusing forces. Figure 10.19*a* illustrates equilibrium of a one-dimensional beam centered on the symmetry axis. All particles have the same betatron wavelength, λ_b—the beam distribution consists of a mixture of amplitudes and phases. Contributions of applied forces, beam-generated forces, and emittance give a force balance at all positions. Suppose we introduce a uniform displacement force slowly compared with the betatron oscillation period. The center of the beam shifts to a new position where the net transverse force equals zero (Fig. 10.19*b*). Note that the applied focusing forces referenced to the new beam center are unchanged. Therefore, the matched beam has the same emittance and radius.

Figure 10.19*c* shows an example of elastic motion. A beam enters a region of linear forces with no displacement force. All particles have the same betatron wavelength. The matched beam has an initial displacement from the symmetry

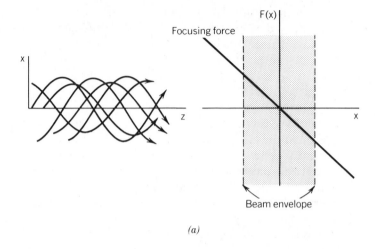

(a)

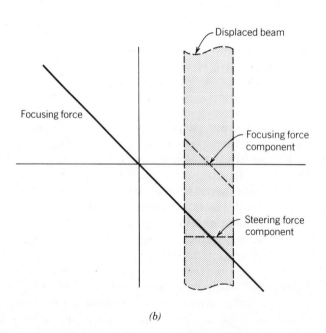

(b)

Figure 10.19   Description of a beam as an elastic body. (*a*) Beam particles in equilibrium in a linear focusing force. Left-hand side: configuration-space orbits, all particles have the same value of λ_b. Right-hand side: spatial variation of focusing force. (*b*) Forces on a displaced beam resolved into a focusing component and a uniform transverse force. (*c*) Configuration-space view of particle orbits in a displaced elastic beam. All particles have the same value of λ_b with oscillation amplitude and phase randomly distributed within a range.

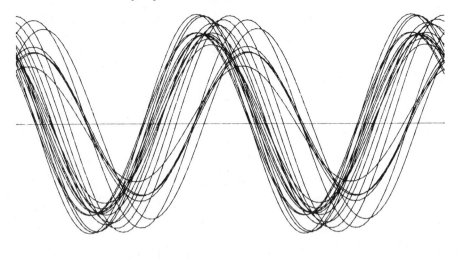

(c)

Figure 10.19 *(Continued)*

axis. The particle motion in the focusing force is the sum of normal oscillations for a matched beam with nonzero emittance with an additional component from the initial displacement. The beam distribution at a downstream location is the sum over particle betatron oscillations. If the beam width is smaller than λ_b, the beam has constant width and exhibits a superimposed oscillation of the centroid. The centroid motion has wavelength equal to the betatron wavelength and an amplitude equal to the initial displacement. We also observe elastic beam motion if a beam propagates in linear-focusing forces in the presence of a displacement force that varies over scale lengths much longer than λ_b. The displacement force shifts the centroid of the beam but does not modify the matched equilibrium.

It is important to recognize that a beam does not behave like an elastic body when the particles have a spread in betatron wavelength. Such spreads result from either nonlinear-focusing forces or a dispersion in kinetic energy. In the presence of nonlinearities, the focusing force referenced to a displaced position differs from the force at the symmetry axis. As a result, a displacement causes modification of the transverse equilibrium. A spread in betatron wavelength affects the transverse oscillations of Fig. 10.19c. Phase mixing of individual particle orbits results in damping of the oscillations with an attendant growth of emittance. Section 13.3 discusses the effect of phase-mix damping and its importance for stabilizing transverse instabilities.

10.7. DRIFTS OF ELECTRON BEAMS IN A SOLENOIDAL FIELD

Solenoidal fields are often used to focus high-current electron beams. We studied equilibria of beams in magnetic fields in Sections 10.2 and 10.3. For those models, we assumed that the beam had cylindrical symmetry and propagated on the axis of the transport system. In Section 10.8, we address the problem of beam guiding in a solenoidal field. We want to find how perturbation forces and system imperfections affect the average beam position. As a preliminary, in this section we shall derive expressions for electron beam drift motion in a solenoidal field and discuss the basis of the drift model for a filamentary beam.

We shall treat the beam in the elastic approximation discussed in Section 10.6. The model is valid when the following conditions hold:

1. Beam electrons have paraxial orbits and the same kinetic energy.
2. Applied focusing forces maintain a matched beam equilibrium with small radius.
3. To calculate macroscopic motions of the beam, we follow the orbit of a test particle at the beam center.

The third condition applies if all electrons have about the same betatron wavelength.

We shall limit attention to the position of the beam centroid in a plane normal to z. In addition to focusing forces with linear variation about the system axis, there are also transverse perturbation forces. To apply the elastic beam model, we assume that these forces are constant over the beam width—they act uniformly on all electrons. As a result, the response of the beam to a perturbing force is identical to the response of individual electrons. We divide perturbing forces into two categories—single-particle forces and collective forces. The effect of a single-particle force is independent of beam properties. For example, the force of a vertical applied magnetic bending field affects all electrons equally and does not depend on the beam current. In contrast, collective forces depend on the density of particles. The wall force described in Section 10.6 is a good example of a collective force—its magnitude is proportional to the beam current. Note that the wall force of Eq. (10.108) is almost constant over the beam width if the beam displacement is small compared with the wall radius, r_w.

Figure 10.20 shows an electron beam in a uniform solenoidal field,

$$\mathbf{B}(x, y) = B_0\mathbf{z}. \tag{10.109}$$

We shall calculate the motion of the beam center in a plane normal to z in the presence of a force:

$$\mathbf{F} = F_x\mathbf{x}. \tag{10.110}$$

We assume that the energy associated with transverse motion is much smaller than the axial kinetic energy. Therefore, the quantity γ is almost unchanged by the transverse motion. Thus we can apply nonrelativistic equations in the transverse directions with an adjusted mass per particle of $m = \gamma m_0$.

We shall concentrate on the orbit of a test electron at the center of the beam.

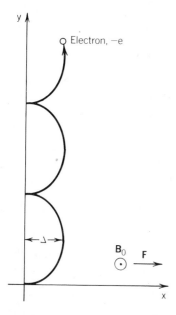

y

○ Electron, $-e$

B_0 F

○⃗

$\leftarrow \!\!\!\Delta\!\!\! \rightarrow$

x

Figure 10.20 Drift motion of electrons in a uniform magnetic field perpendicular to an applied force.

Figure 10.20 shows that the force initially causes the electron to move in the positive x direction. The combined effect of the axial magnetic field and the velocity v_x creates a force that accelerates the electron upward. The $v_y \times B_z$ force, directed to the left, leads to a reversal of v_x. The result is that the test electron follows the scalloped orbit shown. The electron moves in the positive y direction and there is no cumulative displacement in the x direction. The time-averaged motion along y is called a *drift*. Analysis of Fig. 10.20 shows that a particle with charge q has a drift velocity in the direction

$$\mathbf{v}_d \sim q(\mathbf{F} \times \mathbf{B}). \tag{10.111}$$

The drift motion of Eq. (10.111) applies to a single electron or to the center of a filamentary beam.

If we denote the coordinates of the beam center as $[X, Y]$, the equations of motion are

$$\gamma m_0 (d^2 X/dt^2) = -eB_0(dY/dt) + F_x \tag{10.112}$$

and

$$\gamma m_0 (d^2 Y/dt^2) = eB_0(dX/dt). \tag{10.113}$$

With the substitution

$$\xi = Y - (F_x/eB_0)t, \tag{10.114}$$

Eqs. (10.112) and (10.113) become

$$dv_x/dt = -(eB_0/\gamma m_0)v_\xi, \tag{10.115}$$

$$dv_\xi/dt = (eB_0/\gamma m_0)v_x. \tag{10.116}$$

Equation (10.114) represents a coordinate transformation to a frame of reference moving at velocity $(F_x/eB_0)\mathbf{y}$. In the moving frame, Eqs. (10.115) and (10.116) show that the beam center follows a circular path at the relativistic gyro-frequency, $\omega_g = (eB_0/\gamma m_0)$. The beam moves on an orbit that is a composite of a magnetic gyration and the constant drift velocity:

$$\mathbf{v}_d = q(\mathbf{F} \times \mathbf{B})/(qB_0)^2. \tag{10.117}$$

The magnitude of the drift velocity is $|v_d| = F/qB_0$.

We can determine the gyroradius for beam oscillations by noting that the beam has zero transverse velocity at $t = 0$ in the stationary frame. Therefore, the beam moves at speed v_d along the y axis in the transformed frame. The gyroradius is

$$r_g = \gamma m_0 v_d/qB_0. \tag{10.118}$$

Figure 10.20 shows that the amplitude of the oscillatory motion is

$$\Delta = 2r_g = 2\gamma m_0 F/(qB_0)^2. \tag{10.119}$$

In drift orbit theory, we find the approximate position of the beam by assuming drift velocities and neglecting the oscillatory motion. The drift approximation holds when two conditions are satisfied:

1. Over the time of observation, Δt, the beam performs many oscillations in the magnetic field:

$$\omega_g \Delta t = eB_0 \Delta t/\gamma m_0 \gg 1. \tag{10.120}$$

2. The size of the transport system, r_w, is much larger than the amplitude of the beam gyrations:

$$\Delta/r_w = 2\gamma m_0 F/r_w(qB_0)^2 \ll 1. \tag{10.121}$$

The most familiar drift motion occurs when the applied force results from an electric field, $\mathbf{E}$. The drift velocity is

$$\mathbf{v}_d = \mathbf{E} \times \mathbf{B}/B_0^2. \tag{10.122}$$

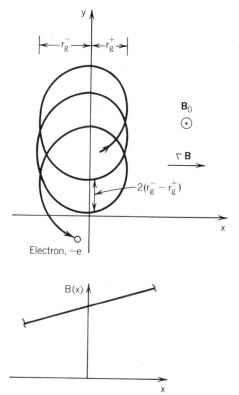

Figure 10.21 Geometry to calculate the drift motion of energetic electrons in a magnetic field with a spatial gradient.

The quantity defined in Eq. (10.122) is the *E-cross-B* velocity. We have already studied this type of motion for electrons in magnetically insulated high-voltage gaps (Section 8.1). Note that the velocity does not depend on the mass or charge of the beam particles.

Drift motions of a beam also result from time-varying electric fields and spatially varying magnetic fields. Section 8.1 described motion in a time-varying electric field, the *polarization drift*. Figure 10.21 illustrates the drift associated with spatial variations of the focusing magnetic field. Suppose that the beam enters the transport region at $x = 0$. At this position, the magnetic field has magnitude B_0 and has a positive gradient (assumed constant) in the x direction, $\nabla B = (\partial B / \partial x) \mathbf{x}$. If electrons in the beam have no initial transverse velocity, the beam does not move normal to the field. On the other hand, if individual electrons in the beam have transverse energy, they gyrate in the magnetic field (Fig. 10.21). The electrons experience a higher magnetic field in the half-plane with $x > 0$—here, their gyroradius is smaller. The opposite effect occurs in the region $x < 0$. The result, illustrated in Fig. 10.21, is that electrons have an oscillatory motion in x and a net drift in y. The electrons move in the $-y$ direction for a positive magnetic field gradient along x. If all electrons have the same transverse energy, the beam center moves downward with no change in the shape of the beam. With a spread in transverse energy, the beam may stretch along the y direction.

We can estimate the magnitude of the drift velocity by noting that an electron moves a distance along y equal to the difference in the average orbit diameters to the left and right of the $x = 0$ in an interval $2\pi/\omega_g$. An electron with transverse velocity $\beta_\perp c$ has an average gyroradius in the region $(x > 0)$ of

$$r_g^+ \cong \gamma m_e \beta_\perp c \left/ e\left[B_0 + \left(\frac{\partial B}{\partial x}\right)\Delta x \right] \right., \tag{10.123}$$

where

$$\Delta x \sim \tfrac{1}{2}(\gamma m_e \beta_\perp c/eB_0). \tag{10.124}$$

The electron has an average gyroradius

$$r_g^- \cong \gamma m_e \beta_\perp c/e[B_0 - (\partial B/\partial x)\Delta x)]$$

in the region $(x < 0)$. The drift velocity is roughly

$$v_d \cong 2(r_g^+ - r_g^-)\omega_g/2\pi. \tag{10.125}$$

Substituting from Eqs. (10.123) and (10.124) gives

$$v_d \cong \frac{\gamma m_e c^2 \beta_\perp^2 (\partial B_z/\partial x)}{\pi q B_0^2}. \tag{10.126}$$

A detailed analysis of the *grad-B* drift[5] gives the velocity as

$$\mathbf{v} = \frac{\gamma m_e c^2 \beta_\perp^2}{2q} \frac{\mathbf{B} \times \nabla \mathbf{B}}{B^3}. \tag{10.127}$$

Equation (10.127) holds in the limit

$$\frac{|\nabla \mathbf{B}|}{B_0} \ll \frac{eB_0}{\gamma m_0 \beta_\perp c}. \tag{10.128}$$

10.8. GUIDING ELECTRON BEAMS WITH SOLENOIDAL FIELDS

We shall apply the results of Sections 10.6 and 10.7 to the problem of guiding high-current electron beams. We have already discussed the role of solenoidal fields for beam focusing. The focusing field confines the beam to a small radius in

---

[5] See, for instance, G. Schmidt, *Physics of High Temperature Plasmas*, Academic Press, New York, 1979, p. 14.

the presence of space-charge and emittance forces. In this section, we turn our attention to the problem of gross beam motion from the axis of an accelerator through force imbalance and instabilities. These effects are particularly important for circular accelerators such as high-current betatrons and recirculating linacs. Although solenoidal focusing is commonly used in these machines, the field geometry has poor stability properties that increase the difficulties of beam transport.

Figure 10.22 illustrates the geometry and the coordinate system for the analysis. A narrow beam propagates through a pipe of radius r_w. We treat the beam as an elastic filament—forces are either uniform over the beam cross section or vary linearly about the symmetry axis of the pipe. The z axis corresponds to the pipe axis. We allow the possibility that the pipe has a gentle bend with radius of curvature much larger than r_w. We define two special directions. The y axis lies along the *vertical direction*—it is perpendicular to the pipe radius of curvature. The x axis lies in the *horizontal direction*—it points outward along the radius of curvature. A focusing magnetic field extends along the z axis—the field may have a constant gradient in the x direction:

$$\mathbf{B} = \left[B_0 + \left(\frac{\partial B}{\partial x} \right) x \right] \mathbf{z}. \tag{10.129}$$

Finally, we assume that motions of the beam center satisfy the drift orbit conditions (Section 10.7). The time scale for transverse beam motion is much longer than the relativistic gyroperiod.

To begin, we shall study a filamentary electron beam in a straight transport

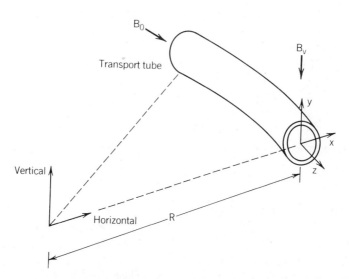

Figure 10.22 Geometry and coordinate system to analyze electron transport in a torus.

a)

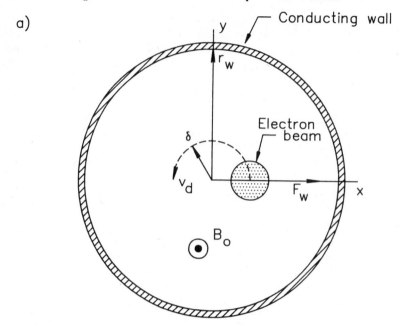

b)

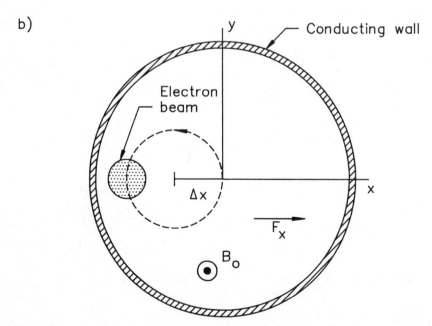

Figure 10.23 Drift motion of an electron beam in a metal transport tube with an applied axial magnetic field. (*a*) Drift resulting from the radial force of induced charge and return current in the wall. (*b*) Drift motion resulting from a lateral force such as centrifugal force.

pipe with a uniform magnetic field, $B_0\mathbf{z}$. There are no other applied forces. The paraxial beam has current I_0. The beam electrons have kinetic energy $(\gamma - 1)m_e c^2$ and axial velocity $v_z = \beta c$. If the beam enters on the pipe axis, there are no transverse forces. On the other hand, a displaced beam experiences a force resulting from the induced charges and currents in the pipe wall (Section 10.6). Figure 10.23a shows that the wall force is radial. Equation (10.108) gives the magnitude of the force for a beam displacement δ. Following Section 10.7, the combined effects of the axial magnetic field and the radial force leads to an azimuthal drift. Substituting the wall force expression [Eq. (10.108)] in the drift velocity equation [Eq. (10.117)], we find that

$$|v_d| = \frac{eI_0\delta}{2\pi\varepsilon_0 r_w^2 \beta c \gamma^2 e B_0}$$

$$= \frac{K\gamma m_e \beta c^2 \delta}{eB_0 r_w^2}. \tag{10.130}$$

The quantity K in Eq. (10.130) is the generalized perveance. Since the drift motion is azimuthal, the beam center follows a circular orbit of radius δ around the axis of the pipe (Fig. 10.23a). The beam rotates at an angular drift frequency $\omega_d = |v_d|/\delta$. Substituting from Eq. (10.130) gives an expression for the drift frequency:

$$\omega_d = \frac{K(\beta c/r_w)^2}{\omega_g}. \tag{10.131}$$

The quantity ω_g is the relativistic gyrofrequency for individual electrons:

$$\omega_g = eB_0/\gamma m_e. \tag{10.132}$$

As an example, suppose a 1-MeV, 1-kA electron propagates in a cylindrical vacuum chamber of radius $r_w = 0.08$ m. The generalized perveance is $K = 5.42 \times 10^{-3}$. The beam electrons have $\gamma = 2.96$ and $\beta = 0.941$. The applied solenoidal field must be strong enough to maintain a small beam equilibrium radius. If the emittance is small, Section 9.4 shows that the radius of a matched beam is

$$r_0 = 2\sqrt{K}(\beta c/\omega_g). \tag{10.133}$$

We take $r_0 = 0.01$ m to satisfy the filamentary beam condition. Equation (10.133) implies a focusing field magnitude of $B_0 = 0.070$ tesla, corresponding to a relativistic electron gyrofrequency of $\omega_g = 4.16 \times 10^9$ sec^{-1}. The rotation caused by the wall force has an angular frequency of $\omega_d = 1.62 \times 10^7$ sec^{-1}. The drift orbit limit gives a good description of beam motion since $\omega_d \ll \omega_g$. The drift motion is slow—it takes 390 nsec to complete a revolution.

Next, we shall apply a lateral perturbation force, $F_x \mathbf{x}$, to a beam in a straight pipe. The force is small enough so that drift orbit theory describes the motion of the beam center. Without a conducting wall, there is no preferred propagation axis. Equation (10.117) implies that the beam drifts without limit in the y direction under the influence of the x-directed force. This behavior results from a property of focusing by a uniform axial magnetic field—there is no stability axis where the beam energy is a minimum. Therefore, a small perturbation force can displace the beam to any transverse location.

Wall forces can stabilize a magnetically focused electron beam in a pipe. Sometimes, the combination of the wall force with a lateral force leads to closed drift orbits that do not intersect the wall. We can express the wall force in the Cartesian coordinates of Fig. 10.23a:

$$\begin{aligned} \mathbf{F}_w &= K\gamma(\beta c)^2 r\mathbf{r} \\ &= K\gamma(\beta c)^2(x\mathbf{x} + y\mathbf{y}). \end{aligned} \tag{10.134}$$

Combining the wall force with the perturbation force, the drift orbit equations for the beam are

$$v_x = -\omega_d y, \tag{10.135}$$

$$v_y = \omega_d x + F_x/eB_0. \tag{10.136}$$

The quantity ω_d is given by Eq. (10.131). With the substitution $\xi = x + F_x/eB_0\omega_d$, Eqs. (10.135) and (10.136) become

$$v_\xi = -\omega_d y, \tag{10.137}$$

$$v_y = \omega_d \xi. \tag{10.138}$$

Equations (10.137) and (10.138) describe circular motion at angular frequency ω_d about a point displaced along the x axis a distance

$$\Delta x = -F_x/eB_0\omega_d. \tag{10.139}$$

If the beam enters on the system axis, the drift orbit passes through the axis (Fig. 10.23b).

Equation (10.139) shows that for given properties of the beam and transport system, there is a maximum tolerable perturbation force. For a beam injected on the axis, the condition $2\Delta x < r_w$ must hold, or

$$F_x < 2r_w eB_0\omega_d. \tag{10.140}$$

Another viewpoint is that for a given perturbation force, there is a minimum value of the generalized perveance for successful beam transport. Here we have an interesting case where the collective force of a high-current beam improves transport. We also note that conducting walls cannot stabilize a neutralized electron beam focused by solenoidal fields. If a high-current electron beam propagates in a plasma so that there is no net current or charge inside a pipe, there is no wall force. Therefore, a small perturbation force will push the beam to the wall.

A beam in a circular accelerator is subject to forces in the horizontal direction. Figure 10.22 shows the geometry for the analysis of a curved transport system. The pipe has minor radius r_w and major radius R. Magnet windings around the pipe create a toroidal magnetic field of average magnitude B_0. A vertical magnetic field, B_v, is necessary to combat the centrifugal force on the beam. The quantity x is the displacement of the beam center from the pipe axis parallel to the radius of curvature, while the y displacement is parallel to the vertical field.

To avoid evaluating complex toroidal field expressions, we assume that $r_w \ll R$. In this limit, the expression for the wall interaction force [Eq. (10.134)] is still almost correct. Ampere's law requires that the magnitude of the focusing field varies in the horizontal direction according to

$$B_z(x) \cong B_0 R/(R + x). \tag{10.141}$$

The focusing field has a horizontal gradient:

$$\nabla \mathbf{B} \cong -(B_0/R)\mathbf{x}. \tag{10.142}$$

To construct a drift equation for the beam, we must list all forces that act on the electrons. In the y direction, the only force is the wall interaction, proportional to the vertical deflection of the beam. There are three forces in the horizontal direction: the wall force, the centrifugal force, and the action of the bending magnetic field. The centrifugal force equals

$$\mathbf{F}_c = \gamma m_0 (\beta c)^2 /(R + x)\mathbf{x} \tag{10.143}$$
$$\cong [\gamma m_0 (\beta c)^2 /R][1 - x/R]\mathbf{x}.$$

We define the vertical bending magnetic field in terms of a scaling field:

$$B_v = \gamma m_0 \beta c/eR. \tag{10.144}$$

The quantity B_v is the field magnitude that gives an electron gyroradius equal to R. Without perturbations and collective effects, a field magnitude B_v would maintain the beam on the symmetry axis. We introduce an error factor ε for the vertical field—the bending force is

$$\mathbf{F}_m = -e\beta c B_v(1 + \varepsilon). \tag{10.145}$$

The quantity ε could represent a mismatch between the bending field and the electron beam energy or an adjustment deliberately introduced to compensate perturbations.

The beam drift velocity in the vertical direction is the sum of the following components:

1. The wall interaction,

$$v_d = \omega_d x. \tag{10.146}$$

2. The magnetic and centrifugal forces,

$$v_d = [\beta c(B_v/B_0)][-\varepsilon + x/R]. \tag{10.147}$$

3. The grad-B drift,

$$v_d = (\beta c/2)(B_v/B_0)(\beta_\perp/\beta)^2. \tag{10.148}$$

The expression for the grad-B drift comes from Eqs. (10.127) and (10.142). In Eq. (10.148), $\beta_\perp c$ equals the average transverse velocity of beam electrons.

The complete drift equations for solenoidal field focusing in a circular accelerator are

$$v_x = -\omega_d y, \tag{10.149}$$

$$v_y = \omega_d x - \varepsilon(\beta c)(B_v/B_0) - (\beta c)(B_v/B_0)(x/R) + (\beta c/2)(B_v/B_0)(\beta_\perp/\beta)^2 \tag{10.150}$$

We can use Eq. (10.150) to find conditions for a beam equilibrium, defined by $v_x = v_y = x = y = 0$. When the electron transverse velocity equals zero ($\beta_\perp = 0$), the beam center lies on the pipe axis if ε equals zero. Usually, the electrons of high-current beams in solenoidal fields have substantial transverse velocity. This results from envelope oscillations if the beam is generated in the field (Section 10.2) or from magnetic deflections in the transition region if the electrons are generated outside the field (Section 10.3). When $\beta_\perp$ is nonzero, the condition for a beam equilibrium is

$$\varepsilon = (\beta_\perp/\beta)^2/2. \tag{10.151}$$

Equation (10.151) shows that the bending field must exceed B_v to counteract the outward force from the interaction between the negative gradient of focusing field and the electron transverse energy. In real beams, the electrons have a spread in $\beta_\perp$. The variation in drift velocities causes distortion of the beam profile.

We can investigate the effects of force imbalances by solving Eqs. (10.149) and (10.150). For simplicity, we take $\beta_\perp = 0$. If we make the substitution

$$\xi = x\left[1 - \frac{\beta c}{R\omega_d}\left(\frac{B_v}{B_0}\right)\right] - \frac{\varepsilon\beta c}{\omega_d}\left(\frac{B_v}{B_0}\right), \tag{10.152}$$

the equations become

$$(d\xi/dt) = -\omega_d[1 - (\beta c/R\omega_d)(B_v/B_0)]y,\tag{10.153}$$

$$(dy/dt) = +\omega_d\xi.\tag{10.154}$$

We can combine Eqs. (10.153) and (10.154) into the single equation,

$$(d^2\xi/dt^2) = -\omega_d^2[1 - (\beta c/R\omega_d)(B_v/B_0)]\xi.\tag{10.155}$$

If the beam enters on the axis ($x = 0, y = 0$) at $t = 0$, Eq. (10.155) has the solution

$$\xi(t) = \zeta_0 \cos \Omega t,\tag{10.156}$$

where

$$\Omega = \omega_d[1 - (\beta c/R\omega_d)(B_v/B_0)]^{1/2}$$

and

$$\zeta_0 = (\varepsilon\beta c/\omega_d)(B_v/B_0).$$

We can substitute Eq. (10.156) into the original equations to give the following expressions for $x(t)$ and $y(t)$:

$$x(t) = \frac{(\varepsilon\beta c/\omega_d)(B_v/B_0)(1 - \cos \Omega t)}{[1 - (\beta c/R\omega_d)(B_v/B_0)]^{1/2}},\tag{10.157}$$

$$y(t) = -(\varepsilon\beta c/\omega_d)(B_v/B_0)\sin \Omega t.\tag{10.158}$$

The beam center sweeps out an ellipse at angular frequency Ω. The major axis of the ellipse lies along the x direction. The maximum distance that the beam center moves from the pipe axis is

$$\Delta x_{max} = \frac{2(\varepsilon\beta c/\omega_d)(B_v/B_0)}{[1 - (\beta c/R\omega_d)(B_v/B_0)]^{1/2}}.\tag{10.159}$$

A necessary condition for the existence of a closed orbit is that the bracketed quantity in the denominator of Eq. (10.159) is positive, or

$$(\beta c/R\omega_d)(B_v/B_0) > 1.\tag{10.160}$$

With some work, we can reduce Eq. (10.160) to the simple stability requirement,

$$K > (r_w/R)^2.\tag{10.161}$$

Given the existance of a closed orbit, a sufficient condition for beam transport is

$$\Delta x_{max} < r_w.\tag{10.162}$$

We shall illustrate the utility of the transport theory with some examples. Suppose we have an accelerator with $R = 1$ m and $r_w = 0.08$. A 1-MeV electron beam with 1-kA current has a generalized perveance of only 5.42×10^{-3}. Equation (10.161) shows that we cannot transport such a beam in the accelerator, even with very high values of B_0. Any field error, ε, drives the beam to the wall. We can achieve a closed drift orbit if we raise the beam current to 5 kA. The generalized perveance of a 5-kA, 1-MeV beam is 2.7×10^2. If the beam has zero emittance, a focusing field of 0.16 tesla maintains the 0.01 m beam radius. The vertical field is $B_v = 4.8 \times 10^{-3}$ tesla, and the wall interaction drift frequency is $\omega_d = 8.1 \times 10^7$ sec^{-1}. Substituting into Eqs. (10.159) and (10.162), we find an upper limit on the error of the vertical field, $\varepsilon < 0.34$. It is easy to adjust the magnetic field to such a tolerance. On the other hand, severe problems arise when the beam accelerates. The generalized perveance drops as $K \sim 1/\gamma^3$—wall forces are small for high-energy electron beams. An accelerating beam quickly reaches a kinetic energy where the condition of Eq. (10.161) does not hold. For example, wall stabilization fails when a 5-kA beam in the given accelerator geometry reaches a kinetic energy of only 1.9 MeV.

10.9. ELECTRON BEAM TRANSPORT IN MAGNETIC CUSPS

A magnetic cusp array consists of a series of solenoid lenses with alternating field polarity. The focusing properties of such a system are similar to a uniform solenoid. The main difference is that the cusp array guides beams more effectively in the presence of perturbations. This property compensates for the higher current that cusp coils require to achieve the same focusing strength. In this section, we discuss several topics related to high-current electron beam propagation in an alternating solenoid lens array.

A. Magnetic Fields

The cusp field geometry has a long history of application to space-charge-dominated beams in devices with permanent magnet focusing. Periodic-permanent-magnet (PPM) arrays consist of a series of annular magnets separated by iron rings (Fig. 10.24a). The magnetization alternates in direction between elements. If the magnet length is comparable to the diameter, a useful fraction of the magnetic flux returns through the inside of the annulus. This portion of the field energy is available for beam focusing.

A series of magnet coils with alternating azimuthal current also generates a series of magnetic cusps. Figure 10.24b shows field lines for a coil array with no ferromagnetic material. The addition of iron flux return structures outside the coils reduces the ampere-turns required to achieve a given on-axis field [CPA]. Figure 10.24c shows numerical solutions for a cusp array with flux return through iron.

We can calculate the magnetic fields of a cusp array directly when there are no

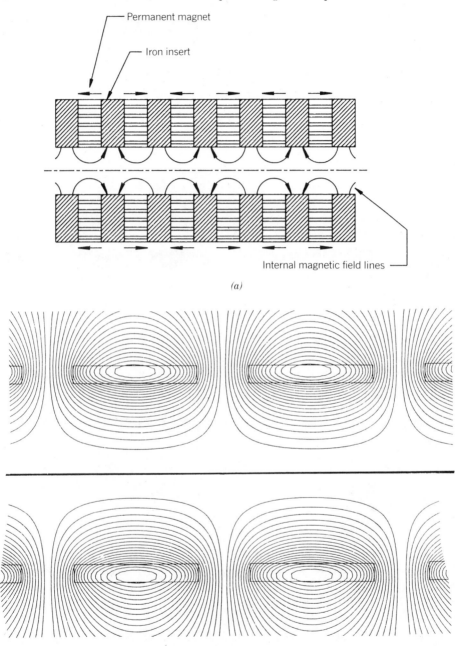

Figure 10.24 Multiple magnetic cusp geometries for electron beam transport. (*a*) Periodic-permanent-magnet (PPM) array. External arrows show direction of magnetization. (*b*) Calculated magnetic field lines, multiple solenoid lenses with alternating polarity. (*c*) Calculated field lines, multiple solenoid lenses with an iron flux-return tubes.

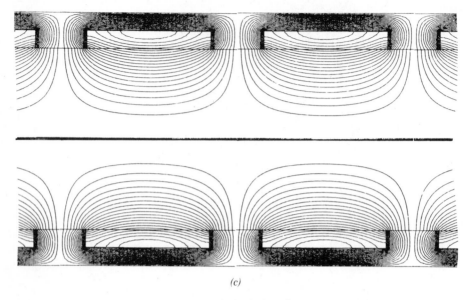

(c)

Figure 10.24 *(Continued)*

nearby ferromagnetic structures. We use the analytic expression for the vector potential, A_θ, from a single current loop [CPA, p. 73] and sum contributions from all coils in the array. Figure 10.25 shows results from such a calculation. The coil current is represented by a series of current loops of radius R_c distributed uniformly in z. The polarity of the current follows a step-function variation with cell length L. Figure 10.25 shows the field on axis, $B_z(0, z)$, for three values of L/R_c.

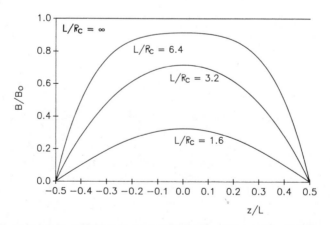

Figure 10.25 Calculated axial field, $B_z(0, z)$, for multiple solenoid lenses (Fig. 10.24b) as a function of the magnet geometry. Uniform magnet windings of radius r_c reversed in direction periodically over a distance L. The magnetic field is normalized to B_0, the field generated by a solenoid winding of infinite length.

The dashed line shows the field of a solenoid with the same value of linear coil current density (ampere-turns per meter). The field in the cusp array is lower because of cancellation between adjacent cells. Field cancellation increases in importance as L/R_c decreases. The figure shows that for moderate values of L/R_c the axial field in a cusp array has the approximate variation:

$$B_z(z) = B_0 \sin(2\pi z/L) = B_0 \sin(k_m z). \tag{10.163}$$

The quantity B_0 depends on the linear current density and L/R_c.

B. Radial Force Balance for a Laminar Beam

We shall next derive the limiting current for a high-energy electron beam in a cusp array. In contrast to the continuous radial focusing force of an extended solenoid, the force is periodic; therefore, it is difficult to derive exact solutions for self-consistent beam equilibria. Instead, we shall use a model based on the average force approximation of Section 9.4. The model holds when the lengths of focusing cells are much shorter than the single-particle betatron wavelength. To simplify the discussion, we limit attention to laminar beam equilibria. Again, we take a paraxial beam with current well below the longitudinal space-charge limit [Eq. (5.77)].

The following equation describes the envelope radius, R, of a laminar high-current electron beam in a cusp array:

$$R'' = \frac{K}{R} - \left[\frac{eB_z(z)}{2\gamma m_e \beta c} \right]^2 R. \tag{10.164}$$

The quantity K is the generalized perveance [Eq. (5.88)]. The first term on the right-hand side of Eq. (10.164) represents the beam-generated forces, while the second term describes the combined effects of focusing and centrifugal forces for electrons with zero canonical angular momentum, $P_\theta = 0$. We find conditions for an axially averaged matched equilibrium by setting $R'' \cong 0$ and averaging the force over the length of a focusing cell. We assume that R remains close to an average value R_0. The equilibrium condition is

$$K \cong e^2 \langle B_z(z)^2 \rangle R_0^2/4\gamma^2 m_e^2 \beta^2 c^2. \tag{10.165}$$

The content of Eq. (10.165) is similar to Eq. (9.56) for a solenoidal field. The difference is that the mean-squared magnetic field appears rather than a uniform field value. If $B_z(z)$ follows the harmonic variation of Eq. (10.163), the radial force balance condition is

$$K \cong e^2 B_0^2 R_0^2/8\gamma^2 m_e^2 \beta^2 c^2. \tag{10.166}$$

Equation (10.165) leads to an expression for the limiting current in a cusp array:

$$I \leqslant \left[\frac{\pi e \varepsilon_0 c}{2m_e}\right](\beta\gamma)\langle B_z(z)^2 \rangle R_0^2 = \left[\frac{\pi e \varepsilon_0 c}{4m_e}\right](\beta\gamma) B_0^2 R_0^2. \qquad (10.167)$$

If B_0 has units of tesla and R_0 is in meters, the current limit for electrons in amperes is

$$I \leqslant 3.67 \times 10^8 (\beta\gamma) B_0^2 R_0^2. \qquad (10.168)$$

As an example, suppose we have a 5-kA electron beam with kinetic energy of 2.5 MeV ($\gamma = 5.89$ and $\beta = 0.986$). If the envelope radius is 0.01 m and the transport pipe radius is 0.04 m, the beam-generated electrostatic potential is only 0.57 MV [Eq. (5.68)]; therefore, electron motion is paraxial. Equation (10.168) implies that the beam is almost in radial equilibrium if the peak field in the cusp array is $B_0 = 0.15$ tesla.

C. Single-Particle Orbits

Although conditions for radial force balance on the beam envelope are similar for a uniform solenoid and a cusp array, the orbits of individual particles are quite different. Figure 10.26a shows particle trajectories projected in the r-θ plane with

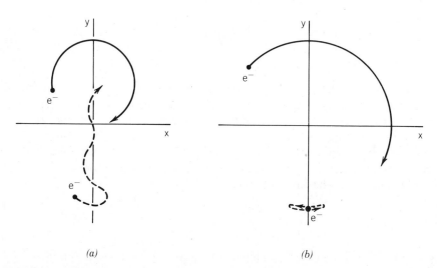

(a)　　　　　　　　　　(b)

Figure 10.26 Electron orbits in magnetic transport systems projected in the r-θ plane. Electrons generated by a nonimmersed cathode. Solid line: Orbit in a uniform solenoidal magnetic field. Dashed line: Orbit in a magnetic cusp array. (a) Single-particle orbits with no beam-generated radial electric field. (b) Orbits in a matched equilibrium for a laminar beam with included space-charge force. Beam in uniform field has a Brillouin equilibrium, while the beam in the cusp array has radial force balance averaged over a cell.

no beam-generated field. In a uniform solenoid, an electron with zero canonical angular momentum follows a circular path that passes through the axis. In a cusp array, electron orbits are more complex. Electrons oscillate in the azimuthal direction because of the reversing field. If the betatron wavelength is much longer than a cell, the azimuthal motion averages to zero over many cells. On the other hand, the radial force points inward in all cells resulting in a cumulative motion. The dashed line of Fig. 10.26a shows a typical orbit. The electron trajectory lies close to a straight line that passes through the axis.

Figure 10.26b plots electron orbits in laminar beams in space-charge equilibrium. In a solenoidal field, the beam has a Brillouin equilibrium (Section 10.3). All electrons follow axis-centered circular orbits with the same rotation frequency (solid line). In contrast, the orbit projections in a cusp array show small oscillations about a point (dashed line).

D. Envelope Oscillations

Although Eq. (10.168) is a necessary condition for a radial beam equilibrium, it is not sufficient to define the current limit in a cusp array. Stability requirements constrain the amplitude of the focusing magnetic field in the periodic system. The instability for a space-charge-dominated beam is the collective equivalent of the familiar single-particle orbital instability [CPA, p. 182]—it occurs when the single-particle vacuum phase advance per focusing cell exceeds 180°. We shall derive the stability criterion for a high-current beam by investigating the behavior of beam envelope oscillations in a harmonically varying focusing force. The derivation also yields validity conditions for the average force model.

We shall study perturbations of a laminar beam about the equilibrium defined by Eq. (10.166) by solving Eq. (10.164) with a harmonic variation of focusing field [Eqs. (10.163)]. We can simplify Eq. (10.164) by introducing dimensionless variables. A good choice for the radial scaling length is R_0, the equilibrium envelope radius. The dimensionless radial coordinate is

$$\rho = R/R_0. \tag{10.169}$$

We also need a characteristic axial scale length. We shall use the reciprocal of the wavenumber for betatron oscillations of a single particle with zero canonical angular momentum in a constant axial magnetic field equal to the root-mean-squared field magnitude, $B_0/\sqrt{2}$. We can find the betatron wavenumber, k_s, by solving the equation:

$$r'' \cong -\left[\frac{e}{2\gamma m_e \beta c}\right]^2 \langle B_z(z)^2 \rangle r = \left[\frac{eB_0}{2\gamma m_e \beta c}\right]^2 \frac{r}{2}. \tag{10.170}$$

Equation (10.170) implies that

$$k_s \cong \frac{eB_0}{2\sqrt{2}\,\gamma m_e \beta c}. \tag{10.171}$$

We can rewrite the equilibrium condition of Eq. (10.166) in terms of the betatron wavenumber as

$$K = (k_s r_0)^2. \tag{10.172}$$

The dimensionless axial variable is

$$Z = k_s z. \tag{10.173}$$

The reduced envelope equation is

$$\rho'' = 1/\rho - 2\rho \sin^2[(k_m/k_s)Z]. \tag{10.174}$$

Equation (10.163) defines the quantity k_m. The single parameter k_m/k_s governs the nature of envelope oscillations. We can solve Eq. (10.174) numerically. A solution for a matched beam has an envelope radius that varies periodically over length π/k_m. Figure 10.27 shows the expected form of the matched beam envelope for the focusing magnetic field of Eq. (10.163). We carry out the numerical solution in the range $0 \leqslant Z \leqslant (k_m/k_s)\,(\pi/2)$. At $Z = 0$, we take $\rho' = 0$ and set $\rho(0) = \rho_0$. The quantity ρ_0 is an initial guess of the radius with magnitude less than unity. We continue solutions with adjusted values of ρ_0 until $\rho' = 0$ at $Z = (k_m/k_s)\,(\pi/2)$.

Figure 10.28 summarizes results of the calculation. The plot shows minimum and maximum values of ρ for a matched beam as a function of k_m/k_s. Envelope oscillations are small when the betatron wavelength is much larger than the

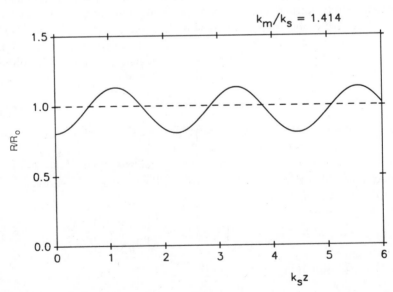

Figure 10.27 Numerical solution to the envelope equation for a high-current electron beam focused by a cusp array.

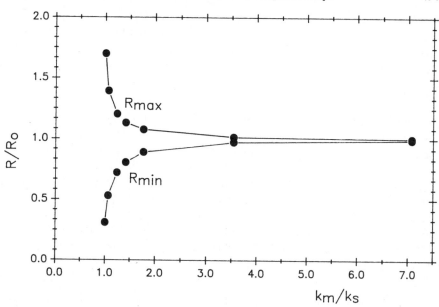

Figure 10.28 Envelope oscillation of a high-current electron beam focused by a cusp array—minimum and maximum values of R/R_0 as a function of k_m/k_s.

focusing cell length, $k_m \gg k_s$. The oscillation amplitude increases when k_m approaches k_s. The solutions show the presence of a beam envelope instability when $k_m/k_s < 1$. In this range, there is no matched beam solution because the value of ρ' at $Z = (k_m/k_s)(\pi/2)$ is less than unity for any choice of ρ_0. The implication is that the amplitude of envelope oscillations grows from cell to cell. The condition for the envelope instability is the same as the condition for a single-particle orbital instability—the phase advance per cell exceeds 180°. Section 13.1 gives detailed discussions of the collective envelope instability in a periodic focusing system.

We can write the stability criterion in terms of a limit on generalized perveance:

$$K \leqslant k_m^2 r_0^2 = (\pi r_0/L)^2. \qquad (10.175)$$

The envelope instability in a cusp array presents little problem for beams commonly used in applications. For example, consider a 5-kA, 2.5-MeV electron beam in an induction linac. The generalized perveance is $K = 3 \times 10^{-3}$. If $r_0 = 0.01$ m, Eq. (10.175) implies that

$$L < \pi r_0/K^{1/2} = 0.57 \text{ m}.$$

The condition on cell length is easy to satisfy. For example, if the beam travels through a tube of radius 0.04 cm, it is straightforward to design cusp coils with

$L \geqslant 0.1$ m. For a choice $L = 0.16$, the quantity k_m/k_s equals 3.6. Inspection of Fig. 10.28 shows that the beam is stable and that the amplitude of envelope oscillations is only $\pm 2\%$.

E.   Radial Force Variation

The radial focusing force in a solenoid of infinite length is linear with displacement from the symmetry axis [CPA, Section 7.5]. The radial variation of focusing force in a cusp array depends on the magnet coil geometry. The force has nonlinear variation when $L \leqslant R_c$. We can make a rough estimate of the radial variation of focusing force in the limit that envelope oscillations are small. Here, the radial position of an electron is almost constant crossing a focusing cell. An axial average gives a good approximation for the force that acts on an electron passing through a lens. If the electrons have zero canonical angular momentum, we can express the combined effect of magnetic and centrifugal forces in terms of the vector potential. The axially averaged force is

$$\langle F_r(r) \rangle = -\frac{1}{L}\int_0^L dz \left[\frac{e^2}{\gamma m_e}\right] A_\theta(r,z)\frac{\partial A_\theta(r,z)}{\partial r}. \tag{10.176}$$

Figure 10.29 shows implications of Eq. (10.176) for numerically calculated values

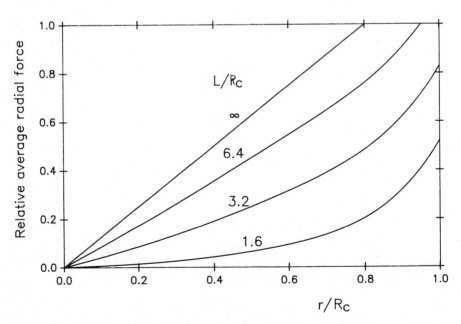

Figure 10.29   Radial variation of axially averaged focusing force in a cusp array as a function of the coil geometry. Uniform magnet windings of radius R_c reversed in direction periodically over a distance L.

of A_θ. The magnet coil geometries are the same as those of Fig. 10.25: $L/R_c = \infty$, 6.4 (*B*), 3.2 (*A*), and 1.6. All geometries have the same value of linear current density. Figure 10.29 shows that the focusing is highly nonlinear when the solenoid lenses are short.

F. Macroscopic Stability of Elastic, Filamentary Beams

Section 10.8 showed that perturbation forces caused drift motions of electron beams in a uniform solenoidal field. A major advantage of the cusp geometry is that drifts do not occur. In the beam rest frame, the direction of the axial magnetic field varies rapidly in time. The drift motions reverse, canceling cumulative displacements. The only forces that affect the beam coherently over many focusing cells are the axi-centered focusing forces. In a cusp array, a transverse perturbation force does not displace a beam without limit. Instead, it moves the beam to a different equilibrium position.

We can extend the elastic beam theory of Section 10.8 to the cusp geometry. Again, we use the coordinate system of Fig. 10.22. To simplify the calculation, assume that $k_m/k_s \gg 1$ and that the radial force varies almost linearly with displacement. For small cell length, we neglect the effect of the azimuthal forces. We express the axially averaged focusing forces in terms of the single-particle betatron wavenumber [Eq. (10.171)]:

$$F_{fx} = -\gamma m_e(\beta c)^2 k_s^2 x, \tag{10.177}$$

$$F_{fy} = -\gamma m_e(\beta c)^2 k_s^2 y. \tag{10.178}$$

The wall forces on an electron beam with current I_0 in a conducting pipe of radius r_w are

$$F_{ix} \cong (eI_0/2\pi\varepsilon_0\beta\gamma^2 c)(x/r_w^2), \tag{10.179}$$

$$F_{iy} \cong (eI_0/2\pi e_0\beta\gamma^2 c)(y/r_w^2). \tag{10.180}$$

Only the focusing and wall forces act in the vertical direction. For beam confinement, the focusing force must exceed the wall force:

$$F_i/F_f = K/(k_s r_w)^2 < 1. \tag{10.181}$$

The limit of Eq. (10.181) usually holds for paraxial beams of practical interest. The focusing magnetic field must be strong enough so that:

$$k_s > (K)^{1/2}/r_w. \tag{10.182}$$

The force-balance expression in the horizontal direction is more complex. For transport in a circular accelerator, we include the possibility of an error in the bending magnetic field. If the main orbit lies on a circle of radius R, we can represent the vertical magnetic field as $B_v(1 + \varepsilon)$, where ε is an error factor and

$$B_v = \gamma m_e \beta c / eR. \tag{10.183}$$

Take x as the displacement of a beam from the ideal radius,

$$x = r - R. \tag{10.184}$$

Combining the centrifugal force, focusing force, and magnetic bending force with an error term, the condition for horizontal force balance is

$$F_x = (e\beta c B_v)\left[-\varepsilon - \frac{x}{R} - (k_s R)^2 \frac{x}{R} + K\left(\frac{R}{r_w}\right)^2 \frac{x}{R} \right]. \tag{10.185}$$

The first term in brackets in Eq. (10.185) represents a vertical field error, the second term a change in centrifugal force associated with a beam displacement, the third term cusp focusing, and the forth term defocusing from wall forces. Equation (10.185) implies that a vertical field error causes a shift in the equilibrium radius of a beam:

$$\Delta x = -\varepsilon R / [1 + (k_s R)^2 - K[R/r_w]^2]. \tag{10.186}$$

In contrast to beam behavior in a uniform solenoidal field, the transverse energy distribution of the electrons has no effect on the beam position—there is no ∇B drift. The beam displacement equals zero when $\varepsilon = 0$.

Vertical field errors in a cusp array are not as critical as those in a uniform solenoid. An equivalent statement is that a circular accelerator with cusp focusing can contain a beam with larger momentum spread. Again, take the example of a 5-kA, 2.5-MeV beam in a circular accelerator with major radius $R = 1$ m and a toroidal vacuum chamber radius of $r_w = 0.04$ m. Suppose the betatron wavelength in the magnetic focusing system is 1.0 m, or $k_s = 6.28$ m^{-1}. Substituting values in Eq. (10.186), we find that a field error of 20% gives a displacement of only 0.005 m.

11

Ion Beam Neutralization

It is more difficult to transport high-current ion beams than electron beams. Nonrelativistic ions move slower than electrons of equal kinetic energy. Therefore, an ion beam has higher space-charge electric fields than an electron beam of the same current. Also, magnetic focusing by beam-generated fields is ineffective for nonrelativistic beams. We must apply neutralization to create and to transport high-flux ion beams. The idea is to mix electrons with the ions to reduce the beam-generated electric field. The process is feasible because of the low mass of the electron. The mobile electrons rapidly enter the beam volume. Low-energy electrons can follow high-energy ions to neutralize a beam propagating into free space. Also, the technology to generate electrons is straightforward compared with the complexity of ion sources (Chapter 7).

There are two ways to neutralize an ion beam with electrons. First, we can direct the beam through a dense plasma. The plasma electrons shift in position to compensate for the added positive charge. Plasma neutralization is an important process for large-area ion extractors that use a gas-injection plasma source. The beam ionizes gas leaking from the source to produce a high-density, low-temperature plasma. Although this neutralization method has practical importance, we shall not discuss it in this chapter. The characteristics of the plasma depend on complex collisional processes. Prediction of the plasma properties and residual electric fields involves applied plasma and atomic physics rather than beam theory.

In this chapter, we shall concentrate on an alternative approach, vacuum neutralization. Here, sources located outside the vacuum beam transport region create the electrons. The electrons join the ions as needed. The resulting neutralized beam has an electron density approximately equal to the beam

501

density. Collisions with the electrons have little effect on the ion trajectories. Therefore, beams neutralized by externally generated electrons can propagate long distances.

Section 11.1 describes longitudinal neutralization where electrons follow an ion beam entering a field-free vacuum. There are two options for the generation of collinear electrons. The first is to accelerate the electrons to the ion velocity by an applied electric field. The second is to allow the space-charge field of the beam to attract electrons. We shall show that the latter method of passive neutralization results in an electron distribution with a density and average velocity equal to that of the ions. Section 11.2 treats a similar process where electrons enter the side of the beam in response to space-charge electric fields. This model applies to ion beams in magnetic quadrupole lenses or bending magnets where the fields prevent the axial motion of electrons. Section 11.3 describes propagation of an ion beam in a bounded field-free region where electrons neutralize space-charge fields but do not cancel the beam current. This effect can be useful for flux measurements of neutralized ion beams.

One motivation for neutralization is to achieve tightly focused ion beams. With complete cancellation of space-charge fields, only emittance limits the focal spot of an intense beam. Section 11.4 shows that an ideal focus does not occur if the neutralizing electrons have a nonzero temperature. The electric fields generated by electron thermal motion can reach high values in a converging beam, defocusing the ions. To conclude the chapter, Section 11.5 reviews the control of neutralizing electrons with applied magnetic fields to guide and to accelerate ion beams.

11.1. NEUTRALIZATION BY COMOVING ELECTRONS

In this section, we study the propagation of neutralized ion beams in free space with no applied electric and magnetic fields. Figure 11.1a shows the ideal neutralized beam. The electron and ion densities are equal, so there is no beam-generated electric field. The electrons move at same velocity as the ions, $v_e = v_i$. The conditions of equal densities and equal velocities imply that the current densities of electrons and ions have equal magnitude but opposite direction. The net current is zero, so there is no beam-generated magnetic field. If T_i is the ion kinetic energy, electrons with equal velocity have kinetic energy

$$T_e = m_e v_i^2/2 = (m_e/m_i)T_i. \tag{11.1}$$

The electron energy is much smaller than T_i. For example, electrons moving with 1-MeV protons have $T_e = 540$ eV. The problem we shall address in this section is how to create a neutralized beam like that of Fig. 11.1a. We limit attention to longitudinal neutralization, where electrons enter the transport region at the same location as the ions and travel in the same direction.

One option for neutralization is to accelerate electrons to kinetic energy T_e

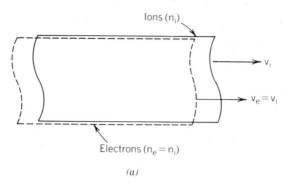

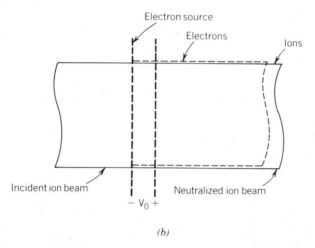

Figure 11.1 Ion beam neutralization. (*a*) Ideal neutralized beam—ions and electrons have equal densities and velocities. (*b*) Active neutralization—acceleration of electrons to match the density and velocity of the ion beam.

and to combine them with the ion beam. This process is called *active neutralization*. We shall analyze the process with a one-dimensional model. Figure 11.1*b* illustrates the geometry. Ions with kinetic energy T_i pass through a set of grids. A cathode grid acts as an unlimited source of electrons. If the voltage difference between the grids is

$$V_0 = (m_e/m_i) T_i/e, \tag{11.2}$$

electrons reach the anode grid with velocity equal to that of the ions. The electric field between the grids has a negligible effect on the velocity of energetic ions. Therefore, the ions have constant density, n_0, and velocity, $v_i = (2T_i/m_i)^{1/2}$, throughout the acceleration and propagation regions.

Ideal neutralization results if the magnitude of the electron current density at the anode equals the ion current density, en_0v_i. We are free to choose the spacing between grids to achieve this condition. Again, we seek a one-dimensional self-consistent equilibrium for electron flow. The main difference from previous analyses is the inclusion of a uniform ion density. The following boundary conditions hold for a steady-state solution:

1. The electrostatic potential of the cathode (at $z = 0$) is $\phi(0) = 0$.
2. The anode at $z = d$ has potential $\phi = V_0 = T_i(m_e/m_i)/e$.
3. Space-charge-limited electron emission reduces the electric field at the cathode to zero, $d\phi(0)/dx = 0$.
4. The electron density equals the ion density at the gap exit, $n_e(x = d) = n_0$.

The one-dimensional Poisson equation that satisfies the boundary conditions is

$$\frac{d^2\phi}{dx^2} = \frac{en_0}{\varepsilon_0}\left[\left(\frac{V_0}{\phi}\right)^{1/2} - 1\right]. \tag{11.3}$$

Following the method of Section 6.4, we can show that the electron current density is given by

$$j_e = \frac{4\varepsilon_0}{9}\left[\frac{2e}{m_e}\right]^{1/2}\frac{V_0^{3/2}}{d^2}G, \tag{11.4}$$

where

$$G = \left[\frac{3}{2\sqrt{2}}\int_0^1 \frac{dX}{(2X^{1/2} - X)^{1/2}}\right]^2 = 1.46.$$

Because of the ion space charge, the electron current density is slightly higher than the single species Child limit [Eq. (5.48)].

As an example of the characteristics of an electron acceleration gap for ion beam neutralization, suppose we have a 100-KeV deuteron beam with current density 1×10^4 A/m². The neutralizing electrons have kinetic energy $T_e = 27$ eV. The acceleration gap must be very narrow to generate the required electron current density at low energy, $d = 0.22$ mm. The fundamental problem of active neutralization is the creation of high-current-density electron flux from a structure that transmits a high-intensity ion beam with little attenuation. Although the one-dimensional mathematical solution is straightforward, the technological realization is quite difficult.

A more practical way to reduce space-charge forces in an ion beam is through *autoneutralization*. In this process the space-charge potential of the ion beam accelerates electrons from a grounded surface. Figure 11.2 illustrates a one-dimensional geometry to describe the process. An ion beam of infinite transverse extent leaves a planar surface at $z = 0$. The region $z > 0$ is a field-free volume. The surface at $z = 0$ supplies an unlimited electron flux. The acceleration of electrons

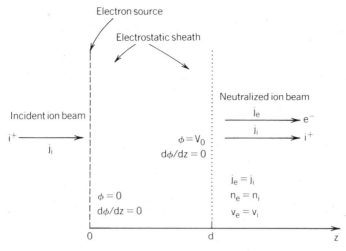

Figure 11.2 Hypothetical desired conditions for ion beam neutralization through space-charge acceleration of electrons.

by the ion space charge is a self-limiting process. We recognize that even a small imbalance of charge in an intense ion beam results in a high value of space-charge potential, $\phi \gg (m_e/m_i)T_i/e$. If the electron density is less than that of the ion beam, the resulting electric fields draw more electrons into the beam. An equilibrium occurs when electrons move into the propagation region at the same rate as the ions.

Figure 11.2 illustrates the desired equilibrium solution for autoneutralization. The space-charge fields accelerate electrons in a thin sheath to match the velocity of the ions. We assume that the sheath occupies the region $0 \leqslant z \leqslant d$. Several conditions constrain the solution of the Poisson equation:

1. The voltage drop across the sheath is $V_0 = (m_e/m_i)T_i/e$.

2. The electron current density at $z = d$ equals the ion current density, $j_e = j_0 = en_0v_i$.

3. The electric field at $z = 0$ must equal zero because the electron flow is space-charge limited, $d\phi(0)/dz = 0$.

4. The electric field at the sheath exit equals the filed inside the propagating beam because there is no charge layer at d. By the assumption of a neutralized beam, $d\phi(d)/dz = 0$.

There are too many boundary conditions for a solution of the Poisson equation. In order to have zero electric field at both sides of the sheath, the potential must follow an S-shaped curve with both positive and negative inflection. The dual inflection occurs only if the ion density exceeds the electron density near $z = d$ and the electron density is larger near $z = 0$. Condition 2 implies that the electron and ion densities are equal at $z = d$; furthermore, we know that $n_e \sim 1/\phi^{1/2}$ in steady state. The implication is that the electron density is higher than the ion density everywhere in the sheath.

We must seek solutions with different boundary conditions to explain how autoneutralization works. One possibility is to look for a steady-state solution where variations of the potential are not contained to a sheath but extend to infinity. For any potential variation, the electron density has the form

$$n_e(z) = A/\sqrt{2e\phi(z)/m_e}. \tag{11.5}$$

We retain the condition of uniform ion density, n_0, and current density, j_0. To ensure that the electron and ion fluxes are equal, the constant A in Eq. (11.5) has the value $(j_0/e)(2eV_0/m_e)$. In the region $z \geqslant 0$, the Poisson equation is

$$\frac{d^2\phi}{dz^2} = \frac{j_0}{\varepsilon_0}\left[\left(\frac{V_0}{\phi}\right)^{1/2} - 1\right]. \tag{11.6}$$

We can simplify Eq. (11.6) by defining the dimensionless variables:

$$\Phi = \phi/V_0. \tag{11.7}$$

$$Z = z/(V_0\varepsilon_0/en_0)^{1/2}. \tag{11.8}$$

The reduced Poisson equation is:

$$\frac{d^2\Phi}{dz^2} = \frac{1}{\sqrt{\Phi}} - 1. \tag{11.9}$$

If emission of electrons from the surface at $Z = 0$ is space-charge-limited, the boundary conditions for the solution of Eq. (11.9) are $\Phi(0) = 0$ and $d\Phi'(0)/dZ = 0$. A dual integration of the equation leads to the solution

$$Z = \sqrt{2}\{\sin^{-1}(\Phi^{1/2} - 1) - [\Phi^{1/2}(1 - \Phi^{1/2}/2)]^{1/2} + \pi/2\}. \tag{11.10}$$

Figure 11.3 illustrates the spatial variation of $\Phi(Z)$. The potential is periodic with values between $\Phi = 0$ and $\Phi = 4$. The distance from the cathode to the first potential maximum is $Z = (2)^{1/2}\pi$. The electrons overneutralize and underneutralize the beam. The electron velocity varies between $v_e = 0$ and $v_e = 2v_i$.

We might conclude from the solution of Eq. (11.10) that effective autoneutralization is impossible. To explain experimental observations, we recognize that the derivation proceeds from two questionable conditions:

1. The model assumes that an equilibrium state exists in the full half plane $z > 0$ for all times. It ignores processes that may occur as ions and electrons fill the propagation region.

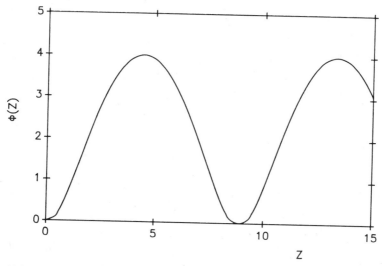

Figure 11.3 Ion beam neutralization by space-charge acceleration of electrons. Steady-state solution when a uniform ion beam occupies the entire region $Z > 0$. $\Phi = \phi/V_0$, $Z = z/(V_0\varepsilon_0/en_0)^{1/2}$.

2. The model takes electron motion as purely one dimensional. The electrons have a delta-function distribution in longitudinal energy.

Regarding the second assumption, we recognize that neutralization is a disordering process where electrons join with ions to form a homogeneous mixture. If we limit motion to one dimension, we may have set an artificial constraint that prevents the electron distribution from attaining thermodynamic equilibrium.

We shall develop a model to show that there are alternative equilibrium solutions that give autoneutralization with a well-defined sheath region. We include time-dependent processes as an ion beam fills the vacuum propagation region, $z \geqslant 0$. The uniform density beam enters the region at $t = 0$ and moves in the z direction at velocity v_i. We resolve the problem of inconsistent boundary conditions by introducing the possibility of low-energy electrons in the propagating beam. These electrons reflect from the moving ion front—Fig. 11.4 illustrates the process. If the potential across the sheath exceeds the voltage V_0, electrons enter the beam with velocity higher than v_i. The electrons try to run ahead of the ion front, but the unbalanced space charge creates a virtual cathode. Electrons reflected from the moving virtual cathode have reduced kinetic energy.

Suppose that the voltage drop across the sheath equals $4V_0$-electrons enter the beam with velocity $2v_i$. Applying conservation of momentum, we find that an electron loses all its kinetic energy when reflected from the moving beam front. As a result, the electron distribution in the propagating beam has two components: stationary reflected electrons and a uniform density of newly injected electrons moving at $2v_i$. Conservation of flux implies that the moving ion front deposits stationary electrons at the same rate as electrons exit the acceleration sheath.

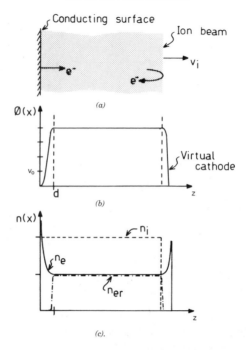

(a)

(b)

(c).

Figure 11.4 Time-dependent solution for ion beam neutralization by the space-charge acceleration of electrons. (a) Geometry, showing electrons reflected by a virtual cathode at the moving ion front. (b) Variation of electrostatic potential with z: $eV_0 = (m_e/m_i)T_i$. (c) Spatial variation of densities of ions (n_i), entering electrons (n_e), and reflected electrons (n_{er}).

Therefore, the density of stationary electrons, n_s, equals the density of injected electrons at the sheath exit, $n_e(d)$. In a neutralized beam, the densities are

$$n_s = n_e(d) = n_0/2. \tag{11.11}$$

By including stationary electrons in the beam, we can solve the Poisson equation in the electron acceleration sheath. The following boundary conditions define the solution:

1. $\phi(0) = 0.$ (11.12)

2. $\phi(d) = 4V_0.$ (11.13)

3. $d\phi(0)/dz = 0.$ (11.14)

4. $d\phi(d)/dz = 0.$ (11.15)

5. $n_e(d) = n_0/2.$ (11.16)

Equation (11.16) implies that the density of electrons near the exit of the sheath is lower than the ion density. By Eq. (11.5), the electron density is higher near the cathode grid. Therefore, it is possible to generate a solution for ϕ that follows an S-shaped curve. The solution of the Poisson equation with conditions (11.12) through (11.16) is identical to Eq. (11.10) in the sheath region, $0 \leqslant z \leqslant d$. The sheath width is

$$d = \pi(2V_0\varepsilon_0/en_0)^{1/2}. \tag{11.17}$$

In the region $z > d$, the additional low-energy electrons give a solution with constant potential, $\phi = 4V_0$, rather than the oscillatory solution of Fig. 11.3. The electric field is confined to the sheath. The propagating beam is field-free. In the beam volume, the net electron density equals n_0 while the *average* electron velocity equals v_i. Figure 11.4 shows a plot of electron density and potential over the sheath and beam.

The modified sheath solution gives an electron distribution in the beam with two discrete velocity components at $v_e = 0$ and $v_e = 2v_i$. In the beam rest frame, the electron components stream through each other with velocity $\pm v_i$. Such a distribution is potentially unstable to the two-stream instability (Section 14.1). This instability randomizes the axial velocity distribution. We can find the actual electron distribution that results from autoneutralization from a one-dimensional computer simulation. Figure 11.5 shows results from a computer program that uses the dimensionless variables of Eqs. (11.7) and (11.8). The figure gives electron phase-space distributions in terms of the dimensionless electron velocity $V = v_e(z)/v_i$. Figure 11.5a shows a distribution at early time when the ion beam has moved only a few sheath widths. The spatial variation of electron velocity closely follows the prediction of Eq. (11.10). As predicted, the peak potential of $4V_0$ occurs at a distance $Z = (2)^{1/2}\pi$ from the source. The solution has an oscillatory component of potential similar to that of Fig. 11.3. Some reflected electrons appear as negative velocity particles.

Figure 11.5b shows the electron distribution at a later time with the ion beam front at $Z = 9$. There is considerable activity near the injection point, but the downstream electrons have settled into an equilibrium with small variations of potential. The average beam potential is V_0 and the average electron velocity equals v_i. Thermalization of the electron distribution results from the two-stream instability. Figure 11.5c plots the electron velocity distribution averaged over the downstream region of Fig. 11.5b. The computer simulation illustrates the main difference between autoneutralization and ideal active neutralization. For the autoneutralization solution, the electrons have a velocity spread of about $\Delta v_e = 0.6v_i$ (full-width at half-maximum).

We can use Eq. (11.17) to find the electron acceleration sheath width. Again, suppose we have a 100-keV deuteron beam with current density 10^4 A/m² — the predicted sheath width is 1.2 mm. This width is much smaller than the width of a high-current ion beam; therefore, the one-dimensional sheath model is a good representation. In many intense ion beam experiments, neutralizing electrons are

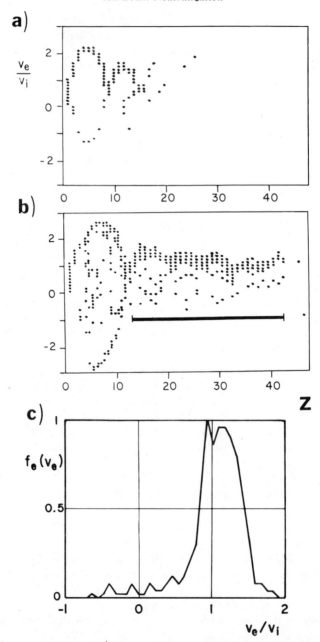

Figure 11.5 Particle-in-cell computer simulation of one-dimensional autoneutralization. A uniform-density ion beam with a sharp front moves into a field-free region. Ion kinetic energy: T_i; ion velocity: v_i; ion beam density: n_0. $V_0 = (m_e/m_i)T_i$, $Z = z/(\varepsilon_0 V_0/en_0)^{1/2}$. (*a*) Axial phase-space plot of the electron distribution with the ion front at $Z = 15$. (*b*) Axial phase-space plot of the electron distribution with the ion front at $Z = 40$. (*c*) Relative electron velocity distribution, averaged over the spatial region marked by the thick line in part (*b*).

generated when the beam passes through a conducting grid or foil. A localized plasma sheet can also supply electrons for autoneutralization.

11.2. TRANSVERSE NEUTRALIZATION

Figure 11.6 shows the geometry for transverse neutralization of an ion beam. The beam passes between conducting boundaries that act as electron sources. The space-charge electric field of the ions pulls electrons from the boundaries. Ideally, the electrons cancel electric fields in the beam. The main differences from the models of Section 11.1 are that the ions propagate through a bounded region and that the electrons need not move with the ion beam. By studying transverse neutralization, we can understand how electrons merge with ion beams in more complex geometries.

The geometry of Fig. 11.6 is a good representation of the transport region near a magnetically insulated ion diode (Section 8.8). The diode magnetic field penetrates into the transport region, inhibiting axial propagation of neutralizing electrons. The electrons can flow only along the magnetic field lines. Under vacuum, the only way to neutralize an intense ion beam is to supply electrons on all magnetic field lines that the beam crosses. Electrons generated on the surfaces shown in Fig. 11.6 flow into the beam. For intense ion beam diodes with pulse lengths less than 0.1 μsec, we must study the response time for this process to determine the success of transverse neutralization.

We shall take an approach similar to that of Section 11.1. We start with a simple equilibrium model that has reasonable assumptions but leads to

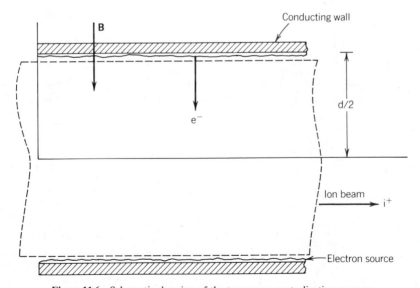

Figure 11.6 Schematic drawing of the transverse neutralization process.

nonphysical results. By analyzing the limitations of the model, we can gain insight into how neutralization occurs in a real system. Finally, to get an accurate description of the disordered collective process, we turn to computer simulations. For the simplified, one-dimensional model, suppose that an ion beam moves in the z direction through a field-free region between two conducting walls at $x = \pm d/2$. The walls can supply a space-charge-limited electron flux. We assume that the maximum space-charge potential energy is much smaller than the ion kinetic energy, $e\phi \ll T_i$. Therefore, we shall concentrate on the electron motion. For simplicity, we let the ions fill the space between the boundaries with a uniform density n_0.

For one-dimensional motion, the density of electrons is inversely proportional to their velocity in the x direction. The density is related to the electrostatic potential by

$$n_e = A/(\phi)^{1/2}. \tag{11.18}$$

We define the wall potential as $\phi(\pm d/2) = 0$. For space-charge-limited electron emission, the wall electric field equals zero, $d\phi(\pm d/2)/dx = 0$. The electric field also equals zero on the symmetry axis, $d\phi(0)/dx = 0$. We can write the Poisson equation for a space-charge equilibrium as

$$\frac{d^2\phi}{dx^2} = -\frac{e}{\varepsilon_0}\left[n_0 - \frac{A}{\phi^{1/2}}\right]. \tag{11.19}$$

We define the quantity ϕ_0 as the potential at the midpoint between the boundaries, $\phi(0) = \phi_0$.

The first integral of Eq. (11.19) is

$$\left(\frac{d\phi}{dx}\right)^2 = -\frac{2e}{\varepsilon_0}[n_0\phi - n_0\phi_0 - 2A\phi^{1/2} + 2A\phi_0^{1/2}]. \tag{11.20}$$

Equation (11.20) satisfies the boundary conditions if $A = n_0\phi_0^{1/2}/2$. Substituting for A and introducing dimensionless variables $\Phi = \phi/\phi_0$, $X = x/x_0$, Eq. (11.20) becomes

$$\int_1^\Phi \frac{d\Phi'}{(\Phi'^{1/2} - \Phi')^{1/2}} = \pm\left[\frac{2en_0d^2}{\varepsilon_0\phi_0}\right]^{1/2} X. \tag{11.21}$$

Equation (11.21) has the multiple solutions

$$\pm X = \left[\frac{\varepsilon_0\phi_0}{2en_0d^2}\right]^{1/2}[2(\Phi^{1/2} - \Phi)^{1/2} - \sin^{-1}(2\Phi^{1/2} - 1) + \sin^{-1}(1)]. \tag{11.22}$$

The maximum electrostatic potential, $\Phi = 1$, occurs at the midplane, $X = 0$. In physical units, the midplane potential is

$$\phi_0 = en_0 d^2/2\pi^2 \varepsilon_0 (2m+1)^2, \qquad (11.23)$$

where $m = 0, 1, 2, \cdots$. The maximum potential from a uniform-density ion beam without neutralizing electrons is $en_0 d^2/8\varepsilon_0$. We define a space-charge potential reduction factor:

$$\phi_0/[en_0 d^2/8\varepsilon_0] = 4/\pi^2(2m+1)^2. \qquad (11.24)$$

Figure 11.7 shows the spatial variation of potential of neutralized and unneutralized beams for $m = 0$ and 1.

One problem with the model is that it does not predict a unique equilibrium state. We expect that a unique set of initial conditions should give a unique final state. We can choose any value of m—the model does not show whether neutralization is effective. For $m = 0$, the reduction factor equals 0.405. This reduction is useless for intense ion beam transport where the electric fields must be less than 10^{-6} of the unneutralized value.

The simplified model is unrealistic for two reasons. First, it takes electron motion as perfectly one dimensional. In this case, conservation of energy implies that an electron that leaves one boundary reaches the other boundary with zero velocity. Therefore, the density of reflexing electrons diverges at the boundaries. For the $m = 0$ solution, electrons spend most of the time near the boundaries and move quickly through the midplane. As a result, the ion beam is overneutralized at the boundary and underneutralized at the midplane. Another problem with the model is that the monoenergetic electron distribution is valid only if the ion

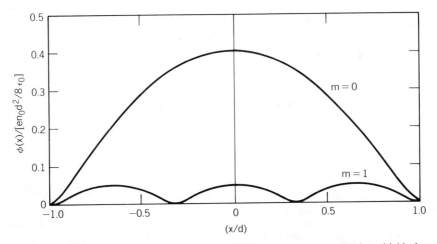

Figure 11.7 Spatial variations of electrostatic potential for transverse neutralization with ideal one-dimensional electron motion and an ion density constant over all time.

density is constant at all times. For a pulsed beam, the ion density and the associated space-charge fields change with time. Therefore, electrons emitted early in the pulse have different orbit properties than those that enter at late times.

Transverse neutralization is effective if electrons in the beam volume do not return to the boundaries. If electrons are trapped in the beam volume, additional electrons can enter from the boundaries until space-charge fields are completely canceled. Relaxing the constraint of one-dimensional motion allows electron trapping. If an electron suffers a deflection normal to the x direction (Fig. 11.8a), conservation of energy implies that it cannot return to the boundaries. A

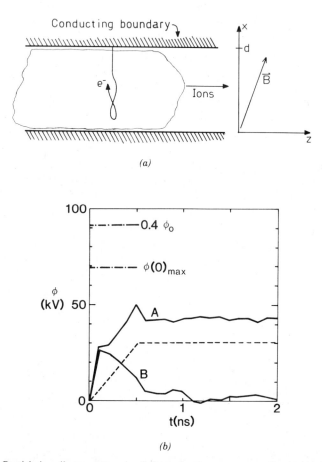

Figure 11.8 Particle-in-cell computer simulation of transverse neutralization. (a) Simulation geometry—the ion density increases with time between grounded conducting electron emitters. The figure shows an electron orbit deflected by a skewed magnetic field. (b) Simulation results—time-variation of electrostatic potential on the midplane. Dashed line shows variation of ion density with peak value $n_0 = 10^{18}\,\text{m}^{-3}$. Quantities $0.4\phi_0$ and $\phi(0)_{max}$ described in text. Curve A: magnetic field inclination angle $0°$—one-dimensional electron motion. Curve B: magnetic field inclination angle $15°$. (Courtesy J. Poukey, Sandia National Laboratories.)

temporal variation of space-charge fields also traps electrons. A rising ion density pulls electrons away from the boundary.

A computer simulation is the best approach to model complex electron orbits. We shall review results from a particle-in-cell simulation for the geometry of Fig. 11.8*a*. Electrons leave conducting boundaries with a space-charge-limited flux. The boundaries and ion density have infinite extent in the *z* direction. The ion density is also uniform in the *x* direction. The simulation model has two main differences from the simple equilibrium model:

1. The ion density rises with time to an equilibrium value.
2. A transverse magnetic field in the transport region influences the electron orbits. When the magnetic field points in the *x*-direction, the electrons have purely one-dimensional orbits as before. On the other hand, tipping the field direction (Fig. 11.8*a*) results in electron velocity components in the *y* and *z* directions.

The skewed magnetic field is an easy way to introduce the effect of geometric variations into the one-dimensional code.

Figure 11.8*b* shows results of the simulation. The graph plots the electrostatic potential at the midplane, ϕ_0, as a function of time. The boundaries are at positions $y = 0$ and $y = d = \pm 0.005$, the ion density rises to a final value $n_0 = 10^{18} \, \text{m}^{-3}$ with a risetime $\Delta t_r = 0.5$ nsec. The dashed line in Fig. 11.8*b* shows the variation of ion density. For the given parameters, the equilibrium model predicts a potential $\phi_0 = [0.405 e n_0 d^2 / 8 \varepsilon_0] = 91 \, \text{kV}$ for $m = 0$. Curve *A* of the figure corresponds to a calculation where the magnetic field lies in the *x* direction. The reduced potential results from the time dependence of space-charge fields. The final potential is lower than the equilibrium model prediction by about a factor of 2. Curve *B* is the result of a simulation with deflected electron orbits— the magnetic field has an inclination of 15°. This small change in geometry results in a dramatic difference in the nature of the solution. After the ion density reaches equilibrium, the potential rapidly drops almost to zero. (The residual potential oscillations at late time result from the finite number of particles in the simulation.) The simulation implies that in real systems with asymmetries, transverse electron neutralization rapidly cancels space-charge electric fields.

Because of their low mass, electrons respond rapidly to changes in the density of an ion beam. Nonetheless, the variation of ion density in pulsed ion diodes is so rapid that the electron response time can result in high values of space-charge potential. The simulation results (Fig. 11.8*b*) show that a nonzero space-charge potential is necessary to draw electrons into the beam volume during the rise of ion density. We can estimate the required potential for the geometry of Fig. 11.9. A sheet ion beam of width $\pm x_b$ and velocity v_i enters a region between electron emitting plates separated by distance $\pm d$. At a given axial location, the ion density varies as

$$n_i(t) = n_0 t / \Delta t_r \qquad (11.25)$$

for $t \leqslant \Delta t_r$, and

$$n_i(t) = n_0$$

for $t > \Delta t_r$. When $v_i \Delta t_r \gg d$, the space-charge electric fields lie predominantly in the x direction.

A transverse magnetic field prevents electron motion in the z direction. We can estimate the electrostatic potential during the ion density rise by invoking global charge balance in a section of the transport system of length Δz. During the rise of ion density, the beam potential must be high enough to pull electrons across the vacuum region from the walls. We assume that field asymmetries are strong enough to trap electrons in the beam; therefore, electrons enter continually from the wall. Finally, we assume that neutralization is effective so that the global integrals of electrons and ion densities over the volume element $2d\Delta z \Delta y$ are almost equal, or

$$\Delta z \Delta y \int_{-d}^{+d} dx \left(\frac{\partial n_i}{\partial t} \right) \cong \Delta z \Delta y \int_{-d}^{+d} dx \left(\frac{\partial n_e}{\partial t} \right). \tag{11.26}$$

Equation (11.26) holds when the space-charge potential is much smaller than that of an unneutralized beam.

The integral of the time variation of ion density over the volume element of Fig. 11.9 equals

$$\Delta z \Delta y \int_{-d}^{+d} dx \left(\frac{\partial n_i}{\partial t} \right) \cong (2x_b \Delta z \Delta y) n_0 / \Delta t_r. \tag{11.27}$$

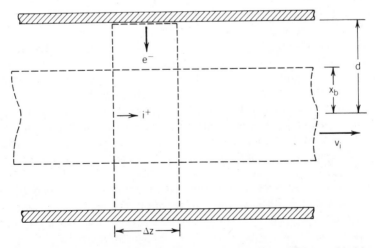

Figure 11.9 Geometry to estimate the peak electrostatic potential in an ion beam with rising density neutralized by the transverse flow of electrons.

If the current density of electrons leaving each boundary is j_e and if the beam traps the electrons, the rate of change of the integrated electron density is approximately

$$\Delta z \Delta y \int_{-d}^{+d} dx \left(\frac{\partial n_e}{\partial t} \right) \cong 2 \left(\frac{j_e}{e} \right) \Delta z \Delta y. \tag{11.28}$$

If ϕ_0 represents the average electrostatic potential of the beam, the electron current density scales as

$$j_e \leqslant (4\varepsilon_0 / 9)(2e/m_e)^{1/2} \phi_0^{3/2} / d^2. \tag{11.29}$$

The inequality depends on the fraction of the transport region filled by the ion beam. Combining Eqs. (11.26)–(11.29) gives the following expression for the beam potential during the ion density rise:

$$\phi_0 \leqslant \left[\frac{9}{8} \frac{e n_0}{\varepsilon_0 \Delta t_r} d^2 x_b \left(\frac{m_e}{2e} \right)^{1/2} \right]^{2/3}. \tag{11.30}$$

When applied to the parameters of the simulation of Fig. 11.8 where the ion beam fills the transport region, Eq. (11.30) overestimates the potential by about a factor of 2.5. The dashed line in the figure shows the estimate. As an application example, consider a neutralized ion beam accelerator (Section 11.5). A Na^+ beam has current density $50 \times 10^4 \, A/m^2$, kinetic energy 20 MeV, and ion density $n_0 = 2.4 \times 10^{17} \, m^{-3}$. The beam has width $x_b = 0.02 \, m$ and propagates between boundaries at $d = \pm 0.04 \, m$. With a beam density risetime of 50 nsec, the predicted midplane electrostatic potential is $\phi_0 \leqslant 30 \, kV$. The transverse electric fields associated with the residual potential can result in beam defocusing, limiting the utility of high-current ion diodes and accelerators.

11.3. CURRENT NEUTRALIZATION UNDER VACUUM

When a high-current ion beam moves into an infinite field-free volume, accompanying electrons provide both space-charge neutralization and current neutralization. We can show that a high-flux ion beam is also current-neutralized if it crosses a finite-length region from a source to an electrically isolated target. Without electron flow, the beam would induce a large voltage drop between the target and source. As an example, suppose an ion beam charges a spherical target of radius 0.1 m. The beam has 10-A current and a 1-MeV kinetic energy. If the distance between target and the source is 0.3 m, the interelectrode capacitance is about $C = 10^{-12} \, F$. The voltage difference is $\Delta V \sim I_i \Delta t / C$, where Δt is the beam pulse length. The deposited charge creates a potential equal to the beam kinetic energy in about 1 μsec. On the other hand, the voltage to accelerate an equal current of electrons is only about 500 V. We expect that if electrons are available at the source electrode, they will flow to the target with the ion beam to cancel the space-charge potential. Figure 11.10a shows the elements of a one-dimensional autoneutralization solution for an ion beam crossing to an isolated target. The net current to the target is zero if it has potential $+eV_0$.

The current of an intense ion beam is not canceled completely when the beam moves through a vacuum region surrounded by conducting boundaries. To describe time variations of the electron current, we shall use the idealized geometry of Fig. 11.10b. A cylindrical beam of ions with kinetic energy T_i travels through a pipe of radius r_w and length d with conducting walls at each end. The beam has current I_i, radius r_0, and velocity $v_i \cong (2T_i/m_i)^{1/2}$. The source plane can supply an electron flux equal to the beam current density. For current

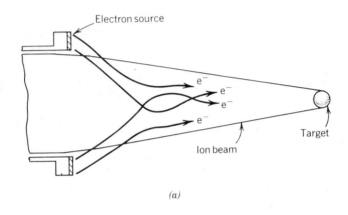

(a)

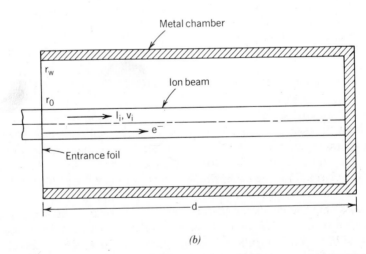

(b)

Figure 11.10 Current neutralization of ion beams by electron flow in vacuum. (a) Intense ion beam focused to an isolated inertial fusion target. (b) Ion beam in a closed pipe neutralized by the axial flow of electrons.

neutralization, the electrons must have kinetic energy; on the other hand, stationary electrons could provide space-charge neutralization. In the static field limit there are no electric fields to give electrons a directed energy because the surrounding walls are all grounded. At late time (compared with the ion transit time, d/v_i), we expect that the neutralizing electrons are stationary. Inside the pipe, there are no electric fields, but there are magnetic fields created by the ion beam current.

Processes are more interesting at early times because the changing magnetic flux inside the pipe can create an axial electric field to accelerate neutralizing electrons. During the initial transit of the beam front through the pipe, electrons follow the beam with velocity v_i. In this phase, the electron distribution is determined by the autoneutralization process described in Section 11.2. When the beam contacts the downstream wall, the flow of neutralizing electrons continues. If the electron flow were to stop immediately, the toroidal magnetic field of the beam would appear instantaneously, creating an infinite electric field. Therefore, the electron flow must decrease gradually. For current neutralization to persist, a continuous flow of electrons from the entrance wall must accelerate to kinetic energy $T_{eo} = (m_e/m_i)T_i$. Changing magnetic flux inside the pipe supplies the accelerating voltage. The small fractional deceleration of ions in the electron acceleration sheath at the entrance wall contributes energy to create the magnetic field. The inductive voltage results from a changing net current. Because the ion current is constant, the electron current must decay.

We can construct a simple model for the decay of neutralizing electron current with the following assumptions:

1. The ion beam current rises rapidly. During the initial ion transit through the pipe, the electron current equals the ion current. At time $t = 0$ when the beam fills the pipe, the net current equals zero. If $I_e(t)$ is the net electron current, then $I_e(0) = -I_i$.

2. Electrons accelerate in a narrow sheath at the entrance wall. The average electron kinetic energy at $t = 0$ is $T_e(0) = T_{eo}$.

3. The beam radius is much smaller than the pipe radius, $r_0 \ll r_w$. As a result, the magnetic field energy from a net chamber current is concentrated in the volume outside the beam. To first order, the inductive voltage acts uniformly on all electrons.

4. The ion and electron densities, n_i and n_e, are uniform over the beam radius.

5. The space charge of the high-intensity ion beam is always neutralized. Thus, the electron density always equals the ion density, $n_e \cong n_i$.

The ion current, I_i, is constant following injection, while the magnitude of the electron current, $I_e(t)$, decreases in time. The net current is $I(t) = I_i - I_e(t)$. The chamber has an inductance L, roughly equal to

$$L = (\mu_0/2\pi)d\ln(r_w/r_0). \tag{11.31}$$

The voltage between the entrance and exit walls is

$$V = L(dI/dt). \tag{11.32}$$

Because the electron and ion densities are always equal, the ratio of ion to electron current is proportional to the ratio of the average particle velocities:

$$I_e/I_i = v_e/v_i, \tag{11.33}$$

where $v_e(t) = [2T_e(t)/m_e]^{1/2}$. We can rewrite Eq. (11.33) as

$$I = I_i(1 - v_e/v_i). \tag{11.34}$$

The electrons gain their velocity in a narrow sheath at the entrance wall. With this condition, Eq. (11.34) becomes

$$I = I_i(1 - \sqrt{eV/T_e}). \tag{11.35}$$

We can combine Eqs. (11.32) and (11.35) into a single equation for the time-variation of total current:

$$L(dI/dt) = (T_{eo}/e)[1 - (I_0/I_i)]^2. \tag{11.36}$$

In terms of the dimensionless variables,

$$\tau = t/(eLI_i/T_{eo}) \tag{11.37}$$

and

$$\mathsf{I} = I/I_i, \tag{11.38}$$

Eq. (11.36) becomes

$$d\mathsf{I}/d\tau = (1 - \mathsf{I})^2. \tag{11.39}$$

With the initial condition that $\mathsf{I} = 0$ at $\tau = 0$, the solution of Eq. (11.39) is

$$\frac{I}{I_i} = 1 - \frac{1}{1 + \tau}. \tag{11.40}$$

Figure 11.11 plots the result of Eq. (11.40). The net current rises to a significant level over the dimensionless interval $\tau = 1$. At late time, the total current approaches the ion current. As expected, stationary electrons provide long-term space-charge neutralization.

In most experiments with pulsed high-current ion beams in bounded chambers, the fractional decay of electron current is small. For example, suppose we have a high-flux beam of 500-keV protons with $j_i = 100 \times 10^4 \ \text{A/m}^2$ and $r_0 = 0.05 \ \text{m}$. The beam travels through a drift chamber of length $d = 1 \ \text{m}$ and

a)

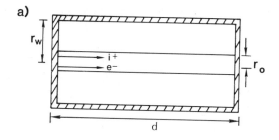

b)

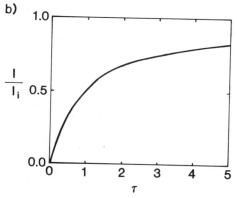

Figure 11.11 Variation of the net current for a neutralized ion beam inside a closed pipe. (a) Geometry of the calculation. (b) Variation of total current, I, as a function of time. I_i equals the constant injected ion current, τ equals $t/(eLI_i/T_{e0})$, and L equals $(\mu_0/2\pi)d \ln(r_w/r_0)$.

radius $r_w = 0.15$ m. The total beam current is $I_i = 7.9$ kA, the chamber inductance is $L \cong 2.2 \times 10^{-7}$ H, and the average electron energy for current neutralization is $T_{e0} = 271$ eV. Inserting the values into Eq. (11.37), the characteristic current decay time is $\tau = LI_i e/T_{e0} = 6.4$ μsec. The quantity τ is much longer than pulse lengths typical of many experiments ($\leqslant 0.1$ μsec).

By adjusting the propagation chamber inductance and the beam width, we can achieve conditions where τ is much shorter than the beam pulse length. In this limit, we can use small propagation chambers for measurements of the current density of energetic, neutralized ion beams. Figure 11.12 shows the geometry of a detector. It consists of a bounded cylindrical chamber with a Rogowski loop [CPA] to measure the net axial current. Ions enter through a foil with thickness less than the ion range. Besides acting as an electron source, the foil provides discrimination against low-energy ions. Each ion in the beam creates several secondary electrons on the inner surface of the foil. We can design the chamber geometry for a rapid decay of electron current. For example, with $d = 0.01$ m, $r_0 = 0.002$ m, and $r_w = 0.015$ m, the chamber inductance is only $L = 4 \times 10^{-9}$ H. For a beam of 500-keV protons with $j_i = 100 \times 10^4$ A/m^2, the

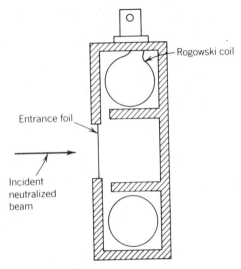

Figure 11.12   A detector to measure the current density of an energetic, high-current ion beam neutralized by electrons. (Courtesy J. Greenly, Cornell University).

decay time is $\tau = LI_i e/T_e \cong 0.2\,\text{nsec}$. As a result, the detector has good time resolution. We can find the ion current density by dividing the net chamber current by the area of the entrance aperture.

11.4.   FOCAL LIMITS FOR NEUTRALIZED ION BEAMS

One motivation to neutralize an ion beam is to focus it to a small spot size. Recent neutralization studies have concentrated on intense ion beam transport to small inertial fusion targets. In Section 5.4, we saw that space-charge forces interfere with focusing. In this section, we shall study processes that limit focusing of neutralized beams under vacuum. Although the focal spot size for a neutralized beam is smaller than that for a bare beam, we shall see that collective effects may cause problems for some applications.

Figure 11.13 shows a pulsed neutralized beam crossing a vaccum region to a target. A current of electrons equal to the ion current enters the beam at the entrance. The electrons almost eliminate electric fields in the beam. Nonetheless, there is a small transverse electric field if the electrons have nonzero transverse temperature, kT_e. Sections 11.1 and 11.2 showed that both the longitudinal and transverse neutralization processes lead to thermal electron distributions. As a first step in the calculation of ion trajectories in a neutralized beam, we shall estimate the magnitude of the thermally generated fields.

We take an ion beam with cylindrical symmetry—changes in the beam dimension take place over axial distances much larger than the beam radius. The

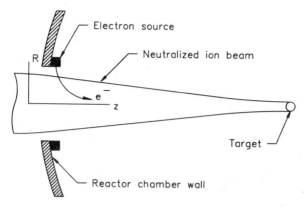

Figure 11.13 Coordinate system to analyze propagation of a pulsed neutralized ion beam to an inertial fusion target in a spherical reactor.

radius variation of high-energy ions is a given function:

$$n_i(r) = n_{io} f(r). \qquad (11.41)$$

The quantity n_{io} is the ion density on axis and $f(r)$ is a normalized function that equals unity at $r = 0$ and drops to zero at large radius. We assume that the electron transverse velocity distribution viewed in the beam rest frame is close to a Maxwell distribution with uniform temperature kT_e. Section 2.11 showed that the electron density is related to the electrostatic potential by

$$n_e(r) = n_{eo} \exp\left[e\phi(r)/kT_e\right]. \qquad (11.42)$$

With the choice $\phi(0) = 0$, the quantity n_{eo} is the electron density on axis. Although Eq. (11.42) applies only to beams in equilibrium, it is useful to estimate the electron density when the ion beam changes slowly compared with the average electron transit time over the beam width.

When a neutralized ion beam propagates in free space, the net beam current is zero. The line charges of ions and electron have equal magnitude, or

$$\int_0^\infty 2\pi r \, dr \, n_i(r) = \int_0^\infty 2\pi r \, dr \, n_e(r). \qquad (11.43)$$

When the neutralizing electrons are cold, the radial distributions of electrons and ions are identical. On the other hand, the density profile of hot electrons may extend radially outside the ion distribution. We can calculate the variation of electron density by substituting Eqs. (11.41) and (11.42) into the cylindrical

Poisson equation:

$$\frac{1}{r}\frac{d}{dr}r\frac{d\phi}{dr} = \frac{e}{\varepsilon_0}\left[-n_{i0}f(r) + n_{e0}\exp\left(\frac{e\phi}{kT_e}\right)\right]. \tag{11.44}$$

We can identify the scaling parameters by rewriting Eq. (11.44) in terms of the following dimensionless variables:

$$\Phi = e\phi/kT_e, \tag{11.45}$$

$$R = r/(kT_e\varepsilon_0/e^2n_{e0})^{1/2} = r/\lambda_d, \tag{11.46}$$

where λ_d is the Debye length [Eq. (6.13)]. The reduced form of Eq. 11.44 is

$$\frac{1}{R}\frac{d}{dR}R\frac{d\Phi}{dR} = \exp(\Phi) - N_{i0}f(R), \tag{11.47}$$

where $N_{i0} = n_{i0}/n_{e0}$. Inspection of Eq. (11.47) shows that Φ changes significantly over a scale length $R \sim 1$.

We can solve Eq. (11.47) numerically by an integration from the axis to large radius. We initiate the calculation with the starting conditions $\Phi(0) = 0$, $d\Phi(0)/dR = 0$, and an assumed value of N_{i0}. Equation (11.43) and Gauss's law imply that the radial electric field outside the beam equals zero. If the choice of N_{i0} is correct, $d\Phi/dR$ approaches 0 at infinite radius. Figure 11.14 illustrates two solutions. In the first (Fig. 11.14a), the ion density is uniform between the axis and a sharp edge. In the second solution (Fig. 11.14b), the ion density drops smoothly to zero. In both cases, the transverse temperature pushes electrons outside the ion distribution. The ion density exceeds the electron density near the axis. As a result, there is a positive radial electric field. The equilibrium solution represents a balance between the radial electric force and the gradient of the electron pressure.

For an ion beam with a well-defined boundary, the electric field is concentrated within a few Debye lengths of the edge. The peak electric field roughly equals

$$E_r \sim kT_e/\lambda_d e. \tag{11.48}$$

For the smooth beam profile, charge separation occurs over the full width of the beam. The radial variation of electric field is almost linear.

We can use Eq. (11.48) to estimate the effects of electron temperature on neutralized ion beam focusing. Figure 11.13 shows the geometry of the calculation. Ions enter a spherical chamber and travel through vacuum to a small target. At the entrance point, the ions draw electrons from a source. From Section 11.1, we expect that the entering electrons have a small but nonzero temperature, kT_{e0}. Near the entrance, electric fields resulting from electron temperature are

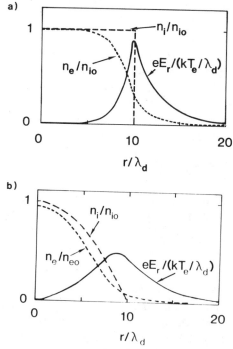

Figure 11.14 Spatial variations of normalized particle density and radial electric field for a cylindrical ion beam neutralized by hot electrons with temperature T_e. $\lambda_d = (kT_e\varepsilon_0/e^2 n_{e0})^{1/2}$. (*a*) Uniform-density ion beam with a sharp boundary. (*b*) Ion beam with a gradual density decrease to zero.

small and have a negligible effect on the ion orbits. On the other hand, electric fields resulting from electron temperature can be very strong near the target. In a short neutralized ion beam, the volume occupied by electrons shrinks substantially as the beam moves toward a focus. Compression of the electrons raises their temperature.

We shall use an envelope equation to describe propagation of a neutralized ion beam. The process of neutralized beam focusing is complex—our model illustrates the application of approximations and the limitations they introduce. The main assumption is that the length of the ion beam pulse is much shorter than the distance from the injection point to the target. As a result, the ions and electrons form an isolated system during propagation. Given the initial electron distribution, we can estimate the final properties by applying the principle of phase-volume conservation (Section 3.8). If the beam is not isolated, an exchange of hot electrons in the beam for cold electrons from the vacuum chamber or target can take place. This process is much more involved, so that we must turn to computer simulations for predictions.

We take the electron distribution at injection as isotropic with temperature kT_{e0}. As the beam travels to the focal point, the electrons compress radially. As a result, the electron temperature varies with the axial position of the beam, $T_e(z) \geqslant T_{e0}$. We can describe the compression of electrons in two special cases:

1. For an ideal radial compression, the transverse energy of electrons increases while the thermal energy in the axial direction remains constant. We apply the theory of Section 3.8 for a two-dimensional compression.

2. If the beam compression is nonuniform in the z direction, some of the transverse energy gain converts to an axial velocity spread. If there is strong coupling between axial and transverse motion, the electron velocity distribution remains isotropic. Here, the beam undergoes a two-dimensional compression with energy shared between three degrees of freedom.

The final electron distribution of a real beam is likely to have properties intermediate between the predictions of the two limiting cases.

The solutions of Eq. (11.47) imply that the radial electric field at the beam roughly equals

$$E_r(z) \cong \eta[kT_e(z)/e\lambda_d]. \tag{11.49}$$

The quantity $T_e(z)$ is the transverse electron temperature and η is a scaling parameter with a value near 0.5. The Debye length of electrons in the beam changes with propagation distance according to

$$\lambda_d(z) = \lambda_{d0}\left[\frac{kT_e(z)}{kT_{e0}}\right]^{1/2}\left[\frac{n_{e0}}{n(z)}\right]^{1/2}. \tag{11.50}$$

For a radial compression, the electron density is related to the envelope radius, R, by

$$n_e(z) = n_{e0}[R_0/R(z)]^{1/2}. \tag{11.51}$$

To construct an envelope equation, we need an expression for the transverse temperature as a function of the beam radius. To begin, consider an ideal two-dimensional compression. From Section 3.8,

$$kT_e(z) = kT_{e0}(R_0/R)^2. \tag{11.52}$$

Substitution of Eq. (11.52) into (11.50) shows that the Debye length is constant during propagation, $\lambda_d(z) = \lambda_{d0}$. For the parameters of inertial fusion beams, the Debye length near the focal point often exceeds the beam radius. Therefore, the radial electric fields are close to those of an unneutralized beam near the target. Combining Eqs. (11.49), (11.50), (11.51), and (11.52) leads to the following expression for the envelope electric field as a function of radius:

$$E_r(R) \cong (\eta kT_{e0}/e\lambda_{d0})(R_0/R)^2. \tag{11.53}$$

The envelope equation for a nonrelativistic ion beam with zero emittance in a field-free region is

$$R'' = [\eta k T_{e0} R_0^2 / m_i v_i^2 \lambda_{d0}] / R^2. \qquad (11.54)$$

We can integrate Eq. (11.54) from the target at $z = 0$ back to the injection point at $z = -L$. Assume that the beam has a waist at the target, so that $R(0) = R_{min}$ and $R'(0) = 0$. At the injection point, $R(-L) = R_0$, $R'(-L) = -\theta_0$. The quantity θ_0 is the envelope injection angle, $\theta_0 \cong R_0/L$. We find that

$$[R'(z)]^2 = \left(\frac{2\eta k T_{e0} R_0}{m_i v_i^2 \lambda_{d0}}\right)\left[\frac{1}{R_{min}} - \frac{1}{R(z)}\right] \qquad (11.55)$$

The quantity v_i is the axial ion velocity. If we drop the term $1/R(z)$ in brackets, Eq. (11.55) leads to a relationship for the minimum beam spot size:

$$\frac{R_{min}}{R_0} \cong \left[\frac{\eta k T_{e0}}{T_i}\right]\left[\frac{R_0}{\lambda_{d0}}\right]\left[\frac{L}{R_0}\right]^2. \qquad (11.56)$$

The quantity T_i is the ion kinetic energy, $m_i v_i^2 / 2$.

We find a different result from Eq. (11.56) if the electron energy growth is uniform in three dimensions. Conservation of phase volume (Section 3.8) implies that

$$k T_e(z) = k T_{e0} (R_0/R)^{4/3}. \qquad (11.57)$$

Using Eq. (11.57) in place of Eq. (11.52), the predicted focal spot size is

$$\frac{R_{min}}{R_0} \cong \left[\frac{3\eta k T_{e0} L^2}{2 T_i R_0 \lambda_{d0}}\right]^{3/2}. \qquad (11.58)$$

We can illustrate the implications of Eqs. (11.56) and (11.58) for beam parameters in a conceptual heavy ion fusion reactor. Suppose that multiple beams of 10-GeV U^+ ions irradiate a target. Each beam carries 5 kA and has a pulse length of 10 nsec. The injection beam radius is $R_0 = 0.06$ m, while the propagation distance is $L = 10$ m. The beam length of 0.9 m is shorter than the propagation length. Neutralizing electrons with the same velocity as the ions have a kinetic energy of 23 keV. Following the discussion of Section 11.1, we assume an initial electron temperature of $k T_{e0} \cong 10$ keV. The initial density of ions and neutralizing electrons is $n_{e0} = 3.1 \times 10^{16}$ m^{-3}. The initial Debye length, $\lambda_{d0} \cong 4.2$ mm, is much smaller than the beam radius. Inserting the parameters into Eq. (11.56) with $\eta = 0.5$ gives a spot size prediction of $R_{min} = 12$ mm for an ideal two-dimensional compression. If the thermal energy is equal in three dimensions, the spot size from Eq. (11.58) is $R_{min} = 5.3$ mm. In both cases, the predicted radius is larger than a fusion target. Therefore, the problem of thermally generated electric fields warrants a more detailed study. Our simplified model may overestimate the electric fields near the target. When the beam contacts the target, hot electrons may exchange with cold electrons from the target surface, short-circuiting the transverse electric fields.

11.5. ACCELERATION AND TRANSPORT OF NEUTRALIZED ION BEAMS

Neutralized ion beams can carry very high power—beam current can exceed 1 kA. Such beams present special problems for acceleration and transport. It is essential to maintain a close balance between ion and electron density throughout the acceleration process. In this section, we shall discuss methods to guide neutralized beams and to increase their kinetic energy. Conventional methods of beam focusing, such as electrostatic lenses or quadrupole magnets, are ineffective. The fields in these devices strongly deflect electrons and may interfere with neutralization of the ion beam. Here, we shall concentrate on alternative focusing methods based on collective effects. High-density ion beams cannot propagate without neutralizing electrons. By inducing small electron displacements, we can generate large space-charge electric fields that can guide energetic ions.

Figure 11.15*a* shows a simple example of electron control for a high-intensity ion beam. A neutralized beam enters a vacuum region through a grounded grid. The entering ion and electron fluxes are exactly equal. The ion beam has nonzero emittance—we represent the spread in velocity by a transverse temperature kT_i. The region has an applied solenoidal magnetic field, B_0. Although the field is too weak to affect the orbits of energetic ions directly, it is strong enough to confine the low-energy neutralizing electrons. In the propagation region, the ions expand while the electrons are confined to a cylindrical volume. The resulting charge separation creates radial electric fields that can focus the ions.

Exact solutions for the electrostatic potential in the electrostatic sheath at the edge of the beam depend on details of the ion distribution. Here, we shall make a rough estimate of the sheath dimension from scaling arguments. A strong magnetic field binds electrons to field lines. In the strong field limit, electrostatic effects determine the sheath width. If the ions have a Maxwell distribution in transverse velocity, we can apply the results of Section 6.2 to find the sheath width. The width is close to an ion Debye length:

$$\lambda_E \sim (kT_i \varepsilon_0 / e^2 n_i)^{1/2}. \tag{11.59}$$

In Eq. (11.59), n_i is the density of the entering ion beam and the subscript E denotes the sheath width from electrostatic effects.

With a weak magnetic field, electron confinement in the presence of space-charge electric fields determines the sheath width. We can apply results from

a)

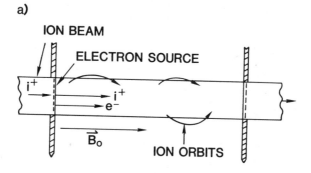

b)

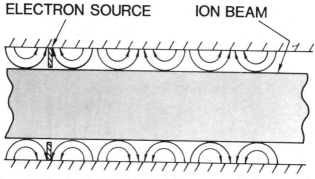

Figure 11.15 Transport of high-intensity neutralized ion beams by the control of electrons. (*a*) Weak solenoidal magnetic field confines electrons, while the space-charge electric field confines ions. (*b*) Multipole magnetic fields concentrated at the boundary confine neutralizing electrons.

Section 8.1. We want to find the size of orbits for electrons in a magnetic field B_0 subject to a voltage of about $V_0 \sim kT_i/e$. From Eq. (8.11), the magnetic sheath width is roughly

$$\lambda_M \sim (2kT_i m_e)^{1/2}/eB_0. \tag{11.60}$$

Figure 11.16 shows plots of the electric and magnetic sheath dimensions as functions of kT_i, n_i, and B_0. The condition $\lambda_M \gg \lambda_E$ defines the weak-magnetic-field regime. In an intermediate regime, the expansion width for the ions equals the larger of λ_E and λ_M.

As an example, suppose we have a neutralized beam of C^+ ions with current density $j_i = 1 \times 10^4 \, \text{A/m}^2$, kinetic energy $= 2 \, \text{MeV}$, and density $n_i = 1.1 \times 10^{16} \, \text{m}^{-3}$. The beam has an angular divergence of $0.5°$, corresponding to a

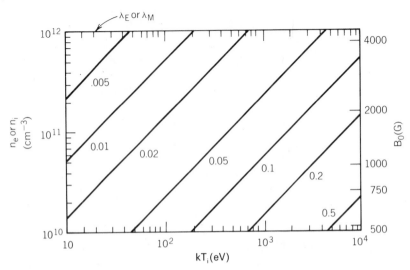

Figure 11.16 Sheaths for electron control of intense neutralized ion beam transport—dimensions in centimeters. λ_M, the scale length for magnetic confinement of electrons, depends on B_0(G) and T_i(eV). λ_E, the scale length for electrostatic confinement of ions, depends on T_i(eV) and n_i (cm^{-3}).

transverse temperature of $kT_i = 150\,\text{eV}$. Equation (11.59) predicts that the electrostatic sheath width is only 0.87 mm. To ensure that $\lambda_M < \lambda_E$, the magnetic field should be $B_0 > 0.048$ tesla. The results show that low magnetic fields can confine energetic ion beams through charge-separation effects.

The system of Fig. 11.15 is impractical for high-flux beams because of the entrance mesh. Also, extraction of the neutralized ion beam from the magnetic field is difficult. Figure 11.15*b* shows an alternative geometry for collective ion beam transport. The applied magnetic field is a cusp array (Section 10.9). Magnetic windings with alternate polarity or permanent magnets line the wall of the transport chamber. The axial lengths of cells are smaller than the coil radius. The resulting cusp field is concentrated at the wall and is small on the axis. Although the weak magnetic fields have little effect on the ions, they strongly focus the neutralizing electrons. The radial electric fields generated by charge separation confine the ions. The cusp array has advantages over the solenoid: (1) the minimum-B field provides stable confinement of the low-energy electrons and (2) the neutralized ion beam emerges from and travels to field-free regions.

The *space-charge lens* is another option for collective focusing of neutralized ion beams. In contrast to the cusp transport system, the space-charge lens is an isolated solenoid lens with linear radial forces that focus a neutralized ion beam toward a point. To analyze its effect, we shall make use of the geometry of Fig. 11.17. An ideal neutralized beam enters a weak solenoid lens. The ions and electrons of the incident beam have only axial velocity. The density and velocity of the electrons exactly equals that of the ions. The magnetic field of the lens has little effect on the energetic ions—without space-charge electric fields, single ions

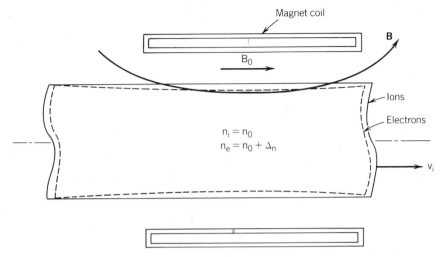

Figure 11.17 Geometry of the space-charge lens.

would pass through with little deflection. On the other hand, the converging magnetic field lines exert a radial force on the electrons. Compression of the electron distribution creates electric fields that point toward the axis.

To calculate electric fields in the space-charge lens, we must find a self-consistent beam equilibrium. For a first-order treatment, we neglect changes in the ion density and consider a long solenoidal field. All electrons have zero canonical angular momentum and equal total energy:

$$T_e = T_i(m_e/m_i). \tag{11.61}$$

The constraints on the electron distribution are similar to those we used to calculate the Brillouin equilibrium—we can apply the results of Section 10.3. The main difference is the presence of the ions. In the theory of Section 10.3, the charge density, n_0, represented a bare electron beam. In the space-charge lens, the negative charge density arises from an excess of electrons near the axis. We shall denote the negative charge enhancement as Δn. For a high-current neutralized beam, we expect that

$$\Delta n \ll n_i, n_e. \tag{11.62}$$

Section 10.3 showed that Δn is uniform in radius in a self-consistent electron equilibrium. Equation (10.33) implies that the charge imbalance has magnitude

$$\Delta n = (eB_0/2m_e)^2 (2\varepsilon_0 m_e/e^2). \tag{11.63}$$

From Eq. (5.26), the resulting radial electric field is

$$E_r = -(\omega_{g0}^2 m_e / 4e) r. \qquad (11.64)$$

The electric field of Eq. (11.64) combined with the centrifugal force balances the focusing force of the magnetic field so that the neutralizing electrons pass through the lens with small change in radius. The electric field varies linearly with r—the field from the perturbed electrons is independent of the radial variation of ion beam density when the fractional charge imbalance is small.

The space-charge electric force on nonrelativistic ions is much larger than the direct magnetic force. Within the solenoid lens, the electric force is

$$F_e = -m_e (eB_0 / 2m_e)^2 r. \qquad (11.65)$$

Equation (9.25) implies that the magnetic force is

$$F_m = -m_i (eB_0 / 2m_i)^2 r. \qquad (11.66)$$

The ratio of forces is

$$F_e / F_i = m_i / m_e. \qquad (11.67)$$

Equation (11.67) shows that the focusing effect from space separation is over a thousand times stronger than the direct action of the magnetic field.

The space-charge lens has the apparent ability to focus intense ion beams with modest magnetic fields. Unfortunately, further analysis shows that a linear force variation is possible only in a restricted parameter regime, limiting application to low-current ion beams. One restriction is that the magnetic field cannot be so strong that it reflects entering electrons. We discussed this process in Section 10.1. Equation (10.11) gives a constraint on the maximum radius of a neutralized beam:

$$r_b < 2m_e v_i / eB_0. \qquad (11.68)$$

To illustrate the implication of Eq. (11.68), suppose we have a 10-MeV C^+ beam with velocity $v_i = 1.26 \times 10^7$ m/sec. For a beam radius of 0.02 m, the equation implies that $B_0 < 7.2 \times 10^{-3}$ tesla. A neutralized beam can penetrate through a lens with higher applied magnetic field by creating axial space-charge fields that pull electrons through the lens. Although the beam crosses the lens, the associated spread in electron total energy violates the conditions of the model and the radial electric field is non-linear.

With a limit on applied magnetic field, we can investigate the constraints imposed by the condition of small fractional change in electron density, Eq.

Figure 11.18 Methods to accelerate high-current neutralized ion beams. (*a*) Low gradient, large-area injector. Biased grids prevent streaming of electrons into the acceleration gaps. (*b*) Radial magnetic field acceleration gap. The magnetic field inhibits electron streaming.

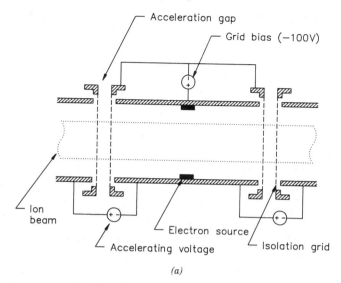

(a)

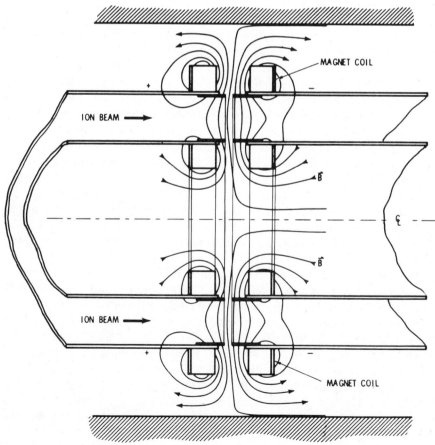

(b)

(11.62). From Eq. (11.63), the ratio of the electron density perturbation to the density in the undisturbed beam is

$$\Delta n/n_e = (eB_0/2m_e)^2 (2\varepsilon_0 m_e/e^2 n_e) \ll 1. \qquad (11.69)$$

We can rewrite Eq. (11.69) in the form that shows that limit on the ion beam current,

$$I \ll (\pi/2)(eB_0^2 \varepsilon_0 v_i/m_e)r_b^2. \qquad (11.70)$$

Inserting parameters for the carbon ion beam with $B_0 = 5 \times 10^{-3}$ tesla and $r_b = 0.02$ m, we find that $I \ll 0.31$. The implication is that space-charge lenses have linear focusing strength only for low-current ion beams.

To conclude this section, we discuss methods to accelerate neutralized ion beams. The problem for achieving a neutralized beam accelerator is the control of electrons in the presence of strong axial electric fields. The electrons cannot cross acceleration gaps with the ions. The neutralizing electrons must be removed from the ion beam at the entrance to a gap and replaced at the downstream side. Also, the drift regions between acceleration gaps must be electrically isolated from the gaps. Accelerating electric fields that penetrate into the drift regions would pull electrons from the neutralized beam and accelerate them backward. Electron loss wastes energy and prevents effective neutralization between gaps.

Figure 11.18*a* shows an accelerator for a high-current ion beam that uses grids for electrical isolation of the acceleration gaps. The gap electric field penetrates into the drift region a small distance comparable to the spacing between grid wires. Electron loss is small if the beam in the drift region has a positive potential relative to the grid. In the drift region, electrons mix with the ion beam through the transverse neutralization process of Section 11.2. A beam with a long pulse length traps a distribution of stationary electrons in the drift spaces between narrow acceleration gaps. Shaped grids create transverse components of electric field to focus the ions. The grid accelerator has the disadvantages of ion loss on the grids and degradation of beam emittance by facet lens effects. Nonetheless, the device can create high-perveance beams of moderate kinetic energy (~ 1 MV) and current density ($\sim 1 \times 10^4$ A/m^2).

The radial magnetic field acceleration gap of Fig. 11.18*b* allows acceleration of high-flux neutralized beams. In contrast to the grid accelerator, the radial field gap has no physical structure to intercept the beam. Transverse magnetic fields prevent electron flow across the acceleration gaps. The radial field gap is similar to the magnetically insulated gap of Section 8.8—a preaccelerated beam replaces the ion source. By conservation of canonical angular momentum, the magnetic fields do not contribute a net azimuthal velocity to accelerated ions. For long-pulse beams, electrons trapped in the acceleration gap allow enhanced ion flux (Section 8.9). Multiple-gap radial field accelerators have generated 3-kA pulsed C^+ beams at 600 keV. Observed current densities of 30×10^4 A/m^2 are well beyond conventional space-charge limits.

12

Electron Beams in Plasmas

We have looked briefly at plasma properties several times in previous chapters. In this chapter, we shall initiate a more detailed study of plasmas with emphasis on their response to pulsed high-current electron beams. There is a large body of literature on the topic because of the possibility of long-distance propagation of electron beams through the atmosphere and the potential use of intense beams for plasma heating. We shall concentrate on the motion of charged particles rather than the atomic processes involved in the creation of plasmas. The models assume preformed plasmas and do not address the involved process of ionization by the beam.

We shall limit attention to infinite plasmas with no applied magnetic field. The ideal plasma responds immediately to a pulsed beam, providing complete neutralization of the beam space-charge and current. Four sections in this chapter review plasma properties that limit their response to changes of charge and current density. Section 12.1 describes how plasma electrons neutralize space charge by shifting their positions. Residual electric fields depend on the temperature of the electrons. The derivations lead to the Debye scale length. Field cancellation may be incomplete over distances smaller than the Debye length. Section 12.2 reviews electrostatic plasma oscillations. These occur when there is a sudden change in charge density, such as the injection of a beam. The oscillations have a characteristic frequency, the electron plasma frequency, ω_{pe}.

Although low-temperature plasmas conduct current, they have a much lower conductivity than metals. Sections 12.5 and 12.6 discusses imperfect conduction in plasmas. Resistive effects modify the distribution of plasma return current for a pulsed beam. Section 12.5 describes how the inertia of electrons delays

535

the plasma response to a rapid change in current. We derive the magnetic skin depth, a characteristic dimension for cancellation of pulsed magnetic fields in plasmas. Section 12.6 reviews the effect of collisional resistivity in plasmas. Resistivity results in spatial spreading of plasma return current over a cross section larger than that of the beam. As a result, the magnetic field in the beam volume can provide a self-confined equilibrium.

Section 12.3 extends the theory of plasma oscillations to the transverse motion of beam electrons about a core of immobile ions. We will apply the results in Section 13.7 to the hose instability of an ion-confined electron beam. Section 12.4 covers the space-charge neutralization of a pulsed electron beam. We shall calculate the time-dependent shift of plasma electrons by a direct numerical solution of the nonlinear moment equations. The treatment illustrates some methods to solve partial differential equations with computers and shows that neutralization is almost complete if the beam current risetime is much longer than $1/\omega_{pe}$.

Section 12.7 describes propagation of a long-pulse-length beam in a resistive plasma. At late time, the plasma return current covers a larger area than the beam current; therefore, there is a nonzero confining magnetic field in the beam volume. The magnetic deflection of electrons in a charge-neutral beam defines an upper limit on the transportable current. The Alfven current, I_A, is a useful scaling parameter for intense beam studies. Section 12.8 reviews the theory of self-contained equilibria for a beam with current less than I_A in a resistive plasma. We shall study the Bennett equilibrium, a self-consistent model for magnetically pinched beams with nonzero emittance. The Bennett equilibrium, based on a Maxwell distribution of transverse energy, is a good representation for collision-dominated beams. Section 12.9 applies the equilibrium results to model long-distance transport of an electron beam through a weakly ionized plasma. Collisions with gas atoms increase the transverse emittance, resulting in an expanding beam envelope. We derive the Nordsieck length, an expression for the collision-limited propagation distance.

12.1 SPACE-CHARGE NEUTRALIZATION IN EQUILIBRIUM PLASMAS

When a high-current beam enters a plasma, plasma particles move to cancel the beam-generated electric field. In this section, we shall calculate the response of a plasma to a steady-state beam and investigate the significance of the plasma Debye length.

To begin, we shall review the quantities we need to characterize a plasma. Plasmas are collections of ions and electrons governed by long-range electromagnetic interactions. Usually, the densities of ions and electrons, n_i and n_e, are almost equal, so that the mixture is space-charge-neutralized. In many experiments on plasma transport of electron beams, the beam density, n_b, is much smaller than the plasma density, $n_b \ll n_e$. The beam drives out some of the plasma electrons to achieve complete space-charge neutralization, where

$$n_i = n_e + n_b. \tag{12.1}$$

Relativistic electron beams generate strong magnetic focusing forces. Such beams can be self-contained with only partial space-charge neutralization. Therefore, the transport of such beams in low-density plasmas, $n_b > n_e$, are of considerable interest. Here, the beam expels all the low-energy plasma electrons.

The kinetic energy of plasma particles affects how they respond to beams. In this chapter, we shall represent the velocity dispersion of plasma ions and electrons as Maxwell distributions with temperatures T_i and T_e. Plasma particles may also have average drift velocities, v_i and v_e. Usually, unconfined plasmas are stable against velocity space instabilities if

$$|v_i|, |v_e| \ll (2kT_e/m_e)^{1/2}. \tag{12.2}$$

The condition of Eq. (12.2) holds when moderate current beams propagate through dense plasmas. On the other hand, high-current pulsed electron beams can induce a large plasma electron drift velocity, leading to a two-stream instability and rapid plasma heating.

Plasmas used for electron beam transport usually have low temperature ($kT_e < 100\,\text{eV}$) and large spatial extent. Methods for beam plasma generation include laser ionization, pulsed plasma guns, low-energy-electron discharges, or collisions of beam particles with a background gas. Beam-generated plasmas are often not fully ionized—neutral atoms are present in the beam volume. Although atoms do not participate in electromagnetic interactions, they may influence the beam and plasma responses through collisions.

We shall use a simple model to calculate the steady-state response of a plasma to an injected beam. The beam enters an unconfined plasma of infinite dimension with no included magnetic field. Low-energy electrons respond rapidly to the presence of the beam, while the massive ions respond slowly. We neglect ion motion and assume that the beam pulse is long enough for plasma electrons to adjust to a modified equilibrium. The electron beam density is uniform in z and has cylindrical symmetry. We write the beam density as

$$n_b(r) = n_{b0} f(r). \tag{12.3}$$

The function $f(r)$ equals unity on the axis and drops to zero on the beam envelope, $f(r_b) = 0$. The plasma ions are spatially uniform with density $n_i(r) = n_0$. If the electrons are in thermal equilibrium, their density is

$$n_e(r) = n_0 \exp[+e\phi(r)/kT_e]. \tag{12.4}$$

The electrostatic potential, $\phi(r)$, is negative over the region of interest. The form of Eq. (12.4) ensures that at large radius the potential drops to zero and the plasma electron density approaches the ion density.

Using Eqs. (12.3) and (12.4), the Poisson equation is

$$\frac{1}{r}\frac{d}{dr}\left[r\frac{d\phi}{dr}\right] = \frac{e}{\varepsilon_0}\left[n_{\mathrm{bo}}f(r) - n_0 + n_0\exp\left(\frac{e\phi}{kT_e}\right)\right].$$ (12.5)

Equation (12.5) is similar to Eq. (11.44) for the neutralization of an ion beam by hot electrons. We again scale the potential in terms of kT_e and the length in terms of the Debye length [Eq. (6.13)]:

$$\Phi = -e\phi/kT_e,$$ (12.6)

$$R = r/\lambda_{\mathrm{d}}.$$ (12.7)

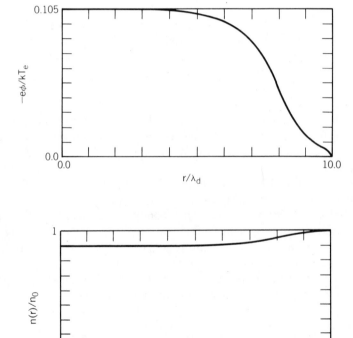

Figure 12.1a Neutralization of a steady-state cylindrical electron beam in a homogeneous plasma. (a) Numerical solutions for a broad beam. Unperturbed plasma electron density, n_0; plasma temperature, T_e; $\lambda_{\mathrm{d}} = (kT_e\varepsilon_0/e^2n_0)^{1/2}$. Beam has uniform density, $n_{\mathrm{bo}} = 0.1n_0$, and a sharp boundary at $r_{\mathrm{b}}/\lambda_{\mathrm{d}} = 8$. Top: spatial variation of electrostatic potential. Bottom: solid line, spatial variation of plasma electron density; dashed line, spatial variation of beam density.

The dimensionless form of Eq. (12.5) is

$$\frac{1}{R}\frac{d}{dR}\left[R\frac{d\Phi}{dR}\right] = -N_{b0}f(R) + [1 - \exp(-\Phi)],\qquad (12.8)$$

where $N_{b0} = n_{b0}/n_0$. Given $f(R)$ and a value for N_{b0}, we solve Eq. (12.8) by integrating from the origin with the symmetry condition $d\Phi/dR = 0$. The proper choice of $\Phi(0)$ gives a solution where $\Phi = 0$ and $d\Phi/dR = 0$ at large radius. This value of $\Phi(0)$ gives an electron density that preserves net neutrality:

$$\int_0^\infty 2\pi r\,dr\,n_e(r) = \int_0^\infty 2\pi r\,dr[n_0 - n_b(r)].\qquad (12.9)$$

Figure 12.1*a* shows a solution of Eq. (12.8) for a broad beam $(r_b/\lambda_d = 8)$ with a sharp boundary. The reduction of the plasma electron density equals the

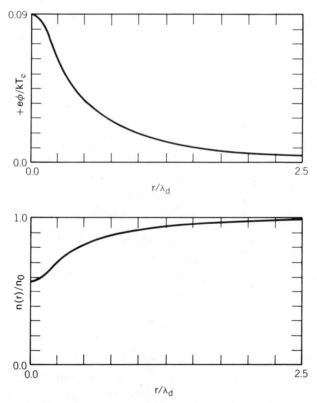

Figure 12.1b Analytic solution for Debye shielding of electric fields in a narrow beam, $r_b/\lambda_d \ll 8$. Top: Spatial variations of electrostatic potential. Bottom: Spatial variation of plasma electron density.

beam density over most of the beam width. Electric fields are concentrated within one Debye length of the beam edge. The on-axis potential energy, $e\phi(0)$, is much smaller than kT_e if the beam density is low, $n_{b0} \ll n_0$.

The nature of Debye shielding in a plasma is often illustrated with the example of the fields of a point test particle with charge q. To find the distribution of plasma electrons, we solve the Poisson equation in spherical coordinates centered on the test particle. We assume that the test charge makes a small perturbation to the plasma distribution, or $\phi(r)/kT_e \ll 1$. The approximate form of the Poisson equation is

$$\frac{1}{r^2}\frac{d}{dr}\left[r^2\frac{d\phi}{dr}\right] \cong \frac{\phi}{\lambda_d^2}. \qquad (12.10)$$

We solve Eq. (12.10) analytically with the boundary condition that the potential approaches the vacuum potential for a charge q as r approaches zero:

$$\phi(r) = (q/4\pi\varepsilon_0 r)\exp(-r/\lambda_d). \qquad (12.11)$$

The expression of Eq. (12.11) is the product of the vacuum electrostatic potential times a term resulting from the plasma, $\exp(-r/\lambda_d)$. The plasma cancels the electric field of the test charge at distances greater than λ_d—the process is called *Debye shielding*. Figure 12.1b shows a numerical solution for a related case, an immersed narrow cylindrical electron beam with $r_b \ll \lambda_d$. The plasma shields the radial electric of the beam over a length scale equal to λ_d.

12.2. OSCILLATIONS OF AN UNMAGNETIZED PLASMA

To study the transport of pulsed beams, we must understand time-dependent plasma processes. Section 12.4 presents numerical calculations of plasma response to a rapidly pulsed electron beam. As a preliminary, this section reviews analytic calculations of plasma oscillations induced by charge imbalances. We limit the discussion to uniform plasmas with infinite extent and no applied magnetic field. Here, all plasma disturbances oscillate at the plasma frequency, ω_{pe}. The quantity $1/\omega_{pe}$ is the characteristic time for plasma electrons to shift position to balance a charge perturbation.

Again, we shall make several approximations to simplify the mathematics and to emphasize the relevant physical processes:

1. The plasma oscillation is a small perturbation about a stationary state.
2. The motion of plasma particles is one-dimensional. Electric fields and particle displacements are in the x direction.
3. The plasma is cold, $kT_e = kT_i = 0$. As a result, the macroscopic electric force determines the motion of electrons. Because all electron orbits at a

position x are identical, the first two moment equations (Section 2.10) gives a complete description.

4. Ions are immobile over time scales for electron motion.

5. Collisions have a negligible effect on electron dynamics. The time between collisions is much longer than $1/\omega_{pe}$.

The steady-state plasma has electron and ion densities that are uniform and equal:

$$n_{e0}(x) = n_{i0}(x) = n_0. \tag{12.12}$$

Equation (12.12) implies that the equilibrium plasma has no electric field. The density of immobile ions remains uniform over time scales of interest:

$$n_i(x, t) = n_0. \tag{12.13}$$

We use moment equations to describe the electron response to a perturbation. The one-dimensional equation of continuity is

$$\frac{\partial n_e}{\partial t} + \frac{\partial}{\partial x}(n_e v_e) = 0. \tag{12.14}$$

The equation of momentum conservation for the nonrelativistic plasma electrons is

$$\frac{\partial v_e}{\partial t} + v_e \frac{\partial v_e}{\partial x} = -\frac{eE_x}{m_e}. \tag{12.15}$$

The function E_x is the electric field resulting from displacement of the electrons. It is related to the electron density through the one-dimensional Poisson equation:

$$\frac{\partial E_x}{\partial x} = \frac{e}{\varepsilon_0}(n_0 - n_e). \tag{12.16}$$

The first term in parentheses is the contribution of the uniform ion density.

Equations (12.14), (12.15), and (12.16) are a coupled set of nonlinear differential equations. We can find an analytic solution in the limit of small changes from equilibrium. We write the variation of electron density as

$$n_e(x, t) = n_0 + \Delta n(x, t), \tag{12.17}$$

where $\Delta n \ll n_0$. In the equilibrium state, there is no average electron velocity or electric field. We denote the small perturbed values of these quantities as Δv

and ΔE. The continuity equation becomes

$$\frac{\partial \Delta n}{\partial t} + n_0 \frac{\partial \Delta v}{\partial x} + \Delta n \frac{\partial \Delta v}{\partial x} = 0. \tag{12.18}$$

We drop the third term in Eq. (12.18), a second-order differential quantity. Later, we shall prove that the term is small. The momentum equation takes the form

$$\frac{\partial \Delta v}{\partial t} + \Delta v \frac{\partial \Delta v}{\partial x} = -\frac{e\Delta E}{m_e}. \tag{12.19}$$

We also eliminate the second term on the left-hand side of Eq. (12.19). The Poisson equation involves only the perturbed density:

$$\frac{\partial \Delta E}{\partial x} = \frac{e\Delta n}{\varepsilon_0}. \tag{12.20}$$

We can combine Eqs. (12.18), (12.19), and (12.20) into a single equation using the chain rule of partial derivatives:

$$\frac{\partial^2 \Delta n}{\partial t^2} = -\left[\frac{e^2 n_0}{m_e \varepsilon_0}\right] \Delta n. \tag{12.21}$$

Equation (12.21) has the general solution

$$\Delta n(x, t) = f_1(x) \cos(\omega_{pe} t) + f_2(x) \sin(\omega_{pe} t). \tag{12.22}$$

The quantities $f_1(x)$ and $f_2(x)$ are arbitrary functions of position—they are constrained by the initial density and velocity variations in the plasma. For any initial state, the plasma response is oscillatory at the characteristic plasma frequency:

$$\omega_{pe} = \left[\frac{e^2 n_0}{m_e \varepsilon_0}\right]^{1/2}. \tag{12.23}$$

The oscillation frequency of Eq. (12.23) is high for plasmas commonly used for electron transport. For example, a plasma with density $10^{19}\,\mathrm{m}^{-3}$ has a plasma angular frequency of $\omega_{pe} = 1.78 \times 10^{11}\,\mathrm{sec}^{-1}$. The oscillation frequency is $f = 28.4\,\mathrm{GHz}$, corresponding to a period of only 0.035 nsec.

To illustrate the physical meaning of Eq. (12.21), suppose we have a plasma with a specified initial density perturbation at $t = 0$ but with no initial electron velocity. Figure 12.2a illustrates the density perturbation, a step function that varies between $n_0 \pm \Delta n_0$ over a spatial period of $2x_0$. For this special case, the function $f_1(x)$ is a step function and $f_2(x) = 0$. Equation (12.22) predicts

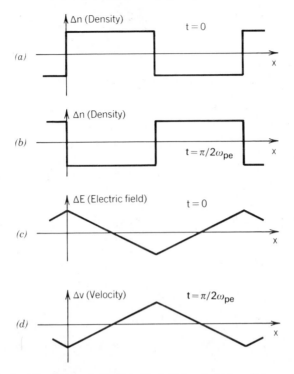

Figure 12.2 Mechanism of plasma oscillations. The initial perturbation of plasma electrons is a step function in density with no velocity change. (*a*) Initial density perturbation. (*b*) Density variation at $t = \pi/\omega_p$. (*c*) Electric field distribution at $t = 0$. (*d*) Velocity distribution at $t = \pi/2\omega_p$.

that the density changes to the form shown in Fig. 12.2*b* at $t = \pi/\omega_{pe}$. The regions with excess electron density become regions of minimum density. A plot of the electric field at $t = 0$ (Fig. 12.2*c*) clarifies the mechanism of the density reversal. From Eq. (12.20), the initial electric field is a sawtooth function with maximum amplitude

$$\Delta E = e\Delta n/2\varepsilon_0. \tag{12.24}$$

Equation (12.19) implies that the spatial variation of velocity is proportional to $-\Delta E(x)$—the temporal velocity variation is 90° out of phase with the electric field. Figure 12.2*d* illustrates the spatial variation of velocity at $t = \pi/2\omega_{pe}$. The direction of the velocity is such that electrons shift from regions of high density toward regions of low density. According to the equation of continuity, the linear spatial variation of velocity means that the density decreases uniformly over a region with $\Delta n > 0$.

We can use Eqs. (12.18) and (12.19) to estimate how far individual electrons move during the density compression and rarefaction. The magnitudes of the perturbed position, velocity, and electric field are related by

$$\Delta x = \Delta v / \omega_{pe} = e\Delta E / m_e \omega_{pe}^2 = 2x_0(\Delta n / n_0). \tag{12.25}$$

In the limit that $\Delta n / n_0 \ll 1$, the amplitude of individual particle oscillations, Δx, is much smaller than the length scale of the disturbance, x_0. Plasma oscillations result from small shifts of large numbers of electrons. We can use Eq. (12.25) to estimate the magnitudes of terms on the left-hand side of Eq. (12.19). The time derivative of a quantity is roughly equal to the magnitude of the quantity multiplied by ω_{pe}—a spatial derivative is comparable to the quantity multiplied by $1/x_0$. A dimensional analysis shows that the first term of Eq. (12.19) has magnitude

$$\frac{\partial \Delta v}{\partial t} \leqslant \omega_{pe} \Delta v, \tag{12.26}$$

while the second term is

$$\Delta v \left(\frac{\partial \Delta v}{\partial x} \right) \leqslant \frac{\Delta v^2}{x_0} \cong \left(\frac{2\Delta v^2}{\Delta x} \right)\left(\frac{\Delta n}{n_0} \right) \cong (\omega_{pe} \Delta v)\left(\frac{\Delta n}{n_0} \right). \tag{12.27}$$

The nonlinear term is smaller by a factor of $\Delta n / n_0$. Hence, we were justified dropping it from the analysis.

The plasma oscillations represented in Fig. 12.2 are stationary. We can generate traveling-wave oscillations if the initial electron distribution includes perturbations in both density and velocity. Because Eqs. (12.18), (12.19), and (12.20) are linear, we can represent any initial variation and subsequent oscillation as a sum of independent harmonic components. Suppose that the density perturbation at $t = 0$ has the form

$$\Delta n(x, 0) = \Delta n_0 \cos(kx) \tag{12.28}$$

and the initial velocity variation equals

$$\Delta v(x, 0) = -\frac{\omega_p}{k} \frac{\Delta n_0}{n_0} \cos(kx). \tag{12.29}$$

Equations (12.28) and (12.29) imply that the spatial functions of Eq. (12.22) are

$$f_1(x) = \cos(kx) \tag{12.30}$$

and

$$f_2(x) = \sin(kx).$$

Combining the trigonometric terms of Eq. (12.22), we can express the density as

$$\Delta n(x, t) = \Delta n_0 \cos(kx - \omega_{pe}t). \tag{12.31}$$

Equation (12.31) is a traveling density perturbation with phase velocity ω_{pe}/k. Such waves are called *plasma waves*. Plasma waves oscillate at frequency ω_{pe}. They may have any phase velocity, depending on the choice of wavenumber. As an example, suppose we inject a 100-keV electron beam into a plasma with density $n_0 = 10^{19}\,\text{m}^{-3}$. The electron beam interacts strongly with plasma waves if the phase velocity of the wave equals the beam velocity, $v_z = \beta c$, where $\beta = 0.548$. Here, beam electrons move with a point of constant phase and see a steady-state electric field. The condition for resonant interaction is

$$\omega/k = \beta c, \qquad (12.32)$$

or

$$\lambda = 2\pi \beta c / \omega_{pe}.$$

For the given plasma density, we expect to observe the growth of plasma waves with $\lambda \cong 5.8 \times 10^{-3}\,\text{m}$.

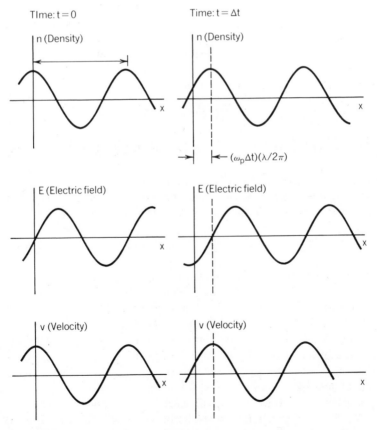

Figure 12.3 Relationships between perturbed electron density, velocity, and electric field for a plasma wave with a positive phase velocity. Left-hand side: $t = 0$; right-hand side: $t = \Delta t$.

For a harmonic disturbance moving in the $+x$ direction, the density, velocity, and electric field are given by

$$\Delta n(x,t) = \Delta n_0 \cos(kx - \omega_{pe}t), \tag{12.33}$$

$$\Delta E = \frac{e\Delta n_0}{\varepsilon_0 k} \sin(kx - \omega_{pe}t), \tag{12.34}$$

and

$$\Delta v = \frac{\Delta n_0}{n_0}\frac{\omega_p}{k} \cos(kx - \omega_{pe}t). \tag{12.35}$$

Figure 12.3 illustrates the relations between plasma quantities in the moving disturbance. The left-hand side of the figure shows the initial plasma state, while the right-hand side shows the quantities after an interval Δt. First, consider the velocity variation. In the region between $x = 0$ and $x = \lambda/2$, the velocity decreases as x increases. Therefore, in this region the velocity dispersion bunches particles (Section 15.3) and the density increases. Conversely, electrons disperse in the region $-\lambda/2 \leqslant x \leqslant 0$, so the density decreases. The combination of increasing and decreasing density causes the point of maximum density to move to the right. Inspection of the spatial variation of electric field shows that electrons accelerate in the region $0 \leqslant x \leqslant \lambda/2$ and decelerate at position $-\lambda/2 \leqslant x \leqslant 0$. The combination of increasing and decreasing velocity causes the point of maximum velocity to move to the right.

12.3. OSCILLATIONS OF A NEUTRALIZED ELECTRON BEAM

In this section, we shall show that a uniform-current-density electron beam in a neutralizing ion background oscillates at a characteristic frequency. The beam oscillation frequency is called the *beam plasma frequency* because it has the same form as the plasma frequency of Section 12.2. We shall apply the results in Section 13.6 to study hose instabilities of ion-confined electron beams.

When shall begin by modeling a nonrelativistic electron beam and add relativistic corrections later. A cylindrical electron beam with uniform density n_e from the axis to radius r_0 moves through an equal density of ions, $n_i = n_e = n_0$. The massive ions are immobile in the transverse direction. In equilibrium, there is no electric field. Electrons move in the axial direction with velocity v_z. For $v_z/c \ll 1$, the effect of beam-generated magnetic forces is small. We shall calculate the response of the beam to a perturbation in the electron rest frame moving at velocity v_z. The condition $v_z \ll c$ means that the particle densities and electric fields are almost the same in the rest frame and laboratory frame.

Suppose that at $t = 0$ we displace all electrons a distance $\delta(z,0)$ from the ions (Fig. 12.4). The charge separation results in an electric field E_x. The electrons subsequently oscillate about the immobile ion core. The quantity $\delta(z,t)$ represents

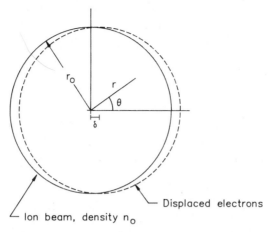

Figure 12.4 Coordinates to analyze transverse oscillations of a cylindrical nonrelativistic electron beam in an ion column.

the time-dependent displacement of the beam center. We can calculate the nature of beam oscillations with the following assumptions:

1. The displacement is small, $\delta \ll r_0$.
2. The electrostatic approximation is valid for the calculation of the field—electromagnetic radiation from the beam is negligible.
3. The axial length for variations of the displacement is much greater than r_0.

The second condition holds if the oscillation period is much longer than r_0/c. With the third assumption, we can approximate local fields with expressions for an infinite-length beam.

As a first step, we shall find the electric field that results from a uniform electron displacement using the polar coordinate system of Fig. 12.4. The figure shows the effect of a small electron beam displacement. Although the core of the beam maintains zero charge density, there is a charge imbalance on the surface. For small displacement, we can represent the charge as a thin surface layer. The magnitude of the surface charge density is proportional to the thickness of the layer—Fig. 12.4 shows that

$$\sigma(\text{coulombs/m}^2) = -en_0\delta\cos\theta. \tag{12.36}$$

Gauss' law implies that the electric fields inside and outside the beam surface are related by

$$E_\theta(\text{out}) = E_\theta(\text{in}), \tag{12.37}$$

$$E_r(\text{out}) = E_r(\text{in}) + \sigma/\varepsilon_0. \tag{12.38}$$

To find the electric field, we can solve the Laplace equation for electrostatic potential inside and outside the surface and match the values of the potential at $r = r_0$ using Eqs. (12.37) and (12.38). The Laplace equation is valid within the beam core because there is no net space charge.

For an axially uniform system, the Laplace equation is

$$\frac{1}{r}\frac{d}{dr}\left[r\,\frac{d\phi(r,\theta)}{dr}\right] + \frac{d^2\phi(r,\theta)}{d\theta^2} = 0. \qquad (12.39)$$

The method of separation of variables gives a general solution for Eq. (12.39). We take the potential as

$$\phi(r,\theta) = R(r)\Theta(\theta). \qquad (12.40)$$

Substitution of Eq. (12.40) into Eq. (12.39) gives individual equations for the radial and azimuthal functions, $R(r)$ and $\Theta(\theta)$. We can write the general solution in terms of cylindrical harmonic functions:

$$\phi(r,\theta) = \sum_{n=0}^{\infty} R_n(r)\Theta_n(\theta), \qquad (12.41)$$

where

$$\Theta_0 = A_0\theta + B_0, \qquad (12.42)$$

$$R_0 = C_0\ln r + D_0, \qquad (12.43)$$

$$\Theta_n = A_n\cos(n\theta) + B_n\sin(n\theta), \qquad (12.44)$$

and

$$R_n = C_n r^n + D_n r^{-n}. \qquad (12.45)$$

The electrostatic potential for the geometry of Fig. 12.4 must have symmetry about the y axis; therefore, we can drop the sine terms in the expansion of Eq. (12.44). The requirement that the potential inside the beam has a finite value at the origin reduces the expansion to the following form:

$$\phi_{\text{in}}(r,\theta) = \sum_{n=1}^{\infty} M_n r^n \cos(n\theta). \qquad (12.46)$$

Similarly, the potential in the region $r > r_0$ approaches zero at large radius. Here, the potential is

$$\phi_{\text{out}}(r,\theta) = \sum_{n=1}^{\infty} N_n r^{-n} \cos(n\theta). \qquad (12.47)$$

We can determine the coefficients in Eqs. (12.46) and (12.47) by matching the solutions at $r = r_0$. Note that the equations must hold at all values of θ. Therefore, all coefficients except those with $n = 1$ equal zero. The constraints on the coefficients for $n = 1$ imply that the electrostatic potential is

$$\phi_{out}(r, \theta) = (\sigma r_0^2/\varepsilon_0)[(\cos \theta)/r], \tag{12.48}$$

$$\phi_{in}(r, \theta) = (\sigma/\varepsilon_0)r \cos \theta = (\sigma/\varepsilon_0)x. \tag{12.49}$$

Figure 12.5 plots electric field lines for the potential of Eqs. (12.48) and (12.49). Outside the beam, the potential is identical to that of an electric dipole located at the origin.

The implication of Eq. (12.49) is that the x directed electric field inside the beam volume is constant,

$$E_x = -en_0\delta/\varepsilon_0. \tag{12.50}$$

A uniform displacement of electrons gives a uniform electric field. The equation of motion for all electrons inside the beam core is

$$m_e(d^2\delta/dt^2) = -(e^2n_0/\varepsilon_0)\delta. \tag{12.51}$$

Equation (12.51) implies that in the beam rest frame, the electrons oscillate at frequency $\omega_{pe} = (e^2n_0/\varepsilon_0m_e)^{1/2}$. We can represent transverse motion of the beam

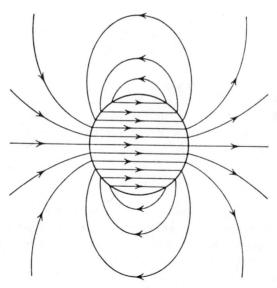

Figure 12.5 Electric field lines resulting from small sideways displacement of a cylindrical electron beam in an ion background. (Adapted from J. D. Jackson, *Classical Electrodynamics*. Used by permission, John Wiley and Sons.)

in the general form

$$\delta(z,t) = \delta(z)\,\mathrm{Re}\exp\left[\omega_{pe}t + \phi(z)\right].\tag{12.52}$$

The quantities $\delta(z)$ and $\phi(z)$ are arbitrary amplitude and phase factors that depend on the initial perturbation. As with any collection of independent oscillators, we can construct standing-wave or traveling-wave disturbances by choosing appropriate forms of $\phi(z)$. With $\delta(z)$ uniform and $\phi(z) = \pm kz$, Eq. (12.52) represents a traveling wave with phase velocity $\pm\omega/k$.

Neutralized relativistic electron beams also can perform transverse electrostatic oscillations. To develop a theory for this regime, we assume that the change of energy in the transverse direction is much smaller than the electron kinetic energy, $(\gamma-1)m_0c^2$. A uniform cylindrical beam of electrons of density n_b travels through a uniform ion cylinder with equal density n_i. In the laboratory frame, there is a beam-generated magnetic force. This force is always symmetric about the center of the beam, so it does not contribute to transverse oscillations. The asymmetric charge distribution of the displaced beam acts like a surface charge layer:

$$\sigma = -en_b\delta\cos\theta.\tag{12.53}$$

Following the previous discussion, the charge layer creates a uniform electric field $E_x = -en_b\delta/\varepsilon_0$. To calculate the electric response to the field, we must remember that they have relativistic mass γm_0 in the laboratory frame. The electron beam oscillates at ω_b, the *beam plasma frequency*:

$$\omega_b = \left[\frac{n_b e^2}{\gamma m_0\varepsilon_0}\right]^{1/2}.\tag{12.54}$$

The beam oscillation frequency differs from Eq. (12.54) if the beam and ion densities are unequal. Generally, n_b is larger than n_i for a neutralized relativistic electron beam in equilibrium. For example, Section 5.5 showed that a uniform-density electron beam with zero emittance in a uniform ion background of the same radius has radial force balance when

$$n_i = n_b/\gamma^2.\tag{12.55}$$

The ions cancel only a fraction of the radial electric field. The ion density for equilibrium must be higher if the electron beam has nonzero emittance. We can symbolize the relation between ion and beam densities by the neutralization fraction [Eq. (5.124)]:

$$f_e = n_i/n_b,$$

where $f_e \geq 1/\gamma^2$.

Suppose an electron beam moves through an ion background of density $f_e n_b$. Figure 12.6a illustrates forces on an equilibrium electron beam in the laboratory frame. With no displacement, there is a balance between the forces of the net radial electric field, emittance, and the beam-generated magnetic field. Figure 12.6b shows forces when the beam moves sideways a small distance δ. The beam-generated magnetic force and the emittance force are unaffected by the displacement. To calculate the electric force, we divide the beam charge into two components with densities $(1 - f_e)n_b$ and $f_e n_b$. The first component creates an radial electric field centered on the beam axis with magnitude equal to the magnetic and emittance forces—it does not contribute to transverse beam oscillation. The second component, when combined with the displaced ion charge, approximates the effect of a charge layer, $\sigma = -e(f_e n_b)\delta \cos\theta$. The result is that the beam oscillates at frequency

$$\omega_\perp = \left[\frac{f_e n_b e^2}{\gamma m_0 \varepsilon_0} \right]^{1/2}. \tag{12.56}$$

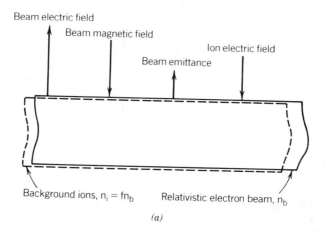

Beam electric field

Beam magnetic field

Ion electric field

Beam emittance

Background ions, $n_i = fn_b$ Relativistic electron beam, n_b

(a)

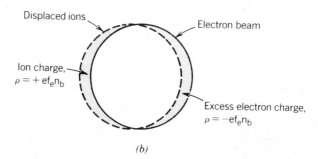

Displaced ions   Electron beam

Ion charge, $\rho = +ef_e n_b$

Excess electron charge, $\rho = -ef_e n_b$

(b)

Figure 12.6 Partial neutralization of a relativistic electron beam by an ion background. (a) Radial forces acting on the beam in equilibrium. (b) Charge distribution for a displaced beam.

12.4. INJECTION OF A PULSED ELECTRON BEAM INTO A PLASMA

In this section, we shall study detailed solutions for the response of a plasma to a rapidly pulsed electron beam using the moment equations of Section 2.10. The model illustrates the excitation of plasma oscillations by a pulsed beam. It also gives us another opportunity to apply numerical methods to the solution of collective problems. We shall solve a set of nonlinear moment equations using the Lax–Wendroff method.

In collective problems, the preliminary analysis and simplifying assumptions are as important as the actual solution. To describe plasma neutralization of a pulsed beam, we want to reduce the problem so that the solutions have generality and do not depend critically on details of the boundary conditions. Nonetheless, we must be certain to include the essential physical processes so that the results reflect the behavior of real systems. Figure 12.7 illustrates the geometry of the calculation. A sheet beam of high-energy electrons enters a uniform, fully ionized plasma enclosed between conducting boundaries at $\pm x_w$. The plasma has an equilibrium electron density of n_{e0}. The beam has half-width x_0.

To reduce the mathematical complexity, we limit attention to solutions where all quantities vary only in the x direction. This assumption is not entirely consistent—the density of a pulsed beam must change along the direction of propagation. In particular, if the beam current at a position on the z axis has risetime Δt_b, then the density varies over the axial length $v_z \Delta t_b$. A one-dimensional calculation gives a good prediction of the local behavior of the plasma if

$$x_w \ll v_z \Delta t_b. \tag{12.57}$$

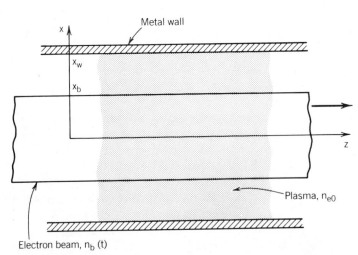

Figure 12.7 Geometry for numerical solutions of fluid equations describing the injection of a pulsed electron beam into a homogeneous, field-free plasma.

We shall adopt the condition of Eq. (12.57) and ignore z derivatives. In the local region of the calculation, the beam density varies with time. We take the beam density as uniform between $\pm x_0$ with a time variation:

$$n_b(t) = n_{b0}g(t). \tag{12.58}$$

The function $g(t)$ varies from 0 to 1 over the beam risetime, Δt_b.

To simplify the model further, we assume that there are no applied magnetic fields and that the beam-generated magnetic fields are small. The calculation of electric fields is easy if we neglect electromagnetic radiation. The electrostatic limit applies if the relaxation time of the plasma is much longer than the time for electromagnetic radiation to cross the system:

$$1/\omega_{pe} \gg x_w/c. \tag{12.59}$$

Finally, we use the approximation of immobile plasma ions—the ion density equals n_{e0} over the duration of the calculation. The validity condition is

$$1/\omega_{pi} \gg \Delta t_b. \tag{12.60}$$

In Eq. (12.60), ω_{pi} is the ion plasma frequency:

$$\omega_{pi} = \left[\frac{e^2 n_{e0}}{m_i \varepsilon_0} \right]^{1/2}. \tag{12.61}$$

The quantity $1/\omega_{pi}$ is the time for ions to respond to an imbalance of space charge.

With a stiff beam and immobile ions, we need solve only moment equations for plasma electrons. We take the plasma electrons as cold. The effects of electron temperature are small if $kT_e \ll e|\phi|$, where $|\phi|$ is the magnitude of electrostatic potential created by the injection of the beam. In the cold electron limit, we can describe the plasma completely with the Poisson equations and the equations of continuity and momentum conservation (Section 12.10). We can write the coupled set of equations in a convenient form by defining dimensionless variables. The dimensionless plasma electron density is referenced to the equilibrium density before injection of the beam:

$$N = n_e/n_{e0}. \tag{12.62}$$

Similarly, the dimensionless maximum beam density is $N_{b0} = n_{b0}/n_{e0}$. As we saw in Section 12.2, there is no inherent length scale associated with plasma oscillations. Therefore, we choose a length scale characteristic of the specific problem, x_w. The dimensionless distance is

$$X = x/x_w. \tag{12.63}$$

The space-charge imbalance created by the beam induces a plasma oscillation. The characteristic time scale is $1/\omega_{pe}$—the dimensionless time is

$$\tau = \omega_{pe}t. \qquad (12.64)$$

We normalize the velocity in terms of the maximum velocity of plasma waves in the system:

$$V = v_e/(x_w\omega_{pe}). \qquad (12.65)$$

Finally, we scale the electric field, E_x, in terms of the maximum field that results if bare plasma electrons fill the entire region between the conducting boundaries. The dimensionless electric field is

$$E = E_x/(-en_{e0}x_w/\varepsilon_0). \qquad (12.66)$$

Recasting a numerical calculation in dimensionless variables is a valuable procedure because the quantities have magnitudes that we can compare to unity. This makes it easier to choose reasonable input parameters, to recognize errors, and to interpret results. For example, an effective plasma neutralization solution has $E \ll 1$. Inserting the dimensionless variables, the equation of continuity [Eq. (2.102)] becomes

$$\frac{\partial N}{\partial \tau} = -N\frac{\partial V}{\partial X} - V\frac{\partial N}{\partial X}. \qquad (12.67)$$

The equation of momentum conservation is

$$\frac{\partial V}{\partial \tau} = -\frac{1}{2}\frac{\partial}{\partial X}(V^2) - E. \qquad (12.68)$$

The first term on right-hand side of Eq. (12.68) represents a change of momentum at a point by electron convection. The second term is the momentum change resulting from space-charge electric fields. The dimensionless form of the Poisson equation is

$$\frac{\partial E}{\partial X} = 1 - N - N_{b0}g(\tau). \qquad (12.69)$$

For a solution on a digital computer, we must convert continuous differential equations to a discrete form. The usual approach is to define approximations to the continuous quantities at a finite set of locations, replacing the spatial and temporal derivatives in the original equations with finite-difference operators. For the present problem, we define the quantities N, V, and E at

Nmesh locations separated by a uniform distance ΔX. To calculate the evolution of the quantities in time, we advance them through a large number of states separated by a uniform interval $\Delta\tau$. Figure 12.8a shows quantities used in the discrete formulation. The set of discrete positions and times is called a *numerical mesh* or *grid*. We specify spatial position with the index j and temporal position by the index n. Discrete quantities defined at *mesh points* represent the original continuous functions

$$N(X,\tau) \rightarrow N(j,n). \qquad (12.70)$$

The difference equations that advance $N(j,n)$, $V(j,n)$, and $E(j,n)$ are correct if the discrete quantities approach the values of the continuous quantities at all positions and times in the limit that ΔX, $\Delta\tau \rightarrow 0$.

We must define the mesh parameters clearly. Mesh quantities along the time axis are simple—time starts at zero and proceeds continuously until the end of the calculation. If we take $n = 0, 1, 2, \ldots$, then $\tau = n\Delta\tau$. The spatial mesh has boundaries. If the spatial index has the range $j = 0, 1, 2, \ldots$, Nmesh-1, Nmesh, then the spatial position is

$$X = j\Delta X. \qquad (12.71)$$

In the present problem, the final mesh point corresponds to the conducting wall at $X = 1$; therefore, the mesh spacing is $\Delta X = 1/\text{Nmesh}$.

The two primary concerns of computer calculations are to preserve numerical stability and to achieve good accuracy. We shall discuss the stability criterion later. Regarding accuracy, we want results that are in close agreement with an exact solution of the partial differential equations. A necessary condition for a valid solution is that the quantities ΔX and $\Delta\tau$ must be smaller than the finest spatial or temporal features of the problem. For example, we want the time step to be smaller than $1/\omega_{pe}$—in our dimensionless units, this is equivalent to $\Delta\tau < 1$. We also want the total computational time to be as short as possible, so $\Delta\tau$ should not be too small. The best approach is to choose a numerical method that gives good accuracy for a moderate number of computational steps. Here, we shall review the two-step Lax–Wendroff method. The errors that result from this finite-difference formulation are on the order of ΔX^2 or $\Delta\tau^2$; therefore, the numerical solutions converge rapidly toward the exact solution as the step-size decreases.

The Lax–Wendroff method[1] is effective for partial differential equations like Eqs. (12.67) and (12.68), where the time derivatives are proportional to the spatial derivations. Section 2.3 stressed the importance of time centering for ordinary differential equations. In this section, we shall extend the two-step method to partial differential equations and seek equations centered in both space and time. Before attacking the coupled nonlinear moment equations, we shall illustrate the method with the simple equation:

$$\partial Y(X,\tau)/\partial\tau = -\partial Y(X,\tau)/\partial X. \qquad (12.72)$$

---

[1] See, for instance, D. Potter, *Computational Physics*, Wiley, New York, 1973, p 67.

Figure 12.8*b* shows the strategy of the numerical solution. Suppose we have a discrete set of values of Y defined over the spatial mesh at time $\tau = n\Delta\tau$. We denote these values as $Y(j,n)$. The object is to advance the values to the next time step, generating a set $Y(j,n+1)$ that agrees with Eq. (12.72). The finite-difference approximation for the spatial derivative of Y at time n is:

$$\partial Y(j + \tfrac{1}{2}, n)/\partial X \cong [Y(j+1,n) - Y(j,n)]/\Delta X. \tag{12.73}$$

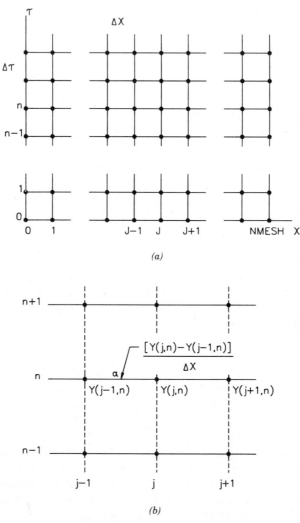

(a)

(b)

Figure 12.8 Elements of the finite-difference solution of fluid equations. (*a*) Definition of a computational mesh in time and a single spatial dimension. Conventions for indices. (*b*) Approximation for the spatial derivative of Y at position $(j - \tfrac{1}{2})/\Delta\chi$ and time $n\Delta\tau$. (*c*) Strategy of the Lax–Wendorff method for a time- and space-centered solution of the convective fluid equation.

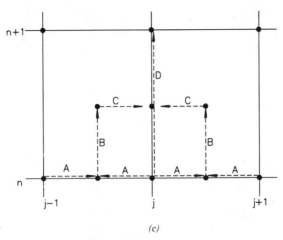

(c)

Figure 12.8 (*Continued*)

Note that the derivative is referenced to a location between two mesh points at time n (point α in Fig. 12.8b). Following the discussion of time centering in Section 2.3, an accurate way to advance Y with Eq. (12.72) is to use the spatial derivative at a spatial mesh point and at a time $n + \frac{1}{2}$ (point β in Fig. 12.8b):

$$Y(j, n+1) \cong Y(j, n) - [\partial Y(j, n + \tfrac{1}{2})/\partial X]\Delta\tau. \qquad (12.74)$$

We can estimate the derivative at the intermediate time point by the two-step process shown schematically in Fig. 12.8c.

In the first step, we calculate quantities at intermediate space and time steps from Eq. (12.72):

$$Y(j + \tfrac{1}{2}, n + \tfrac{1}{2}) = [Y(j + 1, n) + Y(j, n)]/2$$
$$- \{[Y(j + 1, n) - Y(j, n)]/\Delta X\}(\Delta\tau/2). \qquad (12.75)$$

The first term on the right-hand side of Eq. (12.75) is the average value of Y at time n between spatial mesh points j and $j + 1$. The second term is the change in Y in the interval $\Delta\tau/2$ using the spatial derivative at the point $(j + \frac{1}{2}, n)$. Next, we can use the values $Y(j + \frac{1}{2}, n + \frac{1}{2})$ To estimate the derivative at a time-centered position, $\partial Y(j, n + \frac{1}{2})/\partial X$. We then apply the derivatives to advance the set of $Y(j, n)$:

$$Y(j, n+1) = Y(j, n) - [Y(j + \tfrac{1}{2}, n + \tfrac{1}{2}) - Y(j - \tfrac{1}{2}, n + \tfrac{1}{2})](\Delta\tau/\Delta X). \qquad (12.76)$$

Note that second term on the right-hand side approximately equals the desired time-centered derivative of Eq. (12.74).

To show how to apply the algorithm to a real problem, we shall write out all terms to advance the set of nonlinear equations for plasma neutralization (Eqs. 12.67, 12.68 and 12.69):

First Step

A. Continuity Equation
The expression to find intermediate values of the density from the continuity equation is straightforward:

$$N(j + \tfrac{1}{2}, n + \tfrac{1}{2}) = [N(j + 1, n) + N(j, n)]/2$$
$$- (\Delta\tau/2)\{[N(j + 1, n) + N(j, n)]/2\}\{[V(j + 1, n) - V(j, n)]/\Delta X\}$$
$$- (\Delta\tau/2)\{[V(j + 1, n) + V(j, n)]/2\}\{[N(j + 1, n) - N(j, n)]/\Delta X\}. \tag{12.77}$$

B. Momentum Equation
The momentum equation is more involved—we need values of the electric field at intermediate spatial points, $E(j + \tfrac{1}{2}, n)$. Following Eq. (12.69), we can evaluate the electric field by an iterative numerical integration:

$$E(\tfrac{1}{2}, n) = [1 - N_{bo}g(n) - N(0, n)](\Delta X/2),$$
$$E(j + \tfrac{1}{2}, n) = E(j - \tfrac{1}{2}, n) + [1 - N_{bo}g(n) - N(j, n)](\Delta X/2). \tag{12.78}$$

Using the shifted electric field values, the first step to advance the momentum equation is

$$V(j + \tfrac{1}{2}, n + \tfrac{1}{2}) = [V(j + 1, n) + V(j, n)]/2 - (\Delta\tau/2)E(j + \tfrac{1}{2}, n)$$
$$- (\Delta\tau/2)\{[V(j + 1, n)^2 - V(j, n)^2]/(2\Delta X)\}. \tag{12.79}$$

Second Step

With the intermediate quantities $N(j + \tfrac{1}{2}, n + \tfrac{1}{2})$ and $V(j + \tfrac{1}{2}, n + \tfrac{1}{2})$, we proceed to the second step. Again we need to calculate electric field—this time, at the time centered mesh position.

$$E(0, n + \tfrac{1}{2}) = 0,$$

$$E(j, n + \tfrac{1}{2}) = E(j - 1, n + \tfrac{1}{2}) + [1 - N_{bo}g(n) - N(j - \tfrac{1}{2}, n + \tfrac{1}{2})](\Delta X/2). \tag{12.80}$$

The equations to advance N and V through a full time step are

$$N(j, n + 1) = N(j, n) - \Delta\tau\{[N(j + \tfrac{1}{2}, n + \tfrac{1}{2}) + N(j - \tfrac{1}{2}, n + \tfrac{1}{2})]/2\}$$
$$\cdot \{[V(j + \tfrac{1}{2}, n + \tfrac{1}{2}) - V(j - \tfrac{1}{2}, n + \tfrac{1}{2})]/\Delta X\}$$

$$-\Delta\tau\{[V(j+\tfrac{1}{2},n+\tfrac{1}{2})+V(j-\tfrac{1}{2},n+\tfrac{1}{2})]/2\}$$
$$\cdot\{[N(j+\tfrac{1}{2},n+\tfrac{1}{2})-N(j-\tfrac{1}{2},n+\tfrac{1}{2})]/\Delta X\} \tag{12.81}$$

and

$$V(j,n+1)=V(j,n)-\Delta\tau E(j,n+\tfrac{1}{2})$$
$$-\Delta\tau\{[V(j+\tfrac{1}{2},n+\tfrac{1}{2})^2-V(j-\tfrac{1}{2},n+\tfrac{1}{2})^2]/(2\Delta X)\}. \tag{12.82}$$

In the plasma neutralization problem, we must include boundary conditions at the extremities of the mesh, $j=0$ and $j=\text{Nmesh}$. The problem has symmetry about $X=0$. We can carry out the solution in the upper half x plane with the conditions:

$$E(0)=0, \qquad V(0)=0, \qquad \partial N(0)/\partial X=0. \tag{12.83}$$

The boundary conditions at the conducting wall are more subtle. We must assign spatial derivatives of quantities at the wall that are consistent with physical processes in the plasma. One way to define a derivative at $j=\text{Nmesh}$ is to add a virtual mesh point within the conducting wall at $j=\text{Nmesh}+1$. We can think of the wall as a grounded zero-thickness foil—plasma electrons can stream out through the foil. Within the wall, space-charge electric fields are completely canceled. Because there are no forces on electrons inside the wall, there is no difference in the velocity between the last two points, $V(\text{Nmesh},n)=V(\text{Nmesh}+1,n)$—the derivative at the intermediate point always equals zero. Similarly, the continuity equation implies that the electron density at the two points is the same, $N(\text{Nmesh})=N(\text{Nmesh}+1)$. Note that electrons at the virtual mesh point have no effect on electric fields in the plasma. The virtual mesh point resolves some mathematical difficulties but has little effect on the physical results.

We must add other constraints at the wall to represent absorption of the plasma electrons. Because electrons cannot flow inward from the wall, the value of V at $j=\text{Nmesh}$ cannot be negative. Therefore, whenever the density and velocity values in adjacent cells imply a negative value of $V(\text{Nmesh},n)$, we set the quantity equal to zero. We also set $N(\text{Nmesh})$ equal to zero because the wall cannot supply electrons to compensate for electron flow to inner cells.

Table 12.1 lists the body of a computer program to calculate the response of a plasma to a pulsed electron beam. The single-page program, written in BASIC, uses the two-step Lax–Wendroff method to solve Eqs. (12.67), (12.68), and (12.69). The program applies symmetry conditions at $X=0$ and absorbing boundary conditions at $X=1$. The division of the advancing routine into two steps is easy to see. Some other features of the program include:

Lines 132–192: Absorbing wall.
Lines 193–194: Boundary conditions at wall.
Line 200: Boundary condition on symmetry axis.

TABLE 12.1 Solution of One-Dimensional Plasma Neutralization[a]

10 REM—PROGRAM SCNEUT—

30 REM—DEMONSTRATION OF SPACE CHARGE NEUTRALIZATION
40 REM—SHEET ELECTRON BEAM INJECTED INTO A PLASMA

70 REM—MAIN TIME LOOP

80 FOR T = DTAU TO TMAX STEP DTAU

82 GOSUB 1750:REM—ION AND BEAM DENSITY AT NEXT TIME STEP
85 PRINT "——CALCULATING AT T = ";T;"——"
90 REM—STEP 1.   ADVANCE TO INTERMEDIATE POSITION

100 FOR K = 0 TO NMESH:REM K = J + 1/2
110 NI(K) = (N(K) + N(K + 1))/2 − DTAU*(N(K + 1)*V(K + 1) − N(K)*V(K))/(2*DX)
115 IF NI (K) < 0 THEN LET NI (K) = 0
120 VI(K) = (V(K) + V(K + 1))/2 − DTAU*(E(K) + E(K + 1))/2
    − DTAU*(V(K + 1)^2 − V(K)^2)/(4:DX)
130 NEXT K
132 IF VI (NMESH) < 0 THEN LET VI (NMESH) = 0
140 GOSUB 1250:REM EVALUATE EI (K)

150 REM—STEP 2. ADVANCE TO NEXT TIME POSITION
160 FOR J = 1 TO NMESH
170   N(J) = N(J) − DTAU*(NI(J)*VI(J) − NI(J − 1)*VI(J − 1))/DX
175 IF N(J) < 0 THEN LET N(J) = 0
180 V(J) = V(J) − DTAU*(EI(J) + EI(J − 1))/2
    − DTAU*(VI(J)^2-VI(J − 1)^2)/(2*DX)
190 NEXT J

192 IF V (NMESH) < 0 THEN LET V(NMESH) = 0:
    V(NMESH + 1) = V(NMESH)):N(NMESH) = 0
193 V(NMESH + 1) = V(NMESH)
194 N(NMESH + 1) = N(NMESH):E(NMESH + 1) = E(NMESH)
200 N(0) = N(1):V(0) = 0:E(0) = 0

205 GOSUB 1000

206 ND = ND + 1
207 IF ND = NDIAG THEN GOSUB 1500:ND = 0:REM—DIAGNOSTICS
220 NEXT T
230 END

1000 REM—SUBROUTINE TO CALCULATE E(J)
1010 E(0) = 0
1020 FOR J = 1 TO NMESH:
     E(J) = E(J − 1) + DX*((NION(J) + NION(J − 1) − N(J) − N(J − 1))/2):NEXT J

TABLE 12.1 *(Continued)*

```
1030  E(NMESH + 1) = E(NMESH)
1040  RETURN

1250  REM—SUBROUTINE TO CALCULATE EI(K) WHERE K = J + 1/2
1260  EI(0) = DX*(NION(0) − NI(0))/2
1270  FOR K = 1 TO NMESH:
      EI(K) = EI(K − 1) + DX*(NION(K) − (N(K) + N(K − 1))/2):NEXT K
1280  RETURN

1750  REM—SUBROUTINE ION AND BEAM DENSITY
1760  REM—LINEAR RISE TO NB IN TIME TINJ
1765  IF T > TINJ THEN GOTO 1780
1770  FOR J = 0 TO INT (XB*NMESH):NION(J) = 1 − NB*T/TINJ:NEXT J
1780  RETURN
```

[a]Diagnostic, input/output and initialization routines omitted.

Lines 1000–1040 and 1250–1280: Numerical integration to find electric field.
Lines 1750: Subroutine to give beam density as a function of time, $g(\tau)$.

For numerical stability, the spatial and temporal steps must satisfy the constraint

$$\Delta\tau < \Delta X/V(j, n). \tag{12.84}$$

When the calculation goes unstable, small variations in quantities grow in a nonphysical manner. The origin of the numerical instability is easy to understand by inspecting Eq. (12.84). If $V(j, n)\Delta\tau$ is greater than ΔX, electrons travel more than one spatial mesh length in a time step. The interval $\Delta\tau$ is not short enough to fix the mesh location of electrons unambiguously at each time step, leading to nonphysical transport mechanisms.

Figures 12.9 and 12.10 show histories of normalized density, directed velocity, and electric field for a beam density uniform in the range $0 < X < x_0/x_w$ with time variation

$$N_b(\tau) = N_{b0}(\tau/\tau_b) \qquad (0 \leqslant \tau < \tau_b),$$
$$= N_{b0} \qquad\qquad (\tau \geqslant \tau_b). \tag{12.85}$$

The dimensionless form of the equations helps to pick appropriate run parameters. For example, the condition of a fast-rising beam density is

$$\tau_b = \omega_{pe}\Delta t_b \ll 1. \tag{12.86}$$

Figure 12.9 shows results for an instantaneous rise of beam density, $\tau_b = 0$. The beam has $N_{b0} = 0.25$ and $x_0/x_w = 0.25$. The spatial resolution is Nmesh = 25,

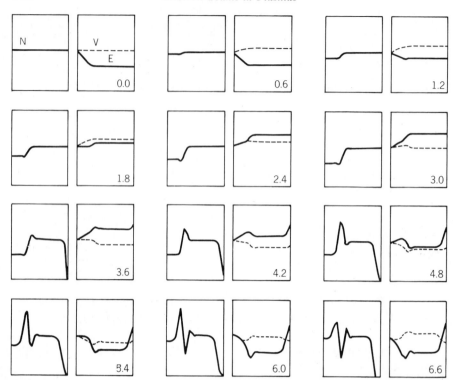

Figure 12.9 History of the normalized plasma electron density (N), directed velocity (V), and electric field (E) as a function of τ following instantaneous injection of a high-energy sheet electron beam. The figure gives plots of the spatial variation of N, V, and E at several values of τ. $N_{b0} = 0.25$, $X_b = 0.25$, $\tau_b = 0$, $\Delta X = 0.04$, $\Delta \tau = 0.05$.

or $\Delta X = 0.04$. The maximum space-charge imbalance in the system at any time is less than 0.25. Therefore, the directed velocity lies in the range $V(j, n) < 0.5$. We take a time step of $\Delta \tau = 0.05$ to guarantee stability. Despite its simplicity, the model gives interesting results. Figure 12.9 confirms the contention of Section 12.2 that a beam with a fast rise time, $\tau \leqslant 1/\omega_{pe}$, creates a large space-charge imbalance. The solution shows high directed plasma electron velocity and large electric fields. Note the inward and outward propagation of a plasma disturbance. As expected, the bounce period in dimensionless time units equals 2π.

Figure 12.10 shows a calculation with a longer beam rise time, $\tau_b = 12$. The directed velocity and electric field are much smaller. By the end of the run ($\tau = 16$), the plasma density has adjusted to the presence of the beam—at the center, it is lower by a factor of 0.75. The large electron density spikes result from the interference of plasma waves—the divergences would disappear if we added effects of electron temperature to the moment equation model.

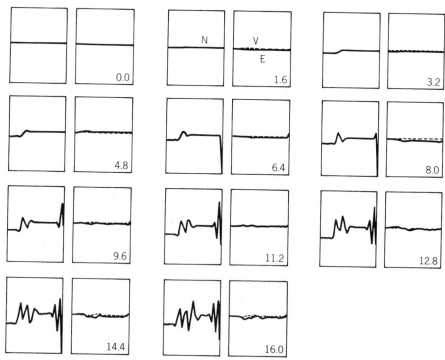

Figure 12.10 History of the normalized plasma electron density (N), directed velocity (V), and electric field (E) as a function of τ following slow injection of a high-energy sheet electron beam. $N_{b0} = 0.25$, $X_b = 0.25$, $\tau_b = 12$, $\Delta X = 0.04$, $\Delta\tau = 0.05$.

12.5. MAGNETIC SKIN DEPTH

A rapidly pulsed electron beam induces current in a conductive plasma. The plasma current opposes the beam current and may interfere with self-pinched propagation of the beam. A pulsed beam entering a plasma creates a changing magnetic field. The resulting inductive electric field accelerates plasma electrons in the direction opposite to that of the beam electrons. If the plasma current density has about the same spatial distribution as the beam current density, the plasma cancels the beam-generated magnetic field—the beam is *current neutralized.*

To understand the response of a plasma to a changing magnetic field, we shall study the simplified geometry of Fig. 12.11. A coil produces a pulsed magnetic field outside the sharp boundaries of a uniform density plasma with infinite extent in the y and z directions. The plasma has width L and equal electron and ion densities, $n_e = n_i = n_0$. Following Fig. 12.11, the magnetic fields are symmetric about the line $x = L/2$. Outside the plasma, the magnetic field is a specified function of time, $B_0(t)$, given by the variation of coil current. We

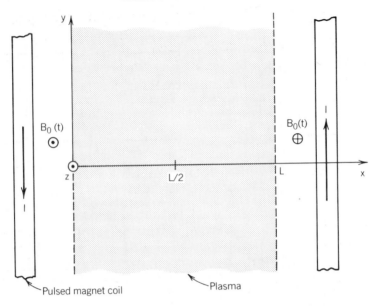

Figure 12.11 Schematic geometry to calculate the magnetic skin depth of a homogeneous plasma.

want to find the distribution of magnetic field inside the plasma as a function of position and time, $B_z(x, t)$.

A perfectly conducting plasma excludes the applied magnetic field for all time. With plasma collisional resistivity, the field ultimately penetrates the plasma (Section 12.6). Even with no collisions, magnetic fields can penetrate a distance into the plasma because of the electron inertia. The penetration distance is called the *magnetic skin depth*.

In calculating the response of a zero-resistivity plasma to a pulsed magnetic field, we initially neglect the motion of massive plasma ions. We shall concentrate on the left-hand boundary of Fig. 12.11. The coil current moves in the $-y$ direction, generating an applied magnetic field in the $+z$ direction. The inductive electric field accelerates plasma electrons in the $-y$ direction—the plasma current flows in the $+y$ direction. The plasma current opposes the applied current, excluding magnetic field from the plasma volume. For a given spatial variation of plasma electron current density at time t, $j_{ey}(x, t)$, Eq. (1.31) determines the spatial variation of magnetic field:

$$\frac{\partial B_z(x, t)}{\partial x} = -\mu_0 j_{ey}(x, t). \qquad (12.87)$$

Electric fields in the plasma result from a changing magnetic flux. With the symmetry of Fig. 12.11, the total magnetic flux enclosed within position x at time t for a unit length in the y direction is the integral over x of $B_z(x, t)$ from

$x = 0$ to $x = L/2$. In the limit of strong-field exclusion at the center of the plasma, the integral can extend to $+\infty$ with little loss in accuracy. Faraday's law is

$$E_y(x, t) = -\int_x^\infty dx \, \frac{\partial B_z(x', t)}{\partial t}. \tag{12.88}$$

To simplify the calculation, we shall temporarily neglect the effect of the magnetic field on the electron orbits—electrons move only in the $-y$ direction. The acceleration is

$$\frac{\partial v_{ey}(x, t)}{\partial t} = -\frac{eE_y(x, t)}{m_e} = -\int_x^\infty dx' \, \frac{e}{m_e} \frac{\partial B_z(x', t)}{\partial t}. \tag{12.89}$$

We can integrate both sides of Eq. (12.89) over time, t'. The integration extends from $t' = 0$ to the time of interest, $t' = t$. If $B_0(0) = 0$, then $v_{ey}(x, 0) = 0$ and $B_z(x, 0) = 0$. With these conditions, the integration gives

$$v_{ey}(x, t) = -j_{ey}(x, t)/en_0 = -(e/m_e) \int_x^\infty dx' B_z(x', t). \tag{12.90}$$

The derivative of Eq. (12.90) with respect to x is

$$\frac{\partial j_{ey}(x, t)}{\partial x} = \frac{-e^2 n_0 B_z(x, t)}{m_e}. \tag{12.91}$$

Combining Eqs. (12.87) and (12.91) gives an equation for the spatial variation of B_z at time t:

$$\frac{\partial^2 B_z(x, t)}{\partial x^2} = \frac{\mu_0 e^2 n_0 B_z(x, t)}{m_e} = \frac{e^2 c^2 n_0 B_z(x, t)}{\varepsilon_0 m_e}. \tag{12.92}$$

With the boundary condition $B_z(0, t) = B_0(t)$, the solution of Eq. (12.92) is

$$B_z(x) = B_0 \exp(-x/\lambda_m). \tag{12.93}$$

The quantity λ_m is the magnetic skin depth:

$$\lambda_m = c/(e^2 n_0/\varepsilon_0 m_e)^{1/2} = c/\omega_{pe}. \tag{12.94}$$

The final form of Eq. (12.94) incorporates the electron plasma frequency, Eq. (12.23). Because of electron inertia, a magnetic field penetrates into zero-resistivity plasma a distance comparable to λ_m. As an example, a plasma with density $10^{19}\,\mathrm{m}^{-3}$ has an electron plasma frequency of $\omega_{pe} = 1.8 \times 10^{11}\,\mathrm{m}^{-3}$. Substitution in Eq. (12.94) gives a collisionless skin depth of $\lambda_m = 1.6\,\mathrm{mm}$.

Figure 12.12 shows the significance of λ_m for the current neutralization of a pulsed electron beam. Usually, the plasma frequency is high enough to guarantee complete space-charge neutralization during the rise of beam current. If the beam propagates in a plasma with no nearby conducting boundaries, the net plasma current has magnitude equal to that of the beam current. If the beam radius is smaller than λ_m, the plasma return current occupies a larger cross-sectional area than the beam (Fig. 12.12a). As a result, the magnetic field

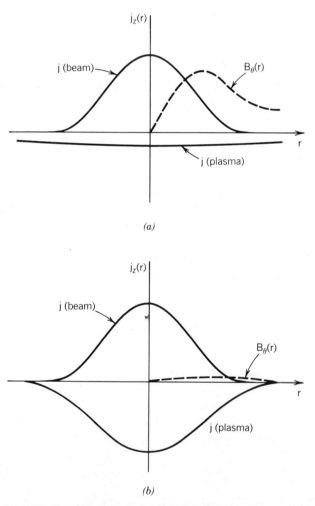

(a)

(b)

Figure 12.12 Spatial profiles of beam current density, plasma current density, and toroidal magnetic field following injection of a cylindrical electron beam into a plasma with zero resistivity. (a) $r_b \ll \lambda_m$. (b) $r_b \ll \lambda_m$.

inside the beam almost equals the value of the beam-generated field without the plasma—the beam can propagate in a self-pinched equilibrium. In the opposite limit of large beam radius, the spatial distributions of beam and plasma current density are almost the same. Figure 12.12b shows that the net magnetic field inside the beam almost equals zero.

We can apply the result of Eq. (12.94) to calculate the magnetic acceleration of a plasma. This process is often applied in pulsed plasma guns for intense ion beam extraction. Here, we must include effects of magnetic bending of electron orbits and acceleration of ions by the resulting charge separation. Figure 12.13 shows the behavior of particles at the boundary between a plasma and a rising magnetic field. The field generates an electron current localized to a surface layer of width λ_m. The magnetic field in the layer also exerts an x-directed force on the electrons. The electrons shift away from the ions in the layer, creating an electric field $E_x(x,t)$. The flow of electrons approaches an instantaneous force equilibrium if the electric field has the value

$$E_x(x,t) \cong v_{ey}(x,t)B_z(x,t). \tag{12.95}$$

The electric field accelerates ions, resulting in motion of the plasma boundary. We can estimate the velocity of the plasma by applying the condition of global force balance over the current sheath. The force per volume in the sheath is $j_{ey}B_z$ (newtons/m^3). The force per area on the plasma equals the integral of the

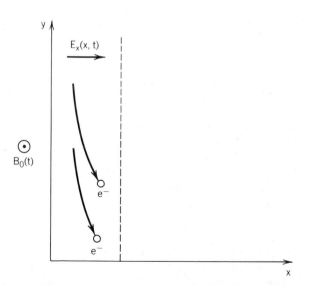

Figure 12.13 Mechanism for the magnetic acceleration of a plasma. The plasma has a boundary at $x = 0$ and extends over the region $x > 0$. A pulsed magnetic field fills the region $x < 0$. Dashed line shows region of induced electron current extending to $x \sim \lambda_m$.

volume force over the sheath width:

$$F_a(t) = \int_0^\infty dx\, j_{ey}(x,t) B_z(x,t) (\text{newtons/m}^2).\qquad(12.96)$$

Equations (12.91) and (12.93) imply that the electron current density has the following variation with depth in the plasma:

$$j_{ey}(x,t) \cong (e^2 n_0 \lambda_m/m_e) B_0(t) \exp(-x/\lambda_m).\qquad(12.97)$$

Substituting Eqs. (12.93) and (12.97) into Eq. (12.97) gives an expression for total force per area on the plasma:

$$F_a = B_0^2(t)/2\mu_0.\qquad(12.98)$$

Equation (12.98) is the familiar expression for the magnetic pressure exerted by the field $B_0(t)$ on a highly conducting body.

The inertia of ions governs the velocity of the plasma front. Suppose that the boundary moves in the $+x$ direction at velocity v_d. In the single-particle limit, we expect that ions gain a velocity $2v_d$ in an elastic reflection from the moving front. On the other hand, experiments show that accelerated ions streaming through a plasma are subject to strong momentum transfer instabilities. As a result, ions are swept up and carried with the moving front at about velocity v_d. If the ion density is n_0, the change of plasma momentum per area per time equals the momentum gain of a single ion multiplied by the number of ions swept up by the moving front per unit time in a unit cross-sectional area:

$$\frac{dp_a}{dt} = (m_i v_d)(n_0 v_d).\qquad(12.99)$$

Equating the rate of change of plasma momentum to the magnetic force gives the following expression for the plasma front velocity:

$$v_d = \left[\frac{B_0^2}{2\mu_0 n_0 m_i} \right]^{1/2}.\qquad(12.100)$$

The quantity v_d of Eq. (12.100) is called the *Alfven velocity*. For a plasma of density $10^{19}\,\text{m}^{-3}$ and an applied field of 0.10 tesla, the Alfven velocity equals 1.41×10^5 m/sec for C^+ ions. The velocity corresponds to a directed ion kinetic energy of 2.5 keV. The available pulsed current density of C^+ ions is $j_i \cong e n_0 v_d = 23 \times 10^4$ A/m^2. This value is much higher than those from steady-state ion sources (Section 7.6).

12.6. RETURN CURRENT IN A RESISTIVE PLASMA

Section 12.5 showed that the magnetic fields of a rapidly pulsed electron beam entering a plasma are almost completed neutralized in the limit $r_0 \gg \lambda_m$. Inductive axial electric fields create a plasma return current—a small reduction of the beam kinetic energy supplies the energy of the plasma current. If the beam current stays roughly constant for $t > 0$, the spatial distribution of return current may change with nonzero plasma resistivity. In this section, we shall study the nature of plasma resistivity. First, we shall derive expressions for resistivity, and then develop equations describing the time variation of plasma current density. The section concludes with a discussion of implications of the diffusion equation, including wake effects.

The cold plasmas that are often used for high-current electron beam propagation are poor conductors compared with metals. Plasma resistivity can lead to significant perturbations of the return current, even for short pulsed electron beams. Resistance results from collisions that interrupt the directed plasma electron motion. Collisions with neutrals are important in weakly ionized plasmas—at high current density, electrons can exchange momentum with plasma ions through collective instabilities. In this section, we shall concentrate on the resistivity of a fully ionized plasma with no applied magnetic field. The current density is moderate—momentum transfer to ions occurs through *Coulomb collisions*. These collisions are electric field deflections that occur when an electron passes close to an ion. The model holds when the electron drift velocity that produces the plasma current, v_d, is smaller than the randomly

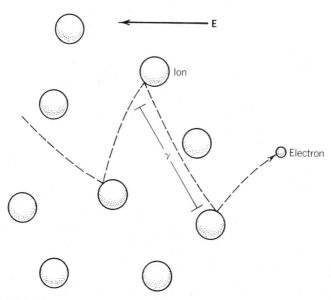

Figure 12.14 Definition of mean free path for collisions of free electrons with background atoms.

directed electron thermal velocity, v_{th}. In this limit, the current density is proportional to the applied electric field and the electron flow is stable against the two-stream instability.

Collisions between electrons cannot change the average momentum of a drifting electron distribution. Momentum transfer takes place between the electrons and stationary ions. We define the mean free path for momentum transfer, λ, as the average directed distance an electron travels before a collision with an ion. In the simplified representation of Fig. 12.14, an applied electric field accelerates electrons. They lose their directed momentum through a collision with an ion after moving an average distance λ. When $v_{th} \gg v_d$, the thermal velocity determines the time between collisions:

$$\Delta t \cong \lambda / v_{th}. \tag{12.101}$$

The collision frequency for momentum transfer to ions is the inverse of Δt,

$$v_{ei} \cong 1/\Delta t = v_{th}/\lambda. \tag{12.102}$$

If electrons accelerate freely between collisions, the average directed velocity superimposed on the random thermal velocity is

$$\mathbf{v}_d = v_{ei} \int_0^{1/v_{ei}} dt' \, \frac{-e\mathbf{E}t'}{m_e} = \frac{-e\mathbf{E}}{2m_e v_{ei}}. \tag{12.103}$$

For a uniform electron density, n_0, the current density associated with the drift velocity is $\mathbf{j}_e = -en_0\mathbf{v}_d$ or

$$\mathbf{j}_e = (e^2 n_0 / 2m_e v_{ei})\mathbf{E}. \tag{12.104}$$

Equation (12.104) shows that plasma electron flow satisfies Ohm's law

$$\mathbf{E} = \rho \mathbf{j}_e, \tag{12.105}$$

if we take the volume resistivity as

$$\rho = 2m_e v_{ei}/e^2 n_0 \, (\Omega - m). \tag{12.106}$$

An analysis of the deflection of electrons traveling past ions leads to an estimate of the collision frequency in a fully ionized plasma.[2] For physically correct results, it is necessary to include plasma shielding effects for the ion charge at distances greater than the Debye length, λ_d. The following equation applies to a plasma with singly charged ions and a Maxwell distribution of electron velocities with temperature T_e:

---

[2] See, for instance, D. J. Rose and M. Clark, *Plasma and Controlled Fusion*, M.I.T. Press, Cambridge, Mass, 1961, p. 167.

$$v_{ei} = \frac{e^4 n_0 \ln(\Lambda)}{32\sqrt{2\pi m_e} \varepsilon_0^2 (kT_e)^{3/2}}.$$ (12.107)

The quantity Λ is the *plasma parameter*:

$$\Lambda = 12\pi \frac{(\varepsilon_0 kT_e/e^2)^{3/2}}{n_0^{1/2}} = 9\left[\frac{4\pi n_0 \lambda_d^3}{3}\right].$$ (12.108)

The final form on the right-hand side of Eq. (12.108) shows that Λ is proportional to the number of particles within a spherical volume of radius λ_d, the *Debye sphere*. A group of charged particles acts like a plasma when there are many particles in a Debye sphere. To illustrate the application of Eqs. (12.106) and (12.107), suppose we have an ion source plasma with a density of $10^{19}\,\mathrm{m}^{-3}$ and an electron temperature of $kT_e = 10\,\mathrm{eV}$. For the Debye length of $7.44\,\mu\mathrm{m}$, there are over 5000 particles in a Debye sphere. The plasma parameter is $\ln\Lambda = 10.8$, giving a collision frequency of $5.8 \times 10^6\,\mathrm{sec}^{-1}$. Finally, the volume resistivity is $\rho = 4.2 \times 10^{-5}\,\Omega\text{-m}$. In comparison, the volume resistivity of pure copper is $\rho = 1.7 \times 10^{-8}\,\Omega\text{-m}$.

Knowing ρ, we can find the variation of plasma return current near a pulsed electron beam. Figure 12.15 illustrates the geometry of the calculation. A cylindrical electron beam of radius r_0 enters a plasma with $\lambda_m \ll r_0$. We neglect the effects of plasma electron inertia—the plasma acts as an ideal resistive material with no phase shift between **j** and **E**. Also, the plasma completely cancels beam-generated electric fields. Our model treats only variations in the radial direction—the following assumptions allow the neglect of axial variation:

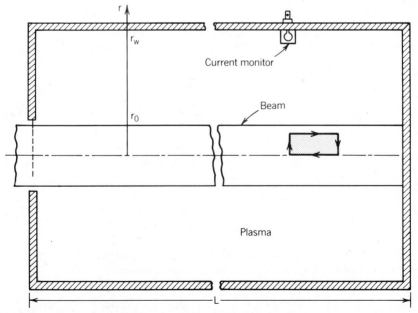

Figure 12.15 Geometry for the calculation of the radial distribution of return current for a cylindrical pulsed electron beam in a plasma-filled chamber.

1. The beam transverses a plasma filled chamber with conducting walls. The chamber has length L and a large wall radius, $r_w \gg r_0$.
2. The time scale for changes in the plasma current density is much longer than the beam transit time through the chamber, $L/\beta c$.
3. The beam is paraxial and axially uniform.

If the electron beam has a fast rise time, the plasma return-current density, $j_e(r)$, initially has the same spatial variation as that of the beam, $j_b(r)$. Therefore, there is no beam-generated magnetic field at $t = 0$.

When ρ is nonzero, an axial electric field must be present to maintain the return current

$$E_z(r) = \rho j_e(r). \tag{12.109}$$

The only possible source of the electric field is a changing magnetic flux because the ends of the chamber are electrically connected. The magnetic field points in the azimuthal direction. To calculate the electric field, we apply Faraday's law to the loop shown in Fig. 12.15. Noting that $E_z(0) = 0$, the electric and magnetic fields are related by

$$\int_0^{r'} dr' \left(\frac{\partial B_\theta(r')}{\partial t} \right) \Delta z = E_z(r)\Delta z. \tag{12.110}$$

The magnetic field results from the sum of beam and plasma currents:

$$B_\theta(r) = \frac{\mu_0}{2\pi r} \int_0^r 2\pi r' \, dr' [\, j_b(r') - j_e(r')]. \tag{12.111}$$

The radial derivative of Eq. (12.110) is

$$\frac{\partial B_\theta(r)}{\partial t} = \frac{\partial E_z(r)}{\partial r} = \rho(r) \frac{\partial j_e(r)}{\partial r}. \tag{12.112}$$

Taking partial derivatives of Eq. (12.111) with respect to radius and time gives the relationship

$$\frac{\partial j_b}{\partial t} - \frac{\partial j_e}{\partial t} = \frac{1}{\mu_0} \frac{1}{r} \frac{\partial}{\partial r} \left(r \frac{\partial B_\theta}{\partial t} \right). \tag{12.113}$$

Combining Eqs. (12.112) and (12.113) leads to an equation that describes the spatial and temporal variation of plasma current density:

$$\frac{\partial j_e(r,t)}{\partial t} = \frac{-1}{\mu_0} \frac{1}{r} \frac{\partial}{\partial r} \left(r\rho(r,t) \frac{\partial j_e(r,t)}{\partial r} \right) + \frac{\partial j_b(r,t)}{\partial t}. \tag{12.114}$$

Equation (12.114) has the form of the familiar diffusion equation. The diffusion constant is $D(r, t) = \rho(r, t)/\mu_0$. The given beam current density acts as a source term for the plasma return current.

We can solve Eq. (12.114) when the resistivity has a constant value in space and time, $\rho(r, t) = \rho_0$, and the beam current density follows a step-function temporal variation. The beam current density rises instantaneously at $t = 0$. Because $j_b(t)$ is constant for $t > 0$, we can remove the source term from Eq. (12.114)—the beam sets the initial condition, $j_e(r, 0) = -j_b(r, 0)$. The equation has the form

$$\frac{\partial j_e}{\partial t} = -\left[\frac{\rho_0}{\mu_0}\right]\nabla^2 j_e. \tag{12.115}$$

The symbol ∇^2 represents the Laplace operator appropriate for the geometry of the problem.

The diffusion equation is one of the most familiar relationships in collective physics. It describes a wide array of phenomena, such as the flow of heat in solids and the diffusion of neutrons. Many analytic solutions to Eq. (12.115) have been tabulated.[3] For example, suppose we have a narrow cylindrical beam that enters a plasma with uniform resistivity ρ_0. The beam current rises instantaneously to I_0 at time $t = 0$. If we can approximate the narrow beam as an on-axis current filament, then the plasma return current outside the beam has the spatial and temporal variations:

$$j_e(r, t) = -\left[\frac{I_0\rho_0}{4\pi\mu_0 t}\right]\exp\left[\frac{-r^2\rho_0}{4\mu_0 t}\right]. \tag{12.116}$$

We can verify the validity of Eq. (12.116) by direct substitution into Eq. (12.115). The integral of j_e over all radii always equals $-I_0$. If we define the average radius of the plasma current as the point where j_e falls to $1/e$ of its peak value, then Eq. (12.116) implies that

$$\langle r \rangle = \sqrt{4\mu_0 t/\rho_0}. \tag{12.117}$$

Expansion of the plasma current is proportional to the square root of time.

We can also solve Eq. (12.115) numerically. This approach is often necessary if the plasma conductivity varies with position and time. Defining the current density on a radial mesh with $r_i = i\Delta r$, the finite difference form of the diffusion equation with a spatially varying resistivity is

$$\frac{\partial j_i}{\partial t} \cong \left[\frac{-1}{\mu_0}\right]\left[\frac{(j_{i+1} - j_i)\rho[(i+0.5)\Delta r](i+0.5) - (j_i - j_{i-1})\rho[(i-0.5)\Delta r](i-0.5)}{i\Delta r^2}\right]. \tag{12.118}$$

[3] See, for example, H. S. Carslaw and J. C. Jaeger, *Conduction of Heat in Solids*, 2nd ed., Clarendon Press, Oxford, 1959.

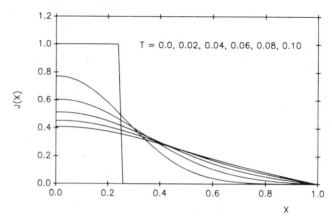

Figure 12.16 Numerical solutions to the equation $\partial^2 J/\partial X^2 = -\partial J/\partial T$ for an initial step function variation of J. Symmetry axis at $X = 0.0$, absorbing wall at $X = 1.0$. Spatial mesh: $dX = 0.02$. Solution time step: $dT = 0.0001$.

We can apply time-centered methods for partial differential equations (Section 12.4) to solve Eq. (12.118). Usually, the time step must be in the range $\Delta t < \Delta r^2/(\rho/\mu_0)$ for numerical stability.

Figure 12.16 shows results of a numerical solution of the diffusion equation. In the calculation, a sheet beam with a rapid rise time has uniform current density, j_b, from the axis to $x = x_0 = x_w/4$. The beam propagates through a uniform plasma with constant resistivity between plates at $x = \pm x_w$. Initially, the plasma current density has the same spatial profile as the beam. Later, the plasma current spreads over a larger area.

The time scale for significant spreading of the plasma current from a cylindrical beam of radius r_0 is the *current decay time*:

$$\tau_d \cong \mu_0 r_0^2/\rho_0. \tag{12.119}$$

The diffusion of the plasma return current involves the interaction between inductive and resistive effects. At $t = 0$, the distribution of return current minimizes the inductance for the plasma current. Later, the plasma current spreads to minimize the resistance. If the plasma current expands to an area significantly larger than that of the beam, the azimuthal magnetic field in the beam volume approaches the value without plasma. The quantity τ_d also represents the time scale for the decay of plasma current in the beam volume. For a resistivity of $\rho = 4.2 \times 10^{-5}\,\Omega$-m and a beam radius of $r_0 = 0.01$ m, the current-neutralization decay time is $\tau_d = 3\,\mu\text{sec}$.

We must be careful analyzing the interactions of intense electron beams ($\gg 1\,\text{kA}$) with plasmas. These beams can deposit energy in the plasma, raising the electron temperature and lowering the local resistivity [Eq. (12.106)]. The region occupied by the beam may have high conductivity, while the surrounding cold plasma is resistive. Intense electron beams induce strong plasma currents. If the plasma electron drift velocity exceeds the thermal velocity, the electrons may exchange momentum with the plasma ions through collective instabilities. For example, suppose an electron beam with $j_b = 1000 \times 10^4 \, A/m^2$ enters a plasma with density $10^{19} \, m^{-3}$ and electron temperature $kT_e = 10 \, eV$. If plasma electrons carry a current density $-j_b$, they travel with drift velocity $6.3 \times 10^6 \, m/sec$. In comparison, the thermal velocity is only $1.06 \times 10^6 \, m/sec$. For these parameters, we expect to observe collective interactions. The resulting

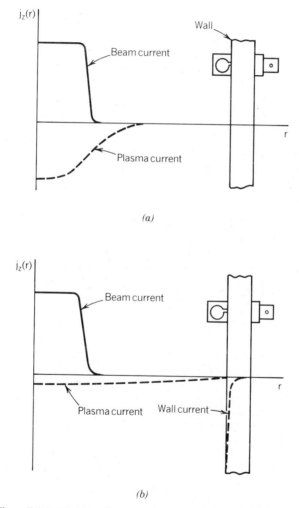

(a)

(b)

Figure 12.17 The radial distribution of return current for a pulsed cylindrical electron beam in a plasma-filled chamber. (a) Immediately after beam injection, the plasma current has almost the same spatial distribution as that of the beam. (b) Later in time, the plasma-current-density distribution is diffuse and the wall carries some of the current.

anomalous resistivity is usually much greater than the single-particle prediction of Eq. (12.106).

Most transport experiments take place inside a conducting pipe with moderate radius. In this geometry, the wall can carry a portion of the return current. Figure 12.15 shows a simplified geometry for such an experiment. A fast-rising electron pulse enters a pipe filled with uniform-resistivity plasma. Immediately after injection, the plasma return current flows in the beam volume (Fig. 12.17a). The net current inside the pipe equals zero—there is no changing magnetic flux to drive axial current along the pipe wall. A beam current probe

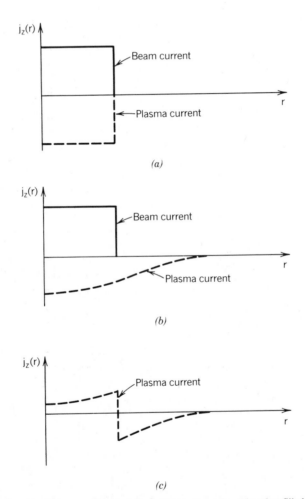

Figure 12.18 Wake fields generated by a pulsed electron beam in a chamber filled with a resistive plasma. Solid line: radial distribution of beam current density. Dashed line: radial distribution of plasma current density. (a) Immediately following injection. (b) Late in the beam pulse. (c) After the beam pulse has passed.

(that senses B_θ) just inside the wall has no signal. After an interval $\sim \tau_d$, the plasma current spreads radially outward toward the pipe (Fig. 12.17*b*). The inside wall of the conducting pipe carries a portion of the return current. The probe senses a net current in the direction of the beam current with magnitude equal to the wall current. At late time ($t \gg \tau_d$), almost all return current flows through the pipe rather than through the resistive plasma—the probe signal is close to that of the net beam current. While the net plasma return current induced by a beam decays inside a conducting chamber, in free space the magnitude of the plasma current always equals the beam current—only the spatial distribution changes.

When an electron beam with a finite pulse length travels through a resistive plasma, it may leave behind a perturbed field distribution. The residual fields may influence electron orbits in following beam pulses. This process may lead to beam instabilities—a perturbation in a lead pulse communicates with and grows in following pulses. Electric or magnetic fields that preserve a memory of a beam pulse are called *wake fields*. Figure 12.18 illustrates the mechanism for wake field formation in a resistive plasma. The current of an electron beam with uniform current density follows a square pulse in time (Fig. 12.18*a*). The pulse length, Δt_b, is comparable to the current decay time, t_d. At time $t = 0^+$, the beam and plasma current densities have the same spatial profile—there is no magnetic field. At the end of the pulse, the plasma current has expanded to a larger radius and there is a net magnetic field within the beam (Fig. 12.18*b*). When the beam current ends at $t = \Delta t_b$, the net plasma current must drop to zero. The falling beam current induces a plasma current density equal in magnitude and direction to that of the beam, j_b. Figure 12.18*c* illustrates the net current density and magnetic field remaining in the plasma just after the beam passes. The opposing plasma currents ultimately diffuse together, annihilating the magnetic field. If a second beam pulse enters before field cancellation, the remaining field may deflect the electrons. The wake field results from the plasma resistivity—in a perfectly conducting plasma, the currents induced by the rise and fall of the beam current cancel exactly, leaving no sign that the beam has passed.

12.7. LIMITING CURRENT FOR NEUTRALIZED ELECTRON BEAMS

The main reason to propagate electron beams through a plasma is to cancel beam-generated electric fields and to allow the transport of very high currents. Section 12.6 showed that sometimes the plasma also reduces the beam-generated magnetic field. If the magnetic field is not completely canceled, the beam electrons oscillate in the radial direction. At a certain value of net current, the magnetic field is high enough to reverse the orbits of some beam electrons. This process sets an upper limit on the net current. The limiting value is the *Alfven current*, I_A. We have already applied the Alfven current as a scaling factor for relativistic beams (Section 10.5).

To calculate the current limit, consider a cylindrical relativistic electron beam in a plasma. The beam pulse length is comparable to or greater than τ_d. Although the plasma cancels the electric field, there is a magnetic field inside the beam volume. We shall not pursue a fully self-consistent model. Instead, we postulate a radial variation of beam current density and investigate the properties of single-particle orbits in the resulting magnetic field. The quantity I is the net current contained within the beam radius, r_0. The function $F(r/r_0)$ specifies the radial variation of current—it equals the fraction of the net current contained within the radius r. By definition, $F(1) = 1$. The toroidal magnetic field is

$$B_\theta(r) = \mu_0 I F(r/r_0)/2\pi r. \tag{12.120}$$

The radial equation of motion for an electron with kinetic energy $(\gamma_0 - 1)m_e c^2$ and zero angular momentum is

$$\frac{d^2 r}{dt^2} = -\frac{ev_z \mu_0 I F(r/r_2)}{2\pi \gamma_0 m_e r}. \tag{12.121}$$

If the magnetic field does not vary in time, the kinetic energy of an electron is constant. As a result, the axial velocity is related to the radial velocity by

$$v_z = c(\beta_0^2 - v_r^2/c^2)^{1/2}. \tag{12.122}$$

The quantity β_0 equals the total velocity divided by c:

$$\beta_0 = (1 - 1/\gamma_0^2)^{1/2}. \tag{12.123}$$

We can reduce Eqs. (12.121) and (12.122) with the dimensionless variables: $R = r/r_0$, $V_R = v_r/\beta_0 c$, $V_Z = v_z/\beta_0 c$, and $\tau = t/(r_0/v_0)$. Substituting $\mu_0 = 1/\varepsilon_0 c^2$, the electron equations of motion are

$$\frac{dR}{d\tau} = V_R, \tag{12.124}$$

$$\frac{dV_R}{d\tau} = -\frac{I}{[4\pi\varepsilon_0 m_e c^3 \beta_0 \gamma_0/e]} \frac{2V_Z F(R)}{R}, \tag{12.125}$$

and

$$V_Z = (1 - V_R^2)^{1/2}. \tag{12.126}$$

The backeted quantity in the denominator of Eq. (12.125) is the Alfven current:

$$I_A = 4\pi\varepsilon_0 m_e c^3 \beta_0 \gamma_0/e. \tag{12.127}$$

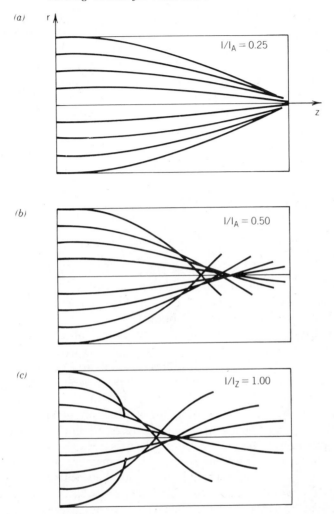

Figure 12.19 Numerical calculations of single-electron orbits in the toroidal magnetic field generated by a uniform current density beam. Beam current, I; Alfven current, I_A. (a) $I/I_A = 0.25$. (b) $I/I_A = 0.50$. (c) $I/I_A = 1.00$.

A single parameter, I/I_A, governs the solutions of Eqs. (12.124)–(12.126). Figure 12.19 illustrates numerical solutions for a uniform current density, $F(R) = R^2$. When $I/I_A \ll 1$, the radial velocity is much smaller than the axial velocity, $V_R \ll V_Z$ (Fig. 12.19a). In this limit, particle orbits at all radii are almost harmonic. At higher values of I/I_A, the oscillations of peripheral electrons are anharmonic (Fig. 12.19c). When the net current equals the value $I = I_A$, the orbits of the outermost particles bend completely inward, reaching $V_Z = 0$ on the axis (Fig. 12.19c). A higher current is impossible because it leads to electrons that travel backward along the axis.

If an electron beam with current I_0 propagates in an equal density of ions, there is no net electric field, but there is a magnetic field because the return current of ions is small. In this instance, the Alfven current is the limit on the beam current, $I_0 \leqslant I_A$. If the beam propagates in a dense plasma, the net current may be smaller than I_0, allowing transport of beam current substantially above I_A. In this case we say that the beam propagates beyond the Alfven limit.

For electrons we can write the Alfven current in terms of practical units as

$$I_A = (17 \times 10^3)\beta_0\gamma_0 \quad \text{[electrons]}. \tag{12.128}$$

As an example, the limiting current for a 1.5-MeV electron beam is about 65 kA. The limiting current for ions is much higher. The Alfven current for an ion with charge state Z and atomic number A is

$$I_A = (31 \times 10^6)(A/Z)\beta_0\gamma_0 \quad \text{[ions]}. \tag{12.129}$$

The Alfven current for 2-MeV protons is 2 MA. The limit is an important concern for space-charge-neutralized light-ion beams leaving magnetically insulated diodes (Section 8.8). These beams can generate high magnetic fields because the applied fields prevent axial electron motion.

The radial distribution of current density, $F(r/r_0)$, affects the value of the limiting current. A beam with an annular cross section can carry higher current than a solid beam. Figure 12.20 illustrates solutions to Eqs. (12.124)–(12.126) for an annular beam. The current density is uniform between r_0 and $r_i = 0.667r_0$. Note that the beam can carry current higher than I_A without reversal of peripheral electron orbits. Regarding the calculation of an annular beam equilibrium, it is clear that the orbits of Fig. 12.20 are not consistent with the assumed annular profile because they cross the axis. We can resolve the problem by including electrons with nonzero angular momentum. Electrons with the proper angular momentum oscillate in the region between r_i and r_0.

The *Budker parameter* is also used to characterize the effect of beam-generated forces on transverse dynamics. It is closely related to the Alfven current. Consider a monoenergetic cylindrical electron beam of radius r_0. The electrons have total kinetic energy $(\gamma_0 - 1)m_ec^2$, average energy $(\gamma - 1)m_ec^2$, total velocity β_0c, and average axial velocity βc. The quantity N is the total number of electrons per unit length in the axial direction, or

$$N = \int_0^{r_0} 2\pi r\, dr\, n_e(r). \tag{12.130}$$

The dimensionless Budker parameter, denoted v, is the product of N times the classical radius of the electron:

$$v = Nr_e, \tag{12.131}$$

where

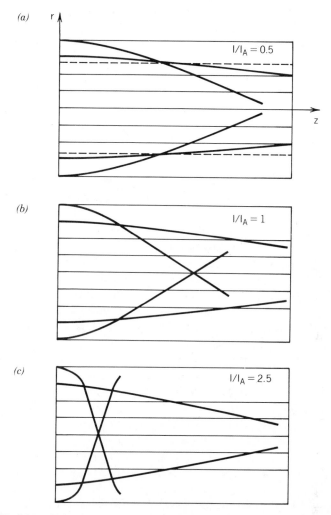

Figure 12.20 Numerical calculations of single-electron orbits in the toroidal magnetic field generated by a beam with an annular radial distribution of current density. Dashed line shows assumed inner radius of annular beam. (a) $I/I_A = 0.50$. (b) $I/I_A = 1.00$. (c) $I/I_A = 2.50$.

$$r_e = e^2/4\pi\varepsilon_0 m_e c^2.$$ (12.132)

If βc is the average axial velocity of electrons in a beam, the beam current equals $I_0 = eN\beta c$. The Budker parameter is related to the Alfven current by

$$\frac{v}{\gamma} = \frac{e^2 N}{4\pi\varepsilon_0 m_e c^2 \gamma} = \left[\frac{I_0}{I_A}\right]\left[\frac{\gamma_0 \beta_0}{\gamma \beta}\right].$$ (12.133)

The second factor in brackets on the right-hand side of Eq. (12.133) is close to unity unless the beam propagates near the Alfven current limit. A beam with high

v/γ carries a current close to I_A. If an unneutralized relativistic electron beam has a value of v/γ close to unity, the difference in electrostatic potential between the axis and the beam envelope is comparable to the average longitudinal kinetic energy:

$$v/\gamma \cong e\Delta\phi/\gamma m_e c^2. \qquad (12.134)$$

The implication of Eqs. (12.133) and (12.134) is that the electrons in beams with high v/γ have substantial transverse energy. Conversely, the condition $v/\gamma \ll 1$ implies that orbits of the beam electrons are paraxial.

12.8. BENNETT EQUILIBRIUM

The well-known Bennett model[4] describes the self-consistent transverse equilibrium of an intense electron beam propagating through a dense plasma background. The model gives an accurate representation of beams observed in experiments because it incorporates a realistic transverse velocity distribution. The Bennet equilibrium is a good starting point for studies of beam stability.

Suppose a high-energy electron beam travels through a plasma with no externally applied forces. If the plasma density is much greater than that of the beam, the beam drives out plasma electrons to achieve almost complete space-charge neutralization. The only radial forces we need consider are the beam-generated magnetic force and the emittance force from the transverse velocity spread of the beam particles. We seek conditions for a *self-pinched electron beam* where the magnetic and emittance forces balance. To simplify the model, assume that there is no plasma return current and that the current of the cylindrical beam is much smaller than the Alfven current,

$$I_0 \ll I_A. \qquad (12.135)$$

The condition of Eq. (12.135) specifies that electron orbits in the beam are paraxial—we can treat transverse and axial motion independently. If all electrons have about the same kinetic energy, $T_e = (\gamma - 1)m_e c^2$, and longitudinal velocity, $v_z = \beta c$, we can use nonrelativistic equations with an adjusted electron mass, $m = \gamma m_e$, to describe transverse motion.

In the model, the electrons have a Maxwell distribution of transverse velocity with uniform temperature. The condition of uniform temperature means that the shape of the velocity distribution is the same at all positions. If the transverse velocity distribution is isotropic, the Maxwell distribution has the form [Eq. (2.59)]

$$g(v_x, v_y) = \left[\frac{\gamma m_e}{2\pi kT}\right] \exp\left[\frac{-\gamma m_e(v_x^2 + v_y^2)}{kT}\right]. \qquad (12.136)$$

[4] W. H. Bennett, *Phys. Rev.*, **45**, 890 (1934).

The temperature in Eq. (12.136) is related to the transverse velocity spread of the beam by

$$kT = \gamma m_{\mathrm{e}} \langle v_x^2 \rangle = \gamma m_{\mathrm{e}} \langle v_y^2 \rangle. \qquad (12.137)$$

We seek a variation of beam density, $n(r)$, that gives force balance at all radii. From the discussion of Section 2.11, the force acting on a Maxwell beam distribution determines the equilibrium density of particles. For a net radial focusing force $F_r(r)$ and a density n_0 on the axis, Eq. (2.131) implies that the density is

$$n(r)/n_0 = \exp\left(\int_0^r F_r(r')dr'/kT \right). \qquad (12.138)$$

The radial force in Eq. (12.138) results from the beam-generated toroidal magnetic field, $B_\theta(r)$. If all electrons have the same value of βc, then the radial magnetic force is

$$F_r(r) = -e\beta c B_\theta(r). \qquad (12.139)$$

If we define $I(r)$ as the total axial beam current contained within the radius r, the toroidal magnetic field is

$$B_\theta(r) = \mu_0 I(r)/2\pi r. \qquad (12.140)$$

For uniform βc, the included current is related to the beam density by

$$I(r) = \int_0^r 2\pi r' \, dr' \, e\beta cn(r'). \qquad (12.141)$$

We can combine Eqs. (12.138)–(12.141) into a single equation that gives the self-consistent equilibrium density of a pinched electron beam:

$$\frac{n(r)}{n_0} = \exp\left\{ \left[\frac{-1}{kT} \right] \int_0^r dr' \frac{e\beta c\mu_0}{2\pi r'} \int_0^{r'} dr'' \, 2\pi r'' e\beta cn(r'') \right\}. \qquad (12.142)$$

We shall see that Eq. (12.142) holds if the density has the radial variation

$$n(r) = \frac{n_0}{[1 + (r/r_0)^2]^2}. \qquad (12.143)$$

The quantity r_0 is a scaling radius. Substituting from Eq. (12.143) and defining the quantity $\chi = r/r_0$, we can rewrite Eq. (12.142) as

$$\frac{1}{[1 + \chi^2]^2} = \exp\left\{ \left[\frac{-e^2\beta^2c^2\mu_0 n_0 r_0^2}{kT} \right] \int_0^\chi \frac{d\chi'}{\chi'} \int_0^{\chi'} \frac{\chi'' d\chi''}{(1 + \chi''^2)^2} \right\}. \qquad (12.144)$$

Carrying out the double integral of Eq. (12.144) leads to the relationship

$$\frac{1}{[1+\chi^2]^2} = \exp\left\{\left[\frac{e^2\beta^2c^2\mu_0n_0r_0^2}{8kT}\right][-2\ln(1+\chi^2)]\right\}. \qquad (12.145)$$

Equation (12.145) holds if the first bracketed quantity on the right-hand side equals unity. This equilibrium constraint is called the *Bennett pinch condition*. We can write the Bennett condition conveniently in terms of the total beam current:

$$I_0 = \int_0^\infty 2\pi r\, dr n(r). \qquad (12.146)$$

Evaluating Eq. (12.146) for the density expression of Eq. (12.143) gives

$$I_0 = (\pi r_0^2)en_0\beta c. \qquad (12.147)$$

A familiar form of the Bennett pinch condition is

$$kT = I_0e\beta c\mu_0/8\pi. \qquad (12.148)$$

We can substitute the Alfven current to derive an alternative form of Eq. (12.148):

$$kT = (I_0/I_A)[\gamma m_e(\beta c)^2/2]. \qquad (12.149)$$

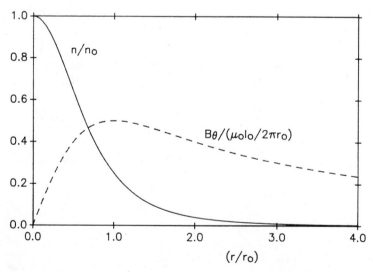

Figure 12.21 Radial variation of the normalized density and toroidal magnetic field for an electron beam in a Bennett equilibrium.

Equation (12.149) implies that the ratio of the transverse beam energy to the longitudinal energy is roughly equal to the ratio of I_0 to I_A.

To this point, we have concentrated on the mathematics of the Bennett model. We can understand the physical meaning of the derivation by verifying that the density of Eq. (12.143) guarantees a balance between the pressure force and magnetic force at all radii. The moment equation for momentum balance [Eq.

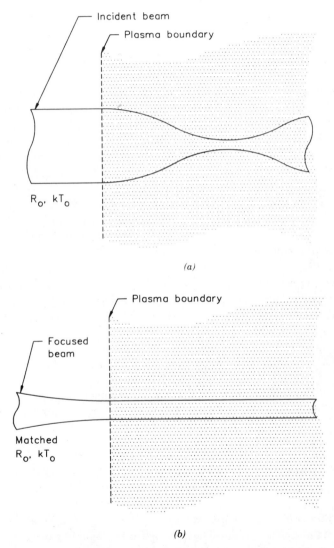

(a)

(b)

Figure 12.22 Matched propagation of a magnetically pinched beam in a plasma. (*a*) The electron transverse energy spread (kT_0) is too low, resulting in overfocusing and envelope oscillations. (*b*) Matching a beam at the plasma boundary by preliminary focusing.

[(2.117)] in an equilibrium beam has the form

$$kT \, \frac{dn(r)}{dr} = - e\beta c n(r) B_\theta(r). \tag{12.150}$$

The left-hand side is the force density associated with a velocity spread while the right-hand side is the magnetic force per volume. We find B_θ by substituting Eq. (12.143) into Eqs. (12.140) and (12.141):

$$B_\theta(r) = (\mu_0 I_0 / 2\pi r)(r/r_0)^2 / [1 + (r/r_0)^2]. \tag{12.151}$$

Using the expressions of Eqs. (12.143) and (12.151), the reader can show that Eq. (12.151) holds at all radii if I_0 and kT satisfy the Bennett condition. Figure 12.21 plots the normalized density and magnetic field as a function of r/r_0. The smooth bell-shaped density profile is a good representation for beams with particles that collide with a background medium.

We have not yet discussed how to calculate the scaling radius, r_0. The Bennett equilibrium condition does not depend on r_0. To define the radius, we must include information on the distribution of the injected beam. Figure 12.22a shows a beam entering a plasma with total current I_0, injection radius R_0, and an emittance characterized by kT_0. If kT_0 equals kT of Eq. (12.148), the beam is in equilibrium and propagates with $r_0 = R_0$. On the other hand, if $kT_0 < kT$, the beam experiences a focusing force in the plasma that leads to decreased radius. In turn, the two-dimensional beam compression leads to increased transverse temperature (Section 3.8). At some value of reduced radius, the pressure force is high enough to reverse the beam compression. The unbalanced nonlinear forces cause envelope oscillations and emittance growth. The best approach for plasma transport is to compress the beam reversibly so that it enters the plasma with a matched distribution

$$kT_0 = kT. \tag{12.152}$$

Figure 12.22b illustrates matching a beam into a plasma transport channel with a linear lens. The quantity R_0 in Fig. 12.22b is the beam radius consistent with Eq. (12.148).

12.9. PROPAGATION IN LOW-DENSITY PLASMAS AND WEAKLY IONIZED GASES

In this section, we shall discuss two examples of high-energy electron beam propagation through a medium. The first topic is laser-guiding of electron beams by selective ionization of a low-density gas. This propagation mode has potential application to high-current electron transport in accelerators. The second topic is the propagation of a self-contained electron beam a long distance through a

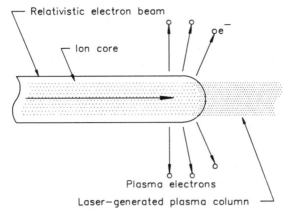

Figure 12.23   Guiding a high-current electron beam by laser generation of a low-density-plasma column.

weakly ionized gas. This propagation mode is the basis of recurring proposals for electron beam weapons. Our derivation has an educational purpose—we shall see how to add collisional effects to the paraxial envelope equation.

Figure 12.23 shows the idea of electron beam guiding by a laser. A high-power dye laser tuned to a resonance of a low-density background gas generates an extended, narrow channel of plasma. The plasma ion density, n_i, is much smaller than the beam density, n_b. The space-charge field of an entering relativistic electron beam ejects the low-energy plasma electrons, leaving a low-density core of ions. When $n_b > n_i$, the beam expels all plasma electrons. Without electrons, there is negligible return current in the beam volume. With sufficient ion density, the electron beam can propagate long distances focused by its own magnetic field. We saw in Section 5.5 that the ion density must be in the range

$$1/\gamma^2 < n_i/n_b < 1 \tag{12.153}$$

for a pinched-beam equilibrium. The exact value of density depends on the beam emittance—for paraxial relativistic beams, the required value of n_i is low.

Laser guiding has been applied to high-current beams in induction linear accelerators. The advantage is that the beam propagates through the accelerator in a tight pinch. The compressed beam has a high value of angular divergence and is less susceptible to emittance growth. The beam-generated focusing force varies nonlinearly with radius. The resulting spread in the betatron wavelength of the beam electrons reduces the growth rate for transverse instabilities (Section 13.3). On the other hand, there are major problems associated with laser guiding in a high-gradient accelerator. The machine must be filled with special gases, such as benzene. Also, plasma electrons may accelerate with the beam. Laser guiding only works for short-pulse beams. Confinement in a channel depends on localization of the ion density. The beam must pass through before the plasma ions expand. If the beam creates additional ions, there is no preferred propagation axis and a perturbation may send the beam sideways; therefore, the pulse length must be shorter than the average time for collisional ionization of the background gas. Electron beams in plasma channels are also subject to a hose instability— Section 13.6 discusses this topic.

We next proceed to electron beam propagation in a weakly ionized gas. In contrast to the previous model, the plasma for beam neutralization is created by the beam through collisional ionization. Usually, the plasma density is much higher than that of the beam. In the dense plasma, the electrons have low temperature and the resistivity is high. Following the discussion of Section 12.6, the plasma return current spreads over a large cross section, allowing self-pinched beam propagation.

Our goal is to modify the paraxial envelope equation to include collisions that change the axial and transverse momentum of electrons. We shall concentrate on the equilibrium properties of beams—the theory of electron beam stability in a dense gas is a complex and open-ended field. We shall adopt some simplifying assumptions:

1. The paraxial beam has cylindrical symmetry.
2. We limit attention to an envelope model that describes the root-mean-square beam radius, $R(z)$, without addressing detailed momentum balance over the cross section.
3. Partial ionization of a background gas provides complete space-charge neutralization of the beam.
4. The magnetic diffusion time is much shorter than the beam pulse length. In the resistive plasma, only a small fraction of the return current flows within the beam volume.
5. There are no applied focusing or acceleration forces.

A set of differential equations describes the beam trace, $R(z)$. Although we shall not derive detailed numerical solutions, it is informative to list the equations along with their physical motivation. The first equation defines the envelope angle:

$$\frac{dR}{dz} = R'. \tag{12.154}$$

The cylindrical envelope equation [Eq. (9.25)] gives the change in R' with z:

$$\frac{dR'}{dz} = -\frac{(d\gamma/dz)R'}{\beta^2\gamma} - \frac{I_0}{I_A R} + \frac{\varepsilon^2}{R^3}. \tag{12.155}$$

The quantities $\gamma(z)$ and $\beta(z)$ are the relativistic parameters of the beam electrons at position z averaged over radius. The first term on the right-hand side of Eq.

(12.155) represents the effect of deceleration. We include it because the beam loses energy in collisions with the gas atoms. Because the gas density is almost independent of the beam density, we can treat γ' as a given function of z:

$$\frac{d\gamma}{dz} = \left[\left(\frac{dE}{dz} \right)_c + \left(\frac{dE}{dz} \right) \right] \bigg/ m_e c^2. \tag{12.156}$$

Section 10.4 gives expressions for the collisional and radiative stopping powers.

The second term in Eq. (12.156) is the focusing force of the beam magnetic field. The quantity I_0 is the total beam current and I_A is the Alfven current [Eq. (12.127)]. With no absorption of beam electrons, I_0 is almost constant in z. If atomic or nuclear collisions cause the loss of a significant fraction of the beam, we can express the current as a given function of z, $I_0(z)$.

The third term represents defocusing by emittance. Although the emittance term has the same form as Eq. (3.42), we recognize that ε varies with z. The rate of emittance increase with z depends on $R(z)$, even though single-particle scattering processes are uniform throughout the homogeneous medium. Because the change in ε depends on the beam geometry, we must develop a separate differential equation. We can write ε^2 in Eq. (12.155) as the product of the mean-squared radius and the mean-squared divergence angle:

$$\varepsilon^2 = R^2 \langle \theta^2 \rangle. \tag{12.157}$$

We define the emittance equation by taking the total derivative of Eq. (12.157), noting that changes in $\langle \theta^2 \rangle$ result from both scattering collisions and beam deceleration:

$$\frac{d(\varepsilon^2)}{dz} = R^2 \langle \theta^2 \rangle \left[\frac{2}{R} \frac{dR}{dz} + \frac{1}{\langle \theta^2 \rangle} \frac{d\langle \theta^2 \rangle}{dz} \bigg|_c + \frac{2\gamma}{\gamma^2 - 1} \frac{d\gamma}{dz} \right]. \tag{12.158}$$

The second term in brackets represents the variation in $\langle \theta^2 \rangle$ from collisions—Eq. (10.62) gives the derivative. The third term states that normalized emittance is conserved (Section 3.4) if there are no collisions.

We can calculate numerical solutions for the variation of beam envelope by advancing Eqs. (12.154), (12.155), and (12.158) simultaneously from given initial conditions. An analytic solution is possible if we neglect collisional energy loss compared to scattering. This assumption often holds for moderate-energy electron beams in air or other gases with intermediate to high atomic number. With constant γ, the envelope equation is

$$\frac{dR'}{dz} = \frac{\varepsilon^2}{R^3} - \frac{I_0}{I_A R}. \tag{12.159}$$

If the beam expansion takes place over a long distance, the envelope angle and its derivative are small—the beam is always close to a radial force equilibrium. The two terms on the right-hand side of Eq. (12.159) are almost equal. Using Eq. (12.157), the angular divergence satisfies the following condition at all values of z:

$$\langle \theta^2 \rangle \cong I_0/I_A. \tag{12.160}$$

For constant I_0, Eq. (12.160) states that the angular divergence of electrons in the beam is constant, despite scattering collisions. In response to collisions, the beam radius expands to maintain a constant value of $\langle \theta^2 \rangle$.

We can describe the expansion of the beam mathematically. In propagating a differential distance Δz, collisions increase the angular divergence of electrons by an amount

$$\Delta \langle \theta^2 \rangle = \left(\frac{d\langle \theta^2 \rangle}{dz} \right)_c \Delta z. \tag{12.161}$$

A beam with enhanced angular divergence is not in a radial force equilibrium. The beam must expand to reduce the divergence by an amount $-\Delta \langle \theta^2 \rangle$. According to Section 3.8, the relationship between divergence angle and beam radius for a differential expansion is

$$\Delta \langle \theta^2 \rangle / \langle \theta^2 \rangle = -2\Delta R/R. \tag{12.162}$$

Substituting from Eqs. (12.160) and (12.161), we find the following equation for the beam radius:

$$\frac{2}{R} \frac{dR}{dz} = \frac{(d\theta/dz)_c}{I/I_A}. \tag{12.163}$$

Integration of Eq. (12.163) gives the Nordsieck equation for expansion of a beam colliding with a background:

$$\frac{R(z)}{R_0} = \exp \left[\frac{\langle \theta^2 \rangle_c^{\vec{}}}{2(I/I_A)} \right]. \tag{12.164}$$

Note that the quantity $\langle \theta^2 \rangle_c$ in Eq. (12.164) is not the angular divergence of the beam. According to Eq. (12.160), the beam divergence, $\langle \theta^2 \rangle$, is constant. Instead, $\langle \theta^2 \rangle_c$ is the mean-squared divergence angle that *single electrons* would gain moving a distance z. The mean-squared divergence angle is a function of z. At $z = 0$, $\langle \theta^2 \rangle_c = 0$ and $R = R_0$.

As an applications example, consider a self-pinched beam with $I_0 = 5\,\text{kA}$ and $(\gamma - 1)m_e c^2 = 50\,\text{MeV}$ that moves through air at a pressure of 1 mtorr. Equation (12.127) shows that $I_A = 1.6\,\text{MA}$. Electron motion is paraxial—the ratio I_0/I_A equals 3×10^{-3}. Following Eq. (10.62), the mean-squared divergence angle for an equilibrium beam is

$$(\langle \theta^2 \rangle)^{1/2} = 55 \text{ mrad } (3°).$$

Suppose the beam enters the medium with small envelope radius, $R_0 = 1$ mm and the envelope radius at the target must be less than 0.1 m. Equation (12.164) implies that the change in single-particle mean-squared divergence angle is

$$\langle \theta^2 \rangle_c \leqslant 2(I/I_A)\ln(100) = 2.76 \times 10^{-2}.$$

Inserting the value into Eq. (10.62) with $Z = 7$ and $N = 7.0 \times 10^{22} \text{ m}^{-3}$ gives a propagation length of 7.5×10^4 m (75 km). Over this length, Eqs. (10.52) and (10.55) predict an energy loss of 18 MeV—the assumption of constant electron energy is marginal.

13

Transverse Instabilities

In this chapter and the next, we shall study instabilities of charged particle beams. An instability is a change in a system where the outcome does not depend on the magnitude of the initial perturbation. The changes feed on themselves and therefore exhibit exponential growth. A familiar example of a system with instabilities is a ball in a region of varying gravitational potential. Suppose we have a complex topography of hills and valleys and the goal is to confine the ball in one spot. First we must find positions of equilibrium with no components of transverse force. One choice is the bottom of a valley, a point of *stable equilibrium*. Small perturbations of the position of the ball result in bounded oscillations about the equilibrium point. The top of a hill is also a position with zero transverse forces, but here the equilibrium is unstable. The slightest displacement causes acceleration of the ball away from the peak. Where the ball ultimately goes depends on the topography of the region, not on the magnitude of the initial perturbation. The potential energy of the ball on the hilltop drives the instability. Any displacement initiates a coupling of the stored potential energy to velocity away from the equilibrium point.

Systems with many particles have absolute stability only if there is no free energy that can change to kinetic energy. A perfectly stable group of particles has a uniform density with infinite extent and a Maxwell velocity distribution. We recognize that charged particle beams are often as far from this ideal state of thermodynamic equilibrium as possible. The beam quality depends on how nonisotropic we can make the particle distribution. Beams are tightly confined in the transverse direction—in RF accelerators, particles are also localized in the axial direction. The distribution is almost monoenergetic, with average momentum predominantly in the longitudinal direction. The creation and transport of

592

charged particles depends on avoiding processes that drive the beam to a state of thermodynamic equilibrium. In beams, the axial motion is a reservoir of free energy. Coupling of even a small part of the beam kinetic energy to transverse motion can result in beam loss.

We have already encountered one instability that affects individual particles — the orbital instability in a periodic focusing system (Section 1.3). When the lens strength is high, the vacuum phase advance may exceed 180°. The resulting overfocusing couples axial motion to transverse particle motion with increasing amplitude. The orbital instability is a single-particle process. In this chapter and the next, we shall concentrate on *collective instabilities*. Here, time-varying fields generated by the beam particles cause coupling of free energy to undersired motions of individual particles or to the entire beam. The growth rate of collective instabilities is usually proportional to the current of the beam.

We divide collective instabilities into two types: transverse and longitudinal. Transverse instabilities, the subject of this chapter, involve energy coupling to beam expansion, to gross sweeping motions, or to increased emittance. Mild instabilities may interfere with applications that call for fine beam steering or focusing. Strong instabilities can cause complete loss of beams to vacuum chamber walls.

We saw in Chapter 1 that the parameters of beams for applications extend over a broad range. Within this range, there are an almost unlimited number of ways that beams seek to avoid confinement. In a single chapter, we cannot hope to cover the entire field of transverse instabilities. Instead, we shall study a few well-known examples in detail. Section 13.1 describes the collective version of the orbital instability. We investigate oscillations of a beam with strong self-fields in a periodic focusing system. The beam exhibits unstable envelope oscillations when $\mu_0 = 180°$. We shall develop a simple but informative particle simulation to model the instability.

Sections 13.5 through 13.8 deal with instabilities that cause transverse sweeping of beams, while Sections 13.2 through 13.4 provide essential background material. Section 13.2 summarizes oscillation modes of a filamentary elastic beam. The approximation applies to thin beams in linear focusing systems where all particles have the same betatron frequency. A uniform orbit displacement at an axial position results in a coherent downstream sweeping motion. Each axial segment of the beam acts as an independent, moving harmonic oscillator. Individual oscillations combine to create traveling betatron waves on the beam. We divide the waves into two types, fast waves and slow waves. A clear understanding of slow betatron waves gives insight into the mechanisms of many transverse instabilities.

Section 13.3 derives expressions for frictional forces on beams that result from external structures such as resistive vacuum chambers or resonant cavities. Although a friction force damps some transverse oscillations, it can augment the amplitude of some slow betatron waves. In contrast, a spread in the betatron frequencies of beam particles always damps transverse oscillations. Such a dispersion can follow from a spread in particle energy or from nonlinear focusing forces. It is sometimes possible to avoid a transverse instability by introducing a spread in betatron frequency.

Section 13.4 reviews the properties of transverse resonant modes, a class of electromagnetic oscillations in cavities. These modes play a critical role in the beam breakup instability because they can extract energy from a sweeping beam. The modes have an oscillating on-axis magnetic field that causes enhanced beam sweeping. Section 13.5 applies the properties of transverse modes to describe the beam breakup instability in induction linear accelerators. This instability is the main limit on beam current in these machines and several other electron accelerators.

Section 13.6 describes the transverse resistive wall instability, a source of problems in high-current storage rings. Here, frictional forces from vacuum chamber wall resistivity amplify slow betatron waves. We shall see that the theory of the resistive wall instability also describes the beam breakup instability when the growth time is much longer than the fill time for transverse modes.

Section 13.7 treats the hose instability of an electron beam confined by an ion channel. The ion-confined propagation mode has application to long-distance electron beam propagation in free space and in accelerators. Hose instabilities occur when a flexible transport system confines a beam. Here, the ion column can move in response to beam-generated forces. The centrifugal force of a deflected beam causes a growing displacement of the ions. Section 13.8 covers the resistive hose instability of a self-pinched electron beam in a homogeneous plasma. The plasma return current can shift sideways in a plasma with nonzero resistivity. The plasma acts like a flexible confinement system, leading to growth of beam sweeping.

Finally, Section 13.9 describes a transverse instability that does not involve changes in beam dimensions or position. The filamentation instability causes increased beam emittance. It results from localized magnetic pinching of a neutralized electron beam. We shall derive growth rates and stability criteria in resistive plasmas and in foil focusing arrays.

13.1. INSTABILITIES OF SPACE-CHARGE-DOMINATED BEAMS IN PERIODIC FOCUSING SYSTEMS

All high-energy particle accelerators use periodic quadrupole arrays for beam focusing. Although quadrupoles provide strong focusing, their periodic nature may also induce transverse beam instabilities. In this section, we shall discuss two instabilities that depend on the space-charge forces of the beam. The first is the resonance instability, important in circular accelerators. Although resonance instabilities are a single-particle effect, the space-charge force determines the onset conditions. The instability sets a limit on the contained current in a storage ring or synchrotron. The second instability, the envelope instability, is a true collective process. The periodic focusing forces drive a growing oscillation of the envelope of a high-current beam. The envelope instability affects both linear and circular accelerators. It sets an upper limit on the strength of focusing lenses, and hence indirectly determines the transportable current.

Sections 8.7 and 15.7 of [CPA] discuss the nature of resonance instabilities. As a review, consider a particle confined by a strong focusing system in a circular accelerator. The strength of the quadrupole lenses determines the wavelengths of betatron oscillations in the horizontal and vertical directions, λ_x and λ_y. We represent the lens strength for a FD array with unequal F and D cells by the parameters

$$\Gamma_1 = \sqrt{\kappa_1}\, l_1, \qquad \Gamma_2 = \sqrt{\kappa_2}\, l_2. \tag{13.1}$$

The quantities l_1 and l_2 are the lens lengths and κ_1 and κ_2 are the lens parameters [Eq. (9.49)]. Assume that lens 1 focuses in the horizontal direction (x), while lens 2 defocuses in the vertical direction (y). If C is the circumference of the circular accelerator, the definition of a strong focusing is that

$$\lambda_x, \lambda_y < C. \tag{13.2}$$

A resonance instability occurs when particles rotating around the machine have an integer number of betatron oscillations:

$$\lambda_x = C/v_x \qquad (v_x = 1, 2, 3, \ldots)$$

or

$$\lambda_y = C/v_y \qquad (v_y = 1, 2, 3, \ldots). \tag{13.3}$$

If either of Eqs. (13.3) holds, then errors in the focusing or bending fields of the accelerator always act at the same phase of a betatron oscillation. As a result, the deflections caused by field errors add coherently, causing a growth in the oscillation amplitude.

The set of quadrupole lenses and bending magnets in a circular accelerator is called a *lattice*. The stability properties of a quadrupole lattice are usually represented by a *necktie diagram*—Fig. 13.1 shows an example. The coordinates are Γ_1^2 and Γ_2^2. The orbital instabilities that we shall discuss later determine the bounded region. To avoid orbital instabilities, the operating point of the lenses must lie inside the necktie-shaped region. The diagram also shows curves of integer values of v_x and v_y. To avoid an integer resonance instability, the operating point must not lie on any of the lines. Figure 13.1 shows an acceptable operating point. The lens settings that determine the operating point are called the *lattice tune*. During the acceleration cycle in a synchrotron, the tune must change so that the operating point always remains within a quadrilateral region on the diagram.

The features of the necktie diagram are different if the lenses have nonlinear components of focusing force. With nonlinear lenses, the particles in a beam have a spread in the values of v_x and v_y. For example, the value of v_x is higher for off-axis particles if the lenses have spherical aberration. We can represent the effect of a

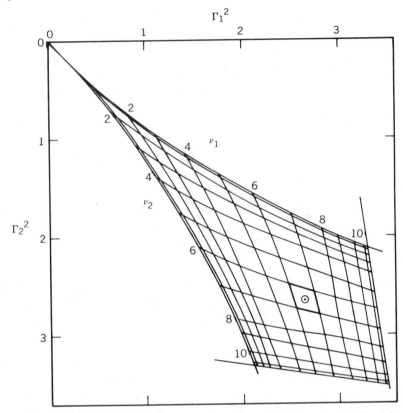

Figure 13.1 Necktie diagram for beam propagation in a circular accelerator with FD quadrupole focusing, 24 focusing cells per revolution. Γ_1 is the focusing strength parameter in the x direction, while Γ_2 describes focusing in the y direction. Interior lines indicate conditions for orbital resonances. The circle shows a possible operating point at $v_1 = v_2 = 6.4$. (From M. S. Livingston and J. P. Blewett, *Particle Accelerators*. Used by permission, McGraw-Hill Book Co.)

spread in betatron wavelength on the orbital resonance conditions by using lines with nonzero thickness in the necktie diagram. With a positive coefficient of spherical aberration (Section 4.4) the resonance lines extend downward and to the right (Fig. 13.2). Nonlinear lenses reduce the size of the allowed regions for stable propagation.

Beam-generated forces have two effects on the tune of a focusing lattice. First, the linear component of the force counteracts the applied focusing force, increasing the single-particle betatron wavelength. Suppose the current trapped in a storage ring rises slowly with time through slow beam injection. If there is no change in the lens setting, space-charge forces shift the lines of constant v_x and v_y to the left and upward. To avoid an integer resonance, the lens strength must change to balance the space-charge effect. The accelerator operator introduces a *tune shift* in the lattice. In principle, a tune shift can compensate the effect of any

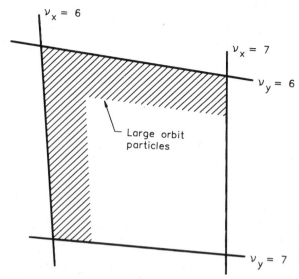

$\nu_x = 6$

$\nu_x = 7$

$\nu_y = 6$

Large orbit
particles

$\nu_y = 7$

Figure 13.2 Modifications of resonance conditions. Close-up view of the marked portion of the necktie diagram in Fig. 13.1. Lens nonlinearities—the shaded area shows the region of the necktie diagram occupied by particles in a finite-width beam.

value of beam current if the space-charge forces are linear. In reality, beams do not have a uniform density with sharp boundaries. The beam profile is often Gaussian or parabolic—the transverse beam-generated force is inevitably nonlinear. Therefore, space-charge forces also increase the thickness of lines of constant ν on the necktie diagram. At high beam current, the region around the operating point shrinks to zero and some beam particles are lost. For injection into a storage ring, the trapped current reaches a saturation level even if beam particles enter continuously.

The single-particle orbital resonance instability determines the boundaries of the necktie region. Chapter 8 of [CPA] discusses this process. The instability occurs when the phase advance per focusing cell of a particle is in the range

$$\mu_0 \geqslant 180°. \tag{13.4}$$

With space-charge forces, we might expect that the stability condition changes. We know that a linear space-charge force reduces the phase advance per cell of a single particle orbit, $\mu < \mu_0$. Is it possible that space-charge forces could counteract the applied focusing force that causes orbit overfocusing? If this were true, we could use stronger lenses in linear accelerators and contain beams with very high current. We shall address this topic in the rest of this section. We shall prove that, unfortunately, beam-generated forces do not change the limit on lens strength set by instabilities. For the condition of Eq. (13.4), the periodic focusing forces couple to an unstable envelope oscillation of a space-charge-dominated beam. The limits of the stability region of the necktie diagram with linear beam-generated forces are the same as those of Fig. 13.1.

To minimize the mathematics of the model, we treat a simple one-dimensional focusing system consisting of a uniform array of thin lenses with focal length f and drift spaces of length L. The lenses focus in the x direction. A paraxial beam with uniform density and zero emittance travels through the array. The symbol X denotes the half-width of the sheet beam. If $X \ll L$, we can approximate transverse electric and magnetic fields of the beam by the expressions for an infinite-length beam.

As an initial condition, we take a laminar beam matched to the focusing array. In a matched equilibrium, X repeats periodically through all focusing cells (Fig. 13.3). The quantity X_0 is the width of the matched beam envelope at the waist midway between two lenses—we take the origin of the z coordinate at a midpoint between lenses. Although the beam density is uniform in x, it varies with z over a focusing cell. The density is related to the envelope width by

$$n(z) = n_0 [X_0 / X(z)].\tag{13.5}$$

We start by calculating the properties of the matched beam equilibrium and then investigate the effect of perturbations. The one-dimensional paraxial ray equation (Section 9.1) describes the change of envelope width between lenses:

$$X'' = K_x,\tag{13.6}$$

where

$$K_x = e^2 n_0 x_0 / \varepsilon_0 m_0 \gamma^3 \beta^2 c^2.\tag{13.7}$$

In a matched beam, the envelope angles at the input and output of a lens are equal

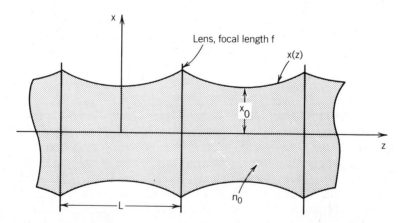

Figure 13.3 Envelope of a laminar sheet beam with space-charge forces in an array of linear thin lenses.

in magnitude but opposite in sign:

$$\Delta X'(L/2^+) = -\Delta X'(L/2^-). \qquad (13.8)$$

Equation (13.8) is equivalent to

$$\Delta X'(L/2^-) = X(L/2)/2f. \qquad (13.9)$$

The solution of Eq. (13.6) for a matched beam with a waist at $z = 0$ is

$$X(z) = X_0 + K_x z^2/2, \qquad X'(z) = K_x z. \qquad (13.10)$$

Substitution in Eq. (13.9) leads to the matched beam condition[1]:

$$K_x = (X_0/Lf)/(1 - L/8f). \qquad (13.11)$$

The maximum values of X and X' occur at the entrance to a lens,

$$\begin{aligned} X_{max} &= X_0/(1 - L/8f), \\ X'_{max} &= (X_0/2f)/(1 - L/8f). \end{aligned} \qquad (13.12)$$

We can write the envelope equation for the region between lenses as

$$X'' = \delta(X_0/Lf)/(1 - L/8f). \qquad (13.13)$$

The quantity δ is a mismatching parameter—for a matched beam, $\delta = 1$.

The plots of Fig. 13.4 illustrate the variation of envelope width for mismatched beams in an infinite lens array. Figure 13.4a shows superimposed envelope trace-space plots for 20 focusing cells. The plots follow a parabolic variation between lenses connected by a straight line to represent the discontinuity in envelope angle at the lens. In Fig. 13.4a, the lens focal length is $f = L/2.15$ (corresponding to a phase advance per cell of $\mu_0 = 94.3°$) and the beam is almost matched ($\delta = 0.975$). Figure 13.4b shows the trace for a strongly mismatched beam ($\delta = 0.75$). Although the envelope width changes considerably between cells, the beam has long-term stability.

The space-charge-dominated beam is subject to an envelope instability when $f < L/4 (\mu_0 = 180°)$. With strong lenses, a small mismatch causes the amplitude of the envelope radius to increase without limit. Figure 13.5 shows envelope

---

[1]A similar beam-matching condition holds for a cylindrical beam in a lens array. If envelope oscillations are small, the matched condition is

$$K \cong (2R_0^2/Lf)(1 - L/4f),$$

The quantity R_0 is the envelope radius and K is the generalized perveance of Eq. (5.88).

a

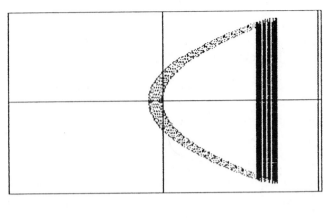

b

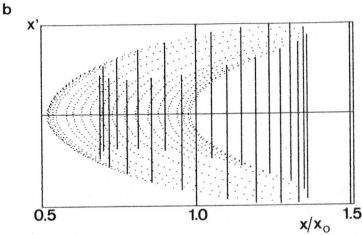

Figure 13.4 Trace-space plots of the variations of envelope width of a nonrelativistic, space-charge-dominated beam propagating in an array of thin lenses. Lenses of focal length f are separated by a distance L. Data for 20 lenses with $f = L/2.15$. Dots show the points of a numerical integration of the envelope equation between lenses. (*a*) Mismatch factor, $\delta = 0.975$. (*b*) Mismatch factor, $\delta = 0.75$.

trace-space plots that illustrate this effect. The solution of Fig. 13.5*a* has a small mismatch ($\delta = 0.98$) and is marginally stable ($L/f = 3.9, \mu_0 = 162°$). A small change in the lens focal length ($L/f = 4.1$) gives the markedly different solution of Fig. 13.5*b*. The periodic lens force excites the natural transverse oscillation frequency of the space-charge-dominated beam. When $L/f > 4$, the beam waist is at a position $z < 0$; therefore, the envelope width at the next lens is larger. This process repeats in the next cell, leading to an envelope oscillation amplitude that grows continuously.

More complex modes of oscillation are possible in a quadrupole array because of coupling between x and y motions. Theory predicts an instability for a beam

a

b

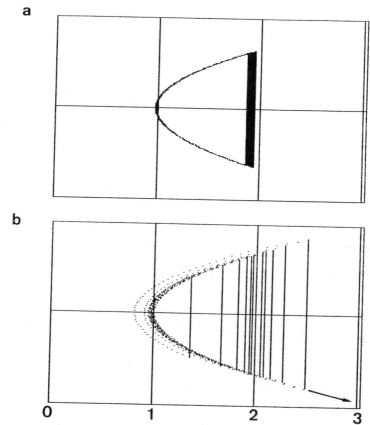

| 0 | 1 | 2 | 3 |

Figure 13.5 Trace-space plots of the variations in the envelope width of a nonrelativistic, space-charge-dominated beam propagating in an array of strong thin lenses. (*a*) Stable beam, 20 lenses, $f = L/3.9$, $\delta = 0.98$. (*b*) Unstable beam, $f = L/4.1$, $\delta = 0.98$.

with a KV distribution at $\mu_0 = 90°$.[2] Excitation of beam oscillations and coupling between x and y motions results in forces with sixfold symmetry. These forces distort the distribution, resulting in a growth of emittance. Figure 13.6 shows the results of a particle simulation for a focusing channel with $\mu_0 = 90°$ and a matched KV beam with nonzero emittance ($\mu = 45°$). The emittance increases by a factor of 2.4 during propagation through 42 FD cells. Emittance growth leads to saturation of the instability. The importance of the $\mu_0 = 90°$ instability is questionable—it would not have occurred if the distribution at cell 42 was used as the input distribution rather than a KV distribution. In a sense, the simulation is a tool to generate stable matched distributions for a given beam current and lens setting. The stability of space-charge-dominated beams in quadrupole arrays is a field of active research. Presently, there is general agreement that beam

---

[2] I. Hofmann, *Nucl. Instrum. Methods*, **187**, 281 (1987).

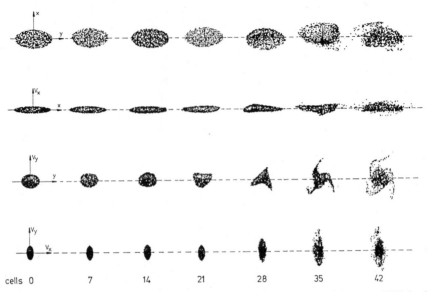

Figure 13.6 Particle-in-cell simulation of beam transport in a quadrupole lens array, FODO with 50% filling. Beam is uniform in z with initial KV distribution. $\mu_0 = 90°$, $\mu = 45°$. Rapid sextupole-type mode in $y - v_y$ leads to emittance growth by a factor of 2.4. (Courtesy of I. Hofmann, *Gesellschaft für Schwerionenforschung*, Darmstadt.)

propagation is stable if

$$\mu_0 \leqslant 60°, \tag{13.14}$$

$$0 \leqslant \mu \leqslant 60°.$$

The problem of beam propagation in a one-dimensional lens array gives us an opportunity to study some techniques of particle simulation codes. The calculation of self-consistent beam-generated forces is particularly simple in the sheet-beam geometry—we can achieve sufficient accuracy with a small number of computational particles. Table 13.1 lists a simulation program, SC_PORT. The program has practical as well as educational value—we can use it to study the emittance growth of space-charge-dominated beams in nonlinear lens arrays. The short program illustrates three essential steps to create an effective particle simulation:

1. Identify the main physical issues and scaling quantities through analytic studies.
2. If possible, recast the relevant equations in dimensionless form to identify free parameters.
3. Choose numerical methods to optimize the accuracy of the results and to minimize the computational time.

TABLE 13.1 SC_PORT: Particle Simulation Program Minus I/O Routines

```
program SC_PORT (input, output);

{Propagation of a space-charge dominated sheet beam through a periodic thin lens
 array.
     1. Matched or mismatched beams.
     2. Linear or non-linear lenses.
     3. Two-step integration, PIC calculation of beam force

 Written in PASCAL.
 -S. Humphries, Jr., 06/88}
{ ---------------------------------------------------------------------------------------------------------- }

const
     NPartMax = 250;
     NMeshMax = 50;

type
     PartArray = array [1..NPartMax] of real;
     MeshArray = array [0..NMeshMax] of real;

var
     {---Indices---}
     k : integer;   {Cell number}
     i : integer;   {Integration step between lenses}

     {---Particle variables---}
     x, xtemp : PartArray; {Transverse position}
     xp, xptemp : PartArray; {Transverse angle}
     NPart : integer; {Number of particles}

     {---Field variable---}
     F : MeshArray; {Normalized transverse force function}
     NMesh : integer; {Number of mesh points}

     {---Program control---}
     NStep : integer; {No. integration steps between lenses}
     NCell : integer; {Number of cells}
     NDiag : integer; {Number of cells between diagnostics}
     XMax : real; {Maximum value of x to terminate program}
     dz : real; {Differential length element for integration}
     dx : real; {Mesh scale length}

     {---Focusing system---}
     LdivF : real; {L/f}
     Delta : real; {Distribution mismatch factor}
     Epsilon : real; {Lens non-linearity factor}
     OrbFact : real; {Space Charge force of beam}
{ ---------------------------------------------------------------------------------------------------------- }

procedure ReadControlParameters
                    (var NStep, NCell, NDiag, NPart, NMesh:integer;
                      var LdivF, Delta, Epsilon, XMax:real);
```

(Continued)

TABLE 13.1 *(Continued)*

```
begin
  {Read input parameters from disk file, system dependent}
end; {ReadControlParameters}

procedure Initialize;
begin
  dz:= 1.0/(1.0*NStep);
  OrbFact:= Delta*LdivF/(1.0 + LdivF/8.0);
end; {Initialize}

procedure MakeInitialDistribution (var x,xp : PartArray);
var
  X0, XP0 : real;
  n : integer;
begin
      {Start with distribution matched to linear lens
      with Delta = 1.0}
  X0:= 1.0 − LdivF/8.0;
  XP0:= -LdivF/(1.0 − LdivF/8.0);
  for n:= 1 to NPart do
    begin
      x[n]:= X0*n/1(1.0*NPart);
      xp[n]:= XP0*n/(1.0*NPart);
    end;
end; {MakeInitialDistribution}

procedure CalculateF (var x : PartArray; var F : MeshArray);
var
  j : integer;
  n : integer;
begin
  for j:= 0 to NMesh do         {Zero function}
    F[j]:= 0.0;
  for n:= 1 to NPart do         {Assign particles to boxes}
    begin
      j:= round(int(x[n]/dx + 1.0));
      F[j]:= F[j] + 1.0;
    end;
    for j:= 1 to NMesh do        {Integrate}
      F[j]:= F[j] + F[j-1];
    for j:= 1 to NMesh do        {Normalize}
      F[j]:= F[j]/(1.0*NPart);
end; {CalculateF}

procedure AdvanceThroughDrift (var x, xp : PartArray);
var
  n : integer; {Particle Index}
  j : integer; {Mesh Index}
  fact : real; {Interpolation factor}
```

TABLE 13.1 *(Continued)*

```
    procedure CheckX (var x : real);
    begin
      if (x < 0.0) then x:= 0.0;
      if (x > XMax) then
        begin
          writeln ('***** Particle out of range, terminating *****');
          halt,
        end;
    end; {CheckX}

begin
        {STEP 1}
  CalculateF (x,F);
  for n:= 1 to NPart do
    begin
      xtemp[n]:= x[n] + xp[n]*dz/2.0;
      CheckX (xtemp[n]);
      j:= round(int(x[n]/dz));
      fact:= x[n]/dz − 1.0*j;
      xptemp[n]:= xp[n] +
          (F[j]*(1-fact) + F[j + 1]*fact)*Orbfact*dz/2.0;
    end;
            {STEP 2}
  CalculateF (xtemp,F);
  for n:= 1 to NPart do
    begin
      x[n]:= xtemp[n] + xptemp[n]*dz;
      CheckX (x[n]);
      j:= round(int(x[n]/dz));
      fact:= xtemp[n]/dz − 1.0*j;
      xp[n]:= xptemp[n] +
          (F[j]*(1-fact) + F[j + 1]*fact)*Orbfact*dz;
    end;
end; {AdvanceThroughDrift}

procedure AdvanceThroughLens (var x,xp : PartArray);
var
  n : integer;
begin
  for n:= 1 to NPart do
    xp[n]:= xp[n] − x[n]*LdivF*(1 + Epsilon*x[n]*x[n]);
end; {AdvanceThroughLens}
procedure MakePlots (x, xp : PartArray; F : MeshArray);
begin
  {System Dependent Plotting Routines}
end; {MakePlots}

{ --------------------------------------------------------------------------------------------------- }
```

(Continued)

TABLE 13.1 *(Continued)*

{*********THE BODY OF THE PROGRAM STARTS HERE ****************}

```
begin
  ReadControlParameters (NStep, NCell, NDiag, NPart, NMesh,
LdivF, Delta, Epsilon, XMax);
  Initialize;
  MakeInitialDistribution (x,xp);
  for k:= 1 to NCell do
    begin
      for i:= 1 to NStep do
              AdvanceThroughDrift (x,xp);
      AdvanceThroughLens (x,xp);
      if ((k mod NDiag) = 0) then MakePlots(x,xp,F);
    end;
end. {SC_ PORT}
```

We have already carried out the first step for the one-dimensional beam in the thin lens array. We shall again use the approximation that $X_0 \ll L$. In this limit, the beam is paraxial. The code follows computational particles in an axial slice of the beam as they move through the array. Two quantities characterize a computation particle: the transverse position within the envelope, $x(z)$, and the angle with respect to the axis, $x'(z)$. We define the following dimensionless variables:

$$\chi = x/X_0, \qquad \chi' = x'/(X_0/L), \qquad \text{and} \qquad \zeta = z/L. \tag{13.15}$$

In the sheet beam geometry, the electric field at the envelope has a constant value, independent of X. By Gauss's law, the field at a position x inside the envelope is proportional to the number of particles between the origin and x. The trace equation for a relativistic particle inside the envelope is therefore

$$\frac{d^2x}{dz^2} = K_x \frac{\text{Computational particles inside } x}{\text{Total number of particles}}, \tag{13.16}$$

where K_x is the generalized perveance. In dimensionless form, Eq. (13.16) becomes

$$\frac{d^2\chi}{d\zeta^2} = \frac{\delta(L/f)}{(1 - L/8f)} F(\chi), \tag{13.17}$$

where

$$F(\chi) = (\text{Particles inside } \chi)/(\text{Total particles}). \tag{13.18}$$

We use the particle-in-cell (PIC) method of Section 2.9 to evaluate the self-consistent force of Eq. (13.17). We define $F(\chi)$ at mesh locations separated by $\Delta\chi$. At the beginning of each time step, we carry out the summation of Eq. (13.18) to find $F_n(n\Delta\chi)$. We then assume that $F(\chi)$ is a smooth function in a real beam. Therefore, we can find the value of F at the position of a particle by interpolation of $F_n(n\Delta\chi)$.

Table 13.1 lists the main portions of SC_PORT in PASCAL. The first section is a list of constants and variables. The main routine is at the end of the listing—it consists of only seven statements. These statements use predefined procedures. The program uses the two-step method to advance the positions and angles of computational particles. Linear interpolation gives the force function at positions between mesh points. Evaluation of the force function consists of counting the particles inside each mesh point. Often, particle simulations use computational particles with a non-zero width (the *cloud-in-cell* method) to smooth statistical noise. This procedure is unnecessary in this problem because the integral involved in the calculation of $F(\chi)$ provides inherent smoothing. The program includes the possibility of beam mismatching and nonlinear lenses with a deflection of the form:

$$\Delta\chi' = -(\chi/f)(1 + \eta\chi^2). \tag{13.19}$$

Figure 13.7 illustrates results on the stability of beams in periodic focusing systems. The runs used 50 computational particles, 20 mesh points, and 6 integration steps between each lens. Figure 13.7a shows a test calculation of a matched laminar beam propagating in a lens array with $L/f = 1.0\,(\mu_0 = 90°)$. The figure includes a trace-space plot of the particle distribution and a plot of the spatial variation of the normalized force, $F_n(n\Delta\chi)$. Except for statistical noise, the distribution does not change traveling through 50 cells. The example verifies that a one-dimensional beam with $\mu_0 = 90°$ and $\mu = 0°$ propagates with little emittance growth.

The second example (Fig. 13.7b) has similar input parameters—the difference is the introduction of a nonlinear lens factor of $\eta = 0.05$. The input beam satisfies the conditions of Eq. (13.12) for a linear lens; therefore, it is initially mismatched to the nonlinear array. Following injection, the beam distribution adjusts in the first few cells to seek a matched force balance. After the adjustment, the beam is stable and has no further emittance growth through 50 cells. The results imply that one-dimensional beams are stable at $\mu = 0$ in both linear and nonlinear focusing systems.

The final example of Fig. 13.8 shows the effects of an envelope instability on the beam distribution. The focusing system has linear lenses with $f = L/4.2$. The beam is matched at injection. The simulation shows rapid local and global distortions of the beam distribution. The program stops after six cells when the growing envelope oscillation carries particles outside the computational region ($XMax = 1.75$).

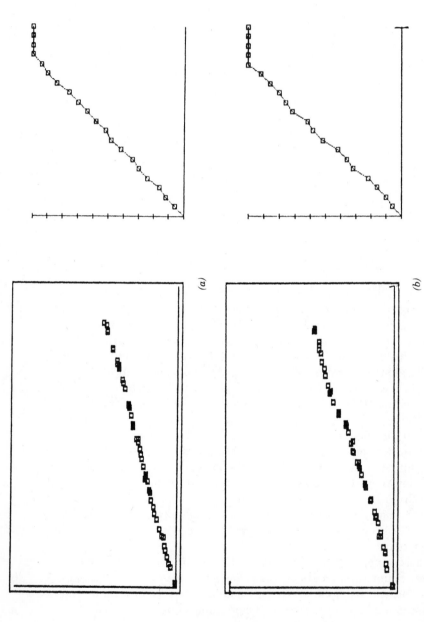

(a)

(b)

Figure 13.7 Results from a one-demensional particle-in-cell simulation of a laminar beam in an array of nonlinear thin lenses. Fifty simulation particles. Variables $\chi = x/x_0$, $\chi' = x'f/x_0$. Lenses impart of a deflection of the form, $\Delta\chi = (L/f)^{1/2}\chi(1 + \varepsilon\chi^2)$. Vacuum phase advance for paraxial particles, $\mu_0 = 57.4°$. Graph on left-hand side shows particle distributions in trace space at lens entrances. Graph on right-hand side shows the normalized electric field variation with radial position. (a) Distribution after propagation through 50 linear lenses, $\varepsilon = 0$. (b) Distribution after propagation through 50 nonlinear lenses, $\varepsilon = 0.05$. Note that beam relaxes into modified matched equilibrium following some emittance growth.

608

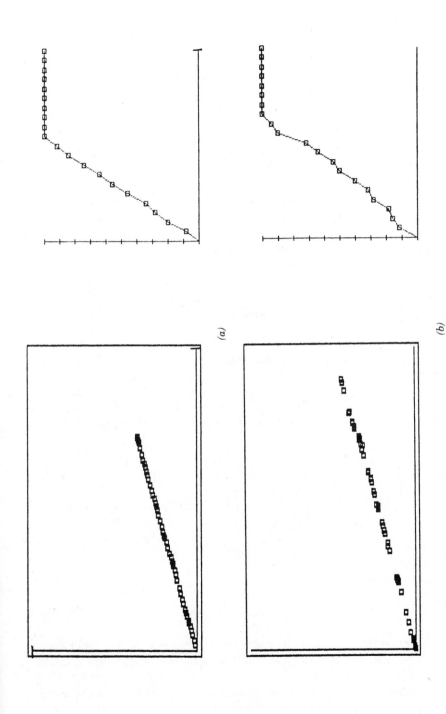

Figure 13.8 Results from a one-dimensional particle-in-cell simulation—envelope instability of a laminar beam in an array of nonlinear thin lenses. $f = L/4.2$ (f is the lens focal length and L is the distance between lenses), $\varepsilon = 0$. (a) Trace-space distribution and electric field variation, beam entering lens 2. (b) Trace-space distribution and electric field variation, beam entering lens 6.

609

13.2. BETATRON WAVES ON A FILAMENTARY BEAM

In this section, we shall study transverse motions of thin beams in linear focusing systems. The model provides a first-order description of beams in most high-energy accelerators. The harmonic components of the beam displacements are called *betatron waves*. We shall find that the waves resolve into two types, *fast* and *slow* waves. A clear understanding of slow waves is an important prerequisite for comprehending a wide variety of transverse beam instabilities. Here, we develop basic expressions for transverse beam motion that will prove useful in following sections.

We shall specify several simplifying conditions. We address only macroscopic motion, ignoring details of beam focusing. The inherent assumption is that the focusing forces are strong enough to maintain a small beam radius. The beam particles are monoenergetic, their motion is paraxial, and the focusing force is linear. With these conditions, all particles have the same betatron wavelength. Following Section 10.6, we can treat the beam as an elastic filament. For motion in the x direction, we can write the focusing force conveniently in the form:

$$F_x = -\gamma m_0 \omega_b^2 x. \qquad (13.20)$$

The quantity ω_b is the betatron oscillation frequency. Although Eq. (13.20) describes a continuous force, it also approximates quadrupole and solenoid lens arrays when the betatron wavelength is much longer than the cell length.

To visualize betatron waves of nonrelativistic beams, we shall calculate oscillations in the beam rest frame and then transform the results to the stationary frame. The nonrelativistic beam particles have $\gamma \cong 1$ and move at axial velocity $v_0 \ll c$. In a reference frame moving at v_0, the particles appear to be stationary in z—they move only in the transverse direction. The force in the rest frame is

$$F_x = -m_0 \omega_b^2 x. \qquad (13.21)$$

We divide the beam into axial segments—let $x(z, t)$ represent the center of the segment at z. The transverse motion of a segment follows the equation

$$d^2x/dt^2 = -\omega_b^2 x. \qquad (13.22)$$

Equation (13.22) has the solution

$$x(z, t) = x_0(z) \sin[\omega_b t + \phi(z)]. \qquad (13.23)$$

All axial segments of the beam oscillate at the same frequency, ω_b. The oscillation of one segment does not depend on the position of others—the segments act as independent harmonic oscillators.

We can define an infinite set of beam oscillations in the rest frame by choosing different values of the amplitude and phase, $x_0(z)$ and $\phi(z)$, for individual segments. We can organize the possibilities by recognizing that a Fourier series can represent any spatial beam displacement. Therefore, we shall study oscillations with a harmonic variation in z. We take x_0 as constant and introduce phase variations of the form $\phi(z) = \pm kz$. The corresponding beam oscillations are traveling waves:

$$x_k(z, t) = x_{0k} \sin(\omega_b t \pm kz). \tag{13.24}$$

We used analogous expressions treating plasma waves in Section 12.2. At a given time, the displacement of Eq. (13.24) varies harmonically in space over the wavelength $\lambda = 2\pi/k$. Although individual segments of the beam are stationary in the rest frame, the points of maximum displacement move with a phase velocity $\pm \omega_b/k$.

The waves described by Eq. (13.24) have a Doppler shift when viewed in the stationary frame. Waves that have phase velocity in the same direction as the beam velocity have an increased transformed oscillation frequency, ω. In the stationary frame, the points of maximum displacement move at a velocity equal to the sum of the rest frame phase velocity and v_0:

$$\omega/k = \omega_b/k + v_0. \tag{13.25}$$

The waves that satisfy Eq. (13.25) are *fast waves*. Conversely, the stationary-frame phase velocity of a wave with a rest-frame phase velocity in the opposite direction from the beam is

$$\omega/k = \omega_b/k - v_0. \tag{13.26}$$

Equation (13.26) describes a *slow wave*.

Equation (13.26) shows that there is a class of slow waves with a stationary-frame phase velocity in the opposite direction from their rest frame velocity. This velocity reversal occurs when the beam velocity is higher than the rest frame phase velocity:

$$v_0 > \omega_b/k. \tag{13.27}$$

For a given ω_b, we can choose a wavenumber such that the wave travels in the same direction as the beam in the stationary frame. When the limit of Eq. (13.27) holds, the slow wave oscillations have the interesting property illustrated in Fig. 13.9—the observed transverse motion of the beam is in the opposite direction from the motion of individual segments. To understand this, consider first the transverse oscillations of a fast wave. Figure 13.9a shows a rest-frame view of beam segment oscillations near the phase of zero displacement for a wave traveling in the $+z$ direction. In the rest frame, the motion of individual segments is synonymous with the motion of the beam—in a region where particles move away from the axis, the beam also moves outward. Figure 13.9b shows the same portion of the beam viewed in the stationary frame. Although individual segments oscillate with angular frequency ω_b, an observer at a location in the stationary frame sees a higher frequency, ω. Nonetheless, Fig. 13.9b shows that the observed motion of the beam in the stationary frame is in the same direction as the motion of individual segments.

The beam behavior is more interesting when it has a slow wave oscillation with $\omega_b/k < v_0$. Figure 13.9c shows the rest-frame displacement of segments near the zero-displacement phase of a slow wave. Again, motion of the beam center is synonymous with motion of the individual segments. Figure 13.9d shows that the view in the stationary frame is markedly different. The beam carries the wave past an observer at a velocity that exceeds the backward wave phase velocity. The

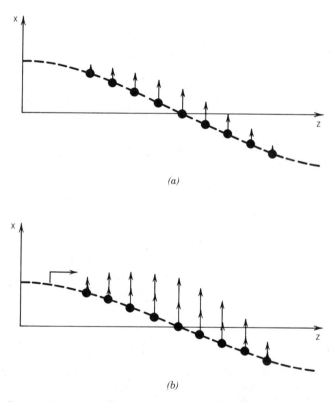

(a)

(b)

Figure 13.9 Betatron waves on a filamentary beam in a linear focusing system. Each particle, represented by a circle, oscillates independently. (a) Beam rest-frame view of a transverse wave with phase velocity in the $+z$ direction. The arrows represent instantaneous velocity vectors. (b) The wave of part a viewed in the stationary frame for a beam moving in the $+z$ direction (fast wave). Total transverse velocity results from the oscillation and convection. (c) Beam rest-frame view of a transverse wave with phase velocity in the $-z$ direction. (d) Stationary-frame view of the wave of part c viewed in the stationary frame (slow wave). The beam appears to move outward at a position where an individual particle moves inward.

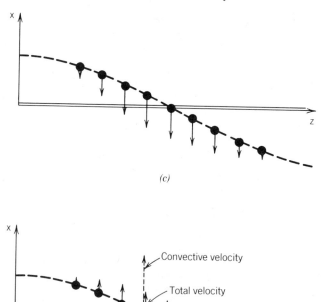

(c)

(d)

Figure 13.9 *(Continued)*

result is that the beam center appears to move in the opposite direction from the motion of the individual segments. Figure 13.9*d* shows that the beam center moves outward at a point where individual segments move inward. To summarize this unusual result, there is a class of betatron oscillations where the motion of the displacement of the beam is 180° out of phase with the motion of individual particles. This type of wave has growing amplitude in the presence of a dissipative force. As an example, suppose a beam in a chamber is subject to a collective frictional force in the direction opposite to the velocity of the beam center. Although frictional forces usually damp oscillatory motion, here they increase the amplitude of a slow wave. The frictional force is directed so that it accelerates individual beam segments. Energy balance arguments can explain the result. A careful analysis shows that the total beam kinetic energy in the axial and transverse directions is *lower* if the beam carries a slow wave. As a result, friction drives the beam to a state with reduced axial energy and enhanced transverse oscillations.

To find relativistic equations for fast and slow waves, it is easier to carry out the calculation entirely in the stationary frame. The derivation yields equations that are useful for the treatment of resistive instabilities (Section 13.5) and the hose instability (Sections 13.6 and 13.7). If the force in the stationary frame follows Eq. (13.20), then individual segments of the beam have harmonic oscillations at the relativistic betatron frequency, ω_b. We shall describe the observed motion of the beam at a point, z, as segments pass.

Let $x(z, t)$ represent the displacement of the beam centroid at axial position z and time t. The variation of x results from two effects:

1. If there is no axial variation of displacement, the quantity x at position z varies in time because of the oscillatory motion of individual beam segments in the focusing force.
2. If there is no focusing force, x at z changes if the beam has an axial variation of displacement. An observer sees different displacements as the beam passes.

We combine both effects to give an equation for the total change of beam displacement observed at a stationary point:

$$\frac{dx}{dt} = \frac{\partial x}{\partial t} - v_0 \frac{\partial x}{\partial z}. \tag{13.28}$$

We shall use harmonic variations of displacement in the stationary frame with the general form

$$x(z, t) = x_0 \exp j(\omega_b t \pm kz). \tag{13.29}$$

The notation of Eq. (13.28) has the following meaning. The total derivative on the left-hand side is the temporal change in x at z from all causes. The first term on right-hand side is the change in x resulting from the focusing force. If the displacement of the beam is uniform in z, the entire beam oscillates at frequency ω_b. In complex notation, we can write the partial derivative as

$$\frac{\partial x}{\partial t} = j\omega_b x. \tag{13.30}$$

The second term on the right-hand side of Eq. (13.28) represents the axial convection of displacement. Suppose the beam has a harmonic variation of displacement with wavelength $\lambda = 2\pi/k$. We can write the second term as

$$-v_0 \frac{\partial x}{\partial z} = \pm jkv_0 x. \tag{13.31}$$

The variation of the beam displacement at z results from two harmonic processes; therefore, the net displacement oscillates harmonically. From Eqs. (13.28), (13.30), and (13.31), the observed frequency is

$$\omega = \omega_b \pm kv_0. \tag{13.32}$$

Equation (13.32) is identical to Eqs. (13.25) and (13.26). To apply Eq. (13.32), we must remember that ω_b and k are measured in the stationary frame.

To conclude, we shall derive an expression for the apparent acceleration of the beam center at a stationary observation point. The result will be useful for the calculation of instability growth rates in the next section. We want to find the rate of change with time of the total derivative, dx/dt. The observed variation of dx/dt combines effects of the temporal change with no spatial variation of velocity and the effects of convection. The total time derivative of dx/dt is

$$\frac{d}{dt}\left(\frac{dx}{dt}\right) = \frac{\partial}{\partial t}\left(\frac{dx}{dt}\right) - v_0 \frac{\partial}{\partial z}\left(\frac{dx}{dt}\right). \tag{13.33}$$

Substitution from Eq. (13.28) gives the following expression:

$$\frac{d^2x}{dt^2} = \frac{\partial^2 x}{\partial t^2} - 2v_0 \frac{\partial}{\partial z}\frac{\partial x}{\partial t} + v_0^2 \frac{\partial^2 x}{\partial z^2}. \tag{13.34}$$

13.3. FRICTIONAL FORCES AND PHASE MIXING

In this section, we shall study processes that change the amplitude of transverse beam oscillations. The first part discusses frictional forces that extract energy from a beam. These forces arise from the resistivity of surrounding structures. Friction stabilizes some beam motions—other oscillations may be subject to the resistive wall instability (Section 13.4). The second part of the section describes the effects of a spread in betatron wavelength on the transverse motion of a beam. Phase-mixing processes always damp transverse oscillations—they play an important role in stabilizing many transverse instabilities.

To begin, consider the transverse forces that act on a beam moving through a cylindrical metal chamber with resistive walls. In Section 10.6, we discussed the forces that result from induced charge and current for a perfectly conducting wall. The section showed that for long-wavelength perturbations the wall forces have an amplitude proportional to the local beam displacement. For perturbations with the form of harmonic betatron waves, the wall force is either in phase or 180° out of phase with the beam displacement, depending on whether the wave is fast or slow (Section 13.2). The force from a perfectly conducting wall affects the amplitude and frequency of transverse oscillations, but does not extract energy from the beam. A frictional force component arises when the wall is resistive. This force is 90° out of phase with the beam displacements.

Figure 13.10 illustrates beam motion in a resistive cylindrical pipe. The pipe

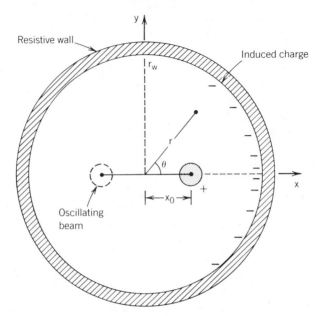

Figure 13.10 Beam propagation in a resistive pipe. Response of the induced wall charge to transverse motion of the beam.

wall has radius r_w and volume resistivity ρ. We assume that the axial wavelength of transverse oscillations is much larger than r_w. At a given axial location, an observer sees the beam center oscillate at frequency ω:

$$x(t) = x_0 \sin \omega t. \tag{13.35}$$

We shall calculate the movement of induced charge caused by the stationary-frame beam displacement of Eq. (13.35). When the beam moves to one side of the pipe, the induced charge also accumulates on that side. Therefore, a beam oscillation causes an oscillatory flow of wall charge. The flow of induced charge in a resistive wall creates a retarding electric field acting on the beam. Because the transverse wall current is proportional to the lateral beam velocity, the resistive electric field is 90° out of phase with the beam displacement. As a result, the frictional force causes a transfer of energy from the beam to the wall. We can write the frictional force on each particle as

$$F_x = -\alpha v_x = -\alpha x_0 \omega \cos \omega t, \tag{13.36}$$

where α is a constant that depends on the system geometry.

It is not difficult to calculate the force of Eq. (13.36) in the limit of high wall resistivity. For this condition, the resistive component of the field generated by the wall is much smaller than the reactive component. The total electric fields and induced charge distribution are close to those of a tube with infinite conductivity. Therefore, we can calculate the flux of image charge for an ideal conductor to find the approximate transverse currents and then use the expression to find the resistive power losses.

In Section 10.6, we derived the electric fields for a displaced beam inside a cylindrical metal pipe. Suppose an infinite-length beam with a charge per unit length of $I_0/\beta c$ moves a distance x from the pipe axis. The total electric field inside the pipe equals the sum of the beam field plus the field from an image line charge $-I_0/\beta c$ located a distance r_w^2/x from the tube axis. In order to calculate resistive wall losses, we must find the distribution of charge induced on the inside surface of the tube wall, $\sigma(x, y, t)$. To calculate $\sigma(x, y, t)$, we note that the surface charge on the wall equals the electric field normal to the wall multiplied by ε_0. We can represent this relationship conveniently in a cylindrical coordinate system referenced to the pipe (Fig. 13.10):

$$E_r(r_w, \theta) = -\sigma(\theta)/\varepsilon_0. \tag{13.37}$$

We can use the image charge construction to find $E_r(r_w, \theta)$. Adding the contributions of the two line charges at the wall position, we find that

$$E_r(r_w, \theta) \cong \frac{I_0}{2\pi\varepsilon_0 r_w \beta c}\left[1 + 2\left(\frac{x}{r_w}\right)\cos\theta\right], \tag{13.38}$$

when $x \ll r_w$.

The transverse current of induced charge in a thin wall per length, $J_\perp(\theta, t)$ A/m, is related to the surface charge density by the law of conservation of charge:

$$\frac{1}{r_w}\frac{\partial J_\perp}{\partial \theta} = \frac{\partial \sigma}{\partial t}. \tag{13.39}$$

Substituting from Eqs. (13.37) and (13.38), we find that

$$J_\perp(\theta, t) = -(I_0 x_0 \omega/\pi\beta c r_w)\cos(\omega t)\sin\theta. \tag{13.40}$$

If the wall is resistive, the transverse current heats the tube wall. The resistive power loss per length to wall in an element $r_w d\theta$ is

$$dP = J_\perp^2(\rho r_w d\theta/\delta). \tag{13.41}$$

The quantity in parentheses on the right-hand side of Eq. (13.41) is the resistance of a wall section of unit axial length and azimuthal length $r_w d\theta$. The quantity δ is the skin depth:

$$\delta = \sqrt{2\rho/\mu\omega}. \tag{13.42}$$

The integral of Eq. (13.41) around the wall gives the total power transfer per length:

$$P(W/m) = [I_0 x_0 \omega \cos(\omega t)/\beta c]^2 (\rho/\pi r_w \delta). \qquad (13.43)$$

The bracketed quantity in Eq. (13.43) has the dimension of current—it is proportional to the transverse velocity of the beam at the observation point. Comparison to Eq. (13.35) shows that the quantity equals the total beam current multiplied by the ratio of the velocity in the x direction to the axial velocity:

$$I_0 x_0 \omega \cos(\omega t)/\beta c = I_0 v_x/v_z. \qquad (13.44)$$

The transverse component of current is the *dipole current*:

$$I_{dx} = I_0(v_x/v_z) = I_0 x_0 \omega \cos(\omega t)/\beta c. \qquad (13.45)$$

Equation (13.43) has form

$$P = I_{dx}^2 Z_\perp, \qquad (13.46)$$

where

$$Z_\perp = (\rho/\pi r_w \delta) \quad (\Omega/m). \qquad (13.47)$$

The quantity $Z_\perp$ is the resistive component of the *transverse impedance*. We shall find that the transverse impedance is a useful parameter to characterize forces on a beam from charges and currents in surrounding structures.[3]

We can relate the resistive part of $Z_\perp$ to the frictional force acting on individual beam particles. Using Eq. (13.36), the power loss per particle resulting from friction is

$$dU/dt = -\alpha v_x^2. \qquad (13.48)$$

Since there are $I_0/e\beta c$ particles per length in beam, the beam power loss per length is

$$P = -(I_0/e\beta c)(\alpha v_x^2). \qquad (13.49)$$

Comparing Eqs. (13.46) and (13.49), we find that

$$Z_\perp = \alpha \beta c/e I_0. \qquad (13.50)$$

---

[3] In literature on the beam breakup instability, the transverse impedance is defined in terms of field integrals over the resonant mode with dimensions Ω/m^2. We shall use the definition in terms of transverse force and the dipole current for consistency with the longitudinal shunt impedance with dimensions Ω/m.

Combining Eqs. (13.36), (13.47), and (13.50), we find an expression for the transverse frictional force per particle from a resistive wall:

$$F_\perp = -(eI_0\rho/\pi\beta cr_w\delta)v_x. \tag{13.51}$$

Note that Eq. (13.51) describes a collective force—the force per particle is proportional to the total beam current.

We recognize that the familiar impedance of circuit theory has dimensions of voltage divided by current. The electric force on a particle multiplied by a scale length and divided by the particle charge has units of volts. Accordingly, we define the transverse impedance, with dimensions of Ω/m, as the transverse force per particle divided by the product of the dipole current and the particle charge:

$$Z_\perp = F_\perp/eI_d. \tag{13.52}$$

We can show that Eq. (13.40) is consistent with the general definition of Eq. (13.52). Multiplying the numerator and denominator of Eq. (13.40) by v_x gives

$$Z_\perp = \alpha v_x/(eI_0 v_x/\beta c) = \alpha v_x/(eI_d). \tag{13.53}$$

We can generalize the definition of transverse impedance by allowing $F_\perp$ in Eq. (13.52) to represent all transverse forces generated by the structure. The generalized impedance may have both reactive and resistive components. As an example, we can use Eq. (10.108) to show that the generalized transverse impedance of a cylindrical pipe is

$$Z_\perp = \left[\frac{\rho}{\pi r_w \delta(\omega)} \right] + j \left[\frac{1}{2\pi\varepsilon_0 \omega r_w^2 \gamma^2} \right].$$

We shall see that resistive interactions with external structures damp some types of betatron oscillations but amplify others. Because most vacuum chamber structures have nonzero resistivity, we might expect that beams are always unstable. Fortunately, there is a process that can stabilize both fast and slow betatron waves. This mechanism is the phase mix damping of oscillations that results from a spread of betatron oscillation wavelength or frequency.

Figure 13.11 illustrates the effect of phase mixing on transverse beam oscillations. Suppose a beam enters a focusing system at $z = 0$ with a displacement x_0. If all particles have the same betatron wavelength, $\lambda_b = 2\pi/k$, the beam has a coherent oscillation of constant amplitude x_0. When the particles have a spread in betatron wavenumber, the coherent oscillation dissipates as particles move downstream. Figure 13.11 shows a plot of harmonic orbits with a spread in k. We can estimate a decay length for coherent oscillations when the collection of particles is distributed uniformly in k between the limits $k - \Delta k/2$ and $k + \Delta k/2$. The beam center at position z equals the average position of all particles in a

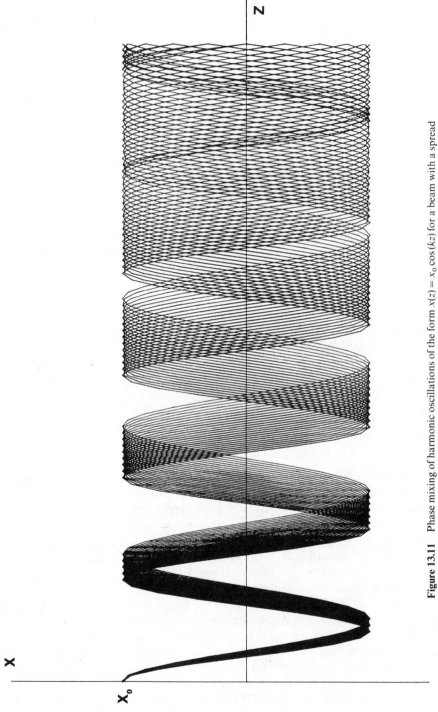

Figure 13.11 Phase mixing of harmonic oscillations of the form $x(z) = x_o \cos(kz)$ for a beam with a spread in k. The figure shows orbits with a uniform distribution of k from 1.00 to 1.25 in increments of 0.01 in the range $0 \le z \le 10\pi$.

segment. Suppose the N particles in a segment enter at $z = 0$ with $x(0, t) = x_0$. For a uniform distribution in k, the average position of the segment as a function of z equals

$$\bar{x}(z) = \sum_{n=1}^{N} x_0 \cos\left\{[k + \Delta k(n)]z\right\}/N,$$ (13.54)

where

$$\Delta k(n) = \Delta k(n - N/2)/N.$$

Expanding the trigonometric function gives

$$\bar{x}(z) = \sum_{n=1}^{N} \left\{x_0 \cos(kz) \cos[\Delta k(n)z] + x_0 \sin(kz) \sin[\Delta k(n)z]\right\}/N.$$ (13.55)

The summation over the second term on the right-hand side of Eq. (13.55) equals zero; therefore, we can rewrite the equation in the form

$$\bar{x}(z) = x_0 \cos(kz)F(\Delta kz).$$ (13.56)

The average beam displacement equals the product of the coherent displacement and a spatial damping factor from phase mixing. For the uniform beam distribution, the damping factor is

$$F(\Delta kz) = \sum_{n=1}^{N} \cos[\Delta kz(n - N/2)/N]/N.$$ (13.57)

Figure 13.12 shows a plot of $F(\Delta kz)$ for large N. The amplitude of the displacement decreases to a value $1/e$ of its initial value at a distance $\Delta z = 4.35/\Delta k$. The decay length normalized to the betatron wavelengths is

$$\Delta z/\lambda_b = (4.35/2\pi)(k/\Delta k).$$ (13.58)

Equation (13.58) applies to spatial damping of a steady-state beam injected with a constant displacement at $z = 0$. We can apply a similar treatment to find the temporal damping of a beam oscillation with a specified displacement at $t = 0$. Suppose a beam has an initial displacement $x_0 \cos(kz)$ at $t = 0$. With a uniform spread in betatron oscillation frequency, $\Delta\omega_b$, the time for the displacement amplitude to decrease by a factor $1/e$ is

$$t_e = 4.35/\Delta\omega_b.$$ (13.59)

The higher the frequency spread, the stronger the damping.

Spreads in the betatron oscillation wavelength or frequency usually arise from two causes: dispersion in the axial momentum of the beam particles and nonlinear focusing forces. In the first case, we recognize that the focal length of

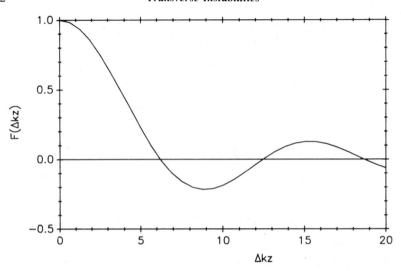

Figure 13.12 Damping factor, $F(\Delta kz)$, as a function of the betatron wavenumber spread of a group of particle orbits, Δkz.

most charged particle lenses depends on the axial momentum of the particle. For example, the focal length of a solenoidal lens varies as $1/p_z^2$ while the focal length of a magnetic quadrupole lens is proportional to $1/p_z$. The betatron wavelength in a periodic lens array is proportional to the lens focal length. If the lenses exert nonlinear forces, the betatron oscillation period also depends on the amplitude of particle oscillations. Although nonlinear forces can lead to some emittance growth, they may help to stabilize transverse oscillations in linear accelerators. Usually, we can estimate the amount of energy spread or nonlinear force component to stabilize a transverse instability such as beam breakup by a simple rule-of-thumb. The damping length from Eq. (13.58) should be shorter than the growth length for the instability.

13.4. TRANSVERSE RESONANT MODES

Induction linacs, coupled-cavity RF linear accelerators, and many other accelerators consist of a series of vacuum cavities connected by a narrow beam transport tube. The cavities can support a wide variety of resonant electromagnetic oscillations. *Transverse modes* have longitudinal electric fields with odd symmetry about a transverse plane passing through the cavity axis. They pose a particular threat to beam stability. Because of their field structure, these modes can extract power from the longitudinal kinetic energy of beams and drive transverse oscillations. We will study the consequences of this process in Sections 13.5 and 13.6. In this section, we shall review the properties of transverse modes in some simple geometries and define scaling parameters to apply in following sections.

    Figure 13.13*a* shows a pillbox structure that interrupts a cylindrical beam tube. The structure could represent a cavity of an RF linac, an acceleration gap of an induction linac, or a vacuum pump port. Figure 13.13b illustrates the electric field of the lowest frequency transverse mode of the pillbox, the TM_{110} mode. The designation TM means that magnetic fields are perpendicular to the beam direction. In a cylindrical cavity, the mode numbers for the TM_{110} mode have the following interpretation:

Azimuthal number (1): The fields vary in the azimuthal direction as $\cos(m\theta)$, where $m = 1$.

Radial number (1): The electric field has a simple variation in the radial direction—there is no field node between the axis and the wall.

Axial number (0): The fields are uniform over the length of the cavity.

Because the TM_{110} mode has a z-directed electric field, it can couple energy from the beam if the beam is not centered on the axis. The electric field magnitude is zero at the cavity axis and has odd symmetry along the x axis. Near the beam, the electric field amplitude varies almost linearly with y. The beam drives the mode if it has a

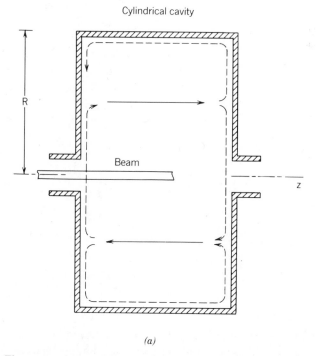

Cylindrical cavity

(a)

Figure 13.13 The transverse mode, TM_{110}, in a cylindrical cavity. (a) Side view of a cavity showing the distribution of axial electric field and wall currents at an oscillation phase with rising E_z. Solid line: electric field; dashed line: wall current. (b) Plot of the spatial distribution of E_z at an oscillation phase of maximum electric field. Top: three-dimensional plot. Bottom: Contour plot of the magnitude of E_z. $(X^2 + Y^2)^{1/2} = r/R$. (c) Schematic equivalent circuit model for the TM_{110} mode.

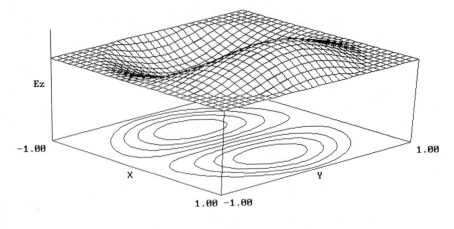

(b)

Outer wall

Axis

(c)

Figure 13.13 *(Continued)*

harmonic displacement along y at the frequency of the mode, ω_{110}. Figure 13.13c shows a representation of the mode in terms of a lumped element model. The displacement currents at the points of maximum electric field are in opposite directions on either side of the $x = 0$ line. The conduction of real current through the walls completes the circuit of the resonant mode. Wall current flows around the outside of the cavity and across the beam axis to connect the two regions of displacement current. Inspection of the current loops in Fig. 13.13a shows that the magnetic field has maximum amplitude at $x = 0$ and points in a direction normal to the beam. This field can deflect charged particles crossing the cavity.

The structure of an induction linac cavity is much more complex than the geometry of Fig. 13.13. Nonetheless, transverse modes in all cavities share common properties. We will study some simple geometries to motivate the definition of scaling parameters to apply to all cavities. The first example is the square cavity, illustrated in Fig. 13.14a. The structure has axial length d and equal sides of length a. Transverse magnetic modes in the square cavity have the designation TM_{nmp}, where m refers to variation along the x axis, n to the y axis, and p to the z axis. We shall limit consideration to modes with uniform axial fields that can interact strongly with a beam. For $p = 0$, the general expressions for the fields of transverse TM modes in a square cavity are

$$E_z(x, y) = E_0 \sin\left[m\pi(a/2 + x)/a\right] \sin\left(n\pi(a/2 + y)/a\right) \exp\left(j\omega_{mn0}t\right), \tag{13.60}$$

$$B_x(x, y) = E_0(n\pi/\omega_{mn0}a) \sin\left[m\pi(a/2 + x)/a\right] \cos\left[n\pi(a/2 + y)/a\right] j \exp\left(j\omega_{mn0}t\right), \tag{13.61}$$

$$B_y(x, y) = E_0(m\pi/\omega_{mn0}a) \cos\left[m\pi(a/2 + x)/a\right] \sin\left[n\pi(a/2 + y)/a\right] j \exp\left(j\omega_{mn0}t\right). \tag{13.62}$$

where $m = 1, 2, 3, \dots$ and $n = 1, 2, 3, \dots$. Equations (13.60)–(13.61) apply in the range $-a/2 \leqslant x \leqslant a/2$, $-a/2 \leqslant y \leqslant a/2$. The resonant frequency of the TM_{mn0} mode is

$$\omega_{mn0} = (\pi c/a)\left[m^2 + n^2\right]^{1/2}. \tag{13.63}$$

In the square cavity, the lowest-order transverse modes are the TM_{120} and TM_{210}. Figure 13.14b and 13.14c give plots of field components for the TM_{120} mode. The following expressions for the axial electric and transverse magnetic fields of the TM_{120} mode are valid near the axis ($x, y \ll a$):

$$\mathbf{E} = E_z\mathbf{z} \cong (2\pi E_0 y/a) \cos\left(\omega_{120}t + \phi\right)\mathbf{z}, \tag{13.64}$$

$$\mathbf{B} = B_x\mathbf{x} \cong -(2\pi E_0/\omega_{120}a) \sin\left(\omega_{120}t + \phi\right)\mathbf{x}. \tag{13.65}$$

The electric field varies linearly with distance from the axis, while the magnetic field is almost uniform across the axis.

In cylindrical cavities, we found that the mode numbers m, n, and p refer to variations in polar coordinates. Transverse modes have $m = 1$. The electric and magnetic fields of TM_{1n0} modes in a cylindrical cavity of radius R are

$$\mathbf{E} = E_0 J_1(x_{1n}r/R) \sin\theta \exp\left(j\omega_{1n}t\right)\mathbf{z}, \tag{13.66}$$

$$\mathbf{B} = +(jE_0/c)\{(R/x_{1n}r)J_1(x_{1n}r/R)\cos\theta\,\mathbf{r} - [J_0(x_{1n}r/R)$$
$$- (R/x_{1n}r)J_1(x_{1n}r/R)]\sin\theta\,\boldsymbol{\theta}\} \exp\left(j\omega_{1n0}t\right). \tag{13.67}$$

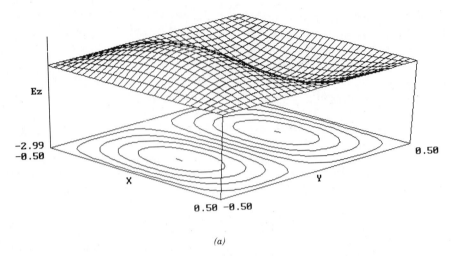

(a)

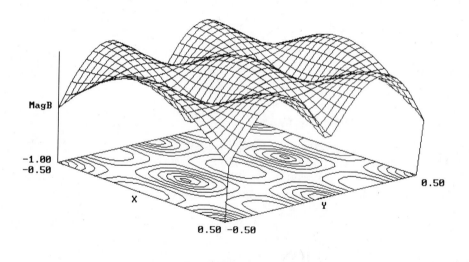

(b)

Figure 13.14 Transverse resonant mode, TM_{120}, in a square cavity. (*a*) Plot of the spatial distribution of E_z at an oscillation phase of maximum electric field. Top: three-dimensional plot. Bottom: Contour plot of the magnitude of E_z. $X = x/a$, $Y = y/a$. (*b*) Plot of the spatial distribution of $|B|$ at an oscillation phase of minimum electric field. Top: three-dimensional plot. Bottom: Contour plot of $|B|$. $X = x/a$, $Y = y/a$.

The functions J_0 and J_1 are Bessel functions. The quantities x_{1n} are the nth zeros of the J_1 function:

$$x_{11} = 3.832, \quad x_{12} = 7.016, \quad x_{13} = 10.173, \quad \ldots \tag{13.68}$$

The resonant frequencies of transverse modes are

$$\omega_{1n0} = cx_{1n}/R. \tag{13.69}$$

The following approximate forms hold near the axis ($r \ll R$):

$$\mathbf{E} = E_z\mathbf{z} \cong (E_0 x_{1n}/2R)y \cos(\omega_{1n0}t + \phi)\mathbf{z}, \tag{13.70}$$

$$\mathbf{B} = B_x\mathbf{x} \cong -(E_0/2c)\sin(\omega_{1n0}t + \phi)\mathbf{x}. \tag{13.71}$$

As with the square cavity, the electric field near the axis is linearly proportional to displacement from the symmetry axis while the magnitude of the magnetic field is almost uniform.

A transverse mode can absorb energy from a beam through the E_z field and exert a transverse force through the B_x field. For a steady-state beam, we can find a simple expression for the transverse force in terms of the beam dipole current. By *steady state*, we mean that the beam properties are constant over an interval equal to the decay time for electromagnetic energy in the cavity. We shall define the decay time later in the derivation. We take an electron beam that moves in the $+z$ direction—the axial current is in the $-z$ direction. Suppose that the beam arrives at a resonant cavity with a harmonic transverse displacement

$$y(t) \sim \cos\omega_\perp t, \tag{13.72}$$

where the frequency $\omega_\perp$ is equal to the resonant frequency of a mode with electric fields of the form of Eqs. (13.64) and (13.70) that vary linearly with y. The transverse velocity of the beam at the cavity is

$$v_y(t) \sim -\sin\omega_\perp t. \tag{13.73}$$

As the beam oscillates across the electric field null on the x axis, it can transfer energy to support the transverse mode. The cavity is a driven harmonic oscillator. In the steady state, we know that the driven cavity oscillation has a phase such that the electric field extracts the maximum energy from the beam. In other words, the electric field points in the $+z$ direction when the beam is in the top half of the cavity ($y > 0$). Comparison to Eq. (13.72) show that the electric field in the upper half plane is

$$E_z(t) \sim \cos\omega_\perp t. \tag{13.74}$$

The axial displacement current density that results from the changing electric field is

$$j_z(t) \sim \frac{\partial E_z}{\partial t} \sim - \sin \omega_\perp t. \tag{13.75}$$

If we apply the integral form of Eq. (1.31) to the current loop formed by the displacement and wall currents, the magnetic field in the beam plane is

$$B_x(t) \sim - \sin \omega_\perp t. \tag{13.76}$$

The force on a particle of the beam resulting from the magnetic field, $- e(v_z \times B_x)$, is proportional to

$$F_y(t) \sim \sin \omega_\perp t. \tag{13.77}$$

The force of Eq. (13.77) is in the direction opposite to the transverse velocity—the magnitude of the force is proportional to that of the beam velocity:

$$F_y(t) \sim - v_y(t). \tag{13.78}$$

Equation (13.78) implies that a driven cavity exerts a frictional force on a continuous beam if the beam displacement varies at the resonant frequency of a transverse mode.

Following the discussion of Section 13.3, the form of Eq. (13.77) suggests that we can assign a transverse impedance to a cavity for a beam oscillation at frequency $\omega_\perp$. Suppose a narrow beam arrives at a cavity having length d and width a. The center of the beam follows the variation $y(t) = y_0 \cos \omega_\perp t$. To apply the arguments to all cavity geometries and transverse modes, we shall not assign specific interpretations to a and d. Rather, we treat them as characteristic lengths and introduce dimensionless geometric scaling parameters. The idea is to express results in terms of known dimensions and scaling constants that are close to unity. Then, we can make good estimates of results before pursuing detailed numerical solutions. Assume the mode of interest has a peak electric field amplitude of E_0. Inspection of the forms of Eqs. (13.64) and (13.70) shows that we can represent the electric field near the axis as

$$E_z(x, y, z, t) \cong E_0(\eta_1 y/a) \cos \omega_\perp t. \tag{13.79}$$

where η_1 is a dimensionless scaling constant.

If the transverse position of the narrow beam is almost constant as it transits the cavity, we can represent the current density in the cavity as

$$j_z(\mathbf{x}, t) = I_0 \delta(y - y_0 \cos \omega_\perp t) \delta(x). \tag{13.80}$$

Equation (13.80) holds if the transit time for beam particles is much smaller than the oscillation period for the transverse mode. To find the time-averaged power transfer from the beam to the transverse mode, we take a spatial integral of the product of beam current density times the electric field:

$$P_{in} = \int_0^{2\pi/\omega} dt \int_{-a/2}^{+a/2} dx \int_{-a/2}^{+a/2} dy \int_0^d dz \frac{j_z(x,y,z,t)E_z(x,y,z,t)}{2\pi/\omega}. \qquad (13.81)$$

Carrying out the integrals, we find that

$$P_{in} \cong (I_0 E_0 d/2)(\eta_1 y_0/a). \qquad (13.82)$$

In a steady state, the balance between the power input from the beam and the power lost to the cavity walls or output couplers determines the amplitude of the transverse mode. The quality factor $Q_\perp$ characterizes the time-averaged power loss, P_{out}, from the transverse mode to the cavity:

$$P_{out} = \omega_\perp U_\perp/Q_\perp. \qquad (13.83)$$

In Eq. (13.83), $U_\perp$ is the stored electromagnetic energy of the transverse mode. If we write the equation in the form

$$dU_\perp/dt = -\omega_\perp U_\perp/Q_\perp, \qquad (13.84)$$

we can identify the time for the decay of electromagnetic energy in the cavity as

$$\Delta t = Q_\perp/\omega_\perp. \qquad (13.85)$$

Equations (13.63) and (13.69) show that we can write $\omega_\perp$ in terms of a dimensionless scaling factor

$$\omega_\perp = \eta_2 c/a. \qquad (13.86)$$

Similarly, the stored electromagnetic energy is proportional to the product of the peak energy density times the cavity volume:

$$U_\perp = \eta_3(\varepsilon_0 E_0^2/2)(a^2 d). \qquad (13.87)$$

Using the definitions of Eq. (13.86) and (13.87), we can write the power loss from the mode as

$$P_{out} = (\eta_2\eta_3)(\varepsilon_0 c E_0^2/2)(ad)/Q_\perp. \qquad (13.88)$$

Equating the input power from the beam to the output power gives an equation for the amplitude of the steady-state electric field as a function of the beam displacement:

$$E_0 = (\eta_1/\eta_2\eta_3)(I_0 Q_\perp/\varepsilon_0 c)(y_0/a^2). \tag{13.89}$$

We introduce a final dimensionless factor to connect the amplitude of the on-axis magnetic field of the transverse mode to the characteristic electric field:

$$B_x(0, 0, z, t) = -(\eta_4 E_0/c) \sin \omega_\perp t. \tag{13.90}$$

Neglecting transit-time corrections, the transverse force on an electron crossing the cavity is

$$F_y(0, 0, z, t) \cong \eta_4 e\beta E_0 \sin \omega_\perp t. \tag{13.91}$$

The dipole current of the beam is

$$I_{dy} = -I_0(y_0\omega_\perp/\beta c) \sin \omega_\perp t. \tag{13.92}$$

Using Eq. (13.52), we find that the magnitude of the transverse impedance for the cavity is

$$Z_\perp = \left[\frac{\eta_1 \eta_4 \beta^2}{\eta_2^2 \eta_3} \right] \left[\frac{Q_\perp}{a} \right] \left[\frac{\mu_0}{\varepsilon_0} \right]^{1/2}. \tag{13.93}$$

The first bracketed quantity on the right-hand side of Eq. (13.93) contains dimensionless quantities—its value is close to unity. For reference, a square cavity with width a and length d has the following scaling constants for the TM_{120} mode:

$$\begin{aligned} \eta_1 &= 2\pi, & \eta_2 &= \pi(5)^{1/2}, \\ \eta_3 &= 1/4, & \eta_4 &= 2/(5)^{1/2}. \end{aligned} \tag{13.94}$$

The second term has the dimension of inverse length. The last term is the familiar free-space impedance, equal to $377\,\Omega$. Therefore, Eq. (13.93) has the proper dimensions of Ω/m.

The impedance of Eq. (13.93) applies only to beam oscillations with frequency within the resonant width of the transverse mode, $\omega \sim \omega_\perp(1 \pm 1/Q_\perp)$. For such oscillations, the cavity has a much stronger influence on the transverse beam dynamics than an equivalent length of resistive wall. Suppose a beam has a betatron oscillation with a stationary-frame frequency of $f = \omega/2\pi = 500$ MHz. Consider, first, the transverse impedance of a stainless-steel transport tube with radius $r_w = 0.04$ m. The resistivity of stainless steel is about $\rho = 81 \times 10^{-8}\,\Omega/m$. Equation (13.42) gives a skin depth of $20\,\mu m$. Substituting in Eq. (13.47), the resistive component of the transverse impedance is $Z_\perp = 0.3\,\Omega/m$. In comparison, assume that the beam passes through a square cavity of width $a = 0.67$ m with

$\omega_{120}/2\pi = 500\,\text{MHz}$. The transverse mode is strongly dampled by selective output couplers so that $Q_{120} = 100$. For $\beta \cong 1$, the dimensionless quantity in Eq. (13.93) is $[\eta_1\eta_4/\eta_2^2\eta_3] \cong 0.46$, giving a transverse impedance of $26.9\,\text{k}\Omega/\text{m}$. Even with strong damping, the cavity exerts a much stronger transverse force than an extended resistive wall for the same value of dipole current. For the parameters of the example, the power extracted from the beam for a given dipole current is more than four orders of magnitude higher for resonant interactions. For this reason, it is imperative to avoid irregularities in the vacuum chambers of storage rings that could support transverse modes. Although the beam current in these devices is low, the storage time usually must be long. Even a slowly growing transverse instability can cause beam loss. At the other end of the scale, transverse resonant instabilities are a source of major problems in high-current linear induction accelerators.

13.5. BEAM BREAKUP INSTABILITY

The *beam breakup (BBU) instability*, also known as the *transverse instability*, is the one of the most damaging collective instabilities in electron accelerators. It limits the transportable current in a variety of machines, including RF linacs, linear induction accelerators, microtrons, and storage rings. Although the mechanism of the instability is the same for all accelerators, the theory varies considerably for different geometries.

The beam breakup instability involves the transfer of particle kinetic energy from longitudinal motion to transverse motion. As a result, the beam has a growing transverse oscillation as it moves through the accelerator. For a weak instability, the oscillations cause emittance growth or aiming errors. A strong beam breakup instability may result in complete beam loss. The instability results from two coupled processes:

1. Excitation of transverse electromagnetic oscillations in an array of cavities by a beam that oscillates in the transverse direction.
2. Displacement of the beam by the magnetic fields of the resonant modes.

Suppose that a constant-current electron beam with a harmonic transverse displacement at frequency $\omega_\perp$ enters a cavity that has a transverse mode with resonant frequency $\omega_\perp$. Following Section 13.4, the beam drives the mode, creating an oscillating magnetic field on the axis. The magnetic field gives the electrons a harmonic angular deflection that leads to sweeping motion at downstream locations. In an array of cavities the amplitude of oscillations grows rapidly in space and time if the magnitude of the beam displacement is higher in cavity $n + 1$ than in cavity n.

We can identify three classes of beam breakup instability that depend on how cavity oscillations couple. The term *cumulative* beam breakup refers to an instability in a linear accelerator where the cavities do not exchange electromagnetic energy. The coupling between the oscillations of individual cavities results entirely from perturbations carried on the beam. The *convective* beam breakup instability occurs if there is an exchange of transverse mode electromagnetic energy between cavities. If the coupled modes of a cavity array have a positive group velocity, the energy of cavity excitations can propagate forward in the accelerator. This effect, similar to the mechanism of a traveling-wave tube, increases the severity of the instability. The *regenerative* instability occurs in coupled cavity arrays with a negative group velocity for transverse mode excitations. Here, electromagnetic energy propagates backward in the accelerator—the mechanism resembles that of a backward wave oscillator. A beam-coupled regenerative instability affects circular electron accelerators such as microtrons. In these devices, electron beams recirculate many times through an RF linac. Transverse beam deflections gained on one pass through the linac contribute to the growth of transverse modes when the beam reenters the accelerator.

Induction linacs consist of a beamline interrupted by isolated acceleration cavities; therefore, they are subject to the cumulative beam breakup instability. Because the machines contain high-current pulsed beams (> 1 kA), the instability can grow even if the transverse modes have low $Q_\perp$. Consider a pulsed electron beam with fast-rising current that enters the first cavity slightly displaced from the symmetry axis. A Laplace transform shows that the dipole current has a broad frequency spectrum—a harmonic component of the current may overlap the resonance width of a transverse mode. As the head of the beam moves downstream, it excites a low level of transverse mode oscillations in all acceleration cavities. The oscillating magnetic field in the first cavity modulates the transverse direction of electrons behind the beam head. These electrons arrive at downstream cavities with a sweeping motion at frequency $\omega_\perp$ that increases the amplitude of transverse mode oscillations. In a frame of reference moving with the beam, the mode amplitude increases exponentially moving back from the beam head. In the stationary frame, the amplitude in a particular cavity increases over a beam pulse until the mode reaches saturation. During the growth of the instability, the oscillation amplitude increases moving away from both the beam head and the injection point. As a result, the peak beam displacement lies at a point between the front of the beam and the injection point. Late in time, when the beam fills the accelerator, the instability approaches a steady state where the mode amplitude grows exponentially with distance from the injection point.

The cumulative beam breakup instability in an RF accelerator with isolated cavities, such as a side-coupled linac, is qualitatively similar to the process in an induction linac. The difference is that the RF accelerator beam has a micropulse structure at the frequency of the fundamental accelerating mode, ω_0. We can use a Fourier analysis to express the current as a steady-state value with harmonic components at integer multiples of ω_0. The longitudinal modulation has little effect on the growth of transverse instabilities if the harmonics of ω_0 do not overlap the resonance width of a transverse mode. With no degeneracy, the instability growth rate depends mainly on the transverse displacement of the average beam current—we can apply an approach similar to that used for the induction linac. Beam breakup is a concern in high-current RF electron linacs because the accelerating cavities have large $Q_\perp$ values. The problem is particularly severe in superconducting linacs.

Complete time-dependent solutions for the beam breakup instability in accelerators with applied focusing forces can be complex. To understand the factors that influence the growth of the instability, we shall develop a simple scaling model. We take a coasting beam in a system of identical, uncoupled cavities. We shall derive a two-cavity instability criterion—a steady-state beam displacement in one cavity causes an amplified sweeping motion in the next cavity. Figure 13.15 shows the geometry of the calculation. Cavity n has characteristic length d and width a. The distance to cavity $n + 1$ equals D. A narrow beam with current I_0 arrives at cavity n with transverse oscillation $y(t) = y_n \cos \omega_\perp t$, where $\omega_\perp$ is the frequency of the transverse mode with the highest growth rate. We seek the amplitude of beam-sweeping in the next cavity, y_{n+1}.

We shall not worry about the phase of the sweeping motion relative to the original displacement or the effects of transverse focusing lenses. A rough criterion for a strong instability is

$$y_{n+1} \geqslant y_n. \tag{13.95}$$

From Section 13.4, we know that the steady-state amplitude of the transverse magnetic field for a beam displacement y_n is

$$B_0 = \left[\frac{\eta_1 \eta_4}{\eta_2 \eta_3} \right] \left[\frac{I_0 Q_\perp}{\varepsilon_0 c^2} \right] \left[\frac{y_n}{a^2} \right]. \tag{13.96}$$

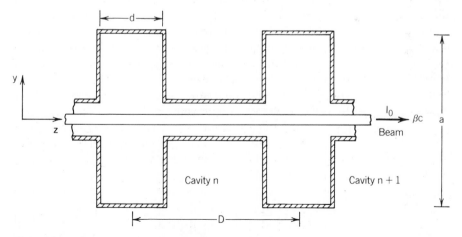

Figure 13.15 Geometry to derive an amplification criterion for the beam breakup instability in uncoupled cavities.

The amplitude of the angular displacement of an electron crossing the cavity equals the transverse impulse divided by the axial momentum, or

$$\Delta\theta = \frac{(eB_0\beta c)(d/\beta c)}{\gamma m_e \beta c} = \frac{eB_0 d}{\gamma m_e \beta c}. \tag{13.97}$$

With no transverse focusing, the amplitude of the sweeping motion in cavity $n+1$ induced by the transverse mode in cavity n is

$$y_{n+1} = \Delta\theta D. \tag{13.98}$$

Combining Eqs. (13.96)–(13.98), we find that

$$y_{n+1}/y_n = \Pi Q_\perp, \tag{13.99}$$

where

$$\Pi = \left[\frac{\eta_1 \eta_4}{\eta_2 \eta_3}\right]\left[\frac{eI_0}{\varepsilon_0 m_e c^3 \gamma \beta}\right]\left[\frac{dD}{a^2}\right]. \tag{13.100}$$

We can rewrite the constant in Eq. (13.100) in terms of the Alfven current:

$$\Pi = 4\pi\left[\frac{\eta_1 \eta_4}{\eta_2 \eta_3}\right]\left[\frac{I_0}{I_A}\right]\left[\frac{dD}{a^2}\right]. \tag{13.101}$$

All the terms in Eq. (13.101) are dimensionless. The magnitude of the first term on the right-hand side depends on the geometry of the cavity. For the TM_{120} mode in a square cavity, the term equals 3.2. The second term is the ratio of the beam current to the Alfven current—for most accelerators, this term is much smaller than unity. The third term includes geometric parameters of the cavity array.

We expect that the beam breakup instability is strong if $\Pi Q_\perp \geqslant 1$. Strong damping of transverse modes reduces the steady-state amplitude of unstable transverse oscillations. For given values of I_0 and $Q_\perp$, small values of d and D reduce the growth of the instability—the accelerating gradient should be as high as possible. The quantity Π is proportional to $1/\gamma$; this is one of the reasons that injectors for induction linacs operate at high voltage.

For a given current, energy, and accelerator geometry, a low value of $Q_\perp$ can reduce the growth of beam oscillations. We can use the condition $\Pi Q_\perp \ll 1$ to give the transverse mode damping required for small growth of the beam breakup instability:

$$Q_\perp \ll \left[\frac{\eta_2 \eta_3}{\eta_1 \eta_4}\right]\left[\frac{I_A}{4\pi I_0}\right]\left[\frac{a^2}{dD}\right]. \tag{13.102}$$

As an example, consider an induction accelerator with a 1-kA, 5-MeV electron beam. We approximate an acceleration gap as a square cavity of width 0.3 m that supports a TM_{120} mode. The gaps have length $d = 0.05$ m with a separation $D = 0.3$ m. Equation (13.102) gives the critical value of the mode quality factor as $Q_\perp < 27$. Because metal cavities usually have values of $Q_\perp$ that exceed 1000, microwave absorbers must be incorporated in the cavities to avoid a strong instability.

Two simple equations developed by Lau and Colombant[4] represent the behavior of the beam breakup instability in a wide variety of linear accelerators. These equations treat the effect of transverse modes in an array of cavities in the spatially continuous approximation. Although we will not perform a detailed derivation of the equations, we can understand the underlying physical principles from an inspection. For the cumulative beam breakup instability, the equations have the form

$$\left(\frac{\partial}{\partial t} + \beta c \frac{\partial}{\partial z}\right)\left[\gamma\left(\frac{\partial}{\partial t} + \beta c \frac{\partial}{\partial z}\right)y(z,t)\right] + \gamma\omega_b^2 y(z,t) = a(z,t) \quad (13.103)$$

and

$$\left(\frac{\partial^2}{\partial t^2} + \frac{\omega_\perp}{Q_\perp}\frac{\partial}{\partial t} + \omega_\perp^2\right)a(z,t) = 2\gamma\omega_\perp^4 \varepsilon y(z,t). \quad (13.104)$$

Equation (13.103) describes the spatial and temporal growth of beam oscillations—the quantity $y(z,t)$ is the transverse beam displacement. The right-hand side of Eq. (13.103) represents the effect of the transverse resonant modes on the beam. The quantity $a(z,t)$ equals the transverse force per particle from a cavity oscillation divided by m_0, the particle rest mass. The force is proportional to the magnetic field of the mode. The quantity ω_b on the left-hand side of Eq. (13.103) is the particle betatron frequency, which may depend on γ. The term $\gamma m_0 \omega_b^2 y$ is the transverse force of a continuous focusing system—the equation holds if all particles have the same betatron wavelength. If the right-hand side of the equation equals zero, the solution gives fast and slow betatron waves (Section 13.2). Inclusion of γ in the second factor on the left-hand side represents the effect of acceleration.

Equation (13.104) describes pumping of transverse cavity oscillations by transverse beam motion. Again, the quantity $\omega_\perp$ is the resonant frequency of a transverse resonant mode. With no beam deflection ($y = 0$), the right-hand side of Eq. (13.104) equals zero. In this case, the equation has the familiar form of a damped harmonic oscillator. A sweeping beam adds a driving term for the mode. The dimensionless constant ε in Eq. (13.104) characterizes the coupling of the beam dipole current to the mode. The constant is related to the transverse impedance [Eq. (13.93)] by

$$\varepsilon = \left(\frac{c}{\omega_\perp}\right)\left(\frac{Z_\perp}{Q_\perp}\right)\left(\frac{I_0}{I_A}\right)\frac{4\pi\beta}{(\mu_0/\varepsilon_0)^{1/2}}. \quad (13.105)$$

[4]D. G. Colombant and Y. Y. Lau, *Appl. Phys. Lett.*, **53**, 2602 (1988).

The quantity I_0 is the average beam current and I_A is the Alfven current. We can rewrite Eq. (13.105) in terms of the dimensionless transverse mode parameters introduced in Section 13.4:

$$\varepsilon = 4\pi \left(\frac{\eta_1 \eta_4}{\eta_2^3 \eta_3} \right) \left(\frac{I_0}{I_A} \right) \beta^4. \tag{13.106}$$

Despite their simple form, Eqs. (13.103) and (13.104) describe the cumulative beam breakup instability in a wide variety of accelerators. As an example, we shall review one solution for a coasting beam with no focusing forces in the limit of a saturated instability. The $Q_\perp$ factor of the cavities limits the amplitudes of the cavity fields and the beam oscillations. Here, we expect a steady-state solution with exponential growth in z. The analysis of Lau and Colombant recovers the familiar growth expression for RF accelerators:[5]

$$|x(z)| \sim \exp\left[1.14(\omega_\perp z/\beta c)(Q_\perp \varepsilon)^{1/2}\right]. \tag{13.107}$$

Substituting from Eqs. (13.86), (13.101), and (13.106), we can rewrite Eq. (13.107) as

$$|x(z)| \sim \exp\left\{ [1.14(\Pi Q_\perp)^{1/2}][\beta z/(dD)^{1/2}] \right\}, \tag{13.108}$$

where Π is the parameter we derived from the two-cavity analysis. If $\Pi Q_\perp \geqslant 1$, Eq. (13.108) shows that there is significant growth of the oscillation amplitude over the distance between two cavities. The square root in the exponent results from summing the beam deflections over cavities with a correct accounting of the oscillation phases.

To model the time-dependent beam breakup instability in an induction linac with discrete cavities, we must include several effects:

1. The short pulses in ferrite-core induction linacs ($\sim 30\,\text{nsec}$) may be comparable to the fill time for transverse modes. We must account for the time-dependent excitation of cavities during the beam pulse.
2. High-current induction linacs have strong lenses to contain the beams—in the low-energy section of the accelerator, the betatron wavelength, λ_b, may be comparable to D so that resonant effects are possible. For example, if $D = \lambda_b/2$, an angular deflection in a cavity results in no displacement in the next cavity, giving a low instability growth rate.
3. The quantities $\omega_\perp$ and $Q_\perp$ may vary between cavities.
4. The electrons gain energy through the accelerator, reducing the effect of the transverse mode magnetic fields and changing ω_b.

It is easy to include most relevant physical processes in a numerical model. We

[5] R. Helm and G. Loew, *Linear Accelerators*, P. M. Lapostolle and A. L. Septier, Eds., North-Holland, Amsterdam, 1970, p. 173.

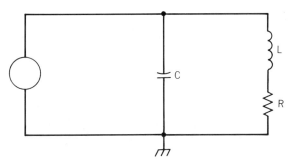

Figure 13.16 Lumped element circuit model used to represent a specified transverse mode in a cavity for numerical calculations of the beam breakup instability.

shall study results from SCAN, a computer program written by the author to calculate power output from devices that use the beam breakup instability to generate microwave energy. The program represents a specific transverse cavity mode as a lumped element LRC circuit. The symmetry of transverse modes about the axis simplifies the model. The program treats a driven oscillation in the upper half cavity driven by half of the beam dipole current. Figure 13.16 shows the circuit model for a cavity. The equations and geometric factors of Section 13.4 lead to expressions for lumped circuit elements to represent half the cavity:

$$C = \eta_3 \varepsilon_0 a^2 / d, \qquad L = \mu_0 d / \eta_2^2 \eta_3, \qquad \text{and} \qquad R = (L/C)^{1/2} / Q_\perp. \qquad (13.109)$$

The code calculates the self-consistent mode amplitude and beam displacement in a series of cavities for a given initial harmonic transverse perturbation. The model divides the relativistic beam into several segments that shift along the accelerator at the speed of light. Three quantities describe a segment: the transverse displacement, the transverse angle, and the kinetic energy $[x(z), x'(z), \gamma(z)]$. A ray transfer matrix for the focusing system advances the displacement and angle between acceleration gaps. The transformation depends on the kinetic energy of a segment. The angle changes within a cavity because of the acceleration and the magnetic field of the transverse mode. Numerical integration of the equations for the circuit of Fig. 13.16 gives the mode amplitude in each cavity. The displaced beam acts as a dipole current source. The program models variations in cavity properties through a spread in values of $Q_\perp$ and $\omega_\perp$.

Figure 13.17 shows results of the program for acceleration of a 1-kA beam in an induction accelerator from a 1-MeV injector to 10 MeV. The accelerator has 36 gaps with an applied voltage of 250 kV—the longitudinal gradient is 0.75 MV/m. The isolated acceleration cavities have identical geometry with length $d = 0.05$ m and transverse dimension $a = 0.3$. We assume a TM_{210} mode with resonant frequency 1.13 GHz. In the run shown, $Q_\perp = 10$ for all cavities. A solenoid magnetic field of magnitude 0.05 tesla focuses the beam.

Figure 13.17a shows the maximum amplitude of the beam displacement as a function of the acceleration gap number at a time 20 nsec after the injection of

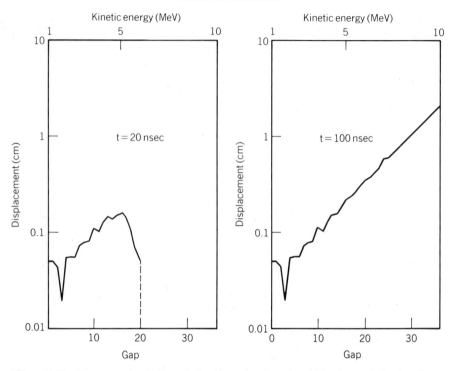

Figure 13.17 Computer simulation of the beam breakup instability in an induction linear accelerator—results from the SCAN code. Electron beam current, 1 kA. Injection energy, 1 MeV. Output energy, 10 MeV. Number of gaps, 36. Voltage per gap, 0.25 MV. Beam injected with harmonic displacement of 0.5 mm at 1.13 GHz. $Q_\perp = 10$. $B_0 = 0.05$ tesla. (*a*) 20 nsec after injection of the beam head. (*b*) 100 nsec after injection of the beam head.

the beam head. At this time, the beam head is halfway through the accelerator. Maximum displacement occurs about four gaps behind the head. At points well behind the head, the displacements have reached their steady-state values—the fill time for the transverse mode is only about $Q/f \cong 7$ nsec. Figure 13.17*b*, at $t = 100$ nsec, shows the saturated instability. The strong variations of displacement amplitude near the injection point result from the focusing forces. At downstream locations, the betatron wavelength is much larger than *D;* therefore, the growth in amplitude is almost exponential. The 2.1-cm displacement at the end of the accelerator is large. The beam gives up 54 MW to transverse resonant modes in the saturated state of the instability, about 0.5% of the output beam power. Other runs of the program show that the maximum beam displacement in saturation depends strongly on $Q_\perp$. The displacement at the accelerator exit drops to 0.7 cm if $Q_\perp = 7.5$.

There are several possible methods to reduce the severity of the beam breakup instability in induction accelerators and RF linacs:

1. The most common technique in induction linacs is to reduce values of $Q_\perp$ for all resonant modes in the acceleration cavities. Damping ferrites in the cavities absorb microwave radiation. The design of the cavity geometry allows transport of most of the electromagnetic energy generated by the beam to the ferrites.

2. The amplitude of beam sweeping at the accelerator exit is proportional to the beam displacement at the entrance. The electron source of a high-current accelerator should be well-aligned with the accelerator axis and should generate a symmetric beam.

3. In high-current induction linacs with a strong focusing system, tuning the lenses can reduce the growth of the instability. The growth is smallest if there is a half integral number of betatron wavelengths between cavities.

4. In both induction and RF linacs, variation of the geometry of accelerating cavities reduces instability growth. Inserts in RF accelerator cavities can modify the resonant frequency of transverse modes while leaving the accelerating mode unchanged.

5. Coupling loops in RF accelerator cavities can transfer the energy of resonant modes to a load absorber. With bandpass filters to reject the

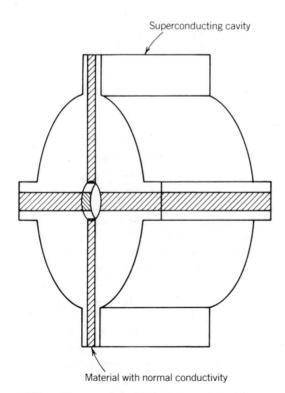

Figure 13.18 A composite cavity to reduce the growth of the beam breakup instability in a superconducting linear accelerator.

accelerating mode, the system gives a low Q value for transverse modes with little reduction in the longitudinal shunt impedance of the accelerator.

6. In systems where some emittance growth is tolerable, nonlinear lenses can be used for beam focusing. A large spread in the betatron oscillation of electrons can reduce the beam breakup instability.

7. Unconventional cavity designs can give high Q for the accelerating mode but a low value for transverse modes. Figure 13.18 shows an example for a high-current superconducting linac. The cavity consists of sections of superconducting material separated by resistive layers. The sectioned construction has little effect on the TM_{010} mode because the wall current flows only in the radial and axial directions. On the other hand, the wall current of transverse modes must cross the absorbing layers, leading to strong attenuation.

13.6. TRANSVERSE RESISTIVE WALL INSTABILITY

Section 15.2 discussed betatron oscillations of a filamentary beam in a linear focusing system. In this section, we will show how the amplitude of slow betatron waves can grow in the presence of resistive forces. We shall treat the transverse resistive wall instability by solving equations of motion for the center of the beam. If the beam is monoenergetic and the focusing forces are linear, the beam acts like an elastic body. We consider transverse displacements in the x direction with a frictional force of the form:

$$F_x = -k\frac{\partial x}{\partial t} \quad \text{[friction]}. \tag{13.110}$$

The quantity $\partial x/\partial t$ is the change in the displacement of the beam at a given point z in the stationary frame of reference. The focusing force on the beam is

$$F_x = -\gamma m_0 \omega_b^2 x, \tag{13.111}$$

where ω_b is the single-particle betatron frequency.

The frictional force results from the interaction between the beam and external structures. From the discussion of Section 13.3, we know that induced charges in the wall of a vacuum chamber with nonzero resistivity create a resistive force. Also, resonant structures can sometimes exert strong frictional transverse forces on beams (Section 13.3). In this section, we shall first study the theory of the resistive wall instability for a homogeneous resistive wall. Then, we shall derive an expression for the growth rate of the instability resulting from interactions with transverse modes in an array of resonant structures.

The model describes a thin beam observed in the stationary frame of reference. The displacement of the center of the beam is $x(z, t)$. The linear focusing force of Eq. (13.111) and the friction force of Eq. (13.110) govern the transverse beam motion. Particle orbits are paraxial and all particles have the same value of γ and axial velocity, $v_z = \beta c$. The friction force is weak—we shall look for harmonic beam oscillations with long-term damping.

The equations we shall derive for the motion of the beam center are deceptively simple. The derivation provides a model for a wide range of transverse instabilities—the reader should carefully note the definitions of quantities and the logic of the equations. We denote the rate of change of x at a constant position z as $\partial x/\partial t$. As we saw in Section 13.2, changes in x observed at a point result from two processes:

1. The beam particles at point z may have a transverse velocity v_x.
2. The beam particles near z may have a gradient in displacement, $\partial x/\partial z$. The beam appears to move sideways at velocity $-\beta c \, \partial x/\partial z$ as new particles move to position z.

The total time rate of change of x at z and t is

$$\frac{\partial x}{\partial t} = v_x - \beta c \frac{\partial x}{\partial z}. \tag{13.112}$$

The quantity $\partial x/\partial z$ is the variation of displacement with position at a constant time.

Similarly, we can write an equation for the change in v_x with time at constant position z, $\partial v_x/\partial t$. The velocity at z changes because of (1) acceleration of beam particles at z by forces and (2) replacement by particles with different velocity. The change in v_x is

$$\frac{\partial v_x}{\partial t} = -\beta c \frac{\partial v_x}{\partial z} - \omega_b^2 x - \kappa \frac{\partial x}{\partial t}. \tag{13.113}$$

The first term on the right-hand side of Eq. (13.113) is the change in v_x from particle convection. The second term is acceleration from the focusing force, proportional to the local displacement of particles. The third term, the friction force, incorporates the constant

$$\kappa = k/\gamma m_0. \tag{13.114}$$

Note that $\partial x/\partial t$ in Eq. (13.113) is the total change of x at a given point, equal to the expression of Eq. (13.112). The solutions of Eqs. (13.112) and (13.113) lead to a variety of beam oscillations, some of which are unstable. We shall proceed to the complete solution in stages.

First, consider a beam uniformly displaced in the axial direction at $t = 0$. We set all spatial derivatives in Eqs. (13.112) and (13.113) equal to zero. Without friction ($\kappa = 0$), the beam oscillates harmonically at the betatron frequency:

$$x(t) = x_0 \exp(-j\omega_b t). \tag{13.115}$$

Next, we add friction and assume a solution of the form

$$x(t), v_x(t) \sim \exp(-j\Omega t), \qquad (13.116)$$

where Ω can have real and imaginary parts. Substituting Eq. (13.116) into Eqs. (13.112) and (13.113) gives

$$\Omega^2 = \omega_b^2 - j\Omega\kappa. \qquad (13.117)$$

Equation (13.117) is a quadratic equation with the solution

$$\Omega = [-j\kappa + (4\omega_b^2 - \kappa^2)^{1/2}]/2 \cong \omega_b - j\kappa/2. \qquad (13.118)$$

The last form of Eq. (13.118) holds for a weak friction force, $\kappa \ll \omega_b$. The corresponding beam displacement is

$$x(t) = x_0 \exp(j\omega_b t)\exp(-\kappa t/2). \qquad (13.119)$$

Oscillations of an axially uniform beam always damp. Solutions with growing amplitude occur only when there are spatial variations of displacement.

We now add harmonic spatial variations. Suppose that perturbations of the beam have the form

$$x(z,t) = x_0 \exp[j(kz - \Omega t)], \qquad (13.120)$$

where k is real and Ω may have an imaginary part. The displacement of Eq. (13.120) has a uniform spatial variation that grows or damps with time. This representation is a good approximation for instabilities with low growth rate in circular accelerators.

First, we review the solution without friction. Substituting the form of Eq. (13.120) into Eqs. (13.112) and (13.113) gives the relationship:

$$(\Omega - k\beta c)^2 = \omega_b^2. \qquad (13.121)$$

Equation (13.121) has the mathematical solutions

$$\Omega/k = \beta c + \omega_b/k, \qquad (13.122)$$

$$\Omega/k = \beta c - \omega_b/k. \qquad (13.123)$$

Equation (13.122) is a fast betatron wave and Eq. (13.123) is a slow wave. For the slow wave, Section 13.2 showed that the motions of individual particles are 180° out of phase with the motion of the beam center at a point when

$$\beta c > \omega_b/k. \qquad (13.124)$$

We can now address the full solution of Eqs. (13.112) and (13.113). With the displacement of Eq. (13.120), we find the following relationship for the complex frequency:

$$(\Omega - k\beta c)^2 = \omega_b^2 - jk\Omega. \tag{13.125}$$

We take the complex square root of both sides of Eq. (13.125). In the limit that $\kappa \ll \omega_b$, the solutions are

$$\Omega \cong k\beta c + \omega_b - jk\Omega/2\omega_b, \qquad \Omega \cong k\beta c - \omega_b + jk\Omega/2\omega_b. \tag{13.126}$$

If the friction force is small, we can replace the values of Ω on the right-hand side of Eqs. (13.126) with the real values from Eqs. (13.122) and (13.123). The result is

$$\Omega \cong k\beta c + \omega_b - jk(k\beta c + \omega_b)/2\omega_b, \tag{13.127}$$

$$\Omega \cong k\beta c - \omega_b + jk(k\beta c - \omega_b)/2\omega_b. \tag{13.128}$$

The fast wave solution of Eq. (13.127) damps for all values of k when friction is present. The slow wave solution of Eq. (13.128) gives the following expression for the beam displacement at a point:

$$x(z,t) = x_0 \exp\left[j(k\beta c - \omega_b)t\right] \exp\left[\kappa(k\beta c - \omega_b)t/2\omega_b\right]. \tag{13.129}$$

The beam center oscillates at the Doppler-shifted frequency $(k\beta c - \omega_b)$. For the slow wave, the friction force results in a growth of the amplitude of oscillations if $k\beta c > \omega_b$. There is a simple interpretation for the instability. When the condition of Eq. (13.124) holds, the motions of individual beam particles are opposite to the macroscopic motion of the beam center observed at a point in the stationary frame. Therefore, the friction force is in a direction to *accelerate* particles, enhancing their transverse oscillations. The *e*-folding time for instability growth is

$$\Delta t_e = 2\omega_b/\kappa(k\beta c - \omega_b). \tag{13.130}$$

The model is valid when $\lambda \gg r_w$ and $r_0 \ll r_w$. As with other transverse collective instabilities, an axial momentum spread or a nonlinear focusing force may prevent the resistive wall instability. The instability cannot grow if phase mixing of transverse oscillations (Section 13.3) is effective on a time scale less than Δt_e.

As an example of a specific frictional force, suppose a beam moves through a circular pipe of radius r_w with nonzero wall resistivity, ρ. Comparison to Eq. (13.51) of Section 13.3 gives the friction constant

$$\kappa = \frac{eI_0\rho}{\gamma m_0 \pi \beta c r_w \delta} \ (\text{sec}^{-1}). \tag{13.131}$$

The quantity δ is the skin depth [Eq. (13.42)], which depends on the frequency of electromagnetic field variations at the wall in the stationary frame:

$$\delta = [2\rho/\mu_0(k\beta c - \omega_b)]^{1/2}. \tag{13.132}$$

The growth time is

$$\Delta t_e = \frac{2\omega_b \gamma m_0 \pi \beta c r_w \delta}{(k\beta c - \omega_b)eI_0\rho}. \tag{13.133}$$

Equation (13.133) shows that the resistive wall instability is a collective effect— the growth time is short for high beam current. We can also write the friction constant in terms of the transverse impedance of the structure (Section 13.3):

$$\kappa = eI_0 Z_\perp/\gamma m_0 \beta c, \tag{13.134}$$

where

$$Z_\perp = \rho/\pi r_w \delta.$$

As a numerical example, consider a high-current electron beam in a betatron with a stainless-steel vacuum chamber ($\rho = 8.1 \times 10^{-7}\,\Omega - $m). We take $I_0 = 100$ A, $\gamma = 10$, and $r_w = 0.08$ m. The growth time scales as $(k\beta c - \omega_b)^{3/2}$. We take a betatron wavelength of 1.5 m, or $\omega_b = 1.3 \times 10^9\,\text{sec}^{-1}$. We expect to observe growing waves at the highest allowed value of k. From the model validity condition $\lambda \gg r_w$, we take $k_{max} \cong 1/r_w$ and $(k\beta c - \omega_b) \sim c/r_w - \omega_b \cong 2.5 \times 10^9$ sec^{-1}. The frequency of the transverse beam oscillation in the stationary frame is $f = 400$ MHz. At this frequency, the skin depth in stainless steel is $\delta = 2.2 \times 10^{-5}$ m. The friction constant is $\kappa \cong 8.6 \times 10^2\,\text{sec}^{-1}$ and the growth time is above $\Delta t_e = 1.2$ msec. The long growth time follows from the low values of transverse impedance of the resistive wall, $Z_\perp = 0.15\,\Omega/\text{m}$. For the parameters of the example, the resistive walls have little effect on the transverse motion of the beam. The period of a betatron oscillation is about a nanosecond—even a small variation in the betatron oscillation frequency of particle orbits results in phase mixing over a time much shorter than 1 msec. In contrast, we shall see that interruptions in the vacuum chamber of the accelerator that support resonant oscillations may have a much shorter growth time.

Metal structures in accelerators have low resistivity while beams usually have high energy and low current. In other words, the driving beam has high impedance while the surrounding walls have low impedance—the direct transfer of energy to the walls is ineffective because of the mismatch. Power transfer to resonant cavities is a more complex process. For a continuous beam, we shall see in Section 15.2 that resonant structures can act as impedance transformers. Some electromagnetic modes have axial electric fields that can extract power from a beam. The beam drives the resonant mode, creating large circulating currents that couple energy to the low resistance of the cavity walls. As a result, the power transfer per length from a beam to resonant structures may be orders of magnitude higher than power to a simple wall resistance.

The resistive wall instability with resonant structures is the limiting case of the saturated beam breakup instability. Section 13.3 showed that the force from a transverse mode on a steady-state beam oscillating at the resonant frequency of the mode has amplitude proportional to the transverse beam velocity and opposite direction. In other words, the transverse impedance of the structure is resistive. If the spacing between resonant structures is much smaller than the beam betatron wavelength, we can represent the interactions with cavities by an axially averaged transverse impedance (Section 13.3):

$$Z_\perp = \left(\frac{\eta_1 \eta_4 \beta^2}{\eta_2^2 \eta_3} \right) \left(\frac{\mu_0}{\varepsilon_0} \right)^{1/2} \frac{Q_\perp f}{a} \left(\frac{\Omega}{\text{m}} \right). \qquad (13.135)$$

The quantity f is the fraction of the beamline occupied by cavities. We can substitute the transverse impedance into Eqs. (13.130) and (13.134) to estimate the growth rate of the instability for a beam perturbation with wavenumber $k = (\omega_\perp + \omega_b)/k\beta$, where $\omega_\perp$ is the resonant frequency of the transverse mode.

As an example, consider a series of square cavities that fills 10% of a beamline. For a relativistic beam, the factor in brackets on the right-hand side of Eq. (13.135) equals 0.46. If $a = 0.2$ m and $Q_\perp$ has a relatively low value of 500, the transverse impedance is 43 $k\Omega$/m. With such high values of transverse impedance, the transverse resistive wall instability can be a major problem in storage rings, even though the beam energy is high and the average current is low. The beams in these devices usually have a small spread in momentum and remain for long times. Extreme care must be exercised to shield vacuum ports and other interruptions of the vacuum chamber that could support transverse resonant modes.

13.7. HOSE INSTABILITY OF AN ION-FOCUSED ELECTRON BEAM

Experiments have shown that it is possible to guide a high-energy electron beam long distances with a preformed low-density ion column. Section 12.9 discussed the *ion-focused regime* (IFR) of electron beam transport. The ion columns are created by forming a low-density plasma in a tenuous medium, either with an electron discharge or with a tuned laser beam. IFR transport is an option for the propagation of high-energy electron beams in space. The idea is to generate an ion column kilometers in length in the upper atmosphere with a laser—the ion column guides a high-power electron beam to a target. For this application, the transverse stability of the beam-column system is a major concern.

In this section, we shall first review general properties of hose instabilities and then study electron beam instabilities in an ion channel. Hose instabilities are possible whenever a flexible transport system confines a particle flow. The simplest example of a flexible confinement system is the firehose, illustrated in Fig. 13.19. The firehose provides radial confinement of a stream of water. In a distorted hose, the stream of water exerts a centrifugal force at the points of maximum curvature. If the flow of water is strong enough, the centrifugal force exceeds restoring forces from the tension of the elastic hose and the amplitude of the perturbation grows.

Neglecting tension forces in the hose, we can calculate the growth rate of the perturbation amplitude by balancing the centrifugal force against the inertia of the hose and the water. Assume the water has a mass per length of M_w and an axial velocity V, while the hose has a mass per length of M_h. If we take a perturbation of the form $y(z, t) = y_0(t) \sin(kz)$, then the centrifugal force per length is

$$F_c = M_w(kV)^2 y_0(t) \sin(kz). \tag{13.136}$$

Balancing the force against the inertia, the amplitude of the perturbation grows according to

$$(M_w + M_h)\frac{d^2 y_0}{dt^2} = M_w(kV)^2 y_0. \tag{13.137}$$

The perturbation grows exponentially, $y_0(t) \sim \exp(\alpha t)$, with growth rate

$$\alpha = kV\sqrt{M_w/(M_w + M_h)}. \tag{13.138}$$

The growth rate is proportional to the water velocity and inversely proportional to the wavelength of the perturbation. In a real firehose, the restoring force of hose tension increases at short wavelength so there is an intermediate wavelength that gives the highest instability growth rate.

For water flowing in a hose or charged particles in a beam, the kinetic energy in the axial direction is much greater than the transverse kinetic energy. A flexible confinement system provides a mechanism to couple axial energy into transverse motion. Because of hose instabilities, a confined beam can acquire more

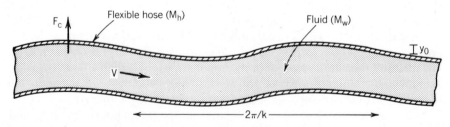

Figure 13.19   Mechanism of the simple firehose instability—a flexible transport system contains a moving fluid.

transverse kinetic energy than a freely expanding beam. If we direct a stream of water into free space, it expands at a rate given by its initial angular divergence. Nonetheless, on the average the water travels in a straight line. In contrast, although a flexible confinement system maintains a uniform beam radius, macroscopic motions of the beam can grow without limit.

We shall study the special case of a relativistic electron beam confined by an ion channel. Again, we seek a simplified model to minimize the mathematics. Figure 13.20a shows a cylindrical electron beam passing through a stationary ion column. The beam and column have uniform densities and equal radii, r_0. The column consists of bare singly charged ions with mass m_i. The paraxial electron beam has kinetic energy $(\gamma - 1)m_e c^2$ and axial velocity βc.

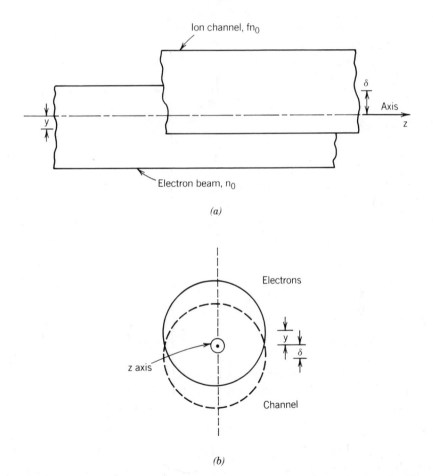

Figure 13.20 Hose instability of an ion-focused electron beam. (a) An electron beam confined by an ion column. The quantity y is the displacement of the electron beam from a symmetry axis while δ is the displacement of the ion column. (b) End view of a displacement between the electron beam and the ion column.

We denote the electron beam density as n_0. The ion channel density is

$$n_c = f n_0. \qquad (13.139)$$

The neutralization fraction, f, is less than unity. We assume that f has a value that guarantees the radial equilibrium of beam electrons. If the electrons have laminar orbits, then $f = 1/\gamma^2$. The neutralization fraction must be higher if the beam has nonzero emittance. We assume that the hose instability results in displacements of both the beam and ion channel from the symmetry axis in the y direction. The quantity $\delta(z, t)$ denotes the center of the flexible ion channel, while $y(z, t)$ is the beam center. The channel is initially centered, while the beam enters at $z = 0$ with a specified displacement, $y(0, t)$.

First, consider the force on the ion channel if it is not collinear with the electron beam (Fig. 13.20b). Section 12.3 discussed the electric fields of displaced charge cylinders. In the limit of small displacement, $(\delta - y) \ll r_0$, the electric field in the area region occupied by the cylinders is uniform with a magnitude linearly proportional to the displacement. The ion space charge cannot exert an average force on itself. Therefore, the force on the ion column results from the space charge of the electron beam. Following Eq. (12.50), the force exerted by the beam on a displaced ion channel is

$$F_y(z, t) = -(e^2 n_0 / \varepsilon_0)[\delta(z, t) - y(z, t)]. \qquad (13.140)$$

The variation of δ at an axial location in the stationary frame follows the equation:

$$\frac{\partial^2 \delta}{\partial t^2} = \frac{\omega_C^2}{f}[y - \delta]. \qquad (13.141)$$

The quantity ω_c in Eq. (13.141) is the plasma frequency of the channel ions:

$$\omega_c = (e^2 n_c / m_i \varepsilon_0)^{1/2}. \qquad (13.142)$$

To complete the solution, we need an equation similar to Eq. (13.141) for the beam transverse acceleration. As in Section 13.6, we must include all contributions to beam motion in the stationary frame. Besides the forces between the beam and ion column, the beam position can also change by convection. The net change in the transverse velocity of the beam is

$$\frac{\partial v_y}{\partial t} = -\beta c \frac{\partial v_y}{\partial z} - \left(\frac{e^2 n_c}{\gamma m_e \varepsilon_0} \right)(y - \delta). \qquad (13.143)$$

Equation (13.143) has following interpretation. The partial derivative on the left-hand side is the change of v_y with time at a constant position. The partial derivative on right-hand side is the axial variation of v_y at a given time. The last term on the right-hand side is the acceleration of the beam resulting from charge separation. We can also write Eq. (13.143) in terms of the convective derivative (Section 13.2):

$$\frac{Dv_y}{Dt} = \left[\frac{\partial}{\partial t} + \beta c \frac{\partial}{\partial z}\right] v_y = (f\omega_b^2)(\delta - y). \tag{13.144}$$

Equation (13.144) incorporates the beam plasma frequency

$$\omega_b = (e^2 n_0/\gamma m_e \varepsilon_0)^{1/2}. \tag{13.145}$$

We can follow similar logic to derive an expression for the change in y at a point:

$$\frac{\partial y}{\partial t} = -\beta c \frac{\partial y}{\partial z} + v_y \tag{13.146}$$

or

$$\frac{Dy}{Dt} = v_y. \tag{13.147}$$

Combining Eqs. (13.144) and (13.147) gives a symbolic expression for changes in the beam displacement:

$$\frac{D^2 y}{Dt^2} = (f\omega_b^2)(\delta - y). \tag{13.148}$$

Expanding the convective derivatives gives the following partial differential equation:

$$\frac{\partial^2 y}{\partial t^2} + 2\beta c \frac{\partial}{\partial t}\frac{\partial t}{\partial z} + \beta^2 c^2 \frac{\partial^2 y}{\partial z^2} = (f\omega_b^2)(\delta - y). \tag{13.149}$$

Equations (13.141) and (13.149) predict the spatial and temporal variations of the centers of the beam and the channel when displacements are small. Before reviewing numerical solutions, we shall investigate the ion hose instability in some limiting cases. To begin, suppose a beam enters an undisturbed channel $[\delta(z,0) = 0]$ that consists of infinitely massive ions. If the electron beam enters with a displacement, it oscillates harmonically about the rigid ion channel as it moves downstream. For a continuous beam, we calculate a steady-state solution by setting time derivatives in Eq. (13.149) equal to zero:

$$\frac{d^2 y}{dz^2} = -f\left(\frac{\omega_b}{\beta c}\right)^2 y. \tag{13.150}$$

Equation (13.150) has the solution

$$y = y_0 \cos(kz),$$ (13.151)

where

$$k = 2\pi/\lambda_b = \sqrt{f(\omega_b/\beta c)}.$$

The electron beam has a betatron oscillation about the rigid ion channel.

Next, we relax the condition of inifinite channel mass. The electric force on the ions is proportional to the electron displacement. We expect that if the electron beam has the displacement of Eq. (13.151), after a time the ion channel displacement is

$$\delta = \delta_0 \cos(kz).$$ (13.152)

The periodic ion channel displacement—we substitute the expression of Eq. (13.152) into Eq. (13.149). Again, we seek a steady-state beam solution with spatial variations. The beam displacement equation is

$$\frac{d^2y}{dz^2} = -k^2 y + \delta_0 k^2 \cos(kz).$$ (13.153)

Equation (13.153) describes a harmonic oscillator with a resonant drive term. The amplitude of oscillations grows as the electron beam moves downstream. In the limit that the change of the channel displacement amplitude is small over a distance $1/k$, Eq. (13.153) has the approximate solution

$$y \cong y_0[1 + (f^{1/2}\,\omega_b/2\beta c)z]\cos(kz) = y_0[1 + kz/2]\cos(kz),$$ (13.154)

When $\delta_0 = y_0$. For a uniform channel displacement, the beam displacement grows linearly with distance from the injection point.

We can now construct a qualitative explanation of the ion hose instability for small displacements. Suppose a displaced beam enters an initially straight ion channel. The beam has constant-amplitude betatron oscillations as it moves downstream. The electric fields from charge separation cause a harmonic displacement of the ion channel at wavelength λ_b. The distorted ion channel results in a beam oscillation amplitude that grows with distance from the injection point. In turn, the enhanced electron motion causes increased channel displacement. The displacements grow in space and time.

To complete the description, we shall estimate the growth time for the channel displacement at a point. Suppose that the charge-separation electric fields are strong—the beam and the channel have about the same displacement, $\delta(z,t) \cong y(z,t)$. A centrifugal force acts on the electron beam at the points of highest inflection. The force couples to the ion channel through the space-charge forces. We can estimate the growth rate for channel displacement by balancing the beam centrifugal force against the channel inertia:

$$(m_i n_c \pi r_0^2) \frac{d^2 \delta}{dt^2} \cong (\gamma m_e n_0 \pi r_0^2)(k\beta c)^2 \delta. \tag{13.155}$$

Equation (13.155) implies that the channel displacement at a point grows exponentially:

$$\delta \sim \exp[(\omega_c / f^{1/2})t]. \tag{13.156}$$

The following example illustrates the growth rate of the ion hose instability. A 10-MeV electron beam ($\gamma = 20.6$) propagates through a channel composed of He^+ ions. The beam has current $I_0 = 1\,\text{kA}$, radius $r_0 = 0.01\,\text{m}$, and density $n_0 = 2.1 \times 10^{17}\,\text{m}^{-3}$. The neutralization fraction must exceed $1/\gamma^2$, or 0.002. We shall take $f = 0.1$. The plasma frequencies are $\omega_b = 5.7 \times 10^9\,\text{sec}^{-1}$ and $\omega_c = 9.6 \times 10^7\,\text{sec}^{-1}$. The length scale for the growth of beam oscillations from Eq. (13.151) is about $2/k = 0.1\,\text{m}$. Equation (13.156) gives the time scale for channel distortion as $f^{1/2}/\omega_c = 3.3\,\text{nsec}$.

The spatial and temporal growth of the ion hose instability for small displacements is rapid for parameters of practical interest. To gauge the significance of the instability, we must study mechanisms that limit the amplitude of oscillations. The main process that leads to saturation of the hose instability is the nonlinear variation of the force between the beam and channel at large values of displacement, $|y - \delta| > r_0$. With a nonlinear force, the wavelength of transverse beam oscillations depends on the amplitude. As a result, the driving term in Eq. (13.153) may not be at resonance with the oscillation of the beam. We can estimate the average force between a beam and channel as a function of their separation by treating two limiting cases. We have already derived an expression for the force of the channel on the beam for small displacements:

$$F_y \cong (fe^2 n_0 / \varepsilon_0)(\delta - y) \qquad [(\delta - y) \ll r_0]. \tag{13.157}$$

At large displacement, the beam does not overlap the channel. Here, the average force on the beam equals the electric force from a line charge with linear density $fn_0 \pi r_0^2$ displaced a distance $(\delta - y)$:

$$F_y = [fe^2 n_0 / 2\varepsilon_0][r_0^2 / (\delta - y)] \qquad [(\delta - y) \gg r_0]. \tag{13.158}$$

The following form approximates the average force on the beam for both small and large displacements:

$$F_y = \left[\frac{fe^2 n_0}{\varepsilon_0}\right]\left[\frac{\delta - y}{1 + 2(\delta - y)^2 / r_0^2}\right]. \tag{13.159}$$

The equations that describe the motions of the beam and channel centers with nonlinear forces are

$$\frac{\partial^2 \delta}{\delta t^2} = \frac{\omega_c^2}{f}\left[\frac{\delta - y}{1 + 2(\delta - y)^2/r_0^2}\right] \tag{13.160}$$

and

$$\frac{\partial^2 y}{\partial t^2} + 2\beta c \frac{\partial}{\partial t}\frac{\partial y}{\partial z} + \beta^2 c^2 \frac{\partial^2 y}{\partial z^2} = (f\omega_b^2)\left[\frac{\delta - y}{1 + 2(\delta - y)^2/r_0^2}\right]. \tag{13.161}$$

Figure 13.21 illustrates numerical solutions to Eqs. (13.160) and (13.161). In the calculation, the electron beam has current of 1 kA and kinetic energy of 3.6 MeV. It propagates through a column of He$^+$ ions with radius $r_0 = 0.02$ m and width $f = 1$. The beam enters with a displacement of 2 mm. Note that the plots of Fig. 13.21 show displacements of the beam and channel following a particular segment of the beam. The top trace shows the displacement of the segment as it propagates away from the injection point. The lower trace shows the channel displacement observed in a frame moving with the beam segment as a function of distance from the injector.

Figure 13.21a shows displacements referenced to a segment that leaves the injection point 4.7 nsec after the start of the beam. For this segment, the quantity $(\delta - y)/r_0$ is much smaller than unity. The oscillation amplitude of the segment grows linearly with distance from the source. The channel displacement seen by an observer who moves with the segment is almost uniform with distance traveled. Note that a plot of the channel displacement as a function of z at a particular time shows an amplitude that grows with distance from the injection point.

Figure 13.21b shows displacements referenced to a segment that enters 7.1 nsec after the beam head. The nonlinear force has a significant effect. Although the channel displacement amplitude is larger than that of Fig. 13.21a, the beam displacement does not grow proportionally. Because of the shift of the beam betatron wavelength, the channel oscillation does not drive the beam at resonance. Figure 13.21c shows the displacement history of a segment entering 21 nsec after the beam head. The beam oscillation amplitude saturates at a level about twice the channel radius. Solutions at late time or long distance show almost a constant beam oscillation amplitude.

Other mechanisms limit the effect of the ion hose instability. The nonlinear confining force introduces a spread in betatron wavelength. The nonlinear force between the beam and channel not only shifts the average betatron wavelength for beam oscillations but also adds to the spread in the wavelength for individual particles. Section 13.3 discussed the stabilizing effect of a spread in betatron wavelength. A rough criterion for stability is that the damping time scale from phase mixing is shorter than the growth time for the instability.

Experiments on the long-distance transport of electron beams through ion channels show that the beam is subject to an instability immediately after injection that causes a growth of emittance. The beam expands to a radius larger

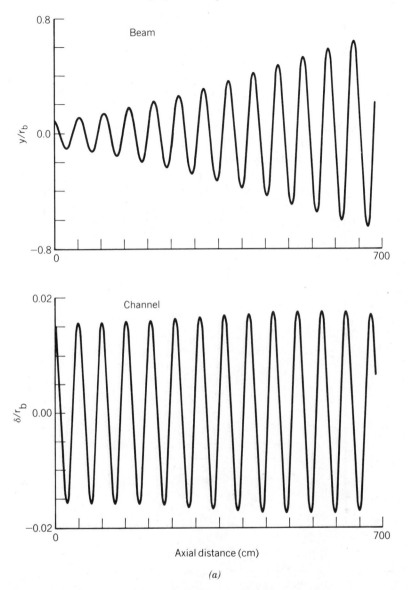

Figure 13.21 Numerical solutions of the instability of an electron beam confined by an ion column. Electron beam current: 1 kA. Electron beam energy: 3.6 MeV. Column of He$^+$ ions with radius $r_0 = 0.02$ m and $f = 1$. Initial beam displacement: 2 mm. (*a*) Top: displacement of a beam segment that enters the plasma 4.7 nsec after the beam head as a function of distance from the injection point. Bottom: Displacement of the ion column at the position of the beam segment. (*b*) Displacement of the beam and ion column for a segment that enters the plasma 7.1 nsec after the beam head. (*c*) Displacement of the beam and ion column for a segment that enters the plasma 21 nsec after the beam head. (Courtesy K. O'Brien, Sandia National Laboratories.)

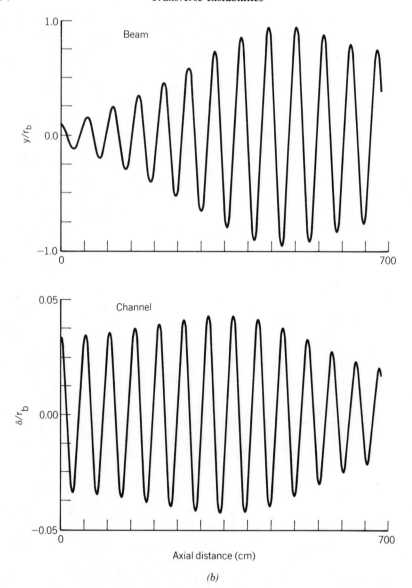

(b)

Figure 13.21 *(Continued)*

than that of the channel. After the beam reaches an equilibrium in the nonlinear force, it can propagate a long distance with little additional emittance degradation. The main limiting process for the propagation length in an IFR channel is radial expansion of the unconfined ions. We can estimate the expansion rate by balancing the transverse pressure force of the high-emittance electron beam against the inertia of the ion column.

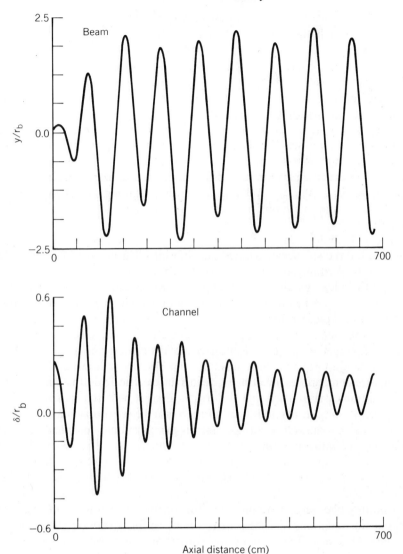

Figure 13.21 *(Continued)*

13.8. RESISTIVE HOSE INSTABILITY

The propagation of intense electron beams through gaseous media has been an active research area for over two decades. The motivation for this work has been the potential application of electron beams for the defense of ships and other large installations. As envisioned, a pulsed high-current electron beam is injected into air to create a region of dense plasma. In principle, following pulses can propagate through the plasma in a self-pinched equilibrium. The goal is to transport beams many kilometers. In present experiments, hose instabilities limit the propagation length to a few meters. The collective forces of the beam interact with the plasma background to cause severe deflections of the beam.

In this section, we shall not attempt a detailed study of electron beam propagation in dense gases. The coupled processes of gas ionization, plasma responses, and beam confinement represent one of the most complex collective problems in beam physics. We shall instead concentrate on the basic mechanisms of hose instabilities in a plasma. There are two main types. The simple firehose instability occurs when a beam travels through a localized plasma column. The resistive hose instability is a much more subtle phenomenon. It can affect self-pinched electron beams in preformed plasmas that extend over a large transverse width.

The simple firehose instability often occurs when an intense electron beam from a pulsed diode enters a neutral gas. The intense electric fields of the beam cause a rapid local breakdown of the gas. The discharge creates a dense plasma column surrounded by neutral gas. The beam propagates through the resistive plasma in a self-pinched equilibrium. The plasma column is a flexible confinement system. The centrifugal force of the beam enhances transverse displacements. Often, the instability deflects the beam and plasma column through 90° within a few meters of the injection point. The points of maximum beam curvature have approximately constant axial positions as the instability grows in time. In time-integrated photographs, beams injected into the atmosphere follow a sinuous trajectory.

The resistive hose instability occurs when a beam propagates through a uniform plasma. The width of the plasma region is much larger than the beam dimension. For this condition, the plasma does not act as a guiding channel as in the ion hose or simple firehose instabilities. Before the onset of instability, the beam propagates in a self-pinched equilibrium in a direction determined by the injection conditions. To describe the instability, we adopt some simplifying assumptions. The electron beam is cylindrical with a uniform current density from the axis to radius r_0. Electrons in the beam are monoenergetic and have paraxial orbits; therefore, their axial velocity, βc, is uniform. The background plasma is uniform and isotropic and of infinite extent. There are no applied magnetic or electric fields. We assume that the plasma density is high enough to neutralize the beam space-charge electric fields and to ensure a small magnetic skin depth, $c/\omega_{pe} \ll r_0$ [Eq. (12.94)]. Finally, the plasma has a scalar resistivity, ρ.

To help understand the instability, we shall review some properties of electron beam equilibrium in a resistive plasma. Suppose a pulsed beam of current I_0 enters a uniform plasma—the beam has transverse displacement X from a symmetry axis. To preserve charge neutrality, the plasma carries a return current—I_0. At the head of the beam, the return current density has the same transverse distribution as the beam current density. Section 12.6 showed that plasma resistivity spreads the return current over larger cross section in latter parts of the beam. Therefore, the distribution of the return current varies as a function of position in the beam pulse.

The most convenient way to describe the interactions of an electron beam and a resistive plasma is to follow individual axial segments of the beam (Fig. 13.22*a*). The treatment differs from those in previous sections of this chapter where we observed displacements of many segments as they passed a constant axial location. To construct equations referenced to the beam, we define a variable τ, equal to the time a beam segment enters the system at $z = 0$. The beam head enters at $t = 0$ and has $\tau = 0$. Each following segment has a characteristic value of τ. If z is the axial position at time t of an axial element parametrized by τ, then

$$z = v(t - \tau). \tag{13.162}$$

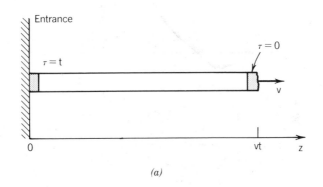

(a)

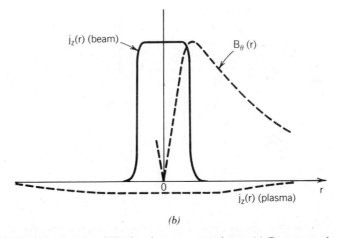

(b)

Figure 13.22 Resistive hose instability in a homogeneous plasma. (*a*) Geometry—the calculation uses the variable τ, the time at which a beam segment passes the injection point ($z = 0$). (*b*) Properties of the beam equilibrium. Radial distribution of beam current density, plasma current density, and toroidal magnetic field at a point well back from the beam head.

Even without an instability, the behavior of the beam head is complex. At the head, there is no net magnetic field in the beam volume. Therefore, electrons at the head are not focused—their orbits expand radially. For simplicity, we shall concentrate on beam segments far behind the head where the plasma current has expanded enough to allow a self-pinched equilibrium. From the discussion of

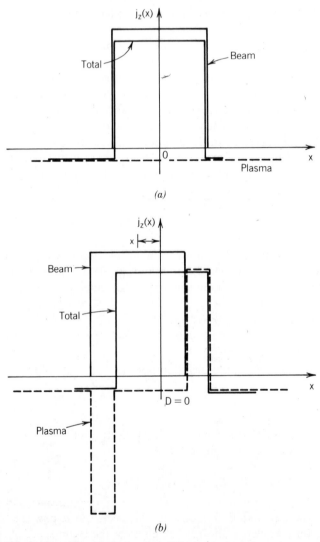

(a)

(b)

Figure 13.23 Mechanism of the resistive hose instability in a homogeneous plasma. (a) Distribution of the beam, plasma, and total current densities in equilibrium at a point well back from the beam head. (b) Changes in current density distributions immediately following a sudden beam displacement. (c) Effect of resistive diffusion on the current density distributions following a displacement. The position of the center of the plasma return current is shifted.

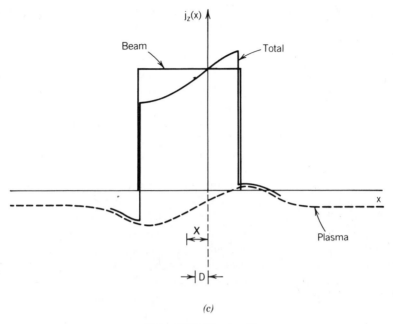

(c)

Figure 13.23 *(Continued)*

Section 12.6, we know that the beam-generated magnetic field has almost its full magnitude in a segment if

$$\tau \gg \tau_d. \tag{13.163}$$

The quantity τ_d in Eq. (13.163) is the magnetic diffusion time,

$$\tau_d = \mu_0 r_0^2 / \rho. \tag{13.164}$$

When the limit of Eq. (13.163) holds, the distributions of beam and plasma current resemble those of Fig. 13.22b.

We shall neglect the fraction of plasma return current within the beam volume. Figure 13.22b shows the variation of the beam-generated toroidal magnetic field as a function of x. With a uniform current density, the field variation is linear. The electrons follow harmonic betatron oscillations with wavenumber k_b. Following Section 12.7, we can express the wavenumber in terms of the beam current, the Alfven current [Eq. (12.127)] and the beam radius

$$k_b = (2I/I_a)^{1/2} / r_0. \tag{13.165}$$

We now have enough information about the equilibrium to explain the response of the plasma to transverse displacements of the beam. Suppose a segment at position z suddenly moves sideways along the x axis (Fig. 13.23a). We denote the center of the displaced segment with parameter τ at z as $X(z, \tau)$. The quantity $D(z, \tau)$ is the average transverse position of the plasma return current at position z before the beam displacement. After the displacement, $X(z, \tau)$ may no longer coincide with $D(z, \tau)$. To the plasma, it appear that there is an instantaneous rise of the beam current density in the direction of the displacement and a drop on the trailing edge. Inductively driven return currents in the highly conducting plasma flow to cancel the change in the beam current density distribution. Figure 13.23a shows the result—the spatial distribution of *total* current is the same before and after the beam displacement. The quantity D does not change immediately after a rapid change in X.

We next consider how X varies after the displacement. To begin, suppose the plasma is a perfect conductor, $\rho = 0$. Here, the plasma provides perfect cancellation of any change in the current density distribution. Therefore, the magnetic axis, D, remains at its initial value independent of the position of the beam—the focusing system is rigid in the transverse direction. For small displacements, $X - D \ll r_0$, all electrons follow coherent betatron oscillations and the beam center oscillates harmonically about D with wavenumber k_b and amplitude $X - D$. We can represent the position of the beam center by the trace equation

$$\left.\frac{\partial^2 X(z, \tau)}{\partial z^2}\right|_\tau = -k_b^2 [X(z, \tau) - D(z, \tau)]. \tag{13.166}$$

The partial derivative symbol on the left-hand side emphasizes that Eq. (13.166) describes the axial variation of displacement for a particular beam element with constant τ. Equation (13.166) gives a harmonic solution with wavenumber k_b.

The position of the magnetic axis is not stationary if the plasma has nonzero resistivity. To complete the solution, we must construct an equation that describes the variation of D. Figure 13.23a showed that sudden beam displacements from D induce positive and negative-going plasma currents on opposite sides of beam. The magnetic diffusion equation [Eq. (12.115)] implies that resistivity spreads these regions of plasma current density in space after the initial perturbation (Fig. 13.23b). Ultimately, the positive and negative plasma current components overlap and cancel each other. The cancellation leads to a shift in the magnetic axis to the new position of the beam center (Fig. 13.23c).

In principle, we can compute the translation rate for the magnetic axis exactly from Eq. (12.115) and the initial distribution of current. The following simple equation gives a good approximation for the motion of the magnetic axis relative to the beam position:

$$\left.\frac{\partial D}{\partial t}\right|_z = -\frac{D - X}{\tau_d}. \tag{13.167}$$

We can justify Eq. (13.167) by noting that the time for the plasma current component to spread over an area comparable to that of the beam is roughly τ_d. Furthermore, the magnetic axis must approach the beam center at long times, $D \cong X$ for $t \gg \tau_d$. We can rewrite Eq. (13.167) using the chain rule of partial derivatives:

$$\frac{\partial D(z, \tau)}{\partial \tau} = -\frac{[D(z, \tau) - X(z, \tau)]}{\tau_d}. \tag{13.168}$$

The quantity $D(z, \tau)$ in Eq. (13.168) is the transverse position of the magnetic axis at a given value of z seen by the beam segment with parameter τ.

Equations (13.166) and (13.168) describe the history of translation of a beam segment in a homogeneous resistive medium. The equations have several solutions, depending on the initial conditions. We choose a simple form of the solution that represents experiments on the resistive hose instability. Suppose that a continuous electron beam enters at $z = 0$. At the injection point, the transverse position of the beam varies at frequency ω,

$$X(0, \tau) = X_0 \cos(\omega \tau). \tag{13.169}$$

The position modulation may result from a beam breakup instability in the accelerator that generates the beam. We seek solutions with a transverse displacement that varies harmonically with time and may grow with distance from the injection point:

$$D(z, \tau) = \mathsf{D} \exp[j(\Gamma z - \omega \tau)], \tag{13.170}$$

$$X(z, \tau) = \mathsf{X} \exp[j(\Gamma z - \omega \tau)].$$

The quantity ω is real, while Γ may have an imaginary part to represent spatial growth.

Substitution of Eqs. (13.170) into Eqs. (13.166) and (13.168) gives

$$\mathsf{D} = \mathsf{X}/(1 - j\omega \tau_d), \tag{13.171}$$

$$\mathsf{X} = \mathsf{D}/(1 - \Gamma^2/k_b^2). \tag{13.172}$$

Eliminating D and X gives the following equation for Γ:

$$\Gamma^2 = k_b^2 \left[1 - \frac{1}{1 - j\omega \tau_d} \right]. \tag{13.173}$$

There are two governing parameters in Eq. (13.173), the unperturbed betatron oscillation wavenumber, k_b, and the quantity $\omega \tau_d$. The second expression equals the ratio of the magnetic diffusion time to the time scale for transverse beam motion. We shall find that the growth of the instability is slow when $\omega \tau_d \gg 1$.

Here, the plasma is almost a perfect conductor. The instability growth is also slow when $\omega\tau_d \ll 1$. In this limit, plasma current components annihilate rapidly so that the magnetic axis is always collinear with the beam axis. Because the plasma provides little restoring force, there is no transverse instability.

To solve Eq. (13.173), assume Γ has real and imaginary parts,

$$\Gamma = \Gamma_r + \Gamma_i. \tag{13.174}$$

We substitute Eq. (13.174) into Eq. (13.173) and set real and imaginary parts separately equal. The resulting equations are

$$\frac{\Gamma_r}{k_b} = \frac{-\omega\tau_d}{(2\Gamma_i/k_b)(1 + \omega^2\tau_d^2)} \tag{13.175}$$

and

$$\left[\frac{\Gamma_r}{k_b}\right]^2 - \left[\frac{\Gamma_i}{k_b}\right]^2 = \frac{\omega^2\tau_d^2}{1 + \omega^2\tau_d^2}. \tag{13.176}$$

Eliminating Γ_r from Eqs. (13.175) and (13.176) and solving the resulting quadratic equation gives the following expression for Γ_i:

$$\left[\frac{\Gamma_i}{k_b}\right]^2 = \frac{1}{2}\left[\frac{-\omega^2\tau_d^2}{1 + \omega^2\tau_d^2} + \left(\frac{\omega^2\tau_d^2}{1 + \omega^2\tau_d^2}\right)^{1/2}\right]. \tag{13.177}$$

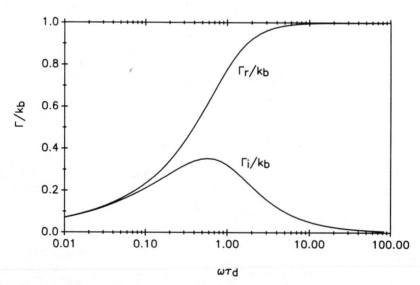

Figure 13.24 Steady-state solution for the resistive hose instability in a homogeneous plasma. Real and imaginary parts of the wavenumber, Γ_r and Γ_i.

Figure 13.24 plots Γ_i and Γ_r as functions of $\omega\tau_d$. The instability growth rate approaches zero when $\omega\tau_d \rightarrow 0$. For low plasma resistivity, $\omega\tau_d \rightarrow \infty$, the growth rate approaches zero and Γ_r approaches k_b. These conditions represent a stable harmonic oscillation at the single-particle betatron wavelength. The following conditions hold at the maximum value of Γ_i:

$$\omega\tau_b = (\tfrac{1}{3})^{1/2} = 0.577,$$

$$\Gamma_i/k_b = (\tfrac{1}{8})^{1/2} = 0.354, \qquad \Gamma_r/k_b = (\tfrac{3}{8})^{1/2} = 0.612. \qquad (13.178)$$

With conditions for maximum growth, the displacement amplitude grows significantly over a distance equal to one betatron wavelength. Figure 13.25 shows the transverse motion of beam element τ and the magnetic axis as the element propagates for $\omega\tau_d = 0.577$. Note that the magnetic axis lags behind the beam to provide a net restoring force. As an applications example, consider propagation of a 10-MeV beam with $I_0 = 10\,\text{kA}$, $r_0 = 0.01\,\text{m}$, and $k_b = 24\,\text{m}^{-1}$. Previous beam pulses have created a plasma channel in air with a density $n_p = 7 \times 10^{22}\,\text{m}^{-3}$ and temperature $T_e = 10\,\text{eV}$. We take the input beam sweeping frequency at a high value typical of accelerator instabilities, $\omega = 2.5 \times 10^9\,\text{sec}^{-1}$. To calculate the resistivity for the nitrogen plasma, we multiply Eq. (12.107) by a factor $Z^2 = 49$. The predicted resistivity is about $\rho = 1.5 \times 10^{-3}\,\Omega\text{-m}$. The field diffusion time in the plasma is $\tau_d = 90\,\text{nsec}$, or $\omega\tau_d = 225$. In the limit of high $\omega\tau_d$, Eq. (13.177) shows that the growth length for the instability is $1/\Gamma_i \cong 2\omega\tau_d/k_b = 20\,\text{m}$.

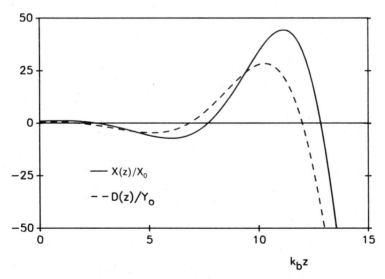

Figure 13.25 Transverse position of the beam and plasma return current at an instant of time for $\omega\tau_d = 0.577$.

   The full implications of the resistive hose instability for long-distance propagation are still the subject of active research and debate. A spread in betatron wavelength reduces the growth of the instability by damping the coherent transverse oscillations of the beam. Such a spread can arise from a distribution of electron energy or from nonlinear focusing forces. For an accurate calculation of the resistive hose instability, we must include effects of the nonuniform transverse variation of beam current density and the nonlinear force between the beam and the plasma magnetic field for displacements comparable to r_0.

13.9. FILAMENTATION INSTABILITY OF NEUTRALIZED ELECTRON BEAMS

Space-charge-neutralized relativistic electron beams often are unstable to filamentation. When no electric field is present, perturbations of the beam-generated magnetic fields can cause local pinching. The magnetic force in regions of enhanced axial current attracts more electrons, amplifying the current. Ultimately, the beam separates into current filaments. The filamentation instability causes no change in the average position of the beam—its main effect is to increase the beam emittance. The magnetic fields couple longitudinal kinetic energy to transverse motion. The instability saturates when the transverse velocity spread is high enough to resist the pinching force.

   Filamentation instabilities occur when intense relativistic beams propagate through a neutralizing plasma. We shall see that filamentation also affects beam focused by a transverse foil array under high vacuum (Section 10.5). In this section, we shall construct models for both cases. The models use the linearized moment equations to predict the onset of an instability. Although they do not describe the saturation state of the instability, we can use physical insights to estimate the outcome.

   To begin, we shall study propagation of an infinite-width beam in a homogeneous plasma. Here, the term "infinite" means large compared with the length scale of filaments. The equilibrium state is simple to describe. The plasma provides both space-charge and current neutralization—there are no steady-state electric or magnetic fields. The beam has a uniform density, n_0, that is much smaller than the plasma density. Furthermore, the beam is paraxial—all electrons have about the same kinetic energy, $(\gamma - 1)m_e c^2$, and axial velocity, βc. We include the possibility of a nonzero beam emittance, represented by the transverse velocity dispersion, $\langle \delta v_x^2 \rangle$.

Figure 13.26 Mechanism for the filamentation instability in a resistive plasma. View along the axis—transverse distribution of beam current density and plasma return current density. (*a*) Immediately following application of a transverse density perturbation to the beam. (*b*) Resistive diffusion causes decay of plasma return current perturbations. Magnetic pinch forces act on the beam particles. (*c*) Growth of beam current-density perturbation.

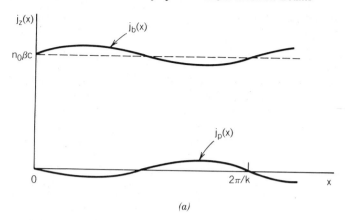

(a)

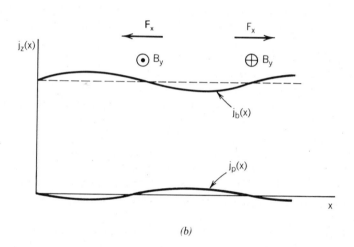

(b)

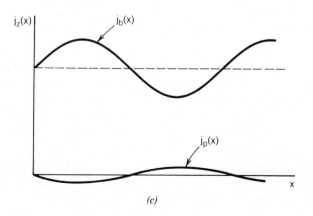

(c)

Suppose the beam has a density perturbation in the x direction equal to

$$n(x) = n_x \cos(kx). \tag{13.179}$$

In a linear analysis, the density perturbation is small, $n_x/n_0 \ll 1$. We shall see that the growth time of the magnetic instability is much longer than $1/\omega_{pe}$, where ω_{pe} is the plasma frequency. Therefore, the rapid flow of plasma electrons ensures that electric fields are always small. In the linear regime, the change in the transverse velocity of electrons has little effect on their axial velocity. With this condition, the axial current density of the beam is proportional to $n(x)$:

$$j_{bz}(x) = - en_x \beta c \cos(kx). \tag{13.180}$$

If the beam current density varies in space and time, the return current density of a perfectly conducting plasma adjusts to seek zero net magnetic field. Field cancellation is effective if the magnetic skin depth (Section 12.5) is much smaller than the wavelength of the perturbation:

$$k \ll \omega_{pe}/c. \tag{13.181}$$

When the condition of Eq. (13.181) holds, the return current of a perfectly conducting plasma is

$$j_{px}(x) = + en_x \beta c \cos(kx). \tag{13.182}$$

If there is no net magnetic field, there is no driving force for the instability. In this case, a small transverse beam velocity spread is sufficient to damp density perturbations.

The filamentation instability can occur in a plasma with nonzero resistivity ρ. Figure 13.26 illustrates the sequence of events. Suppose we perturb the beam density instantaneously, as in Fig. 13.26a. In response, the plasma generates return current components that flow in the opposite direction to the modified beam current. Because of resistive diffusion, the return current density decays to a uniform distribution in a time roughly equal to

$$\tau_d = \mu_0/k^2\rho. \tag{13.183}$$

After a magnetic diffusion time, the beam-plasma system approaches the state of Fig. 13.26b. The beam-generated magnetic fields cause pinching and enhancement of the density perturbations (Fig. 13.26c). The combination of the beam inertia, the magnetic pinch force, and the plasma magnetic diffusion time control the growth of beam filaments.

We can use linearized moment equations for a mathematical description of filamentation. Figure 13.27 shows the coordinate system of the analysis. The perturbed quantities are the beam density, n_x, the directed beam transverse

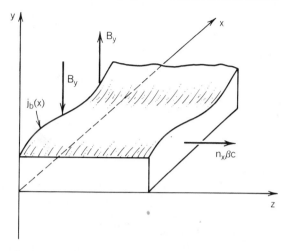

Figure 13.27 Coordinate system to analyze magnetic pinching of an electron beam in a resistive plasma. Also shown is a schematic view of the variation of beam current density for the assumed perturbations.

velocity, v_x, the magnetic field, B_y, and the axial plasma return current density, j_p. The plasma electrons carry the return current—we assume plasma ions are immobile. Because pinching is a slow process, we can use the static form of Ampere's law:

$$\frac{\partial B_y}{\partial x} = \mu_0(-en_x\beta c + j_p).$$ (13.184)

The beam continuity equation is

$$\frac{\partial n_x}{\partial t} = -n_0\frac{\partial v_x}{\partial x}.$$ (13.185)

The beam momentum equation has the form

$$\frac{\partial v_x}{\partial t} = -\langle\delta v_x^2\rangle\frac{\partial n_x}{\partial x} + \frac{e\beta c B_y}{\gamma m_e}.$$ (13.186)

The quantity $\langle\delta v_x^2\rangle$ is the mean-squared beam transverse velocity spread. The diffusion equation for beam current density from Section 12.6 completes the set of equations. We modify Eq. (12.114) to include the possibility of continuous changes in beam current density:

$$\frac{\partial j_p}{\partial t} = \frac{\rho}{\mu_0}\frac{\partial^2 j_p}{\partial x^2} + e\beta c\frac{\partial n_x}{\partial t}.$$ (13.187)

First-order variables vary as $\exp(ikz)\exp(\alpha t)$—the quantity α is the anticipated instability growth rate. Substitution in Eqs. (13.184) through (13.187) gives
a set of four algebraic equations in four unknowns. Elimination of variables leads
to an equation for α as a function of k and the characteristics of the equilibrium
beam–plasma system:

$$\alpha^2 = -k^2 \langle \delta v_x^2 \rangle + \beta^2 \omega_b^2 \left[1 - \frac{1}{1 + k^2 \rho/\mu_0 \alpha} \right]. \tag{13.188}$$

The quantity ω_b in Eq. (13.188) is the beam plasma frequency

$$\omega_b = (e^2 n_0/\gamma m_e \varepsilon_0)^{1/2}, \tag{13.189}$$

while $\beta \omega_b$ is a characteristic magnetic oscillation frequency. A positive value of α^2
gives growing perturbations. Equation (13.188) shows that a transverse velocity
spread has a stabilizing effect, while the beam-generated magnetic force
contributes to an instability. Note that the magnetic force contribution drops to
zero when $\rho = 0$. There is no filamentation instability in a perfectly conducting
plasma.

If we neglect the velocity spread term, Eq. (13.188) gives the following
expression for the growth rate:

$$\left[\frac{\alpha}{\beta \omega_b} \right]^3 + \left[\frac{k^2 \rho}{\mu_0 \beta \omega_b} \right] \left[\left(\frac{\alpha}{\beta \omega_b} \right)^2 - 1 \right] = 0. \tag{13.190}$$

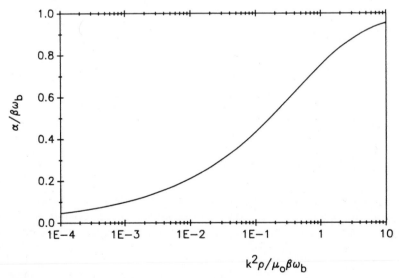

Figure 13.28 Normalized growth rate for the pinching instability of an electron beam with no
transverse velocity spread in a resistive plasma.

Figure 13.28 shows a plot of $\alpha/\beta\omega_b$ as a function of $k^2\rho/\mu_0\beta\omega_b$. When $k^2\rho/\mu_0\beta\omega_b \ll 1$, magnetic field diffusion in the plasma controls the instability growth rate. When $k^2\rho/\mu_0\beta\omega_b > 1$, the interaction between the beam inertia and magnetic force determine α. In this limit, the growth rate approaches its maximum value:

$$\alpha/\beta\omega_b \cong 1. \tag{13.191}$$

For relativistic beams, the assumption of good space-charge neutralization is valid because $\alpha \leqslant \beta\omega_b \ll \omega_{pe}$. The filamentation instability does not occur if there is a sufficient transverse velocity spread. We can estimate the stability requirement by setting α equal to zero in Eq. (13.188). A beam does not break into filaments if

$$\langle \delta v_x^2 \rangle \geqslant (\beta\omega_b/k)^2. \tag{13.192}$$

The following example illustrates some implications of the results. A 5-MeV, 5-kA electron beam of radius 0.02 m propagates in a hydrogen plasma with electron temperature $kT_e = 10\,\text{eV}$. The beam density is $8.3 \times 10^{16}\,\text{m}^{-3}$ and the beam plasma frequency is $\omega_b = 5 \times 10^9\,\text{sec}^{-1}$. If the plasma density is 10^3 times the beam density, the plasma frequency is $\omega_{pe} = 5 \times 10^{11}\,\text{s}^{-1}$. Following Section 12.6, the plasma resistivity is $\rho = 4.6 \times 10^{-5}\,\Omega\text{-m}$. To estimate the maximum growth rate, we must find the reasonable value of perturbation wavenumber. Our model holds only if the quantity $1/k$ exceeds the collisionless skin depth. For the assumed plasma parameters, $c/\omega_{pe} = 6 \times 10^{-4}\,\text{m}$; therefore, we shall take $k = 1/(2 \times 10^{-3}\,\text{m}) = 500\,\text{m}^{-1}$. This wavenumber corresponds to a perturbation wavelength of 0.013 m, small enough to justify use of the infinite-beam model. Combining all quantities, we find that

$$k^2\rho/\mu_0\beta\omega_b \cong 1.8 \times 10^{-3}. \tag{13.193}$$

The low value in Eq. (13.193) shows that magnetic diffusion in the plasma governs the filamentation growth rate. Figure 13.28 shows that

$$\alpha = (\beta\omega_b)(0.121) = 6 \times 10^8. \tag{13.194}$$

The growth time is 1.7 nsec. During this time, the beam travels 0.5 m in the axial direction. The instability causes enhancement of the beam emittance. We expect that the instability saturates when the transverse velocity spread satisfies Eq. (13.192), or

$$(\langle v_x^2 \rangle)^{1/2}/\beta c = (\omega_b/ck) = 0.033. \tag{13.195}$$

An angular divergence of 2° stabilizes the beam against further filamentation.

Filamentation instabilities also act on beams that propagate under high vacuum through a foil array (Section 10.5). The transverse foils neutralize the space-charge electric fields of the beam; therefore, it is not surprising that magnetic filamentation may occur. An exact analysis of magnetic filamentation of a cylindrical beam in a foil array is a difficult problem—the fields and beam density vary in three dimensions. Instead, we shall construct a simple model for the sheet beam geometry of Fig. 13.29. The continuous beam moves in the z direction. It has a narrow width in the y direction and extends infinitely in x. In equilibrium the beam density is uniform in the x direction—we shall find conditions where perturbations along x lead to magnetic pinching. In a foil transport system, the sheet beam travels through closely spaced transverse meshes that reduce electric fields but have little effect on magnetic fields. This geometry, with variations in y and z, is difficult to describe with an analytic model. As an alternative, we shall use the simplified model of Fig. 13.29. The beam moves between symmetric planar boundaries. To represent the effect of partial space-charge neutralization, we locate the induced charge boundary closer to the sheet beam than the induced current boundary. The charge boundary, at positions $\pm y_e$, is a surface of constant electrostatic potential with $E_x = 0$. The return current boundary at positions $\pm y_m$ is a surface of constant vector potential with $B_y = 0$.

If electrons in the sheet beam have paraxial orbits, we can use nonrelativistic equations of motion with an adjusted mass, γm_e. Focusing forces confine the beam in the y direction—electrons move only in the x direction. In equilibrium, the beam has a uniform charge density ρ_0 over a half-thickness Δy. A thin sheet beam satisfies the condition $\Delta y \ll y_e, y_m$. We take perturbations in the x direction with a wavelength much longer than the beam thickness:

$$k \leqslant 1/\Delta y. \tag{13.196}$$

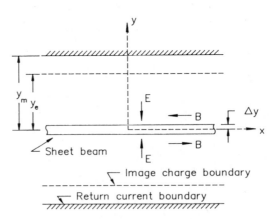

Figure 13.29 Sheet beam geometry to analyze the filamentation of an electron beam in a foil focusing array.

Finally, we compute electric and magnetic fields from static equations—the static approximation is valid if the instability growth times are much longer than $1/kc$.

In equilibrium the electric and magnetic fields have the following values at the upper and lower edges of the beam:

$$E_{y0}(\pm \Delta y) = \pm \rho_0 \Delta y/\varepsilon_0, \qquad B_{x0}(\pm \Delta y) = \pm \mu_0 \rho_0 \Delta y \beta c. \qquad (13.197)$$

We take a charge density perturbation of the form

$$\rho(x) = \rho_0 + \Delta \rho \cos (kx). \qquad (13.198)$$

For the density variation of Eq. (13.198), the beam current density varies as

$$j_z(x) = \rho_0 \beta c + \Delta \rho \beta c \cos (kx). \qquad (13.199)$$

First, we shall calculate the electrostatic potential variations caused by the perturbed density. The solution to the Laplace equation outside the beam that satisfies the boundary condition at the induced charge boundary is

$$\Delta \phi(x, y) = 2\Delta \phi_k \cos(kx)\sinh [k(y - y_e)]. \qquad (13.200)$$

We apply a matching condition for the electric field at the beam boundary. When $k\Delta y \ll 1$, the normal component of electric field approaches the value

$$\Delta E_y(0^+) \cong - \partial \Delta \phi(x, 0)/\partial y = \Delta \rho \Delta y \cos (kx)/\varepsilon_0. \qquad (13.201)$$

Equation (13.201) states that the change in the perturbed normal electric field across the sheet beam almost equals the net beam charge per area divided by ε_0. Equation (13.201) implies that

$$\Delta \phi_k = - (\Delta \rho \Delta y/4\varepsilon_0)\cosh (ky_e). \qquad (13.202)$$

The electric field component ΔE_x affects the growth of filamentation instabilities. Taking $\Delta E_x = - \partial \Delta \phi(x, y)/\partial x$, we find the following expression:

$$\Delta E_x(x, y) = (\Delta \rho \Delta y/2\varepsilon_0) \tanh (ky_e) \sin (kx). \qquad (13.203)$$

We can apply a similar treatment to calculate the perturbed magnetic field. Solving for the perturbed vector potential leads to the expression

$$\Delta B_y(x, y) = (\Delta \rho \Delta y \beta/2\varepsilon_0 c) \tanh (ky_m) \sin (kx). \qquad (13.204)$$

The net force in the x direction resulting from the fields of the beam is

$$F_x = (e\Delta\rho\Delta y/2\varepsilon_0)\sin(kx)[\tanh(ky_e) - \beta^2\tanh(ky_m)]. \qquad (13.205)$$

The linearized moment equations for beam motion in the x direction are

$$\frac{\partial\Delta\rho}{\partial t} + \frac{\partial}{\partial x}(\rho_0\Delta v_x) = 0 \qquad (13.206)$$

and

$$\gamma m_0\rho_0\frac{\partial\Delta v_x}{\partial t} + \gamma m_0\langle\delta v_x^2\rangle\frac{\partial\Delta\rho}{\partial x} = \frac{\rho_0 e\Delta y}{2\varepsilon_0}\Delta\rho\sin(kx)[\tanh(ky_e) - \beta^2\tanh(ky_m)].$$

$$(13.207)$$

The quantity Δv_x is the directed transverse velocity while $\langle\delta v_x^2\rangle$ is a random velocity spread in the x direction. If perturbed quantities vary as $\exp(\alpha t)\sin(kx)$, Eqs. (13.206) and (13.207) lead to an expression for the linear instability growth rate:

$$\alpha^2 = -k^2\langle\delta v_x^2\rangle - (eJ_z/2\gamma m_0\beta c\varepsilon_0)k\tanh(ky_e)$$
$$+ \beta^2(eJ_z/2\gamma m_0\beta c\varepsilon_0)k\tanh(ky_m), \qquad (13.208)$$

The quantity J_z is the current of the equilibrium beam per unit length along x:

$$J_z = en_0\Delta y\beta c. \qquad (13.209)$$

The transverse velocity spread and the space-charge electric force inhibit filamentation, while the beam-generated magnetic force drives the instability. The beam is stable if $y_e = y_m$. Here, the repulsive electric forces are stronger than the magnetic force. Taking $y_e < y_m$ simulates the effect of transverse focusing foils — the proximity of the induced charge boundary reduces the electric field in the x direction. Filamentation raises the beam emittance. Ultimately, the velocity spread grows to a level sufficient for stability. Equation (13.208) implies that the root-mean-squared beam angular divergence for stability equals

$$\Delta\theta \cong \frac{(\langle\delta v_x^2\rangle)^{1/2}}{\beta c} = \left[\frac{(qJ_z)[\beta^2\tanh(ky_m) - \tanh(ky_e)]}{(2\varepsilon_0\gamma m_0\beta^3 c^3)k}\right]^{1/2}. \qquad (13.210)$$

We can use Eq. (13.208) to predict a beam propagation length for the growth of filaments:

$$L_g \sim \beta c/\alpha. \qquad (13.211)$$

To estimate L_g, assume that the foils almost completely cancel space-charge electric fields. This condition implies that $\tanh(ky_e) \cong 0$ in Eq. (13.208). Furthermore, the distance from the beam to the return current surface is much larger than $1/k$, or $\tanh(ky_m) \cong 1$. With these approximations and the assumption of small initial emittance, the filamentation propagation length is

$$L_g \geq (2\gamma m_0 \beta c^3 \varepsilon_0 / eJ_z k)^{1/2}. \tag{13.212}$$

As an example, consider a 2-kA, 500-kV sheet electron beam of width $\Delta y = 0.005\,\text{m}$ and length 0.1 m. The linear current density is $J = 2 \times 10^4\,\text{A/m}$. The fastest instability growth corresponds to the maximum value of k. Our model holds when $k < 1/\Delta y$. If we take $k = 2/\Delta y = 200\,\text{m}^{-1}$, Eq. (13.212) predicts a growth length of only $L_g = 0.034\,\text{m}$. The high growth rate is consistent with observations of strong filamentation of beams from relativistic electron guns with anode grids. The best approach to propagation in a foil transport array is to compress the electron beams to an emittance-dominated equilibrium before injection into the focusing system.

14

Longitudinal Instabilities

A longitudinal instability causes randomization of the axial velocity distribution of a beam. Energy from the directed axial motion couples to a longitudinal velocity spread. Longitudinal instabilities involve acceleration and deceleration of particles—they are driven by axial electric fields generated by beam perturbations. In this chapter, we shall concentrate on a few examples of longitudinal instabilities to illustrate the techniques of analysis.

A beam in field-free vacuum or in an infinitely long, uniform pipe is always stable. Here, we can transform to the beam rest frame without changing the nature of boundaries. In this frame, the beam consists of particles at rest with no free energy. The kinetic energy of the beam can couple to instabilities only if there is a reference in the stationary frame to facilitate energy transfer. For example, longitudinal instabilities may occur when a beam moves through a stationary plasma. Other references include pipes with resistive force components that depend on the absolute beam velocity and nonuniform vacuum chambers where the moving beam excites electromagnetic oscillations.

Section 14.1 reviews the two-stream instability for a beam in a plasma. We limit attention to infinitely wide electron beams—the approximation is useful for many applications where the unstable wavelengths are much smaller than the beam width. A two-stream instability involves coupling to plasma electrons. Beam perturbations drive growing plasma waves that transfer energy to the background electrons and increase the beam energy spread. Two-stream instabilities may also indirectly affect electron beam propagation in a plasma. Momentum coupling between drifting plasma electrons and stationary ions results in a resistivity for return current flow that is much higher than the collisional prediction (Section 12.6).

We shall study two other instabilities of finite-width beams in structures. For reference, Section 14.2 derives general expressions for axial electric fields in a

674

perturbed cylindrical beam. The results are applied in Section 14.3 to derive equations for the negative-mass instability. This process affects beams with strong space-charge fields in circular accelerators. The term *negative mass* refers to an unusual property of highly relativistic particles in circular machines. Particles with increased energy take longer to circulate around the machine because of their larger radius. An accelerating force reduces the average circulation or angular velocity as though the particle had a negative value of mass. For this reversed response, the repulsive axial space-charge forces in a region of enhanced beam density cause growth of the perturbation.

Section 14.4 describes the longitudinal resistive wall instability. The process may occur in storage rings and has application in the resistive wall microwave amplifier. Frictional forces from the wall amplify longitudinal slow waves. Wave growth or damping rates are sensitive to details of the beam axial velocity distribution. Here, we cannot use simple moment equation models. The resistive wall instability provides an opportunity to follow an analysis based on a direct solution of the Vlasov equation.

14.1. TWO-STREAM INSTABILITY

Two-stream instabilities may occur when particle beams with different velocities flow through each other. Free energy is available to cause axial bunching of the beam particles—the space-charge fields of the bunched beams convert part of the directed kinetic energy to an axial velocity dispersion. A mild two-stream instability causes longitudinal emittance growth—a strong instability leads to destruction of the beam. Two-stream instabilities are a major concern when beams propagate through plasmas. Examples include long-distance propagation of high-energy electron beams, transport of low-energy ion beams emerging from high-current injectors, and the propagation of heavy ion beams in an inertial fusion reaction chamber. The two-stream instability may be desirable in applications such as plasma heating by intense electron beams.

In this section, we shall use simple models to understand the mechanism of the two-stream instability. We take beams and plasmas with infinite extent and apply perturbations in a single direction. We treat the plasma as a collection of cold ions and electrons and model beams with a small velocity spread. For these conditions, we can apply moment equation models to find expressions for instability growth rates. A Vlasov equation model is necessary to include the effects of plasma temperature and beam velocity dispersion—we will develop such a model when we study the longitudinal resistive wall instability in Section 14.4.

As an introduction, consider the simplest geometry for a two-stream instability: two equal and opposite electron streams flow through an immobile ion background. Figure 14.1 illustrates the geometry and velocity distribution. Although symmetric streams seldom appear in applications, we will study this model because it leads to perturbations that are stationary in the ion rest frame.

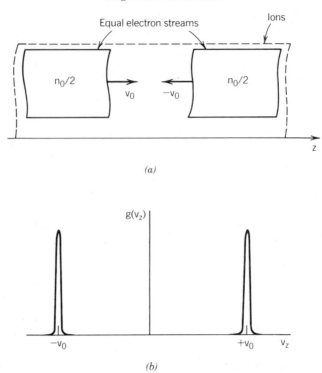

Figure 14.1 Instability of symmetric, infinite-width streams of electrons in an ion background. (*a*) Geometry of calculation. (*b*) Initial axial velocity distributions of the streams.

This makes it easy to explain the instability mechanism. The ions have a uniform density n_0. Initially, the two streams of nonrelativistic electrons have velocities $v_z = + v_0\mathbf{z}$ and $v_z = - v_0\mathbf{z}$. The density of both electron streams is $n_0/2$. The total electron and ion densities are equal—there is no equilibrium electric field. We shall investigate perturbations that grow in time with a uniform rate at all positions. Often, beams in experiments enter a region at a given point over an extended period of time. Here, perturbations grow both in time and with distance from the injection point. The analysis to describe the convective growth of the two-stream instability is more involved, although the mechanism is the same.

To understand the principle of the two-stream instability, imagine a small local enhancement of electron density in the beams of Fig. 14.1. The excess of negative charge could result from a local depression of the ion density. The electrostatic potential in the region is negative. Electrons would be driven from the region of negative potential to restore charge balance. The response is different for electrons with a directed velocity. They can travel across the negative potential region. By conservation of energy, the electrons move more slowly when the potential is negative. If the electron motion is one dimensional, the condition of flux conservation (Section 2.10) implies that the density of the flowing electrons is higher in the negative potential region. Electrons that follow sense an enhanced negative potential and move through the region even more slowly. For certain conditions, the amplitude of the potential grows. Ultimately, the space-charge electric field may reach a level that severely distorts the electron velocity distribution.

We shall describe electron bunching with moment equations. Each stream satisfies equations of continuity and momentum conservation. By symmetry, there is no net current at any position, so there are no beam-generated magnetic fields. We assume that the electron current density is much higher than the displacement current density that results from the growing electric field. With this condition, we can calculate the electric field in the static approximation. The one-dimensional Poisson equation relates the electric field to the perturbed particle density. Let N^+ represent the density of electrons moving in $+z$ direction and V^+ denote their directed velocity. The equation of continuity for the electrons is

$$\frac{\partial N^+}{\partial t} + \frac{\partial}{\partial z}(N^+ V^+) = 0. \tag{14.1}$$

The momentum equation is

$$\frac{\partial V^+}{\partial t} + V^+ \frac{\partial V^+}{\partial z} = \frac{-eE_z}{m_e}. \tag{14.2}$$

Similar equations hold for the density and velocity of electrons traveling in the opposite direction, N^- and V^-. With the Poisson equation

$$\frac{\partial E_z}{\partial z} = \frac{e}{\varepsilon_0}[n_{i0} - N^+ - N^-], \tag{14.3}$$

we have a set of five equations in five unknowns.

Moment equations hold only during the initial stages of the instability before randomization of the axial velocity. Therefore, we can simplify the mathematics by recasting equations in terms of small perturbations about the equilibrium. We define the following perturbed quantities:

$$\begin{aligned} n^+ &= N^+ - n_{i0}/2, & v^+ &= V^+ - v_0, \\ n^- &= N^- - n_{i0}/2, & v^- &= V^- - v_0. \end{aligned} \tag{14.4}$$

The quantity E_z is a small quantity because there is no equilibrium electric field. To reduce the equations, we substitute Eqs. (14.4) into Eqs. (14.1)–(14.3) and eliminate terms that involve the product of variational quantities. The result is

$$\frac{\partial n^+}{\partial t} = -v_0 \frac{\partial n^+}{\partial z} - \left(\frac{n_{i0}}{2}\right)\frac{\partial v^+}{\partial z}, \tag{14.5}$$

$$\frac{\partial v^+}{\partial t} = -v_0 \frac{\partial v^+}{\partial z} - \frac{eE_z}{m_e}, \tag{14.6}$$

$$\frac{\partial E_z}{\partial z} = -\frac{e}{\varepsilon_0}[n^+ + n^-]. \tag{14.7}$$

Equations (14.5)–(14.7) are called the *linearized moment equations*—all terms are linearly proportional to perturbation quantities or their derivatives.

We can convert the differential equations to algebraic equations by taking harmonic variations in space and time. Assume all first-order quantities are proportional to $\exp[j(kz - \omega t)]$. With the restriction that perturbations grow in time throughout space, k is a real number while ω may have both real and imaginary parts. The moment equations take the form

$$-j\omega n^+ = -jkv_0 n^+ - j(kn_{i0}/2)n^+, \qquad -j\omega v^+ = -jkv_0 v^+ - eE_z/m_e,$$

$$-j\omega n^- = +jkv_0 n^- - j(kn_{i0}/2)n^-, \qquad -j\omega v^- = +jkv_0 v^- - eE_z/m_e, \tag{14.8}$$

$$jkE_z = -(e/\varepsilon_0)[n^+ + n^-].$$

We can eliminate variables from Eqs. (14.8) to derive a single equation with all terms linearly proportional to one variable such as F_z. Canceling E_z gives an equation that relates the frequency, ω, to the wavenumber, k:

$$1 - \frac{\omega_{pe}^2/2}{(\omega - kv_0)^2} - \frac{\omega_{pe}^2/2}{(\omega + kv_0)^2} = 0. \tag{14.9}$$

Equation (14.9) incorporates the electron plasma frequency:

$$\omega_{pe} = (e^2 n_0/\varepsilon_0 m_e)^{1/2}. \tag{14.10}$$

Rearranging Eq. (14.9) gives a quadratic equation in ω

$$\omega^2 = (\omega_{pe}^2/2)\{1 + 2(kv_0/\omega_{pe}) \pm [1 + 8(kv_0/\omega_{pe})^2]^{1/2}\}. \tag{14.11}$$

The solution with the plus sign in Eq. (14.11) always gives a real value of ω. On the other hand, the solution of ω with a minus sign may be imaginary. By inspection of Eq. (14.11), we can see that the right-hand side is negative if

$$kv_0 < \omega_{pe}. \tag{14.12}$$

When the condition of Eq. (14.12) holds, ω is purely imaginary—the magnitude of ω gives the growth rate for the two-stream instability. Figure 14.2 shows a plot of $j\omega$ as a function of kv_0/ω_{pe}. The maximum growth rate of $j\omega = \omega_{pe}/2\sqrt{2}$ occurs at a wavenumber of $(kv_0/\omega_{pe}) \cong \sqrt{3/8}$.

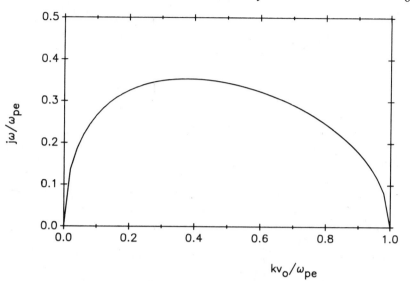

Figure 14.2 Two-stream instability growth rate—symmetric, infinite-width streams of electrons in an ion background.

The physical meaning of Eq. (14.12) is clear if we square both sides and multiply by $m_e/2$:

$$m_e v_0^2/2 < e^2 n_{i0}(\lambda/2\pi)^2/\varepsilon_0. \tag{14.13}$$

The left-hand side of Eq. (14.13) is the kinetic energy of electrons in equilibrium. The right-hand side equals an electrostatic potential energy, $e\phi_{max}$. The quantity ϕ_{max} is the electrostatic potential that would occur if there were no electrons in a region of width $(2)^{1/2}/k$, roughly equal to one-fourth the perturbation wavelength:

$$\phi_{max} \sim e n_{i0}(\lambda/2\pi)^2/\varepsilon_0. \tag{14.14}$$

Electron bunching depends on how much the space-charge electric fields decelerate electrons. Equation (14.14) states that an instability grows for a perturbation wavelength where the resulting space-charge imbalance significantly changes the electron velocity. The growth time for the instability is very short—for most laboratory plasmas, the plasma frequency is high.

We must turn to computer simulations to get information about the advanced stages and saturation of the two-stream instability. Figure 14.3 shows results from a one-dimensional simulation of the equal stream problem. The program applies periodic boundary conditions in space over a distance z_0. The run of Fig. 14.3 uses the parameters $n_0 = 1 \times 10^7 \, \text{m}^{-3}$ and $v_0 = 4.2 \times 10^7 \, \text{m/sec}$ (5-keV electrons). The plasma frequency is $\omega_{pe} = 1.8 \times 10^{10} \, \text{sec}^{-1}$. From the previous analysis, we expect a maximum growth rate of about $Im(\omega) \sim 6.4 \times 10^9 \, \text{sec}^{-1}$, or a growth time of 0.08 nsec. The perturbation wavelength for maximum growth is 0.025 m. With a computational region of width $z_0 = 0.059$ m, we expect to observe about two wavelengths of the perturbation. To start the run, the program randomly distributes 2000 computation particles over a 30-cell mesh. The instability grows from the initial statistical noise.

Figure 14.3 shows the electron phase-space distribution at five times after the simulation starts: 0.00, 0.56, 1.12, 1.68, and 2.24 nsec. The distribution at 0.56 nsec is a good representation of the linear instability regime. The growth of electron clumps results in harmonic distortions of the velocity distribution at the expected wavelength. At 1.12 nsec, the instability has progressed past the linear regime and the distortion of the velocity distribution is highly anharmonic. At 1.68 nsec, the electrostatic potential is strong enough to turn electrons. Positive- and negative-going streams mix in phase space. At 2.24 nsec the electron distribution is almost completely thermalized. With saturation of the instability, the electric field amplitude decreases. At late time, the streams are stable, although with a considerable loss of beam quality.

Next, we consider an example that better represents experiments, propagation of a low-density electron beam through a plasma. Again, we limit perturbations to a single dimension. We shall use nonrelativistic equations to describe motion of the beam electrons and add relativistic corrections later. There are no applied or beam-generated magnetic fields. In equilibrium the beam has density n_{b0} and velocity v_0. The stationary cold plasma has electron density n_{e0} and a density of immobile ions, n_{i0}. In equilibrium, there is no electric field, or $n_{i0} = n_{e0} + n_{b0}$. Finally, the beam density is low, $n_{b0} \ll n_{e0}$.

We shall first derive mathematical results from the linearized moment equations and then discuss their physical implications. There are continuity and momentum equations for both the beam and plasma electrons. If perturbed quantities have variation $\exp[j(kz - \omega t)]$, solution of the linearized moment equations gives

$$1 - \frac{\omega_{pe}^2}{\omega^2} - \frac{\omega_{pb}^2}{(\omega - kv_0)^2} = 0, \qquad (14.15)$$

where $\omega_{pb} = (e^2 n_{b0}/\varepsilon_0 m_e)^{1/2}$ and $\omega_{pe} = (e^2 n_{e0}/\varepsilon_0 m_e)^{1/2}$.

The two-stream instability results in a transfer of the stored kinetic energy of the beam to the plasma electrons. In the limit of low beam density, Section 12.2 implies that excitation of the plasma creates a wave with frequency near the

---

Figure 14.3 Particle-in-cell simulation of the nonlinear evolution of the two-stream instability. Symmetric, infinite-width streams of electrons in an ion background. $n_{i0} = 10^{17} \, \text{m}^{-3}$. $v_0 = 4.2 \times 10^7$ m/sec. Length of region shown: 0.059 m; 2000 particles, 30 cell mesh. (a) 0.56 nsec. (b) 1.12 nsec. (c) 1.68 nsec. (d) 2.24 nsec.

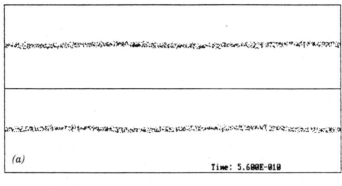

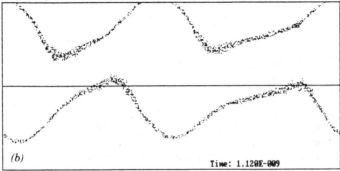

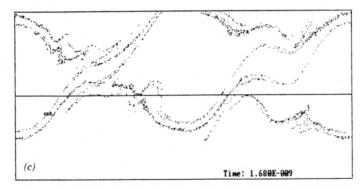

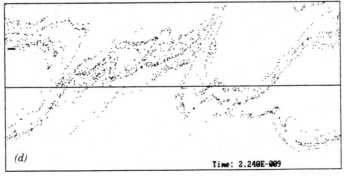

plasma frequency:

$$\omega \cong \omega_{pe}. \tag{14.16}$$

The beam couples energy to a wave strongly only if the wave phase velocity is close to v_0, or $\omega/k \cong v_0$. We shall look for solutions where

$$\omega/k = v_0 + \delta/k. \tag{14.17}$$

The quantity δ is a small frequency difference, $|\delta| \ll \omega_{pe}$. Substituting Eqs. (14.16) and (14.17) into Eq. (14.15) gives

$$1 - \omega_{pe}^2 \left[\frac{1}{(kv_0)^2} - \frac{2\delta}{(kv_0)^3} \right] - \frac{\omega_{pb}^2}{\delta^2} \cong 0. \tag{14.18}$$

We shall solve for δ as a function of the wavenumber k. A detailed analysis of Eq. (14.18) shows that the imaginary component of δ has its maximum value for wavenumber $k = \omega_{pe}/v_0$. With this condition, the equation has the simple form

$$\delta^3 = \omega_{pb}^2 \omega_{pe}/2. \tag{14.19}$$

Equation (14.19) is a cubic equation with three solutions:

$$\delta = \left[\frac{\omega_{pb}^2 \omega_{pe}}{2} \right]^{1/3}, \tag{14.20}$$

$$\delta = \left[\frac{\omega_{pb}^2 \omega_{pe}}{2} \right]^{1/3} \left[-\frac{1}{2} + j \frac{\sqrt{3}}{2} \right], \tag{14.21}$$

and

$$\delta = \left[\frac{\omega_{pb}^2 \omega_{pe}}{2} \right]^{1/3} \left[-\frac{1}{2} - j \frac{\sqrt{3}}{2} \right]. \tag{14.22}$$

The first solution has only a real part—it represents a stable oscillation. The second has negative imaginary part, giving a damped oscillation.

The third solution corresponds to the two-stream instability. Note that the real part of the frequency shift is negative. This means that the frequency is less than ω_{pe} and that the phase velocity of the disturbance, $[\omega_{pe} + \mathrm{Re}(\delta)]/k$, is less than the beam velocity. The imaginary part of the frequency shift is the instability growth rate:

$$\mathrm{Im}(\omega) = \frac{\sqrt{3}}{2} \left(\frac{\omega_{pb}^2 \omega_{pe}}{2} \right)^{1/3}. \tag{14.23}$$

The growth rate is proportional to $n_{b0}^{1/3}$. The rate has a high value, even for low beam density. For example, $\text{Im}(\omega) = 0.1\omega_{pe}$, when $n_{b0}/n_{e0} = 0.001$.

The mechanism for a two-stream instability of a beam in a dense plasma is easy to understand if we review the dielectric properties of a plasma. Remember, the relative dielectric constant of a linear and homogeneous medium, $\varepsilon/\varepsilon_0$, has the following definition. For an applied electric field $\mathbf{E}_a$, the total electric field within the medium is

$$\mathbf{E} = \mathbf{E}_a/(\varepsilon/\varepsilon_0). \tag{14.24}$$

In a plasma, the total electric field is the sum of external fields and the contribution from the plasma, $\mathbf{E} = \mathbf{E}_a + \mathbf{E}_p$. Suppose we apply a small z-directed electric field,

$$E_a \exp[j(kz - \omega t)], \tag{14.25}$$

to an infinite plasma with immobile ions. The linearized moment equations that describe the plasma response are

$$-j\omega n_e + jkn_{e0}v_e = 0, \tag{14.26}$$

$$-j\omega v_e = -eE, \tag{14.27}$$

$$jkE = jkE_a - en_e/\varepsilon_0. \tag{14.28}$$

Equation (14.28) is the Poisson equation with electric field contributions from the applied field and the plasma electrons. Combining Eqs. (14.26)–(14.28) gives the following relationship between the applied field and the field inside the plasma:

$$E = E_a/(1 - \omega_{pe}^2/\omega^2). \tag{14.29}$$

Comparing Eqs. (14.24) and (14.29), we see that the plasma has a frequency-dependent dielectric constant:

$$\varepsilon/\varepsilon_0 = 1 - \omega_{pe}^2/\omega^2. \tag{14.30}$$

At high frequency, $\omega > \omega_{pe}$, the total electric field points in the same direction as the applied field. In contrast to ordinary solids or liquids, the high-frequency dielectric constant of a plasma is less than unity. Plasmas exhibit very unusual behavior at low frequency, $\omega < \omega_{pe}$. The total field within the plasma points in the *opposite* direction from the applied field. This effect arises because of overshoot by the oscillating plasma electrons. It is the negative dielectric constant of a plasma that leads to the two-stream instability.

Consider a high-energy, low-density beam propagating through a plasma. Suppose that beam has a given density perturbation with wavenumber k. The variation of beam density creates an applied electric field in the plasma that oscillates at frequency $\omega \cong kv_0$. If

$$\omega = kv_0 < \omega_{pe}, \tag{14.31}$$

then the total electric field in the plasma is in the opposite direction from the electric field created by the beam. As a result, the electric force on a beam acts to enhance the bunching, leading to the growth of density perturbations. Equation (14.31) is the criterion for the growth of a two-stream instability of a low-density beam in a plasma.

Applications of the two-stream instability to plasma heating involve high-power-density relativistic beams. We have found that the growth of the two-stream instability depends on changes in the axial velocity of beam particles. Because a relativistic beam travels near the speed of light, variations of kinetic energy cause only small changes of axial velocity. Therefore, bunching takes place slowly for a relativistic beam. To describe such beams, we must correct the moment equations to account for this effect.

The beam continuity equation expresses conservation of flux, independent of the relationship between force and velocity. Therefore, the continuity equation does not change for a relativistic beam. We must modify the equation of momentum conservation. For the moment, we shall ignore changes in momentum density resulting from convection. The relativistic momentum equation is

$$\frac{dp_z}{dt} = \frac{d(\gamma m_e \beta c)}{dt} = eE_z. \tag{14.32}$$

Expanding the expression in parentheses gives

$$\frac{d(\gamma m_e \beta c)}{dt} = (m_e c)\left[\gamma\left(\frac{d\beta}{dt}\right) + \beta\left(\frac{d\gamma}{dt}\right)\right]. \tag{14.33}$$

Using the expression $\gamma = 1/(1 - \beta^2)^{1/2}$ we can rewrite the bracketed term in Eq. (14.33) as $[\gamma^3(d\beta/dt)]$. The modified momentum equation is

$$\frac{dv_z}{dt} = \frac{eE_z}{\gamma^3 m_e}. \tag{14.34}$$

For small perturbations, we can take γ in Eq. (14.34) as approximately constant. Then, the equation looks like a nonrelativistic equation with the rest mass replaced by the quantity $\gamma^3 m_e$. The adjusted mass for small perturbations is called the *longitudinal mass*. In the linear limit, we can apply Eq. (14.15) to a relativistic beam, taking the beam plasma frequency as

$$\omega_{pb} = (e^2 n_{b0}/\varepsilon_0 \gamma^3 m_e)^{1/2}. \tag{14.35}$$

As an example, consider a 1-kA, 5-MeV electron beam of radius 0.025 m that travels through a plasma of density $n_{e0} = 10^{20} \, \text{m}^{-3}$. The beam has $\gamma = 10.8$, $\beta = 0.996$ and $n_{b0} = 1.1 \times 10^{16} \, \text{m}^{-3}$. The plasma frequency is $\omega_{pe} = 5.6 \times 10^{11} \, \text{sec}^{-1}$, while the beam plasma frequency with corrected longitudinal mass is $\omega_{pb} = 1.7 \times 10^{8} \, \text{sec}^{-1}$. The wavelength corresponding to the fastest growth rate is $\lambda = 2\pi\beta c/\omega_{pe} = 1.1 \times 10^{-2} \, \text{m}$. Note that the short wavelength justifies the use of the infinite beam model. The growth rate for the two-stream instability is $\text{Im}(\omega) = 10^{8} \, \text{sec}^{-1}$. We expect a significant beam disturbance after a propagation distance of $\sim 1.6 \, \text{m}$. Experimentally observed instabilities that disorder highly relativistic electron beams usually have a growth more rapid than the prediction of our simple model. Because of the high longitudinal mass, waves that propagate at an angle with respect to the beam direction have a higher growth rate. The prevalent instability of a relativistic beam involves an electromagnetic wave that causes both longitudinal and transverse bunching. A general treatment

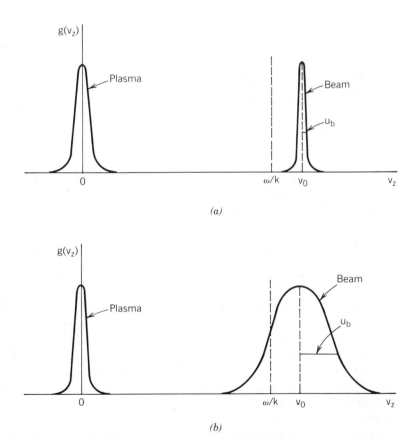

(a)

(b)

Figure 14.4 Velocity distributions for a low-density electron beam in a plasma. (*a*) Velocity distributions for the plasma and a cold beam. (*b*) Velocity distributions for the plasma and a hot beam—the wave phase velocity is at the point of the maximum beam distribution slope.

gives the two-stream and filamentation instabilities (Section 13.9) as limiting cases.

Mechanisms that reduce the growth rate of the two-stream instability are important for applications. Stabilizing effects can help beam propagation or hinder plasma heating. One quantity that affects the growth rate is the longitudinal temperature of the beam. For two equal streams of electrons, we found that the velocity spread for stability is large, comparable to the directed velocity. On the other hand, the stabilizing velocity spread for a low-density beam in a plasma can be much smaller than v_0. Figure 14.4 illustrates a velocity-space map for a low-density beam and plasma electrons. The beam has a nonzero velocity spread, u_b.

For the two-stream instability, the growing plasma wave has a phase velocity less than the beam velocity, $\omega/k < v_0$. If the velocity difference between the wave and beam is much larger than u_b (Fig. 14.4a), then results that we already derived for a cold beam are approximately correct. The validity criterion for the cold beam expressions is

$$|\omega/k - v_0| \gg u_b. \tag{14.36}$$

We can explain the result of Eq. (14.36) with the following argument. If the growth time of the instability is shorter than the time for particles with a velocity spread to disperse over distance $1/k$, the cold-beam results are adequate. The mathematical expression of this condition is

$$1/\mathrm{Im}(\omega) \ll 1/ku_b. \tag{14.37}$$

Equation 14.37 is equivalent to Eq. (14.36) if we remember from Eq. (14.22) that $\mathrm{Im}(\omega) \sim |\omega - kv_0|$. Substituting for the growth rate from Eq. (14.23), the cold beam expressions hold when

$$\frac{u_b}{v_0} \ll \left[\frac{n_{b0}}{n_{e0}}\right]^{1/3}. \tag{14.38}$$

A high beam velocity spread reduces the growth rate of the two-stream instability. When $u_b/v_0 \gg (n_{b0}/n_{e0})^{1/3}$, only a fraction of the beam electrons with velocity close to ω/k transfer energy to the wave. Beam electrons that move slower than ω/k extract energy from the wave, while fast moving particles contribute to wave growth. As a result, the phase velocity of a growing wave shifts to a location in velocity space that corresponds to the maximum positive slope of the beam distribution function (Fig. 14.4b). A detailed calculation of the hot-beam growth rate involves plasma kinetic theory beyond the scope of this section. Instead, we shall quote the main results of the theory. For a beam with a displaced Maxwell distribution, the wavenumber for maximum growth is

$$k = \omega(v_0 - u_b). \tag{14.39}$$

The instability growth rate for a hot beam is

$$\text{Im}(\omega) \cong \left[\frac{\sqrt{3}}{2} \left(\frac{\omega_{pb}\omega_{pe}}{2} \right)^{1/3} \right] \left[\left(\frac{n_{b0}}{n_{c0}} \right) \frac{v_0}{u_b} \right]^2. \tag{14.40}$$

In Eq. (14.40), the first bracketed term is the cold-beam growth rate, while the second term is a beam temperature correction. The growth rate decreases with beam temperature as $\text{Im}(\omega) \sim 1/u_b^2$.

A nonzero beam temperature reduces the growth rate of the two-stream instability, but does not provide complete stability in a lossless plasma. The instability may be inhibited in a collisional plasma if the momentum exchange rate for plasma electrons, v_c, satisfies the condition

$$\frac{v_c}{\omega_{pe}} \geq \left(\frac{\pi}{2} \right)^{1/2} \frac{\omega_{pb}^2}{\omega_{pe}^2} \frac{v_0^2}{u_b^2}. \tag{14.41}$$

As we add more effects, the descriptions of the interactions between a beam and a plasma rapidly become more complex. For example, theory shows that the addition of a collisional term to the plasma dielectric response can reduce the two-stream instability growth rate, even for a cold beam. On the other hand, a resistive plasma can support growing slow waves. Here, the beam may be unstable to the distributed equivalent of the resistive wall instability (Section 14.4). We can also add effects of the plasma electron temperature. Although the plasma temperature has little effect on the growth rate of relativistic electron beam instabilities, it is important in calculating two-stream interactions of ions passing through a neutralizing plasma. The velocity of the high-energy ions may fall within the velocity spread of the plasma electrons. The analysis is considerably more involved than the moment equation models we have studied. An accurate description must include interactions between beam and plasma ions and Landau damping of unstable waves by the plasma electrons. Additional complications arise if we add effects of finite beam width and applied magnetic fields.

14.2. BEAM-GENERATED AXIAL ELECTRIC FIELDS

In most accelerators, a narrow beam propagates in a metal vacuum chamber. In the following three sections, we shall study longitudinal instabilities of beams in surrounding structures. To describe longitudinal dynamics, we must find the axial electric field of finite-width beams in the presence of induced wall charges. In this section, we shall derive expressions for the axial electric field of narrow beams in uniform cylindrical pipes. Figure 14.5 shows the geometry of the models. A beam of radius r_0 moves along the axis of a metal cylinder with radius r_w. We shall limit the treatment to beam-density variations in the axial direction, neglecting

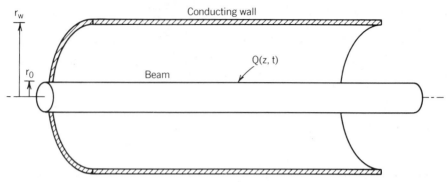

Figure 14.5 Narrow cylindrical beam in a pipe. The beam line charge varies in space and time, $Q(z, t)$.

transverse beam displacements or changes in width. The magnetic field of a beam with cylindrical symmetry equals zero on the axis; therefore, changes in the axial velocity of particles result solely from axial electric fields.

In equilibrium, a beam with current I_0 and average axial velocity v_0 has a uniform line charge density:

$$Q_0 = I_0/v_0 \quad \text{(coulombs/m)}. \tag{14.42}$$

The uniform density does not contribute to axial electric fields. Instabilities cause axial variations of the line charge. In turn, the density perturbations lead to axial electric field components that can further modify particle dynamics. In the following discussions, we shall concentrate on harmonic components of line charge density with the form

$$Q_k(z, t) = Q_k \exp\left[j(kz - \omega t)\right]. \tag{14.43}$$

We can describe any function $Q(z)$ by a Fourier synthesis of harmonic components. Equation (14.43) describes a charge-density variation in the stationary frame of reference. In this frame, the disturbance has wavelength $\lambda = 2\pi/k$ and moves at a phase velocity of ω/k.

We shall study two analytic calculations that give field expressions useful for stability analyses. The first model describes a narrow nonrelativistic beam. Here, we can apply an electrostatic model because of the low beam velocity. The results have application to bunching instabilities of heavy ions in linear induction accelerators (Section 14.5). The line charge density of bunched beams often has a broad range of harmonic components. For this reason, we shall derive field expressions that hold for both short wavelengths ($\lambda \leqslant r_w$) and long wavelengths ($\lambda \gg r_w$). The second model is appropriate for relativistic beams. We shall calculate the complete electromagnetic fields, including the effect of wall resistance. We shall apply the results to the negative mass instability (Section 14.3) and the resistive wall instability (Section 14.4). The model treats only long-wavelength perturbations ($\lambda \gg r_w$).

First, we shall derive the axial electric fields of a perturbed nonrelativistic beam in a perfectly conducting pipe. The beam is narrow,

$$r_0 \ll r_w, \tag{14.44}$$

with a velocity much smaller than the speed of light:

$$v_0 \ll c. \tag{14.45}$$

We shall see that the phase velocity of the charge perturbation is also small, $\omega/k \ll c$. The condition of Eq. (14.45) justifies the neglect of beam-generated magnetic forces. Also, the charge density and axial electric fields measured in the stationary frame are almost equal to those in the rest frame of the perturbation. As a result, we can carry out an electrostatic field calculation in the perturbation rest frame and then use the axial electric field values in the stationary frame.

In the perturbation rest frame moving at velocity ω/k, the line charge density equals

$$Q_k(z) = Q_k \cos{(kz)}. \tag{14.46}$$

To find the axial electric field resulting from the line charge of Eq. (14.46), we shall solve the Poisson equation to find $\phi_k(r, z)$ and calculate $\partial \phi_k / \partial z$ near the axis. The effects of charge in the vacuum chamber wall enter through the boundary condition:

$$\phi_k(r_w, z) = 0. \tag{14.47}$$

For a narrow beam, we apply the condition

$$kr_0 \ll 1. \tag{14.48}$$

When the limit of Eq. (14.48) holds, the radial electric field on the beam envelope is close to the value for an infinite-length beam:

$$E_{rk}(r_0, z) \cong Q_k(z)/2\pi\varepsilon_0 r_0. \tag{14.49}$$

To simplify the calculation, we compute the electrostatic potential only in he vacuum region between r_0 and r_w. The effect of the beam space charge enters through the boundary condition near the axis:

$$\frac{\partial \phi_k(r_0, z)}{\partial r} \cong -\frac{Q_k \cos{(kz)}}{2\pi\varepsilon_0 r_0} \qquad (r_0 \to 0). \tag{14.50}$$

Using the method of separation of variables, we seek a solution to the cylindrical Poisson equation with the form

$$\phi_k(r, z) = R_k(r) \cos(kz). \tag{14.51}$$

Substitution in the Poisson equation leads to the following expression for the radial function $R_k(r)$:

$$\frac{1}{r} \frac{d}{dr} r \frac{dR_k(r)}{dr} - k^2 R_k(r) = 0. \tag{14.52}$$

Equation (14.52) has the general solution

$$R_k(r) = A_k K_0(kr) + B_k I_0(kr), \tag{14.53}$$

where K_0 and I_0 are zero-order modified Bessel functions.[1] Near the axis, $kr \ll 1$, the functions have the asymptotic form

$$K_0(kr) \cong -\ln(kr/2) - 0.5772, \qquad I_0(kr) \cong 1. \tag{14.54}$$

The boundary conditions of Eqs. (14.47) and (14.50) give values for the constants A_k and B_k. We can find the derivatives of the potential near the axis from the asymptotic forms of Eq. (14.54). The resulting expression for electrostatic potential is

$$\phi_k(r, z) = \frac{Q_k}{2\pi\varepsilon_0} \left[K_0(kr) - \frac{K_0(kr_w)I_0(kr)}{I_0(kr_w)} \right] \cos(kz) \tag{14.55}$$

In the narrow-beam limit, the radial electric field changes considerably over the beam cross section, while the axial field is almost constant. Therefore, the average axial electric field within the beam is almost equal to the derivative of the potential at the envelope:

$$E_z(r_0, z) \cong -\frac{\partial\phi(r_0, z)}{\partial z}. \tag{14.56}$$

Inserting Eq. (14.55) into Eq. (14.56) gives the expression

$$E_{zk}(r_0, z) \cong \frac{kQ_k}{2\pi\varepsilon_0} \left[K_0(kr_0) - \frac{K_0(kr_w)I_0(kr_0)}{I_0(kr_w)} \right] \sin(kz). \tag{14.57}$$

We can check the validity of the model by writing Eqs. (14.55) and (14.57) in the long-wavelength limit. Using the asymptotic forms for the modified Bessel

[1] See, for instance, M. Abramowitz and I. A. Stegun, Eds., *Handbook of Mathematical Functions*, Dover, New York, 1970, Chapter 9.

functions, we find

$$\phi_k(r_0, z) \cong (Q_k/2\pi\varepsilon_0) \ln(r_w/r_0) \cos(kz) \tag{14.58}$$

and

$$E_{zk}(r_0, z) \cong (Q_k/2\pi\varepsilon_0) \ln(r_w/r_0) k \sin(kz). \tag{14.59}$$

when $kr \ll 1$. Comparison of Eq. (14.58) with Eq. (5.68) shows that $\phi_k(r_0, z)$ approaches the value of potential on the edge of an infinite-length beam with line charge density $Q_k \cos(kz)$. The electric field is simply the axial derivative of Eq. (14.58). We can write Eq. (14.59) in a useful form that describes the axial electric field for arbitrary charge perturbation, $Q(z)$, in the rest frame:

$$E_z(r_0, z) \cong -\frac{dQ}{dz} \left[\frac{\ln(r_w/r_0)}{2\pi\varepsilon_0} \right]. \tag{14.60}$$

Equation (14.60) is valid if all components in the Fourier decomposition of $Q(z)$ satisfy the condition $kr_w \ll 1$.

Figure 14.6 shows the ratio of $\phi_k(r_0, z)$ and $E_k(r_0, z)$ for the complete model [Eqs. (14.55) and (14.57)] divided by the values for an infinite-length beam [Eqs. (14.58) and (14.59)]. The figure plots the quantity

$$\left[K_0(kr_0) - \frac{K_0(kr_w)I_0(kr_0)}{I_0(kr)} \right] \left[\frac{1}{\ln(r_w/r_0)} \right] \tag{14.61}$$

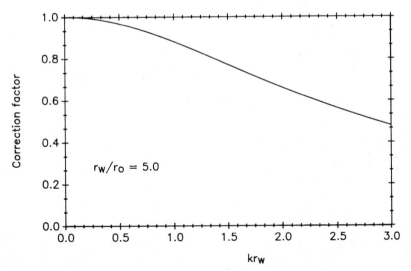

Figure 14.6 Deviation of the electrostatic potential and axial electric field of a beam with a harmonic line charge perturbation from the long-wavelength predictions.

as a function of kr_w for choice $r_w/r_0 = 5$. The error in the infinite-beam model exceeds 10% when $\lambda \leqslant 7r_w$.

As an applications example, we shall use Eq. (14.60) to find the properties of axial space-charge waves on a narrow beam. These waves are similar to the plasma waves discussed in Sections 12.2 and 14.1. The main difference is that the flow of particles extends over a finite cross section. We shall calculate the wave properties in the beam rest frame and then make a transformation to the stationary frame. We assume that perturbed quantities are small and use linearized equations. In the rest frame, the charge continuity equation is

$$\frac{\partial Q}{\partial t} = -\frac{I_0}{v_0}\frac{\partial v}{\partial z}. \tag{14.62}$$

The quantities $Q(z,t)$ and $v(z,t)$ are small changes in the line charge density and average axial velocity. Following Eq. (14.42), I_0/v_0 is the equilibrium line charge density. For long-wavelength perturbations, Eq. (14.60) leads to the following momentum equation for a cold beam:

$$\frac{\partial v}{\partial t} = -\frac{q}{m_0}\left[\frac{\ln(r_w/r_0)}{2\pi\varepsilon_0}\right]\frac{\partial Q}{\partial z}. \tag{14.63}$$

The quantities q and m_0 are the charge and rest mass of beam particles.

With perturbed quantities that vary as $Q, v \sim \exp[j(kz \pm \omega t]$, Eqs. (14.62) and (14.63) give the dispersion relationship

$$\omega/k = \pm v_0\sqrt{K\ln(r_w/r_0)}. \tag{14.64}$$

The quantity K is the generalized perveance of the nonrelativistic beam,

$$K = qI_0/2\pi\varepsilon_0 m_0 v_0^3. \tag{14.65}$$

Equation (14.64) shows that the phase velocity of a wave with $\lambda \gg r_w$ is independent of the wavelength. In this respect, the waves differ from space-charge waves in an infinite-width beam (Section 12.2). Figure 14.7 illustrates the nature of axial space-charge waves on a filamentary beam in a pipe for arbitrary values of λ. The plot shows the dimensionless frequency, $\omega r_0/v_0 K^{1/2}$, as a function of kr_0. A small value of kr_0 means that the wavelength is much larger than the beam radius. Figure 14.7 shows that waves with $\lambda \gg r_0$ have a frequency-independent phase velocity—following the dispersion relationship of Eq. (14.64). For $kr_0 \geqslant 1$, the perturbation wavelength is comparable to or less than the beam radius. In this limit, the effect of induced wall charge is small and the waves behave like those in an infinite-width beam. The angular frequency approaches the constant value ω_b, the beam plasma frequency of Section 12.3.

The Doppler-shifted phase velocity of long-wavelength space-charge waves in

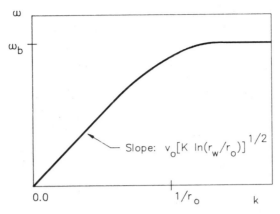

Figure 14.7 Dispersion relationship for axial space-charge waves on a beam.

the accelerator frame is

$$\omega/k = v_0[1 \pm \sqrt{K \ln (r_w/r_0)}\,]. \tag{14.66}$$

Equation (14.66) shows that there are fast and slow space-charge waves. For most beams, the generalized perveance is much smaller than unity, $K \ll 1$. Therefore, the phase velocities of the waves are close to the beam velocity. The slow wave is a negative energy wave—the longitudinal energy of the beam decreases as wave amplitude increases. Therefore, dissipative forces cause growth of the slow-wave amplitude. Section 14.4 describes this effect.

To complete the section, we will derive expressions for the axial electric field of relativistic beams. The approach is different from the preceding electrostatic calculation. We must include the full set of Maxwell equations to represent the effects of time variation of the beam magnetic field. To prepare for the treatment of resistive wall interactions in Section 14.4, we include a low resistivity in the vacuum chamber wall.

Again, consider a narrow beam with equilibrium line charge density $I_0/\beta c$. In the stationary frame, the perturbed line charge density equals

$$Q_k(r, z, t) = Q_k \exp [j(kz - \omega t)]. \tag{14.67}$$

In the stationary frame, there are significant values of both the beam-generated electric and magnetic fields. The perturbed beam generates a toroidal magnetic field, B_θ and electric fields E_r and E_z. The magnetic field contributes an inductive component to the on-axis E_z.

We limit perturbations to long wavelength and use field expressions for an infinite-length beam. For relativistic beams, we must be careful when we define the long-wavelength limit. In the perturbation rest frame where there are only beam-generated electric fields, the infinite-beam approximation holds if the axial electric field is much smaller than the radial field. In other words, the perturbation wavelength in the rest frame is much larger than the wall radius,

$$\lambda' \gg r_w. \tag{14.68}$$

We apply a Lorentz transformation to express the condition of Eq. (14.68) in terms of stationary-frame quantities:

$$\lambda = \lambda'/\gamma \gg r_w/\gamma. \tag{14.69}$$

When the condition of Eq. (14.69) holds, the radial component of electric field outside the beam is

$$E_r(r, z, t) \cong (Q_k/2\pi\varepsilon_0 r) \exp[j(kz - \omega t)] \qquad (r > r_0). \tag{14.70}$$

The toroidal magnetic field results from the beam current and the displacement current of changing electric fields. The relevant Maxwell equation is

$$\nabla \times \mathbf{B} = \left(\frac{1}{c^2}\right)\frac{\partial \mathbf{E}}{\partial t} + \mu_0 \mathbf{j}. \tag{14.71}$$

The radial component of Eq. (14.71) is

$$\frac{-\partial B_\theta}{\partial z} = \left(\frac{1}{c^2}\right)\frac{\partial E_r}{\partial t}. \tag{14.72}$$

If all quantities vary as $\exp[j(kz - \omega t)]$, the toroidal magnetic field is related to the radial electric field by

$$B_\theta = \omega E_r/kc^2. \tag{14.73}$$

The boundary condition for the axial electric field at the vacuum chamber is $E_z = 0$ if the wall is a perfect conductor. With resistivity, the flow of return current creates an axial voltage drop along the wall. Because the wall excludes high-frequency magnetic fields, the surface current is proportional to $B_\theta(r_w, z, t)$. We can find the boundary condition on electric field at a resistive wall in most introductory texts on electromagnetism.[2] The relationship in MKS units for a wall with volume resistivity ρ is

$$E_z(r_w, z, t) = -(\omega\partial/2)(j - 1)B_\theta(r_w, z, t). \tag{14.74}$$

The quantity δ in Eq. (14.74) is the skin depth for penetration of magnetic fields into the conductor,

$$\delta = (2\rho/\mu_0\omega)^{1/2}. \tag{14.75}$$

---

[2] See, for instance, D. K. Cheng, *Field and Wave Electromagnetics*, Addison-Wesley, Reading, Mass, 1989, p. 369.

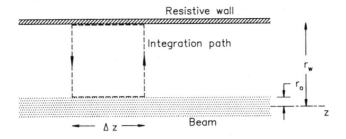

Figure 14.8 Path of the circuital integral to derive Eq. (14.76).

We can combine Eqs. (14.73) and (14.74) with Faraday's law to yield the axial electric field at the edge of a thin beam. We carry out a circuital integral of electric field around the path of Fig. 14.8 and equate the result to the rate of change of included toroidal magnetic field flux. The path consists of a radial segment from the wall to the beam at z, an axial segment of length dz along the envelope of the beam, a radial segment out to the wall, and then a return to the starting point along the wall. The mathematical expression of Faraday's law is

$$\int_{r_w}^{r_0} E_r(r', z)\,dr' + \int_{r_0}^{r_w} E_r(r', z + \Delta z)\,dr' + E_z(r_0, z)\Delta z + \frac{j-1}{4\pi\varepsilon_0}\frac{\delta\omega^2}{kc^2}\frac{Q_k}{r_w}\Delta z$$
$$= \frac{j\omega^2}{kc^2}\int_{r_0}^{r_w} E_r(r', z)\,dr'\,dz. \tag{14.76}$$

The third term on the left-hand side is the quantity we seek. The wall integral of the fourth term uses Eq. (14.74). Equation (14.73) was applied to rewrite B_θ in terms of E_r in the fourth term and in the expression on the right-hand side. If we divide all terms of Eq. (14.76) by dz and take the limit $dz \to 0$, the first two terms on the left-hand side reduce to

$$\int_{r_0}^{r_w} \frac{\partial E_r}{\partial z}\,dr\,\Delta z. \tag{14.77}$$

Performing the integrals, we find the following approximate expression for E_z on the beam envelope:

$$E_z(r_0, z) \cong \frac{Q_k}{2\pi\varepsilon_0}\exp\left[j(kz - \omega t)\right]\left[jk(1 - \beta_w^2)\ln\left(\frac{r_0}{r_w}\right) - (1 - j)\frac{\omega\delta\beta_w}{2cr_w}\right]. \tag{14.78}$$

The quantity β_w is the ratio of the phase velocity of the perturbation to the speed of light,

$$\beta_w = \omega/kc. \tag{14.79}$$

Most vacuum chambers for beam transport have low resistivity; therefore, the second term in brackets of Eq. (14.78) is much smaller than the first. Dropping the imaginary part of the second term, the electric field expression has the simple form

$$E_z(r_0, z) \cong -\frac{dQ(z)}{dz}\frac{1-\beta_w^2}{2\pi\varepsilon_0}\ln\left(\frac{r_w}{r_0}\right) - Q(z)\frac{\omega\delta\beta_w}{4\pi\varepsilon_0 c r_w}. \tag{14.80}$$

Equation (14.80) has a straightforward physical interpretation. The first term on the right-hand side is the electric field that arises from the combined effects of space charge and wall charge in a perfectly conducting chamber. The expression is the same as Eq. (14.60) except for the relativistic factor $1 - \beta_w^2 = 1/\gamma_w^2$. The origin of the relativistic correction is evident if we consider a coordinate transformation between the rest frame of the perturbation and the stationary frame. A perturbation with charge density $Q(z)$ and wavenumber k in the stationary frame has charge density $Q(z)/\gamma_w$ and wavenumber k/γ_w when observed in a frame moving at velocity $\beta_w c$. If we write the rest-frame expression for electric field in Eq. (14.60) in terms of the stationary-frame quantities, we must include a factor of $1/\gamma_w^2$. In transforming the fields back to the stationary frame, the magnitude of the axial electric field does not change [Eq. (1.51)].

The second term on the right-hand side of Eq. (14.80) represents the contribution of wall resistivity to E_z. We can derive this term directly from physical arguments. The condition that there is no beam-generated magnetic field outside the vacuum chamber means that the magnitude of the wall return current equals that of the beam current in the long-wavelength limit. The return current associated with a line charge perturbation is

$$I_i = -Q(z, t)\beta_w c. \tag{14.81}$$

We can write the resistive component of the axial field in Eq. (14.80) as

$$E_z(r_0, z) \cong I_i(\omega\delta/4\pi\varepsilon_0 c^2). \tag{14.82}$$

An alternative form for the quantity in parentheses on the right-hand side of Eq. (14.82) is

$$\omega\delta/4\pi\varepsilon_0 c^2 = \rho/2\pi r_w\delta. \tag{14.83}$$

We recognize that the right-hand side of Eq. (14.83) equals the resistance per unit axial length of the cylindrical chamber. Therefore, the resistive electric field component equals the wall current multiplied by the wall resistance per unit length.

14.3.   NEGATIVE MASS INSTABILITY

The negative mass instability is a major concern for high-current electron beams contained in recirculating accelerators. Figure 14.9 illustrates the physical mechanism of the instability. Electrons follow closed orbits defined by a bending magnetic field and focusing forces. The presence of the negative mass instability depends on the time for electrons to circulate around the machine, τ. The value of τ depends on the electron momentum, p. For nonrelativistic electrons, higher momentum corresponds to increased velocity and a lower value of τ. In contrast, the velocity of relativistic electrons varies little with momentum. For these particles, the change in rotation period with momentum results mainly from the change in the average orbit radius in the bending magnetic field. Relativistic electrons with higher momentum followed a longer path and therefore have a higher value of τ.

Suppose the beam density in a circular accelerator is higher at one azimuthal location, the perturbed space charge produces electric fields along the beam axis that accelerate electrons on the forward side of the bunch and decelerate them on the trailing side. The circulation time of relativistic electrons on the leading edge

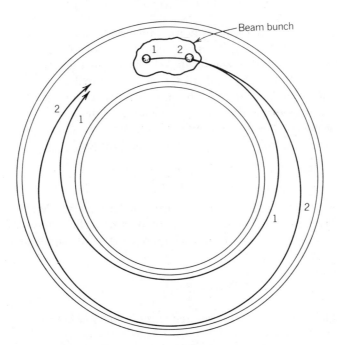

Figure 14.9   Mechanism of the negative mass instability.

increases; therefore, these electrons move backward in the beam. The space-charge fields cause an increase in the amplitude of the density perturbation as the beam circulates. We apply the term *negative mass* because the electrons appear to move in the direction opposite the force. The instability breaks the beam into azimuthal clumps, resulting in emittance growth or loss of particles.

To develop a mathematical model for the negative mass instability, we must review some properties of particle orbits in circular accelerators [CPA, Chapter 15]. Suppose the beam consists of relativistic particles with charge q, average energy $(\gamma - 1)m_0c^2$, average momentum p, and average velocity βc. The particles move in closed orbits around an accelerator or storage ring. For simplicity, we treat circular orbits with average radius R and neglect particle acceleration over the growth time of the instability. Momentum is related to the orbit radius by

$$R = \gamma m_0 \beta c / e B_0 = p/eB_0. \tag{14.84}$$

The quantity B_0 is the vertical bending field that defines the circular orbits.

A variation of momentum, δp, about p causes a change in the orbit radius, δR. To represent this change, we shall introduce the parameter γ_t, the *transition gamma*. It equals the square root of the ratio of the relative momentum change to the relative change in gyration radius:

$$\gamma_t^2 = \frac{\delta p/p}{\delta R/R}. \tag{14.85}$$

The physical meaning of the transition gamma will become apparent later in the derivation. The value of γ_t depends on the properties of the focusing system. For example, a uniform vertical magnetic field focuses particles in the horizontal direction. From Eqs. (14.84) and (14.85), we can see that $\gamma_t = 1$.

As another example, suppose the beam moves through a strong focusing system, such as quadrupole lens array. The forces of the lenses point toward the main beam axis. If δR is the horizontal distance from the axis, we can represent a general linear focusing force in the horizontal direction as

$$F_r = -\gamma m_0 \omega_r^2 \delta R, \tag{14.86}$$

where ω_r is the betatron oscillation frequency in the lens array. Making a balance between the force of Eq. (14.86), the force of the bending field, and the centrifugal force, we find the following expression for the change of average orbit radius with a change of momentum:

$$\delta R/R = (\delta p/p)(1 + \omega_r^2/\omega_0^2). \tag{14.87}$$

The quantity ω_0 is the circulation frequency in the accelerator,

$$\omega_0 = \beta c/R. \tag{14.88}$$

From Eq. (14.87), we can identify γ_t for a strong focusing system as

$$\gamma_t^2 = 1 + (\omega_r/\omega_0)^2 = 1 + v_r^2. \tag{14.89}$$

The quantity v_r is the number of betatron oscillations in the focusing system per revolution. The condition,

$$v_r > 1, \tag{14.90}$$

defines *strong focusing*. The opposite limit holds for *weak focusing*. We can carry out a similar derivation to show that the transition gamma for a betatron-type field [CPA] with field gradient, n, is

$$\gamma_t^2 = 1 - n. \tag{14.91}$$

To describe the negative mass instability, we must find an expression that gives the transit time around the accelerator as a function of the momentum. The circulation time equals

$$\tau = 2\pi R/\beta c. \tag{14.92}$$

The differential of Eq. (14.92) is

$$\delta\tau/\tau = \delta R/R - \delta\beta/\beta. \tag{14.93}$$

The momentum equals

$$p = \gamma\beta m_0 c. \tag{14.94}$$

The differential of Eq. (14.94) is

$$\delta p/p = \delta\gamma/\gamma + \delta\beta/\beta = \delta\beta/\beta/(1 - \beta^2) = \gamma^2(\delta\beta/\beta). \tag{14.95}$$

Combining Eqs. (14.85), (14.93), and (14.95), we find that

$$\frac{\delta\tau}{\tau} = \frac{\delta p}{p}\left[\frac{1}{\gamma_t^2} - \frac{1}{\gamma^2}\right]. \tag{14.96}$$

Equation (14.96) illustrates the physical meaning of γ_t. When $\gamma < \gamma_t$, the main result of an increased momentum is an increased velocity—the electron takes less time to complete a circuit. A positive value of δp gives a negative value of $\delta\tau$. Above transition, $\gamma > \gamma_t$, particles with higher momentum take longer to complete a revolution because their orbit radius is larger. Relativistic beams with kinetic energy above the transition energy, $(\gamma_t - 1)m_0 c^2$, are subject to the negative mass instability.

We can develop a moment equation model that describes the linear regime of the negative mass instability with some limiting assumptions:

1. A beam of radius r_0 propagates in a toroidal vacuum chamber with minor radius r_w. The beam is thin, $r_0 \ll r_w$ and the vacuum chamber has small curvature, $R \gg r_w$.
2. The wavelength of density clumps along the propagation direction is much larger than r_w, or $kr_w \ll 1$.
3. The beam displacement, δR, is much smaller than r_w.
4. The growth time for the instability is much longer than the circulation time. We can characterize the relative velocity of an axial segment by taking an average over many transits.

The model describes the line density and average velocity of the beam observed at a point in the accelerator frame of reference. In this frame, the beam has line density $N_0 \, \mathrm{m}^{-1}$ in equilibrium. We take small perturbations about this value:

$$n(z, t) = N(z, t) - N_0, \tag{14.97}$$

where $n \ll N_0$. The z axis lies along the direction of beam propagation. The velocity $V(z, t)$ refers to the relative motion of a segment within the beam averaged over many transits. It varies about the equilibrium value βc:

$$v(z, t) = V(z, t) - \beta c. \tag{14.98}$$

The linearized one-dimensional continuity equation is

$$\frac{\partial n}{\partial t} + \beta c \frac{\partial n}{\partial z} + N_0 \frac{\partial v}{\partial z} = 0. \tag{14.99}$$

The equation for changes in axial velocity involves contributions from convection, longitudinal velocity dispersion, and space-charge electric fields. We are familiar with the first two effects. To derive the equation, we must relate changes in the time-averaged velocity in the circular accelerator to the axial electric field. Following Section 14.2, the electric field along the direction of propagation is proportional to the derivative of the line density. With the conditions that (1) $kr_w \ll 1$, (2) the perturbation moves at same velocity as the beam, and (3) the wall is highly conductive, we can adopt Eq. (14.80):

$$E_z \cong -\left(\frac{q}{2\pi\varepsilon_0}\right)\left[\frac{\ln(r_w/r_0)}{\gamma^2}\right]\frac{\partial n}{\partial z}. \tag{14.100}$$

The average circulation velocity around the accelerator depends on changes in both the actual velocity and the orbit radius. Suppose that we mark particular segment of the beam that passes the observation point at $t = 0$. Then, we note the time that elapses for one revolution around the machine, $\tau + \delta\tau$. The average circulation velocity is

$$V + \delta v = 2\pi R/(\tau + \delta\tau). \tag{14.101}$$

If $\delta\tau$ is positive, the segment moves backward compared with the rest of the beam. The differential change in circulation velocity is

$$\delta v = -\beta c(\delta\tau/\tau). \tag{14.102}$$

From Eqs. (14.96) and (14.102), we can relate the change in the circulation velocity to momentum changes. Dividing both sides by a small interval leads to the equation

$$\frac{dv}{dt} = \frac{dp}{dt}\left[\frac{1}{\gamma m_0}\right]\left[\frac{1}{\gamma_t^2} - \frac{1}{\gamma^2}\right]. \tag{14.103}$$

With a space-charge electric field, the momentum of a relativistic particle changes according to

$$\frac{dp}{dt} = qE_z. \tag{14.104}$$

Incorporating Eq. (14.104), we find the following equation for the time rate of change of average beam velocity observed at a point:

$$\frac{\partial v}{\partial t} = -\beta c\frac{\partial v}{\partial z} - \frac{\langle \delta v^2 \rangle}{N_0}\frac{\partial n}{\partial z} + \frac{qE_z}{\gamma^3 m_0}\left[\frac{\gamma^2}{\gamma_t^2} - 1\right]. \tag{14.105}$$

Following the usual procedure of linear analysis, we take variations of n, v, and E_z of the form $\exp[j(kz - \omega t)]$. Combining Eqs. (14.99), (14.100), and (14.103) gives the following dispersion relationship:

$$(\omega - k\beta c)^2 = k^2\langle \delta v^2 \rangle - \frac{k^2 q^2 N_0}{4\pi\varepsilon_0 m_0\gamma^5}\left[2\ln\left(\frac{r_w}{r_0}\right)\right]\left[\frac{\gamma^2}{\gamma_t^2} - 1\right]. \tag{14.106}$$

We can simplify Eq. (14.106) by introducing the Budker parameter, (Section 12.7), $v = e^2 N_0/4\pi\varepsilon_0 m_0 c^2$ and limiting k to values that give an integral number of wavelengths around the machine circumference:

$$k = M/R. \tag{14.107}$$

The quantity M is an integer.

For a laminar beam ($\langle \delta v^2 \rangle = 0$), the negative mass dispersion relationship is

$$(\omega - M\omega_0)^2 = -\frac{M^2 c^2 v}{R^2 \gamma^5}\left[2\ln\left(\frac{r_w}{r_0}\right)\right]\left[\frac{\gamma^2}{\gamma_t^2} - 1\right]. \qquad (14.108)$$

In recirculating accelerators, perturbations grow in time rather than space. Therefore, we assume that k is a real number and seek imaginary components of ω. The solution of Eq. (14.108) for $v \ll 1$ is

$$\omega = M\omega_0 \pm \frac{jMc}{R}\left\{\frac{v}{\gamma^5}\left[2\ln\left(\frac{r_w}{r_0}\right)\right]\left[\frac{\gamma^2}{\gamma_t^2} - 1\right]\right\}^{1/2}. \qquad (14.109)$$

Equation (14.109) has the following interpretation. The first term on the right-hand side is the real part of the frequency. Density perturbations pass a stationary observation point at frequency $M\omega_0$. The second term has an imaginary value when the beam kinetic energy exceeds the transition energy, $\gamma > \gamma_t$. The imaginary solution with a negative sign corresponds to a growing perturbation.

To illustrate the result, consider the growth rate for the negative mass instability in a high-current betatron experiment. To contain intense electron beams, these devices use strong focusing lenses. Suppose the properties of the focusing system give a transition energy of 5 MeV ($\gamma_t = 10.8$). We shall calculate the instability growth rate for 6-MeV beam ($\gamma = 12.7$) at a current of 1 kA. The line density is $N_0 = 2.1 \times 10^{13}$ m^{-1}, giving a value of the Budker parameter $v = 0.059$. We take $R = 1$ m, $r_w = 0.05$ m, and $r_0 = 0.01$ m. The fastest growth occurs at the maximum value of M. The long-wavelength model holds only if $k < 1/r_w$. Therefore, a good guess for the maximum value of azimuthal mode number is

$$M \sim R/2r_w = 10.$$

Substituting values into Eq. (14.109) gives $\omega_i = 1.4 \times 10^6$ sec^{-1}. The linear growth time is 710 nsec, equal to 34 beam revolutions of 21 nsec each.

The example illustrates that the negative mass instability can grow rapidly. It results in increased emittance in both the longitudinal and transverse directions. For high-current beams, the spreads of momentum and radius can be large, leading to beam loss on the walls. The success of high-current circular electron accelerators hinges on suppressing the instability. One stabilizing effect is a longitudinal velocity spread. Equation (14.106) shows that the growth rate is zero when

$$\langle \delta v^2 \rangle \geq \frac{vc^2}{\gamma^5}\left[2\ln\left(\frac{r_w}{r_0}\right)\right]\left[\frac{\gamma^2}{\gamma_t^2} - 1\right]. \qquad (14.110)$$

For relativistic beams with velocity close to the speed of light, it is more useful to write a criterion in terms of the momentum spread rather than the velocity spread. Applying Eq. (14.95), we find that

$$\langle \delta v^2 \rangle = \frac{\langle \delta p^2 \rangle}{(\gamma^3 m_0)^2}. \tag{14.111}$$

Substituting in Eq. (14.110), the momentum spread for stability is

$$\frac{\delta p}{p} \geqslant \left\{ \frac{v}{\gamma} \left[2 \ln \left(\frac{r_w}{r_0} \right) \right] \left[\frac{\gamma^2}{\gamma_t^2} - 1 \right] \right\}^{1/2}. \tag{14.112}$$

For the parameters of the example, the fractional momentum spread is $\delta p/p = 0.076$, corresponding to a kinetic energy spread of about 0.45 MeV. Although the value is high, Eq. (14.87) shows that the beam could be contained

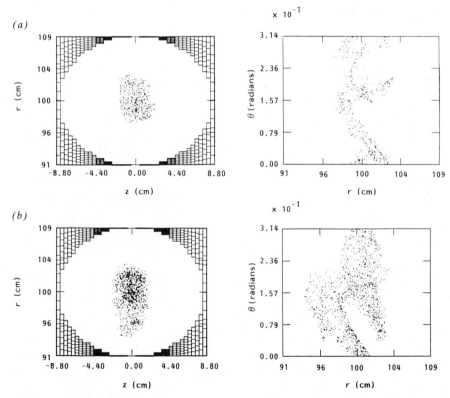

Figure 14.10 Particle-in-cell computer simulation of the negative mass instability in a high-current betatron with alternating-polarity solenoidal lenses. Calculation for the $M = 20$ mode for a ring with 20 lenses. Initial beam radius, 2 cm. Figures show particle distributions late in time when the beam is stabilized by axial velocity spread. Instability causes an enhanced energy spread and an increased RMS beam radius. (a) Drifting beam, $\gamma = 7$. (b) Drifting beam, $\gamma = 12$. (Courtesy T. Hughes, Mission Research Corporation.)

within an acceptable horizontal width by a strong-focusing system. An alternative method to suppress the negative mass instability is to neutralize the beam with a plasma. As described in Section 12.4, the plasma electrons move rapidly to cancel electric fields that result from the nonuniform beam density. Plasma stabilization has been observed in experiments on high-current electron rings.

For a comprehensive treatment of the negative mass instability, we must turn to computer simulations. Figure 14.10 shows a calculation for a betatron with a strong-focusing system consisting of an array of solenoid lenses (Section 10.9). Additional stabilizing effects include a spread in the gyration period from nonlinear focusing forces and the finite-beam width. The simulation includes the three-dimensional variation of transverse focusing forces in the toroidal geometry. The accelerator has 20 lenses around a ring with major radius of $R = 1$ m. The initial radius of the 10-kA beam is 0.02 m. The simulation follows the fastest growing mode with $M = 20$. The figure shows the nonlinear growth and saturation of the instability for a beam segment occupying one-twentieth of the circumference. The instability saturates when the longitudinal momentum spread reaches a value consistent with Eq. (14.112).

14.4. LONGITUDINAL RESISTIVE WALL INSTABILITY

We saw in Section 14.2 that filamentary beams in a pipe may carry slow space-charge waves. In the presence of resistive dissipation, these negative-energy waves grow in amplitude. In this section, we shall study longitudinal instabilities of a beam in a resistive pipe or a cavity array with a resistive component of impedance. We shall develop two models: (1) a moment equation description for nonrelativistic beams that illustrates the mechanism of the instability, and (2) a complete solution of the Vlasov equation for relativistic beams. The second model includes contributions of the beam velocity distribution. It yields practical information and also gives us an introduction to advanced methods of beam theory.

To develop the moment equation model, we shall adopt several simplifying assumptions. The beam consists of nonrelativistic particles that initially have the same axial momentum. We shall neglect the effects of axial velocity spread and space-charge forces. These conditions apply to electron beams in resistive wall microwave amplifiers and to ions in induction linear accelerators. In both cases, resistive loading by the structure is high. The space-charge electric fields in the beam are much smaller than the electric fields generated by the flow of return current in the wall.

Figure 14.11 shows the geometry of the calculation. A narrow cylindrical beam of radius r_0 propagates in a pipe of radius r_w. In equilibrium, the beam has current I_0 and all particles move at axial velocity v_0. The particle line density is

$$N_0 = I_0/qv_0. \tag{14.113}$$

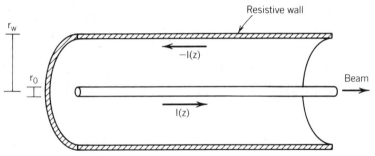

Figure 14.11 Geometry to describe the resistive wall instability of a cylindrical, nonrelativistic beam in a linear pipe.

We shall concentrate on long-wavelength perturbations, $\lambda \gg r_w$. For long λ, the distribution of return current in the wall has equal magnitude but opposite direction to that of the beam. If we allow axial variations of beam current, $I(z)$, then the wall current is approximately $-I(z)$. The resistance per length of a pipe with volume resistivity ρ, is

$$R = \pi/2\pi r_w \delta(\omega). \qquad (14.114)$$

The quantity δ is the skin depth of Eq. (13.42), a function of frequency. Note that ω is the frequency of current variations observed in the stationary frame of reference. Equation (14.114) holds if the wall thickness is larger than δ. The flow of return current through the wall creates an axial electric field. In the long-wavelength limit, the axial electric field is almost uniform over the cross section of pipe:

$$E_z(r, z) \cong -I(z)R. \qquad (14.115)$$

We shall use linear equations, valid for small amplitude oscillations. Consider small variations of line density and axial velocity, n and v, where $n \ll N_0$ and $v \ll v_0$. The beam current varies as

$$I(z) = I_0 + i(z), \qquad (14.116)$$

where $i \ll i_0$. For a stability analysis, we need to find only the time-varying portion of the axial electric field:

$$e_z = -i(z)R. \qquad (14.117)$$

The linear equations of continuity and momentum conservation for the cold beam are

$$\frac{\partial n}{\partial t} = -\frac{1}{q}\frac{\partial i}{\partial z} \qquad (14.118)$$

and

$$\frac{\partial v}{\partial t} = -v_0 \frac{\partial v}{\partial z} + \frac{q e_z}{m_0}. \tag{14.119}$$

A complete set of equations requires an expression for i in terms of n and v. The total current is

$$I_0 + i(z) = q[n_0 + n(z)][v_0 + v(z)]. \tag{14.120}$$

Because n and v are small quantities, the expanded form of Eq. (14.120) is

$$i(z) \cong q n_0 v(z) + q v_0 n(z). \tag{14.121}$$

We shall find a solution of Eqs. (14.118), (14.119), and (14.121) for a linear transport system with a steady-state beam. The solution, appropriate for a resistive wall amplifier, oscillates harmonically in time but may grow with distance from the injection point. All first-order quantities vary as

$$n, v, i \sim \exp[j(\Gamma z - \omega t)]. \tag{14.122}$$

In Eq. (14.122), ω is a real number, while Γ may be complex. Substitution in Eqs. (14.118) and (14.121) gives an expression for the perturbed current:

$$i = \left[\frac{q n_0 k_0}{k_0 - \Gamma} \right] v = \left[\frac{I_0 k_0}{v_0 (k_0 - \Gamma)} \right] v. \tag{14.123}$$

Equation (14.123) incorporates the scaling wavenumber, k_0:

$$\omega / k_0 = v_0. \tag{14.124}$$

A perturbation with wavenumber k_0 and frequency ω moves at the same velocity as the beam.

Combining Eqs. (14.117), (14.119), and (14.123), we find the following dispersion relationship:

$$-\omega + \Gamma v_0 = j \left[\left(\frac{k_0}{k_0 - \Gamma} \right) \left(\frac{q I_0 R}{m_0 v_0} \right) \right]. \tag{14.125}$$

Reviewing the derivations of previous sections, we expect that the phase velocity of the perturbation is close to the beam velocity. We shall define a small quantity, ζ, to represent the deviation in wavenumber:

$$\zeta = \Gamma - k_0 \qquad (|\zeta| \ll k_0). \tag{14.126}$$

The quantity ζ may have real or imaginary parts. The real part represents a difference between the beam velocity and the perturbation phase velocity, corresponding to fast or slow waves. The imaginary part determines whether the waves damp or grow in space.

Substituting from Eq. (14.126), Eq. (14.125) reduces to

$$\zeta^2 = -j\left[\frac{qI_0k_0R}{m_0v_0^2}\right]. \tag{14.127}$$

The complex square root gives two solutions for Eq. (14.127):

$$\zeta = (C/\sqrt{2})(1-j), \tag{14.128}$$

$$\zeta = (C/\sqrt{2})(-1+j). \tag{14.129}$$

The quantity C is the growth constant,

$$C = (qI_0k_0R/m_0v_0^2)^{1/2}. \tag{14.130}$$

Substituting the first solution for ζ in Eq. (14.122) gives a wave that grows in space. The waves have phase velocity:

$$\omega/\Gamma_r \cong v_0(1 - C/\sqrt{2}k_0). \tag{14.131}$$

The velocity of Eq. (14.131) is less than v_0; therefore, the growing solution is a slow wave. The growth length is

$$\lambda_g = (2)^{1/2}/C. \tag{14.132}$$

The growth constant has a straightforward physical interpretation. We can rewrite C/k_0 as

$$\frac{C}{k_0} = \left[\frac{qI_0R/k_0}{4(m_0v_0^2/2)}\right]^{1/2} \ll 1. \tag{14.133}$$

The quantity qI_0R/k_0 is the average kinetic energy lost by the beam to the wall resistance over a distance $1/k_0$. In a practical microwave amplifier, the energy loss must be small compared with the total kinetic energy of beam, $m_0v_0^2/2$. The implication is that $C/k_0 \ll 1$. Equation (14.131) shows that the phase velocity of the slow wave is close to the beam velocity for a small value of C/k_0.

The results of the cold-beam model are valid if the beam has a small axial velocity spread:

$$\Delta v_z/v_0 \ll C/k_0. \tag{14.134}$$

The limit of Eq. (14.134) usually holds for beams from electrostatic injectors. We can also find conditions where it is valid to neglect space-charge forces. In the long-wavelength limit, the axial electric field resulting from a current perturbation i is [Eq. (14.59)]

$$e_z \cong (k_0 i / 2\pi\varepsilon_0 v_0) \ln (r_w / r_0). \tag{14.135}$$

The electric field component from wall resistance is

$$e_z = - iR. \tag{14.136}$$

Space-charge forces are small when the expression of Eq. (14.135) is much smaller than that of Eq. (14.136):

$$R \gg (k_0 / 2\pi\varepsilon_0 v_0) \ln (r_w / r_0). \tag{14.137}$$

As an example, suppose a 10-A, 10-MeV carbon beam accelerates in an induction linear accelerator. Equation (14.137) implies that the contribution from space-charge fields is small if $R > 20\,\mathrm{k\Omega/m}$. A typical acceleration gap has a voltage of 100 kV and an AC shunt impedance to ground of $5-10\,\mathrm{k\Omega}$. Depending on the number of gaps per meter, the neglect of space-charge effects is marginally valid.

We need a more detailed model to treat resistive wall instabilities of high-energy beams in storage rings. We must include relativistic effects in calculations of the particle dynamics and electromagnetic fields. Furthermore, the resistance per unit length is low; therefore, space-charge forces play a dominant role. Finally, we must include a detailed description of the longitudinal velocity distribution. The resistive wall instability is different from instabilities that we have already studied in Chapters 13 and 14. The instability depends on the growth of a space-charge wave with a phase velocity less than the beam velocity. Growth or damping of the wave depends sensitively on the axial velocity distribution of particles near the phase velocity. This effect is called *Landau damping*. To address the process, we must carry out a direct solution of the Vlasov equation.

Before beginning a study of the resistive wall instability, it is useful to review why we did not need information about the axial velocity distribution in previous sections. For transverse instabilities such as the hose instabilities of Sections 13.7 and 13.8, the spread in axial momentum had little effect on the transport of kinetic energy from the axial to the transverse direction—the stabilizing effect was indirect through phase-mix damping of betatron oscillations (Section 13.3). As a result, the growth rate of transverse instabilities depends on the mean-square momentum spread with only a weak dependence on the shape of the axial momentum distribution. In the simple derivation of the negative mass instability in Section 14.3 with perfectly conducting walls, the perturbation of the line-charge density moved at the same velocity as the beam. In a velocity space map,

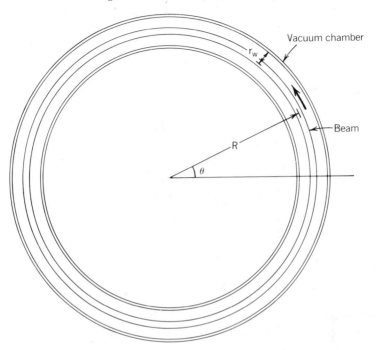

Figure 14.12 Coordinate system to analyze the resistive wall instability for a relativistic beam in a circular accelerator.

the phase velocity was centered in the beam distribution. The perturbations of the negative mass instability were not slow waves with a growth rate dependent on local variations of the velocity distribution. For this reason, the axial velocity spread enters only as a global pressure force that resists bunching.

The understand stabilization of the resistive wall instability, we shall follow the theory of Neil and Sessler.[3] The model applies to a continuous beam in a storage ring. Figure 14.12 shows a polar coordinate system for a generic circular accelerator. The toroidal transport tube has a major radius R and a wall radius r_w. The vacuum chamber has a low wall resistivity and small curvature:

$$R/r_w \gg 1. \tag{14.138}$$

In the limit of Eq. (14.138), we can adopt the field expressions for a straight pipe beam (Section 14.2).

If the instability growth time is much longer than the particle circulation time, we can take perturbations that have a uniform harmonic variation in position but grow in time. All perturbed quantities are proportional to

$$\exp[j(kR\theta - \Omega t)], \tag{14.139}$$

[3] V. K. Neil and A. Sessler, *Rev. Sci. Instrum.*, **36**, 429 (1965).

where Ω can have both real and imaginary parts. The allowed beam perturbations in a circular machine are periodic over angle 2π. The periodic condition limits the possible values of k:

$$2\pi/k = 2\pi R/M, \qquad M = 1, 2, 3, \ldots . \tag{14.140}$$

Variations of the beam line charge have the form

$$Q(\theta, t) = Q_1 \exp[j(M\theta - \Omega t)]. \tag{14.141}$$

From Eq. (14.80), the amplitude of the azimuthal electric field on the axis of the vacuum chamber resulting from the line charge variation of Eq. (14.141) is

$$RE_\theta = -\frac{jMQ_1(1 - \beta_w^2)}{2\pi\varepsilon_0} \ln\left(\frac{r_w}{r_0}\right) - \frac{\omega\delta\beta_w RQ_1}{4\pi\varepsilon_0 c r_w}. \tag{14.142}$$

The quantity β_w is the phase velocity of the perturbation, equal to the ratio of the real part of the frequency to the wavenumber. Note that Eq. (14.142) holds only for long-wavelength perturbations in systems with small curvature. We assume that horizontal displacements of particles from the pipe axis are small; therefore, Eq. (14.142) gives an approximation for the axial field acting on all particles in the beam cross section.

We will combine Eq. (14.142) with the Vlasov equation to find a dispersion relationship that gives the complex frequency, Ω, as a function of the mode number M and the properties of the beam and accelerator. In the one-dimensional analysis, we characterize particle orbits by their angle, θ, and momentum in the azimuthal direction. We must choose variables for the Vlasov equation carefully. Remember that the derivation of the equation followed from the principle of phase area conservation for a collisionless distribution. The principle holds only if the phase-space coordinates are *canonical variables*.[4] For example, x and p_x are canonical variables in a Cartesian coordinate system.

If we use polar coordinates with position variable θ, then the canonical momentum is the *canonical angular momentum*, P_θ, defined by

$$P_\theta = \gamma m_0 r v_\theta + q r A_\theta. \tag{14.143}$$

The first term on the right-hand side of Eq. (14.143) is the product of the orbit radius and the ordinary momentum in the azimuthal direction, $p_\theta = \gamma m_0 v_\theta$. If displacements in the storage ring are much smaller than the major radius, then $r \cong R$. The second term on the right-hand side involves the azimuthal vector potential, A_θ. We can write the related vector potential to the flux of vertical field contained within the particle orbit:

[4] See, for instance, H. Goldstein, *Classical Mechanics*, Addison-Wesley, Reading, Mass, 1965, Section 2.6.

$$A_\theta(R,0) = \int_0^R 2\pi r'\,dr'\,B_z(r',0)/2\pi R. \tag{14.144}$$

In Eq. (14.144), B_z arises mainly from the vertical magnetic that bends particles in a circular orbit. Changes of magnetic flux cause betatron acceleration—this process has a negligible effect in high-energy storage rings.

We shall study distributions with small variations of particle canonical angular momentum about a mean value, P_0. The variable W represents deviations from the mean:

$$W = 2\pi(P_\theta - P_0). \tag{14.145}$$

Equation (14.145) includes the factor of 2π for compatibility with the notation of the Neil and Sessler model. The beam distribution function, $f(\theta, W, t)$, depends on the azimuth, the error in canonical angular momentum, and the time. The Vlasov equation for azimuthal motion is

$$\frac{\partial f}{\partial t} + \left(\frac{d\theta}{dt}\right)\frac{\partial f}{\partial \theta} + \left(\frac{dW}{dt}\right)\frac{\partial f}{\partial W} = 0. \tag{14.146}$$

Equation (14.146) has the following meaning: the value of f remains constant along the orbit of a particle in (θ, W) space defined by $(d\theta/dt)$ and (dW/dt).

We shall solve the Vlasov equation in the limit that the distribution function has small variations about an equilibrium. We divide the function into parts representing the equilibrium and a perturbed part, $f = f_0 + f_1$, where $|f_1| \ll |f_0|$. In equilibrium, the beam is uniform in azimuth. With no instability, applied forces do not depend on θ, and W is a constant of particle motion. We know that any function of W is a valid equilibrium distribution. Combining this fact with the assumed form for beam perturbations [Eq. (14.139)], we can write the distribution function as

$$f(\theta, W, t) = f_0(W) + f_1(W)\exp[j(M\theta - \Omega t)]. \tag{14.147}$$

Substituting Eq. (14.147) in Eq. (14.146) gives

$$\left[j\Omega f_1(W) - jM\left(\frac{d\theta}{dt}\right)f_1(W)\right]\exp[j(M\theta - \Omega t)]$$

$$= \left(\frac{dW}{dt}\right)\left\{\left[\frac{\partial f_0(W)}{\partial W}\right] + \left[\frac{\partial f_1(W)}{\partial W}\right]\exp[j(M\theta - \Omega t)]\right\}. \tag{14.148}$$

Dropping the small second term in braces on right-hand side of Eq. (14.148), the linearized Vlasov equation takes the form

$$f_1(W)\exp\left[j(M\theta - \Omega t)\right] = \frac{-j(dW/dt)}{[\Omega - M(d\theta/dt)]}\frac{\partial f_0(W)}{\partial W}.$$ (14.149)

Note that the quantity $d\theta/dt$ in the denominator depends on W—the circulation time for a particle is a function of its canonical angular momentum.

For a self-consistent solution we must relate dW/dt to the azimuthal electric field created by the perturbed particle distribution. Taking the time derivative of Eq. (14.145) and substituting from Eq. (14.143), we find

$$\frac{dW}{dt} = 2\pi R\left(\frac{dp_\theta}{dt} + \frac{qdA_\theta}{dt}\right) = 2\pi qRE_\theta + 2\pi Rq\left(\frac{dA_\theta}{dt}\right).$$ (14.150)

Changes in the ordinary momentum, p_θ, result from the azimuthal electric force, eE_θ. The second term on the right-hand side represents betatron acceleration from a changing magnetic flux. We assume that the applied magnetic field is constant during the growth of the instability. Also, we note that the vertical field created by the circulating beam current is much smaller than the local fields created by perturbation. As a result, we drop the second term on the right-hand side of Eq. (14.150).

We integrate the Vlasov equation over W to obtain a dispersion relationship. Following the normalization convention introduced in Section 2.5, the integral over W of f_0 gives the total number of particles in the accelerator. If the equilibrium line density is N_0, then

$$2\pi RN_0 = \int_{-\infty}^{+\infty} dWf_0(W).$$ (14.151)

To simplify the form of results, we define a normalized equilibrium distribution function, $g_0(W)$, that has an integral of unity:

$$g_0(W) = f_0(W)/2\pi RN_0.$$ (14.152)

We can express the perturbed line charge density as an integral over the perturbed part of the distribution function:

$$Q_1 = e\left[\int dWf_1(W)\right]\Big/2\pi R \qquad \text{(coulombs/m)}.$$ (14.153)

Integrating both sides of Eq. (14.149) over W and substituting from Eqs. (14.152) and (14.153), we find the following expression for the linearized Vlasov equation:

$$Q_1\exp\left[j(M\theta - \Omega t)\right] = -2j\pi q^2RE_\theta N_0\int_{-\infty}^{+\infty}\frac{dW}{[\Omega - M(d\theta/dt)]}\frac{\partial g_0(W)}{\partial W}.$$ (14.154)

Substitution of Eq. (14.142) for the electric field leads to the dispersion relationship

$$-1 = (U - jV)I,\qquad(14.155)$$

where

$$I = \int_{-\infty}^{+\infty} \frac{dW}{[\Omega - M(d\theta/dt)]} \frac{\partial g_0(W)}{\partial W},\qquad(14.156)$$

$$U = q^2 N_0 M(1 - \beta_w^2)\ln(r_w/r_0)/\varepsilon_0,\qquad(14.157)$$

and

$$V = q^2 R N_0 \delta\Omega\beta_w/2c\varepsilon_0 r_w.\qquad(14.158)$$

Equation (14.155) holds for any uniform transport tube. We can calculate the functions U and V for alternative wall geometries, such as a rectangular vacuum chamber. The functions have positive values for any geometry. For low wall resistivity, we will find that the phase velocity of perturbations is almost equal to that of the beam, $\beta_w \cong \beta$. Therefore, the real part of the perturbation frequency observed in the stationary frame is $\mathrm{Re}(\Omega) \cong M\omega_0$, where ω_0 is the circulation frequency of the beam, $\omega_0 = \beta c/R$. Another implication of low resistivity is that the real part of the frequency is much larger than the imaginary part. As a result, we can approximate the functions U and V as constants:

$$U \cong q^2 N_0 M(1 - \beta^2)\ln(r_w/r_0)/\varepsilon_0,\qquad(14.159)$$

$$V \cong q^2 R N_0 \delta(M\omega_0)(M\omega_0)\beta/2c\varepsilon_0 r_w.\qquad(14.160)$$

The symbol $\delta(M\omega_0)$ denotes the wall skin depth evaluated at a frequency of $M\omega_0$. In the limit of small ρ, $|U| \gg |V|$.

As a final modification, we shall write the term $d\theta/dt$ of the function I in terms of W and the properties of the machine. The angular frequency for motion around the ring depends on the particle energy, $E = (\gamma - 1)m_0 c^2$. Increasing the energy of nonrelativistic particles causes them to circulate faster because of their increased velocity. In contrast, relativistic particles circulate slower because they have larger orbit radii. The dividing line between the types of behavior is the transition energy, $(\gamma_t - 1)m_0 c^2$. If f is the particle circulation frequency in the ring, we can write a Taylor series expansion about the mean beam energy, E_0:

$$f(E) \cong f_0(E_0) + \left(\frac{\partial f}{\partial E}\right)\Delta E,\qquad(14.161)$$

where $\Delta E = E - E_0$. The time variation of θ averaged over many transits is

$$\frac{d\theta}{dt} \cong \omega_0 + 2\pi \left(\frac{\partial f}{\partial E} \right) \Delta E = \omega_0 + 2\pi f \left(\frac{\partial f}{\partial E} \right) W. \qquad (14.162)$$

The final form in Eq. (14.162) comes from the relationship

$$\Delta E = qE_0\beta c\Delta t = 2\pi\Delta P_\theta(\beta c/2\pi R) = Wf, \qquad (14.163)$$

where f is the circulation frequency of particles with canonical angular momentum $P_0 + 2\pi W$. Substitution from Eq. (14.162) gives a modified expression for I:

$$I = \int_{-\infty}^{+\infty} \frac{dW}{[\Omega - M\omega_0 - M\kappa_0 W]} \frac{\partial g_0(W)}{\partial W}, \qquad (14.164)$$

where

$$\kappa_0 = 2\pi f \left(\frac{\partial f}{\partial E} \right). \qquad (14.165)$$

Above transition, particles with increased energy take longer to circulate around machine, or $\kappa_0 < 0$.

We can now investigate the implications of Eq. (14.155) for different equilibrium distributions. The simplest choice is a cold beam where all particles have the same value of canonical angular momentum in equilibrium:

$$g_0(W) = \delta(W). \qquad (14.166)$$

The expression for I becomes

$$I = \int \frac{dW(d\delta/dW)}{(\Omega - M\omega_0 - M\kappa_0 W)}. \qquad (14.167)$$

To evaluate Eq. (14.167), we can use the following property of the delta function,

$$\int f(x) \frac{d[\delta(x-a)]}{dx} = -\frac{df(a)}{dx}. \qquad (14.168)$$

The result is

$$I = M\kappa_0/(\Omega - M\omega_0)^2. \qquad (14.169)$$

Substitution in Eq. (14.155) gives the dispersion relationship

$$(\Omega - M\omega_0)^2 = M\kappa_0(U - jV). \tag{14.170}$$

Despite its simplicity, Eq. (14.170) has several interesting implications. First consider a case where the wall has no resistivity, or $V = 0$. Below transition, $\kappa_0 > 0$, Eq. (14.170) implies that Ω is real—the beam is stable. Above transition, Ω has an imaginary part and the beam may be unstable. The solution with $\kappa_0 < 0$ is

$$\Omega = M\omega_0 \pm j(M\kappa_0 U)^{1/2}. \tag{14.171}$$

The instability represented by Eq. (14.171) is the negative mass instability of Section 14.3. Note that there is no shift in the real part of the frequency—the perturbation phase velocity equals the beam velocity, consistent with the results of the moment equation model of Section 14.3.

We can show that the content of Eq. (14.171) is identical to that of Eq. (14.109)—both models predict the same growth rate. In the relativistic limit, Eq. (14.96) implies that

$$\frac{df}{f} = -\frac{d\tau}{\tau} \cong -\frac{dE}{E}\left[\frac{1}{\gamma_t^2} - \frac{1}{\gamma^2}\right]. \tag{14.172}$$

Comparing Eqs. (14.165) and (14.172), we find that

$$\kappa_0 \cong -\frac{2\pi(\beta c/2\pi R)^2}{\gamma^3 m_0 c^2}\left[\frac{\gamma^2}{\gamma_t^2} - 1\right]. \tag{14.173}$$

Substituting for U and κ_0 in Eq. (14.171) from Eqs. (14.159) and (14.173), after some algebra we find that Eqs. (14.171) and (14.109) are identical.

Next, consider solutions with $V > 0$. Above transition, the growth rate of the negative mass instability is much higher than that of the resistive wall instability; therefore, the resistive contribution is of little concern. Solutions below transition, $\kappa_0 > 0$, are important because the resistive wall effect is the only source of instability. If we take κ_0 as a positive number equal to the magnitude of $2\pi f (\partial f/\partial E)$, the dispersion relationship is

$$(\Omega - M\omega_0)^2 = M\kappa_0(U - jV) = M\kappa_0 U(1 - jV/U). \tag{14.174}$$

The second term in parentheses is much smaller than unity—we can use the binomial theorem to find the square root:

$$\Omega \cong M\omega_0 \pm (M\kappa_0 U)^{1/2}(1 - jV/2U). \tag{14.175}$$

The positive sign in Eq. (14.175) corresponds to a fast wave with phase velocity greater than the beam velocity. The wall resistivity damps this wave. The minus sign gives a slow wave, with a growth time of

$$\tau_r = (2/V)(U/M\kappa_0)^{1/2}. \qquad (14.176)$$

Although Eq. (14.175) gives us a mathematical answer, the physical implications are not obvious. We can rewrite the equation in a form that displays scaling relationships more clearly. Substituting from Eqs. (14.157), (14.158), and (14.173), introducing the Budker parameter [Eq. (12.131)], and carrying out extensive algebra, we find the resistive wall instability growth time is

$$\tau_r \cong \frac{\sqrt{8r_w/[M\omega_0\delta(M\omega_0)\beta]}}{(\nu/\gamma)^{1/2}(1-\gamma^2/\gamma_t^2)^{1/2}} = \frac{\sqrt{8r_w/[k\beta^2c\delta(M\omega_0)]}}{(\nu/\gamma)^{1/2}(1-\gamma^2/\gamma_t^2)^{1/2}}. \qquad (14.177)$$

Note that major radius, R, does not appear in Eq. (14.177). Although R influences the growth rate for the negative mass instability, it has no affect on the growth rate of resistive wall instabilities. This result follows from our use of the straight pipe expression for the axial electric field. The only constraint introduced by the circular geometry is that the wavenumber has a discrete set of values, $k = M/R$. As in Eq. (14.133), the instability growth is inversely proportional to the square root of the beam current. The growth time is long for a low-current, high-energy beam in a metal chamber. As with transverse instabilities, resonant structures in the beamline can present a greatly enhanced impedance. If the perturbation frequency, $M\omega_0$, overlaps the resonance width for a TM_{0n0} mode, a cavity array has a high value of resistance per length. This fact follows from the properties of resonant cavities as impedance transformers.

The main reason to pursue the Vlasov equation solution is to investigate the stabilizing effects of an axial momentum spread. We shall use a Gaussian distribution in W to represent a spread in canonical angular momentum:

$$g_0(W) = \frac{1}{\Delta W(\pi)^{1/2}} \exp\left(\frac{-W^2}{\Delta W^2}\right). \qquad (14.178)$$

To incorporate the distribution of Eq. (14.178), we need a modified form for the dispersion relationship. Removing a factor of $M\kappa_0$ and multiplying both sides of Eq. (14.155) by $(U + jV)$ gives

$$\frac{M\kappa_0(U + jV)}{U^2 + V^2} = \int dW \frac{dg_0}{dW} \frac{1}{W - W_1}, \qquad (14.179)$$

where

$$W_1 = (\Omega - M\omega_0)/M\kappa_0. \qquad (14.180)$$

We insert Eq. (14.178) into Eq. (14.179) and express the result in terms of the normalized variable $\xi = W/\Delta W$. The final form of the dispersion relationship results from an integration by parts:

$$\frac{M\kappa_0 \Delta W^2 (U + jV)}{(U^2 + V^2)} = \frac{1}{\pi^{1/2}} \int_{-\infty}^{\infty} \frac{\exp(-\xi^2)d\xi}{(\xi - \xi_0)^2} = D(\xi_0), \qquad (14.181)$$

where

$$\xi_0 = (\Omega - M\omega_0)M\kappa_0 \Delta W. \qquad (14.182)$$

The function $D(\xi_0)$ is the *plasma dispersion function*.[5] We must exercise care in evaluating the definite integral, particularly at the discontinuity at $\xi = \xi_1$. The portion of the integration near ξ_1 is important because it involves those particles in the distribution that move at the phase velocity of the slow wave. These particles cause growth or damping of the wave.

We want to find conditions for beam stability, so we assume *a priori* that Ω and ξ_0 are real numbers and look for consistent conditions. We can find the integral of Eq. (14.181) along the real axis from $-\infty$ to $+\infty$ by taking a contour integral in the complex plane, noting the pole at around $\xi = \xi_0$. We expect that $U \gg V$, or

$$\text{Re}\,[D(\xi_0)] \gg \text{Im}\,[D(\xi_0)]. \qquad (14.183)$$

Equation (14.183) implies that ξ_0 is much larger than unity; in this limit, the plasma dispersion function has the approximate form

$$D(\xi_0) \cong 1/\xi_0^2 - 2j(\pi)^{1/2}\xi_0 \exp(-\xi_0^2). \qquad (14.184)$$

For a valid solution with real ξ_0, the real and imaginary parts on both sides of Eq. (14.181) must be separately equal when we substitute Eq. (14.184) on the right-hand side. The resulting equations are

$$\frac{1}{\xi_0^2} = \frac{M\kappa_0 \Delta W^2 U}{U^2 + V^2} \qquad (14.185)$$

and

$$-2j(\pi)\xi_0 \exp(-\xi_0^2) = \frac{jM\kappa_0 \Delta W^2 V}{U^2 + V^2}. \qquad (14.186)$$

Equation (14.185) gives he following value for the real frequency of the slow wave:

$$\Omega \cong M\omega_0 - (M\kappa_0 U)^{1/2}. \qquad (14.187)$$

---

[5] Properties of the plasma dispersion function are tabulated in B. D. Fried and S. D. Conte, *The Plasma Dispersion Function* (Academic Press, New York, 1961).

Equation (14.187) is identical to Eq. (14.175) derived for a delta function distribution—the shift in the phase velocity of the slow wave does not depend on the form of the distribution.

Analysis of Eq. (14.186) gives a sufficient condition for stability, equivalent to a real value solution for ξ_0. Substituting from Eq. (14.185), the stability criterion is

$$\xi_0^3 \exp(-\xi_0^2) = -V/2U(\pi)^{1/2}. \tag{14.188}$$

In principle, given values of U and V we can use Eqs. (14.185) and (14.188) to calculate the spread in angular momentum. ΔW, for stability. This approach, although exact for the particular distribution, does not lend itself to a straightforward physical explanation.

As an alternative, Neil and Sessler propose the following approximate rule for stability. The criterion follows from an analysis of the properties of the complex integral for $D(\xi_0)$. The necessary stability condition for well-behaved distributions is that $g_0(W)$ extends over a range of canonical angular momentum that includes W_0. In other words, the spread in circulation frequency must be comparable to $\Omega - M\omega_0$. If the frequency spread is smaller, the cold-beam results apply. From Eq. (14.165), the circulation frequency spread is $\Delta\omega = 2\kappa_0\Delta W$, so that the stability criterion is

$$M\Delta\omega > 2(M\kappa_0 U)^{1/2}, \tag{14.189}$$

or

$$\frac{\Delta W}{2\pi P_0} > \left[\frac{U}{M\kappa_0}\right]^{1/2} \frac{1}{2\pi\gamma\beta m_0 cR}. \tag{14.190}$$

We can simplify Eq. (14.190) to

$$\frac{\Delta W}{2\pi P_0} > \frac{1}{\beta^2}\left[\frac{v}{\gamma}\frac{\ln(r_w/r_0)}{2[1-\gamma^2/\gamma_t^2]}\right]^{1/2}. \tag{14.191}$$

The required spread in canonical angular momentum does not depend on the wall resistivity, the mode number M, or the radius of the circular accelerator R. A comparison to Eq. (14.112) shows that the same stability criterion holds for the negative mass instability above transition if we replace the factor in the denominator by $\gamma^2/\gamma_t^2 - 1$.

At this point, you may wonder why we followed such a difficult path to arrive at a conclusion that we could have found from the moment equations and some common sense. One reason is that other times we might not be so lucky. Some instabilities and methods for radiation generation depend sensitively on the velocity distribution—they saturate at levels quite different from projections based on a displaced Maxwellian. Another reason is that rules of thumb like Eq. (14.189) must ultimately be supported by detailed analyses. Although intuition is valuable in collective physics, it can often lead to incorrect predictions when processes are so complex that we cannot construct a simple physical picture. The resistive wall instability is a good example of the synergism between intuition and detailed mathematical treatments in collective physics. The mathematics guides us to conclusions that we can justify with physical insights.

15

Generation of Radiation with Electron Beams

In this final chapter, we shall review the generation of coherent electromagnetic radiation by charged particle beams. An understanding of this application demands a comprehensive knowledge of beam physics. All the devices we shall discuss use electron beams because these beams are easy to generate and to transport at high power levels.

A multitude of methods have been developed to generate microwave radiation—we shall address only a few of the main types. All beam-driven radiation sources share common features. In every device, an electron beam with directed kinetic energy interacts with a traveling or standing electromagnetic wave. The wave has an electric field that oscillates at angular frequency ω. The beam electrons must move against the electric field to perform work on the wave. A steady-state beam does not contribute a time-averaged energy to the wave—the electric field of the wave alternately accelerates and decelerates the particles. We must use a *modulated* beam to drive the wave. The term modulation implies a synchronized time variation of beam current or position at frequency ω. We can distinguish three types of beam modulation used in microwave sources:

1. For a standing wave that varies in time at an axial location, we can switch the beam on and off so that current flows only during the decelerating phase of the wave. The two-cavity klystron illustrates this modulation method.
2. For a traveling wave with phase velocity equal to the beam velocity, we can bunch electrons along the axial direction. The beam amplifies the wave if there is an excess of electrons in the decelerating phase of the wave. The traveling-wave tube illustrates such an interaction.

720

3. If the amplitude of the wave electric field varies in the direction normal to the beam, we can sweep the beam position in the transverse direction to achieve net power transfer. The magnetron is a familiar example of transverse beam modulation.

The free-electron laser combines two modulation approaches. Wave amplification depends on synchronized beam sweeping combined with axial bunching of the beam density.

We can classify beam-driven microwave sources as either oscillators or amplifiers. Oscillators create microwave energy through resonant instabilities. Here, an electromagnetic disturbance grows from noise at a favored frequency determined by a resonant circuit. Oscillators usually operate in the saturation state of the instability. In contrast, an amplifier operates in the linear regime of a beam instability. The beam-driven output is proportional to the amplitude of an input wave. A single oscillator can drive several amplifiers in parallel to generate phased microwave energy at high power.

Section 15.1 describes the inverse diode, the simplest method to extract the energy of a modulated beam. The beam energy changes directly to electrical energy in a deceleration gap. We can use inverse diodes to generate RF energy from a beam with axial modulation or a sweeping motion. The beam drives current through a load connected to the inverse diode collector. For good efficiency, the load impedance should be close to the beam impedance, equal to V_0 divided by the peak beam current. Transmission lines and waveguides that transport electromagnetic energy have characteristic impedances in the range 30–300 Ω. An inverse diode driven by an intense electron beam matches such a line. For example, a 200-keV, 1-kA beam has an impedance of 200 Ω.

Steady-state electron beams used in conventional microwave sources are poorly matched to electromagnetic lines and loads. For example, a 1-A, 20-keV beam has an impedance of 20 kΩ. We use resonant cavities to match a high-impedance beam to an electromagnetic load. Section 15.2 describes the function of resonant cavities as impedance transformers. This matching technique has important applications in klystrons and magnetrons.

Section 15.3 summarizes axial modulation of electron beams by velocity bunching. This method has application to high-frequency sources. Beam chopping is inefficient, while beam gating using the grid of an electron gun has limited frequency range because of transit-time effects of low-energy electrons. Velocity bunching occurs after a continuous beam passes through a resonant cavity with an axial electric field at frequency ω. The beam emerges with a harmonic velocity shift—after a drift distance, the velocity dispersion changes to a modulation of the axial density.

We apply the theory of velocity bunching to the klystron in Section 15.4. In this device, an electron beam interacts with two or more discrete resonant cavities. An input signal drives the first cavity, the buncher. The modulated beam drives electromagnetic oscillations in one or more downstream load cavities. Klystrons have high gain, good stability, and high pulsed-power capability—

they are often used to power RF accelerators. A single klystron can produce a pulsed output of 50 MW in the gigahertz range.

The traveling-wave tube, discussed in Section 15.5, illustrates the continuous interaction of a beam with a wave over an extended distance. We review some properties of slow-wave structures. These devices support electromagnetic waves with an axial electric field component and a phase velocity close to that of the beam. Because of these properties, beam electrons undergo a long-term average acceleration or deceleration. In the traveling-wave amplifier, an input wave bunches the electrons of an initially continuous beam. Bunching causes an enhanced density in the decelerating regions of the wave. The wave amplitude grows from energy extracted from the beam. The result is an exponential growth of wave amplitude and beam modulation. With correct design of the slow-wave structure, traveling-wave tubes amplify radiation over a broad frequency range.

Section 15.6 reviews the principles of magnetrons. These crossed-field devices have perpendicular applied magnetic and electric fields. The electric field drives a sheet electron beam that drifts normal to the magnetic fields. The beam moves through a structure that supports a slow wave with an electric field component along the electron drift direction. The wave field causes a transverse drift of electrons—in decelerating regions of the wave the electrons shift to locations where the axial electric field is stronger. This displacement results in a net transfer of energy to the wave. The magnetron is usually used as a high-power oscillator. Magnetrons achieve high energy-conversion efficiency because electrons lose most of their kinetic energy before exiting.

The final two sections review an area of intense recent interest, free-electron lasers. These devices generate short-wavelength radiation—depending on beam energy, the output can extend to the optical regime. They have the potential ability to operate as tunable sources at high power and high efficiency. In free-electron lasers, a relativistic electron beam drives a transverse-electromagnetic (TEM) wave. An alternating polarity applied magnetic field causes a harmonic transverse displacement of the beam. The dipole current component of the beam acts on the transverse electric fields of the wave. The magnetic field of the TEM wave bunches the beam toward regions of wave deceleration. Because of the properties of the relativistic Doppler shift, the electrons in a free-electron laser can drive electromagnetic radiation with a wavelength much shorter than the scale of magnetic field variations. Section 15.7 describes the synchronization mechanism between transverse single-particle oscillations and the TEM wave. Section 15.8 concentrates on axial bunching and phase dynamics of groups of electrons in the free-electron laser.

15.1. INVERSE DIODE

An *inverse diode* converts the kinetic energy of a charged particle beam to electrical energy. The device performs the complementary function to a beam injector. In an inverse diode, a charged particle beam crosses a deceleration gap to

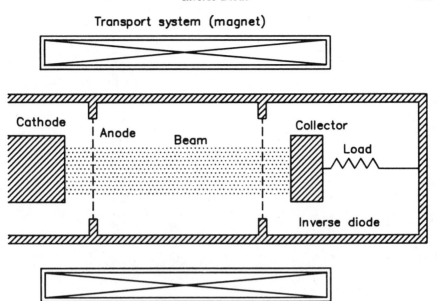

Figure 15.1 Components of an inverse diode to recover the energy of an electron beam: electron gun, transport system, inverse diode, load circuit.

a collector at elevated potential. The collected beam current can drive an external load at high voltage. The inverse diode is the simplest method for energy extraction from a directed flux of charged particles. The device has potential applications to high-power RF radiation generators. At the end of the section, we shall discuss two examples of such generators, the scanned beam switch and the relativistic beam oscillator.

Figure 15.1 shows an inverse diode. An electron beam from an electrostatic acceleration gap moves through a transport system to a collector. In an ideal system, the collector can recover the full beam energy if the following conditions hold:

1. The beam is laminar. This condition holds only if electrons from the source have no random components of longitudinal or transverse energy.
2. Focusing forces in the transport system are linear. The lenses generate an image of the source beam at the collector.
3. The electric and magnetic fields in the inverse diode are identical to those in the injector.

When the conditions hold, electron motion in the inverse diode mirrors motion in the injector. With a voltage equal to that of the injector, the collector recovers the complete beam kinetic energy.

In reality, the inverse diode is not an exact reflection of the injector. Beams

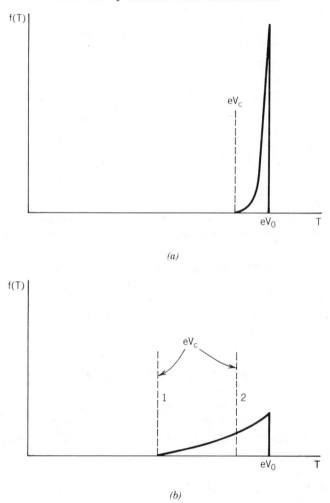

Figure 15.2 Effect of beam energy spectrum on the efficiency of an inverse diode. Collector voltage V_c; peak beam kinetic energy, eV_0. (a) With a small energy spread, the collector captures the full beam current with V_c close to V_0. (b) With a broad energy spread, the inverse diode must operate at reduced voltage to capture the full beam current (1) or to collect a reduced beam current near the full voltage (2).

have spreads of transverse and axial momentum. Figure 15.2 illustrates the effect of kinetic energy spread, T, in a beam of electrons incident on a one-dimensional inverse diode. In Figure 15.2a, all electrons have almost the same kinetic energy, eV_0, and velocity in the z direction. All electrons reach the collector if its voltage is slightly below V_0. Figure 15.2b shows a more realistic beam distribution—there is a spread of kinetic energy below eV_0. If the collector voltage equals V_0, no electrons reach the surface—the conversion efficiency of kinetic energy to electrical energy equals zero. With lower voltage, the collector captures some of the electron beam. On the other hand, the collected electrons arrive with extra kinetic energy that is wasted heating the surface. For a given beam distribution, there is a value of collector voltage that gives the highest conversion efficiency.

A complete treatment of inverse diode physics could include effects of finite beam width, spatial variations of applied fields, and beam-generated magnetic fields. A problem of this complexity requires a numerical simulation. Instead, we shall study a one-dimensional model to understand the basic principles. We shall use values of electrostatic potential appropriate for an electron beam, although inverse diodes can also recover the energy of ion beams. The approach is similar to the space-charge flow calculations of Chapters 5 and 6. We apply conservation of particle flux in equilibrium to express the particle density as a function of the electrostatic potential. We then solve the Poisson equation to find a self-consistent flow solution.

Figure 15.3 shows the geometry of the calculation. The inverse diode consists of a grounded input mesh at $z = d$ and a collector with bias voltage $-V_c$ at $z = 0$. The diode has spacing d. Electrons with current density j_0 enter from the positive z direction. If the electrons are monoenergetic, the space-charge solution is the mirror image of the solution of Section 5.2. Instead, we allow a spread in the z-directed kinetic energy, T. Suppose that we know the distribution of electron flux in terms of T to an endpoint energy eV_0. We define the flux spectrum, $f(T)$, so that $j_0 f(T) dT$ equals the fraction of the incident current density carried by electrons with longitudinal kinetic energy between $T - dT/2$ and $T + dT/2$. The flux function satisfies the normalization equation:

$$\int_0^{eV_0} dT\, F(T) = 1. \tag{15.1}$$

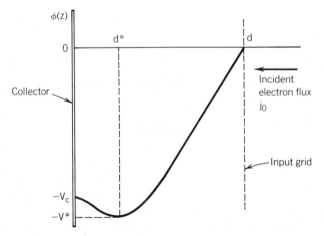

Figure 15.3 Quantities to calculate self-consistent space-charge equilibria for electron flow in a planar inverse diode.

Figure 15.3 illustrates the nature of a self-consistent solution. An external circuit maintains the collector voltage at $-V_c$, absorbing the energy of incident electrons. The electrostatic potential decreases moving from the entrance grid to the collector. At each position in the gap, some low-energy electrons turn while higher-energy particles move forward. For a given flux function and values for j_0, V_c, and V_0, we can find a unique equilibrium. Assume that the solution has a point where the electric field goes to zero. We denote the potential and position of this point as V^* and d^*. The value of V^* determines the fraction of incident electrons that pass the point. Beyond d^*, the magnitude of the potential must decrease because the Poisson equation implies the $d^2\phi/dz^2$ is positive everywhere in the gap. Therefore, all particles that enter the inverse diode with $T > eV^*$ pass d^* and continue to the collector. All other particles turn back.

We can show that for a given fraction of collected electrons, the best position for the collector is the $E = 0$ point. Suppose we start with a solution with $d = d^*$ and $V_c = V^*$. If we keep V^* constant and move the collector downstream ($d > d^*$), the fraction of electrons that passes d^* remains the same. To ensure that all these electrons reach the collector, we must set V_c below V^*. With the same collected current density but lower collector voltage, the efficiency of the inverse diode is lower. Conversely, if we maintain the collector voltage at $V_c = V^*$ but move the electrode closer to the entrance mesh ($d < d^*$), the transported current is the same but the electric field on the collector has a positive value. With a positive field, secondary electrons generated on the collector surface flow back to the entrance grid, reducing the efficiency of the system. In the following discussion, we shall take solutions where $E_z(0) = 0$.

In equilibrium, conservation of charge implies that the product of particle density and velocity is uniform with position for each energy group of electrons. We can express the velocity of relativistic electrons that enter the inverse diode with kinetic energy T as a function of the electrostatic potential

$$v_z(\phi) = \beta c = c(\gamma^2 - 1)^{1/2}/\gamma, \tag{15.2}$$

where

$$\gamma = 1 + (T + e\phi)/m_e c^2.$$

The equations that determine the electron equilibrium are conveniently written in terms of dimensionless variables:

$$X = T/eV_0, \tag{15.3}$$

the input kinetic energy,

$$\Phi = -e\phi/eV_0, \tag{15.4}$$

the electrostatic potential, and

$$Z = z(ej_0/2\varepsilon_0 ceV_0)^{1/2}, \tag{15.5}$$

the normalized position. We also define two dimensionless constants:

$$\alpha = eV_0/m_e c^2,\qquad(15.6)$$

a distribution relativistic factor, and

$$\Phi_c = -eV_c/T_0,\qquad(15.7)$$

the collector voltage. The normalized flux function is

$$F(X)\,dX = F(T/T_0)\,d(T/T_0).\qquad(15.8)$$

The distance scale of Eq. (15.5) is the gap width of an ultrarelativistic space-charge-limited diode [Eq. (6.5)] for a current density, j_0, of monoenergetic electrons with $T = eV_0$.

Substituting the dimensionless quantities, Eq. (15.2) becomes

$$v = \frac{c[2\alpha(X-\Phi)+\alpha^2(X-\Phi)^2]^{1/2}}{1+\alpha(X-\Phi)}.\qquad(15.9)$$

Electrons with incident energy X in dX make a contribution to the density at a position with potential Φ equal to their incident flux divided by their local velocity:

$$\Delta n(\Phi, X) = \frac{[j_0 F(X)dX][1+\alpha(X-\Phi)]}{c[2\alpha(X-\Phi)+\alpha^2(X-\Phi)^2]^{1/2}}.\qquad(15.10)$$

The total density at the position is the integral of Eq. (15.10) over all electrons with sufficient kinetic energy to overcome the potential Φ. In doing the integral, we must separate the contributions from electrons that pass the position once and continue to the collector ($X > \Phi_c$) and from electrons that pass through the region twice because they reflect at higher potential ($\Phi < X < \Phi_c$). The correct expression for the particle density is:

$$n(\Phi) = (j_0/ec)N(\Phi),\qquad(15.11)$$

where

$$N(\Phi) = \int_\Phi^{\Phi_c} \frac{2F(X)\,dX[1+\alpha(X-\Phi)]}{[2\alpha(X-\Phi)+\alpha^2(X-\Phi)^2]^{1/2}} + \int_{\Phi_c}^1 \frac{F(X)\,dX[1+\alpha(X-\Phi)]}{[2\alpha(X-\Phi)+\alpha^2(X-\Phi)^2]^{1/2}}.$$

We can find $N(\Phi)$ for a given flux choice of Φ_c and substitute the result into the one-dimensional Poisson equation to find the gap spacing:

$$\frac{d^2\Phi}{dZ^2} = -2N(\Phi).\qquad(15.12)$$

We can solve Eq. (15.12) numerically for any $N(\Phi)$, starting at the collector with the conditions $\Phi(0) = \Phi_c$, $\Phi'(0) = 0$. The position where Φ equals zero defines the normalized gap width.

For given $F(T)$, the current density that reaches the collector is

$$j = j_0 \int_{eV_c}^{eV_0} F(T)\,dT = j_0 \int_{\Phi_c}^{1} F(X)\,dX. \tag{15.13}$$

The conversion efficiency equals the electrical energy produced per unit area of

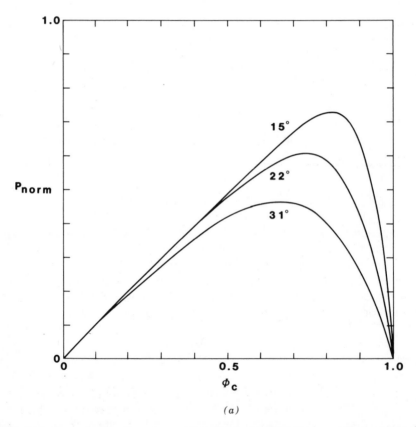

(a)

Figure 15.4 Properties of a planar inverse diode for a nonrelativistic electron beam with kinetic energy eV_0 and a Gaussian distribution of transverse angle with a mean-squared value $\langle\theta^2\rangle$. (a) Collected power per area normalized to the incident beam power density as a function of V_c/V_0 for three values of $\langle\theta^2\rangle$. (b) Normalized gap spacing for zero electric field on collector, $D^* = d^*/[(4\varepsilon_0/9j_0)^{1/2}(2e/m_e)^{1/4}(T_0/e)^{3/4}]$.

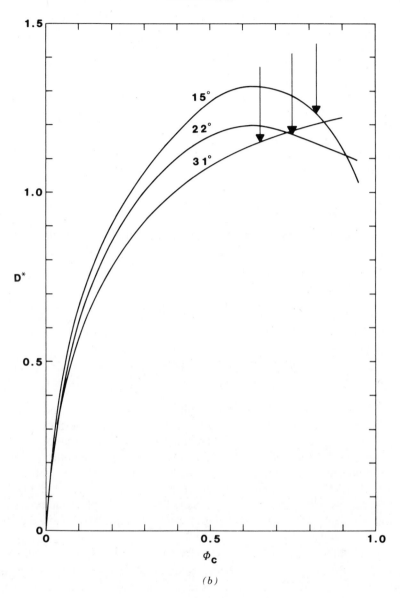

(b)

Figure 15.4 *(Continued)*

the collector divided by the incident beam power flux:

$$\varepsilon_e = eV_c \int_{eV_c}^{eV_0} F(T)\,dT \Big/ \int_0^{eV_0} TF(T)\,dT$$

$$= \Phi_c \int_{\Phi_c}^1 F(X)\,dX \Big/ \int_0^1 XF(X)\,dX. \qquad (15.14)$$

We can illustrate application of the theory with the example of a nonrelativistic scattered beam distribution. Following Section 10.4, we take a Gaussian distribution in transverse angle, θ:

$$P(\theta) \sim \exp\left(-\theta^2/\langle\theta^2\rangle\right). \qquad (15.15)$$

The quantity $\langle\theta^2\rangle$ is the mean-squared scattering angle. For small angles, the longitudinal kinetic energy is related to θ by

$$T/T_0 = X \cong 1 - \theta^2. \qquad (15.16)$$

An expression for the spectral function follows from the relationship $F(X)\,dX \cong P(\theta)\,d\theta$:

$$F(X) = \frac{\exp\left[-(1-X)/\langle\theta^2\rangle\right]}{\langle\theta^2\rangle[1 - \exp(-1/\langle\theta^2\rangle)]}. \qquad (15.17)$$

Figure 15.4 shows results of numerical solutions of Eqs. (15.12) and (15.14) for three different values of $\langle\theta^2\rangle$. Figure 15.4a plots the energy efficiency, ε_{norm}, as a function of the collector voltage, Φ_c. At a high value of Φ_c, the fraction of current collected is low. At low Φ_c, most of the current reaches the collector but the inverse diode recovers only a fraction of the kinetic energy. As a result, the efficiency reaches a maximum at an intermediate value of Φ_c. Note that ordered beams with low $\langle\theta^2\rangle$ give higher efficiency. Figure 15.4b plots the normalized gap width as a function of Φ_c and the input spectrum. The arrows show the points of peak efficiency.

As an example, suppose an incident electron beam has $j_0 = 100 \times 10^4\,\text{A/m}^2$, $eV_0 = 250\,\text{keV}$ and $\sqrt{\langle\theta^2\rangle} = 15$. With a beam radius of $r_0 = 0.05\,\text{m}$, the incident beam power is $1.96\,\text{GW}$. Figure 15.4a shows that the maximum power transferred to the inverse diode is $1.4\,\text{GW}$ at a collector voltage of $200\,\text{kV}$. Figure 15.4b gives a normalized gap spacing of $D^* = 1.2$, or a physical gap width of $d = 0.02\,\text{m}$. Because $d \ll r_0$, the one-dimensional model should give a good approximation.

Inverse diodes have application to devices for the generation of high-power pulsed microwave radiation. Here, a diode can capture most of the power of a modulated electron beam and transfer it to a high-voltage vacuum transmission line. The anticipated power levels ($> 1\,\text{GW}$) and frequency range ($100\text{–}500\,\text{MHz}$)

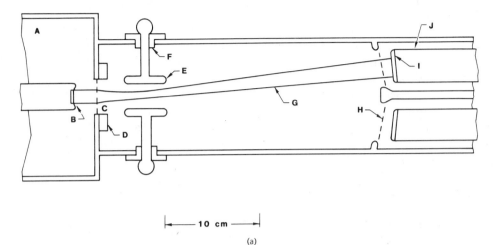

(a)

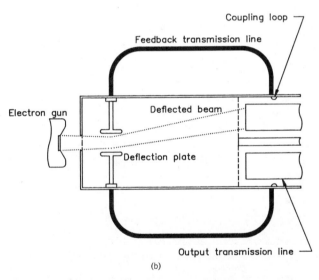

(b)

Figure 15.5 Pulsed-power switching by beam scanning. (*a*) Geometry of a scanned-beam-switch amplifier, side view. (*A*) Vacuum housing of electron beam generator. (*B*) Extended linear cathode. (*C*) Electron gun anode. (*D*) Matching magnetic lens. (*E*) Deflection plate. (*F*) Vacuum insulator and voltage feedthrough. (*G*) Deflected sheet beam in vacuum. (*H*) Inverse diode entrance grid. (*I*) Inverse diode collector. (*J*) High power strip transmission line. (*b*) Scanned beam switch with feedback acts as a high-power RF oscillator.

are well suited to high-gradient particle accelerators. Figure 15.5 shows one such device, the scanned beam switch. A relativistic electron gun generates an extended sheet beam—the results of Section 9.1 show that such beams can propagate long distances under vacuum. The beam passes between electric field deflection plates and continues to one of two inverse diodes. The collectors of the diode form the ends of vacuum transmission lines. Operation of the scanned beam switch as an amplifier is easy to understand. A bipolar square wave on the deflection plates sweeps the beam between the collectors, generating a high-power square wave on the output lines. The sheet beam geometry is critical to the operation of the scanned beam switch. It leads to low source current density, small beam expansion, and small system dimensions along the deflection direction.

Figure 15.5*b* shows a scanned beam switch configured as an oscillator. Feedback transmission lines connect the deflection plate to the output lines. The transmission lines act as quarter-wave resonators excited by the coupling loop in the output lines. The feedback lines have almost open-circuit terminations at the deflection plates. Let τ equal the period of the resonant mode. Depending on the orientation of the loop, the device oscillates if the net elapsed time for propagation of the electron beam and the electromagnetic radiation in the output line from the deflection plates to the coupling loop equals

$$\Delta t = \tau(1/4 + m) \qquad \text{or} \qquad \Delta t = \tau(3/4 + m), \tag{15.18}$$

where $m = 0, 1, 2, \ldots$

Figure 15.6 illustrates another device that uses an inverse diode to generate microwave power, the relativistic feedback oscillator. The relativistic oscillator is the high-power equivalent of a familiar vacuum triode circuit. It consists of a high-power electron diode with three electrodes and a feedback circuit. The

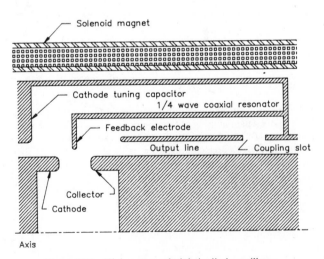

Figure 15.6   High-power relativistic diode oscillator.

cathode generates an annular electron beam in a magnetically insulated diode. The beam crosses to the collector of an inverse diode at the entrance of a high-power vacuum transmission line. The anode of the magnetically insulated diode connects to a high-Q resonant circuit excited by a coupling loop in the output line. The anode acts as a control grid to modulate the electron beam. The

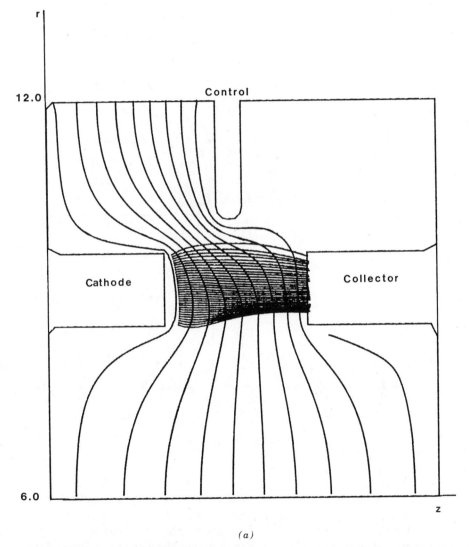

(a)

Figure 15.7 Numerical computation of electron flow in a foilless diode for a relativistic diode oscillator using the EGUN code. Cathode voltage, $-600\,\mathrm{kV}$; collector voltage, $0\,\mathrm{kV}$; applied axial magnetic field, 0.2 tesla. Distances in centimeters. (*a*) Control voltage, $0\,\mathrm{kV}$; predicted current, $28.9\,\mathrm{kA}$. (*b*) Control voltage, $-450\,\mathrm{kV}$; predicted current, $10.0\,\mathrm{kA}$.

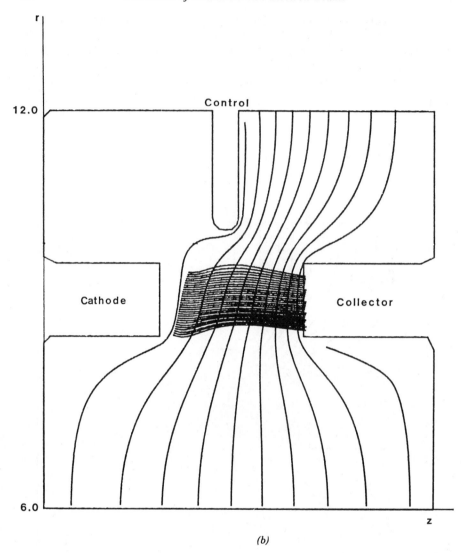

(b)

Figure 15.7 *(Continued)*

modulated electron beam drives the output transmission line at the frequency of the feedback oscillator with the proper choice of transit time from the diode to the coupling loop. The parallel inductance and capacitance on the output line is a resonant termination—the combined elements have high impedance only for the fundamental oscillation frequency. The termination keeps the DC voltage of the collector low, increasing the efficiency for converting beam energy to radiation at the fundamental frequency.

Figure 15.7 shows predictions from a ray tracing code for the foilless diode in a relativistic oscillator. The cathode voltage is $-600\,\text{kV}$, the collector voltage is near ground potential. The uniform solenoidal magnetic field is 0.2 tesla. Simulations at two values of the anode potential show that the anode voltage controls the current, even though the electrode is well outside the flow. In Fig. 15.7a, the anode potential equals zero and the current is 28.9 kA. Dropping the anode voltage to $-450\,\text{kV}$ reduces the current to 10.0 kA (Fig. 15.7b). Figure

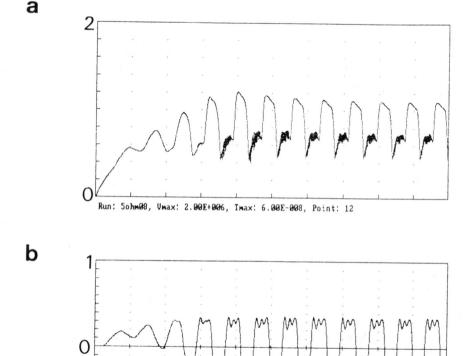

a

Run: 5ohm08, Vmax: 2.00E+006, Imax: 6.00E-008, Point: 12

b

Run: 5ohm08, Vmax: 1.00E+006, Imax: 6.00E-008, Point: 19

0   60

Figure 15.8 Circuit simulation of a high-current relativistic diode oscillator. Three-electrode magnetically insulated diode with matched current of 30 kA at 400 kV; 12-Ω output transmission line (1.5 nsec) connected to 50-Ω feedback line with 200-Ω termination (1.5 nsec). Termination in output line, DC short circuit, 12-Ω at 200 MHz. Pulse generator, 100-nsec transmission line with 800-kV charge, 30-nH series switch inductance. (a) Cathode voltage (ordinate: 0 to 2 MV, abscissa: 0 to 60 nsec). (b) Voltage at output termination (ordinate: 0 to 2 MV, abscissa: 0 to 60 nsec). Output power: 7.5 GW at 200 MHz.

15.8 shows the results of a circuit simulation of a relativistic oscillator driven by a 1-MV pulse generator. The figure shows the voltage on the cathode and output transmission line following initiation of the voltage pulse. The code predicts a net power output of 7.5 GW at 200 MHz with a 24% energy efficiency.

Power output coupling through inverse diodes and vacuum transmission lines has application to a variety of beam-driven RF sources including klystrons. To study the limits on power flux in a vacuum transmission line, suppose that a sheet beam with linear current density J and kinetic energy eV_0 drives a strip transmission line of width D. The impedance of such a line is

$$Z_0 = (\mu_0/\varepsilon_0)^{1/2}(d/2D). \tag{15.19}$$

The quantity d is the gap between the center and outer conductor. The voltage resulting from complete capture of the beam is

$$V_{line} = (JD)Z_0 = (\mu_0/\varepsilon_0)^{1/2}(dJ/2). \tag{15.20}$$

Under ideal conditions, the voltage of the inverse diode collector approaches V_0. Combining Eqs. (15.19) and (15.20) with the condition $V_{line} = V_0$, we find that the spacing between the transmission line of the electrodes must be

$$d \geqslant 2V_0/J(\mu_0/\varepsilon_0)^{1/2}. \tag{15.21}$$

The electric field in the strip transmission line is

$$E = V_0/d \leqslant J(\mu_0/\varepsilon_0)^{1/2}/2. \tag{15.22}$$

If we assume a maximum allowed electric field of $E = 15$ MV/m, Eq. (15.22) predicts a linear current density limit of $J = 7.95 \times 10^4$ A/m, independent of beam energy. As an example, suppose a 600-keV electron beam of width 0.5 m drives a strip transmission line. The maximum allowed beam current is about 40 kA, giving an output power of 24 GW.

15.2. DRIVING RESONANT CAVITIES WITH ELECTRON BEAMS

Resonant cavities are often used to extract energy from charged particle beams for microwave radiation generation. In this section, we shall study some features of resonant cavities and develop lumped circuit element models that give a simplified description of beam interactions with resonant modes. We shall concentrate on electron beams because they are easier to generate, accelerate, and transport than ion beams of equal power. In comparison with direct conversion devices like the inverse diode, resonant cavities have two important capabilities: frequency selection and impedance transformation.

Regarding frequency selection, we shall find that modulated beams often carry a broad range of harmonic components. While an inverse diode responds to all frequencies, energy transfer from a beam to a resonant cavity is strong only at specific frequencies. As a result, resonant cavities can generate highly mono-chromatic radiation from a modulated beam.

Impedance transformation is the second function of a resonant cavity. The charged particle beams in many devices have high kinetic energy and low current. For such beams, the impedance of an inverse diode and output transmission line would have to be very high for efficient energy conversion. Such a transmission line is impractical to build. On the other hand, most microwave loads have low AC impedance. We shall see that the resonant cavity can act as a transformer, converting beam energy at high voltage and low current to electromagnetic energy at low voltage and high current.

To begin, consider interaction between a single charged particle and a harmonic field variation. Figure 15.9 shows single electrons with kinetic energy $(\gamma_0 - 1)m_0c^2$ and velocity $v_z = \beta_0 c$ crossing a planar gap with width d and applied voltage

$$V(t) = V_0 \sin(\omega_0 t). \tag{15.23}$$

In cases of practical interest, the applied electric field is almost uniform in z across the gap. Therefore, the voltage of Eq. (15.23) produces an electric field:

$$E_z(t) = (V_0/d)\sin(\omega_0 t). \tag{15.24}$$

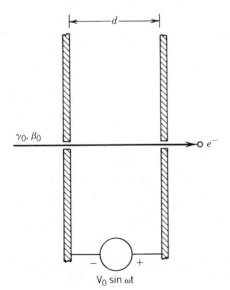

Figure 15.9 Electron beam crossing a planar gap with a harmonic applied voltage.

The kinetic energy of an electron that crosses the gap changes by an amount

$$\Delta T_e = \int dz\, eE_z. \tag{15.25}$$

To evaluate Eq. (15.25), we must account for changes in the electric field during the particle transit. To estimate the effect, we assume that $eV_0 \ll (\gamma_0 - 1)m_e c^2$ so that the electron velocity is almost constant in the gap. The expression of Eq. (15.25) becomes an integral over the transit time with the replacement $dz \cong \beta_0 c\, dt$. We represent the time at which the electron crosses the gap midplane as t_0. The *phase* of the particle with respect to the oscillating gap field is

$$\phi = \omega_0 t_0. \tag{15.26}$$

An electron of phase ϕ crosses the gap in the interval $(\phi/\omega_0 - d/2\beta_0 c) \leqslant t \leqslant (\phi/\omega_0 + d/\beta_0 c)$. We can rewrite Eq. (15.25) as

$$\Delta T_e \cong \frac{eV_0 \beta_0 c}{d} \int_{\phi/\omega_0 - d/2\beta_0 c}^{\phi/\omega_0 + d/2\beta_0 c} \sin(\omega_0 t)\, dt. \tag{15.27}$$

Evaluating the integral of Eq. (15.27) and expanding the resulting trigonometric functions gives the result

$$\frac{\Delta T_e}{eV_0} \cong \left[\left(\frac{2\beta_0 c}{\omega_0 d} \right) \sin \left(\frac{\omega_0 d}{2\beta_0 c} \right) \right] \sin\phi. \tag{15.28}$$

The quantity $V_0 \sin\phi$ is the voltage across the acceleration gap at the time the electron crosses the midplane. The factor in brackets is the *transit time factor*, T. We can also write T in terms of the electron transit time, $\Delta t = d/\beta_0 c$:

$$T = (2/\omega_0 \Delta t) \sin(\omega_0 \Delta t/2). \tag{15.29}$$

The quantity T is always less than or equal to unity. Equation (15.28) also shows that peak particle acceleration occurs when electrons cross at a phase of $\phi = \pi/2$. The gap decelerates electrons with phase in the range $-\pi \leqslant \phi \leqslant 0$.[1]

We next consider energy interchange between the gap and a beam of monoenergetic electrons with current $I(t)$. With the polarity definitions of Fig. 15.9, the energy transferred from the cavity to the beam over an interval t_0 is

$$\Delta E = \int_0^{t_0} dt (V_0 \sin\omega_0 t) I(t) T. \tag{15.30}$$

[1] Note that the phase definition of Eqs. (15.23) and (15.26) agrees with the common convention of particle phase used in RF accelerators [CPA, Chapter 13].

There is no time-averaged energy transfer to the cavity if the beam current is constant. The average energy exchange is also zero if the beam current has a harmonic variation at a frequency not equal to ω_0. The gap accelerates the beam or the beam drives electromagnetic oscillations only if the current varies at frequency ω_0.

In most devices, the beam current always flows in the same direction. The current of a modulated beam turns on and off rather than reversing direction. One possible form for the current of a modulated beam is

$$I(t) = (2I_0)[1 + \cos(\omega_0 t - \phi)]/2. \tag{15.31}$$

We can resolve the expression of Eq. (15.31) into a steady-state component of current that has no average energy exchange with the gap and a harmonic component at ω_0 with a strong interaction. We shall ignore the steady-state current and concentrate on the fundamental harmonic component, $I_0 \cos(\omega_0 t - \phi)$.

The time-averaged power transferred from the cavity to the harmonic component of an electron beam is

$$\langle P \rangle = \int_0^{2\pi/\omega_0} \frac{TV_0 \sin(\omega_0 t)I_0 \cos(\omega_0 t - \phi)dt}{(2\pi/\omega_0)}$$
$$= (V_0 I_0 T/2) \sin \phi. \tag{15.32}$$

In the phase range $0 \leqslant \phi \leqslant -\pi$, the beam drives voltage oscillations in the gap. If we add an external circuit to absorb energy from the gap voltage, we have the basis for a microwave generator. The power extracted from a beam with a given I_0 depends on the amplitude of gap voltage, the transit-time factor, and the phase between the beam current and the gap voltage.

Many cavities and slow-wave structures used for microwave generation and particle acceleration have geometries that are similar to a right circular cylinder. Therefore, we shall concentrate on the resonant modes of the *pillbox* cavity of Fig. 15.10a. The cavity has radius r_0 and length d. The most important resonant mode for charged particle acceleration or deceleration is the TM_{010} mode. The zeros in the subscript show that the electric and magnetic fields of the mode are uniform in the θ and z directions. The number 1 shows that the electric field has a simple radial variation with no nodes. The electric field of the TM_{010} mode is purely axial and has a maximum value at $r = 0$. As a result, the mode interacts strongly with a modulated, on-axis beam. The electric field is

$$\mathbf{E}(r, \theta, z, t) = E_0 J_0(2.405r/r_0) \sin(\omega_0 t)\mathbf{z}, \tag{15.33}$$

where J_0 is the zero-order Bessel function and the mode frequency is

$$\omega_0 = 2.405c/r_0. \tag{15.34}$$

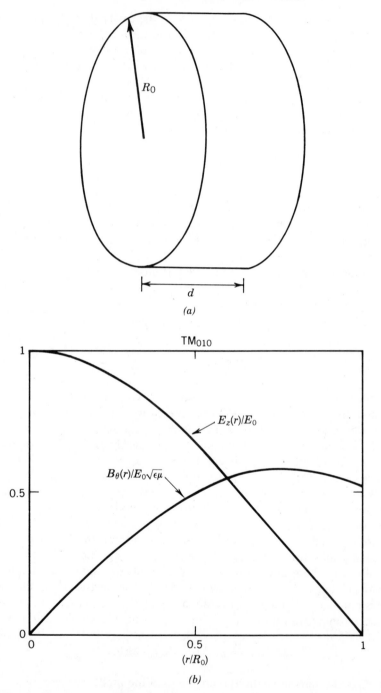

Figure 15.10 Cylindrical or pillbox resonant cavity. (*a*) Geometry. (*b*) Radial variation of normalized electric and magnetic fields.

The magnetic field points in the θ direction. It results from the axial displacement current from the oscillating electric field. The magnetic field is 90° out of phase with the electric field:

$$\mathbf{B}(r, \theta, z, t) = -(E_0/c)J_1(2.405r/r_0)\cos(\omega_0 t)\theta. \tag{15.35}$$

Figure 15.10*b* shows the radial variations of electric and magnetic fields.

The nature of the TM_{010} mode is easy to understand. The oscillating voltage across the center of the cavity creates axial displacement current. The flow of real current around the outside of the cavity completes the circuit. The changing magnetic field created by the current flow supports the voltage between the cavity faces. The electromagnetic energy is alternately exchanged between electric and magnetic fields. The total field energy in the cavity at any time is

$$U = (\pi r_0^2 d)(\varepsilon_0 E_0^2/2)J_1^2(2.405). \tag{15.36}$$

With no drive current, a resonant oscillation damps because of resistive losses associated with real current flow in the walls. We represent resistive power loss in a resonant cavity by the dimensionless quantity Q:

$$Q = \omega_0 U/\langle \text{Power loss to walls}\rangle = \omega_0 U/(-dU/dt). \tag{15.37}$$

Resonant cavities fabricated of copper usually have Q values in the range $> 10^4$. The theoretical Q value for a cylindrical cavity oscillating in the TM_{010} mode is

$$Q = (d/\delta)/(1 + d/r_0). \tag{15.38}$$

The quantity δ is the skin depth in the cavity walls. If the inside wall of the cavity has a volume resistivity ρ, then

$$\delta = (2\rho/\mu_0\omega_0)^{1/2}. \tag{15.39}$$

In most practical devices for microwave radiation generation, the driving beams are either continuous or vary over long times compared with the RF period. In this limit, we can represent a given mode of a resonant cavity with a lumped circuit element model. The model is useful for a localized beam interaction region. In this case, we only need to know the electric field at the beam position—we do not need a detailed description of fields throughout the cavity. We can represent any mode by an equivalent *RLC* circuit. We shall see how to choose the values of elements and how to define the mode voltage and current for the TM_{010} mode of a cylindrical cavity.

Figure 15.11 shows an equivalent circuit to represent the TM_{010} mode. The capacitor represents the concentration of electric fields near the axis, while the inductor represents the magnetic fields at large radius created by the current of the mode and the beam. The beam acts as an on-axis current source in

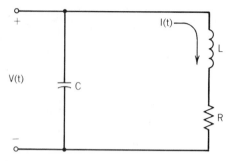

Figure 15.11   Lumped element circuit model to represent the TM_{010} mode in a cylindrical resonant cavity.

parallel with the cavity capacitance. The resistor in series with the inductor gives power loss from current flow in the cavity walls. We must choose L and C to give the correct resonant frequency:

$$LC = 1/\omega_0^2. \tag{15.40}$$

The quantity ω_0 is given by Eq. (15.34) if the cavity is a right circular cylinder of radius r_0. Note that the LC circuit has only one resonant frequency, while a cavity has many. The model of Fig. 15.11 applies only to a single specified mode.

The main point of interest is interaction of a beam with the on-axis electric field; therefore, a good definition of the cavity voltage is the integral of the axial electric field at $r = 0$:

$$V(t) = \int_0^d E_z(z, t)\, dz = E_0 d \sin(\omega_0 t). \tag{15.41}$$

We shall define the cavity current, $I(t)$, as the net current that flows through the cavity wall at $r = r_0$. This quantity equals the area integral of the axial displacement current:

$$j_{dz}(r, t) = \varepsilon_0 \frac{\partial E_z(r, t)}{\partial t}. \tag{15.42}$$

Substituting from Eq. (15.33), the displacement current is

$$j_{dz}(r, t) = \varepsilon_0 \omega_0 E_0 J_0(2.405 r/r_0) \cos(\omega_0 t). \tag{15.43}$$

A radial integral gives the net current

$$I(t) = \int_0^{r_0} (2\pi r\, dr)\varepsilon_0(2.405 c/r_0) E_0 J_0(2.405 r/r_0) \cos(\omega_0 t)$$

$$= 2\pi \varepsilon_0 E_0 r_0 c J_1(2.405) \cos(\omega_0 t). \tag{15.44}$$

If we choose the mode capacitance to satisfy the equation

$$C = I(t)/[dV(t)/dt],$$ (15.45)

then, Eqs. (15.41) and (15.44) imply that

$$C = \left[\frac{\varepsilon_0 \pi r_0^2}{d} \right] \left[\frac{2J_1(2.405)}{2.405} \right].$$ (15.46)

By Faraday's law, the voltage across the cavity axis is proportional to the spatial integral of the time derivative of magnetic flux between the axis and the outer wall. Applying Eq. (15.35), we can write this integral as

$$V(t) = \frac{\omega_0 E_0 d}{c} \int dr J_1 \left(\frac{2.405r}{r_0} \right) \sin(\omega_0 t).$$ (15.47)

The inductance satisfies the equation

$$L = V(t)/[dI(t)/dt].$$ (15.48)

The combination of Eqs. (15.44) and (15.47) gives the following equation for the inductance:

$$L = \left[\frac{\mu_0 d}{2\pi} \right] \left[\frac{1}{2.405 J_1(2.405)} \right].$$ (15.49)

Equations (15.46) and (15.49) give expressions for L and C that satisfy Eq. (15.40).

Resistive losses of energy result from the flow of real current around the outside of the cavity. We can model this power transfer by adding a resistor, R, in series with the inductor. We choose the value of R to be consistent with the Q value of the TM_{010} mode. If I_0 is the maximum value of wall current, then the stored electromagnetic energy in the mode is $LI_0^2/2$. The time-averaged power absorbed by the resistor is $I_0^2 R/2$. Substituting the expressions in Eq. (15.37) gives

$$R = (L/C)^{1/2}/Q.$$ (15.50)

The quantity $(L/C)^{1/2}$ has the units of ohms and is sometimes called the *cavity impedance*.

As an applications example, consider a beam-bunching cavity with $r_0 = 0.12$ m, $d = 0.04$ m, and a peak voltage of $V_0 = 50$ kV. The resonant frequency of the TM_{010} mode is $\omega_0 = 6.0 \times 10^9$ sec^{-1} ($f = 0.96$ GHz). If the cavity has copper walls, the theoretical value of Q from Eqs. (15.38) and (15.39) is 3×10^4. In practice, effects of wall roughness reduce the Q value—we shall take $Q = 1.5 \times 10^4$. Inserting values in Eqs. (15.46), (15.49), and (15.50) gives

$C = 4.3 \times 10^{-12}$ F, $L = 6.4 \times 10^{-9}$ H, and $R = 2.57 \times 10^{-3} \Omega$. The net wall current is $I_0 = 1.3$ kA, while the stored energy in the cavity is $U = 5.4 \times 10^{-3}$ J. The time-averaged power loss to the walls is 2.2 kW. A microwave generator must supply 2.2 kW to support the 50-kV peak voltage across the cavity.

We now turn to the problem of beam–cavity interactions. To begin, suppose that the beam is the only source of power for the cavity and that the cavity walls are the only power sink. We model the beam as an ideal current source. The time variation of the current of a beam does not depend on the cavity voltage; rather, it depends on the previous acceleration history. We add a current source to the circuit element model in parallel with the capacitor to model the beam. The component of beam current at ω_0 is $I_0 \sin(\omega_0 t)$.

For a steady-state beam, we seek an equilibrium solution. It is easy to calculate the cavity voltage for a given beam current using the formalism of complex impedance [CPA, Section 12.1]. The AC impedance of a two-terminal circuit element is the ratio of the voltage across the terminals to the current through the device:

$$Z(\omega) = V(\omega)/I(\omega). \tag{15.51}$$

The impedance may vary with frequency. If it has a complex part, the current and voltage differ in phase. Common passive circuit elements have the following impedances.

Resistor, R: $Z = R$,

Capacitor, C: $Z = -j/\omega C$, $\qquad\qquad$ (15.52)

Inductor, L: $Z = j\omega L$.

Note that the voltage and current of capacitors and inductors are 90° out of phase. We can combine impedances in series and parallel following the same rules used for resistors.

For the circuit model of Fig. 15.11, the cavity presents the following impedance to the beam for the TM_{010} mode:

$$Z(\omega) = \frac{R + j\omega L}{[1 - \omega^2 LC] + j\omega RC}. \tag{15.53}$$

When $\omega = \omega_0$, the bracketed quantity in the denominator of Eq. (15.53) equals zero and the impedance takes the following form:

$$Z(\omega_0) = [R + j(L/C)^{1/2}](L/C)^{1/2}/jR. \tag{15.54}$$

If the cavity has high Q, Eq. (15.50) shows that the first term in brackets in Eq. (15.54) is small, $R \ll (L/C)^{1/2}$. Dropping the term, we find that the cavity

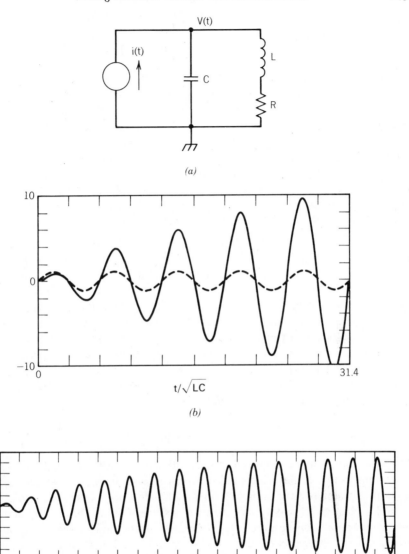

(a)

(b)

(b)

Figure 15.12   Numerical solutions for driven oscillations of a damped harmonic oscillator. (*a*) Early time variation of beam current (dashed line) and cavity voltage (solid line). (*b*) Expanded time axis showing growth and saturation of cavity voltage oscillations.

impedance is resistive:

$$Z(\omega_0) \cong (L/C)/R. \tag{15.55}$$

Equation (15.55) has two important implications. First, with the polarity conventions of Fig. 15.9, the cavity voltage in steady state is 180° out of phase with the beam current. In other words, the cavity oscillation assumes a phase to extract the maximum power from the beam. Second, the impedance, $Z(\omega_0)$, in Eq. (15.55) is much higher than the load resistance, $(L/C)R \gg R$. This result shows that the cavity acts as a transformer. A small beam current transfers energy at high voltage to the large circulating current of the resonant mode.

It is informative to observe the full time-dependent solution for electromagnetic oscillations in a driven cavity. Figure 15.21a shows numerical solutions of the circuit model of Fig. 15.11 for a beam with current:

$$i(t) = 0 \qquad (t < 0), \tag{15.56}$$
$$i(t) = \sin(t) \qquad (t \geqslant 0).$$

The figure plots the voltage of a cavity with $L = 1$, $C = 1$, $R = 0.05$, and $Q = 20$. Note that the cavity voltage approaches a phase of 180° relative to the beam current after an initial transient lasting a few periods. Figure 15.12b shows an extended view of the voltage oscillations. The time to saturation is called the cavity *fill* time. The fill time is approximately Q/ω_0, equal to 20 for the parameters of the example. If we stop the current at time t_0, Eq. (15.37) predicts that the stored electromagnetic energy of the mode decays as

$$U(t) = U(t_0)\exp[-\omega_0(t - t_0)/Q]. \tag{15.57}$$

The saturation voltage amplitude is $V_0 = 20$, consistent with Eq. (15.55).

To this point we have included only cavity wall losses. In devices for the generation of microwave radiation, only a small fraction of the available power should be lost to the cavity. We want to remove most of the electromagnetic power for applications. Figure 15.13 illustrates one method for energy extraction from a cavity. The center conductor of a transmission line penetrates the cavity near the outer radius, where it forms a loop that connects to the cavity wall. The area outlined by the loop is normal to the magnetic field of the TM_{010} mode. The oscillating field induces a voltage on the transmission line. Power travels from the cavity to a load through the line.

It is difficult to calculate the exact combined fields of the loop and the resonant mode. Fortunately, we can find simple approximations for power coupled to the transmission line in a parameter range of practical interest. We assume that the current that flows in the coupling loop has a small effect on the total fields in the cavity. In other words, the magnetic field near the loop is close to the field of an unperturbed TM_{010} mode. In this limit, we can estimate the transmission input voltage by applying Faraday's law assuming that the

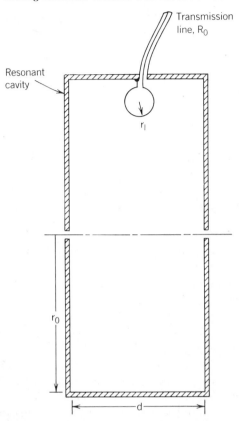

Figure 15.13 Energy extraction from a resonant cavity by a coupling loop and a transmission line.

net magnetic field is close to that of the resonant mode at $r = r_0$:

$$V_{\text{line}}(t) \cong (\pi r_1^2)\omega_0(E_0/c)J_1(2.405)\sin(\omega_0 t). \tag{15.58}$$

The quantity r_1 is the loop radius. If the loop drives a transmission line with matched termination R_0, then the time-averaged power leaving the cavity on the transmission line is

$$P_{\text{out}} = V_{\text{line}}^2/2R_0. \tag{15.59}$$

Equation (15.59) is valid if the following conditions hold:

1. The magnetic field generated by the loop current is much smaller than that of the resonant mode.
2. The energy removed by the transmission line in one cycle of the cavity oscillation is much smaller than the stored energy in the cavity, U.

We shall see that the two conditions are equivalent. Suppose a loop with azimuthal length l connects to a resistor, R_0. The mode magnetic field at the loop is

$$B_{\text{mode}} = (E_0/c)J_1(2.405)\exp(j\omega_0 t). \qquad (15.60)$$

A current $I_1\exp(j\omega_0 t)$ flows in the loop in response to the mode field. The loop current creates a diamagnetic field

$$B_{\text{loop}} = -\mu_0 I_1 \exp(j\omega_0 t)/l. \qquad (15.61)$$

The quantity I_1 may have a complex component to represent a phase difference. We can equate the loop voltage to the rate of change of the net enclosed magnetic flux:

$$I_1 R_0 = j\omega_0[(E_0/c)J_1(2.405) - \mu_0 I_1/l]. \qquad (15.62)$$

Equation (15.62) shows that there are two regimes of steady-state loop operation. In the limit

$$(\mu_0\pi r_1^2/l)/R_0 \ll (1/\omega_0), \qquad (15.63)$$

the field inside the loop is close to the applied field and Eq. (15.59) holds. At the opposite extreme, the loop current is high and the magnetic field inside the loop is much smaller than the field of the resonant mode.

We can write the power absorbed through the coupling loop in terms of a quality factor for the resonant mode:

$$Q_1 = \omega_0 U/P_{\text{out}}. \qquad (15.64)$$

Equation (15.59) gives the average output power, P_{out}. Substitution from Eq. (15.58) gives the following expression:

$$Q_1 = (r_0^2 d)R_0/\pi\mu_0 r_1^4\omega_0. \qquad (15.65)$$

The quantities r_0 and d are the cavity dimensions. Substituting for the output transmission line impedance, R_0, from the condition of Eq. (15.63) gives the following relationship:

$$Q_1 \gg (\pi r_0^2 d)/(\pi r_1^2 l). \qquad (15.66)$$

For small loop perturbation, Eq. (15.66) shows that Q_1 must be much larger than the ratio of the cavity volume to the volume enclosed by the loop. When $Q \gg 1$, the loop extracts a small fraction of the stored cavity energy in each cycle.

We can find the transformer ratio of the cavity when the limit of Eq. (15.66) holds. The voltage amplitude for power extraction from the beam is $E_0 dT$. Equation (15.58) gives the voltage on the output transmission line. The transformer step-down ratio equals the ratio of the voltages:

$$N = \frac{E_0 dT}{V_{\text{line}}} = \frac{Tr_0 d}{2.405 J_1 (2.405) \pi r_1^2}.$$

(15.67)

Equation (15.67) shows that a beam-driven cavity is an effective step-down transformer. If the loop area is much smaller than the cross-section area of the cavity, $2r_0 d$, large values of N result. With no transformer losses, the current amplitude in the transmission line is a factor of N times the amplitude of the beam current component at ω_0. Wall losses reduce the energy efficiency of the cavity as a transformer. When there are both wall losses and output coupling, we can find R in the lumped circuit element model by using a total quality factor

$$Q = (1/Q_w + 1/Q_l)^{-1}.$$

(15.68)

The quantity Q_w is the wall loss factor of Eq. (15.38).

15.3. LONGITUDINAL BEAM BUNCHING

We found in Section 15.3 that a charged particle beam interacts strongly with a resonant cavity only if it has a current modulation at the resonant frequency. One way to create electron beams with a low-frequency modulation is to apply an oscillating voltage to a grid electrode close to the cathode—the grid turns the flow on and off. This method is not practical at microwave frequencies because of transit-time limitations on the acceleration of slow electrons in the cathode-grid gap.[2] For high-frequency modulation, the usual approach is to modify the velocity distribution of the full energy beam.

In this section, we shall discuss methods to modulate high-energy electron and ion beams by adding variations in longitudinal energy. The longitudinal compression of a steady-state beam into periodic pulses is called *beam bunching.* Longitudinal bunching has application to the klystron microwave amplifier (Section 15.4) and to beam conditioning for injection into RF linacs. Axial compression of a pulsed beam amplifies the instantaneous beam power; therefore, bunching is a major component of inertial fusion driver systems using ion beams.

We shall concentrate initially on beam bunching by impressing periodic variations of longitudinal velocity on a beam. This method is useful only for nonrelativistic beams because relativistic particles travel close to the speed of light, independent of their energy. We shall start with a simple model that neglects the effects of space-charge fields and longitudinal velocity spread.

---

[2] See, for instance, C. K. Birdsall and W. B. Bridge, *Electron Dynamics of Diode Regions*, Academic Press, New York, 1966.

Suppose an incident monoenergetic beam has kinetic energy T_0 and uniform current I_0. The velocity of all particles is $v_0 = (2T_0/m_0)^{1/2}$. The beam crosses a gap with a voltage waveform, $V(t)$, periodic over time $1/f_0$. For nonrelativistic beams, the energy shift from the buncher results in a variation of axial velocity. The gap accelerates some particles and decelerates others. As the beam propagates, fast particles overtake slower ones. The beam current is higher at points where the particle orbits converge along z. At a downstream point, the beam has a modulated current with a component at frequency f_0.

To calculate the current variations of a drifting beam, let the variable t represent the time a group of particles crosses the buncher. We want to find the time-dependent current at a distance L from the buncher (Fig. 15.14). The variable t' represents the time that the group of particles reaches the point. The variation of axial velocities causes a variation of current at the observation point, $I(t')$.

All particles that cross the buncher ultimately pass the observation point. Consider the charge that passes through the buncher in a small interval, $I_0 \, dt$. Depending on the axial velocity variation, the charge may take a longer or shorter time, dt', to pass the observation point. Conservation of charge implies that $I_0 \, dt = I \, dt'$, or

$$I(t) = I_0(t)/(dt'/dt). \tag{15.69}$$

To find $I(t)$, we must calculate t' as a function of t. Particles that cross the buncher at t arrive at the observation point after a drift delay:

$$t' = t + L/v(t). \tag{15.70}$$

The quantity $v(t)$ is the exit velocity from the buncher gap:

$$v(t) = v_0[1 + eV(t)/T_0]^{1/2}. \tag{15.71}$$

In limit that $eV(t) \ll T_0$, Eq. (15.71) reduces to

$$t' \cong t + (L/v_0)[1 - eV(t)/2T_0]. \tag{15.72}$$

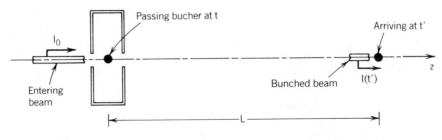

Figure 15.14 Axial bunching of a nonrelativistic beam by an applied voltage in a cavity.

Equation (15.72) implies that

$$dt'/dt = 1 - (L/v_0)(e/2T_0)(dV(t)/dt). \qquad (15.73)$$

We can generate a harmonic energy shift by directing a beam along the axis of a resonant cavity oscillating in the TM_{010} mode. For a harmonic voltage, the energy shift at the cavity exit is

$$eV(t) = eV_0T\sin(\omega_0 t), \qquad (15.74)$$

where T is the transit-time factor of Section 15.2. The ratio of the interval at the observation point to that at the buncher is

$$dt'/dt = 1 - (L\omega_0/v_0)(eV_0T/2T_0)\cos\omega_0 t. \qquad (15.75)$$

Inserting Eq. (15.75) into Eq. (15.69), gives the following expression for beam current at the observation point:

$$I(t) = I_0/[1 - \chi\cos\omega_0(t)]. \qquad (15.76)$$

The quantity χ is the *harmonic bunching parameter*:

$$\chi = (L\omega_0/v_0)(eV_0T/2T_0). \qquad (15.77)$$

Note that $I(t)$ is the current at the observation point for particles that leave the injector at time t. In Section 15.4 we shall use Eq. (15.76) to find the net current as a function of time at the observation point, $I(t')$. Figure 15.15 illustrates results of this calculation—the graphs show $I(t')$ for different values of χ. A high value of χ corresponds to increased distance from the bunching cavity. At $L = 0$, the current is constant. Moving downstream, the current increases for particles that pass through the buncher during the rising portion of the voltage waveform. The density divergence at $\chi = 1$ occurs when particles that cross the buncher near $t = 0$ overtake one another. By inspection of Fig. 15.15, we expect good compression between $\chi = 1$ and $\chi = 2$. Section 15.4 shows that the best value of the bunching parameter for beam modulation is $\chi = 1.84$.

The harmonic buncher produces a beam with a strong modulation component at frequency f_0—it is useful for microwave generation (Section 15.4). For other applications, such as beam conditioning for injection into RF linacs, the goal is to gather the beam particles into narrow bunches. Here, the ideal bunching voltage waveform is the sawtooth:

$$V(t) = V_0t/(1/2f_0) \qquad (-1/2f_0 < t \leqslant 1/2f_0). \qquad (15.78)$$

We can find the location of maximum particle convergence by substituting the waveform of Eq. (15.78) into Eqs. (15.69) and (15.73). Bunching occurs when

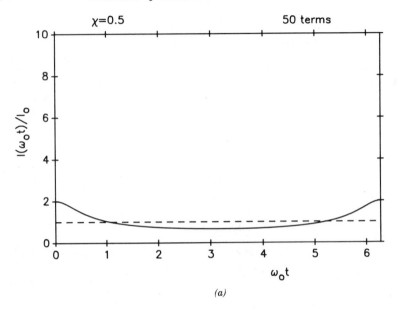

(a)

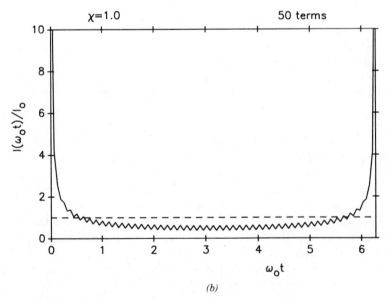

(b)

Figure 15.15 Current as a function of time downstream from a harmonic buncher with frequency ω_0 and different values of bunching parameter χ [using 50-term series expansion, Eq. (15.105)]. (a) $\chi = 0.5$. (b) $\chi = 1.0$. (c) $\chi = 1.5$. (d) $\chi = 2.0$.

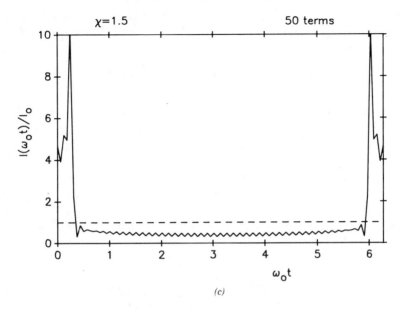

(c)

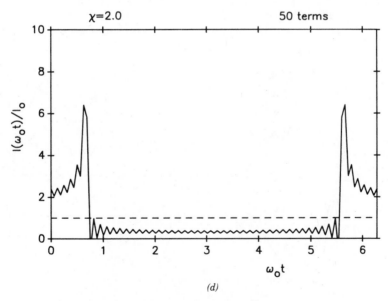

(d)

Figure 15.15 *(Continued)*

the denominator equals zero. We can derive the same result from the following argument. Ignoring transit-time effects, particles that cross the buncher at the extremes of the voltage variation have velocities

$$v \cong v_0(1 \pm eV_0/2T_0) = v_0 + \Delta v. \tag{15.79}$$

A particle at the bunch tail must travel a distance L in the reduced time $(L/v_0 - 1/2f_0)$ to catch up to the particles that cross the buncher when $V(t) = 0$, or

$$\Delta v \left(\frac{L}{v_0} - \frac{1}{2f_0} \right) = \frac{v_0}{2f_0}. \tag{15.80}$$

In the limit that $eV_0 \ll T_0$ and $(1/2f_0) \ll L/v_0$, the bunching length is

$$L = (v_0/f_0)(T_0/eV_0). \tag{15.81}$$

We can illustrate the application of Eq. (15.81) with an example relevant to inertial fusion with intense light ion beams. Consider a proton beam driven by a pulsed power generator with a ramped voltage that rises from 4 to 5 MV over a 100-nsec pulse length. We can resolve the voltage pulse into a sawtooth superimposed on a 4.5-MV square pulse. The parameters imply that $V_0 = 0.5\,\text{MV}$, $f_0 = 1.0 \times 10^7\,\text{sec}^{-1}$, and $v_0 = 2.94 \times 10^7\,\text{m/sec}$. Substituting in Eq. (15.81) gives a bunching distance of $L = 26\,\text{m}$. The long distance implies that

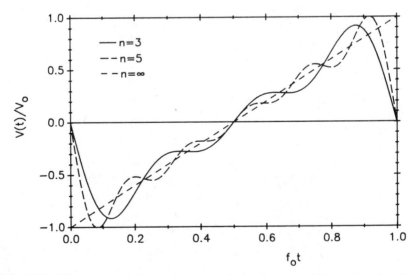

Figure 15.16 Effective cumulative voltage waveforms in a multiple cavity buncher— approximations to a sawtooth. Results included for three and five cavity bunchers compared with the ideal waveform.

longitudinal bunching of intense ion beams is possible only with an effective method for extended transverse confinement.

At high frequency, it is difficult to generate sawtooth voltage waveforms in a nonresonant structure. An alternate approach useful for accelerator bunchers simulates the sawtooth waveform with multiple resonant cavities. The beam crosses several bunching cavities with frequencies $f_0, 2f_0, 3f_0, \ldots$. The total energy shift is the sum of cavity voltages. The strategy is to adjust the phases and amplitudes of the cavity oscillations so that the voltages at the time of the particle transit constitute the initial terms of a Fourier series expansion of a sawtooth. Figure 15.16 illustrates the summed waveform for a three-cavity buncher. Bunchers with multiple resonators involve complex technology. They are used for beam conditioning when the available flux from the particle source limits the accelerator output.

The Applegate diagram is an effective method to visualize the bunching process. The diagram plots several particle orbits in the drift space beyond the buncher with axial distance on the ordinate and time on the abscissa. The slope of the drift orbits depends on the velocity at the exit of the buncher. Figure 15.17 is

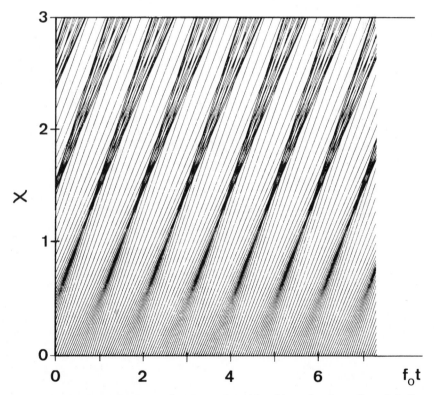

Figure 15.17 Applegate diagram—plot of several particle orbits as functions of χ and f_0t for an applied voltage at $\chi = 0$ of the form $\sin(2\pi f_0 t)$.

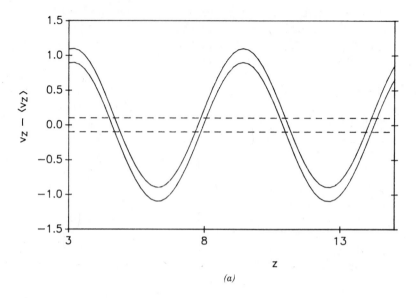

(a)

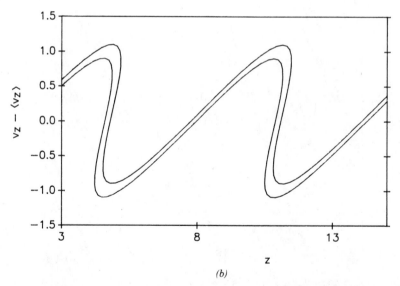

(b)

Figure 15.18 Effect of a small axial velocity spread on beam bunching with an harmonic applied voltage. Axial phase-space plots. (*a*) Beam distribution emerging from the buncher. (*b*) Beam distribution near the point of peak bunching.

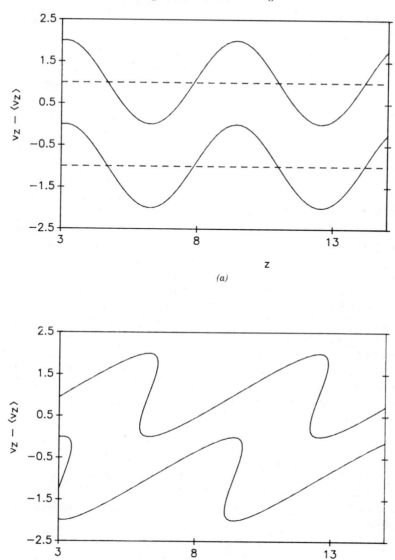

(a)

(b)

Figure 15.19 Effect of a large axial velocity spread on beam bunching with an harmonic applied voltage. Axial phase-space plots. (*a*) Beam distribution emerging from the buncher. (*b*) Beam distribution near the point of peak bunching.

an Applegate plot for a harmonic voltage applied at $z = 0$. Note the convergence of orbits at downstream locations.

The divergence of beam current to an infinite value predicted by Eq. (15.76) does not occur in real beams. The effects of space charge and axial velocity spread limit the peak longitudinal compression. A longitudinal velocity spread limits axial compression in the same way that a transverse velocity spread limits focusing. Figure 15.18 illustrates the effect of velocity spread for a harmonic buncher. The figure shows an axial phase-space plot of a nonrelativistic beam with a uniform distribution function. The incident beam has an axial velocity spread of $\pm \delta v$. The buncher impresses a harmonic velocity variation with amplitude $\pm \Delta v_0$. Figure 15.18 shows a case where $\Delta v_0 \gg \delta v$. From Section 2.5, the configuration space beam density is proportional to the width of the distribution along the velocity axis. When $\Delta v_0 \ll v_0$, the current, $I(z)$, is proportional to the density. Figure 15.18a shows that the beam distribution emerging from the buncher has uniform width in v_z although it is strongly distorted. The distribution changes after propagation. Figure 15.18b shows that the width of the distribution along v_z is large at positions where fast particles overtake slow ones. The resulting density exhibits strong bunching. Figure 15.19 illustrates a similar sequence of events for a beam with a large initial velocity spread, $\delta v \approx \Delta v_0$. Here, the width of the beam along v_z is not a strong function of z—bunching is less pronounced.

We can derive a quantitative expression for the effect of axial velocity spread most easily for the sawtooth bunching waveform. Figure 15.20a shows a beam segment of length $2\Delta z$ and velocity width $2\delta v$ entering a sawtooth buncher. The distribution leaving the buncher (Fig. 15.20b) has a linear velocity displacement with peak amplitude Δv_0—we assume that

$$\Delta v_0 \gg \delta v. \tag{15.82}$$

Peak bunching occurs after the beam drifts for a time $\Delta z/\Delta v_0$ (Fig. 15.20c). Because of the velocity spread, the beam segment has a nonzero width $2\Delta z'$ at the point of peak bunching. The amplification of current at the bunching point is approximately

$$I/I_0 \cong \Delta z/\Delta z'. \tag{15.83}$$

Conservation of emittance implies that

$$\Delta z' \Delta v_0 \approx \Delta z \, \delta v. \tag{15.84}$$

Therefore, the *bunching ratio* is

$$I/I_0 \cong \Delta v_0/\delta v. \tag{15.85}$$

Equation (15.85) shows that a small initial velocity spread is necessary for strong bunching.

We can estimate the effect of axial space-charge electric fields by viewing the bunching process in the rest frame of the beam. In this frame, the buncher creates a flow of particles toward a common point. Strong space-charge fields slow the converging particles and ultimately reflect them. This effect determines a distance of closest approach and hence a peak value for the bunched current. A complete self-consistent treatment of beam bunching with space-charge must account for the nonlinear axial forces from the beam and from wall charges in conducting boundaries. Here, we shall use a simplified model to estimate bunching limits for a nonrelativistic beam. A narrow cylindrical beam of radius r_0 propagates in a

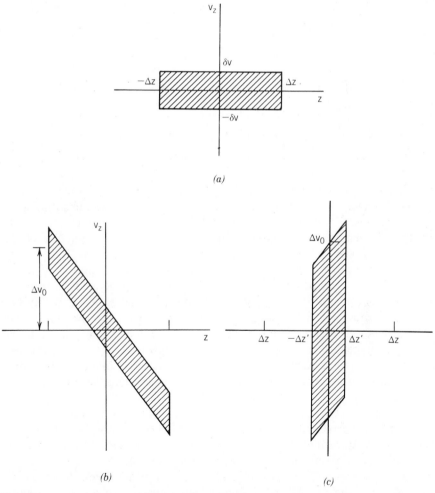

Figure 15.20 Axial phase-space plots for the bunching of a rectangular beam distribution with a sawtooth applied voltage. (*a*) Beam distribution entering the buncher. (*b*) Beam distribution leaving the buncher. (*c*) Beam distribution at the point of peak bunching.

circular pipe of radius r_w. The beam has a small initial velocity spread that satisfies the condition of Eq. (15.82). We assume that the beam length at all axial positions is much smaller than r_w, justifying the use of field expressions for an infinite-length beam.

Particles emerging from the buncher converge toward a common point. The compression results in an increase of the electrostatic potential in the beam. The process stops when $e\phi$ reaches a level comparable to the kinetic energy of convergence. To derive a bunching limit, we must estimate the kinetic energy of particles in the beam rest frame. In the stationary frame, the particles emerge from the buncher with a kinetic energy variation $\pm eV_0$ and a corresponding shift of axial velocity, $\Delta v_0 \cong \pm eV_0/m_0 v_0$. In the rest frame, the kinetic energy of convergence is

$$\Delta T \cong m_0 \Delta v_0^2/2 = (eV_0/2)(eV_0/m_0 v_0^2)$$
$$= (eV_0/4)(eV_0/T_0). \tag{15.86}$$

Equating ΔT to the space-charge potential energy of Eq. (5.68), we find the space-charge limit on the bunching ratio:

$$\frac{I}{I_0} \leqslant \left[\frac{\beta V_0}{240}\right]\left[\frac{eV_0}{T_0 I_0}\right]\left[\frac{1}{\ln(r_w/r_b)}\right]. \tag{15.87}$$

with V_0 in volts. To illustrate application of Eq. (15.87), consider bunching a 1-A, 100-keV electron beam. We take $r_w/r_0 = 4$ and $V_0 = 25\,\text{kV}$. Substituting in Eq. (15.87) gives $I/I_0 \leqslant 10$. Space-charge forces also modify the point of peak bunching. Depending on the beam geometry, bunching occurs at smaller values of χ.

Relativistic electron beams have advantages for the generation of high-power microwave radiation. The cancellation of beam-generated electric and magnetic fields makes confinement of high-power beam easier at relativistic energies. Because the beam particles all have velocity $v_z \cong c$, we must find alternate methods of modulation. Given a successful bunching method, relativistic electron beams have the important advantage that they preserve the modulation structure over long distances. The beam could interact with a long array of resonant structures—in principle, a single beam could drive an entire RF accelerator.

Figure 15.21 shows one method to bunch a relativistic beam. A constant-current beam emerges from a bunching cavity with a harmonic variation of kinetic energy. Instead of a linear drift space, the beam enters a helical transport region with a vertical magnetic field. The electrons are bunched when they emerge because they follow different path lengths through the bending field. High-energy particles have longer paths and therefore take more time to reach the output. After a 360° revolution, the particle orbits converge to a common axis at different times.

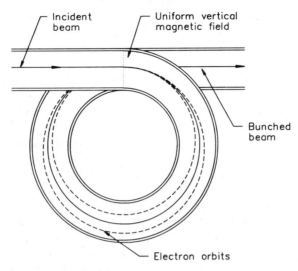

Figure 15.21 Bunching a relativistic electron beam by introduction of an energy modulation and transport through a uniform transverse magnetic field.

A mathematical description of the bunching process is easy in the ultrarelativistic limit. We can modify Eq. (15.70) so that the transit time from the buncher to the observation point depends on the path length, $L(t)$, rather than on the particle velocity:

$$t' = t + L(t)/c. \tag{15.88}$$

The path length is the sum of the distance through the bending field. In a uniform axial magnetic field, B_0, the electrons follow circular orbits with an energy-dependent radius:

$$r_B(t) \cong \gamma(t)m_0c/eB_0 = (m_0c/eB_0)[\gamma_0 + eV(t)/m_0c^2]. \tag{15.89}$$

The quantity $eV(t)$ is the buncher voltage and γ_0 is the entrance kinetic energy. The total path length is

$$L(t) \cong 2\pi r_g(t) = 2\pi r_{B_0}[1 + eV(t)/\gamma_0m_0c^2]. \tag{15.90}$$

where $r_{B_0} = \gamma_0m_0c/eB_0$. Combining Eqs. (15.69), (15.88), and (15.90) gives the following prediction for the time-dependent current emerging from the helix:

$$I(t') = I_0/[1 + \chi_r \cos \omega_0(t' - L/c)]. \tag{15.91}$$

The quantity χ_r in Eq. (15.91) is the *relativistic harmonic bunching parameter* for a magnetic field,

$$\chi_r = (2\pi\omega_0 r_{B_0}/c)(eV_0/\gamma_0 m_0 c^2). \qquad (15.92)$$

Equation (15.91) is similar to Eq. (15.76) except for the plus sign in the denominator. The difference arises because bunching occurs for particles that cross the buncher during the fall of the voltage. The calculation in an electric field buncher is much more complex because the field and the total electron energy vary with radius. For a rough estimate of the field magnitude, we can find an average value that gives a circular orbit on the equilibrium axis:

$$r_{B_0} \sim \gamma m_0 c^2/eB_0. \qquad (15.93)$$

As an applications example, assume a harmonic buncher oscillates at frequency $\omega_0/2\pi = 1$ GHz with a peak voltage of $V_0 = 50$ kV. The input electron beam has kinetic energy 2 MeV. The beam relativistic factors are $\gamma_0 = 4.91$ and $\beta_0 = 0.979$. Substituting parameters in Eq. (15.92) with $\chi_r^2 = 1.84$ gives $r_{B_0} = 0.7$ m. The field for magnetic bending is $B_0 = 0.012$ tesla.

15.4. KLYSTRON

In this section, we shall begin our study of beam devices to generate microwave radiation. In the microwave regime of the electromagnetic spectrum, wavelengths are comparable to the size of structures—they lie in the range of 1 mm to 10 cm. As a result, closed resonant cavities and waveguides are usually used to transport the radiation. For historical reasons, microwave radiation is divided into the frequency bands listed in Table 15.1.

The klystron is the most widely used microwave source for high-energy particle accelerators. For this application, klystrons have several advantages. They function as high-gain amplifiers with excellent frequency stability. A master oscillator can drive several klystrons in parallel to provide the high peak-power levels required for RF linacs.

Figure 15.22 shows the two-cavity klystron amplifier. A steady state or pulsed electron beam with current I_0 and kinetic energy T_0 enters the first cavity, the buncher. An input signal at frequency f_0 supports a TM_{010} cavity oscillation. The buncher imparts a harmonic velocity shift to the beam. Following the discussion of Section 15.3, the velocity shift results in a modulated current downstream. The load cavity, located at the point of peak modulation, usually has a TM_{010} resonance at frequency f_0. Section 15.2 showed how the modulated beam drives a resonant oscillation. Microwave energy is extracted from the load cavity by a loop or other coupling device to drive low-impedance loads. A beam dump at the end of the klystron absorbs the unused components of beam energy.

Many klystrons have reentrant cavities like that of Fig. 15.22. Compared with a pillbox cavity of the same diameter, a reentrant cavity has a lower TM_{010}

TABLE 15.1   Microwave Bands

Band	Frequency Range (GHz)	Wavelength (m)	Band	Frequency Range (GHz)	Wavelength (mm)
UHF	0.30	1.000	*Ka*	26.5	11.32
	1.12	0.268		40	7.50
L	1.12	0.268	*Q*	33	9.09
	1.70	0.176		50	6.00
LS	1.70	0.176	*U*	40	7.50
	2.60	0.115		60	5.00
S	2.60	0.115	*M*	50	6.00
	3.95	0.076		75	4.00
C	3.95	0.076	*E*	60	5.00
	5.85	0.051		90	3.33
XC	5.85	0.051	*F*	90	3.33
	8.20	0.037		140	2.14
X	8.20	0.037	*G*	140	2.14
	12.40	0.024		220	1.36
Ku	12.40	0.024	*R*	220	1.36
	18.00	0.017		325	0.92
K	18	16.67			
	26.5	11.32			

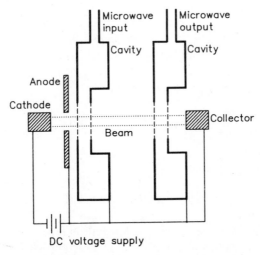

Figure 15.22   Schematic drawing of a two-cavity klystron amplifier.

frequency because of the added capacitance on axis. As a result, reentrant cavities are more compact for a given operating frequency. Another advantage of the geometry is that the narrow gap gives a transit-time factor (Section 15.2) close to unity. The gaps in klystrons are often defined by grids. The grids limit the axial extent of cavity fields and confine oscillations to a single cavity. We shall see that the klystron operates as an oscillator rather than as an amplifier if microwave energy couples back from the load cavity to the buncher.

To find the best geometry for a klystron, we must analyze the frequency content of bunched electron beams. Figure 15.3 shows that the beam downstream from a harmonic buncher has a modulation at the fundamental frequency, f_0, and a steady-state component. The sharp peaks suggest the presence of high-frequency components. We can find the harmonic content of the bunched beam current by a Fourier analysis. Section 15.3 gives the current at a position corresponding to bunching parameter χ as

$$I(t) = I_0/(1 - \chi \cos \omega_0 t). \tag{15.94}$$

We seek a Fourier decomposition of the downstream current as a function of t', the arrival time

$$I(t) = a_0 + \sum_{n=0}^{\infty} [a_n \cos(n\omega_0 t') + b_n \sin(n\omega_0 t')]. \tag{15.95}$$

Following standard techniques, we can find the coefficients in Eq. (15.95) in terms of integrals over the current waveform:

$$a_0 = \int_{-\pi}^{+\pi} I(\omega_0 t') d(\omega_0 t')/2\pi, \tag{15.96}$$

$$a_n = \int_{-\pi}^{+\pi} I(\omega_0 t') \cos(n\omega_0 t') d(\omega_0 t')/\pi, \tag{15.97}$$

and

$$b_n = \int_{-\pi}^{+\pi} I(\omega_0 t') \sin(n\omega_0 t') d(\omega_0 t')/\pi. \tag{15.98}$$

Equation (15.69) implies that

$$I(\omega_0 t') d(\omega_0 t') = I_0 d(\omega_0 t). \tag{15.99}$$

Substitution of Eq. (15.99) simplifies the integrals in Eqs. (15.96)–(15.98). We can then evaluate the integrals over the variable t, the particle crossing time at the buncher. The expression for a_0 becomes

$$a_0 = \int_{-\pi}^{+\pi} I(\omega_0 t') d(\omega_0 t')/2\pi$$

$$= \int_{-\pi}^{+\pi} I_0 d(\omega_0 t')/2\pi = I_0. \tag{15.100}$$

We can modify the higher-order terms by using the expression

$$t' = t + (L/v_0) - (eV_0/2T_0)(L_0/v_0)\sin(\omega_0 t). \tag{15.101}$$

Substitution of Eqs. (15.99) and (15.101) in Eqs. (15.97) and (15.98) leads to the equations

$$a_n = \int_{-\pi}^{+\pi} I_0 \cos[n(\omega_0 t + L\omega_0/v_0 - \chi \sin \omega_0 t)] d(\omega_0 t)/\pi \tag{15.102}$$

and

$$b_n = \int_{-\pi}^{+\pi} I_0 \sin[n(\omega_0 t + L\omega_0/v_0 - \chi \sin \omega_0 t)] d(\omega_0 t)/\pi. \tag{15.103}$$

Equations (15.102) and (15.103) have the analytic solutions[3]:

$$a_n = 2I_0 J_n(n\chi) \cos(n\omega_0 L/v_0), \tag{15.104}$$

$$b_n = 2I_0 J_n(n\chi) \sin(n\omega_0 L/v_0),$$

where J_n is a Bessel function of order n. Combining Eqs. (15.95), (15.100), and (15.104) gives the downstream current:

$$I(t') = I_0 \left\{ 1 + \sum_{n=1}^{\infty} 2J_n(n\chi) \cos\left[n\left(\omega_0 t' - \frac{L\omega_0}{v_0} \right) \right] \right\}. \tag{15.105}$$

The expression of Eq. (15.105) consists of a series of phase-shifted harmonic components at frequency $n\omega_0$ multiplied by the coefficients $2J_n(n\chi)$. Figure 15.23 plots the coefficients for different harmonics as a function of the bunching parameter χ. The proportion of high-order modes reaches a peak at $\chi = 1$ where the density diverges to infinity. The amplitude of the fundamental mode component reaches a maximum value at $\chi = 1.84$—at this point, $I_1 = 1.16 I_0$. To generate radiation at frequency f_0, the best location for the load cavity of a klystron is

$$L = 1.84(v_0/\omega_0)(eV_0/2T_0). \tag{15.106}$$

[3]See, for instance, H.B. Dwight, *Tables of Integrals and Other Mathematical Data*, Macmillan, New York, 1947.

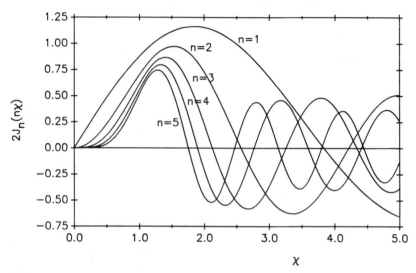

Figure 15.23   Harmonic content of a beam with a sinusoidal energy modulation as a function of the bunching parameter χ.

Figure 15.23 shows that the beam carries significant components of high-order modes at different values of χ. With some sacrifice of energy efficiency, a klystron can function as a frequency multiplying amplifier. Table 15.2 lists the best values of bunching parameter to generate high-frequency output radiation along with the ratio of the harmonic current component to the steady-state beam current. Figure 15.24 shows a klystron frequency doubler.

The harmonic analysis gives an upper limit for the energy efficiency of a klystron. Although the beam carries a broad frequency spectrum, energy extraction occurs only from the current component in resonance with the output cavity. If the load cavity has a resonance at the frequency of the buncher cavity and if $\chi = 1.84$, then the cavity extracts energy from the component of beam current $1.16 I_0 \cos(\omega_0 t)$. Section 15.2 showed that the impedance of the load cavity at resonance is resistive. If R_0 is the on-axis impedance, then the voltage

TABLE 15.2   Frequency Multiplication in a Klystron

Harmonic number, n (f/f_0)	Bunching Parameter for Maximum Output	I_n/I_0 (maximum)
1	1.84	1.16
2	1.52	0.97
3	1.40	0.87
4	1.33	0.80
5	1.28	0.75
10	1.18	0.61

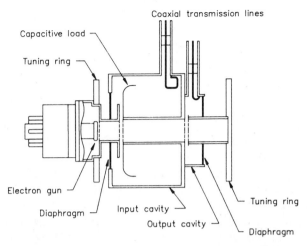

Figure 15.24 Components of a klystron frequency multiplier.

across the load cavity is $V(t) = 1.16 I_0 R_0 \cos(\omega_0 t)$. The time-averaged power extracted from the beam is

$$P = (1.16 I_0)^2 R_0/2. \tag{15.107}$$

The power rises as the square of the beam current. There is an upper limit on $I_0 R_0$—the amplitude of the cavity voltage cannot exceed T_0/e. A higher voltage results in reflection of electron orbits. This process reduces the average beam current and may cause feedback to the buncher cavity. The limiting value of the on-axis cavity impedance is

$$R_0 \leqslant (T_0/e)/(1.16 I_0). \tag{15.108}$$

For the highest value of R_0, the extracted power from the beam is

$$P_{max} = 1.16 I_0 T_0/2e = 0.58 I_0 T_0/e. \tag{15.109}$$

The beam power incident at the buncher is $I_0 T_0/e$. Therefore, the maximum efficiency for the two-cavity klystron is 0.58. The efficiency of actual devices is considerably lower for several reasons. The cavity impedance must be less than the limit of Eq. (15.108). There are power losses to the load cavity walls and output waveguides. Also, the thermionic cathode of the electron gun requires continuous power input. A low-power ($\sim$ 10-W) two-cavity klystron may have an efficiency of 10% and a power gain of only 10 decibels (dB).

The power gain of a klystron amplifier equals the extracted output power divided by the input power to excite the buncher. For a gain value, G, in dB the power multiplication factor is

$$\frac{P_{\text{output}}}{P_{\text{input}}} = 10^{(G/10)}. \tag{15.110}$$

Higher gains and efficiencies are possible with multicavity klystrons. Figure 15.25 shows a four-cavity klystron. The voltage in the first cavity is low, insufficient to provide complete beam modulation in the second cavity. Nonetheless, the resonant component of current at the second cavity is high enough to drive the cavity to a high voltage. The second cavity acts as a buncher, producing an enhanced current modulation in the third. With proper design, the distance between the third and forth cavities satisfies Eq. (15.106)—at the fourth cavity, the beam has the highest possible harmonic current component at f_0. In the limit where a klystron has many cavities over the bunching length, we can describe the growth of current variations by the theory of Section 14.4. Beam dynamics in a multicavity klystron is the discrete, nonlinear analogy of a resistive wall instability. A commercial five-cavity klystron may generate 25-MW microwave pulses at 3 GHz with a 50-dB gain. Typical beam parameters for such a large device are $T_0 = 250$ keV and $I_0 = 250$ A, giving an energy efficiency of 40%.

Klystrons also function as oscillators if there is feedback from the load cavities to the buncher. Either the beam or a portion of the electromagnetic output energy can carry feedback information. The simplest method to operate a two-cavity klystron as an oscillator is to connect the cavities with a transmission line terminated by coupling loops. A fraction of the power from the load cavity maintains the resonant oscillation in the bunching cavity. The steady-state amplitudes of electromagnetic fields in the cavities are determined by the beam power and the loaded Q values. It is also possible to achieve feedback through the beam modulations. This approach is the basis for the *reflex klystron*. The device requires only a single resonant cavity—by changing the geometry of the cavity, the reflex klystron can provide a reference signal over a wide frequency range.

Figure 15.26 shows the geometry of the reflex klystron. A single resonant cavity acts as both a buncher and a load. Incident electrons pass through the cavity in the $+z$ direction. If the cavity has a TM_{010} mode oscillation, the electrons emerge with a shift in axial kinetic energy at frequency f_0. A repeller electrode creates a retarding axial electric field in the region downstream from the cavity. The field is strong enough to stop the electrons and return them to the cavity. Electrons with high energy penetrate farther into the region of electric field; therefore, they take a longer time to return to the cavity. The energy dependence of the reflex time can lead to bunching. With the correct choice of parameters, the returning beam is strongly modulated.

We can use the model of Section 15.3 to describe beam bunching in the reflex klystron. Again, we shall neglect the effects of space charge and axial velocity spread. If the cavity has a voltage $V(t)$, electrons moving in the $+z$ direction

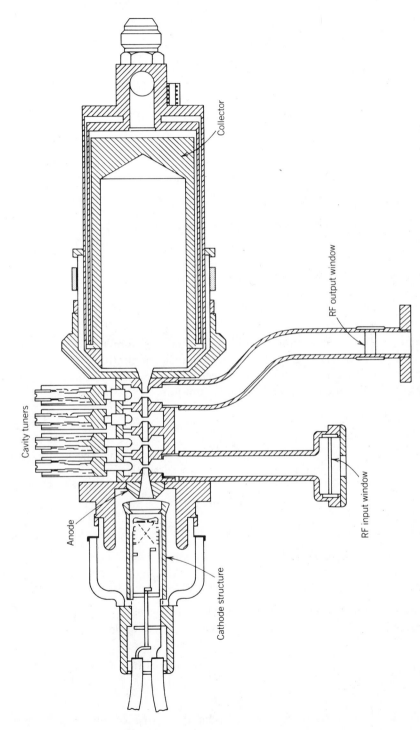

Figure 15.25 Cut-away view of the Varian VA-849 four-cavity klystron. (Courtesy of Varian Associates.)

769

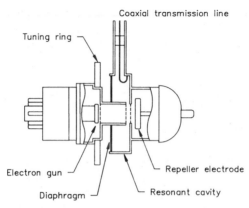

Figure 15.26 Components of a reflex klystron.

emerge with axial velocity

$$v(t) = v_0[1 + eV(t)/T_0]^{1/2}, \qquad (15.111)$$

where v_0 and T_0 are the velocity and kinetic energy of nonrelativistic electrons from the gun. The variable t is the time when incident electrons cross the gap. Suppose that there is a uniform axial electric field, E_0, downstream from the cavity. The following equation describes the motion of electrons in this region:

$$m_e(dv'/d\tau) = -eE_0. \qquad (15.112)$$

The variable τ is the elapsed time after leaving the cavity. The solution of Eq. (15.112) with the initial condition that $v'(0) = v(t)$ is

$$z'(\tau) = v(t)\tau - eE_0^2\tau^2/2m_e. \qquad (15.113)$$

Setting $z(\tau) = 0$ in Eq. (15.113), we find that the interval for an electron to return to the cavity is

$$\tau = 2m_e v(t)/eE_0. \qquad (15.114)$$

Finally, we can use Eq. (15.111) to find the time when electrons cross the cavity in the reverse direction, t', as a function of the first crossing time:

$$t' = t + \tau \cong t + \frac{2m_e v_0}{E_0} + \left[\frac{m_e v_0}{eE_0}\right]\left[\frac{eV(t)}{T_0}\right]. \qquad (15.115)$$

Setting $V(t) = V_0 \sin(\omega_0 t)$, taking the time derivative of Eq. (15.115), and substituting in Eq. (15.69) gives the current of the returning electrons:

$$I(t) = -\frac{I_0}{1 + \chi_0 \cos(\omega_0 t)}. \tag{15.116}$$

The bunching parameter for the reflex klystron oscillator is

$$\chi_0 = 2\omega_0 V_0 / E_0 v_0. \tag{15.117}$$

The highest oscillator output occurs when $\chi_0 = 1.84$.

There are two differences between Eq. (15.94) and Eq. (15.116) for the two-cavity klystron. First, the plus sign in the denominator shows that high-energy electrons take longer to return to the gap. Second, $I(t)$ has a negative value because returning electrons move in the opposite direction from incident electrons. For a strong oscillation, electrons should return to the cavity with the correct phase to drive the voltage oscillation. Suppose a group of incident electrons crosses the cavity near $t = 0$. Because of the rising voltage, these particles will be bunched when they return to the cavity. The bunch drives the cavity oscillation if $V(t)$ has a maximum negative value when they return. We can write this condition mathematically as

$$\tau = (m + 3/4)(1/f_0), \qquad m = 0, 1, 2, \ldots. \tag{15.118}$$

Equation (15.118) is the elapsed return time for electrons that cross the gap when $V(t) = 0$. Setting the equation equal to Eq. (15.114) with $v(t) = v_0$ gives

$$2m_e v_0 / eE_0 = (m + 3/4)(1/f_0). \tag{15.119}$$

The best operation of the reflex klystron occurs when $\chi_0 = 1.84$ and the condition of Eqs. (15.119) holds. For a given cavity tune, f_0, the voltages on the electron gun and repeller electrode are adjusted for maximum power output.

As an application example, suppose we want to design a 3-GHz reflex klystron with a 30-keV electron beam ($v_0 = 1.0 \times 10^8$ m/sec). If $m = 1$, Eq. (15.119) gives a value $E_0 = 2 \times 10^6$ V/m. If $\chi_0 = 1.84$, then the cavity voltage is $V_0 = 10^4$ V. We can find conditions where V_0 equals the optimum value by applying power balance for a given value of net cavity Q. The quality factor includes contributions from wall loading and output coupling. For a given beam power, we can achieve a desired value of V_0 by adjusting the output power coupling.

The available power from a single klystron is limited, in part, by beam optics. The 250-keV, 250-A beam that we used as an example has a perveance of 2 μperv, near the limit of conventional gun design. The strong focusing magnets necessary for high-perveance beams are dominant contributions to the mass and size of high-power klystrons. A potential method to extend the pulsed power output of klystrons to the gigawatt range is the use of relativistic driving beams. We discussed techniques to bunch such beams in Section 15.3. One advantage of relativistic beams is that modulations persist over long distances because the electrons have a longitudinal mass of $\gamma^3 m_e$ (Section 14.1). A high-power modulated beam can drive an extended array of load cavities. Also, relativistic beams reduce demands on the focusing system. For example, suppose a solenoidal magnetic field focuses the beam (Section 9.4). For a given beam radius and field magnitude, the matched current scales as

$$I \sim \beta\gamma. \tag{15.120}$$

The corresponding beam power equals the product of the current and kinetic energy:

$$P \sim (\gamma - 1)(\gamma^2 - 1)^{1/2}. \tag{15.121}$$

For a relativistic beam, the allowed power scales approximately as γ^2. As an example, a magnetic field less than 0.1 tesla can contain a 5-GW electron beam with $T_0 = 5\,\text{MeV}$ and $I_0 = 1\,\text{kA}$.

15.5. TRAVELING-WAVE TUBE

The traveling-wave tube is a widely used microwave amplifier. Although it does not achieve the power levels of a klystron, it has the advantage that it amplifies over a broad frequency range. Wide-band traveling-wave tubes used for FM communications have high gain (35–50 dB) over a full octave bandwidth (2:1 frequency shift). Traveling-wave tubes are also used in tunable radars. A pulsed S-band tube can generate 10 MW of power at an efficiency of about 40%. We shall study the traveling-wave tube in detail because it illustrates an important area of beam physics. The equations we shall derive describe a broad class of devices that rely on longitudinal coupling of beams to slow wave structures.

Figure 15.27 shows a schematic view of a high-power traveling-wave tube. A Pierce gun (Section 7.1) generates a space-charge-dominated nonrelativistic electron beam that propagates through a solenoidal magnetic field in a Brillouin equilibrium (Section 10.3). The beam passes through a structure that supports a slow electromagnetic wave with phase velocity close to that of the beam. A low-level input excites the slow-wave structure at the beam entrance. The electric field of the wave bunches the beam. The electrons shift in phase relative to the wave so that the beam gives up energy. The wave amplitude increases with a consequent increase in beam bunching. The result is that the power flux of the wave is much larger at the end of the tube while the beam energy is lower.

We shall begin the analysis by reviewing some properties of slow-wave structures, particularly the helical transmission line. The helical line in used in many low-power tubes because it has low dispersion—the phase velocity of waves is almost independent of frequency. This characteristic gives the tube its

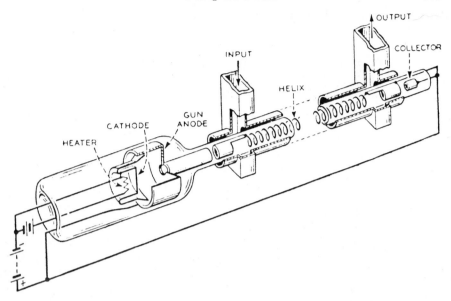

Figure 15.27   Schematic view of traveling-wave tube with a helical transmission line. (From J.R. Pierce, *Traveling Wave Tubes*. Used by permission, Van Nostrand Reinhold Company.)

broad bandwidth. Figure 15.28*a* shows the geometry of the helical transmission line. The main difference from a standard coaxial line is that the center conductor is a helical winding rather than a cylinder. We can understand wave transport on the helical line by comparing the device to a standard coaxial line. Figure 15.28*b* illustrates the familiar lumped circuit element model for a coaxial line with inner radius r_i and outer radius r_o. The model divides the line into differential elements of length Δz. The quantity $\mathscr{C}$ is the capacitance per unit length between the inner and outer conductor of the coaxial line:

$$\mathscr{C} = 2\pi\varepsilon_0/\ln(r_o/r_i). \tag{15.122}$$

and $\mathscr{L}$ is the inductance per unit length:

$$\mathscr{L} = (\mu_0/2\pi)\ln(r_o/r_i). \tag{15.123}$$

The quantity $V(z,t)$ is the voltage of the inner conductor relative to the grounded outer conductor and $I(z,t)$ is the axial current flowing along the inner conductor. For harmonic waves that flow in the $+z$ direction, the voltage and current have the form

$$V(z,t) = V_0 \exp[j(\omega t - kz)], \tag{15.124}$$

$$I(z,t) = (V_0/Z_0)\exp[j(\omega t - kz)],$$

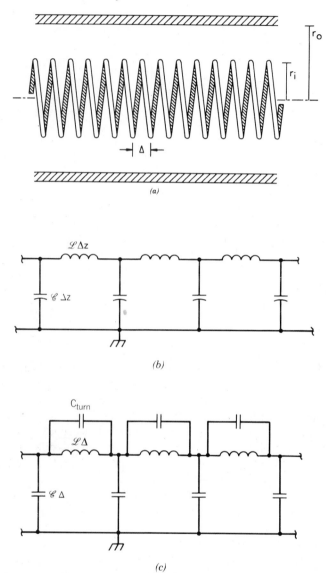

Figure 15.28 Coaxial transmission line with a helical center conductor. (*a*) Geometry. (*b*) Lumped circuit element model of a conventional coaxial transmission line with a cylindrical center conductor. (*c*) Modified lumped circuit element model for a coaxial transmission line with a helical center conductor.

where

$$Z_0 = (\mathscr{L}/\mathscr{C})^{1/2}. \tag{15.125}$$

The quantity Z_0 is the *characteristic impedance* of the line. The phase velocity of the waves,

$$\omega/k = 1/(\mathscr{L}\mathscr{C})^{1/2}, \tag{15.126}$$

is independent of frequency. In a vacuum coaxial line, the phase velocity equals the speed of light.

We cannot use waves on a coaxial line to extract energy from a beam for two reasons. First, the phase velocity of the waves is always greater than the beam velocity. Second, waves on a coaxial line are transverse—they have no electric field component along the propagation direction. The helical transmission line solves these problems—it slows the waves and creates an axial electric field on the axis. To analyze the structure, we take r_i as the radius of the helical windings and r_o as the radius of the grounded outer conductor. The windings advance an axial distance Δ per turn. The quantity ψ is the pitch angle of the helix. The distance between turns is

$$\Delta = 2\pi r_i \tan \psi. \tag{15.127}$$

We assume that the distance between turns is much smaller than the distance between the helix and outer conductor:

$$\Delta \ll r_o - r_i. \tag{15.128}$$

When the condition of Eq. (15.128) holds, the capacitance per length between the helix and outer conductor is almost equal to the value for the coaxial transmission line, Eq. (15.122). The main effect of the helix is to increase the inductance per length. Current flow through the winding creates an internal axial magnetic field that supplements the external toroidal field. Inspection of Eq. (15.126) shows that increased $\mathscr{L}$ results in reduced phase velocity.

We can apply the transmission line model of Fig. 15.28*b* with a modified value of $\mathscr{L}$ if the capacitance between turns of the helix is small. The validity condition is that the axial flow of real current along the helix is much larger than the axial displacement current. If we denote the axial displacement current as i_d and the real current as i_r, then the following condition defines a line with low dispersion:

$$i_d/i_r \ll 1. \tag{15.129}$$

We can derive an alternative form of Eq. (15.129) in terms of the properties of the line and the transmitted wave. The capacitance between two turns of the helix is roughly $C_{turn} \sim 2\pi\varepsilon_0 r_i$. The capacitance between one turn and the outer conductor is about $C_{out} \sim \mathscr{C}\Delta$, where Eq. (15.122) defines $\mathscr{C}$. The ratio of the capacitances is

$$C_{out}/C_{turn} \sim \Delta/(r_o - r_i). \tag{15.130}$$

The inductance per turn is $\mathscr{L}\Delta$, where $\mathscr{L}$ is the inductance per length of the helix. Suppose that a wave with frequency ω creates a voltage difference ΔV between two turns. The ratio of axial displacement current to real current on the center conductor is

$$\frac{i_d}{i_r} \sim \frac{(\mathscr{C}\Delta)[\Delta/(r_o - r_i)]\omega\Delta V}{[\Delta V/(\mathscr{L}\Delta)\omega]}. \tag{15.131}$$

If i_d is small, then k of the wave satisfies Eq. (15.126). We can rewrite the condition of Eq. (15.129) as

$$i_d/i_r \sim (k\Delta)^2[\Delta/(r_o - r_i)] \ll 1 \tag{15.132}$$

or

$$\Delta \ll (1/k)[(r_0 - r_i)/\Delta]^{1/2}. \tag{15.133}$$

A typical value of the helix spacing is $\Delta < (r_o - r_i)/5$. Equation (15.133) implies that the helical line acts like a transmission line for long wavelengths, $\lambda \gg \pi\Delta$.

When $\lambda \gg r_o$, we can make a simple estimate of the total inductance per length by adding the contribution from an infinite-length solenoid:

$$\mathscr{L} = \frac{\mu_0}{2\pi} \ln\left(\frac{r_o}{r_i}\right) + \frac{\mu_0 \pi r_i^2}{(2\pi r_i \tan\psi)^2}. \tag{15.134}$$

The characteristic impedance of the helical line is

$$Z_0 = \left[\frac{\mu_0}{\varepsilon_0}\right]^{1/2} \frac{\ln(r_o/r_i)}{2\pi} \left[1 + \frac{1}{2\tan^2\psi \ln(r_o/r_i)}\right]^{1/2}. \tag{15.135}$$

Inserting Eq. (15.134) into Eq. (15.126) gives the following expression for the phase velocity:

$$\frac{\omega}{k} \cong \frac{(\mu_0 \varepsilon_0)^{-1/2}}{\{1 + 1/[2\tan^2\psi \ln(r_o/r_i)]\}^{1/2}} \cong c \tan\psi [2\ln(r_o/r_i)]^{1/2}. \tag{15.136}$$

The final form of Eq. 15.136 is valid for $\psi \ll 1$.

Note that phase velocity of Eq. (15.136) is approximately equal to the length Δ divided by the time for an electromagnetic disturbance to follow a turn of the helix, $2\pi r_i/c$. One interpretation of the effect of the helix is that it forces electromagnetic waves to follow a longer path. As an example of parameters, consider a helical line with phase velocity equal to the velocity of a 5-keV electron beam, $v_o = 0.14c$. For the choice $r_i = 0.01$ and $r_o = 0.03$ m, Eq. (15.136) predicts a pitch angle of $\psi = 5.4°$. The distance per turn is $\Delta = 0.006$ m. The helix acts like transmission line for $\lambda \gg 0.02$ m, or $f \ll 2\,\text{GHz}$.

For a given slow-wave structure, we can derive equations for the complete traveling-wave tube circuit. Suppose an electron beam of kinetic energy eV_0 and current I_0 travels inside the helix. The beam interacts with the axial electric field generated by voltage differences between the windings. To understand the device, we shall add the effects of the beam to the lumped circuit element model of Fig. 15.28b. For a helical line, a natural choice for the length of differential elements is the distance between turns, Δ. Let the index k represent the number of turns on the winding from the device entrance. The quantity I is axial component of real current in the helix—the current at turn k is I_k. Similarly, V_k is the average voltage between the outer conductor and the kth turn.

Coupling between the beam and the helix is capacitive because the beam does not contact the wires. Figure 15.28c shows an element of the circuit with the contribution from the beam. We distinguish beam node points on the axis from the circuit node points on the helix. The charge at beam node point k is $q_k \Delta$, where q_k is the beam line charge density. The charge at k changes with time because of the axial flow of beam current. The principle of charge conservation implies that the total time-varying beam current entering the beam node point flows through the coupling capacitance to the circuit node point. As a result, the beam deposits a charge $q_k \Delta$ at the circuit node point k.

If i_k is the beam current at point k, the conservation of charge implies that

$$\frac{\partial q_k}{\partial t} = -\frac{\partial i_k}{\partial z}. \tag{15.137}$$

Modifying the standard analysis of the lumped circuit element model [CPA, Chapter 9], the transmission line equations are

$$\frac{\partial I(z,t)}{\partial z} = -\mathscr{C}\frac{\partial V(z,t)}{\partial t} - \frac{\partial i(z,t)}{\partial z}, \tag{15.138}$$

$$\frac{\partial V(z,t)}{\partial z} = -\mathscr{L}\frac{\partial I(z,t)}{\partial t}. \tag{15.139}$$

The second term on the right-hand side of Eq. (15.138) is the beam contribution. Taking harmonic perturbations, we can convert Eqs. (15.138) and (15.139) to algebraic form. In the traveling-wave tube, we seek a steady-state solution with waves that grow with distance from the injection point. Therefore, we shall use the form

$$V, I, i \sim \exp[j(\gamma z - \omega t)]. \tag{15.140}$$

The frequency ω is a real number, while the wavenumber, γ, may have a complex part to represent spatial growth. Substituting into Eq. (15.138) gives following circuit equations:

$$\gamma I = \omega \mathscr{C} V - \gamma i, \tag{15.141}$$

$$\gamma V = \omega \mathscr{L} I.$$

The solution for the circuit voltage in terms of beam current is

$$V = i(\omega \mathscr{L} \gamma)/(\omega^2 \mathscr{L} \mathscr{C} - \gamma^2). \tag{15.142}$$

We can express Eq. (15.142) in a convenient form using the characteristic impedance of the transmission line [Eq. (15.125)]:

$$V = \frac{k\gamma}{k^2 - \gamma^2} Z_0 i. \tag{15.143}$$

The quantity k in Eq. (15.143) is the wavenumber of a wave on the transmission line with frequency ω when there is no beam:

$$k = \omega (\mathscr{L} \mathscr{C})^{1/2}. \tag{15.144}$$

To complete the solution, we need an equation that describes the effect of the transmission line voltage on the beam. The line circuit couples to the beam through its axial electric field. In the limit that $\lambda \gg r_i$, the electric field is approximately

$$E_{z(k+1/2)} \cong -(V_{k+1} - V_k)/\Delta. \tag{15.145}$$

Equation (15.145) holds when the beam is closely coupled to the transmission line. We can add correction factors to account for contributions from the beam space charge and other effects. The following equation describes small changes in the axial velocity of a cold, nonrelativistic electron beam:

$$\frac{\partial v_z}{\partial t} = -\frac{e E_z}{m_e} - v_0 \left(\frac{\partial v_z}{\partial z} \right). \tag{15.146}$$

Equation (15.145) implies that $E_z \cong -\partial V/\partial z$. Taking velocity and voltage variations of the form of Eq. (15.140), Eq. (15.146) becomes

$$v_z(\omega - \gamma v_0) = (e\gamma/m_e)V. \tag{15.147}$$

To simplify the notation, we shall define an effective wavenumber, k_0, that relates the wave frequency to the beam velocity,

$$\omega/k_0 = v_0. \tag{15.148}$$

Equation (15.147) becomes

$$v_z = [(e\gamma/m_e v_0)/(\gamma - k_0)]V. \tag{15.149}$$

To complete the beam equation, we must relate changes in the axial velocity to changes in the beam current. For a linear analysis, we assume that the current consists of a steady-state value and a small perturbation, $I_0 + i$. The beam current at a point depends on both the average beam velocity $(v_0 + v_z)$ and the line density $(n_0 + n)$:

$$(I_0 + i) = -e(n_0 + n)(v_0 + v_z). \tag{15.150}$$

The steady-state line density is

$$n_0 = I_0/ev_0. \tag{15.151}$$

Dropping second-order terms, Eq. (15.150) implies that

$$i \cong -(I_0/v_0)v_z - ev_0 n. \tag{15.152}$$

The line density of the beam is also related to the axial current by the condition of conservation of charge:

$$e\left(\frac{\partial n}{\partial t}\right) = -\frac{\partial i}{\partial z}. \tag{15.153}$$

Combining equations and applying the form of Eq. (15.140), we find the following relationship between the axial velocity of beam and the current:

$$i = -[(k_0 I_0/v_0)/(k_0 - \gamma)]v_z. \tag{15.154}$$

Equations (15.149) and (15.154) imply the following relationship for the beam current in terms of the circuit voltage:

$$i = -[k_0\gamma/(k_0 - \gamma)^2](I_0/2V_0)V. \tag{15.155}$$

In Eq. (15.155), eV_0 is the kinetic energy of the beam,

$$eV_0 = (m_e v_0^2/2)^{1/2}. \tag{15.156}$$

We have enough information to construct a complete equation for wave behavior in the traveling-wave tube. The combination of Eqs. (15.143) and (15.155) gives the following dispersion relationship:

$$-\left[\frac{Z_0 I_0}{2V_0}\right]\left[\frac{k_0\gamma^2 k}{(k^2 - \gamma^2)(k_0 - \gamma)^2}\right] = 1. \tag{15.157}$$

Equation (15.157) specifies the complex propagation constant, γ, in terms of the properties of the beam and transmission line.

We can derive a simple solution of Eq. (15.157) by invoking a condition from a more advanced analysis. The maximum growth rate for unstable solutions occurs when

$$k_0 = k. \tag{15.158}$$

Equation (15.158) means that the beam velocity equals the phase velocity of an unperturbed wave in the transmission line. We seek a solution where the phase velocity of the wave is close to the beam velocity:

$$\gamma = k_0 + \xi. \tag{15.159}$$

The quantity ξ in Eq. (15.159) is small, $|\xi| \ll k_0$. Substituting in Eq. (15.157), we find

$$\xi^3 = -(Z_0 I_0 / 4 V_0) k_0^3. \tag{15.160}$$

Equation (15.160) is similar to the dispersion relationship for the two-stream instability (Section 14.1). It has three solutions:

$$\xi_1 = -C k_0, \tag{15.161}$$

$$\xi_2 = -C k_0 [\tfrac{1}{2} - j(3)^{1/2}/2], \tag{15.162}$$

$$\xi_3 = -C k_0 [\tfrac{1}{2} + j(3)^{1/2}/2]. \tag{15.163}$$

Equations (15.161)–(15.163) incorporate the gain parameter, C:

$$C = (Z_0 I_0 / 4 V_0)^{1/3}. \tag{15.164}$$

The gain parameter is proportional to the cube root of the ratio of the circuit impedance to the beam impedance, V_0/I_0. When the circuit impedance is high and the beam impedance is low, variations of the beam current induce large voltages on the transmission line. In turn, the voltage causes strong perturbations of the electron velocity.

In the solution of Eq. (15.161), ξ is a real number. For this mode, the presence of the beam causes a shift in the wavenumber but no growth of the wave. The second solution [Eq. (15.162)] has a positive imaginary part—the corresponding wave decays. The third solution is the most interesting—the wave grows in the presence of the beam. The voltage of waves on the helical transmission line varies as

$$V(z,t) \sim \exp\{j[k_0(1 - C/2)z - \omega t]\} \exp\{[(3)^{1/2} C k_0 / 2]z\}. \tag{15.165}$$

Note that the phase velocity of the wave of Eq. (15.165) is less than the beam velocity—the growing solution is a slow wave. The difference between the wave and beam velocities is proportional to C. The amount of energy extracted from the beam also scales with C. When the condition for the best coupling between the beam and transmission line holds [Eq. (15.158)], the growth length of the wave amplitude is

$$L_e = 2/(3)^{1/2} C k_0. \tag{15.166}$$

The traveling-wave tube is an amplifier—the frequency and amplitude of the output signal follow the input signal. It is essential that there is no feedback from the high-power output to the low-power input. Feedback occurs if there is an imperfect impedance match at the output, reflecting a fraction of the wave energy toward the input. One solution to the problem is to coat the helix with a resistive material at a point about halfway along the tube. Although the resistive coating damps waves that travel in both directions, it has little affect on the gain of the tube. Although the layer almost eliminates the forward-directed waves, the beam crosses unimpeded. The partially bunched beam carries information about the input wave across the resistive layer. A resistive coating that attenuates waves on the transmission line by 70 dB may reduce the tube gain by only 3 dB. Like the klystron, the traveling-wave tube can also function as an oscillator. In this application, it is desirable for wave energy to propagate back toward the beam input. An oscillating traveling-wave tube is called a *backward wave oscillator*.

We should note that there are alternative slow-wave structures that have application to traveling-wave tubes besides the helical transmission line. High-power devices often incorporate a coupled cavity structure similar to those used on linear accelerators. Chapter 14 of [CPA] discusses coupled cavity arrays for charged particle acceleration. In an accelerator, the phase velocity of the traveling wave increases with distance along the structure. In contrast, the phase velocity in a traveling-wave tube is slightly less than the beam velocity and decreases with distance. Another difference is that the beam in a traveling-wave tube is continuous at injection—bunching results from the interaction with the slow wave. The coupled cavity structure has two advantages over the helical line: (1) it is mechanically stronger and (2) the geometry allows higher RF electric fields without breakdown. The disadvantage is that a cavity array has dispersion. The wavenumber, and hence the wave phase velocity, depends on the frequency. Tubes with a coupled cavity structure generally operate with a single frequency input.

15.6. MAGNETRON

The magnetron was the first successful source of high-power microwave radiation. It is still a widely used microwave oscillator, with applications in industrial processing, microwave ovens, and radar. The magnetron achieves the highest energy efficiency of any source in common use. Conversion efficiencies of 50–80% are typical. Magnetrons can generate high output power. In pulsed mode, single magnetrons can produce over 1 MW of microwave power. Average power levels of several kilowatts are possible.

The magnetron converts the energy of a sheet electron beam to microwave radiation. An applied electric field normal to a magnetic field drives the beam. Figure 15.29 shows a multicavity traveling-wave magnetron, the most common type. It consists of coaxial cathode and anode electrodes immersed in a uniform axial magnetic field. A radial electric field between the electrodes causes an azimuthal drift of electrons emitted from the cathode. The anode is a complex structure, interrupted periodically by resonant cavities. The cavities support electromagnetic oscillations—the electric field of the modes points in the azimuthal direction and has maximum amplitude where the cavities connect to the anode–cathode gap. The oscillations in individual cavities interact through the shared region of the anode–cathode gap—the structure is a closed coupled-cavity array. The coupled electromagnetic oscillations resolve into slow electromagnetic waves that propagate in the azimuthal direction. An electron beam that travels at the same velocity as the wave phase velocity receives periodic displacements in the radial direction. The concentration of electron density at different radii results in a net transfer of energy from the drifting electrons to the wave. The resulting microwave power is extracted from the magnetron through one or more coupling devices in the resonators.

The self-consistent motion of electrons in the electromagnetic fields of the magnetron is a complex collective problem. As a result, development of the magnetron has been largely empirical, guided by conceptual models. This section reviews the principles of radiation generation by drifting electrons in crossed fields. After studying electron equilibria, we shall develop models for coupled electromagnetic oscillations in magnetrons. Finally, we shall discuss qualitative features of the interaction of electrons and oscillating fields.

We have already studied self-consistent equilibria of electrons in related crossed-field systems. We derived Brillouin flow solutions for conical, planar, and

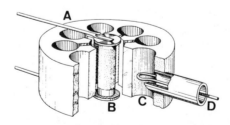

Figure 15.29 Schematic diagram of a multicavity traveling-wave magnetron with output coupled to a coaxial transmission line. (*A*) Anode block. (*B*) Cathode with end hats to inhibit axial electron loss. (*C*) Resonant cavity. (*D*) Coupling loop and coaxial transmission line. (From H. J. Reich, P. F. Ordung, H. L. Krauss, and J. G. Skalnik, *Microwave Theory and Techniques.* Used by permission, Van Nostrand Reinhold Company.)

cylindrical geometries. It is easy to extend the results to the coaxial geometry of Fig. 15.30. Nonrelativistic equations usually provide a good description of electron motion. The applied voltage between the cathode and anode, V_0, is almost always less than 100 kV; furthermore, we shall see that the electron drift energy is much smaller than eV_0. The diamagnetic field of the nonrelativistic rotating electron distribution is much smaller than the applied field (Section 10.3). Therefore, we take a uniform magnetic field with value B_0.

With a cylindrical anode and no microwave fields, the forces on the electrons are symmetric in the azimuthal direction. The canonical angular momentum of electrons, P_θ, is a conserved quantity. If all electrons leave the cathode with zero azimuthal velocity, the distribution is a delta function in canonical angular momentum, $\delta(P_\theta)$. If we normalize the vector potential so that it equals zero on the cathode, $A_\theta(r_c) = 0$, then the electron velocity is

$$v_\theta(r) = eA_\theta(r)/m_e. \tag{15.167}$$

The vector potential and axial magnetic field are related by

$$B_z = (1/r)\partial(rA_\theta)/\partial r = B_0. \tag{15.168}$$

Intergrating Eq. (15.168) from the surface of the cathode to radius r, we find that

$$A_\theta(r) = B_0(r^2 - r_c^2)/2r. \tag{15.169}$$

Inserting Eq. (15.169) into Eq. (15.167) gives the following expression for the

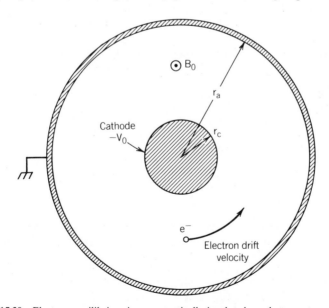

Figure 15.30 Electron equilibrium in a magnetically insulated gap between two cylinders.

angular velocity of electrons as a function of radius:

$$\frac{d\theta}{dt} = \left[\frac{eB_0}{2m_e}\right]\left[1 - \frac{r_c^2}{r^2}\right] = \left[\frac{\omega_g}{2}\right]\left[1 - \frac{r_c^2}{r^2}\right]. \qquad (15.170)$$

In contrast to the result for a solid cylindrical beam [Eq. (10.29)], Eq. (15.70) shows that the angular frequency varies with radius.

The constraint of Eq. (15.167) permits a variety of equilibrium solutions. The type of solution depends on how the radial electron velocity, v_r, varies with radius. The Brillouin flow equilibrium corresponds to the choice $v_r(r) = 0$. The corresponding distribution is a delta function in both canonical angular momentum and total energy. We expect that electrons have such a laminar-flow equilibrium if the magnetron voltage rises to its final value over a time much longer than $1/\omega_g$ and if electrons remain in the gap indefinitely. Clearly, these conditions cannot apply to the magnetron because electrons continually escape by the generation of microwave radiation. In reality, electrons follow the scalloped orbits discussed in Section 8.1. Nonetheless, all introductory texts discuss the magnetron in terms of the Brillouin equilibrium for the following pragmatic reasons. First, there is an infinite set of possible beam distributions, and we have little guidance on how to choose the right one. Second, the derivation with scalloped electron orbits is too difficult for an analytic solution. We should view the Brillouin flow solution in the following treatment as a limiting case that gives physical insight and useful scaling relationships.

When electron flow is laminar, the equation of energy conservation has the form

$$-e\phi(r) + m_e r^2 (d\theta/dt)^2 = 0. \qquad (15.171)$$

Equation (15.171) holds for the following boundary conditions on the electrostatic potential:

$$\phi(r_c) = 0, \qquad \phi(r_a) = V_0. \qquad (15.172)$$

Combining Eqs. (15.170) and (15.171), the self-consistent radial variation of potential for a Brillouin equilibrium is

$$\phi(r) = \left[\frac{eB_0^2}{8m_e}\right]\left[\frac{r^2 - r_c^2}{r}\right]^2. \qquad (15.173)$$

Suppose that the electron cloud extends almost to the anode. Setting $r = r_a$ in Eq. (15.173) gives a minimum value of magnetic field for magnetic insulation of the gap:

$$B_0 \geqslant \left[\frac{8m_e V_0}{e}\right]^{1/2}\left[\frac{r_a}{r_a^2 - r_c^2}\right] = B_H. \qquad (15.174)$$

Equation (15.174) states the *Hull cutoff condition*—we denote the cutoff value of magnetic field as B_H. When B_0 exceeds B_H, the electrons extend partly across the gap. The region filled by electrons is often called the *Brillouin cloud*.

We shall see that the strongest interaction between the electrons and the electromagnetic fields occurs on the surface of the Brillouin cloud at radius r_0. We must be sure that this surface is far enough from the cathode so that the electrons experience strong microwave fields. On the other hand, the surface should be far enough from the anode so that electrons interact coherently with the traveling wave over an extended distance. The radius of the Brillouin cloud surface is usually in the range

$$(r_0 - r_c)/(r_a - r_c) \sim 0.1\text{--}0.2.$$

We can express r_0 in terms of the applied voltage and magnetic field. Differentiation of Eq. (15.173) gives the radial electric field within the Brillouin cloud, while the Laplace equation gives the field between r_0 and r_a. The solutions must match at r_0. Integrating the electric field between the cathode and anode and setting the result equal to V_0 gives the following equation:

$$\left[\frac{(r_0/r_a)^2 - (r_c/r_a)^2}{(r_0/r_a)} \right]^2 + \left[\frac{(r_0/r_a)^4 - (r_c/r_a)^4}{(r_0/r_a)^2} \right] \ln\left(\frac{r_a}{r_0} \right)^2$$
$$- \left(\frac{B_H}{B_0} \right)^2 \left[1 - \left(\frac{r_c}{r_a} \right)^2 \right]^2 = 0. \tag{15.175}$$

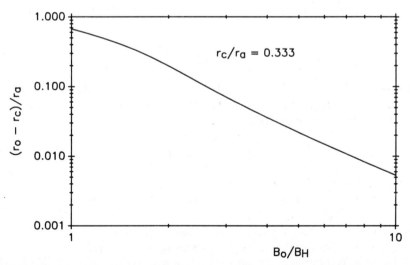

Figure 15.31 Normalized width of the Brillouin cloud between cylindrical electrodes as a function of applied field. r_c, cathode radius; r_a, anode radius; r_0, Brillouin cloud surface radius; B_0, applied magnetic field; B_H, Hull cutoff magnetic field.

A numerical solution of Eq. (15.175) yields r_0 as a function of B_0. Figure 15.31 plots the normalized width of the Brillouin cloud, $(r_0 - r_c)/r_a$, as a function of B_0/B_H for the choice $r_c/r_a = 0.333$. Following the discussion of Section 8.1, the width scales approximately as $1/(B_0/B_H)^2$.

The most important property of the electron distribution for the operation of the magnetron is the azimuthal velocity of electrons on the Brillouin cloud surface. The velocity should closely match the phase velocity of the slow wave supported by the resonant structure. Equations (15.170) and (15.174) imply that

$$\frac{v_\theta}{c} = \frac{(2eV_0/m_ec^2)^{1/2}}{(1 - r_c^2/r_a^2)} \left[\frac{B_0}{B_H} \frac{r_0}{r_a} \left(1 - \frac{r_c^2/r_a^2}{r_0^2/r_a^2} \right) \right]. \tag{15.176}$$

As an example, for $V_0 = 70\,\text{kV}$, $r_c/r_a = 0.5$, and $B_0/B_H = 2$, then $r_0/r_a = 0.67$ and $v_\theta/c = 0.41$.

To understand the magnetron, we must be familiar with electromagnetic oscillations in complex structures. We shall concentrate on unperturbed modes of magnetron resonators with no electrons. The resonant cavities in the anode block may have different shapes to suit specific applications—the simplest geometry to analyze is that of Fig. 15.32a—the cavities are slots of uniform width. The metal extensions that define the cavities are called *vanes*. We assume that the structure has infinite length in the direction out of the page.

To begin, consider a single anode slot resonator in isolation. The slot has width d and length l. We can treat the slot as a parallel-plate transmission line [CPA, Section 9.8]. The line has a short-circuit termination at the outer radius and an open-circuit termination where it meets the anode–cathode gap. The isolated shorted line is a quarter-wave resonator—the fundamental mode has frequency

$$f_0 = \omega_0/2\pi = c/4l. \tag{15.177}$$

The cavity of Fig. 15.32a has $l = 0.02\,\text{m}$; therefore, $f_0 = 3.75\,\text{GHz}$. Figure 15.32b shows calculated field variations. The electric field is strongest at the open end—the field varies with radius as

$$E(r) = E_0 \cos\left[\pi(r - r_a)/2l\right]. \tag{15.178}$$

The mode magnetic field has maximum amplitude at the shorted end.

For a given mode, Section 15.2 showed that we can represent a resonant cavity by an equivalent LC circuit. Figure 15.32c shows the circuit for the single-slot resonator with an arbitrary length Δz in the z direction. We define the mode voltage as the integral of electric field across the open-circuit end—the strength of this electric field determines the amplitude of a traveling wave in the gap. We take the oscillator current, i, equal to the total displacement current across the slot. Following the method of Section 15.2, the equivalent capacitance and inductance are

$$C = (2/\pi)\varepsilon_0 l\Delta z/d, \qquad L = (2/\pi)\Delta z \mu_0 dl. \qquad (15.179)$$

For the dimensions in Fig. 15.32a ($d = 7.32 \times 10^{-3}$ m, $l = 0.02$ m), Eq. (15.179) implies that $C/\Delta z = 1.54 \times 10^{-11}$ F/m, $L/\Delta z = 1.17 \times 10^{-10}$ H-m.

In the magnetron, the resonators are not independent because adjacent cavities share a vane. The set of N resonators forms a coupled-cavity array. Figure 15.32d shows an equivalent circuit model for a set of slots ignoring the effect of the magnetron cathode. Resonant oscillations in individual cavities generate differences in the relative voltages of the connecting vanes. The phased vane voltage oscillations define a traveling-wave component of azimuthal electric field. If all cavities have the same resonant frequency, ω_0, then we can write the voltage on the nth vane as:

$$V_n(t) = V_n \exp(j\omega_0 t). \qquad (15.180)$$

The resonant frequency of an isolated cavity is $\omega_0 = (LC)^{1/2}$. Figure 15.32e shows the lumped circuit element model with the effect of the cathode included. There is an extra capacitance, C_c, between each vane and the cathode. Displacement current flows across the anode–cathode gap when the cathode and vanes are at different voltages. We take the cathode as a reference equipotential plane with $V = 0$ because the azimuthal electric fields of the traveling wave equal zero on its surface.

The solution of the equivalent circuit of Fig. 15.32e gives insight into the operation of the magnetron and the properties of coupled-cavity arrays. Because all voltages and currents vary harmonically at angular frequency ω, we can apply the method of AC impedances discussed in Section 15.2. The quantity I_n is the total current that crosses cavity n from vane n to $n + 1$—the quantity equals the sum of currents through the inductor and capacitor. The difference in voltage between two vanes is the product of the net current times the AC impedance of the resonator, Z_r:

$$V_{n-1} - V_n = I_{n-1} Z_r. \qquad (15.181)$$

The resonator impedance is the parallel combination of impedances for the inductor and the capacitor (Eqs. (15.52)]:

$$Z_r = j\omega L/(1 - \omega^2 LC). \qquad (15.182)$$

Similarly, the vane voltage difference across the next cavity is

$$V_n - V_{n+1} = I_n Z_r. \qquad (15.183)$$

We can find another relationship for the voltage at vane n. The voltage equals the

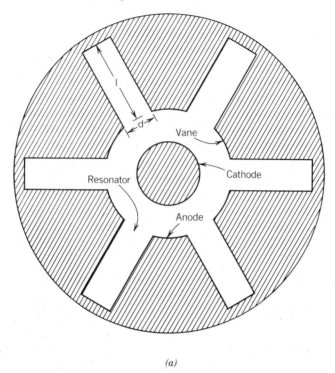

(a)

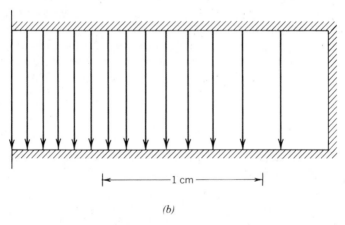

|← 1 cm →|

(b)

Figure 15.32 Properties of azimuthal traveling waves in the magnetron. (*a*) Cross section of magnetron with slot resonators. (*b*) Variation of electric field for the fundamental mode in a single-slot resonator. (*c*) Lumped element circuit model for an independent single-slot resonator. (*d*) Array of slot resonators coupled by shared vanes with an anode of infinite diameter. (*e*) Array of slot resonators coupled by anode vanes with capacitance between the vane tips and the cathode.

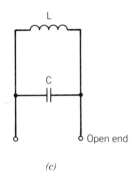

(c)

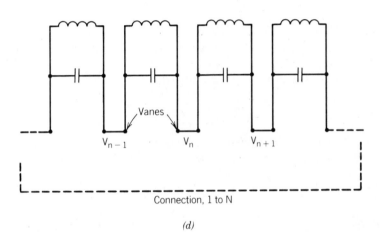

Connection, 1 to N

(d)

(e)

Figure 15.32 (*Continued*)

product of the total current flowing to the vane times the shunt impedance to the cathode, Z_s:

$$V_n = (I_{n-1} - I_n)Z_s. \tag{15.184}$$

The shunt impedance is:

$$Z_s = 1/j\omega C_c. \tag{15.185}$$

The combination of Eqs. (15.181), (15.183), and (15.184) gives the following difference equation for the amplitudes of the vane voltages:

$$V_{n+1} - 2[1 + Z_r/2Z_s]V_n + V_{n-1} = 0. \tag{15.186}$$

Equation (15.186) is a familiar relationship in accelerator physics. It also describes the propagation of particles through periodic focusing systems. Chapter 8 of [CPA] discusses the solution of the difference equation in detail. The amplitudes of the vane voltages satisfy the relationship

$$V_n = V_0 \cos(n\mu). \tag{15.187}$$

The quantity μ is a phase difference in the voltage between adjacent vanes. Substituting Eq. (15.187) into Eq. (15.186), we find that

$$\cos \mu = 1 + Z_r/2Z_s. \tag{15.188}$$

Substitution from Eqs. (15.182) and (15.185) gives the following equation for the frequency of a coupled mode with phase advance μ:

$$\omega^2 LC_c/2(1 - \omega^2 LC) = 1 - \cos \mu. \tag{15.189}$$

We can rewrite Eq. (15.189) in terms of the dimensionless quantities

$$\Omega = \omega/\omega_0 = \omega LC, \tag{15.190}$$

$$\kappa = C_c/C. \tag{15.191}$$

The quantity Ω is the frequency divided by the frequency of the uncoupled cavity mode, while κ is the normalized value of the shunt capacitance that couples the oscillators together. The frequency equation is

$$\Omega = \left[\frac{1 - \cos \mu}{1 - \cos \mu + \kappa/2}\right]^{1/2}. \tag{15.192}$$

The values of μ allowed for a magnetron with N cavities and vanes must satisfy a periodic condition—the voltage level must repeat after traversing N cavities:

$$\cos(N\mu) = 1. \tag{15.193}$$

Equation 15.193 implies that

$$\mu = m(2\pi/N), \qquad m = 1, 2, 3, \ldots. \tag{15.194}$$

Values of m higher than N are redundant; for example, the voltage variation for $m = N + 1$ is the same as that for $m = 1$. We conclude that there are N unique modes of oscillation for N values of μ. Figure 15.33 shows the frequency as a function of μ_n for a six-cavity magnetron. When $m = N$, the phase advance is $\mu_N = 2\pi$. The corresponding oscillation is called the 2π *mode*. This mode is of little interest because it has zero frequency. If the magnetron has an even number of cavities, then there is a mode with $\mu = \pi$ for $m = N/2$. The π *mode* is the standard oscillation mode for a magnetron—it has the highest frequency. In the π mode, the voltage polarity reverses between adjacent cavities. Figure 15.34 shows the π-mode electric fields. The frequency is

$$\Omega_{n/2} = \omega/\omega_0 = 1/(1 + \kappa/4)^{1/2}. \tag{15.195}$$

Figure 15.35 shows a calculation of the π-mode resonant frequency with the SUPERFISH program. The computer code solves the electromagnetic equations on a triangular mesh fitted to the boundaries. Figure 15.35a shows the mesh for the magnetron problem—Fig. 15.35b illustrates computed

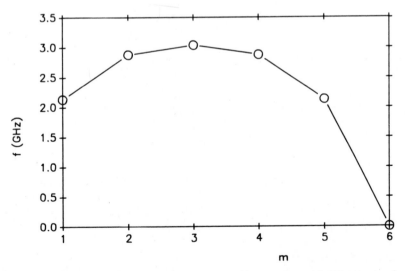

Figure 15.33 Predicted resonant frequencies for a six-cavity magnetron with slot resonators for the dimensions of Fig. 15.35.

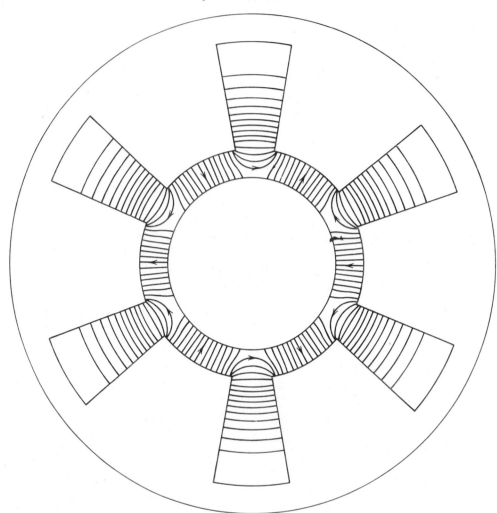

Figure 15.34   Electric-field distribution for the π mode of a six-cavity magnetron using the SUPERFISH code. Cathode radius, 1.58 cm; anode radius, 2.11 cm; outer radius of resonator, 4.11 cm. Predicted resonant frequency, 2.416 GHz.

electric field lines. To avoid redundant calculations and to achieve the highest possible accuracy, the code calculates fields in $1/12^{th}$ of the six-cavity magnetron and applies symmetry conditions at the boundaries. In the π mode, the electric field lines are normal at the bottom boundary that bisects a cavity and parallel to the top boundary that bisects a vane. The code predicts a resonant frequency for the coupled cavity array of 2.87 GHz, 25% lower than the 3.75-GHz resonant frequency of the isolated cavity. We estimate that the coupling capacitance to the cathode per vane is roughly one-sixth of the capacitance per length of a coaxial

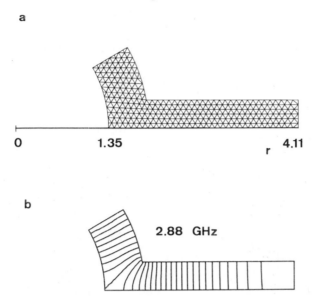

a

0	1.35	4.11

r

b

2.88 GHz

Figure 15.35 Calculation of the π-mode resonant frequency of a six-cavity magnetron with slot resonators using the SUPERFISH code. Predicted resonant frequency, 2.88 GHz. (*a*) Geometry of 1/12th of the device with a triangular mesh for the relaxation solution. (*b*) Electric fields of the π mode.

capacitor with outer radius 0.021 m and inner radius 0.016 m, or $C_c \cong 3.2 \times 10^{-11}$ F/m. With a value of $\kappa = C_c/C = 2.1$, Eq. (15.195) predicts a resonant frequency of 3.04 GHz, about 6% higher than the code solution.

For the best power extraction, the drift velocity of electrons at the surface of the Brillouin cloud should match the phase velocity of the π mode fields. Electrons experience a nonzero average deceleration (or acceleration) if they move a distance $2\pi r_0/N$ in a time $1/2f_0$. The phase velocity of the traveling wave near the cathode is approximately

$$\omega/k(\pi \text{ mode}) \cong 4\pi r_c f_0/N. \tag{15.196}$$

For the geometry of Fig. 15.35, Eq. (15.196) predicts a phase velocity of 9.5×10^7 m/sec, corresponding to an electron kinetic energy of 26 keV. The figure is higher than typical electron drift velocities. For a practical device with the given cathode radius, we must modify the magnetron geometry to reduce the wave phase velocity, either by adding more cavities or by reducing the resonant frequency by elongating the cavities. The frequency is also reduced by end strapping, a process used on many π mode magnetrons.

End-strapping is used on magnetrons to ensure that the available beam energy drives only the π mode. Without precautions, a magnetron could generate multiple-frequency output, useless for most applications. The problem of multimode excitation occurs because the electrons in realistic distributions have a spread in azimuthal velocity. They can drive modes with frequency and phase velocity close to those of the π mode. One approach to improve mode purity is to increase the frequency separation between the π mode and competing modes. Figure 15.36 illustrates how end-strapping modifies the resonant frequencies. Two concentric straps connect the ends of alternate vanes. For π-mode excitation, the straps have opposite voltage polarity. The capacitance

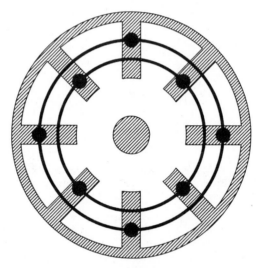

Figure 15.36　Magnetron end-strapping to shift the resonant frequency of the π mode.

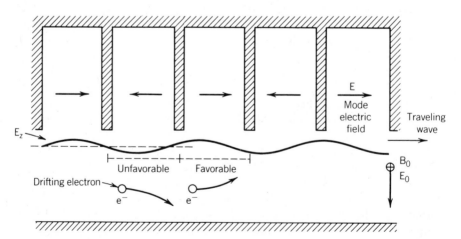

Figure 15.37　Mechanism for resonant interaction of drifting electrons with a traveling wave in a magnetron.

between the straps loads the π mode, pulling down the frequency. Loading is less pronounced for other modes. A similar strapping method is used in some RFQ (radiofrequency quadrupole) linacs to separate the quadrupole mode of oscillation from the undesirable dipole mode [CPA, Section 14.6].

We can now address the topic of electron interaction with electromagnetic waves in a magnetron. Figure 15.37 illustrates the mechanism. The figure shows drifting electrons on the surface of the Brillouin cloud. For clarity, the figure displays the circular system as an extended straight line. With the correct choice of parameters, the cavities support a traveling wave with a phase velocity equal to the electron velocity. Some electrons give up energy to the wave, while others absorb energy. Figure 15.37 shows the regions of acceleration and deceleration. If electron-wave interactions are the same in both regions, there is no amplification of the wave. On the other hand, amplification takes place if the electromagnetic wave modifies the properties of the electron cloud.

Electrons moving at the same velocity as the traveling wave experience a constant azimuthal electric field. This field superimposes a radial $\mathbf{E} \times \mathbf{B}$ drift (Section 8.1) on the azimuthal motion. In the region of decelerating electric field, the electrons drift radially outward. Conversely, electrons drift toward the cathode in regions of accelerating electric field. The electrons that yield energy to the wave move to a position where they interact more strongly—these electrons are called *favorable phase electrons*. Electrons that absorb energy from the wave toward the cathode where the azimuthal electric field is weak—they are called *unfavorable phase electrons*.

Because of the radial electron displacement, the drifting beam gives up more energy to the wave than it absorbs. The favourable phase electrons drift outward in rotating helical spokes until they reach the anode. In the π mode, there are $N/2$ spokes. We should note that the mechanism that concentrates electrons in phase does not depend on longitudinal bunching—instead, it results from the radial electron drift and the variation in wave amplitude across the anode–cathode gap. In contrast to klystrons, it is straightforward to operate magnetrons at relativistic voltages for high-power pulsed microwave generation. The energy of the unfavorable phase electrons is not wasted. In magnetrons with thermionic cathodes, the kinetic energy lost by these electrons helps to heat the cathode. In steady-state magnetrons, it is often possible to turn off the cathode heater after a warm-up time.

For high energy conversion efficiency, the favorable phase electrons should give up most of their potential energy to microwave fields as they cross the gap. If this condition holds, the electrons arrive at the anode with little kinetic energy. An equivalent statement is that the kinetic energy of electron drift motion should be much smaller than eV_0. To derive a rough criterion, we can estimate the drift velocity using the unperturbed radial electric field at the anode. Using Eqs. (10.122) and (15.174), we find that

$$\frac{m_e v_d^2 / 2}{e V_0} \cong \left(\frac{B_H}{B_0} \right)^2 \left\{ \frac{[1 - (r_c/r_a)^2]^2}{16[\ln(r_a/r_c)]^2} \right\}. \tag{15.197}$$

The geometric factor in braces is less than 0.2 for common geometries. Because magnetrons usually operate well above the Hull cutoff, (15.197) implies that drift energy is small.

15.7. MECHANISM OF THE FREE-ELECTRON LASER

The free-electron laser has catalyzed renewed interest in the generation of coherent radiation by electron beams. The device is a tunable source of high-intensity radiation at short wavelengths. Present experiments span the wavelength range from 0.1 to 1000 μm. In principle, free-electron lasers can achieve the efficiency for converting beam energy to photons.

The free-electron laser is not actually a laser. We can explain its operation completely within the framework of classical physics. Free electrons do not occupy well-defined quantized energy levels; therefore, population inversions are undefined. The free-electron laser is most closely related to beam-driven microwave sources such as the traveling-wave tube. The principle of the free-electron laser was discovered and investigated experimentally as early as 1960 in the Ubitron, a microwave generator.[4] Despite the misnomer, we shall use the terms free-electron laser and FEL to agree with current usage.

The FEL has some unique characteristics compared with the other beam-driven radiation sources discussed in this chapter:

1. Relativistic electron beams can drive FELs. Because high-current relativistic beams are easy to transport, projections show that FELs can achieve very high output power.
2. Most beam-driven microwave sources generate radiation with a wavelength comparable to the physical size of the surrounding structures. In contrast, the wavelength of the radiation from an FEL can be much shorter than the size of the surrounding cavities.
3. Electromagnetic waves in conventional microwave sources are usually transverse magnetic (TM) waves. Here, the beam exchanges energy by interaction with axial electric fields. On the other hand, the FEL generates transverse electromagnetic (TEM) waves with no axial electric field components.
4. In many microwave devices, the beam drives slow waves. The TEM waves in the FEL travel at the speed of light.

In this section, we shall learn how single electrons can exchange energy resonantly with a TEM wave. Section 15.8 describes how a continuous beam of electrons with a spread of axial kinetic energy can transfer power to an electromagnetic wave over long distances. To begin, we shall review some properties of TEM waves. These excitations propagate in free space or in a

[4] R. M. Phillips, *Trans. IRE Elec. Devices*, **7**, 231 (1960).

chamber with boundaries much larger than the radiation wavelength. Neglecting the effects of diffraction and amplification, the electric field of a monochromatic, linearly polarized TEM wave with frequency ω_0 is

$$\mathbf{E}(z,t) \cong \mathbf{E}_0 \sin (k_0 z - \omega_0 t - \phi). \tag{15.198}$$

The quantity $\mathbf{E}_0$ is a vector normal to the z direction. We will use the notation that all parameters that refer to the electromagnetic wave have the subscript "0." The constant k_0 is the wavenumber, related to the radiation wavelength by

$$k_0 = 2\pi/\lambda_0. \tag{15.199}$$

The phase velocity of the wave is

$$v_{\text{phase}} = \omega_0/k_0 = c. \tag{15.200}$$

The quantity ϕ is a phase factor that is important in the theory of phase dynamics in the FEL (Section 15.8). For simplicity, we consider a linearly polarized wave with electric field in the x direction. The transverse components of electric and magnetic fields are

$$E_x(z,t) = E_0 \sin (k_0 z - \omega_0 t + \phi), \tag{15.201}$$

$$B_y(z,t) = (E_0/c) \sin (k_0 z - \omega_0 t + \phi). \tag{15.202}$$

Two conditions must hold for energy exchange between the wave and an electron:

1. The electron must oscillate in the x direction to perform work on the wave electric field.
2. The electron oscillations must maintain synchronism with the wave electric field over an extended distance.

If the final condition is not true, energy moves back and forth between the beam and the wave—the time-averaged power transfer equals zero. There are several methods to impart oscillatory transverse motion on an electron that moves in the z direction. The most common approach is to direct the beam through a region of spatially varying magnetic field. The field region is called a *wiggler*.

Figure 15.38 shows a wiggler consisting of an array of permanent magnets with alternating polarity. The magnets generate a field in the y direction that varies periodically along z. If we pick the origin of the axial coordinate at the boundary between a positive and a negative polarity magnet, the axial variation of magnetic field is

$$B_{yw}(z) \cong B_w \sin (k_w z). \tag{15.203}$$

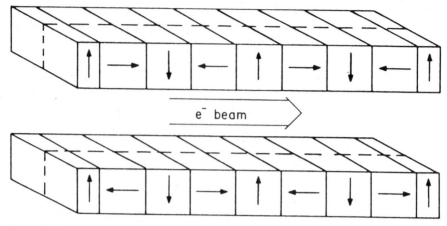

Figure 15.38 Permanent magnet wiggler for a free-electron laser. (From T. Marshall, *Free Electron Lasers.* Used by permission, MacMillan Book Co.)

We shall denote all quantities related to the wiggler with the subscript "w." The constant k_w is the wavenumber of the wiggler,

$$k_w = 2\pi/\lambda_w. \tag{15.204}$$

The wiggler wavelength, λ_w, equals the length of two magnets. Permanent magnet arrays allow small values of λ_w.

With no electromagnetic wave, the equations of motion for an electron with kinetic energy $T_e = (\gamma - 1)m_0c^2$ that passes through the wiggler field are

$$\gamma m_0(dv_x/dt) = (-e)(-v_zB_y), \tag{15.205}$$

$$\gamma m_0(dv_z/dt) = (-e)(v_xB_y). \tag{15.206}$$

Equations (15.205) and (15.206) incorporate the condition of constant γ in a region with no applied electric field. Electrons oscillate in the x direction. The limit

$$v_x \ll v_z \tag{15.207}$$

holds for all practical devices. Equation (15.207) implies that the axial velocity is almost constant. Therefore, we can replace the axial velocity in Eq. (15.205) with an average value taken over one wavelength of the wiggler:

$$v_{z0} = \langle v_z \rangle. \tag{15.208}$$

The wiggler frequency, ω_w, is the angular frequency in the stationary frame for magnetic field variations experienced by a particle moving at velocity v_{z0}:

$$\omega_w = v_{z0} k_w. \tag{15.209}$$

Using Eqs. (15.203) and (15.207), we can rewrite Eq. (15.205) as

$$\gamma m_0 \frac{dv_x}{dt} = e v_{z0} B_w \sin(\omega_w t). \tag{15.210}$$

Equation (15.210) has the solution

$$v_x = -\frac{e v_{z0} B_w}{\gamma m_0 \omega_w} \cos(\omega_w t), \tag{15.211}$$

$$x = -\frac{e v_{z0} B_w}{\gamma m_0 \omega_w^2} \sin(\omega_w t). \tag{15.212}$$

Equations (15.211) and (15.212) hold if (1) the average electron velocity in the x direction equals zero and (2) the origin of the x coordinate is at the position where the electron has maximum transverse velocity.

An expression for the longitudinal velocity follows from the conservation of energy. If v_0 is the total velocity of the injected electron, then

$$v_z = (v_0^2 - v_x^2)^{1/2} \cong v_0 - \frac{v_x^2}{2v_0}. \tag{15.213}$$

Using Eq. (15.211), the average axial velocity is

$$v_{z0} \cong v_0 - \frac{\langle v_x^2 \rangle}{2v_0} \cong v_0 - \frac{v_{z0}}{4} \left(\frac{e B_w}{\gamma m_0 \omega_w} \right)^2. \tag{15.214}$$

After some algebra, we can combine Eqs. (15.213) and (15.214) to show that the axial velocity of electrons in the wiggler varies as

$$v_z(t) = v_{z0} - \frac{v_{z0}}{4} \left(\frac{\omega_{gw}}{\gamma \omega_w} \right)^2 [2\cos^2(\omega_w t) - 1]$$

$$= v_{z0} - \frac{v_{z0}}{4} \left(\frac{\omega_{gw}}{\gamma \omega_w} \right)^2 \cos(2\omega_w t). \tag{15.215}$$

Equation (15.215) incorporates the nonrelativistic gyrofrequency for electrons in a field of magnitude B_w, $\omega_{gw} = e B_w / m_0$.

The equations for electron velocity and position are usually written in terms of the wiggler strength parameter, a_w. The parameter equals the electron γ

multiplied by the ratio of the peak transverse velocity to the average longitudinal velocity:

$$a_w = \gamma v_x(0)/v_{z0} = (eB_w/m_0)/\omega_w. \tag{15.216}$$

The final form of Eq. (15.216) shows that a_w equals the ratio of ω_{gw} to the wiggler frequency. The following equations describe electron motion:

$$v_x(t) = v_{z0}(a_w/\gamma)\cos(\omega_w t), \tag{15.217}$$

$$x(t) = (v_{z0}/\omega_w)(a_w/\gamma)\sin(\omega_w t), \tag{15.218}$$

$$v_z(t) \cong v_{z0}[1 - (a_w^2/4\gamma^2)\cos(2\omega_w t)]. \tag{15.219}$$

To find resonant conditions for FEL operation, we must calculate the total distance an electron travels passing through one period of the wiggler. The differential element of pathlength, ds, is related to distance along the axis by

$$ds = (dz^2 + dx^2)^{1/2} \cong dz[1 + (dx/dz)^2/2]. \tag{15.220}$$

Modification of Eq. (15.217) gives the relationship $dx/dz \cong (a_w/\gamma)\sin(k_w z)$. The total distance an electron travels while moving an axial distance λ_w is

$$s \cong \lambda_w(1 + a_w^2/4\gamma^2). \tag{15.221}$$

Figure 15.39 illustrates the mechanism by which an electron in the wiggler resonantly transfers energy to a TEM wave. The figure represents a typical case where the wavelength of the radiation is much shorter than the wavelength of the wiggler, $\lambda_0 \ll \lambda_w$. We shall follow an electron that enters the system at $z = 0$ and $t = 0$. Figure 15.39 shows the electric field at the time the electron enters—the wave field equals the expression of Eq. (15.198) with $\phi = -\pi/2$. Hence, we say that the particle shown has a phase of $-\pi/2$ relative to the wave.

At the injection point, $z = 0$, the wiggler magnetic field deflects the electron in the x direction. The electron moves against the force of the wave electric field. As a result, the electron gives up energy to the wave near the entrance. With an arbitrary relationship between v_{z0} and ω_0/k_0, electron motion downstream is uncorrelated with the electric field of the wave—the electron crosses regions of positive and negative transverse electric field with zero time-averaged energy exchange. On the other hand, there is a special condition, illustrated in Fig. 15.39, where the electron drives the wave over an extended distance. The wave travels a distance $\lambda_w + \lambda_0$ in the same time it takes for the electron to move through a wiggler period λ_w. For given values of λ_0 and λ_w, there is a particular value of v_{z0} that satisfies this condition. Figure 15.39 shows the motions of the electron and wave at resonance for times $t = 0$, $\pi/2\omega_w$, and π/ω_w. Note the relationship between the direction of the wave electric field and

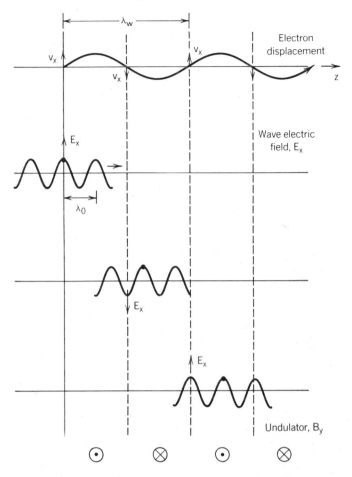

Figure 15.39 Resonant interaction of a single oscillating electron with a transverse electromagnetic wave. Top line: configuration-space view of an electron orbit in a wiggler. Lower lines: Spatial distribution of the wave transverse electric field at points of maximum electron velocity. Bottom: Polarity of the vertical magnetic field in the wiggler.

the direction of the particle velocity—the electron continuously transfers energy to the wave.

Electrons that move at the resonant velocity may enter the system at different times and may have different values of phase with respect to the wave. The phase determines whether a particle amplifies the wave or whether the wave accelerates the particle. For example, consider an electron that enters with phase $\phi = \pi/2$. We can make a drawing similar to Fig. 15.39 to show that the wave performs work on the electron, amplifying its oscillatory motion. For phase values $\phi = 0, \pi, 2\pi, \ldots$, there is no time-averaged energy transfer between the particle and the wave.

We can derive a mathematical statement of the condition illustrated in Fig. 15.39 by comparing the distances of travel for the electron and wave. In the interval $\Delta t = 2\pi/\omega_w$, the electron travels a distance $\lambda_w(1 + a_w^2/4\gamma^2)$ at velocity v_0. For resonant interaction, the wave should move a distance $\lambda_w + \lambda_0$ at velocity c in the same interval. Equating the transit times of electron and wave gives the condition

$$\frac{\lambda_w + \lambda_0}{c} = \frac{\lambda_w(1 + a_w^2/4\gamma^2)}{v_0}. \tag{15.222}$$

We can rewrite Eq. (15.222) as

$$\frac{\lambda_0}{\lambda_w} \cong \left[\frac{1}{\beta} - 1\right] + \frac{a_w^2}{4\gamma^2}. \tag{15.223}$$

The quantity β is the relativistic velocity factor v_0/c. Expanding the quantity in brackets,

$$[(1/\beta) - 1] \cong 1/2\gamma^2, \tag{15.224}$$

we find the familiar form of the FEL resonance condition:

$$\frac{\lambda_0}{\lambda_w} = \frac{1 + a_w^2}{2\gamma^2}. \tag{15.225}$$

Equation (15.225) implies that energetic electrons with high γ can transfer energy to radiation with a wavelength much shorter than that of the wiggler. For example, a 100-MeV relativistic beam in a wiggler with wavelength $\lambda_w = 0.02$ m resonantly drives electromagnetic radiation with wavelength $\lambda_0 = 0.5\,\mu m$. This wavelength corresponds to radiation in the visible range. Tuning of the FEL, changing the resonant λ_0, is accomplished by varying either the amplitude of the wiggler field or the electron kinetic energy.

We can derive the resonance condition of Eq. (15.225) from an alternative viewpoint based on the relativistic Doppler shift. For simplicity, we take a weak wiggler with $a_w \ll 1$. The condition means that transverse oscillations contribute a small enhancement to the total electron path length. Electrons move in the axial direction with velocity $\beta c\mathbf{z}$ and kinetic energy $(\gamma - 1)m_0c^2$. In the electron rest frame, the wiggler appears to move at velocity $-\beta c\mathbf{z}$. Because of the Lorentz contraction, the wiggler cell length observed in the electron rest frame has the reduced value of λ_w/γ. The relativistic transformation of fields [see Section 1.3] implies that the transformed wiggler magnetic field creates a component of transverse electric field in the electron rest frame that oscillates at frequency

$$\omega' = \gamma\omega_w. \tag{15.226}$$

The electric field from the time-varying wiggler magnetic field displaces electrons in the x direction. The oscillating electrons emit dipole radiation at frequency

ω'. In the rest frame, electrons radiate primarily in the x direction. After transformation back to the stationary frame, the radiation pattern is strongly peaked within an angle $1/2\gamma$ of the z direction.

To find the frequency of the radiation in the stationary frame, we must apply the relationships for the relativistic Doppler shift. Consider a TEM wave with wavenumber $k'_0 = \omega'/c$. In a frame moving at velocity $-\beta c$ relative to the direction of the wave, the observed wavenumber is[5]:

$$k_0 = \gamma(k'_0 - \beta k'_0). \tag{15.227}$$

Equation (15.227) implies the frequency transformation

$$\omega_0 = \gamma \omega'(1 - \beta). \tag{15.228}$$

The quantity ω_0 is the frequency of the forward-directed radiation from the oscillating electron observed in the stationary frame. Substituting from Eq. (15.226) and noting that $1 - \beta \cong 1/2\gamma^2$, we arrive at the following expression:

$$\frac{\omega_w}{\omega_0} = \frac{\lambda_0}{\lambda_w} = \frac{1}{2\gamma^2}. \tag{15.229}$$

Equation (15.229) is identical to Eq. (15.225) when a_w equals zero.

15.8. PHASE DYNAMICS IN THE FREE-ELECTRON LASER

Our discussion of the FEL mechanism in the previous section concentrated on isolated electrons. The generation of a high-power photon beam demands a beam with many electrons. The following processes complicate the description of the collective interaction of an electron beam with TEM waves:

1. Electrons in the driving beam lose energy to the wave. With no adjustment of the wiggler properties, the beam soon falls out of resonance because of the decreasing γ.

2. Particles in beams always have a distribution of orbit parameters. We cannot assume that all electrons precisely meet the resonant condition of Eq. (15.225).

3. Even if all electrons have an energy that satisfies the resonant condition, the wave does not grow if electrons have a uniform distribution in phase, ϕ. With a uniform distribution, half of the electrons contributes energy to the electromagnetic wave, while the other half absorbs energy.

---

[5] J. D. Jackson, *Classical Electrodynamics*, 2nd ed., Wiley, New York, 1975, p. 521.

We can apply the familiar theory of longitudinal phase oscillations to understand the amplification of a TEM wave in a free electron laser over an extended interaction length. ([CPA, Chapter 13] discusses phase dynamics in detail.) In this section, we shall derive the phase equations for the FEL and use them to estimate requirements on electron beam quality. We know that the phase of a transverse oscillation of an electron determines whether it amplifies the wave or absorbs energy. The definition of the phase of an electron orbit in Section 15.7 maintains a close connection with the theory of phase dynamics in RF linacs. A positive value of phase corresponds to particle acceleration (wave damping), while a negative value gives wave amplification and particle deceleration. To clarify the definition, Fig. 15.40 plots particle orbits with different phases.

To pump a wave, the electrons in a beam must be concentrated in the phase range

$$0 \leqslant \phi \leqslant -\pi. \tag{15.230}$$

We shall follow two steps to describe electron containment. First, we shall calculate conditions where an ideal electron moves at constant phase relative to the electromagnetic wave. Given the properties of this ideal orbit, we shall then investigate the stability of other electrons with similar orbits. An electron orbit is stable in phase if it oscillates about the phase of the ideal orbit.

To begin, we shall calculate the orbit of an ideal electron that moves in the combined fields of the wiggler and wave at constant phase. The electron is called the *synchronous particle*—the phase of the synchronous particle is the *synchronous phase*, ϕ_s. We take ϕ_s in the range of Eq. (15.230). At a particular time, a wiggler contains a set of synchronous particles spaced a distance λ_0 apart. As synchronous electrons move through the wiggler, they lose axial energy through interaction with the wave. The characteristics of the wiggler must vary in the axial direction to maintain a synchronous particle.

At first glance, we might expect that the best choice for the synchronous phase is $\phi_s = -\pi/2$. At this value, electrons clustered about the synchronous particle transfer energy to the TEM wave at the maximum rate. Later, we shall find that neighboring particles cannot remain synchronized with the wave if $\phi = -\pi/2$. Therefore, we allow the possibility that ϕ_s may have any value in the range of Eq. (15.230).

We shall first calculate the time-averaged work performed on the electromagnetic field by the synchronous particles as a function of ϕ_s. The instantaneous power loss from a synchronous electron at position z_s to a TEM wave with wavenumber k_0 and frequency ω_0 is

$$P(z_s, t) = [-v_x(z_s)][-eE_0 \sin(k_0 z_s - \omega_0 t - \phi_s)]. \tag{15.231}$$

We adopt the convention of Fig. 15.40 and let the synchronous particle enter

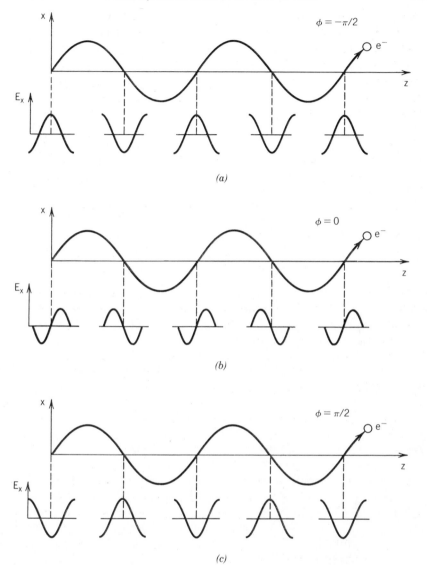

Figure 15.40 Phase of resonant electrons relative to a transverse electromagnetic wave. Plots of the electron orbit in a wiggler and the wave transverse electric field at points of maximum electron velocity. (a) $\phi = -\pi/2$. (b) $\phi = 0$. (c) $\phi = \pi/2$.

the system at $t = 0$. The position of the electron is

$$z_s = v_{z0}t. \qquad (15.232)$$

Substituting Eqs. (15.217) and (15.232) into Eq. (15.231), the time-dependent power transfer is

$$P(t) = \left[\frac{eE_0 v_{z0} a_w}{\gamma_s} \right] \cos(\omega_w t) \sin[(kv_{z0} - \omega_0)t - \phi_s]. \qquad (15.233)$$

The quantity v_{z0} is the average axial velocity of the synchronous particle with kinetic energy $(\gamma - 1)m_e c^2$.

By definition, the synchronous particle fulfills the resonance condition of Eq. (15.225):

$$\frac{\lambda_w + \lambda_0}{c} = \frac{\lambda_w}{v_{z0}} \qquad (15.234)$$

or

$$k_0 v_{z0} - \omega_0 = -\omega_w.$$

We can rewrite Eq. (15.233) as

$$P(t) = \left[\frac{eE_0 v_{z0} a_w}{\gamma_s} \right] \cos(\omega_w t) [\sin(-\omega_w t)\cos(-\phi_s) + \cos(-\omega_w t)\sin(-\phi_s)].$$

$$(15.235)$$

The power averaged over the time for the synchronous electron to move through one period of the wiggler is

$$P(t) = (eE_0 v_{z0} a_w / 2\gamma_s) \sin(-\phi_s). \qquad (15.236)$$

Equation (15.236) confirms the physical picture of Fig. 15.39. Maximum wave pumping occurs when the synchronous electron has $\phi_s = -\pi/2$; there is no time-averaged interaction between the wave and particle if $\phi_s = 0, \pi, \ldots$.

An electron loses kinetic energy when it pumps a TEM wave, because the wave exerts an axial force on the electron. The TEM wave has field components E_x and B_y, while the electrons have velocity components v_z and v_x. The only axial component of force is $-ev_x B_y$. Substituting expressions from Eqs. (15.202) and (15.217), the axial retarding force is

$$F_z(t) = -ev_x(t)B_y(t)$$
$$= -e(v_{z0} a_w / \gamma_s)\cos(\omega_w t)(E_0/c)\sin(-\omega_w t - \phi_s). \qquad (15.237)$$

Averaging Eq. (15.237) over one period of the wiggler gives the long-term force of the wave:

$$F_z(z) = \frac{dp_z}{dt} = \left[\frac{ev_{z0} a_w E_0}{2\gamma_s c} \right] \sin(\phi_s). \qquad (15.238)$$

The axial force of Eq. (15.238) is called the *ponderomotive force* of the wave. Equation (15.238) is consistent with the power equation, Eq. (15.236). The ponderomotive force decelerates electrons with synchronous phase in the range $0 < \phi_s < -\pi$.

We can write Eq. (15.238) for highly relativistic particles as

$$\frac{d\gamma_s}{dz} \cong \frac{ea_w E_0(z)}{2\gamma_s m_0 c^2} \sin \phi_s = \left[\frac{k_0 a_0(z) a_w}{2\gamma_s} \right] \sin \phi_s. \qquad (15.239)$$

In Eq. (15.239), the amplitude of the wave electric field is written as a function of axial position to include the possibility of wave pumping. The final form of Eq. (15.239) incorporates a new parameter, a_0, the wave constant:

$$a_0(z) = eE_0(z)/k_0 m_0 c^2. \qquad (15.240)$$

Electromagnetic waves with a high value of a_0 have either high intensity or long wavelength. The wave constant is the ratio of eE_0/k_0 to the electron rest energy. The quantity eE_0/k_0 equals the change in the potential energy of an electron moving a distance $1/k_0$ in an electric field E_0.

For a known variation $E_0(z)$, Eqs. (15.240) and (15.239) imply conditions for the existence of a synchronous electron. For a self-consistent calculation, we need another equation that gives the variation of $E_0(z)$ that results from wave pumping by the beam. We can apply conservation of wave energy. Let A_0 be the cross-sectional area of both the electron and photon beams. The average electromagnetic wave energy per length is $U_0 = \varepsilon_0 E_0^2 A_0/2$. We assume that the wave traps a fraction of the beam electrons in stable orbits near the synchronous phase—they exchange energy with the wave at about the same rate as the synchronous particle. We denote the time-averaged excess current carried by trapped particles as $I_t(z)$. We shall discuss methods to estimate I_t later in this section. Setting the increase in U_0 equal to the change in energy per unit length of the trapped electrons, we find that

$$\frac{dE_0(z)}{dz} \cong -\frac{I_t(z) a_w \sin \phi_s}{2\gamma_s c \varepsilon_0 A_0}. \qquad (15.241)$$

We can introduce correction factors into Eq. (15.241) if the radiation and electron beams do not overlap perfectly. For constant $I_t(z)$, a simultaneous solution of Eqs. (15.239) and (15.241) gives expressions for $\gamma_s(z)$ and $E_0(z)$. The solutions depend on the injection properties of the beam and the incident TEM wave. Advanced computer simulations are necessary to predict changes in I_t as the wave amplitude grows.

Throughout the wiggler, the energy of a synchronous electron must satisfy the equation

$$\gamma_s = \left[\frac{\lambda_w(1 + + a_w^2)}{2\lambda_0} \right]^{1/2}. \tag{15.242}$$

Neglecting refractive effects and the active optical properties of the beam, the quantity λ_0 is constant over the length of the device. Equation (15.242) implies that the characteristics of the wiggler must vary along z to maintain resonance. Either the wiggler wavelength (λ_w) or the amplitude of the wiggler field (a_w) must decrease to compensate for the reduction in γ_s. Because the wiggler is usually constructed with the shortest possible cell size and highest possible entrance magnetic field, the common approach is to lower a_w by increasing the transverse distance between magnet poles. A wiggler with a varied transverse profile along the axial direction is called a *tapered wiggler*. An alternative method to maintain resonance is to accelerate the electrons in a uniform wiggler to maintain constant γ_s. Although this approach has attractive features, the associated technology is complex.

We shall now discuss the orbits of electrons close to the synchronous particle. We seek conditions where these electrons have *phase stability*—they perform stable oscillations about the position of the synchronous particle. The trapped particle current, I_t, is nonzero only if many electrons are contained stably in a region of phase near ϕ_s. Without a stabilizing mechanism, there would be no hope of adjusting and compensating electron velocity with such precision that the resonance condition holds over the full length of a wiggler. As an example of the required accuracy, in a short-wavelength FEL experiment the position of electrons must be accurate to within microns over a wiggler length of several meters.

To describe phase oscillations, we take the synchronous particle properties, ϕ_s and $\gamma_s(z_s)$, as known functions. To characterize the orbits of electrons with properties close to those of the synchronous particle, we define small quantities

$$\Delta\gamma = \gamma - \gamma_s \tag{15.243}$$

and

$$\Delta\phi = \phi - \phi_s. \tag{15.244}$$

If an electron has phase ϕ with respect to the wave, modification of Eq. (15.239) implies that the change in γ averaged over a wiggler cell is

$$\frac{d\gamma}{dz} = \frac{d(\gamma_s + \Delta\gamma)}{dz} \cong \left[\frac{k_0 a_0 a_w}{2\gamma_s} \right] \sin\phi. \tag{15.245}$$

Substitution from Eq. (15.239) gives the following equation for the energy error of a particle in terms of the phase difference

$$\frac{d\Delta\gamma}{dz} = \left[\frac{k_0 a_0 a_w}{2\gamma_s}\right][\sin\phi - \sin\phi_s].$$ (15.246)

We can find another equation to relate the phase difference of a nonsynchronous particle to its energy error. Suppose an electron is a distance Δz in front of the synchronous particle. The phase difference is proportional to the ratio of the position error to the wavelength of the TEM wave, or

$$\Delta\phi = -2\pi(\Delta z/\lambda_0).$$ (15.247)

The change in the position error as the nonsynchronous particle moves through the wiggler is related to the error in the longitudinal electron velocity, $\Delta\beta_z c$:

$$\frac{d(\Delta z)}{dz} = \frac{\Delta\beta_z c}{v_{z0}}.$$ (15.248)

If the transverse kinetic energy is much smaller than the longitudinal energy, we can write the longitudinal velocity error in terms of the variation in γ:

$$\frac{\Delta\beta_z}{v_{z0}c} \cong \frac{\Delta\gamma}{\gamma_s^3}.$$ (15.249)

Collecting Eqs. (15.247)–(15.249) gives the following equation for the axial derivative of the phase error:

$$\frac{d(\Delta\phi)}{dz} = -\frac{2\pi}{\lambda_0}\frac{\Delta\gamma}{\gamma_s^3} \cong -2k_w\frac{\Delta\gamma}{\gamma_s}.$$ (15.250)

The last form follows from Eq. (15.229).

Equations (15.246) and (15.250) are the FEL phase equations—their form is similar to the equations of axial motion in RF accelerators. The equations describe the displacements of particles relative to the synchronous particle as a function of position in the wiggler. In the limit that the phase oscillations of nonsynchronous electrons are rapid compared with changes in the properties of the synchronous particle, we can combine Eqs. (15.246) and (15.250) into the familiar nonlinear differential equation

$$\frac{d^2\Delta\phi}{dz^2} \cong -\left[\frac{k_w a_w k_0 a_0}{\gamma_s^2}\right][\sin\phi - \sin\phi_s].$$ (15.251)

To derive Eq. (15.251), we dropped axial derivatives of ϕ_s.

In the limit of small phase oscillations ($\Delta\phi \ll \phi_{\mathrm{s}}$), the phase equation is

$$\frac{d^2\Delta\phi}{dz^2} \cong -\Omega_z^2 \Delta\phi, \tag{15.252}$$

where

$$\Omega_z = k_{\mathrm{w}}[2a_{\mathrm{w}}a_0 \cos\phi_{\mathrm{s}}/(1 + a_{\mathrm{w}}^2)]^{1/2}. \tag{15.253}$$

Equation (15.252) has bounded solutions for synchronous phase in the range

$$0 \leqslant \phi_{\mathrm{s}} \leqslant -\pi/2. \tag{15.254}$$

When the condition of Eq. (15.254) holds, the solution of Eq. (15.252) is

$$\Delta\phi(z) = \Delta\phi_0 \cos(\Omega_z z). \tag{15.255}$$

Equation (15.255) may confuse readers familiar with phase dynamics in RF accelerators—note that the independent variable is z rather than t. Figure 15.41 illustrates the physical meaning of Eq. (15.255)—the figure shows phase relationships among a group of trapped electrons as a function of z. Note that the length scale in the figure is much larger than the wavelength of the RF radiation, λ_0. The graphs in (a) show the synchronous particle and a particle with phase error $\Delta\phi_0$ and zero energy error, $\Delta\gamma$. Here, the test particle has the same energy is the synchronous particle but has a position error smaller than the electromagnetic wavelength. The graphs in (b) show the particle positions after the group travels an axial distance $\pi/2\Omega_z$ in a time $\pi/2\Omega_z v_{z0}$. Here, the test particle has a phase error $\Delta\phi = 0$ and an energy error $\Delta\gamma_0$. We can calculate the magnitude of the energy error in terms of the phase error from Eq. (15.246):

$$\frac{\Delta\gamma_0}{\gamma_{\mathrm{s}}} \cong \frac{\Delta\phi_0 \Omega_z}{2k_{\mathrm{w}}} = \Delta\phi_0 \left[\frac{a_{\mathrm{w}}a_0 \cos\phi_{\mathrm{s}}}{2(1 + a_{\mathrm{w}}^2)}\right]^{1/2}. \tag{15.256}$$

We define the axial distance the group travels in the wiggler during the time for one-quarter of a phase oscillation as the bounce distance:

$$z_{\mathrm{b}} = \pi/2\Omega_z. \tag{15.257}$$

Analysis of Eq. (15.251) gives limits on the properties of electron orbits for stable phase oscillations. As in RF accelerators, the wave traps only electrons that are close to the synchronous particle in position and energy. The region of allowed $\Delta\phi$ and $\Delta\gamma$ is called the FEL bucket. The size of the trapped region shrinks to zero when $\phi_{\mathrm{s}} = -\pi/2$. Although the size of the stable region is maximum when $\phi_{\mathrm{s}} = -\pi$, there is no time-averaged wave pumping at this value

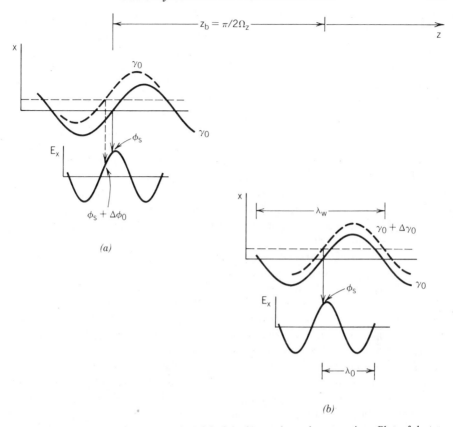

Figure 15.41 Illustration of the meaning of the free-electron-laser phase equations. Plots of electron orbits and the variation of the transverse wave electric field as a function of axial position, z. Synchronous electron orbit: solid line. Nonsynchronous electron orbit: dashed line. Left-hand side: Nonsynchronous electron has the same kinetic energy as the synchronous electron, but a phase different from ϕ_s. Right-hand side: After the beam propagates a distance $\pi/2\Omega_z$, the phase of the nonsynchronous particle equals ϕ_s, but its energy differs from that of the synchronous particles.

of synchronous phase. In free-electron-laser experiments, it is impossible to control the beam distribution in $\Delta\phi$. The beam current from conventional accelerators is almost constant over the time scale of an electromagnetic wave period. Therefore, only a fraction of the injected electrons close to the synchronous phase are initially trapped. The experimenter can control the spread of beam kinetic energy. It is essential that the initial $\Delta\gamma$ is small enough so that the injected wave captures some electrons. We can estimate the energy spread from Eq. (15.256)—we take both ϕ_s and $\Delta\phi_0$ equal to 1 rad. The allowed spread in injected energy for significant beam capture is

$$\frac{\Delta\gamma_0}{\gamma_s} \cong 0.5 \left[\frac{a_w a_0}{1 + a_w^2} \right]^{1/2}. \qquad (15.258)$$

Equation (15.258) shows that the energy spread for beam trapping is larger at long wavelengths because of the $1/k_0$ dependence in a_0. For experiments on the generation of microwave radiation by the FEL mechanism, it is possible to use pulsed high-current electron beams with poor emittance and moderate kinetic energy. On the other hand, the generation of short-wavelength radiation demands high-energy beams with very low emittance and energy spread.

It is interesting to note the implications of the axial confinement of electrons in bunches for the scattered radiation model of Section 15.7. In that viewpoint, the wiggler fields accelerate electrons in the transverse direction—the electrons emit dipole radiation. When the distribution of radiating electrons is uniform along the axis, the radiation from individual dipoles interferes destructively. As a result, the beam emits almost no radiation. In contrast, the dipole radiators in a confined beam are space a distance λ_0 apart in the stationary frame. For this geometry, the radiators constitute a moving phased array, giving strong emission of a narrow photon beam in the forward direction.

As an example of the parameters of a free-electron-laser experiment, we shall consider the Palladin experiment at Livermore National Laboratory. The beam source is a pulsed induction linac. The machine generates a 1-kA beam at 45 MeV. The available beam power is 45 GW at $\gamma = 89.1$. The laser amplifies radiation at $\lambda_0 = 10.6\,\mu$m. For this wavelength, a pulsed CO_2 laser provides an input wave strong enough to trap a portion of the beam. We assume a laser input power of 50 MW (50 J in a 1 μsec pulse). The peak magnetic field strength in the permanent magnet wiggler is $B_w = 0.3$ tesla. To find the wiggler wavelength, we can substitute values in Eq. (15.225). The result is $\lambda_w = 0.0514$ m, a convenient length for fabrication. The 25-m Palladin wiggler consists of 500 cells. The associated wiggler constant is $a_w = 1.44$.

To estimate the electric field of the TEM wave at injection, we take electron and photon beams with 2.5 mm radii. The cross-sectional area is $A_0 = 2 \times 10^{-5}\,\text{m}^{-2}$. The wave electric field, E_0, is related to the power flux by

$$P \cong (c\varepsilon_0 A_0/2)E_0^2. \tag{15.259}$$

Equation (15.259) implies that $E_0 = 4.3 \times 10^7$ V/m. Substitution in Eq. (15.240) gives the wave constant as $a_0 = 1.3 \times 10^{-4}$. If $\phi_s = -45°$, the bounce distance is about 16 cells of the wiggler. Because z_b is much shorter than the total wiggler length, Eq. (15.251) provides a good description of beam phase dynamics. Equation (15.258) gives a value of allowed energy spread for trapping at the injection point of less than 1%. We expect that the wave traps a fraction of the beam at the entrance—the proportion of trapped electrons increases with distance through the wiggler as the wave amplitude grows. To estimate the growth length for the wave, consider a position in the wiggler with 10% of the beam electrons trapped, $T_t = 100$ A. The other electrons move incoherently and have no time-averaged interchange of energy with the wave. Equation (15.241) gives a radiation growth length of

$$E_0/(dE_0/dz) \sim 2\gamma c\varepsilon_0 A_0 E_0/a_w I_t \sin\phi_s = 4\,\text{m}. \tag{15.260}$$

In this section and the previous one, we reviewed the basic mechanism of the free-electron laser. Although we shall not discuss other material in detail, it is important to recognize that there are many outstanding areas of concern for the design of efficient devices for short-wavelength radiation generation. Numerical computer models are essential to describe axial dynamic processes such as initial beam trapping, wave amplification, and capture of a substantial fraction of a beam in a tapered wiggler. For high-current beams, we must include the effects of axial space-charge forces and space-charge waves on the beam. Transverse motions of electrons are also important. Emittance contributes to the kinetic energy spread of the beam and also makes it more difficult to guide narrow beams through long wigglers. Confinement of the TEM wave is another important problem—long wigglers exceed the diffraction length for the photons beams in free space. A important area of recent research is *gain guiding* of the wave. Here, the collective interaction between the wave and the beam confines the radiation over distances much longer than the diffraction length.

Introductory works on the free-electron laser often emphasize the theory of the free-electron-laser oscillator. In the oscillator, a uniform electron beam in a wiggler creates and pumps a TEM wave from initial noise. Oscillator theory treats the self-consistent processes where the ponderomotive force of a weak wave causes a small concentration of the beam in the phase region of Eq. (15.230). Amplification of the wave results in adiabatic trapping of more particles and modulation of the electron beam. We shall not discuss this theory in detail—the process is similar to the growth of radiation in the traveling-wave tube (Section 15.5).

Bibliography

Banford, A. P., *The Transport of Charged Particle Beams,* Spon, London, 1966.

Beck, A. H. W., *Space-Charge Waves and Slow Electromagnetic Waves,* Pergamon, London, 1958.

Birdsall, C. K., and Bridges, W. B., *Electron Dynamics of Diode Regions,* Academic Press, New York, 1966.

Birdsall, C. K., and Langdon, A. B., *Plasma Physics via Computer Simulation,* McGraw-Hill, New York, 1985.

Brau, C. A., "Free-electron lasers," in *Physics of Particle Accelerators (AIP 184),* Vol. 2, Month, M., and Dienes, M. (Eds.), American Institute of Physics, New York, 1989.

Brewer, G. R., "High-Intensity Electrons Guns," in *Focusing of Charged Particles,* Vol. 2, Septier, A. (Ed.), Academic Press, New York, 1967.

Brewer, G. R., "Focusing of high-density electron beams," in *Focusing of Charged Particles,* Vol. 2, Septier, A. (Ed.), Academic Press, New York, 1967.

Briggs, R. J., *Electron Stream Interaction with Plasmas,* MIT Press, Cambridge, MA, 1964.

Briggs, R. J., and Toepfer, A. J. (Eds.), *Proc. Fifth Intl. Conf. on High-Power Particle Beams (CONF-830911),* Lawrence Livermore National Laboratory, Livermore, CA, 1983.

Brown, I. G., "The plasma physics of ion sources," in *The Physics and Technology of Ion Sources,* Brown, I. G. (Ed.), Wiley, New York, 1989.

Brown, I. G. (Ed.), *The Physics and Technology of Ion Sources,* Wiley, New York, 1989.

Button, K. J. (Ed.), *Infrared and Millimeter Waves,* Academic Press, New York, 1979.

Camarcat, N., Doucet, H. J., and Buzzi, J. M., "High-power pulsed electron beam transport," in *Applied Charged Particle Optics,* Part A, Septier, A. (Ed.), Academic Press, New York, 1983.

Chao, A. W., "Coherent instabilities of a relativistic bunched beam," in *Physics of High-Energy Particle Accelerators (AIP 105),* Month, M. (Ed.), American Institute of Physics, New York, 1983.

Chodorow, M., and Susskind, C., *Fundamentals of Microwave Electronics,* McGraw-Hill, New York, 1964.

Collin, R. E., *Foundations for Microwave Engineering,* McGraw-Hill, New York, 1966.

814

Colonias, J. S., *Particle Accelerator Design–Computer Programs*, Academic Press, New York, 1974.

Coslett, V. E., *Introduction to Electron Optics*, Oxford University Press, Oxford, 1950.

Dahl, P., *Introduction to Electron and Ion Optics*, Academic Press, New York, 1973.

Duggan, J. L., and Morgan, I. L. (Eds.), *Application of Accelerators in Research and Industry 86*, North-Holland, Amsterdam, 1987.

Duggan, J. L., Morgan, I. L., and Martin, J. A. (Eds.), *Application of Accelerators in Research and Industry 84*, North-Holland, Amsterdam, 1985.

Gandhi, O. P., *Microwave Engineering and Applications*, Pergamon Press, New York, 1981.

Gareyte, J., "Beam observation and the nature of instabilities," in *Physics of Particle Accelerators (AIP 184)*, Vol. 1, Month, M., and Dienes, M. (Eds.), American Institute of Physics, New York, 1989.

Gewartowski, J. W., and Watson, H. A., *Principles of Electron Tubes*, Van Nostrand, Princeton, NJ, 1965.

Gillespie, G. H., Yu-Yun Kuo, Keefe, D., and Wangler, T. P. (Eds.), *High-Current, High-Brightness and High-Duty-Factor Ion Injectors (AIP 139)*, American Institute of Physics, New York, 1986.

Granatstein, V. L., and Alexeff, I. (Eds.), *High-Power Microwave Devices*, Artech House, Boston, 1988.

Grivet, P., and Septier, A., *Electron Optics*, Pergamon Press, London, 1972.

Hawkes, P. W., *Quadrupole Optics*, Springer-Verlag, Berlin, 1966.

Hawkes, P. W., *Electron Optics and Electron Microscopy*, Taylor and Francis, London, 1972.

Hendrickson, R. (Ed.), *Proc. 1979 Particle Accelerator Conf., IEEE Trans. Nuclear Science* (NS-26, 2 volumes), Institute of Electrical and Electronics Engineers, Piscataway, NJ, 1981.

Hofmann, I., "Transport and focusing of high-intensity unneutralized beams," in *Applied Charged Particle Optics*, Part A, Septier, A. (Ed.), Academic Press, New York, 1983.

Hohler, G. (Ed.), *Collective Ion Acceleration*, Springer-Verlag, Heidelberg, 1979.

Holmes, A. J. T., "Beam transport," in *The Physics and Technology of Ion Sources*, Brown, I. G. (Ed.), Wiley, New York, 1989.

Humphries, S., Jr., *Principles of Charged Particle Acceleration*, Wiley, New York, 1986.

Humphries, S., Jr., and Lockner, T. R., "High-power pulsed ion beam acceleration and transport," in *Applied Charged Particle Optics*, Part A, Septier, A. (Ed.), Academic Press, New York, 1983.

Hutter, R., "Beams with space-charge," in *Focusing of Charged particles*, Vol. 2, Septier, A. (Ed.), Academic Press, New York, 1967.

Hyder, A. K., Rose, M. F., and Guenter, A. H. (Eds.), *High-brightness Accelerators*, Plenum Press, New York, 1988.

Jameson, R. A., and Taylor, L. S. (Eds.), *Proc. 1981 Linear Accelerator Conf. (LA-9234-C)*, Los Alamos National Laboratory, Los Alamos, NM, 1982.

Kapchinskii, I. M., *Dynamics in Linear Resonance Accelerators*, Atomizdat, Moscow, 1966.

Katsouleas, T., and Dawson, J. M., "Plasma acceleration of particle beams," in *Physics of Particle Accelerators (AIP 184)* (3 volumes), Month, M., Dienes, M. (Eds.), American Institute of Physics, New York, 1989.

Keil, E., "Computer programs in accelerator physics," in *Physics of High-Energy Particle Accelerators (AIP 105)*, Month, M. (Ed.), American Institute of Physics, New York, 1983.

Keller, R., "Ion extraction," in *The Physics and Technology of Ion Sources*, Brown, I. G. (Ed.), Wiley, New York, 1989.

Kim, K. J., "Characteristics of synchrotron radiation," in *Physics of Particle Accelerators (AIP 184)*. Vol. 1, Month, M., and Dienes, M. (Eds.), American Institute of Physics, New York, 1989.

Kirstein, P. T., Kino, G. S., and Waters, W. E., *Space-Charge Flow*, McGraw-Hill, New York, 1967.

Klemperer, O., and Barnett, M. E., *Electron Optics*, Cambridge University Press, London, 1971.

Kuswa, G. W., Quintenz, J. P., Freeman, J. R., and Chang, J., "Generation of high-power pulsed ion beams," in *Applied Charged Particle Optics*, Part A, Septier, A. (Ed.), Academic Press, New York, 1983.

Lambertson, G., "Dynamic devices—Pickups and kickers," in *Physics of Particle Accelerators (AIP 153)* (2 volumes), Month, M., and Dienes, M. (Eds.), American Institute of Physics, New York, 1987.

Lapostolle, P. M., and Septier, A. (Eds.), *Linear Accelerators*, North-Holland, Amsterdam, 1970.

Lawson, J. D., "Space-charge optics," in *Applied Charged Particle Optics*, Part A, Septier, A. (Ed.), Academic Press, New York, 1983.

Lawson, J. D., *The Physics of Charged-particle Beams*, Clarendon Press, Oxford, 1988.

Lehnert, B., *Dynamics of Charged Particles*, North-Holland, Amsterdam, 1964.

Lejeune, C., "Extraction of high-intensity ion beams from plasma sources: Theoretical and experimental treatments," in *Applied Charged Particle Optics*, Part A, Septier, A. (Ed.), Academic Press, New York, 1983.

Lejeune, C., and Aubert, J., "Emittance and brightness: Definitions and measurements," in *Applied Charged Particle Optics*, Part A, Septier, A. (Ed.), Academic Press, New York, 1983.

Lichtenberg, A. J., *Phase-Space Dynamics of Particles*, Wiley, New York, 1969.

Lindstrom, E. R., and Taylor, L. S. (Eds.), *Proc. 1987 IEEE Particle Accelerator Conference* (3 volumes) (IEEE 87CH2387-9), Institute of Electrical and Electronics Engineers, Piscataway, NJ, 1987.

Livingston, M. S., and Blewett, J. P., *Particle Accelerators*, McGraw-Hill, New York, 1962.

Loew, G. A. (Ed.), *Proc. 1986 Linear Accelerator Conf.*, Stanford Linear Accelerator Center, Stanford Univ., Stanford, CA, 1986.

Marshall, T. C., *Free-Electron Lasers*, Macmillan, New York, 1985.

Meleka, A. H., *Electron-Beam Welding—Principles and Practice*, McGraw-Hill, New York, 1971.

Michelotti, L., "Phase space concepts," in *Physics of Particle Accelerators (AIP 184)*, Vol. 1, Month, M., and Dienes, M. (Eds.), American Institute of Physics, New York, 1989.

Miller, R. B., *Physics of Intense Charged Particle Beams*, Plenum Press, New York, 1982.

Month, M. (Ed.), *Physics of High-Energy Particle Accelerators (AIP 105)*, American Institute of Physics, New York, 1983.

Month, M., and Dienes, M. (Eds.), *Physics of Particle Accelerators (AIP 153)* (2 volumes), American Institute of Physics, New York, 1987.

Month, M., and Dienes, M. (Eds.), *Physics of Particle Accelerators (AIP 184)* (3 volumes), American Institute of Physics, New York, 1989.

Nagy, G. A., and Szilagyi, M., *Introduction to the Theory of Space-Charge Optics*, Macmillan, London, 1974.

Nation, J. A., "High-power relativistic electron beam sources," in *Applied Charged Particle Optics*, Part A, Septier, A. (Ed.), Academic Press, New York, 1983.

Nation, J. A., and Sudan, R. N. (Eds.), *Proc. Second Intl. Conf. on High-Power Particle Beams*, Laboratory of Plasma Studies, Cornell University, Ithaca, NY, 1977.

Northrup, T. J., *The Adiabatic Motion of Charged Particles*, Interscience, New York, 1963.

Okress, E. (Ed.), *Crossed-Field Microwave Devices*, Academic Press, New York, 1961.

Olson, C. L., "Collective acceleration of ions by an intense relativistic electron beam," in *Applied Charged Particle Optics*, Part A, Septier, A. (Ed.), Academic Press, New York, 1983.

Pantell, R. H., "Free-electron lasers," in *Physics of Particle Accelerators (AIP 184)*, Vol. 2, Month, M., and Dienes, M. (Eds.), American Institute of Physics, New York, 1989.

Paul, A. C., and Neil, V. K., "High-current relativistic electron guns," in *Applied Charged Particle Optics*, Part A, Septier, A. (Ed.), Academic Press, New York, 1983.

Pierce, J. R., *Travelling-Wave Tubes*, Van Nostrand, Princeton, NJ, 1950.

Pierce, J. R., *Theory and Design of Electron Beams*, Van Nostrand, Princeton, NJ, 1954.

Placious, R. C. (Ed.), *Proc. 1981 Particle Accelerator Conf., IEEE Trans. Nuclear Science* (NS-28, 2 volumes), Institute of Electrical and Electronics Engineers, Piscataway, NJ, 1981.

Reimer, L., *Scanning Electron Microscopy*, Springer, Berlin, 1985.

Rose, H., and Spehr, R., "Energy broadening in high-density electron and ion beams: The Boersch effect," in *Applied Charged Particle Optics*, Part A, Septier, A. (Ed.), Academic Press, New York, 1983.

Rosenblatt, J., *Particle Accelerators*, Methuen, London, 1968.

Rostoker, N., and Reiser, M. (Eds.), *Collective Methods of Acceleration*, Harwood Academic Publishers, Chur, Switzerland, 1979.

Rowe, J. E., *Non-Linear Electron–Wave Interaction Phenomena*, Academic Press, New York, 1965.

Scharf. W., *Particle Accelerators and Their Uses*, Harwood Academic Publishers, Chur, Switzerland, 1986.

Scharlemann, E. T., and Prosnitz, D., (Eds.), *Free-Electron Lasers*, North-Holland, Amsterdam, 1986.

Schiller, S., Heisig, U., and Panzer, S., *Electron Beam Technology*, Wiley, New York, 1982.

Schriber, S. O., and Taylor, L. S. (Eds.), *Charged Particle Optics*, North-Holland, Amsterdam, 1987.

Septier, A., "Production of ion beams of high intensity," in *Focusing of Charged Particles*, Vol. 2, Septier, A. (Ed.), Academic Press, New York, 1967.

Septier, A. (Ed.), *Focusing of Charged Particles* (3 volumes), Academic Press, New York, 1967.

Septier, A. (Ed.), *Applied Charged Particle Optics* (3 volumes), Academic Press, New York, 1983.

Servranckx, R., and Brown, K. L., "Circular macine design techniques and tools," in *Physics of Particle Accelerators (AIP 153)* (2 volumes), Month, M., and Dienes, M. (Eds.), American Institute of Physics, New York, 1987.

Shevchik, V. N., Shvedov, G. N., and Soboleva, A. V., *Wave and Oscillatory Phenomena in Electron Beams at Microwave Frequencies*, Pergamon Press, Oxford, 1966.

Slater, J. C., *Microwave Electronics*, Van Nostrand, Princeton, NJ, 1950.

Steffen, K. G., *High Energy Beam Optics*, Wiley, New York, 1965.

Stix, T. H., *The Theory of Plasma Waves*, McGraw-Hill, New York, 1962.

Strathdee, A. (Ed.), *Proc. 1983 Particle Accelerator Conf., IEEE Trans. Nuclear Science* (NS-32, 2 volumes), Institute of Electrical and Electronics Engineers, Piscataway, NJ, 1987.

Stuhlinger, E., *Ion Propulsion for Space Flight*, MacGraw-Hill, New York, 1964.

Sturrock, P., *Static and Dynamic Electron Optics*, Cambridge University Press, London, 1955.

Talman, R., "Resonances in accelerators," in *Physics of Particle Accelerators (AIP 153)* (2 volumes), Month, M., and Dienes, M. (Eds.), American Institute of Physics, New York, 1987.

Talman, R., "Multiparticle phenomena and Landau damping," in *Physics of Particle Accelerators (AIP 153)* (2 volumes), Month, M., and Dienes, M. (Eds.), American Institute of Physics, New York, 1987.

Tollestrup, A. V., and Dugan, G., "Elementary stochastic cooling," in *Physics of High-Energy Particle Accelerators (AIP 105)*, Month, M. (Ed.), American Institute of Physics, New York, 1983.

Trivelpiece, A. W., *Slow-Wave Propagation in Plasma Waveguides*, San Francisco Press, San Francisco, 1967.

Vlasov, A. D., *Theory of Linear Accelerators*, Atomizdat, Moscow, 1965.

Weng, W. T., "Space-charge effects—Tune shifts and resonances," in *Physics of Particle Accelerators (AIP 153)* (2 volumes), Month, M., and Dienes, M. (Eds.), American Institute of Physics, New York, 1987.

Wilson, P. B., "Introduction to wake fields and wake potentials," in *Physics of Particle Accelerators*

(AIP 184), Vol. 1, Month, M., and Dienes, M. (Eds.), American Institute of Physics, New York, 1989.

Wilson, R. G., and Brewer, G. R., *Ion Beams With Applications to Ion Implantation*, Wiley, New York, 1973.

Wollnik, H., *Optics of Charged Particles*, Academic Press, New York, 1988.

Wollnik, H. (Ed.), *Charged Particle Optics*, North-Holland, Amsterdam, 1981.

Yamanaka, C. (Ed.), *Proc. Sixth Intl. Conf. on High-Power Particle Beams*, Institute of Laser Engineering, Osaka Univ., Osaka, Japan, 1986.

Yonas, G. (Ed.), *Proc. First Intl. Conf. on High-power Particle Beams* (*SAND76-5122*), Sandia National Laboratories, Albuquerque, NM, 1976.

Index

HYSIC
351   ll   3122

 MONTH